Biomechanics and Biology of Movement

Biomechanics and Biology of Movement

Benno M. Nigg, PhD
Director, Human Performance Laboratory
University of Calgary

Brian R. MacIntosh, PhD
Professor, Faculty of Kinesiology
University of Calgary

Joachim Mester, PhD
Director, Institute of Training and Movement
German Sport University

Editors

Human Kinetics

Library of Congress Cataloging-in-Publication Data

Biomechanics and biology of movement / [edited by] Benno M. Nigg, Brian R. MacIntosh, Joachim Mester.

    p. cm.

    Includes bibliographical references and index.

    ISBN 0-7360-0331-2

    1. Human mechanics. 2. Human locomotion. 3. Exercise--Physiological aspects. 4. Energy metabolism. I. Nigg, Benno Maurus. II. MacIntosh, Brian R., 1952- III. Mester, J. (Joachim)

QP303 .B56836 2000

612.7'6--dc21   99-059795

ISBN 0-7360-0331-2

Acquisitions Editor: Loarn D. Robertson, PhD; **Managing Editor:** Cynthia McEntire; **Assistant Editor:** John Wentworth; **Copyeditor:** Judy Peterson; **Proofreader:** Pamela S. Johnson; **Indexer:** Marie Rizzo; **Permission Manager:** Cheri Banks; **Graphic Designer:** Nancy Rasmus; **Graphic Artist:** Kathleen Boudreau-Fuoss; **Cover Designer:** Jack W. Davis; **Printer:** Sheridan Books

Printed in the United States of America

10 9 8 7 6 5 4 3 2 1

Human Kinetics
Web site: http://www.humankinetics.com/

United States: Human Kinetics
P.O. Box 5076
Champaign, IL 61825-5076
1-800-747-4457
e-mail: humank@hkusa.com

Canada: Human Kinetics
475 Devonshire Road Unit 100
Windsor, ON N8Y 2L5
1-800-465-7301 (in Canada only)
e-mail: humank@hkcanada.com

Europe: Human Kinetics, P.O. Box IW14
Leeds LS16 6TR, United Kingdom
+44 (0)113-278 1708
e-mail: humank@hkeurope.com

Australia: Human Kinetics
57A Price Avenue
Lower Mitcham, South Australia 5062
(08) 82771555
e-mail: liahka@senet.com.au

New Zealand: Human Kinetics
P.O. Box 105-231, Auckland Central
09-523-3462
e-mail: humank@hknewz.com

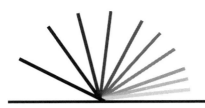

Contents

Part IV Fatigue and Exercise

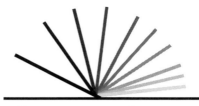

Contributors

Adams, Douglas J.
Department of Orthopaedic Surgery
University of Connecticut Health Center
Farmington, CT

Alexander, R. McNeill
University of Leeds
Department of Pure and Applied Biology
Leeds, England

Allen, David G.
Department of Physiology
University of Sydney
Sydney, New South Wales, Australia

Alt, Wilfried
Sportmedizinisches Institut
Frankfurt am Main, Germany

Anderson, Donald D.
Biomechanics Laboratory
Minneapolis Sports Medicine Center
Minneapolis, MN

Bobet, Jacques
Assistant Professor, Physical Therapy
University of Alberta
Edmonton, AB, Canada

Denoth, Jachen
Laboratorium fur Biomechanik
ETH Zurich
Schlieren, Switzerland

Fischer, Kenneth J.
Musculoskeletal Research Center
University of Pittsburgh
Pittsburgh, PA

Gollhofer, Albert
Universität Stuttgart
Institut für Sport und Sportwissenschaft
Stuttgart, Germany

Hale, Joseph E.
Biomechanics Laboratory
Minneapolis Sports Medicine Center
Minneapolis, MN

Hay, James G.
Department of Sport and Exercise Science
University of Aukland
Aukland, New Zealand

Herzog, Walter
Faculty of Kinesiology
University of Calgary
Calgary, AB, Canada

Holash, R. John
Faculty of Kinesiology
University of Calgary
Calgary, AB, Canada

Hoppeler, Hans
University of Bern
Bern, Switzerland

Komi, Paavo V.
Kinesiology Laboratory
Department of Biology of Physical Activities
University of Jyvaskyla
Jyvaskyla, Finland

Lemon, Peter W.R.
3M Centre
University of Western Ontario
London, ON, Canada

Lohrer, H.
Orthopäd. Abteilung im Sportmedizinischen Institut
Frankfurt am Main, Germany

MacIntosh, Brian R.
Faculty of Kinesiology
University of Calgary
Calgary, AB, Canada

Mester, Joachim
Institute of Training and Movement
German Sport University
Cologne, Germany

Mikulcik, Edwin C.
Department of Mechanical Engineering
University of Calgary
Calgary, AB, Canada

Minetti, Alberto E.
Biomechanics Research Group
Department of Exercise and Sport Science
Crewe + Alsager Faculty
Manchester Metropolitan University
Hassall Road, Alsager, United Kingdom

Neptune, Richard R.
Rehabilitation R & D Center (153)
VA Palo Alto Health Care System
Palo Alto, CA

Nicol, Caroline A.
Université Aix-Marseille II
Marseille Cedex, France

Nigg, Benno M.
Faculty of Kinesiology
University of Calgary
Calgary, AB, Canada

Orizio, Claudio
Department of Biomedical Sciences and Biotech-
   nologies
Institute of Human Physiology
University of Brescia
Brescia, Italy

Stefanyshyn, Darren J.
Faculty of Kinesiology
University of Calgary
Calgary, AB, Canada

Stein, Richard
Department of Physiology
University of Alberta
Edmonton, AB, Canada

van den Bogert, Anthony J.
Department of Biomedical Engineering
Lerner Research Institute
Cleveland Clinic Foundation
Cleveland, OH

Weibel, Ewald R.
Anatomisches Institut der Universität Bern
Bern, Switzerland

Yeadon, Maurice R.
Department of Sports Science
Loughborough University
Loughborough, United Kingdom

Zehr, E. Paul
Faculty of Physical Education and Recreation
University of Alberta
Edmonton, AB, Canada

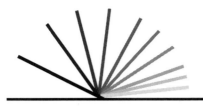

Introduction

B.M. Nigg, B.R. MacIntosh, and J. Mester

The life of an average member of civilized society has changed substantially over the last two centuries. About 100 to 200 years ago, most daily activities were associated with movement and physical activity. Today, at the beginning of the third millennium, most professional occupations are sedentary. Most people in the developed countries have minimal daily physical activity, and computers and television dominate the lives of many members of today's society. Most people have a substantial amount of leisure time available. Some use it for physical activities. They enjoy the outdoors or are engaged in daily workout routines. However, the majority of people in the developed countries are not physically fit. Many people are overweight and have only limited mobility. With the increasing life expectancy afforded by health care, mobility and longevity have become some of the most precious aspects of life. Thus, human movement, exercise, and sport have developed in the 20th century into important lifestyle options.

Parallel to this development, science that deals with human movement, exercise, and sport has become increasingly important. The number of scientists from many different disciplines concentrating on researching movement, exercise, and sport is still increasing. These scientists attempt to understand (a) the functioning of the human body as it relates to movement, exercise, and sport, (b) biological responses to force stimuli, (c) how to prolong mobility for all age groups, and (d) how to improve performance, whether in sport, in the work place, or during a walk. They work together to solve these important questions, which are relevant for the well-being of humankind.

Scientific disciplines dealing with human movement, exercise, and sport include anatomy, biochemistry, biomechanics, neurosciences, and physiology. Textbooks dealing with aspects of movement, exercise, and sport typically discuss discipline-related aspects. Titles such as *The Biomechanics of Sports Techniques* (Hay 1978), *The Physiology of Joints* (Kapandji 1970), *Functional Anatomy in Sports* (Weineck 1986), or *Biomechanics of Sports* (Vaughan 1989) are typical examples of such publications.

They are discipline driven and lack approaches where the multidisciplinary question of interest drives the method(s) of inquiry. However, human life is not discipline oriented. Human life is exposed daily to practical questions, which ask for an answer. Topics that are important for movement, exercise, and sport include work and energy, balance and control of human movement, load and excessive load during movement and exercise, and fatigue during exercise.

Work and energy aspects are important for an athlete running a marathon, a soccer player participating in a three-week tournament, a mountaineer who wants to climb Mount Everest, or a speed skater attempting to break a world record. Additionally, work and energy questions are important for nonathletes interested in healthy nutrition as well as for people who want to maintain their health and achieve an adequate body weight. Furthermore, work and energy are equally important for a person with an artificial leg (e.g., a below-the-knee amputee) or an elderly person who wants to play her or his daily round of golf or to walk to maintain physical fitness. Work and energy questions are relevant to all sectors of the human population.

Balance and control of movement are important for many sport activities such as balance beam exercises in gymnastics, shooting, somersaulting and twisting in trampolining, and activities on the trapeze. However, balance and control are equally important for children as they learn to move correctly and control their movements and for elderly people who may experience impaired control. Impaired mobility can result in exclusion from a large segment of life activities and people so affected will experience a decrease in the quality of life.

Excessive load during sport and exercise is specifically important for competitive sport activities. Injuries resulting from excessive repetitive forces are speculated to be the cause of early arthritis and disability. Therefore, it is important to understand the factors contributing to excessive loading and the strategies that can be used to avoid inappropriate loading situations. Furthermore, it is important to understand when a load on the musculoskeletal system is beneficial and contributes to the development of strong and healthy biological structures.

Fatigue is a consequence of repeated use of muscles or other tissues and limits the duration a given activity can be performed. Fatigue may be the limiting factor, which when minimized permits winning or when evident results in losing a competition. Fatigue may impair mobility or general physical capability during daily activities. Furthermore, fatigue may be a contributing factor to the development of acute or chronic injuries. The ability to assess the presence and magnitude of fatigue becomes an important objective in the quest to understand the circumstances and consequences associated with muscle fatigue. Fatigue is also important in the work place and in daily leisure activities. If fatigue can be reduced, work performance may improve or leisure activities may become more enjoyable.

These questions and problems cannot be solved with the methods and approaches of only one scientific discipline. To understand the effects of excessive load, for instance, one needs contributions from biomechanics, biochemistry, and neuroscience. Answering questions related to fatigue requires contributions from physiology, biochemistry, neuroscience, and muscle mechanics. Balance and control questions can be answered only with contributions from neuroscience, biomechanics, anatomy, and physiology. Work and energy questions require contributions from physiology, biochemistry, biomechanics, and thermomechanics. Consequently, when studying movement, exercise, and sport-related questions one should attempt to understand the many facets contributing to the question and synthesize them into a comprehensive analysis.

The editors of this book identified the primary biological and physical knowledge associated with work and energy, balance and motor control, load and excessive load, and fatigue. The editors defined the important components for each topic and invited world-renowned experts in these areas to contribute from their viewpoint and wealth of understanding to the identified topic. Thus, the different sections of this book attempt to discuss important questions using input from the many disciplines of movement, exercise, and sport sciences that belong to the physical and biological sciences. The authors of the various chapters are experts in their fields. To integrate the presented knowledge, the editors added a synthesis to each topic. The contributions were organized in the form of a concise and comprehensive textbook in the hope that this text may provide a new approach to the exciting field of movement, exercise, and sport sciences.

The book is aimed at students and professionals working in kinesiology (e.g., biomechanics, physiology, physiotherapy, athletic therapy, and ergonomics). Readers who already have a basic knowledge of biology, physiology, and biomechanics will obtain the greatest benefits from the use of this text. Thus, the book is expected to be used for advanced undergraduate or early graduate level courses. Furthermore, the book will be a useful resource for research-oriented undergraduate and graduate students.

Each part starts with selected historical highlights for the interested reader. However, the book can be used without reading these historical highlights. This initial section is followed by basic and applied discussions of the topic of interest, followed by a synthesis of the discussed topic. Each part ends with a summary and a list of the most important definitions used. The reader is not expected to read through these definitions. They are collected to provide the reader a place to quickly find definitions if needed. Furthermore, the reader of this book is not required to use all the chapters. It is possible to choose a selection of contributions to a given topic and to make this selection based on the basic knowledge and the specific interest of the reader or student.

Over the last two decades the quality of students in kinesiology and the exercise sciences has improved substantially. Their expertise in biological and mechanical methods has vastly improved. In many universities, students in kinesiology and exercise sciences are among those with the highest entrance averages of all university faculties. The field of kinesiology is expanding and proves to be attractive for many brilliant young students and scientists. A strong indication for this development in kinesiology and the general field of exercise sciences is the establishment of the IOC Olympic Prize, a research award of $500,000 (U.S.) given for excellence in research on movement, exercise, and sport. It illustrates that intellectual leaders of human society are convinced that science dealing with movement, exercise, and sport is extremely important and should be recognized. This book takes this phenomenon into account. The concepts are presented in a relatively simple way. However, readers will be challenged with the intellectual depth achieved in each section of the text.

References

Hay, J.G. 1978. *The Biomechanics of Sports Techniques.* Englewood Cliffs, NJ: Prentice Hall.

Kapandji, I.A. 1970. *The Physiology of the Joints.* Edinburgh: Churchill Livingstone.

Vaughan, C.L. 1989. *Biomechanics of Sports.* Boca Raton, FL: CRC Press.

Weineck, J. 1986. *Functional Anatomy in Sports.* Chicago: Year Book Medical.

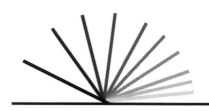

PART I

Work and Energy

Work and Energy Historical Highlights

Around 500 B.C.

Believing that it would enhance muscle strength and size, Greek athletes begin to consume large quantities of meat instead of the predominantly vegetarian diet of the time.

480 B.C.

Hippocrates recognizes the importance of a balance between food intake (energy) and physical activity (energy expenditure).

300s

Aristotle uses the term *energy* to denote something "in action."

1695

Leibniz formulates explicitly the principle of conservation of mechanical energy for a particle in terms of "live force" (kinetic energy) and "dead force" (potential energy).

1700s

Lavoisier makes major contributions in the areas of metabolism and respiration.

1744

Following a suggestion of his teacher (D. Bernoulli, 1700–1782), **Euler** uses a definition of the "potential force" (strain energy) of an elastic bar to solve the buckling problem using variation calculus.

1788

Lagrange adopts the principle of virtual work as the basis for his analytical mechanics.

1824

Carnot publishes *Réflexions sur la Puissance Motrice du feu* in which he outlines the foundations for the second law of thermodynamics.

1840s

Believing that protein is the major fuel for exercising muscle, **von Liebig** recommends that physically-active individuals consume large quantities of meat.

1851

Kelvin synthesizes the works of Carnot and Joule into a precise formulation of the first and second laws of thermodynamics.

1865

Blix recognizes that a linear relationship exists between energy intake and the intake of many nutrients.

1896

Chaveau theorizes that carbohydrate is the only fuel for exercise.

1901

Zuntz and **Schumburg** recognize that both carbohydrate and fat can be used as fuel for exercise and develop tables relating carbohydrate and fat use to CO_2 production and O_2 consumption.

1913

Hill starts his famous experiments on heat measurements of contracting frog skeletal muscle.

1920s

Cathcart suggests that protein plays only a minor role as an exercise fuel.

1920s

Levine observes that carbohydrate intake before and during a marathon run enhances performance.

1930s

Christensen, Hanson, and **Dill** verify the importance of high-carbohydrate diets for endurance-performance athletes especially when exercise is intense.

1935

Fenn finds an exponential relation between muscular force and contraction velocity and proposes that energetics dictates this relationship.

1936
Schoenheimer and **Rittenburg** develop the stable isotope methodology for metabolic studies.

1938
Hill finds a hyperbolic relation between muscular force and contraction velocity and relates this to the thermodynamics of muscle contraction.

1941
Lipman discovers the role of ATP as a kind of common currency for metabolic energy.

1943
Keys suggests that, although vitamin deficiency impairs exercise performance, additional vitamin intake does not improve performance.

1946
Using a radio frequency coil, **Bloch** obtains a proton signal from his finger.

1947
Adolph recognizes that water intake needs to match sweat loss or exercise performance and eventually health deteriorate.

1957
A.F. Huxley proposes the *cross-bridge theory,* which describes the mechanisms of muscular contraction. This allows a comprehensive view of mechanical and energetic properties of skeletal muscle.

1960s
Cade develops glucose- and electrolyte-replacement beverages.

1960s
Holloszy introduces biochemical techniques to assess muscle metabolism.

1962
Bergström and **Hultman** reintroduce the needle biopsy technique, making possible the study of muscle metabolism during exercise.

1967
Saltin and **Hermansen** develop strategies for carbohydrate-loading diets.

1968
Cavagna describes *positive work* done by a previously stretched muscle.

1970s
Hoult, Gadian, and **Radda** use magnetic resonance spectroscopy to assess muscle metabolism noninvasively.

1970s–1990s
Waterlow, Wolfe, and **Young** use the metabolic tracer technique to assess nutritional needs.

1980s–present
Booth uses the molecular biology approach to address metabolic questions.

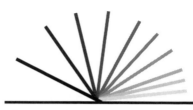

Chapter 1

Mechanical Considerations of Work and Energy

B.M. Nigg, D. Stefanyshyn, and J. Denoth

Work, energy, and performance during physical activities depend on biochemical, physiological, thermodynamic, and mechanical aspects. For example, among other factors, the performance of an athlete during a running competition depends on

- the biochemical composition of the blood providing the metabolic components for muscle work,
- the transformation of oxygen into muscle work,
- the production of heat during the running task, and
- the mechanical efficiency of the athlete, shoe, and surface system.

These factors themselves are influenced by the psychological state of an athlete and can change with a changing psychological state. The resulting work, energy, and performance can be determined by defining the system of interest and by applying the laws of conservation of energy.

This chapter discusses mechanical work and energy and the methods for evaluating them. It discusses the law of conservation of energy and presents selected examples for the study of mechanical work and energy. This chapter attempts to provide the background for understanding mechanical work and energy in locomotion and illustrates selected

mechanical possibilities to influence the amount of energy an athlete has available for a given task.

The Concept of Mechanical Work and Energy

Work and energy aspects during locomotion, physical exercise, and sport have been discussed in many studies over the time span of the 20th century (Aleshinsky 1986; Cavanagh and Williams, 1982; Cavagna, Thys, and Zamboni 1976; Clement et al. 1982; di Prampero 1986; Elftman 1939; Fenn 1930a; Frederick, Howley, and Powers 1982; Frederick et al. 1983; Nigg and Anton 1995; Stefanyshyn and Nigg 1998a; Williams 1985; Winter 1978).

An athlete has three major strategies available to improve the work-energy balance during locomotion:

1. To store and return energy
2. To minimize the loss of energy
3. To optimize muscle functions

The possibility of storing and returning energy in muscle-tendon units or in equipment has been addressed frequently in the scientific literature (e.g., Alexander and Bennet-Clark, 1977; Alexander and Vernon, 1975; Asmussen and Petersen, 1974;

Cavagna, Dusman, and Margaria 1968; Stefanyshyn and Nigg, 1997; van Ingen Schenau 1984). This possibility is discussed in detail later in this chapter.

Examples of storage and return of energy in muscle-tendon units specifically include animals. Camels, kangaroos, and horses have been described as excellent examples with a substantial amount of energy stored and returned during a ground contact.

Examples of storage and return of energy in sport equipment include elastic poles for pole-vaulting, springboards for diving, tuned surfaces for indoor track and field surfaces (McMahon and Greene 1979), and springfloors for tumbling and gymnastics. The pole, the springboard, the track surface, or the springfloor surface are deformed. In this process energy is stored in the deformed structure. This energy can be returned to the athlete in the second half of the movement. It is a fascinating possibility to improve performance. The ability to store and return energy was instrumental in changing and improving performance in many sport activities. The world record in pole-vaulting was increased substantially by about 20% when athletes changed from aluminium to fiberglass poles. Indoor track and field world records were broken when tuned indoor sport surfaces that allowed a vertical deformation of the running surface were installed. Triple and maybe even quadruple somersaults are now possible in gymnastics, moves that would not have been possible on the old, conventional sport surfaces with no storage and release of energy.

The strategy of minimizing the loss of energy to improve performance has only rarely received the attention of researchers or practitioners (Nigg 1997; Stefanyshyn and Nigg 1998b). It suggests that performance can be improved by minimizing the waste of energy for purposes that do not contribute to the improvement of performance. For example, an unstable speed- or in-line skate requires the athlete to spend energy stabilizing the ankle joint. This energy does not contribute to the actual performance and is, therefore, wasted. Similarly, muscle work performed to avoid soft tissue wobbling during impact activities, for example, landing in running, does not contribute to the actual running performance and thus corresponds to lost energy.

The strategy to optimize muscle function with the goal of improving performance has been addressed in only a few publications. A recently published study attempted to optimize performance by maximizing power output (Yoshihuku and Herzog 1990). These approaches attempted to determine a general optimal force-velocity and length-tension relationship for the total system to improve muscle work and, therefore, performance. The position of an athlete on a bicycle, for instance, can be adjusted in order to optimize the ranges in which muscles work. By making this adjustment, performance can be improved substantially.

The advantages, disadvantages, and possibilities of these three strategies are discussed in chapter 4. The following paragraphs concentrate on basic aspects of work and energy. An interested reader may enjoy these sections. A reader less interested in the mechanical aspects may want to skip these sections and start at "The Laws of Conservation of Energy" (page 11). However, the authors suggest that a thorough understanding of the energy chapter is a must for any person dealing with energy and performance questions in human or animal movement.

Mechanical Work

The mechanical work, W, performed by a force vector, **F**, acting on a particle is defined as the line integral

$$W = \int \mathbf{F} \bullet d\mathbf{r}. \qquad\qquad 1.1$$

Unit of work: $[W] = [\text{force}] \cdot [\text{distance}] = N \cdot m = \text{Joule} = J$.

Using the rules of vector algebra, equation 1.1 can be written as

$$W = \int \mathbf{F} \bullet d\mathbf{r} = |\mathbf{F}| \cdot |d\mathbf{r}| \cdot \cos(\alpha) \qquad 1.2$$

where

W = work performed by the force **F**,

F = force vector acting on the particle,

d**r** = infinitesimal small displacement vector,

• = sign for the scalar (dot) product of the two vectors,

· = sign for a normal multiplication, and

α = angle between the force vector, **F**, and the instantaneous displacement vector d**r**.

In some examples a very special case is discussed, that is, the case where (a) the force vector **F** is constant and (b) the force **F** and displacement d**r** vectors act in the same direction. For this special case the term $\cos(\alpha)$ is equal to 1 and the formula for work (equation 1.2) can be written as

$$W = |\mathbf{F}| \cdot |d\mathbf{r}| \qquad\qquad 1.3$$

which can be written as

$$W_{11} = F_{const} \cdot d_{11} \qquad\qquad 1.4$$

where

W_{11} = work performed by the constant force, F_{const}, over the displacement in the direction of the acting force,

F_{const} = | **F** | = constant force acting on the particle, and

d_{11} = | d**r** | = displacement parallel to the acting force.

Example #1

In the first example, an athlete travels to a competition. He stands at the train station and holds a suitcase in each hand. Determine the mechanical work the athlete performs.

Equation 1.2 can be used to solve this example.

$$W = \int \mathbf{F} \bullet d\mathbf{r} = |\mathbf{F}| \cdot |d\mathbf{r}| \cdot \cos(\alpha). \qquad 1.5$$
$$W = |\mathbf{F}| \cdot |d\mathbf{r}| \cdot \cos(\alpha).$$

For each of these three variables we have specific information:

| **F** | = body weight + weight of the two suitcases.

| d**r** | = 0 m because the suitcases don't move.

α = unknown, because there is no movement.

Therefore,

$W = 0 \, N \cdot m.$
$W = 0 \, J.$

Example #2

In the second example, determine the work a person must perform to lift a mass of 0.5 kg by 0.2 m. Again, equation 1.2 can be used to solve this example.

$$W = |\mathbf{F}| \cdot |d\mathbf{r}| \cdot \cos(\alpha) \qquad\qquad 1.6$$

where | **F** | = –5 N equals the force acting on the mass, which corresponds to the weight of the mass but acts in positive axis direction (note that, in a first approximation, the earth acceleration has been assumed to be 10 m/s^2).

The minus sign is present because the weight is pointing downward while the positive axis direction has been assumed to be upward.

| d**r** | = +0.2 m. The plus sign is present because the change in position is pointing upward.

$\alpha = 0°.$

$\cos(\alpha) = 1.$

Thus,

$W = (5 \, N) \cdot (0.2 \, m) \cdot 1.$
$W = 1 \, N \cdot m.$
$W = 1 \, J.$

The term *work* has different meanings when used in mechanics, in physiology, or in daily conversation. A person standing at the gate of an airport carrying two suitcases performs no mechanical work. However, physiologically one would measure a different oxygen uptake with and without the suitcases. Therefore, this person performs work in a physiological but not in a mechanical sense. A person sitting in a chair in a downtown office developing new marketing strategies "works" in the sense of the daily use of the term "work" (at least she gets paid for it). If one would measure the muscle activity during this "work" one would register a low level of activity. Consequently, the physiological work is not zero. However, the mechanical work is zero, as long as she does not move. Thus, it is important to distinguish between these different forms of work and to understand their importance in the context of a specific question.

Mechanical work appears in different forms. Work might be performed against gravity when lifting a mass. Work might be performed when stretching or compressing a spring. Work might also be performed when accelerating a mass. Specifically, the most important forms of mechanical work and their definitions are

work against gravity =   1.7
$W_{gr} = m \cdot g \cdot \Delta H$

work to deform a (linear) spring =   1.8
$W_{sp} = \frac{1}{2} \cdot k \cdot \Delta x^2$

work to accelerate a mass (translation) =   1.9
$W_{acctr} = \frac{1}{2} \cdot m \cdot v^2$

work to accelerate a mass (rotation) =   1.10
$W_{accrot} = \frac{1}{2} \cdot I \cdot \omega^2$

where

m = mass of the particle,

g = acceleration due to gravity,

ΔH = change in height,

k = (linear) spring constant,

Δx = deformation of a (linear) spring,

v = speed of the mass,

ω = angular velocity of the mass about its center of mass, and

I = moment of inertia of the mass.

Mechanical Energy

Work and energy are two terms that are closely related. Energy is defined as the ability to do work.

The same general comments can be made for energy as have been made for work. The term *energy* is differently (and often loosely) used in physics, physiology, or daily life.

Mechanical energy can appear in different forms. The energy a mass possesses because it is in motion is called kinetic energy. This energy may have translational or rotational components. The energy a mass possesses because of its position is called potential energy. Examples of potential energy include energy due to a certain height of a mass with respect to a baseline and energy of a stretched or compressed spring. Specifically, the most important forms of mechanical energy and their definitions are

potential energy due to gravity =   1.11
$$E_{potgr} = m \cdot g \cdot \Delta H$$

deformation energy of a (linear) spring =   1.12
$$E_{potspr} = \tfrac{1}{2} \cdot k \cdot \Delta x^2$$

kinetic energy for translation =   1.13
$$E_{kintr} = \tfrac{1}{2} \cdot m \cdot v^2$$

kinetic energy for rotation =   1.14
$$E_{kinrot} = \tfrac{1}{2} \cdot I \cdot \omega^2.$$

Calculation of Mechanical Energy During Human Movement

Historically, mechanical energy of human movement has been determined using two distinct methods, the segmental energy calculation and the joint energy calculation.

The Segmental Energy Calculation

Fenn (1930a and 1930b) was the first scientist to use the segmental energy calculation. This approach determines the mechanical energy of a body by summing the translational kinetic energy E_{kintr}, the rotational kinetic energy E_{kinrot}, and the potential energy E_{pot} of one or several segments. The total energy of segment i, E_{toti}, is the sum of the three contributing terms:

$$E_{toti} = \tfrac{1}{2} m_i v_i^2 + \tfrac{1}{2} I_i \omega_i^2 + m_i g H_i \qquad 1.15$$

where

m_i = mass of segment i,

v_i = translational speed of the center of mass of segment i,

I_i = moment of inertia of segment i,

ω_i = angular velocity of segment i,

g = acceleration due to gravity, and

H_i = height of the center of mass of segment i with respect to an arbitrarily chosen reference frame.

The segmental energy method can be used to determine net increases or decreases in energy of a system of segments (the entire body) by summing the increases or decreases of each individual segment. The mathematical procedure to determine these mechanical energy components is straightforward. However, the relationship between this mechanical energy and the actual physiological or even mechanical energy expenditure is not always trivial (Aleshinsky 1986; Williams 1985; Winter 1978). It is difficult if not impossible to determine

- the transfer of mechanical energy within one segment (change of energy form),
- the transfer of mechanical energy between segments,
- the possible storage and re-use of energy, and
- the meaning of "positive" and "negative" work.

The segmental energy calculation is used to calculate the *transfer of energy* or the *energy flow* from one segment to a neighboring segment. A change of energy of one specific segment must be the result of energy produced by muscles or energy transferred from an adjacent segment. An increase of energy of a segment must be the result of a net inflow and a decrease the result of a net outflow of energy from proximal and distal muscles and joints (Winter 1978). Again, this method is mathematically straightforward. However, this method does not allow determining whether the energy flow was a result of muscle activity or energy transfer from neighboring joints and whether it occurred at the proximal or distal end. Additionally, the method does not account for energy transfer due to biarticular muscles (Bobbert and van Ingen Schenau 1988; Pandy and Zajac 1991; van Ingen Schenau and Cavanagh 1990). Consequently, the segmental energy approach can be used in determining the flow of mechanical energy. However, it cannot be used to determine the source of the energy flowing from one segment to another one (van Ingen Schenau and Cavanagh 1990; van Ingen Schenau et al. 1990).

Muscle-tendon units can *store energy*. This energy can be used at a later time. However, the segmental energy approach is not able to account for this possibly important aspect of locomotion. This may or may not be critical for human movement. It is, however, extremely critical for some forms of animal locomotion (e.g., horse, camel).

Work performed during *concentric and eccentric muscle contractions* is often referred to as "positive" and "negative" work. This approach is not in agreement with the understanding of work and energy in physics. In concentric and in eccentric force application physiological work is performed. However, the use of the concept of "negative work" suggests that energy is gained and lost in these processes, which is obviously not the case. This inappropriate use of the term "work" resulted, for example, in large discrepancies of the calculated mechanical power during running. The reported values range between 273 W and 1775 W (Williams and Cavanagh 1983).

In summary, the segmental energy method can be used to determine net increases or decreases in energy of a system of segments (the entire body) by summing the increases or decreases of each individual segment. However, it cannot be used to determine the source of the energy flowing from one segment to another one or to account for storage and re-use of energy in muscle-tendon units, and it may be misleading when calculating work during concentric and eccentric force applications.

When using the segmental energy approach one must be aware of the limitations of this approach. It has often been used inappropriately, and results from such calculations have often been overinterpreted.

Joint Power and Joint Energy Calculation

The joint power or joint energy approach was first used by Elftman (1939). He quantified joint power and energy for the ankle, knee, and hip joints during walking. The joint power and joint energy approach uses *resultant joint forces* and *resultant joint moments*. The mechanically equipollent resultant joint forces and resultant joint moments replace the (often) complicated force distributions at the distal and proximal end of a segment.

The resultant joint force and joint moment with respect to a point A are the mechanically equipollent force and moment replacing all forces with lines of action crossing the joint with respect to this point A.

The calculated resultant joint forces and moments do not correspond in any way to actual forces and moments that would be measured with appropriate measuring devices. The resultant joint forces and moments are theoretical (resultant) quantities that can be used for further calculations. The resultant joint moments have also been named *net muscle moments* (Winter 1978), *tendon moments* (Quanbury, Winter, and Reimer 1975), or *control moments* (Aleshinsky 1986).

The *total power* flowing to (+) or from (–) a segment i is defined as

$$P_{tot}^i = \sum_{j=1}^{n} \left(P_{Fj}^i + P_{Mj}^i \right) \qquad 1.16$$

where

P_{tot}^i = total power flowing to or from segment i,

P_{Fj}^i = power due to the resultant joint force at joint j with respect to segment i,

P_{Mj}^i = power due to the resultant joint moment at joint j with respect to segment i,

j = index for the various joints of the segment of interest,

n = number of joints of the segments of interest, and

i = number of segment.

The power due to the resultant joint force for the joint j for a given segment is defined as

$$P_{Fj} = \mathbf{F}_j \bullet \mathbf{v}_j \qquad 1.17$$

or

$$P_{Fj} = |\mathbf{F}_j| \cdot |\mathbf{v}_j| \cdot \cos \alpha_j = F_j \cdot v_j \cdot \cos \alpha_j \qquad 1.18$$

where

P_{Fj} = power due to the resultant joint force for the joint j,

$\mathbf{F}_j$ = resultant joint force vector at the joint j,

$\mathbf{v}_j$ = velocity vector of joint j,

α_j = angle between the force and velocity vectors, and

• = sign for scalar (dot) product.

The joint power for two adjacent segments contacting at joint j is equal in magnitude but opposite in sign (see figure 1.1). This indicates that the joint reaction force can only transfer energy between neighboring segments (Winter and Robertson 1978). A positive value of the power for one segment indicates that this power is being transferred to a segment. A negative sign indicates that the power is being transferred from this segment.

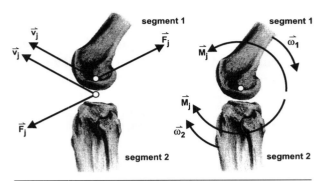

Figure 1.1 Joint reaction force and joint velocity for two adjacent segments (left) and resultant joint moment and angular velocity (right).

The power due to the resultant joint moment for the joint j for a given segment is defined as

$$P_{Mj} = M_j \cdot \omega_s \qquad 1.19$$

where

P_{Mj} = power due to the resultant joint moment for the joint j,

M_j = resultant joint moment at joint j, and

ω_s = angular velocity of the segment of interest (segment i).

Energy flows into a segment if the resultant joint moments and the segmental angular velocities act in the same direction. Energy flows out from a segment if the resultant joint moments and the segmental angular velocities act in opposite directions.

Sometimes, the energy flow is determined using the formula

$$P_j = M_j \cdot \omega_j = M_j \, (\omega_{(i+1)} - \omega_i) \qquad 1.20$$

where

P_j = mechanical power generated (+) or absorbed (–) at joint j,

M_j = resultant joint moment of joint j,

ω_j = angular velocity at the joint j due to the movement of segment i + 1 with respect to segment i,

$\omega_{(i+1)}$ = angular velocity of segment i + 1 with respect to a reference frame, and

ω_i = angular velocity of segment i with respect to a reference frame.

When the resultant joint moment and the angular velocity are in the same direction—such as during a concentric contraction—positive power results.

Positive power is also referred to as *power generation*. When the resultant joint moment and the angular velocity are in opposite directions—such as during an eccentric contraction—negative power results. Negative power is also referred to as *power absorption*.

Equation 1.19 describes information on which joints provide the mechanical energy generation or absorption during locomotion. The mechanical energy at a joint j, E_j, is determined by the time integral of the joint power:

$$E_j = \int P_j \, dt = M_j \cdot \omega_j \, dt \qquad 1.21$$

where

E_j = mechanical energy generated (+) or absorbed (–) at joint j,

P_j = mechanical power at joint j,

M_j = resultant joint moment of joint j, and

ω_j = relative angular velocity of the two segments meeting at joint j.

When concentric and eccentric contractions occur the integral must be divided into separate parts to determine the positive and negative energy (Winter 1978).

Total joint power is a result of

- muscle power (as indicated in equation 1.18),
- elastic energy stored in and released (with a time delay) from muscle-tendon units,
- heat production by eccentrically contracting muscles, and
- heat production due to internal (joints, fibers, etc.) friction.

The exact origin of positive power and the exact destination of negative power cannot be determined from the joint power calculation. Thus, joint power cannot be related to metabolic power due to muscle activity. The relationship of metabolic energy and mechanical work is not well understood. Metabolic energy does not directly produce external mechanical work. Tasks that produce a similar amount of mechanical work may require substantially differing amounts of metabolic energy and vice versa (Epstein 1994).

In summary, the joint energy calculation can be used to determine the energy generation and energy absorption in a joint. However, the joint energy calculation does not provide insight into the reason for this energy generation or absorption. When using the joint energy calculations one must be aware of the limitations of this approach. Results from joint energy calculations have often been overinterpreted.

The Laws of Conservation of Energy

Work, energy, and performance during physical activities depend on the laws of conservation of energy.

The First Law of Thermodynamics

Energetics of an animal or human system during locomotion is an extremely complex phenomenon (Epstein 1994). In most applications dealing with locomotion or physical activity the uptake of oxygen plays a major role in the necessary work and the resulting performance. Thus, work and energy considerations for human locomotion require a law that includes conservative and nonconservative forces. The first law of thermodynamics provides this general law. It describes the behavior of the internal energy U. This internal energy is composed of all possible forms of energy:

$$U = E_{therm} + E_{mech} + E_{electr} + E_{magn} \qquad 1.22$$
$$+ E_{chem} + E_{other}$$

where

U = the internal energy of the system of interest,

E_{therm} = the thermodynamic energy of the system of interest,

E_{mech} = the mechanical energy of the system of interest,

E_{electr} = the electric energy of the system of interest,

E_{magn} = the magnetic energy of the system of interest,

E_{chem} = the chemical energy of the system of interest, and

E_{other} = any other form of energy of the system of interest.

Equation 1.22 indicates that there are many different forms of energy present in a system. For the internal energy, U, the first law of thermodynamics states the following:

> The change of the internal energy of a system is (for any possible process) equal to the sum of the work and heat that is added to the system during this process.

Or as a mathematical formula,

$$\Delta U = \Delta W + \Delta Q \qquad 1.23$$

where

ΔU = change of internal energy,

ΔW = work added to or lost from the system, and

ΔQ = heat added to or lost from the system.

If the system of interest consists of the human body this equation must be expanded to account for processes such as O_2 consumption, CO_2 exhaustion, intake of nutrition, and removal of waste. In this case, the corresponding equation is

$$\Delta U = \Delta W + \Delta Q + \Delta E_{O2/CO2} + \Delta E_{nutrition} \qquad 1.24$$
$$+ \Delta E_{removal}.$$

In the human body there are many different processes that influence the internal energy. For instance, during muscle activation ATP is broken down into ADP, Ca^{2+} is released from the sarcoplasmic reticulum, and when stimulation stops, Ca^{2+} is transported back into the sarcoplasmic reticulum. All these processes are associated with exchange of energy.

The Mechanical Aspect of the Work-Energy Principle

When concentrating on the mechanical aspect of the laws of conservation of energy and the work-energy principle one uses the term *conservative forces*. Conservative forces are defined as follows:

> A force acting on a system is called conservative if the work needed to move the system from point A to point B is independent of the selected path.

The expression *conservative force* derives from the fact that energy is *conserved* when only conservative forces act on a system. Examples of conservative forces include the force of gravity or the force of an ideal linear spring (for which the force only depends on the deformation, following the law of Hook). For such conservative forces the law of conservation of energy is as follows:

> The sum of the external kinetic and the external and internal potential energy of a system is constant if the external and internal forces acting between the mass points are conservative.

Examples of nonconservative forces include the force of air resistance, the frictional forces in gases, fluids, and rigid bodies, and the frictional resistance in muscles and tendons. For nonconservative forces the mechanical energy does not remain constant as a transfer of mechanical energy into another form of energy occurs. An obvious example of such an energy transfer is the development of heat due to friction of a shoe sole with the ground during a game of tennis or squash.

The work-energy principle states (for conservative forces) the following:

> The work performed by the resultant force acting on a particle is equal to the particle's change in kinetic and potential energy from one instant in time, t_1, to a later instant in time, t_2.

The work-energy principle is a mechanical concept that may be derived from Newton's second law and can be formulated mathematically as follows:

$$W = \Delta E_{kin} + \Delta E_{pot}. \qquad 1.25$$

Assuming that the four energy forms described in equations 1.9 to 1.12 are relevant for the question at hand the equation can be written as

$$W = \Delta E_{kintr} + \Delta E_{kinrot} + \Delta E_{potgr} + \Delta E_{potspr} \qquad 1.26$$
$$W = [\tfrac{1}{2} m\, v_2{}^2 - \tfrac{1}{2} m\, v_1{}^2]$$
$$+ [\tfrac{1}{2} \cdot I \cdot \omega_2{}^2 - \tfrac{1}{2} \cdot I \cdot \omega_1{}^2]$$
$$+ [m \cdot g \cdot H_2 - m \cdot g \cdot H_1]$$
$$+ [\tfrac{1}{2} \cdot k \cdot x_2{}^2 - \tfrac{1}{2} \cdot k \cdot x_1{}^2].$$

The work-energy principle provides a different approach to solve specific practical problems. Whenever differences in displacements or velocities are known, the work-energy principle is typically better suited to solve these problems than Newton's second law.

Conservation of Energy

Energy of a system is its capacity to perform work. The law of *conservation of energy* states the following:

> If no external agent exerts force on a system, then the total energy of the system remains constant.

The law of conservation of energy is important for the understanding of nature in general and human locomotion in particular. Two examples should illustrate the importance of the law of conservation of energy for exercise and sport activities.

For the first example, consider a system consisting of the human body. An athlete is able to perform work that corresponds to about 6 MJ (Megajoules) without additional energy input into his body from the outside. For a marathon, an athlete needs about 10 MJ of energy. Consequently, a marathon runner must add energy from the outside to his body. Energy from the outside for this example can be provided through food and beverages. This is an example to illustrate that energy can be added to a system from the outside.

For the second example, consider a system consisting of a trampolinist and a trampoline (assuming that air resistance and friction can be neglected). The trampolinist jumping down into the trampoline changes potential energy into kinetic energy. This kinetic energy just before touching the trampoline is used to deform (stretch) the trampoline surface and energy is stored in the elastic fabric (energy of a spring). This energy is returned to the trampolinist and transferred into kinetic energy, which, in turn, is transferred into potential energy. The athlete can perform additional work during the contact with the trampoline (e.g., by bending and extending the knees) that will be stored in the trampoline and will be returned as additional potential energy at an increased height of the flight phase.

Examples

Further examples are added and discussed in more detail in further sections to illustrate various aspects of the concept of work and energy during exercise and sport.

Example: Application of the Work-Energy Principle for Running With and Without Shoes

The total work performed during a physical activity can be quantified by measuring the oxygen consumption of an athlete during the activity of interest (Cavanagh and Williams 1982; Fukuda, Ohmichi, and Miyashita 1983). Measurements of oxygen consumption for running barefoot and running with running shoes show typically a difference of about 4 to 5% in favor of running barefoot. Running barefoot requires less oxygen (and less work) than running with running shoes. This finding is at first glance surprising and the magnitude of the difference is substantial. If one assumes a linear relationship between the relative work and the relative time differences in a race (which is probably not completely correct), an athlete running a marathon without shoes would gain about 6 to 7 minutes over an athlete with shoes. For a sprint of 100 meters a difference of 5% would correspond to about half a second! Even if the work-time analogy is not completely correct it seems worthwhile to think about the various possibilities where this energy was lost.

Question to be answered.

Determine the additional mechanical work that an athlete wearing shoes must perform against gravity and to accelerate the additional shoe mass. (Note: These two forms of mechanical work are only part of the total additional work. There is additional work, for instance, due to the deceleration of the shoe! Furthermore, the protection aspect may have an effect on the work balance.)

Assumptions.

The following information can be assumed:

- The total (physiological) work done by a runner during a marathon is about 2000 to 2500 kcal or about 10^7 J (10 MJ).
- The additional shoe mass is $\Delta m = 100$ g. (Note that this mass is less than the mass of a running shoe. However, at the end of this section results will be presented for different shoe masses.)
- Each foot (and shoe) is lifted during each step by $\Delta H = 0.2$ m.
- The maximal speed of the swing leg during the swing phase is 10 m/s corresponding to middle or long-distance running.

- The step length (left toe to right toe) is about 2 m, which corresponds to n ≈ 20,000 steps during a marathon.

Solution.

Additional work due to gravity is

$$\Delta W_{gr} = n \cdot \Delta m \cdot g \cdot \Delta H \qquad\qquad 1.27$$
$$= 20{,}000 \cdot 0.1 \text{ kg} \cdot 10 \text{ m/s}^2 \cdot 0.2 \text{ m}$$
$$= 4{,}000 \text{ J}$$
$$\Delta W_{gr} = 0.4 \cdot 10^4 \text{ J}.$$

Additional work due to the acceleration of the additional shoe mass during each step is

$$\Delta W_{acc} = n \cdot \tfrac{1}{2} \cdot \Delta m \cdot \Delta v^2 \qquad\qquad 1.28$$
$$= 20{,}000 \cdot 0.5 \cdot 0.1 \text{ kg} \cdot 100 \text{ m}^2/\text{s}^2$$
$$= 100{,}000 \text{ J}$$
$$\Delta W_{acc} = 10^5 \text{ J}.$$

The solution leads to the following observations:

- The mechanical work to accelerate the additional shoe mass is about 25 times larger than the additional work to lift the additional mass.
- The additional work due to the additional mass of a running shoe mass of 0.1 kg corresponds to about 1% of the total physiological work. In terms of time, this additional work could correspond to about 1 to 1.5 minutes of a marathon time per 100 g additional shoe mass. The results for different masses and different maximal foot speeds are illustrated in figure 1.2.

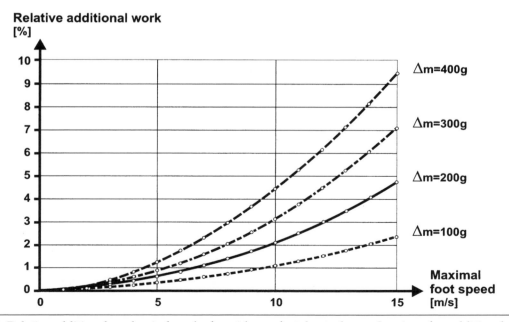

Figure 1.2 Relative additional mechanical work of an airborne foot due to the acceleration of an additional shoe mass as a function of the maximal speed for running.

- Based on these estimates one may rethink the times and victories of barefoot runners! Or one may think about shoes that mimic barefoot running.

Example: Storage and Return of Energy for a Linear Spring

The classic example of an energy-storing mechanism is a spring. If the spring is linear the force that is required to deform (stretch or compress) the spring is

$$F = k \cdot \Delta x \qquad 1.29$$

and the maximum amount of energy stored in an ideal linear spring is

$$E_{sp} = \tfrac{1}{2} k \, \Delta x^2 \qquad 1.30$$

where

E_{sp} = energy stored in the spring,

k = spring constant or spring stiffness, and

Δx = spring deformation.

Question to be answered.

Compare the ability to return energy for two elastic materials with different stiffness.

Assumptions.

The following information can be assumed:

General: F = 1,000 N, k = 10^5 N/m.

Material A: k_A = k = 10^5 N/m.

Material B: k_B = 2k = $2 \cdot 10^5$ N/m.

Solution.

Material A (softer)

$F = k_A \cdot \Delta x$. Thus, $\qquad 1.31$

$\Delta x = \dfrac{1}{k} \cdot F.$

Substituting,

$E_{sp} = \dfrac{1}{2} k \cdot \dfrac{1}{k^2} \cdot F^2$, or

$E_{sp} = \dfrac{1}{2} \cdot \dfrac{F^2}{k}.$

Numerically, E_{sp} = 5 J.

Material B (stiffer)

$F = k_B \cdot \Delta x$. Thus, $\qquad 1.32$

$\Delta x = \dfrac{1}{2k} \cdot F.$

Substituting,

$E_{sp} = \dfrac{1}{2} \cdot 2k \cdot \dfrac{1}{4k^2} \cdot F^2$, or

$E_{sp} = \dfrac{1}{4} \cdot \dfrac{F^2}{k}.$

Numerically, E_{sp} = 2.5 J.

Under the influence of the same force, the stiffer material will have half the displacement of the more compliant material. For a given force the stiffer material is able to store half the energy of the more compliant material. The energy that can be returned by a system depends on its deformation for a given force application. Thus, the more compliant the material, the greater its potential for energy storage due to the greater deformation it allows (as long as it does not bottom out!).

The relationship between the stiffness k, the deformation Δx, and the energy stored for a linear spring (see figure 1.3) shows that the possibly stored energy increases linearly with increasing stiffness. However, the amount of energy that can be stored increases quadratically with increasing deformation.

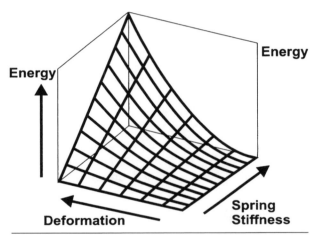

Figure 1.3 Relationship between the stiffness k, the deformation Δx, and the energy stored for a linear spring, illustrating that the possibly stored energy increases linearly with increasing stiffness.

The model of a linear elastic spring is the simplest example; however, any elastically deformable material is capable of storing energy. Examples of materials or constructions that can store and return energy during physical activities include trampolines, diving boards, poles in pole-vaulting, and athletic surfaces.

Example: Work to Deform a Spring or Energy Return From a Shoe Sole or a Sport Surface

Work to deform a spring is a form of work that is quite common in sport activities. A diver deforms the diving board; a gymnast deforms the floor for the free exercise. In both cases these athletes use the energy of the deformed board or surface to improve their jumping ability—the energy is returned to the athlete. Another example is the *tuned track* (McMahon and Greene 1979), a track that returns energy to the runners and, consequently, improves performance. Estimations of energy return for such examples are discussed in detail in the next section.

Storage of energy in a shoe sole.

In this example, the claim that sport shoe soles can return energy is discussed. During each ground contact the shoe sole is deformed, corresponding to the deformation of a spring. Ideally, one would like to return this absorbed (hopefully "stored") energy to the athlete. In order to do this correctly the following conditions must be fulfilled (Nigg and Segesser 1992):

- Energy returned must be substantial enough to make a difference.
- Energy must be returned at the right time.
- Energy must be returned with the right frequency.
- Energy must be returned at the right location.

Question to be answered.

Determine the stiffness of a shoe sole that should return 1% energy per ground contact for running and for jumping. Discuss the feasibility of the concept.

Assumptions.

The following information can be assumed:

- The magnitude of the maximally possible returned energy can be estimated assuming the behavior of a perfect linear spring. Thus,

$$W_{sp} = \tfrac{1}{2} \cdot k \cdot \Delta x^2 \qquad 1.33$$

and

$$F = k \cdot \Delta x. \qquad 1.34$$

- The work during marathon running has a strong physiological component. The total physiological work performed during a single foot contact in marathon running is about

$$W(\text{one step}) \approx \frac{W(\text{marathon})}{20,000 \text{ steps}}. \qquad 1.35$$

Therefore,

$$W(\text{one step}) \approx \frac{10^7 \text{ J}}{20,000}, \text{ or}$$

$$W(\text{one step}) \approx 500 \text{ J}.$$

- The mechanical work performed by an athlete of 80 kg mass who jumps (lifts his center of mass) 1.25 m high (for a high jump of about 2.20 m) is

$$W(\text{jump}) \approx m \cdot g \cdot \Delta H. \qquad 1.36$$

Substituting values,

$$W(\text{jump}) \approx 80 \text{ kg} \cdot 10 \text{ m/s}^2 \cdot 1.25 \text{m, or}$$

$$W(\text{jump}) \approx 1000 \text{ J}.$$

- The average ground reaction force in marathon running is about 1000 N (Cavanagh and Lafortune 1980).
- The average ground reaction force in high jumping is about 2000 N.

Solution.

Energy returned must be substantial enough to make a difference. Equations 1.33 and 1.34 provide equation 1.37.

$$k = \frac{F^2}{2 \cdot W}. \qquad 1.37$$

Using formula 1.37, one can determine the shoe sole stiffness needed for a 1% return of energy. Two examples are calculated, one for marathon running and one for high jumping.

In marathon running, the ground reaction forces acting are assumed to be 1000 N. The energy needed for one ground contact is assumed to be 5 J. Therefore,

$$k = \frac{10^6 \text{ N}^2}{2 \cdot 5 \text{ J}}, \text{ or} \qquad 1.38$$

$$k_{run} = 10^5 \text{ N/m}.$$

In high jumping, the ground reaction forces acting are assumed to be 2000 N. The energy needed for one ground contact is assumed to be 10 J. Therefore,

$$k = \frac{4 \cdot 10^6 \text{ N}^2}{2 \cdot 10 \text{ J}}, \text{ or} \qquad 1.39$$

$$k_{jump} = 8 \cdot 10^5 \text{ N/m}.$$

The mechanical sole characteristics of marathon running shoes and high jumping shoes are typically between 10^6 N/m and 10^7 N/m. The compression stiffness of 10^5 N/m corresponds to a rather soft shoe sole (i.e., a very soft running shoe). Such shoes are typically not used for competition. However, the calculations suggest that (assuming that everything else remains constant) a soft shoe sole would energetically be advantageous. The major reason that the strategy of a soft shoe sole to improve performance has not been used yet is probably the additional weight when the shoe sole becomes softer and thicker. Tubular sole construction may provide an improvement in this context.

The results indicate that a compressed soft shoe sole can ideally return about 1% of the total energy for marathon running and about 0.5% for a good high jump. This is always for the case that all the other conditions outlined above are fulfilled. For this reason these other conditions are briefly discussed.

Energy Must be Returned at the Right Time

An example of a diver may illustrate this idea. The diving board moves up and down. While moving down, energy can be stored in the board. While moving up, energy can be returned from the board to the athlete. It is obvious that the board should first be deformed (in about the first 50% of contact) and that the board can return the energy in the second half of contact.

The application of these considerations to the shoe sole example shows that the shoe sole must be deformed (energy must be stored in the shoe sole) in the first 50% of ground contact, and energy must be returned from the shoe to the athlete in the second half of the contact phase.

These considerations have implications for the actual sport activities of running and jumping. In running, during the return from the heel, let's assume that the ground contact of the foot in marathon running lasts about 200 ms and the ground contact of the heel lasts about 100 ms. Heel takeoff occurs about 100 ms after first ground contact. To act at the right time, the energy stored in the heel of the shoe should be returned between about 100 and 200 ms after first ground contact. However, this energy return from the heel occurs early in the stance phase (between about 50 to 100 ms). The energy is returned at the wrong time.

During the return from the forefoot, energy return should occur toward the end of ground contact. The return of energy to the forefoot should, therefore, occur in the second half of ground contact (between 100 and 200 ms after first contact).

In jumping, takeoff occurs from the forefoot. Contact time in jumping is about 200 ms. The "spring" (shoe sole) should expand (return energy) between 100 and 200 ms.

Energy Must be Returned With the Right Frequency

The example of the diver may again illustrate this concept: Whenever a diver prepares to dive from

the 3 m board, she works on that big wheel on the side of the board. Turning the wheel changes the support point (fulcrum) under the board. When the support point is shifted toward the front of the diving board, the board is shortened and this shortens the time of oscillation of the board by changing the natural frequency of the board. A light diver typically would move the fulcrum to the front; a heavy diver would move it to the back. The example of the diver has similarities with the example of the shoe. If a shoe should return energy with the right frequency the natural loading-unloading cycle should be exactly as long as the ground contact time. The shoe sole should store energy during the first 50% of contact and release this energy during the second 50% of ground contact.

These considerations also have implications for the actual sport activities of running and jumping. For a marathon runner with, for instance, a contact time of 250 ms, the compression of the shoe sole should occur during the first 125 ms of contact, and the release of energy should occur during the second 125 ms of contact. Thus, one half of a sine wave lasts 250 ms. This corresponds to a frequency of 2 Hz (two hertz or two full oscillations per second). Thus, the loaded natural frequency of the shoe sole for a long distance running shoe should be about 2 Hz.

For a sprinter with the contact time of 100 ms, the storage of energy should occur during the first 50 ms of ground contact and the release of energy during the second 50 ms of ground contact. This corresponds to a frequency of 5 Hz (five hertz or five full oscillations per second). Thus, the loaded natural frequency of the shoe sole for sprinting should be about 5 Hz.

For a high jumper with the contact time of 200 ms, the storage of energy should occur during the first 100 ms of ground contact and the release during the second 100 ms of ground contact. This corresponds to a frequency of 2.5 Hz. Thus, the loaded natural frequency of the shoe sole for a high-jump shoe should be about 2.5 Hz.

Energy Must be Returned at the Right Location

Compression of the heel does not help the takeoff of the forefoot with one exception: if the compression of the heel could be transferred to the forefoot in some way and released there at the right time. Construction of shoe soles that would allow this seems, however, not possible.

Energy Return in Current Sport Shoes

The natural frequency of the unloaded shoe sole is typically between 100 and 200 Hz. The natural frequency of the loaded shoe sole is not reported in the literature. A conservative estimate of the natural frequency of the loaded shoe sole using the currently available shoe construction techniques is around 20 to 30 Hz. That means that the energy is not returned with the right frequency and would typically not be returned at the right time.

In order to return energy with the correct frequency, shoe sole constructions must be found that allow this. The current constructions do not allow this. Sport surfaces are able to return energy (e.g., gymnastics surfaces or "tuned tracks") because they can be tuned to these frequencies of about 2 Hz! For sport shoes it seems not possible to achieve relevant return of energy using the conventional shoe construction materials in an open or closed cell construction. It is speculated that the only theoretical possibility to achieve some energy return is a construction that uses a tubular construction strategy. Such a construction may provide the right frequencies and has the advantage that it can be compressed to about 80% of its sole thickness, which would increase the available deformation distance and, therefore, the potentially stored and released energy.

In principle it is possible to store energy in a shoe sole by deforming the sole. In principle it is possible to return energy from a shoe sole by expanding the compressed shoe sole. However, technically there are difficulties that seem rather large. The maximal returned energy from a shoe sole one can reasonably expect is typically smaller than 1% of the energy needed for the task at hand. Also, it is technically difficult to return the energy from a shoe sole at the right time, with the right frequency, and at the right location. It is, however, much easier to fulfill the outlined requirements for energy return through a sport surface since its area-elasticity can be used to tune the return properly.

Summary

Work, energy, and performance during physical activities depend on biochemical, physiological, thermodynamic, and mechanical aspects. The performance of an athlete during a running competition depends on, among other factors, the biochemical composition of the blood providing the metabolic components for muscle work, the transformation of oxygen into muscle work, the production of heat during the running task, and the mechanical efficiency of the athlete, shoe, and surface system. Those factors themselves are influenced by the psychological state of an athlete. The resulting work, energy, and performance can be determined by defining the system of interest and by applying the laws of conservation of energy. An athlete has three major strategies available to improve the work-energy balance during locomotion: to store and return energy, to minimize the loss of energy, and to optimize muscle functions. The most important forms of mechanical work and their definitions are

$$\text{work against gravity} = \qquad\qquad 1.40$$
$$W_{gr} = m \cdot g \cdot \Delta H$$
$$\text{work to deform a (linear) spring} =$$
$$W_{sp} = \tfrac{1}{2} \cdot k \cdot \Delta x^2$$
$$\text{work to accelerate a mass (translation)} =$$
$$W_{acctr} = \tfrac{1}{2} \cdot m \cdot v^2$$
$$\text{work to accelerate a mass (rotation)} =$$
$$W_{accrot} = \tfrac{1}{2} \cdot I \cdot \omega^2.$$

Work and energy are two terms that are closely related. Energy is defined as the ability to do work. Mechanical energy can appear in different forms. The energy a mass possesses because it is in motion is called kinetic energy. This energy may have translational or rotational components. The energy a mass possesses because of its position is called potential energy. Examples of potential energy include energy due to a certain height of a mass with respect to a baseline and energy of a stretched or compressed spring. Mechanical energy of human movement has been determined using the segmental energy and the joint energy calculation.

The first law of thermodynamics, in general, governs work and energy considerations. The change of the internal energy of a system is (for any possible process) equal to the sum of the work and heat that is added to the system during this process. When concentrating on the mechanical aspects the work-energy principle governs all aspects of work and energy. It states that the work performed by the resultant force acting on a particle is equal to the particle's change in kinetic and potential energy from one instant in time, t_1, to a later instant in time, t_2. For the specific case where no external forces act on the system of interest, the total energy of the system remains constant.

References

Aleshinsky, S.Y. 1986. An energy "sources" and "fraction" approach to the mechanical energy expenditure problem. I. Basic concepts, descriptions of the model, analysis of a one-link system movement. *J. Biomech.* 19: 287–93.

Alexander, R.McN., and Bennet-Clarke, H.C. 1977. Storage of elastic strain energy in muscles and other tissues. *Nature* 265: 114–17.

Alexander, R.McN., and Vernon, A. 1975. The mechanics of hopping by kangaroos *(Macropodidae). J. Zoology (London)* 17: 265–303.

Asmussen, E., and Petersen, B. 1974. Apparent efficiency and storage of elastic energy in human muscles during exercise. *Acta Physiol. Scand.* 92: 537–45.

Bobbert, M.F., and van Ingen Schenau, G.J. 1988. Coordination in vertical jumping. *J. Biomech.* 21: 249–62.

Cavagna, G.A., Dusman, B., and Margaria, R. 1968. Positive work done by a previously stretched muscle. *J. Appl. Physiol.* 24: 21–32.

Cavagna, G.A., Thys, H., and Zamboni A. 1976. The sources of external work in level walking and running. *J. Physiol. (London).* 262:639-657.

Cavanagh, P.R., and Lafortune, M.A. 1980. Ground reaction forces in distance running. *J. Biomech.* 13: 397–406.

Cavanagh, P.R., and Williams, K.R. 1982. The effect of stride length variation on oxygen uptake during distance running. *Med. Sc. Sports Exerc.* 14: 30–35.

Clement, D.B., Taunton, J.E., Wiley, J.P., Smart, G.W., and McNicol, K.L. 1982. Investigation of metabolic efficiency in runners with and without corrective orthotic devices. *Int. J. Sports Med.* 2: 14–15.

di Prampero, P.E. 1986. The energy cost of human locomotion on land and in water. *Int. J. Sports Med.* 7: 55–72.

Elftman, H. 1939. Forces and energy changes in the leg during walking. *Am. J. Physiol.* 125: 339–56.

Epstein, M. 1994. Energy considerations. In *Biomechanics of the musculoskeletal system,* ed. B.M. Nigg and W. Herzog, 492–550. Chichester, UK: Wiley.

Fenn, W.O. 1930a. Frictional and kinetic factors in the work of sprint running. *Am. J. Physiol.* 92: 583–611.

Fenn, W.O. 1930b. Work against gravity and work due to velocity changes in running. *Am. J. Physiol.* 93: 433–62.

Frederick, E.C., Clarke, T.E., Larsen, J.L., and Cooper, L.B. 1983. The effects of shoe cushioning on the oxygen demands of running. In *Biomechanical aspects of sport shoes and playing surfaces,* ed. B.M. Nigg and B.A. Kerr, 107–14. Calgary: University Printing.

Frederick, E.C., Howley, E.T., and Powers, S.K. 1982. Lower O_2 cost while running on air cushion type shoe. *Med. Sc. Sports Exerc.* 2: 14–15.

Fukuda, H., Ohmichi, H., and Miyashita, M. 1983. Effects of shoe weight on oxygen uptake during submaximal running. In *Biomechanical aspects of sport shoes and playing surfaces,* ed. B.M. Nigg and B.A. Kerr, 115–19. Calgary: University Printing.

McMahon, T.A., and Greene, P.R. 1979. The influence of track compliance on running. *J. Biomech.* 12: 893–904.

Nigg, B.M. 1997. Impact forces in running: Current opinion. *Orthopedics* 8: 43–47.

Nigg, B.M., and Anton, M. 1995. Energy aspects for elastic and viscous shoe soles and playing surfaces. *Med. Sc. Sports Exerc.* 27: 92–97.

Nigg, B.M., and Herzog, W., ed. 1998. *Biomechanics of the musculoskeletal system.* 2d ed. Chichester, UK: Wiley.

Nigg, B.M., and Segesser, B. 1992. Biomechanical and orthopedic concepts in sport shoe construction. *Med. Sc. Sports Exerc.* 24: 595–602.

Pandy, M.G., and Zajac, F.E. 1991. Optimal muscular coordination strategies for jumping. *J. Biomech.* 24: 1–10.

Quanbury, A.O., Winter, D.A., and Reimer, G.D. 1975. Instantaneous power and power flow in body segments during walking. *J. Hum. Movement Studies* 1: 59–67.

Stefanyshyn, D.J., and Nigg, B.M. 1997. Mechanical energy contribution of the metatarsophalangeal joint to running and sprinting. *J. Biomech.* 30: 1081–85.

Stefanyshyn, D.J., and Nigg, B.M. 1998a. Dynamic angular stiffness of the ankle joint during running and sprinting. *J. Appl. Biomech.* 14: 292–99.

Stefanyshyn, D.J., and Nigg, B.M. 1998b. Contribution of the lower extremity joints to mechanical energy in running, vertical jumps, and running long jumps. *J. Sports Sciences* 16: 177–86.

van Ingen Schenau, G.J. 1984. An alternative view to the concept of utilization of elastic energy. *J. Hum. Movement Science* 3: 301–35.

van Ingen Schenau, G.J., and Cavanagh, P.R. 1990. Power equations in endurance sports. *J. Biomech.* 23: 865–81.

van Ingen Schenau, G.J., van Woensel, W.W., Boots, P.J., Snackers, R.W., and de Groot, G. 1990. Determination and interpretation of mechanical power in human movement: Application to ergometer cycling. *Eur. J. Appl. Physiol.* 61: 11–19.

Williams, K.R. 1985. The relationship between mechanical and physiological energy estimates. *Med. Sc. Sports Exerc.* 17: 317–25.

Williams, K.R., and Cavanagh, P.R. 1983. A model for the calculation of mechanical power during distance running. *J. Biomech.* 16: 115–28.

Winter, D.A. 1978. Calculation and interpretation of mechanical energy of movement. *Exerc. Sport Sc. Rev.* 6: 183–201.

Winter, D.A., and Robertson, D.G. 1978. Joint torque and energy patterns in normal gait. *Biol. Cybernetics* 29: 137–42.

Yoshihuku, Y., and Herzog, W. 1990. Optimal design parameters of the bicycle-rider system for maximal muscle power output. *J. Biomech.* 23: 1069–79.

Chapter 2

Storage and Release of Elastic Energy in the Locomotor System and the Stretch-Shortening Cycle

R.McN. Alexander

This chapter describes the elastic properties of tendons and ligaments and shows how these properties are important for human movement. A stretched spring (or a stretched tendon or ligament) stores elastic strain energy, which can be returned in an elastic recoil. We see how energy stored in stretched tendons and ligaments is released very rapidly in the final stages of takeoff for a jump, improving jumping performance by the principle of the catapult. We also see how, in running, energy is stored in tendons and ligaments at each footstep and returned at takeoff for the next step, so that energy is carried forward from one step to the next by the principle of the bouncing ball. This reduces the work that the muscles have to do to maintain the same speed in successive strides and so saves metabolic energy, enabling us to run at much less energy cost than if we had no springs in our bodies. Finally, we see how the elastic properties of the fatty pads under our heels (and of the heels of running shoes) cushion impacts of the heel with the ground.

Principles of Elasticity

Figure 2.1 is designed to explain some of the basic concepts of elasticity. Figure 2.1a shows a cylinder of an elastic material, of length l_0 and cross-sectional area A_0. In figure 2.1b, forces F act on the ends of the cylinder, stretching it to a greater length l. Stretching it has made it more slender, and its cross-sectional area is reduced to A. The greater the force F, the more the cylinder is stretched, and figure 2.1c is a graph of the force against the extension. The graph is shown as a straight line, though graphs of force against extension for real biological materials are curved, as we will see shortly. In this ideal case, in which the graph is straight

$$F = k(l - l_0). \qquad 2.1$$

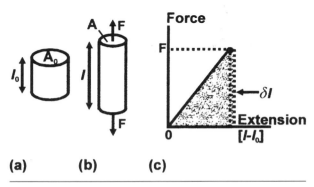

Figure 2.1 A cylinder of an elastic material (a) that is stretched (b) by a force F. (c) A graph of this force against the extension of the cylinder.

Reprinted with the permission of Cambridge University Press from R.McN. Alexander, 1988, *Elastic mechanisms in animal movement*.

19

The slope k of the graph is the stiffness of the cylinder, and the reciprocal of the slope (1/k) is the compliance.

Suppose (still looking at figure 2.1c) that the cylinder were stretched a tiny bit more, by an additional amount δl. The work needed for this extra stretch would be the force (which increases by only a tiny amount if δl is very small) multiplied by the extra stretch; it is $F \times \delta l$. This equals the area of the narrow strip of height F and width δl, at the right-hand edge of the graph. By an extension of this argument, the stippled area under the graph equals the work done in stretching the cylinder from its initial length l_0 to length l. This method of calculating the work takes account of the fact that the force increases gradually as the cylinder is stretched. In this ideal case in which the graph of force against extension is a straight line, the stippled area is a triangle and the work W needed to stretch the cylinder from its initial length l_0 to length l is given by

$$W = \tfrac{1}{2}F \cdot (l - l_0) \qquad\qquad 2.2$$

(see also chapter 3).

The elastic properties of a piece of material depend on its dimensions and on the material of which it is made. Not only is a strip of steel stretched less than a strip of rubber of the same dimensions by the same force, but a long strip of rubber stretches more than a short one, and a stout strip of rubber stretches less than a slender one. These effects are handled by means of the concepts of tensile stress, tensile strain, and Young's modulus, which will now be explained.

The *tensile stress* in a specimen that is being stretched is the force divided by the cross-sectional area. In figure 2.1b, the *true tensile stress* is F/A, but this definition is an inconvenient one to use because additional measurements are needed to find out how much more slender the specimen gets as it is stretched. It is much more convenient to divide the force by the initial cross-sectional area, obtaining the *nominal tensile stress* F/A_0. In any case, there is very little difference between the true and nominal stresses if the specimen is stretched only a little. Tendon can be stretched by no more than about 10% before it breaks (see later discussion), and such small extensions have little effect on its cross-sectional area. Stress is measured in newtons per square meter (N/m^2), sometimes called pascals (Pa).

The *nominal tensile strain* in a specimen that is being stretched is the extension divided by the initial length, $(l-l_0)/l_0$. There is no need, in this book, to explain how true tensile strain is defined (and you would be amazed at the explanation, unless

you have a subtle mathematical mind). Strain is the ratio of one length to another, so it has no units.

Two different-sized specimens of the same material, subjected to the same tensile stress, suffer the same tensile strain. *Young's modulus* is (tensile stress/tensile strain) and is the same for all specimens of a given material, irrespective of their size. It is large (of the order of $10^{11}\ N/m^2$) for steel and much smaller (of the order of $10^6\ N/m^2$) for rubber.

Too large a force will break a specimen. The tensile stress at which a material breaks is the *ultimate tensile strength*.

The Properties of Body Parts

After that outline of basic principles, we are ready to consider some real materials, starting with tendons. Tendons taken from a cadaver or a dead animal seem inextensible when handled, but they are stretched appreciably by the large forces that act on them in the living body. Figure 2.2a shows how the elastic properties of tendon have been investigated using a dynamic testing machine. This is the type of machine that engineers use to measure the strength and compliance of engineering materials such as metals and plastics. The machine has a heavy metal frame, which is necessary to withstand the large forces that it can exert. At the top is a load cell, an electrical device that measures force. At the bottom is an actuator, which can be made to move up and down by hydraulic power. The tendon that is to be tested is held at its ends by clamps, one connected to the load cell and the other to the actuator. The actuator is driven down and up, stretching the tendon and allowing it to recoil. Electrical outputs from the machine indicate the force and the actuator position and can be used to generate graphs of force against extension.

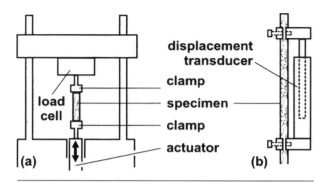

Figure 2.2 Diagrams of *(a)* a dynamic testing machine and *(b)* an extensometer.

Reprinted with the permission of Cambridge University Press from R.McN. Alexander, 1988, *Elastic mechanisms in animal movement.*

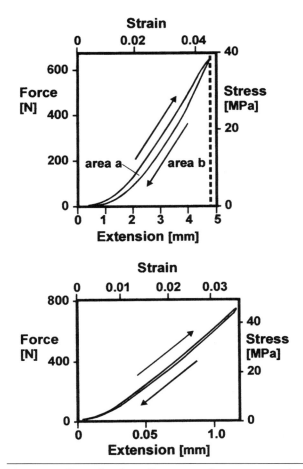

Figure 2.3 Records of tensile tests on the gastrocnemius tendon of a wallaby. Results obtained by using (*a*) the position of the actuator, and (*b*) an extensometer to show length changes.

Reprinted with the permission of Cambridge University Press from R.McN. Alexander, 1988, *Elastic mechanisms in animal movement.*

Figure 2.3a shows the result of an experiment on a tendon from the leg of a wallaby, but very similar results are obtained with human tendon. The actuator was moving up and down sinusoidally with a frequency of 2.2 cycles per second, approximately the frequency of the wallaby's hopping strides. The peak force of 700 N is approximately equal to the peak force that would have acted on the tendon in a typical hop. The machine was allowed to run for a few seconds, stretching the tendon and allowing it to recoil, before the record was taken; this was done because tendons and other biological materials take a few cycles of stretching and recoil before reaching a steady state. Thus the test simulated what would have happened to the tendon when the living animal was hopping.

The unstretched length of the tendon (between the clamps) was measured so the strain could be calculated from the extension: it is shown along the top of the graph. The cross-sectional area of the tendon had also been measured by an indirect but convenient method. A measured length of the tendon was weighed so that cross-sectional area could be calculated as (mass)/(length × density). The density of tendon is 1120 kg/m³ (Ker 1981). Stress was calculated by dividing the force by the cross-sectional area and is shown on the right-hand edge of the graph. There are two major differences between this empirical graph and the schematic one shown in figure 2.1c: the lines are not straight, and there is a loop rather than a single line because the force is slightly less for any given extension when the tendon is recoiling than when it is being stretched.

Areas under the graph represent work, as in figure 2.1c. The work the machine did to stretch the tendon is represented by the area under the upper line, that is, by the area (*a+b*). The work the tendon did as it shortened on the machine is represented by the area *b* under the lower line. Thus the area of the loop *a* represents energy lost (as heat) in the cycle of stretching and recoil.

The method used to obtain figure 2.3a was simple but had serious shortcomings. Each clamp gripped the tendon over a length of about 20 mm so there is some uncertainty about the length of tendon being stretched, making it difficult to calculate strain confidently. Energy was lost by slight movement of the regions of the tendon gripped within the clamps, so measured energy losses (area A in figure 2.3a) will tend to be too large. Also, to hold the tendon firmly enough, the clamps had to be done up so tightly that the tendon was seriously distorted in and near the clamps, resulting in uneven stresses there; consequently, the tendon may break at lower loads than if it were stressed evenly. These sources of error can be avoided by using an extensometer (see figure 2.2b). This is a displacement transducer (an instrument that measures length changes) attached to a short length of the tendon, well away from the clamps. The extensometer's own clamps grip the tendon only lightly, which is adequate because no large forces have to be transmitted between tendon and extensometer. Figure 2.3b is a record of an experiment on the same tendon as figure 2.3a, showing the force sensed by the load cell plotted against the length changes sensed by the extensometer. The length changes are much smaller than in figure 2.3a, although the forces are about the same, because only the length within the extensometer was being recorded. Notice that the loop is relatively narrower than in figure 2.3a, showing that energy losses were relatively smaller.

In figure 2.1c, Young's modulus is stress/strain, equal to the gradient of the graph. Notice that the lines in figure 2.3b are curved at strains below 0.01 and more or less straight at higher strains. The figure usually given as Young's modulus for tendon is the gradient of the straight part of the graph. It has been determined for tendons from the arms, legs, and tails of humans and various other mammals, including wallabies, sheep, deer, camels, dogs, cats, monkeys, and dolphins. It generally lies in the range 1.0 to 1.5 GPa ($1.0–1.5 \times 10^9$ N/m^2) (Bennett et al. 1986; Pollock and Shadwick 1994). The ultimate tensile strength of tendon is difficult to measure because uneven stresses in the clamps are apt to result in failure, starting where the stress is locally high, although the stress in the main body of the tendon may be much lower. Like Young's modulus, ultimate tensile stress seems to be uniform for tendons from all parts of the body of all mammals, at least 100 MPa (1.0×10^8 N/m^2) (Bennett et al. 1986; Wang and Ker [1995] obtained a mean value of 140 MPa for wallaby tail tendon). Tendon can be stretched to a strain of about 0.10 before it breaks.

Another property of tendon that is important for studies of human movement is *energy dissipation*. Remember that figure 2.3a shows work ($a + b$) being used stretching the tendon, a being lost and only b being returned in its elastic recoil. Energy dissipation is defined as $a/(a + b)$, that is, as the energy lost as a fraction of the work done stretching the tendon. It lies in the range 0.05 to 0.12 for most of the mammalian tendons for which reliable measurements have been made (Bennett et al. 1986; Pollock and Shadwick 1994). The same information is sometimes conveyed by giving the *percentage recovery*, $100\,b/(a + b)$, which is 88 to 95% in the case of tendon.

It is important for tendon function that its energy dissipation should be low. Tendons would work less well as energy-saving springs in running, if less energy were returned in their elastic recoil and more dissipated as heat. Also, the heat raises the temperature of the tendon and may damage it (Birch, Wilson, and Goodship 1997). Temperatures up to 45°C have been measured in leg tendons of galloping horses.

All the tests described above have been performed on external tendons (see figure 2.4), tendons that are clear of the bellies of their muscles. However, tendons generally continue as aponeuroses on the surfaces of muscle bellies or (in the case of multipennate muscles) as tendon sheets within the muscle belly. When the muscle and tendon exert

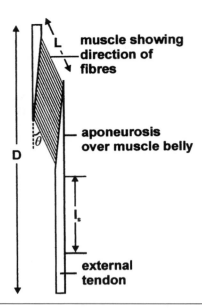

Figure 2.4 A diagram of a muscle and its tendons.
Reprinted, by permission, from R.F. Ker, R.McN. Alexander, and M.B. Bennett, 1988, "Why are mammalian tendons so thick?" *Journal of Zoology* (Cambridge University Press) 216:309-324.

force, the aponeuroses and internal tendons are stretched and store strain energy, as well as the external tendons. Figure 2.4 shows that, for a typical muscle with fibers of uniform length, the force exerted by every muscle fiber is routed through the same total length of tendon. Using the symbols shown in the figure, this is (D – L cos θ). The angle of pennation (θ) of mammalian muscles is generally less than 30°, so cos θ is generally between 0.87 and 1.00, close enough to 1 for the effective length of tendon to be estimated as (D – L) for most practical purposes. For example, the overall length D of the human gastrocnemius muscle is about 450 mm, and the muscle fibers are around 50 mm long, so the effective length of tendon is 400 mm (Yamaguchi et al. 1990). We will see later that the peak stress in this tendon in running is about 50 MPa. This would cause a strain of 0.04 (see figure 2.3), making the tendon 16 mm longer. Though they seem inextensible when handled, tendons can stretch enough to allow appreciable joint movement.

As well as the tendons, the muscle fibers have elastic properties, but these seem to be much less important for the functions described in this chapter than the elastic properties of tendons (Alexander and Bennet-Clark 1977). If a muscle fiber that is contracting isometrically (holding constant length) is suddenly allowed to shorten, its tension falls, just as the tension in a stretched rubber band falls if it is allowed to shorten (but the tension in the muscle fiber recovers if it is held at its new length for a while, whereas the tension in a rubber band does not). To make the tension in the muscle fiber fall briefly to zero, its length must be reduced by at least 1.3%. This shows that when the fiber is contracting isometrically, it is stretched elastically by 1.3% (to a strain of 0.013). There is evidence that the elastic compliance is mainly in the cross bridges (Huxley and Simmons 1971). Muscle fibers can exert larger forces when being forcibly stretched than when contracting isometrically, and Flitney and Hirst (1978) showed that in these circumstances elastic strains up to 3% can occur. Muscles can, of course, be stretched by much more than 3%, but when they are, cross bridges detach and reattach. When the cross bridges detach, any strain energy stored in them is lost.

This tells us that 50 mm muscle fibers in the gastrocnemius can stretch elastically by up to 1.5 mm, and the shorter muscle fibers of the soleus will stretch less. We have already estimated that the Achilles tendon, the tendon shared by these muscles, stretches elastically in running by 16 mm. A muscle and its tendon exert equal forces, so by equation 2.2 the strain energies stored in them are proportional to their extensions, and far more energy can be stored in the Achilles tendon than in its muscles. The quadriceps muscles have longer muscle fibers and shorter tendons than gastrocnemius and soleus, but their tendons are nevertheless much longer than their muscle fibers (Yamaguchi et al. 1990), and more strain energy is probably stored in the tendons than in the muscle fibers. Some of the other thigh muscles have muscle fibers that are longer than their tendons (Yamaguchi et al. 1990), and they may store more strain energy than their tendons, but the possible importance for running of strain energy storage in these muscles does not seem to have been assessed. More generally, there seems to be little evidence that muscle fibers are important strain energy stores in humans or in other mammals.

Though tendon elasticity seems to be more important than muscle elasticity, tendon is not the only material whose elastic properties are important for human movement. We will have to consider the ligaments of the arch of the foot and the fatty pad under the heel.

Films of barefoot runners show that the arch of the foot is perceptibly flattened by the forces that act when the foot is on the ground. Alexander (1987) compares stills from film of a man resting his foot lightly on the ground (with most of his weight on the other foot) and the same man at the stage of a running stride at which the force on the ground is

greatest. The ankle is 10 mm nearer the ground in the latter picture. Ker and coworkers (1987) set out to discover whether this flattening of the foot under load was an elastic phenomenon.

Figure 2.5a shows forces on the foot of a 70 kg man at the stage of a running stride when these forces are highest, based on force plate records with simultaneous film. The force records show that at this stage the ground exerts on the foot a force of about 2.7 times body weight (in this case, a force of 1.9 kN), which acts vertically upwards. The force is distributed over the whole area of contact of the foot with the ground, but the records show that its center of pressure is on the ball of the foot, as indicated by the arrow. The moment of this force about the ankle must be balanced by the force in the Achilles tendon, which is calculated by the principle of levers to be 4.7 kN. Vector addition of these forces shows that the reaction at the ankle joint must be 6.4 kN, in the direction shown. (The weight of the foot and inertial forces acting on it have been ignored in this analysis because they are much smaller than the forces shown.) Thus two upward forces and one downward force act on the foot, flattening the arch of the foot and stretching its ligaments. The ligaments in question are the plantar aponeurosis, the long and short plantar ligaments, and the spring ligament.

Ker and coworkers (1987) performed the experiment shown in figure 2.5b to investigate the elastic properties of amputated feet. It simulates the pattern of forces that act in running. The foot is mounted in the dynamic testing machine used for the tests on tendon, but, whereas the tendons were stretched, the foot is compressed. The ball of the foot rests on a steel block that pushes on it when the actuator is raised, simulating the force exerted on the living foot by the ground. The fatty heel pad has been dissected off, exposing the calcaneus, which rests on another steel block that pushes upward on it, simulating the upward pull of the Achilles tendon. Both steel blocks are mounted on rollers to accommodate the slight lengthening of the foot that occurs when the arch flattens. The tibia is attached to the load cell, which pushes down on it, simulating the reaction at the ankle joint. The actuator is moved up and down through a range of 10 mm (corresponding to the degree of compression seen in the living runner's foot) at a frequency chosen to simulate running.

A typical record from an experiment is shown in figure 2.5c. It shows that the foot behaved as a spring, being compressed and recoiling as the actuator moved up and down. However, the loop formed by the graph is relatively wider than the loop in the tendon graph (see figure 2.3b). This tells

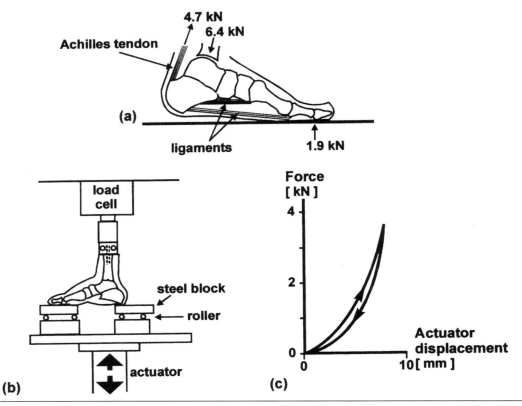

Figure 2.5 *(a)* A diagram of a foot, showing the peak forces that act on it in running. *(b)* An experiment in which the forces that would act on a foot in running are simulated, and *(c)* a record from the experiment.

Reprinted, by permission, from R.McN. Alexander, 1990b. "The spring in your step," *Proceedings of the Royal Institution.* 62:1-14.

us that the foot (at least as mounted for this experiment) is a less good spring than tendon, dissipating as heat a larger proportion of the work done on it. The energy dissipation was found to be 0.22, three times as high as for tendon. Experiments in which the ligaments of the arch of the foot were cut in turn showed that all of them play a part in giving the foot its springlike properties.

The feet used in the experiments described by Ker and coworkers (1987) had been amputated on account of irreparable vascular disease, and it was feared that the disease might have made them deteriorate. However, we were later able to repeat the experiment on a healthy foot that had had to be amputated after a traffic accident. There was no apparent difference in elastic properties between this foot and the feet with vascular disease.

Now we will estimate how much elastic strain energy is stored in the Achilles tendon and foot at the instant shown in figure 2.5a. The cross-sectional area of the tendon in adult men is about 90 mm², so the 4.7 kN force on it would exert a stress of 50 MPa. From measurements of the areas under the graphs in figure 2.3, it can be calculated that a tendon with

the dimensions of the Achilles tendon, stressed to 50 MPa, would store 35 joules strain energy. Similarly, from figure 2.5c, it can be calculated that the forces shown in figure 2.5a would store 17 J strain energy in the ligaments of the arch of the foot.

Next, we consider the properties of the heel pad. This is a pad of fatty tissue and collagen fibers, interposed between the calcaneus and the sole of the foot. Its properties have been investigated in two different ways, which gave very different results. Bennett and Ker (1990) removed heel pads from amputated feet and squeezed them in their dynamic testing machine, with the actuator moving up and down sinusoidally. These experiments gave energy dissipations of about 0.3. Other investigators (see review, Aerts and De Clercq 1993) performed experiments in which a heavy pendulum struck the heel of a living subject. The height to which the pendulum rose on its rebound was measured and used to calculate the energy dissipation. These experiments gave dissipations ranging from 0.76 to 0.95, far higher than those measured with the dynamic testing machine. The discrepancy was resolved by Aerts and associates (1995), who performed both kinds of experiments on the same isolated heel pads. Three principal effects combined to explain it.

• Pendulum experiments with isolated heel pads gave mean dissipations (after problems with an insufficiently rigid support had been eliminated) of 0.66, substantially less than in the experiments with living subjects. The difference seems to be due to energy dissipation in other parts of the subjects' legs, as well as in the heel pad. Such losses seem inevitable, as a living subject's foot cannot be rigidly fixed.

• Much more energy is dissipated in the first cycle of a test in a dynamic testing machine than in later cycles. Mean dissipation fell from 0.48 in the first cycle to 0.32 after 10 or more cycles, by which time a steady state had been attained. The first cycle is the one that should be compared to pendulum tests, but steady state results are more relevant to running.

• Energy is lost in pendulum tests on irregularly shaped specimens such as heel pads, due to vibrations superimposed on the planar swing of the rebounding pendulum.

From this we may conclude that after the first few strides, during which a runner's heels settle down to a steady state, the heel pads behave as viscoelastic pads with an energy dissipation of about 0.3. The peak force on the heel at impact with the ground in barefoot running is 1.5 to 2 kN, compressing the heel by about 9 mm (De Clercq, Aerts, and Kunnen 1994). The strain energy stored in the pad is much less than the 7 to 9 J calculated by applying equation 2.2 to these data because the pad is a highly nonlinear spring.

The iliosacral joint and the vertebral column also have elastic compliances, but rough calculations by Alexander (1997a) indicate that the quantities of strain energy that they store in running are too small to be important.

In addition to the elastic compliances of the body itself, the compliances of shoes and of the ground surface may have some significance for human movement. They are discussed in chapter 4.

Springs Functioning as Catapults

When a child uses a catapult, he or she stretches the rubber, doing work on it quite slowly. When released, it recoils very rapidly, projecting the missile much faster than the unaided hand could do. The energy returned in the recoil equals the work done to stretch the rubber (less a small fraction dissipated as heat), but it is returned at a greatly increased rate. In other words, the catapult amplifies power, the rate of doing work. This section shows how tendons and other elastic structures in the body function as power amplifiers in actions such as jumping.

The faster a muscle fiber is shortening, the less force it can exert, until at its maximum shortening speed (v_{max}) it exerts no force at all (see chapter 11). An individual muscle may be composed of fibers with a wide range of values of v_{max}, for example, from 0.3 to 3 fiber lengths per second in the soleus muscle of a horse (Rome, Sosnicki, and Goble 1990), but every muscle has a maximum shortening speed determined by the properties of its fastest fibers. In contrast, the elastic recoil of an unloaded tendon can be so fast as to impose no practical limit on the speed of human movement. If a muscle stretches its tendon in the early part of a movement starting from rest, and the tendon recoils in the later part, the final speed may be greater than would have been possible by muscle shortening alone.

Jumping is the most thoroughly studied example of this effect. The principles involved have been illustrated by means of the simple mathematical model shown in figure 2.6a, which was designed to be applicable to jumpers of all sizes, from fleas to humans (Alexander 1995). It consists of a rigid trunk and two, two-segment legs. The jump is powered by knee extensor muscles that have realistic force-velocity properties and are connected to the skeleton by compliant tendons. The model is started from rest and jumps vertically by extending its knees as rapidly as possible. The model can be made to represent animals of different sizes with different body proportions and muscle properties. For the simulations illustrated, the parameters of the model were adjusted to represent an adult human.

Figure 2.6 shows a simulation of a squat jump, that is a jump that starts from rest with the knees bent. Initially the muscles are inactive and the knee is prevented from bending further by a passive stop. The ground force equals body weight. At time zero the muscles are fully activated and begin to shorten, stretching their tendons and extending the knees. In the early stages the knees are extending very slowly, but the tendons are being stretched as force builds up. The muscle fibers have to shorten quite fast to stretch the tendons and so exert forces well below their isometric force (the force they could exert if they were prevented from shortening). As movement continues, the tendons stretch more slowly, but the knees extend at an increasing rate. Muscle force peaks without reaching the isometric level

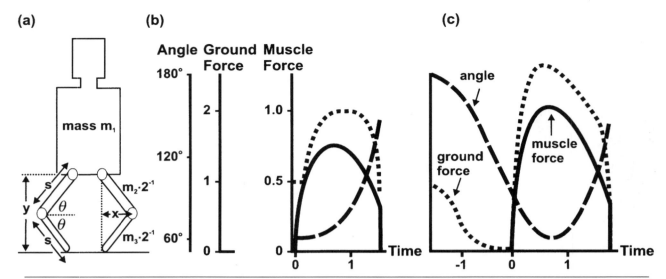

Figure 2.6 *(a)* A simple model of jumping. *(b)* A simulated squat jump and *(c)* a simulated countermovement jump by the model. Knee angle, ground force/body weight, and muscle force/isometric force are plotted against a dimensionless time parameter.

Reprinted, by permission, from R.McN Alexander, 1995. "Leg design and jumping technique for humans, other vertebrates and insects," *Philosophical Transactions of the Royal Society* B347:235-248.

because the muscle fibers are still shortening. As the rate of extension of the knees increases, the muscle fibers shorten faster and exert less force, allowing the tendons to recoil. Eventually the feet lose contact with the ground, at which stage the knees are extending very rapidly under the combined influence of muscle shortening and tendon recoil. Just before takeoff in the example shown, the muscle is exerting twice as much force as it could do at the same angular velocity of the knee if there were no tendon compliance (0.4 times its isometric force instead of 0.2).

The computer program that generated the graph continues after the feet have left the ground, following the movement of the model and calculating how high its trunk rises. In the case illustrated, the hip joints rose to a height of 2.59 s (s is leg segment length; see figure 2.6a). When the simulation was repeated with no tendon compliance the jump was less high, to only 2.44 s. The power amplifying effect of tendon compliance increases the height of the jump.

The squat jump is not the best possible technique. Even more effective use can be made of tendon compliance by making a countermovement. Whereas a squat jump starts with the athlete stationary with knees bent, a countermovement jump starts from a standing position. The knees bend and immediately reextend. Komi and Bosco (1978) found that healthy young men could jump 50 mm higher in countermovement jumps than in squat jumps.

Figure 2.6c shows a simulation of a countermovement jump. From the starting position with the knees almost straight, the trunk is allowed to fall freely with the muscles relaxed. (The force on the ground does not quite fall to zero because of the forces required to give angular acceleration to the leg segments.) At time zero, the muscles are activated and the fall of the trunk is halted and reversed. (The time for which the trunk was allowed to fall before the muscles were activated was chosen to make the minimum knee angle match the angle from which the squat jump started.) When the muscles are first activated, they are stretched rapidly by the falling body, raising the force they exert above the isometric level and stretching the tendons more than in a squat jump. Consequently, when the body starts to rise, there is more strain energy stored in the tendons than at the corresponding stage of a squat jump. This enables the model to jump higher than it did in the squat jump, raising the hip joints to a height above the ground of 2.68 segment lengths compared to 2.59 in the squat jump. The increased height of the jump is dependent on the elastic compliance of the tendons; with no elastic compliance the model rises to the same height in both styles of jump.

However, this simple mathematical model omits a property of muscle that is important in this context—stretch activation. Not only does stretching an activated muscle increase the force that the muscle exerts while being stretched, but it has a force-enhancing effect that persists for a short while after stretching has ceased. This effect is not easily distinguished from the effect of tendon compliance in experiments with intact muscles (Cavagna and Citterio 1974), but it has been demonstrated clearly in experiments with isolated fibers (Cavagna et al. 1985; Edman, Elzinga, Noble 1982). It must contribute to the advantage of the countermovement jump.

Bobbert and coworkers (1996) stressed another effect that is probably important. A countermovement increases the time available for muscle activation, enabling the muscles to be fully activated before knee extension starts.

Thus the advantage of performing a countermovement before a standing jump seems to depend partly on the power-amplifying effect of elastic compliance in series with the muscle fibers, partly on stretch activation, and partly on the time available for muscle activation.

Springs Saving Energy

This section shows how the springlike structures we have found in the legs save energy in running by the principle of the bouncing ball. A falling ball has kinetic energy that it loses when it is halted by hitting the ground. The force of the impact deforms it, storing elastic strain energy in it; the kinetic energy is converted to strain energy. In the elastic recoil, the strain energy is converted back to kinetic energy, and the ball leaves the ground with, ideally, the speed it had when it first hit the ground. Once it had been thrown, a perfectly elastic ball in a frictionless world would go on bouncing forever, with no need for any fresh input of energy.

Force plate records of human running (or of kangaroo hopping; see figure 2.7) show that, whenever a foot is on the ground, the force it exerts is more or less aligned with the leg. When the foot is first set down it is in front of the trunk, and the force on it slopes in such a way as to decelerate the body. Later in the step, the foot is behind the trunk and the force on it slopes in such a way as to accelerate the body. The body slows down and speeds up again in each step, and its kinetic energy is at a minimum as the trunk passes over the supporting foot. Also, running is a series of leaps; the body is rising when a foot leaves the ground and falling when the other foot hits the ground. The body is lowest and has least gravitational potential energy at the stage when the trunk passes over the supporting foot. Kinetic and gravitational potential energy pass through minima simultaneously.

The force exerted by the ground on a runner's foot is greatest at midstance. The moments about the knee and ankle and, consequently, the forces that the triceps surae and quadriceps muscles must exert, are also greatest at this stage (for joint moments calculated for human runners, see McCaw and DeVita 1995; for measurements of triceps surae forces in hopping wallabies, see Biewener and Baudinette 1995). As a result, the strain energy stored in the tendons of these muscles is greatest at the stage of the stride when kinetic energy plus gravitational potential energy is least. The deformation of the arch of the foot, and so the strain energy stored in its ligaments, is also greatest at this stage of the stride, when the forces on the foot are at their maxima.

For a 70 kg runner at a middle distance speed, the kinetic energy plus gravitational potential energy lost and regained at each foot-fall is about 100 J (see Ker et al. 1987). If this energy were removed by muscles acting as brakes, with no elastic tendons, it

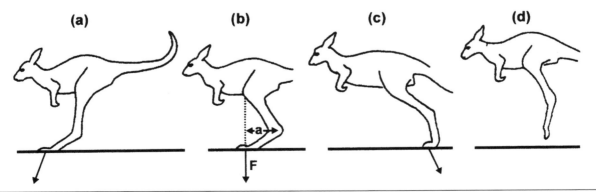

Figure 2.7 Force directions and kinetic and potential energy fluctuations are similar in human running, in kangaroo hopping, and in the bouncing of a ball. These outlines were traced from a film of a kangaroo hopping. The arrows show the directions of the forces exerted by the feet on the ground.

would be degraded to heat and lost, and the muscles would have to do work to restore the lost energy. However, about half of the energy is stored as strain energy and returned in an elastic recoil, an estimated 35 J in the Achilles tendon and 17 J in the ligaments of the arch of the foot. Only half of the kinetic plus potential energy has to be supplied afresh by muscular work. Indeed, less work even than this may be needed; we have not estimated the savings that must occur by storage of strain energy in the tendons of the quadriceps muscles. There will be a further small saving if the runner wears shoes with compliant soles, such as trainers (see chapter 4).

Muscles use metabolic energy at a low rate all the time to maintain the processes of life. They use additional metabolic energy not only when they shorten, doing work, but also when they contract isometrically without shortening, and so without doing work. (Remember that work is force times distance moved; if the point of application of the force does not move, no work has been done.) It has been argued that the metabolic energy cost of running is primarily the cost of exerting force, rather than of doing work (Kram and Taylor 1990). Compliant tendons do not affect the forces our leg muscles have to exert in running, so if metabolic cost depended only on force, the tendon's effect in reducing the work the muscles have to do actually would not save any metabolic energy. However, the metabolic rate of a muscle does not depend only on the force but also on the rate of shortening and so of doing work (Woledge, Curtin, and Homsher 1985). By means of a simple mathematical model that took account of this, Alexander (1997b) confirmed the potential of elastic mechanisms for saving energy. His model was not specifically of running but of oscillatory movements in general, including animal flight and swimming as well as running.

Springs Cushioning Impacts

It is much less dangerous to drive a car into a pile of mattresses than into a brick wall. The reason is that the mattresses will decelerate you over a longer distance, with less force. Similarly, the fatty pad under the calcaneus and the compliant heels of running shoes cushion the impact of the foot with the ground (Valiant 1990).

Conclusion

The elastic properties of tendon and of the ligaments of the arch of the foot play an important role in human movement, much more important than the elastic properties of muscle. The basis of this role is the capacity to store mechanical energy and to return it in an elastic recoil. We exploit this capacity in jumping, using the principle of the catapult: energy that has been stored by relatively slow muscle contraction is released in a very rapid elastic recoil to enable us to attain high speeds at takeoff. Muscle forces and energy storage can be enhanced by a countermovement, but this is only one of several effects that give countermovement jumping an advantage over squat jumping. We also exploit the energy storing capacity of tendons and foot ligaments in running, when they enable us to carry mechanical energy forward from one step to the next, by the principle of the bouncing ball. Thus metabolic energy is saved. Another important function is served by the elastic properties of the fatty pad under the heel, which cushions impacts of the foot with the ground.

References

Aerts, P., and De Clercq, D. 1993. Deformation characteristics of the heel region of the shod foot during a simulated heel strike: The effect of varying midsole hardness. *J. Sport Sci.* 11: 449–61.

Aerts, P., Ker, R.F., De Clercq, D., Ilsley, D.W., and Alexander, R.McN. 1995. The mechanical properties of the human heel pad: A paradox resolved. *J. Biomech.* 28: 1299–308.

Alexander, R.McN. 1987. The spring in your step. *New Scientist* 114(1558): 42–44.

Alexander, R.McN. 1988. *Elastic mechanisms in animal movement.* Cambridge: Cambridge University Press.

Alexander, R.McN. 1990a. *Animals.* Cambridge: Cambridge University Press.

Alexander, R.McN. 1990b. The spring in your step. *Proc. Roy. Inst.* 62: 1–14.

Alexander, R.McN. 1995. Leg design and jumping technique for humans, other vertebrates, and insects. *Phil. Trans. Roy. Soc.* B347: 235–48.

Alexander, R.McN. 1997a. Elasticity in human and animal backs. In *Movement stability and low back pain,* ed. A. Vleeming, V. Mooney, T. Dorman, C. Snijders, and R. Stoeckart, 227–30. New York: Churchill Livingstone.

Alexander, R.McN. 1997b. Optimum muscle design for oscillatory movements. *J. Theor. Biol.* 184: 253–59.

Alexander, R.McN., and Bennet-Clark, H.C. 1977. Storage of elastic strain energy in muscle and other tissues. *Nature* 265: 114–17.

Bennett, M.B., and Ker, R.F. 1990. The mechanical properties of the human subcalcaneal fat pad in compression. *J. Anat.* 171: 131–38.

Bennett, M.B., Ker, R.F., Dimery, N.J., and Alexander, R.McN. 1986. Mechanical properties of various mammalian tendons. *J. Zool. London A* 209: 537–48.

Biewener, A.A., and Baudinette, R.V. 1995. In vivo muscle force and elastic energy storage during steady speed hopping of tammar wallabies *(Macropus eugenii). J. Exp. Biol.* 198: 1829–41.

Birch, H.L., Wilson, A.M., and Goodship, A.E. 1997. The effect of exercise-induced localized hyperthermia on tendon cell survival. *J. Exp. Biol.* 200: 1703–8.

Bobbert, M.F., Gerritsen, K.G.M., Litjens, M.C.A., and van Soest, A.J. 1996. Why is countermovement jump height greater than squat jump height? *Med. Sci. Sports Exerc.* 28: 1402–12.

Cavagna, G.A., and Citterio, G. 1974. Effect of stretching on the elastic characteristics and the contractile component of frog striated muscle. *J. Physiol.* 239:1–14.

Cavagna, G.A., Mazzanti, M., Heglund, N.C., and Citterio, G. 1985. Storage and release of mechanical energy by active muscle: A non-elastic mechanism? *J. Exp. Biol.* 115: 79–87.

De Clercq, D., Aerts, P., and Kunnen, M. 1994. The mechanical characteristics of the human heel pad during foot strike in running: An in vivo cinematographic study. *J. Biomech.* 27: 1213–22.

Edman, K.A.P., Elzinga, G., and Noble, M.I.M. 1982. Residual force enhancement after stretch of contracting frog single muscle fibers. *J. Gen. Physiol.* 80: 769–84.

Flitney, F.W., and Hirst, D.G. 1978. Cross-bridge detachment and sarcomere "give" during stretch of active frog's muscle. *J. Physiol.* 276: 449–65.

Huxley, A.F., and Simmons, R.M. 1971. Mechanical properties of the cross bridges of frog striated muscle. *J. Physiol.* 218: 59P–60P.

Ker, R.F. 1981. Dynamic tensile properties of the plantaris tendon of sheep *(Ovis aries). J. Exp. Biol.* 93: 282-302.

Ker, R.F., Alexander, R.McN., and Bennett, M.B. 1988. Why are mammalian tendons so thick? *J. Zool.* 216: 309–24.

Ker, R.F., Bennett, M.B., Bibby, S.R., Kester, R.C., and Alexander, R.McN. 1987. The spring in the arch of the human foot. *Nature* 325: 147–49.

Komi, P.V., and Bosco, C. 1978. Utilization of stored elastic energy in leg extensor muscles by men and women. *Med. Sci. Sports* 10: 261–65.

Kram, R., and Taylor, C.R. 1990. Energetics of running: A new perspective. *Nature* 346: 265–67.

McCaw, S.T., and DeVita, P. 1995. Errors in alignment of center of pressure and foot coordinates affect predicted lower extremity torques. *J. Biomech.* 28: 985–88.

Pollock, C.M., and Shadwick, R.E. 1994. Relationship between body mass and biomechanical properties of limb tendons in adult mammals. *Am. J. Physiol.* 266: R1016–21.

Rome, L.C., Sosnicki, A.A., and Goble, D.O. 1990. Maximum velocity of shortening of three fiber types from horse soleus muscle: Implications for scaling with body size. *J. Physiol.* 431: 173–85.

Valiant, G.A. 1990. Transmission and attenuation of heel strike accelerations. In *Biomechanics of distance running,* ed. P.R. Cavanagh, 225–47. Champaign, IL: Human Kinetics.

Wang, X.T., and Ker, R.F. 1995. The creep rupture of wallaby tail tendons. *J. Exp. Biol.* 188: 831–45.

Woledge, R.C., Curtin, N.A., and Homsher, E. 1985. *Energetic aspects of muscle contraction.* London: Academic Press.

Yamaguchi, G.T., Sawa, A.G.U., Moran, D.W., Fessler, M.J., and Winters, J.M. 1990. A survey of human musculotendon actuator parameters. In *Multiple muscle systems,* ed. J.M. Winters and S.L-Y. Woo, 717–73. New York: Springer-Verlag.

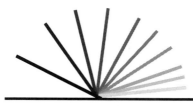

Chapter 3

Length Changes of Muscle-Tendon Units During Athletic Movements

J.G. Hay

Success in athletics (i.e., competitive sport) depends largely on the techniques used in the execution of the relevant motor skills. Such techniques have frequently been studied using cinematography and videography (to determine the kinematics of the motions involved) and various force- and pressure-measuring devices (to determine the external kinetics of these motions). The activity of the muscles responsible for the externally observed kinematics and kinetics has often been determined using electromyography (to establish which muscles are active, and to what level of activation). In addition, the lengths of muscle-tendon units have occasionally been determined using cinematography or videography to add another dimension to our understanding of how athletic movements are produced and controlled.

This chapter considers studies in which the lengths of muscle-tendon units have been determined during the performance of motor skills commonly used in athletics. It describes the modes of action of the primary muscles involved in the performance, considers how well joint actions predict the modes of action of the muscles, and evaluates the role of the stretch-shorten cycle in the performance.

Elaboration of the Concepts

It has long been known that the forced lengthening of an active skeletal muscle before allowing it to shorten leads to an enhanced response during shortening (for review, see Cavagna 1977). This enhancement during the shortening phase of a lengthening-shortening sequence of muscle actions, commonly known as the *stretch-shorten cycle* (SSC), has frequently been studied using the motions of one or more segments of the intact human body. Thus, for example, Cavagna, Dusman, and Margaria (1968) used elbow flexion motions of the forearm-plus-hand segment on a fixed upper-arm segment; and, at the other extreme, Asmussen and Bonde-Petersen (1974) used squat, countermovement, and drop jumps using primarily trunk, thigh, lower leg, and foot segments.

There are two obvious and important limitations in many of the studies of such motions. The first is methodological in nature and arises from the common practice of assuming the actions of the muscles on the basis of the observed motions of the segments. Thus, it is frequently assumed that flexion motions at a joint reflect a lengthening of the extensor muscles and a shortening of the flexor muscles crossing that joint and, conversely, that extension motions reflect a shortening of the extensor muscles and a lengthening of the flexor muscles crossing the joint. It is also frequently assumed that a flexion or extension motion at one joint is indicative of the changes in the lengths of the muscles crossing other joints. Thus, the instant at which maximum knee flexion is observed in a jumping skill is often taken as an indication of the instant at which the leg extensors of the lower extremity—only some of which actually cross the knee joint—change from a lengthening to a shortening mode of action (see, e.g., Lees, Fowler, and Derby 1993; Luhtanen and Komi 1978). These assumptions seem entirely reasonable when used with reference to uniarticular muscles that shorten and lengthen in accord with the flexion and extension motions of the joints they cross. Whether they are similarly reasonable when used with reference to biarticular or multiarticular muscles is another matter. If all the involved joints move so that they shorten a muscle then, of course, it shortens. Alternatively, if they all move so that they lengthen the muscle, then it lengthens. On the other hand, if they do not move to such consistent effect, the muscle presumably shortens, lengthens, or maintains a constant length according to the relative influences of the joints concerned.

The second limitation arises from the relatively limited number of motions studied and from their artificial nature and, thus, their doubtful relevance to the motor skills encountered in athletics. Some, involving motions at just one to three joints (for example, seated elbow flexion, seated knee extension and seated leg thrusts against a pendulum), while often seen in strength training programs, are well removed from the motor skills commonly used in athletics. And even the vertical jumps frequently used in studies of the SSC differ considerably from the jumping skills used in athletics. The squat jump, which involves only an extension of the hip, knee, and ankle joints, is rarely, if ever, seen in athletics. The only possibility that comes to mind is the upward jump used in ski jumping, and even here there is some question as to whether the final extension is preceded by an initial flexion of these joints. The countermovement jump, involving an initial flexion of the joints of the lower extremity followed by a final extension of these same joints, appears at first glance to be more closely related to the motor skills seen in athletics. These latter skills, however, are almost invariably performed from a run and with full use of the arms. The drop or depth jump is an interesting motor skill that has often been used in studies of the SSC, but it too has little apparent

similarity to the motor skills used in athletics. Indeed, it appears to have been first developed as an experimental task and then, with the widespread embracing of plyometric exercise, to have been subsequently adopted as a staple of training regimes for athletes in sports and events requiring explosive leg speed.

These limitations in previous studies of the SSC can be overcome by studying the actions of the muscles involved rather than assuming them from the actions of the joints, and by studying the performance of motor skills commonly encountered in athletics rather than those that are rarely, if ever, seen in that context.

The importance of the SSC to the performance of athletic motor skills depends, first, on whether active muscles actually experience a lengthening-shortening sequence of actions and, second, on whether any such lengthening-shortening sequence that does exist results in an observable improvement in the performance. The present consideration of these matters focuses on four motor skills involving predominantly muscles of the lower extremity—the running long jump, the volleyball block performed following approach runs of varying length, sprinting, and cycling.

Long Jump

Hay, Thorson, and Kippenhan (1999) asked eleven elite female long jumpers to perform six jumps with maximum effort from a full-length approach. High speed motion picture cameras were used to gather kinematic data, and a force platform was used to record the vertical forces exerted on the ground during the takeoff phase of each jump.

The trial in which each subject recorded her best jump was selected for inclusion in a cross-sectional analysis. In addition, the availability of six analyzable trials by one subject permitted a separate longitudinal analysis of her technique. The film from the two cameras for each of the selected trials was digitized, starting just before the takeoff into the last step and ending when the subject left the field of view after the takeoff from the board. Body landmarks defining a 14-segment model of the subject were digitized in each frame, and the resulting coordinate data were used to obtain 3-D coordinates using the direct linear transformation (DLT) procedure (Abdel-Aziz and Karara 1971).

A three-dimensional, four-segment model of the trunk and the jumping leg was developed to estimate muscle-tendon unit lengths (referred to here as muscle lengths, for the sake of brevity). Four muscles and two muscle groups (all referred to here as muscles) were included in the model. These muscles were the gluteus maximus, hamstrings, rectus femoris, vasti, gastrocnemius, and soleus. Each of these muscles was represented by one or more straight lines. The attachment points of each muscle were determined from (a) digitized coordinates defining segment endpoint locations and (b) literature values for attachment locations. Using this model, the length of each muscle was calculated throughout the duration of the takeoff. The length of each muscle was then plotted as a function of time for each of the 11 subjects (cross-sectional analysis)

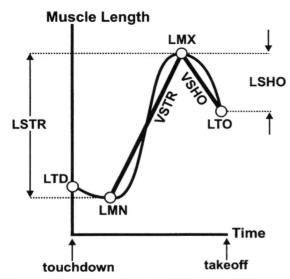

Figure 3.1 Variables used to characterize changes in the lengths of muscles that lengthened and then shortened immediately before takeoff. LTD and LTO = the lengths of the muscle at the instants of touchdown and takeoff; LMN and LMX = the minimum and maximum lengths of the muscle during the takeoff; LSTR = the length by which the muscle was lengthened (or stretched) before it reached LMX; LSHO = the length by which the muscle was shortened after it reached LMX; and VSTR (the stretch velocity) = the average rate at which the muscle was lengthened with respect to time before it reached LMX. VSTR was computed by dividing the LMX – LMN difference by the elapsed time between the instants at which these two lengths were attained. Similarly, VSHO (the shortening velocity) = the average rate at which the muscle was shortened with respect to time after it reached LMX. VSHO was computed by dividing LTO – LMX by the elapsed time between the instants at which these two lengths were attained. Variables used to characterize changes in the lengths of muscles that lengthened or shortened throughout the takeoff were defined in a similar manner.

Reprinted, by permission, from J.G. Hay, E.M. Thorson, and B.C. Kippenhan, 1999, "Muscle-tendon length changes during the takeoff of a running long jump," *Journal of Sports Science* 17:159-172.

and each of the six trials (longitudinal analysis). The form of each of these curves was then classified visually according to the sequence of lengthening and shortening actions observed. Variables were defined to characterize the length vs. time histories of the muscles as shown in the example of figure 3.1.

A muscle's contribution to the vertical forces that propel the body into the air depends primarily on whether it is active. How its length changes with time is secondary; it is important only if the muscle is active. For the purpose of the analysis, it was assumed that the muscles of interest were active throughout the takeoff. Although no attempt was made in this study to establish if and when these muscles were active, there is support for this assumption in the research literature (Carpentier et al. 1989; Ito et al. 1989; Kakihana, Yamanouchi, and Suzuki 1995; Kyrolainen et al. 1988). A visual examination of the sample EMG tracings presented in these studies revealed that most of the muscles of interest in the Hay, Thorson, and Kippenhan (1999) study were active for almost the entire duration of the takeoff.

Modes of Action

A representative example of the results obtained when muscle lengths were plotted against time for one trial is presented in figure 3.2.

Gluteus Maximus and Hamstrings

For almost all subjects (cross-sectional analysis) and for all trials (longitudinal analysis), the length of the gluteus maximus remained nearly constant for the first 30 to 50% of the takeoff and then decreased steadily until the instant of takeoff. In subtle contrast, the length of the hamstrings decreased in linear or nearly linear fashion from touchdown to takeoff.

Vasti

With one exception, the form of the muscle length vs. time curve for the vasti was consistent for all subjects and trials. The muscle lengthened during the first half of the takeoff and shortened during the second half. This is exactly as expected from this muscle, passing over a single joint that was itself flexing and extending. The extent of the lengthening and subsequent shortening was very small. This is also as expected, given that the range of motion observed at the knee joint was modest.

Rectus Femoris

For 6 of the 11 subjects in the cross-sectional analysis, the length of the rectus femoris increased in a nearly linear fashion throughout the takeoff. For

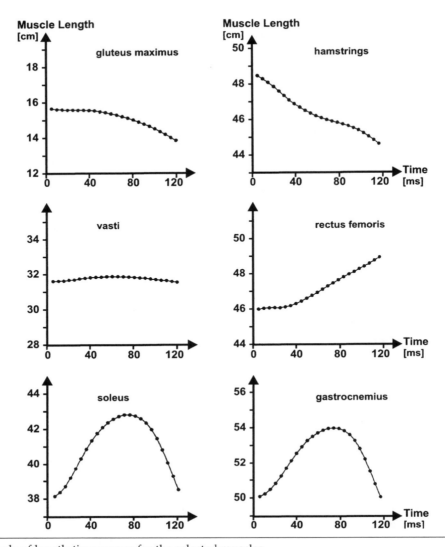

Figure 3.2 Example of length-time curves for the selected muscles.
Reprinted, by permission, from J.G. Hay, E.M. Thorson, and B.C. Kippenhan, 1999, "Muscle-tendon length changes during the takeoff of a running long jump," *Journal of Sports Science* 17:159-172.

four of the remaining subjects and for all six of the trials in the longitudinal analysis, a predominant lengthening phase, occupying the final 55 to 90% of the duration of takeoff, was preceded by a period of isometric action. In short, with one exception, the action of the rectus femoris was either solely or predominantly a lengthening action.

Soleus and Gastrocnemius

The plots for the length of the soleus vs. time and for the length of the gastrocnemius vs. time had two characteristic forms—a form in which the muscle was seen to shorten, lengthen, and then shorten again, and an otherwise identical form in which the initial shortening phase was not in evidence (see figure 3.3). The three-phase form was seen in 7 of the

11 subjects of the cross-sectional and in all six trials of the longitudinal analysis. The obvious explanation for the initial shortening of the triceps surae muscles in some cases and not in others lies in the manner in which the takeoff foot made contact with the ground at touchdown. If it landed flat-footed, the shank of that leg then rotated upward about the ankle joint—a dorsiflexion action resulting in a lengthening of soleus and gastrocnemius. On the other hand, if it landed heel first, the ankle immediately plantarflexed until the sole of the foot was flat on the ground, at which time the shank then began to rotate upward about the ankle joint. This sequence of plantarflexion followed by dorsiflexion produced an initial shortening followed by a lengthening of the soleus and gastrocnemius.

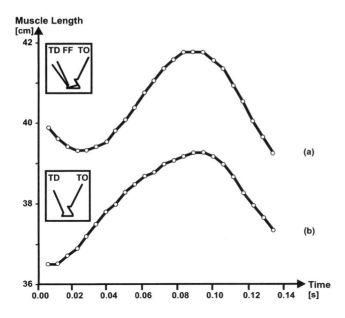

Figure 3.3 Typical length-time curves for soleus (shown here for two subjects) and for gastrocnemius: *(a)* subject DD with heel landing and three-phase (shortening-lengthening-shortening) sequence, and *(b)* subject TD with flat-footed landing and two-phase (lengthening-shortening) sequence. (Hay, J.G., Thorson, E.M., and Kippenhan, B.C. 1996. Why have our women long jumpers been much less successful than our men? Report to U.S. Olympic Committee and U.S.A. Track and Field. p. 44.) Note: The insets show the position of the foot at touchdown ending the last step (TD), at the instant the foot is flat on the ground for the first time following a heel-first touchdown (FF), and at the instant of takeoff (TO).

Joint Angles vs. Time

For the vasti, the instant at which maximum knee flexion was recorded was a valid indicator of the instant at which the muscle ceased to lengthen and began to shorten or, in other words, of the instant at which it reached its maximum length. For the soleus and gastrocnemius, the instant at which maximum knee flexion was recorded underestimated by 10 to 21% of the time of takeoff the instant at which the muscles reached their maximum lengths. For the gluteus maximus, hamstrings, and rectus femoris, the instant at which maximum knee flexion was recorded was completely inappropriate as an indicator of modes of muscle action because none of these muscles exhibited a lengthening-shortening sequence of actions. In light of these findings, one can only conclude that the instant at which maximum knee flexion is recorded is a poor indicator of when extensor muscles of the lower extremity other than the vasti change from eccentric activity to concentric activity.

Relationships With Change in Vertical Velocity

Correlations were computed between the measures of muscle length changes (see figure 3.1) and the change in vertical velocity of the center of gravity (CG) during the takeoff. For the cross-sectional analysis, none of these correlations was found to be significantly greater than zero, and for the longitudinal analysis, seven were found to be significantly greater than zero.

The large difference in the number of significant correlations in the two analyses (0 out of 16 in the cross-sectional case and 7 out of 16 in the longitudinal) suggests that differences in the physical characteristics of the subjects in the former case may have concealed otherwise important relationships and that, in the future, studies involving multiple longitudinal analyses might be more productive than further studies with a cross-sectional design.

The following account is confined to a consideration of the seven significant findings for the one subject of the longitudinal analysis. It is obvious that the results obtained in an analysis of the actions of one subject cannot be generalized to those of other subjects, even if they are also members of a specified population—for example, elite female long jumpers. At best, the results might be generalized to all jumps performed by this subject under similar conditions. Nonetheless, these results provide some useful initial insights into the relationships that exist between muscle actions and the vertical velocity generated during the takeoff of one elite female long jumper. They also show something of the potential of a longitudinal approach to the study of such issues. By experimentally controlling the influence of anthropometric, physiological, and at least some psychological factors, a longitudinal approach permits the analysis to focus on the muscle actions, technique, and performance.

The significant correlations obtained for those muscles that exhibited a lengthening-shortening sequence of muscle actions (or SSC) warrant some additional comment. Studies of enhancement due to the use of a SSC have often involved a comparison of the results obtained when a motor skill is performed with and without a countermovement. The difference in the dependent variable in the two cases (for example, the height achieved in a standing vertical jump) is measured and attributed to enhancement. An experimental design like this is well suited to the study of the benefits of using a SSC in some motor skills (e.g., vertical jumps, bench presses, and single-joint motions) but not in others (e.g., running jumps and throws). In the latter case, the contrasting of with- and without-countermovement variants of the skill of interest is difficult to contemplate because of the obvious difficulty in devising a without-countermovement version.

The analysis employed here was not designed to determine the magnitudes of the enhancements obtained from use of SSCs, but rather to establish whether significant relationships existed between measures of the mechanical activity of those muscles that experienced a lengthening-shortening sequence of actions (SSC) and the change in vertical velocity (the dependent variable). Because enhancement occurs during the shortening phase of a SSC, evidence of its contribution to the change in vertical velocity was sought from the correlations involving measures of the muscle action during that phase. A significant correlation between one of these measures and the change in vertical velocity was taken to indicate that the use of a SSC made a useful contribution to the change in vertical velocity. Conversely, the absence of a significant correlation was taken to indicate that it did not.

Gluteus Maximus and Hamstrings

The magnitude of one of the correlations for these muscles was significantly greater than zero. This correlation ($r = -.85$) between the change in length of the hamstrings and the change in vertical velocity during the takeoff indicated that the more the hamstrings were shortened, the larger was the gain in vertical velocity during the takeoff.

Vasti

One significant correlation was found among the correlations for the measures used to characterize the lengthening-shortening activity of the vasti and the change in vertical velocity. This correlation ($r = .81$) indicated that the longer the distance over which the muscle was stretched, the larger was the gain in vertical velocity.

There are at least two possible explanations for this finding. It may be that the longer the distance over which the muscle was stretched, the greater the mechanical work that was done on the muscle and the larger the change in the vertical velocity that was generated while it was being stretched. Alternatively, it may be that the longer the distance over which the muscle was stretched, the greater the elastic energy stored in the muscle and the greater the change in vertical velocity that was realized when the muscle was subsequently allowed to shorten. In the first of these cases, one might expect to see a strong relationship between the distance

over which the muscle was stretched and the change in the vertical velocity before it reached its maximum length. In the second, one might similarly expect a strong relationship between the distance over which the muscle was stretched and the change in vertical velocity after the muscle reached its maximum length. Correlations computed for the two cases yielded coefficients of $r = .57$ and $r = .01$, respectively. Although neither of these coefficients was significantly greater than zero they did suggest a preference for the first of the two explanations.

Rectus Femoris

No significant correlation was found between any of the four measures of the activity of rectus femoris and the change in vertical velocity.

Soleus and Gastrocnemius

Five of the correlations between the measures of soleus and gastrocnemius muscle actions and the change in vertical velocity were found to be significantly greater than zero. These significant correlations indicated that

- the shorter the length of the soleus at touchdown and the shorter the minimum length of the soleus, the larger was the gain in vertical velocity;
- the longer the distance over which the soleus and gastrocnemius were stretched, the larger was the gain in vertical velocity; and
- the faster the velocity at which the soleus was stretched, the larger was the gain in vertical velocity.

In short, the results suggest that having these muscles short at the outset and then stretching them rapidly over a long distance was consistent with generating large vertical velocities during the takeoff.

Additional correlations were computed between the lengths over which the muscles were stretched and the corresponding velocities at which they were stretched. These correlations were significantly greater than zero in both cases and suggested that the longer the distance over which the muscles were stretched, the faster the velocity of stretching that could be attained.

The obvious way in which the stretch lengths and velocities of the triceps surae muscles might have a causal relationship with the gain in vertical velocity is through use of the stretch-shorten cycle. If the magnitude of the enhancement due to the use of this mechanism increased with stretch length and stretch velocity, it would be reasonable to expect that the vertical forces exerted against the ground as the ankle was plantarflexed at a given velocity of shortening would also increase and that this increase in vertical forces would lead to an increased gain in vertical velocity.

There is at least one major obstacle to this line of argument. The concentric action of the triceps surae muscles began very late in the support phase of the takeoff. In the longitudinal analysis, for example, the maximum lengths of the muscles—signaling the start of the concentric phase—were recorded, on average, after 72% (soleus) and 78% (gastrocnemius) of the duration of the takeoff had elapsed. This means that variations in enhancement due to variations in stretch length and stretch velocity could only have had an effect on the change in vertical velocity during the final 22 to 28% of the takeoff. (Note: These latter values are means for the six trials. The corresponding ranges were 20 to 36% for the soleus and 15 to 31% for the gastrocnemius.)

The vertical ground reaction force vs. time curve for the best trial by the subject of the longitudinal analysis is shown in figure 3.4. Visual inspection of this typical curve is sufficient to show that the magnitude of the resultant change in vertical velocity during the shortening of the triceps surae is relatively small. It is possible, despite this relatively small change in vertical velocity, that trial-to-trial variations in the change in vertical velocity (due to variations in enhancement) led to the observed significant correlations between stretch length and stretch velocity (on the one hand) and the change in vertical velocity (on the other). However, given that the trial-to-trial variations in the change in vertical velocity during the first three-fourths of the takeoff were probably much greater, it seems more likely that these significant correlations were due to relationships between the independent variables and the change in vertical velocity during the first three-fourths of the takeoff. This argument was supported by the correlations among the relevant variables, which showed that the stretch lengths and velocities of the triceps surae muscles were much more closely related with the changes in vertical velocity that took place before their maximum lengths were reached than with the changes that took place subsequently.

In total, these various results suggest that

- the shorter the muscles were at the outset and the longer the distance over which they were subsequently stretched, the faster they were stretched;
- in accord with the force-velocity relationship for muscle, the faster they were stretched, the

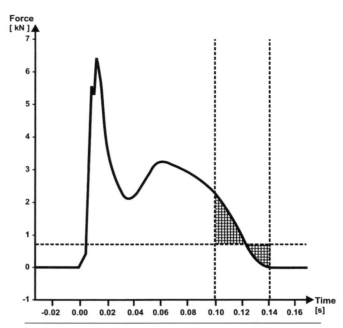

Figure 3.4 The vertical ground reaction force vs. time curve for the best trial by the subject of the longitudinal analysis. The horizontal line represents the weight of the subject and the two vertical lines indicate the times at which the maximum lengths of the triceps surae muscles were recorded and the time of takeoff. The upper, marked area represents the gain in vertical velocity obtained during the initial concentric action of the triceps surae muscles. The lower area represents the loss in vertical velocity during the final concentric action of these same muscles. The vector sum of this gain and loss in vertical velocity is equal to the change in vertical velocity during the concentric action of the muscles. This change in velocity is clearly very small compared to the change represented by the area above the weight line and to the left of the line indicating the maximum lengths of the muscles.

Reprinted, by permission, from J.G. Hay, E.M. Thorson, and B.C. Kippenhan, 1999, "Muscle-tendon length changes during the takeoff of a running long jump," *Journal of Sports Science* 17:159-172.

larger the forces they generated and the larger was the resulting gain in vertical velocity; and

- enhancement due to use of the stretch-shorten cycle did not make a significant contribution to the development of vertical velocity via the actions of the gastrocnemius and soleus muscles.

Summary

Assuming that the muscles were active throughout the takeoff, the observed muscle length vs. time changes were consistent with the gluteus maximus acting isometrically at first and then concentrically;

the hamstrings acting concentrically throughout; the rectus femoris acting isometrically at first and then eccentrically, or eccentrically throughout; and the vasti, soleus, and gastrocnemius acting eccentrically at first and then concentrically. In the case of the soleus and gastrocnemius, the modes of action described were preceded in some cases by a short period of concentric activity.

The instant at which maximum knee flexion is recorded is a very poor indicator of when extensor muscles of the lower extremity other than the vasti change from eccentric activity to concentric activity.

Only one of the seven measures of muscle action found to be significantly related with the change in vertical velocity of the CG—the change in length of the hamstrings—was a measure of the concentric activity of a muscle. All of the others were measures of the eccentric activity.

There was no evidence to suggest that enhancement due to use of the stretch-shorten cycle was associated with the gain in vertical velocity of the CG during the takeoff. For the three muscles that experienced an eccentric-concentric sequence of muscle action, the only measures that were significantly related with the change in vertical velocity were measures of the eccentric phase, and these measures were much more closely related with the change in the vertical velocity before the final concentric phase than with the change in vertical velocity during that final phase. Indeed, rather than supporting the notion that eccentric-concentric actions facilitated the generation of vertical velocity during the final concentric phase, the evidence seemed to suggest that it was fast eccentric actions earlier in the takeoff that enabled the muscles to exert large forces and thus generate large gains in vertical velocity.

Volleyball Block

Grosvenor (1994) analyzed vertical jumps performed from an initial standing position and following approach runs of one, three, five, and seven steps. An experienced male volleyball player served as the subject and performed these jumps with a takeoff from two feet to reach with both hands toward a volleyball suspended directly above a marked takeoff area. Although this skill is rarely performed with an approach run of more than three steps, the jumps were otherwise similar to those performed in the execution of a volleyball block. Three trials were performed under each of the five conditions and the subject's performances were recorded in side view using a motion picture camera. For each of the

conditions, the trial in which the highest vertical velocity of the CG at takeoff was recorded was analyzed. Although Grosvenor's study was not concerned with muscle action, her digitized coordinate data were subsequently used by the present author to compute muscle-tendon length changes during the takeoff for the same six muscles (and using the same computational procedures) as used in the Hay, Thorson, and Kippenhan (1999) study.

Modes of Action

In the *standing vertical jump*, although there was some variation among muscles with respect to the mode of muscle action early in the takeoff, five of the six muscles had a lengthening-shortening sequence of actions immediately prior to the instant of takeoff. The rectus femoris, which exhibited a shortening-constant-lengthening sequence of actions, was the sole exception.

For the *running vertical jump*, with two exceptions in a total of 24 muscle × condition combinations (soleus and gastrocnemius in one-step condition), the modes of action observed were the same as reported for the running long jump. That is, the gluteus maximus and hamstrings shortened; the rectus femoris lengthened; the vasti lengthened and then shortened; and the soleus and gastrocnemius shortened, lengthened, and then shortened again. This last sequence was the same as that recorded in the running long jump for those subjects who landed heel first at the end of the last step of the approach and was exactly as expected, given that the subject here landed on his heels at the end of the last step of his running vertical jumps.

Summary

Standing vertical jumps and running vertical jumps performed from a two-feet takeoff involve different modes of muscle action. For the standing vertical jump, a lengthening-shortening sequence of muscle actions was evident in the final part of the takeoff for five of the six muscles analyzed. The existence of this characteristic sequence of actions means it is possible that use of the stretch-shorten cycle led to an enhanced contribution of muscles crossing the hip, knee, and ankle joints to the generation of vertical velocity at the instant of takeoff. For the running vertical jump, the same characteristic sequence was in evidence only in the records for the vasti and triceps surae muscles. The use of the stretch-shorten cycle in this case may also have led to an enhanced contribution of these muscles to the generation of vertical velocity at the instant of takeoff. Alternatively, it may be that, as in the case of the running long jump, the vertical forces exerted as the muscles shortened were too small and were exerted too late in the takeoff to have a significant effect on the outcome (that is, on the vertical velocity of the athlete at the instant of takeoff).

A running long jump and a running vertical jump involve almost identical modes of action of the major extensor muscles of the takeoff leg (or legs). This suggests that the results obtained in the Hay, Thorson, and Kippenhan (1999) study of the running long jump may well be applicable to running jumps in general and not only to long jumps performed by elite female long jumpers.

Sprinting

Simonsen, Thomsen, and Klausen (1985) analyzed the performances of two male sprinters (best 100 m times of 10.7 and 11.1 s) "to confirm or disconfirm earlier suggestions and theories about the mechanics behind muscular activity during sprint running" (p. 524). Cinematography was used to determine the lengths of the muscle-tendon units of nine muscles of the right leg over one complete cycle of sprinting action, and telemetered electromyography was used to determine the electrical activity.

For the purposes of analysis, the running cycle was divided into four phases (renamed here), which were the primary focus of the study, to describe the actions of the right leg:

1. A *support phase* with the right foot in contact with the ground
2. An *early recovery phase* with the body in flight following right-foot takeoff
3. A *midrecovery phase* with the left foot in contact with the ground
4. A *late recovery phase* with the body in flight following left-foot takeoff

Modes of Action

The data for one of the subjects running at maximum speed between 30 to 60 m from the start are shown in figure 3.5. These data were gathered, three muscles at a time, during three separate trials. The graphs of EMG activity show that, with the exception of the rectus femoris, all nine muscles were active during the latter part of the midrecovery phase and throughout the late recovery phase—presumably in preparation for the heavy loading of the lower extremity at landing—and into the first part of the support phase of the right leg. The rectus femoris was active briefly during the late recovery phase and was the only muscle active during the

latter half of the support phase. In this instance, it became active at the same time as the hamstrings became inactive, as might have been expected.

Considered in conjunction with the records of EMG activity, the data for muscle-tendon unit lengths during the late recovery and support phases showed the following:

- The gluteus maximus and hamstrings were active eccentrically early in the late recovery phase and then concentrically until the middle, or a little beyond the middle, of the support phase.

- The rectus femoris contracted eccentrically and then concentrically during the first half of the late recovery phase and then eccentrically again during the second half of the support phase.

- The vastus lateralis and medialis—treated as a single muscle—contracted concentrically almost throughout the late recovery phase and then toward the end of that phase and into the first part of the support phase acted eccentrically.

- The gastrocnemius acted eccentrically through most of the late recovery phase and then isometrically for the remainder of that phase and the first half of the support phase.

- The soleus acted eccentrically, concentrically, and then eccentrically again during the late recovery phase and then, contrary to the gastrocnemius, eccentrically and briefly concentrically during the first part of the support phase.

- The tibialis anterior acted concentrically, eccentrically, and, finally, concentrically again during the late recovery phase and at the very beginning of the support phase.

The role of the stretch-shorten cycle in sprint running was discussed by Simonsen, Thomsen, and Klausen (1985), who stated, "The hamstring muscles and the gluteus maximus are able to perform . . . eccentric work during flight, thereby storing elastic energy to be released during the [support] phase. This is a remarkable ability compared to the monoarticular muscles, which in turn are restricted to perform both eccentric and concentric work during the [support] phase" (p. 530).

There are several aspects of these statements that appear deserving of comment. First, because the maximum lengths of the hamstrings and gluteus maximus are attained (and the subsequent shortening of these muscles begins) well before the end of

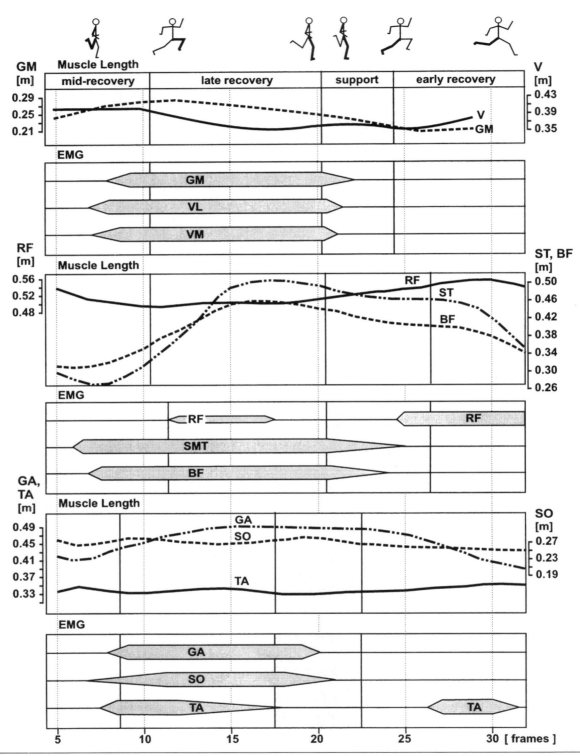

Figure 3.5 Muscle length and electromyographic (EMG) activity vs. time (in number of frames) for one running cycle of the right leg (thickened in stick figures at top) for subject CM. The muscles for which data are shown are the gluteus maximus (GM), vastus lateralis (VL) and vastus medialis (VM)—or VL and VM combined (V), rectus femoris (RF), semitendinosus (ST)—or semimembranosus (SM) and ST combined (SMT), gastrocnemius (GA), soleus (SO), and tibialis anterior (TA).

Adapted, by permission, from E.B. Simonsen, L. Thomsen, and K. Klausen, 1985, "Activity of mono- and biarticular leg muscles during sprint running," *European Journal of Applied Physiology* 54:524-532.

the late recovery phase, whatever elastic energy may have been stored in them is likely to be largely depleted before the right foot touches the ground at the beginning of the support phase. This seems to be the case for the gluteus maximus, which, in the example of figure 3.5, has a maximum length of 28.5 cm reduced to 24.5 cm by the time of touchdown and decreases only an additional 3.0 cm during the takeoff (all measures visually estimated from the original figure). Second, if one assumes that these muscles are far from being maximally active during the late recovery phase—a seemingly reasonable assumption given that the load on them is surely fairly low—the elastic energy stored is probably also much less than could be stored in other circumstances. Finally, the comment about the uniarticular muscles needs some clarification. The articular muscles referred to here are the vastus medialis, vastus lateralis, and soleus, all of which lengthened in the first part of the support phase and shortened in the second part. The comment concerning the uniarticular muscles being "restricted to perform" in this way was clearly made in reference to their roles in sprinting and not intended as a comment on the properties of uniarticular muscles in general.

In a further study requiring measurement of length changes in lower-extremity muscles during sprinting, van Don (1998) used video cameras to record in side view the left-leg actions of 15 college sprinters (10 female, 5 male) and telemetered, surface electromyography to record the electrical activity of the four hamstring muscles—semitendinosus (ST), semimembranosus (SM), biceps femoris long head (BFL), and biceps femoris short head (BFS)—during one full cycle of sprinting. She then digitized the video records and used the coordinates of the relevant landmarks as input to a model to compute the instantaneous muscle fiber lengths of the four muscles as a function of the normalized time for one stride cycle. Three of the four muscles were represented by straight lines joining origin and insertion and the fourth (ST) was represented by two straight lines. The thickness of each muscle (i.e., the perpendicular distance between the tendons) and the length of its tendon were assumed constant, and the instantaneous pennation angles for each muscle were computed using initial values obtained from the research literature and the digitized coordinate data.

Figure 3.6 shows the absolute muscle fiber lengths (measured in meters) plotted vs. time (expressed as a percentage of the time for the complete cycle of leg actions) for a typical subject. The curves shown here for ST and BFL are very similar in form to the corresponding curves of Simonsen, Thomsen, and Klausen (1985)—see figure 3.5— even though the former are measures of fiber length with pennation angle taken into account, and the latter are measures of the length of the entire muscle-tendon unit with no heed taken of pennation angle. In both cases, the two muscles shorten in an approximately linear fashion during the support phase, continue to shorten during the early recovery phase, lengthen during the midrecovery phase and early part of the late recovery phase, and then shorten again just prior to the subject regaining ground support. (Note: In the van Don study, the form of the curve for the SM was almost identical to those for the ST and BFL; it was also similar in form, therefore, to the curve for the ST in the Simonsen, Thomsen, and Klausen study.)

The form of the curve for the BFS differed considerably from that for the other three muscles studied. It showed that the muscle shortened and lengthened slightly during support and then shortened and lengthened markedly over approximately equal parts of the recovery. The curve for the BFS was a rough mirror image of the curve for the vasti in the Simonsen, Thomsen, and Klausen (1985) study, as

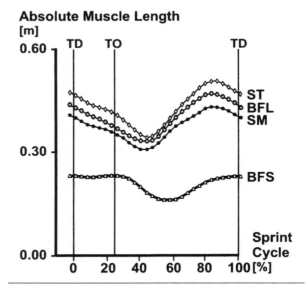

Figure 3.6 Muscle fiber length vs. time (in percentage of time for one leg cycle) for semimembranosus (SM), semitendinosus (ST), long head of biceps femoris (BFL), and short head of biceps femoris (BFS). The vertical lines labeled TD and TO refer to the instants of touchdown beginning the support phase and takeoff ending the support phase.

Adapted, by permission, from Don, B. van. 1998. Hamstring Injuries in Sprinting. Unpublished PhD Dissertation, University of Iowa.

one might have expected given that these muscles are uniarticular and act on opposite sides of the same joint.

Injury Mechanisms

The primary focus of the van Don (1998) study was not the modes of action of the hamstring muscles, nor the role of the stretch-shorten cycle in sprint running. It was, instead, the strains and strain rates to which the hamstring muscles are exposed in sprinting and the association between these measures and the incidence of hamstring injury. To these ends, van Don determined the strain (defined as the change in muscle fiber length with respect to its length when the muscle is in the anatomical position), strain rate, and muscle activation level (defined as the percentage of the maximum value of the rectified and filtered raw electromyographic signal) for each of the four hamstring muscles throughout one cycle of sprinting. She found that the BFL was the hamstring muscle that experienced the greatest maximum strain during the sprint cycle and suggested that this was a potential reason why the BFL experienced a greater incidence of injury in sprinters than any of the other hamstring muscles. She also found that the maximum strain rates recorded for the BFL and the SM were similar and that these maximum rates were greater than those for the ST and BFS; and this led her to conclude that "it cannot be concluded that the greater incidence of injury of the BFL compared to the other hamstring muscles is due to greater strain rates" (p. 137).

Although hamstring injuries have most frequently been reported to occur during the "late forward swing phase," the instants of touchdown and of takeoff have also been mentioned in this context. Van Don's investigation led her to the conclusion that the biarticular hamstring muscles experience their greatest lengthening during the late forward swing phase and this "could be a possible cause for injury."

In an interesting and useful aside, van Don also reported that sensitivity analyses showed errors of the order of 5 to 10% "in the estimated locations of the origin of the biarticular muscles, pennation angle used, and the use of a two-dimensional approach instead of a three-dimensional approach, would not have changed the results of . . . the study" (p. 148).

Summary

Muscular activity generally stopped before the end of the support phase, a phenomenon that Simonsen, Thomsen, and Klausen (1985) considered was due to the relaxation time of the muscles—"although electrical activity ceases about 50 ms before takeoff, there is still mechanical tension in the muscles" (p. 529).

At touchdown and during the first part of the support phase, the vasti and the soleus muscles functioned as "shock absorbers" by performing eccentric work; or, to state it another way, the uniarticular muscles were the only ones studied that played a role in shock absorbing.

Sprinters swing the shank forward before the whole leg is moved toward the ground and this provides a means for the biarticular hamstring and gluteus maximus muscles to perform eccentric work. (Note: Simonsen, Thomsen, and Klausen noted that the superficial part of the gluteus maximus, which is about 2/3 of the whole muscle, is connected to the iliotibial tract and thereby to the tibia. They therefore characterized it as a biarticular muscle that might be expected to act as a hamstring muscle.) Furthermore, since flexion of the hip and extension of the knee take place at the same time, a large lengthening of the hamstring muscles takes place.

The long head of the biceps femoris experienced the greatest maximum strain and, with the semimembranosus, the largest maximum strain rates of all the hamstring muscles during a sprint cycle. It was concluded, therefore, that the greater rate of injury of the long head of the biceps (compared with the other three hamstring muscles) was due to the greater strains it experienced.

Cycling

Hull and Hawkins (1990) developed a general procedure for determining the presence of stretch-shortening cycles in muscles "through identification of regions of positive and negative work in individual muscles" (p. 623) and, because of conflicting claims in the literature concerning the presence of negative work in cycling, applied their procedure to the study of this activity. (Note: "In positive work, the muscle shortens while under contraction, whereas in negative work, the muscle lengthens during contraction. Thus, in positive work the muscle does work on the environment whereas in negative work the environment does work on the muscle" [p. 621].) To this latter end, they analyzed the performance of one subject over one complete cycle of leg actions during a ride on a stationary bicycle. Ten muscles of the right leg were studied. These were the gluteus maximus, rectus femoris, the long head of biceps femoris, semimembranosus, vastus medialis and lateralis, tibialis anterior, the medial and lateral heads of gastrocnemius, and soleus.

Potentiometers fixed to the right crank and pedal were used to measure the orientations of these two parts of the bicycle as a function of time. The data obtained were then used as input to a five-bar linkage model of the lower limb to determine the included joint angles, the muscle-tendon unit lengths, and, by differentiation, the muscle-tendon unit velocities. For these purposes, it was assumed that

- the leg motion occurred in a single plane,
- there was no relative motion between the pelvis and the seat of the bicycle, and
- the axes of rotation did not shift relative to their respective joints.

Surface electromyography was used to determine the period over which each muscle was active and, through normalizing "to the highest EMG signal recorded for [the] particular muscle over the crank cycle" (p. 632), the level of that activity.

Joint Angles vs. Crank Angle

The crank angle was defined equal to 0° when the right foot was at its highest point (top dead center, TDC), 180° when it was at its lowest point (bottom dead center, BDC), and 360° when it was again at its highest point. The included angles for the right hip, knee, and ankle joints increased from a low value at

TDC to a peak value somewhere between 120° and BDC and decreased thereafter through the remainder of the crank cycle. As figure 3.7 shows, these joints did not attain their maximum and minimum values simultaneously: "the extremes of knee motion [occurred] about 20° earlier in the crank cycle than the hip and about 20° later than the ankle" (p. 631). Furthermore, comparison of the data in figures 3.7 and 3.8 revealed that the extreme values for the joint angles did not coincide with those for the muscle lengths of the biarticular muscles but did do so for the muscle lengths of the uniarticular muscles. In short, the conclusions reached here for cycling were very like those reached for the long jump (Hay, Thorson, and Kippenhan 1999); that is, the joint angles were not good indicators of the lengthening and shortening actions of the biarticular muscles but were good indicators of these actions of the uniarticular muscles.

Presence of Stretch-Shorten Cycles

Figure 3.9 summarizes the findings of Hull and Hawkins (1990) with respect to the presence or absence of SSCs in cycling. In this figure, the thick black lines indicate those crank angles through which the muscles decreased in length (cf. figure 3.8) and positive work may, therefore, have been done. The thin black lines indicate crank angles through which

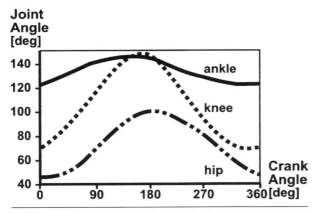

Figure 3.7 Joint angles of the right leg over one crank cycle. The hip angle is the angle between the pelvis and the thigh (with full extension = 180°), the knee angle is the angle between the shank and the thigh (with full extension = 180°), and the ankle angle is the angle between the shank and a line connecting the ankle joint axis to the pedal spindle axis.

Reprinted, by permission, from M.L. Hull, and D.A. Hawkins, 1990, Analysis of work in multisegmental movements: Application to cycling. In *Multiple muscle systems*, edited by J.M. Winters and S.L.Y. Woo. (New York: Springer Verlag), 621-638.

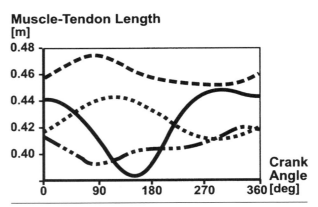

Figure 3.8 Sample plots of muscle-tendon unit length vs. crank angle over one crank cycle for four biarticular muscles: rectus femoris (▬ ▬), biceps femoris (▪▪▪▪), semimembranosus (▬▬), and gastrocnemius (▬ ▪▪ ▬).

Reprinted, by permission, from M.L. Hull, and D.A. Hawkins, 1990, Analysis of work in multisegmental movements: Application to cycling. In *Multiple muscle systems*, edited by J.M. Winters and S.L.Y. Woo. (New York: Springer Verlag), 621-638.

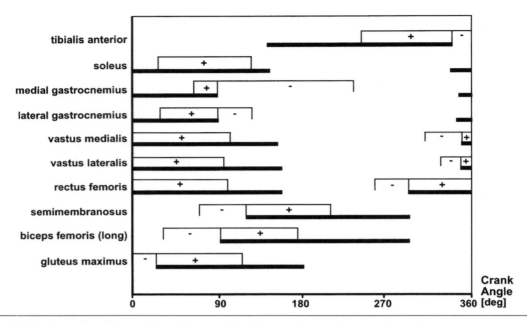

Figure 3.9 Regions of positive and negative work for ten muscles and one crank cycle.

Reprinted, by permission, from M.L. Hull, and D.A. Hawkins, 1990, Analysis of work in multisegmental movements: Application to cycling. In *Multiple muscle systems*, edited by J.M. Winters and S.L.Y. Woo. (New York: Springer Verlag), 621-638.

the muscles were active at a level of activation greater than 30% of the maximum recorded electromyographically for that muscle. For each muscle, crank angles for which there is an overlap of thick and thin lines are those angles through which positive work was actually done. The positive signs serve to reinforce this point. Those angles indicated in a similar fashion with a negative sign are angles through which negative work was done. Finally, a left to right sequence showing a negative sign followed by a positive one is indicative of a lengthening-shortening sequence of actions of an active muscle. Such sequences are indicative, in other words, of the presence of a SSC.

Hull and Hawkins concluded from the evidence of figure 3.9 that all of the muscles except those crossing the ankle joint (that is, all but the top four listed in figure 3.9) exhibited a SSC. They also noted that the negative work was done over ranges of crank angle from 25 to 65° and concluded that with such values "comes the possibility of developing substantial levels of force during stretch. Consequently, the stretch-shortening cycle with its concomitant increase in both efficiency and power appears to be an important mechanism in cycling mechanics" (p. 634).

Summary

As in the case of the long jump, joint angles were not good indicators of the lengthening and shortening

actions of the biarticular muscles but were good indicators of these actions of the uniarticular muscles.

The actions of the gluteus maximus, the hamstrings (biceps femoris long head and semimembranosus), and the quadriceps (vastus medialis and lateralis and rectus femoris) all included a lengthening-shortening sequence when the muscles were active—or, in other words, a stretch-shorten cycle.

Summary and Potential Applications

The reservations regarding methodology and generality mentioned in the introduction to this chapter appear to have been dealt with satisfactorily in the studies cited here. However, these studies also had some important limitations that should be recognized. First, most of the conclusions reached were based on data gathered on just a single subject, and some reservation regarding the generality of the findings is clearly warranted. (The one apparent exception was the study of van Don, in which data were gathered on 15 subjects. However, due to technical problems with the telemetry system used to collect electromyographic data, complete data were gathered on just four of these subjects.) Second, the results obtained and the conclusions reached were no doubt influenced by the experimental conditions under which the data were collected. For example, Hull and Hawkins (1990) suggested that the fast pedaling rate (85 rpm) used in their study increased the negative work regions shown in figure 3.9, compared with the slower pedaling rates (50 rpm and 65 rpm) used in two previous studies, and that this might be the reason they found SSCs for the muscles crossing the hip and knee joints while the previous investigators did not. Third, the activation levels of the muscles examined were handled in different ways in the studies cited. Some investigators relied on evidence from the research literature, some confined their analyses to whether the muscles were active or inactive (or, in common parlance, "on" or "off"), and some also included measures of normalized activation levels. Confidence in the results obtained must be tempered to some degree by the method used in a given case. Finally, all the models used to compute muscle-tendon unit lengths and muscle-fiber lengths assumed constant tendon lengths and, with one exception, a 0° angle of pennation. Unfortunately, little is known about the validity of these assumptions or, more important, the effect that a lack of validity may have had on the conclusions reached.

Joint Angles

The common use of joint angles as an indication of modes of action of muscles (that is, concentric, eccentric, and, occasionally, isometric) appears to be ill advised, except in those rare instances when interest is focused solely on one-joint muscles. Joint actions were found to be poor indicators of the actions of biarticular muscles in both the long jump and cycling—the two cases in which this issue was examined.

Role of Stretch-Shorten Cycle

Stretch-shorten cycles are often assumed to exist and to have a significant effect on the results obtained in the performance of a motor skill. The methods used to determine muscle-tendon length changes in the studies cited provide a means of determining whether these assumptions are well founded. In this regard, there was some contradiction of conventional wisdom relative to the role of the SSC in the motor skills considered here. In particular, enhancement due to the use of the SSC appeared to be unimportant in the takeoff of a running long jump and a running volleyball block and appeared to be important in cycling. The reverse has previously been stated or implied in both cases.

Future Research

The research on length changes in muscle-tendon units in the takeoff to jumping skills and in a single cycle of sprinting and cycling discussed here have produced useful insights into the roles of the lower extremity muscles in these motor skills and into a method by which they might be studied. These have important implications for future research in this area.

In most of the studies cited, the conclusions reached were based on the data for just one subject. In the Simonsen, Thomsen, and Klausen (1985) study of sprinting, a small amount of additional data was presented to demonstrate substantial differences in muscle length changes for a second subject of lesser ability. And in the Hay, Thorson, and Kippenhan (1999) study of the long jump, a longitudinal analysis (involving multiple trials by a single subject) yielded numerous significant results while those for a cross-sectional analysis (involving one trial by each of eleven subjects) yielded none—presumably because constitutional differences (anthropometry, strength, flexibility, etc.) among subjects masked the influence of differences in the techniques or movement patterns they used. Collectively, these several "findings" suggest that, in the future, studies involving multiple longitudinal analyses might be more productive than further studies with just one subject or with a cross-sectional design.

The research discussed here has repeatedly been concerned with determining if and when SSCs are present in the actions of muscles that are central to the performance of a motor skill. More research needs to be done to expand our knowledge in this regard—specifically, to confirm or reject what has already been reported and to extend this knowledge to include other relevant motor skills. This, however, is just a start. The presence and temporal location of an SSC is only of interest if the SSC contributes in a significant way to the performance of the motor skill. Its mere existence is of little interest. In this regard, we need to continue our efforts to develop methods to evaluate the contribution of an SSC to the performance of athletic skills and especially to those involving explosive movements.

The study of the mechanisms of muscle injury has long been hampered by a lack of research methods equal to the task. As a result, some of the most obvious and central of questions relating to such injuries still want satisfactory answers. While the merits of the approach are still to be established, the recent use of muscle fiber length changes in sprint running to determine the strains and strain rates experienced by the individual muscles of the hamstring group is deserving of further thorough examination.

Applications in Athletics

Given that research on muscle-tendon unit length changes during the performance of athletic movements is sparse indeed, it is clearly premature to project the practical consequences or benefits of such research. One can only examine some of the possibilities.

"Recognizing the possible variability in positive and negative work regions between subjects and the impact that the stretch-shortening phenomenon has on human performance" (p. 635), Hull and Hawkins (1990) have suggested that diagrams like figure 3.9 might ultimately be used to identify athletes with the potential to excel in cycling and in other sports. They have also suggested two approaches that might be taken to attain such a goal. The first involves the collection of data on groups of elite and non-elite athletes who are comparable physiologically and the identification of those characteristics of the muscle length and muscle activity vs. crank angle records that distinguish between the groups. The second involves extending the model they have developed to determine muscle-tendon unit length changes so that it also determines the forces exerted by the individual muscles. It also involves the development of "a suitable biomechanical model to assess performance." It is clear, however, that there are major obstacles to overcome before either of these possibilities can become a reality.

The use of plyometric exercises in the training of athletes in explosive sports and events has increased dramatically over the last decade. This development has no doubt been due primarily to the large volume of research showing the performance benefits of the SSC in selected motor skills amenable to experimental control and, in particular, to those involving vertical jumping. If further work supports the initial indications that the SSC does not play a significant role in the development of vertical velocity during the takeoff to a running long jump (and perhaps also during the takeoff to running jumps in general) and that the characteristics of the stretching of the triceps surae muscles do have an important role in this respect, such findings may have important practical consequences. It may come to be recognized, for example, that plyometric exercises are beneficial training exercises not because they increase the enhancement obtained from the use of SSCs, but because they develop an athlete's ability to benefit from the stretching that precedes the shortening phase of an SSC. Or, to put this in a different way, it may be recognized that coaches and athletes have been doing the right thing (using plyometric exercises) for the wrong reason. It may be, then, that the current emphasis on plyometric training shifts to, or is shared with, training in which the emphasis is on the stretching phase of a movement alone and not on the entire SSC and that exercises like drop or depth jumps (in which athletes step or jump down from a platform, land, and then immediately jump upward) are replaced by exercises limited to the first two parts of this three-part sequence—that is, to the initial drop and landing. But this is, of course, mere speculation. What must be established first is whether the initial indications are supported by further research.

References

Abdel-Aziz, Y.I., and Karara, H.M. 1971. Direct linear transformation from comparator coordinates into object space coordinates in close range photogrammetry. In *ASP Symposium on Close Range Photogrammetry.* Falls Church, VA: American Society of Photogrammetry.

Asmussen, E., and Bonde-Petersen, F. 1974. Storage of elastic energy in skeletal muscles in man. *Acta Physiol. Scand.* 91: 385–92.

Carpentier, A., Balestra, C., Guissard, N., and Duchateau, J. 1989. EMG activity during long jump takeoff from different heights [Abstract 289]. In *Congress proceedings: XII International Congress of Biomechanics,* ed. R.J. Gregor, R.F. Zernicke, and W.C. Whiting. Los Angeles: University of California.

Cavagna, G.A. 1977. Storage and utilization of elastic energy in skeletal muscle. In *Exercise and sport sciences reviews,* Vol. 5, ed. R.S. Hutton. Santa Barbara, CA: Journal Publishing Affiliates. 89–129.

Cavagna, G.A., Dusman, B., and Margaria, R. 1968. Positive work done by a previously stretched muscle. *J. Appl. Physiol.* 24: 21–32.

Grosvenor, J. 1994. The velocity of approach, joint actions, and the range of motion of the rotating lever mechanism in jumping. BS (Honors) paper, University of Iowa.

Hay, J.G., Thorson, E.M., and Kippenhan, B.C. 1999. Muscle-tendon length changes during the takeoff of a running long jump. *J. Sports Sci.* 17: 159–72.

Hull, M.L., and Hawkins, D.A. 1990. Analysis of work in multisegmental movements: Application to cycling. In *Multiple muscle systems,* ed. J.M. Winters and S.L.Y. Woo, 621–38. New York: Springer-Verlag.

Ito, T., Azuma, T., Nishijima, Y., and Tokuyama, H. 1989. Electromyographic study on the triple jump. In *Environment of sports performance,* ed. K. Watanabe, 229–36. Hiroshima: Organizing Committee of Japanese Society of Biomechanics.

Kakihana, W., Yamanouchi, T., and Suzuki, S. 1995. Electromyographic activity and biomechanics of long jumps with different takeoff angles. Paper presented at the American College of Sports Medicine annual meeting, Minneapolis, MN.

Kyrolainen, K., Avela, J., Komi, P.V., and Gollhofer, G. 1988. Function of the neuromuscular system during the last two steps in the long jump. In *Biomechanics XI-B*, ed. G. de Groot, A. P. Hollander, P.A.Huijing, and G.J. van Ingen Schenau, 557–60. Amsterdam: Free University Press.

Lees, A., Fowler, N., and Derby, D. 1993. A biomechanical analysis of the last stride, touchdown, and takeoff characteristics of the women's long jump. *J. Sports Sci.* 11: 303–14.

Luhtanen, P., and Komi, P.V. 1978. Mechanical factors influencing running speed. In *Biomechanics VI-B*, ed. E. Asmussen and K. Jorgensen, 23–29. Baltimore: University Park Press.

Simonsen, E.B., Thomsen, L., and Klausen, K. 1985. Activity of mono- and biarticular leg muscles during sprint running. *Eur. J. Appl. Physiol.* 54: 524–32.

van Don, B. 1998. Hamstring injuries in sprinting. PhD diss., University of Iowa.

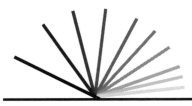

Chapter 4

Work and Energy Influenced by Athletic Equipment

D.J. Stefanyshyn and B.M. Nigg

Athletic equipment is designed to protect athletes, provide comfort, and enhance performance. A recreational athlete may use athletic equipment to improve comfort (e.g., by using a specific backpack), to improve safety aspects (e.g., by using a helmet for rock climbing), or to minimize pain (e.g., by using a shoe insert to prolong the pain-free walking distance). Helmets, shin guards, ankle braces, ice hockey equipment, and many other pieces of athletic equipment are used to protect specific body parts during contact with an opponent, an implement, or the environment.

A competitive athlete attempts to use athletic equipment to enhance performance. Athletic equipment that has been developed to enhance performance is constructed to optimize the energy transfer between athlete and equipment. As mentioned earlier, this is generally achieved by maximizing the (conservative) energy that is returned, minimizing the (nonconservative) energy that is lost, and optimizing the musculoskeletal system.

To optimize the use of athletic equipment it is necessary to understand how athletic equipment influences an athlete's performance. The following section concentrates on this aspect, enhancing performance using athletic equipment. It addresses the first two of the aspects that were previously presented: how athletic equipment can store and re-

utilize (conservative) energy and how (nonconservative) energy loss to the environment can be minimized.

Energy Return

The relationship between stiffness k, deformation Δx, and stored energy has been discussed earlier. Briefly, energy return increases linearly with increasing material stiffness and quadratically with increasing deformation. The approximate ranges of maximal possible energy storage for different athletic surfaces, assuming that the material behaves like an ideal linear spring, are summarized in table 4.1.

Storage and Return of Energy in Sports Equipment

Any elastic material is capable of storing energy. Some examples of materials or equipment that can store and return energy during physical activities include trampolines, diving boards, poles in pole-vaulting, surfaces, and shoes. The next few paragraphs discuss the potential of sport equipment to store and return energy during actual physical activity.

Trampoline and Diving Board

A trampoline is capable of storing large amounts of energy due to its rather low stiffness (approximately

Table 4.1 Energy Return From Elastic Sports Equipment

Equipment	k (N/m)	Δx (m)	E_{sp} (J)	Height (m)*
Trampoline	5,000	0.800	1600	2.30
Tumbling floor	50,000	0.100	250	0.36
Gymnastic floor	120,000	0.050	150	0.22
Running track	240,000	0.010	12	0.02
Gymnasium floor	400,000	0.005	5	0.01

*The calculated height indicates the height a mass of 70 kg could be raised using the maximal possible returned energy.

Approximate values for spring constants (k (N/m)), maximal deformations (Δx (m)), and maximal energy storage (E_{sp} (J)) of different athletic equipment assuming a linear elastic spring behavior.

5000 N/m) and the large deformation this allows. A diving board is slightly stiffer than a trampoline (k ≈ 6000–7000 N/m) (Kooi and Kuipers 1994). However, it does not store more energy than a trampoline because the deformation that it allows is smaller than the deformation of the trampoline springs and fabric. While in contact with the trampoline or the diving board, the athlete works to further deflect the apparatus. Thus, the deflection of the equipment results from the kinetic energy of the landing athlete and the work of this athlete during the first half of contact. The athlete also performs additional work during the second half of the contact phase in order to accelerate his or her center of mass upward relative to the surface. The combination of the energy return from the trampoline and the work performed by the athlete when in contact with the apparatus allows greater heights to be achieved.

Gymnastics Floors

Gymnastics and tumbling make extensive use of energy return from equipment. The gymnastics floor routine is performed on a springy surface that is specially designed to store and return energy to the athletes. The additional energy allows for spectacular tumbling routines that would not be possible without the additional energy from the surface. A competition surface for the floor exercises in gymnastics has a stiffness of approximately 120 kN/m. A competition surface for tumbling has a stiffness of approximately 50 kN/m. Assuming an ideal elastic spring, these material characteristics correspond to an additional jumping height of about 0.20 m for gymnastics and about 0.30 to 0.40 m for tumbling. This additional height corresponds to an additional time in the air (0.20 s for gymnastics and 0.30 s for tumbling), which allows for more somersaults and twists. In analogy to the trampolining and diving examples, the gymnasts or tumblers work during the short ground contact to increase their performance. Additionally, gymnasts also make use of energy storage and return when they compress the springboard during vaults.

Track and Field Surfaces

Indoor track and field surfaces can be tuned to return energy. McMahon and Greene (1978, 1979) studied track surfaces with different compliance. They used a theoretical model with experimental input. Their theoretical model predicted

- that runners would increase their step length and ground contact time for increased compliance of a running surface that would result in slower running speeds,

- that the fastest running surface would be infinitely stiff without any deformation if the human body was an ideal elastic system consisting of only mass and stiffness without any damping,

- that the stiffest or hardest surfaces were not necessarily the surfaces where the athletes would be the fastest,

- that there is some intermediate value of stiffness for a track surface that maximizes running speed when assuming that the musculoskeletal system includes some internal damping that dissipates energy,

- that this optimal stiffness is two to four times an athlete's lower leg stiffness, and

- that this optimal stiffness leads to a decrease in ground contact time and an increase in step length.

Using this surface stiffness, the researchers predicted a 1 to 3% increase in running speed. Experimental results from a running track constructed with a stiffness of approximately three times man's lower leg stiffness ($k \approx 80$ kN/m) showed speed enhancements of approximately 2%. A 2% increase is certainly substantial in any track and field running discipline. It represents approximately 0.2 s in a fast 100 m sprint and about 80 s in a marathon.

Tennis Racquets

The type and tension of tennis racquet strings can have an influence on the efficiency of energy return during impacts of tennis racquets with tennis balls. Nylon strings have been found to be slightly inferior to gut strings in their ability to store and return energy to the ball (Ellis, Elliot, and Blanksby 1978). Lower string tensions were also found to have higher rebound coefficients (velocity of the ball before impact compared to after impact) (Baker and Wilson 1978; Ellis, Elliot, and Blanksby 1978; Bosworth 1981; Elliot 1982). It has been shown that the strings can be considered as purely elastic (Brody 1979; Leigh and Lu 1992). Therefore, the increase in energy return is due to less energy being lost in the ball. For example, Leigh and Lu (1992) report that as string tension decreases from 200 N to 98 N, the compression of the tennis ball is decreased by 36% while the impact times are only decreased by 18%. The result is a decrease in the rate of compression of the tennis ball and less energy being lost. One drawback of flexible strings is that control is sacrificed (Brody 1979).

Vaulting Poles

The ultimate goal in pole-vaulting is to obtain a maximal height jump. In a rough approximation, this is equivalent to the athlete obtaining maximal potential energy. The kinetic energy of the athlete during the run-up is converted to potential strain energy in the pole as the pole flexes. Then as the pole extends, the strain energy is converted to gravitational potential energy as the athlete's center of mass is raised. The ultimate performance is substantially influenced by the ability of the pole to store and return strain energy. Theoretical models have been developed to study the influence of pole stiffness on performance (Braff and Dapena 1985; Ekevad and Lundberg 1997). Both studies indicate that there is an optimal stiffness that maximizes performance. A pole that is too stiff straightens before the athlete is in the maximal vertical position, pushing the athlete horizontally back away from the bar. A pole that is not stiff enough straightens too slowly and the athlete is horizontally past the bar before the maximal vertical position is reached. The optimal stiffness is dependent on the mass and

strength of the athlete. Similar to the trampoline example, the athlete performs additional work by extending his or her arms just before releasing the pole to maximize performance.

Sport Shoes

In the past two decades, several unsuccessful attempts have been made to produce energy return with the help of sport shoes (Alexander and Bennet 1989; McMahon 1987; Turnball 1989). There are several factors that represent possible reasons for these unsuccessful attempts (see chapter 1). The main reason, however, is that the deformation of the shoe sole is generally small.

The maximal deformation of the shoe sole under the forefoot, where most takeoff movements are initiated, is typically small. Currently, sport shoe manufacturers do not want to produce high performance sport shoes with soft forefoot soles because they would produce instability during stance and takeoff. Three different possibilities of energy return during running are calculated to illustrate the current situation and the theoretically possible solutions, using the following assumptions:

1. The peak takeoff force in all examples is $F = 2000$ N.
2. The first example uses a deformation of $d_1 = 2$ mm = 0.002 m.
3. The second example uses a deformation of $d_2 = 5$ mm = 0.005 m.
4. The third example uses a deformation of $d_3 = 10$ mm = 0.010 m.

The results of the calculations, with the corresponding sole stiffness and the maximal possible energy return, are summarized in table 4.2.

These energy return values (see table 4.2) should be compared with the total energy spent during one ground contact. Based on oxygen consumption measurements the total energy spent during a marathon corresponds to about 10^7 J. Assuming about 20,000 steps during a marathon, the total energy spent during one ground contact in marathon running can be estimated as about 500 J (Nigg and Segesser 1992). Using this approximation, the maximal returned energy is about 0.4% of the total energy spent during one ground contact for a deformation of about 2 mm, 1% for a deformation of about 5 mm, and 2% for a deformation of about 10 mm. Thus, existing shoes do not deform enough in the forefoot to utilize energy return. A construction that would allow a deformation of 10 mm while still being stable could provide an estimated maximal

Table 4.2 Maximal Energy Return for Three Shoe Conditions

Force (F) (N)	Stiffness (k) (N/m)	Deformation (Δx) (mm)	Maximal returned energy ($E_{returned}$) (J)
2000	1,000,000	2	2
2000	400,000	5	5
2000	200,000	10	10

Energy return assumes a linear spring behavior with no energy dissipation.

increase in performance, in running, for example, of about 2%, a substantial increase.

In addition to small deformations, other factors that play a role in limiting the energy return from sport shoes include the following:

- The shoes do not act like ideal linear springs.
- The stored energy is not returned at the right location, with the right frequency, and at the right time.

The importance of these aspects to energy return in athletic footwear has been discussed earlier. However, these limitations are not restricted only to sport shoes. They are important for energy return in all types of sport equipment and are discussed in the following paragraphs.

Consequently, one should work in two directions to improve the performance enhancement in sport shoes. First, one should develop shoe soles that allow a deformation under the forefoot of at least 10 mm. Second, one should use materials and constructions that act like ideal springs.

Nonlinearity and Energy Dissipation

The model of a linear elastic spring is the simplest example of energy storage. However, materials used for sport equipment typically do not behave linearly and often dissipate some of the stored energy. Nonlinearity and dissipation of energy are common material properties of such equipment. A typical force deformation diagram (see figure 4.1) demonstrates this. It can be seen that the curves for increasing and decreasing force are not coincidental. This is known as hysteresis. The area under the increasing force curve represents the energy that is put into or the work that is done on the material. The area under the decreasing force curve represents the energy that is returned by or the work that is done by the material. The difference in force is the area between the two curves, which is also known as the

hysteresis loop. This is the energy that is dissipated within the material. Therefore, in returning to its original shape, the work done by the equipment on the athlete is less than the work done by the athlete to deform the equipment. The result is that a piece of athletic equipment will only be able to return a percentage of the energy that is put into the equipment. The fraction of the input energy returned is called the efficiency of the equipment and is determined by

$$\eta = \frac{E_{output}}{E_{input}} \qquad 4.1$$

where

η = efficiency,

E_{output} = energy the equipment returns to the athlete, and

E_{input} = energy the athlete puts into the equipment.

The efficiency of a piece of equipment can range from zero to one. A zero efficiency indicates that a material does not return any of the input energy while an efficiency of one indicates that a material returns all of the input energy. An efficiency of one would only be true for a perfectly elastic material. An efficiency of greater than one can never be attained as it would indicate that the material is generating energy.

A trampoline does not behave like an ideal spring. Some of the energy put into the trampoline is dissipated as the trampoline returns to its original position (see figure 4.2). A trampolinist falling from a height H_0 into the trampoline will bounce back to the height H_1, which is only about 80% of its original height H_0 (Vaughan 1980). About 20% of the original energy is dissipated in the trampoline as frictional and heat energy.

Running shoe midsole materials are not 100% efficient. It has been shown that running shoe soles

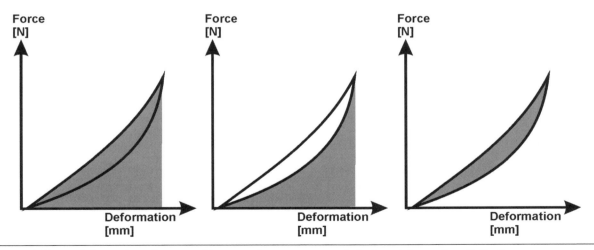

Figure 4.1 Force deformation diagrams illustrate the energy stored during deformation (left), the energy returned (center), and the total loss of energy (right) for a nonlinear system.

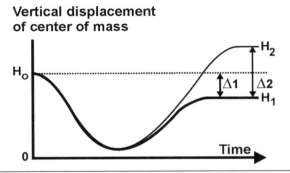

Figure 4.2 Vertical displacement of the center of mass of an athlete bouncing on a trampoline. The thick line represents the path of the center of mass if the athlete performs no additional work when in contact with the trampoline. The difference between the initial height and the final height, $\Delta 1$, is a result of the energy dissipated by the trampoline. The thin line represents the path of the center of mass if the athlete performs additional work when in contact with the trampoline. The increase in height, $\Delta 2$, is a result of the additional work performed.

are only 60 to 70% efficient (Alexander and Bennet 1989; Shorten 1993). The values of potential enhancement of performance discussed previously should, therefore, be reduced by 30 to 40% for those materials, to account for the energy dissipation. However, even those potential enhancements (e.g., 1.4%) are substantial.

Return of Energy at the Right Time

The return of stored energy by athletic equipment must occur at the appropriate time in an athlete's performance as outlined earlier. Pole-vaulters use

their poles to store energy as they flex. When the pole reaches its maximum deformation and begins to return the stored energy, the athlete must be in exactly the right position. The vaulter must have his center of mass close to or behind the line of the pole so that the pole exerts a force on the vaulter that results in an increase in height of the athlete. Furthermore, the vaulter must tuck his knees toward his chest to minimize his moment of inertia, which further facilitates the raising of the athlete (Angulo-Kinzler et al. 1994). The timing is critical. If the athlete's technique is just slightly off and the pole returns the energy a moment too soon or a moment too late, the ultimate performance will be compromised.

Return of Energy at the Right Location

Energy stored in athletic equipment can be of use to an athlete only if it can be returned at the right location. Diving boards, trampolines, and gymnastics surfaces are good examples where energy is returned at the right location, the location of the athlete at takeoff. In all of these cases, the location of maximal deformation, and therefore maximal possible energy return, is the location where the returned energy can have the most influence on performance. This is due to the fact that the athlete who caused the deformation remains in the same position on the equipment during both the energy storage and return phases. However, if the athlete compresses the equipment at one location during the storage phase and then moves to another location during the return phase, the energy return will not be at the right location to maximize performance.

An example of where this may occur is in sport shoes and surfaces.

During a running stride, a heel-toe runner lands on the heel, attains a foot flat position, and then rolls off the forefoot during takeoff. Therefore, the heel is not the location where effective use can be made of returned energy (Nigg and Segesser 1992). For energy return to have an optimal effect on performance, the energy must be returned in the forefoot region during takeoff. However, manufacturers of sport shoes who have attempted to return energy with their shoes have almost exclusively been concerned with the heel. Although the midsole of the forefoot is capable of storing and returning energy (Shorten 1993), the thickness in the forefoot region is quite small (approximately 1 cm), which limits the amount of energy that can be stored in that region.

Return of Energy With the Right Frequency

A piece of equipment (e.g., a sport surface or a diving board) stores energy when deformed and releases this energy when the deformation restores. The storage occurs typically in about the first half and the release in about the second half of the total time of contact. The deformation and release correspond roughly to one-half of a sine wave. The total duration of a full sine wave allows the determination of the frequency of this deformation and release process. In the loaded condition, sport equipment should match this frequency. In other words, the loaded natural frequency of sport equipment should correspond to the actual frequency of the deformation and restoration process.

To illustrate this concept with an example, the ground contact time in sprinting is about 100 ms. One-half sine wave takes about 100 ms. One full sine wave lasts about 200 ms. This corresponds to a frequency of 5 Hz. Therefore, an optimal sprinting surface should have a loaded natural frequency of 5 Hz.

A diver on a diving board has a natural frequency in the range of 2 to 5 Hz. When the diver exerts a downward force on the board, the board deforms. Then as the board returns in an upward direction, it exerts a force propelling the diver upward. If the diver wants to perform a second bounce on the board he or she should match the frequency of the vibrating board to maximize the deflection and the energy stored in the board. The diver utilizes the frequency of the system to enhance the amount of energy that can be stored and, therefore, returned. In fact, it appears that skilled divers contact the board when it is near its maximal downward velocity (Jones, Pizzimenti, and Miller 1993), thus matching the frequency of the vibrating board.

The natural frequency of a system or piece of equipment is dependent on the mass and the stiffness of the system:

$$f_n = \frac{\sqrt{\dfrac{k}{m}}}{2\pi} \qquad\qquad 4.2$$

where

f_n = the natural frequency of the system,

m = the mass of the system, and

k = the stiffness of the system.

The natural frequency changes with changes in either the mass or the stiffness of the system. For example, taking the system consisting of a diver and a diving board, the system has a natural frequency that is dictated primarily by the stiffness of the board and the mass of the athlete. When a diver of different mass is on the board, the system has a different natural frequency. If the new diver is heavier, the natural frequency of the new system is lower, and if the new diver is lighter, the natural frequency of the system is higher. Therefore, it is in the best interest of the diver to be able to adjust the board to match the frequency associated with the diver's mass. The frequency of vibration of the diving board is dependent on the length of the board and is modified by adjusting the position of the fulcrum under the board.

Despite the difficulties of storing and returning energy in sport equipment that have been presented, there are several instances where the concept of energy return in sport equipment is successfully applied. As was mentioned previously, trampolines, diving boards, vaulting poles, tennis racquets, and athletic surfaces are examples where energy storage and return have a large influence on performance.

Minimizing the Loss of Energy

An athlete performing a given task spends energy for aspects that are directly related to performance and energy for aspects that are not directly related to performance of the task (see figure 4.3). The concept of minimizing the loss of energy attempts to minimize the use of energy that is not related to performance but would be available to enhance performance if it would not have been spent unnecessarily. Such energy loss includes loss due to

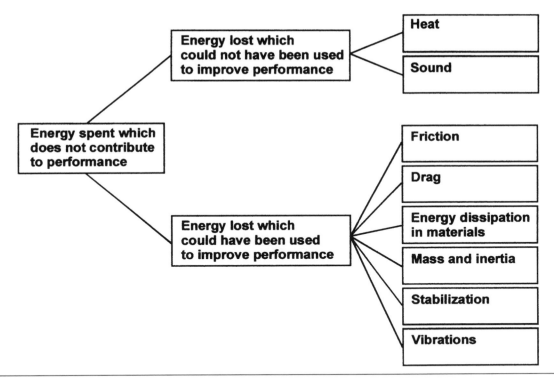

Figure 4.3 Schematic illustration of the different forms of energy used during a physical activity specifically listing those energy aspects that could be used for an enhancement of performance if they would not be spent unnecessarily.

- friction,
- drag,
- mass and inertia,
- energy dissipation in materials,
- stabilization, and
- vibrations.

Many athletic activities can be characterized as physical endeavors to overcome external resistance forces. Swimming and speed skating are prime examples where an athlete works against forces from the environment such as the resistance of the water, air, and ice. Energy is used to overcome these resistive forces. If the magnitudes of these external resistive forces can be reduced, the amount of energy that is lost combating these forces can be decreased. The net result is that an athlete can perform a given task while expending less energy or use the saved energy to enhance performance. The following paragraphs discuss possibilities to reduce the expenditure of unnecessary energy.

Friction

Friction forces occur when two contacting surfaces slide with respect to each other. The frictional force is called a *static friction force* if the two objects are not

moving with respect to each other. The frictional force is called a *dynamic friction force* if the two objects are moving with respect to each other. The friction force that exists between two objects is directly proportional to the normal force between the two objects.

The simplest mathematical description of frictional behavior is Coulomb friction. Coulomb friction assumes that the frictional behavior of the two contacting objects is independent of contact area and relative velocity of the two surfaces. Coulomb friction will be assumed for the following examples. However, one should be aware that this is a simplification of many real athletic situations.

For the dynamic case, the friction force is determined by

$$\mu_{dyn} = \frac{F_{tan}}{N} \qquad 4.3$$

where

μ_{dyn} = dynamic friction coefficient,
F_{tan} = frictional force tangential to the two contacting surfaces, and
N = normal force.

For the static case, the friction force is determined by

$$\mu_{stat} \leq \frac{F_{tan}}{N} \qquad 4.4$$

where

μ_{stat} = static friction coefficient,

F_{tan} = frictional force tangential to the two contacting surfaces, and

N = normal force.

For objects where the shape of the two contacting surfaces does not change (e.g., two hard surfaces), the friction coefficient should not depend on the mass of the two objects. A reduction in mass will reduce the absolute friction force but the relative magnitude stays the same.

If the two objects are not moving with respect to each other, the force of friction is equal to or less than the product of the static friction coefficient and the normal force. For example, if a force less than $\mu_{stat} \cdot$ N is applied to the object, the object will not move. In this case, the opposing friction force will be equal to the force applied. Once the applied force exceeds the product of the static friction coefficient and the normal force, the objects will begin to move with respect to one another. At the instant the object begins to move, the friction force will decrease slightly. Thus, the dynamic friction coefficient is smaller than the static friction coefficient. Therefore, the force that is required to overcome friction and initiate movement is greater than the force that is required to sustain movement.

The absolute magnitude of the frictional force is directly proportional to both the normal force and the coefficient of friction. Thus, the frictional resis-

tance can by reduced by reducing either the normal force or the coefficient of friction. For most athletic activities, the normal forces are dictated, as they depend on the movement being performed. Therefore, attempts to reduce the frictional resistance during athletics have concentrated on reducing the friction coefficient associated with different equipment. Typical friction coefficients for different athletic equipment are shown in table 4.3.

Friction coefficients between ice and skate blades have been reported to lie between 0.003 and 0.007 (de Koning, de Groot, and van Ingen Schenau 1992; Jobse et al. 1990; Kobayashi 1973). These values are dependent on the ice conditions, especially ice temperature, as well as the material and structure of the skate blades. If a normal force during speed skating of 700 N is assumed (approximately the weight of a skater with a mass of 70 kg), the force due to friction is between 2.1 and 4.9 N. Therefore, the energy lost due to friction during a 500 m sprint is approximately 1050 to 2450 J. It has been estimated that the energy expended by a skater during a 500 m sprint is approximately 16,500 J (van Ingen Schenau, de Boer, and de Groot 1989). Therefore, energy lost to ice friction accounts for approximately 6 to 15% of the total energy generated by the athlete. Consequently, skates with a friction coefficient of 0.003 rather than 0.007 will provide a major advantage to an athlete. In fact, reducing the coefficient of friction by just 0.001 will reduce the amount of energy the athlete requires by 2%.

Ice friction is also extremely important in bobsledding. Differences in the sled runners such as shape and material and differences in the track such as slope, temperature, and ice composition have a dramatic effect on the friction forces. Through advances in these variables, the ice friction force for a

Table 4.3 Approximate Translational Friction Coefficients for Athletic Equipment on Different Surfaces

Equipment	Surface	Friction coefficient	Reference
Skates	Ice	0.003–0.007	Jobse et al. 1990
Bobsled runners	Ice	0.01–0.05	van Valkenburg 1988
Skis	Snow	0.05–0.20	Frederick and Street 1988
Tennis balls	Wood	0.25	Brody 1984
	Artificial surfaces	0.50–0.60	Brody 1984
Tennis shoes	Artificial grass	1.3–1.8	Nigg 1986
Basketball shoes	Wooden floor	1.0–1.2	Valiant 1994
	Wooden floor (dusty)	0.3–0.6	Valiant 1994
Cleated shoes	Astroturf	1.2–1.7	Valiant et al. 1985

standard bobsled run averages about 1.5 to 5% of the normal force (Van Valkenburgh 1988). Similar to speed skating, small changes in the friction forces can have a dramatic influence on reducing the energy lost to the environment and ultimately on performance.

The friction between snow and a gliding ski is the most important mechanical factor limiting performance in cross-country skiing (Frederick and Street 1988). Snow friction depends on the characteristics of the snow, the characteristics of the ski, and the preparation (waxing) of the skis. The coefficient of friction between ski and snow varies between 0.05 and 0.2. Frederick and Street (1988) found that the characteristics of the skis had a larger influence than the ski preparation on friction between the ski and the snow. Stiff skis glided better on hard snow because more pressure was distributed to the tips and tails. Conversely, more flexible skis glided better on softer snow due to a more even pressure profile. They estimated that ski design and waxing could influence performance by about 8 to 10%, which corresponds to about eight minutes in a 30 km cross-country race. Friction between ski and snow also plays an important role in alpine skiing. A reduction in the snow-ski friction will result in higher speeds and shorter times for races.

Friction is also intrinsic in various other sport equipment. Friction in the bearings of in-line skates is a limiting factor in achieving top speeds during in-line skating. The higher the friction in the bearings, the greater the resistance to rolling, the harder the athlete has to work to achieve the same speed. Similarly, friction exists in bicycle chains and gears. The friction force in bicycle chains and gears can account for up to 5% of the force required to propel the bicycle and rider (Faria and Cavanagh 1978). However, of greater importance in cycling is the energy that can be conserved if drag forces are reduced.

Drag

Any object passing through a viscous medium experiences a resistive force opposite to the direction of motion. This resistive force is known as a *drag force* or simply as *drag*. Air and water are the two most common viscous mediums that athletes encounter. Swimmers spend energy to travel through the water due to the drag of the water. Similarly, runners, speed skaters, skiers, and cyclists must work against air resistance.

There are two main aspects when considering drag forces associated with athletic equipment and athletic activities: viscous or friction drag and pressure drag.

Viscous or friction drag results from surface friction between the medium, such as air or water, and the athlete or athletic equipment. Viscous drag depends on the surface properties of the material over which the medium moves.

Pressure drag results from the pressure in front of a moving object being greater than the pressure behind the object. Pressure drag depends on the projected surface area of the object.

Mathematically, friction drag and pressure drag are combined to determine the overall drag force, which is described by

$$F_d = \frac{1}{2} \rho \, v^2 \, A \, C_d \qquad\qquad 4.5$$

where

F_d = drag force,

ρ = density of the medium,

v = speed of the object,

A = frontal area of the object, and

C_d = drag coefficient.

Pressure drag is represented by the first part of the equation, $\frac{1}{2} \rho \, v^2 \, A$. The drag coefficient, C_d, is included in the equation to account for friction drag.

Air density can change depending on the height above sea level. Through theoretical modeling, Ward-Smith (1985) estimated that a 100 m sprint run at Mexico City would be 0.2 s faster than an identical sprint at sea level due to the lower air density at altitude. Water density can change depending on the mineral content of the water. For example, salt water is denser than fresh water. However, the density of air or water is predetermined for any athletic competition. Thus, an athlete can reduce the drag during a competition by decreasing the velocity, the frontal area, or the drag coefficient.

The largest influence in reducing the drag coefficient is achieved by decreasing the velocity (see figure 4.4). It is obvious from equation 4.5 that if the velocity is zero, the drag force is zero and that the drag force increases with the square of the velocity. Most athletic competitions, however, require the velocity to be maximized. Therefore, reducing the velocity is generally not a useful method of reducing drag. However, it has been shown that the work required to overcome drag can be reduced by reducing fluctuations in velocity (Nigg 1983, 1984). The theoretical models used in these studies on swimming and rowing showed that the least amount of

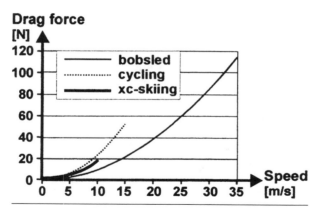

Figure 4.4 Drag forces associated with three sports as a function of the speeds obtained in the different sports.

energy was required when the velocity was constant. For rowing, changes in velocity of the boat occur due to the forward and backward movement of the rower on the sliding seat. The model predicted that changes in velocity of 15 to 25% would lead to an increase of 3 to 10% in the work required by the rower.

Numerous athletic activities involve high speeds where the friction coefficient of apparel can have a large influence on the drag forces. Downhill and cross-country skiing, ski jumping, speed skating, swimming, cycling, bobsled, and luge events are all activities where apparel design can have a substantial influence on performance. High performance athletes in these sports wear tight fitting neoprene or spandex suits to minimize the drag coefficient. Wind tunnel tests on current cross-country suits have shown reductions in drag forces of 6 to 10% in comparison to traditional wool suits and caps (Frederick and Street 1988). Similar tests have shown speed-skating suits to be 2 to 3% faster than traditional woolen suits at high velocities (van Ingen Schenau 1982). However, the woolen suit was faster than the tight suit at velocities below 6 or 7 m/s. The fact that the wool suit was faster at lower velocities than the neoprene or spandex suit is related to the drag resistance for laminar and turbulent flow.

Laminar flow is classified as regular flow where individual fluid particles follow paths that do not cross those of neighboring particles (Massey 1968). Turbulent flow is irregular flow characterized by intermingling fluid particles. For nonstreamlined bodies, laminar flow is associated with larger aerodynamic drag forces than turbulent flow. Thus, the wool suit worn while speed skating reinforces the turbulent flow at lower velocities, resulting in lower drag forces at these low speeds. Introducing turbulent flow by means of small anomalies on athletic

suits has been proposed as a possible means of further reducing the drag coefficient (van Ingen Schenau, de Boer, and de Groot 1989; van Valkenburgh 1988).

Apparel for downhill skiing has been closely regulated for safety reasons. As a result, current suits actually have about 5% higher drag than the first stretch suits introduced about 30 years ago (Holden 1988). The original neoprene suits were hazardous due to the extremely low friction coefficient on snow. If a skier fell, he or she would actually accelerate rather than decelerate in the fallen position. Current suits have placed the neoprene on the inside of the suit and are required to have a minimum porosity to allow the body to breathe. The required porosity has led to the increase in drag forces.

Overcoming drag has been estimated to account for 4 to 8% of the total energy cost of running, depending on the running speed (Davies 1980; Pugh 1970). Due to the higher speeds, drag has been estimated to account for 8 to 13% of the total energy cost of sprinting (Davies 1980; Frohlich 1985; Pugh 1971). Although the specific influence of athletic apparel on drag forces during running has not been addressed, apparel design will have an influence on the drag coefficient. Again due to the larger velocities, apparel will have a larger influence on drag during the sprinting events. Recently sprinters have moved toward a one-piece tracksuit, which minimizes friction drag.

In addition to apparel, other athletic equipment has been modified to decrease drag coefficients. The dimples in a golf ball are essential in reducing drag of the ball, allowing the ball to be hit for distances that cannot be obtained with smooth balls. The layer of air that is in contact with the ball is known as the boundary layer. The boundary layer can be classified as either laminar or turbulent flow. If the air in the boundary layer travels in streams parallel to one another, the flow is laminar. If the air in the boundary layer is not in parallel streams (e.g., random, swirling, and crossing paths), the flow is turbulent. As the air flows around the ball, it eventually separates from the boundary layer causing a turbulent region behind the ball (see figure 4.5a). The drag coefficient depends on the location where this separation occurs. The earlier the separation occurs, the larger the drag coefficient. The dimples in golf balls are designed to produce a turbulent boundary layer, which results in separation of the layers occurring farther back on the ball (see figure 4.5a). The final effect is that the drag coefficient is reduced substantially by using a dimpled golf ball.

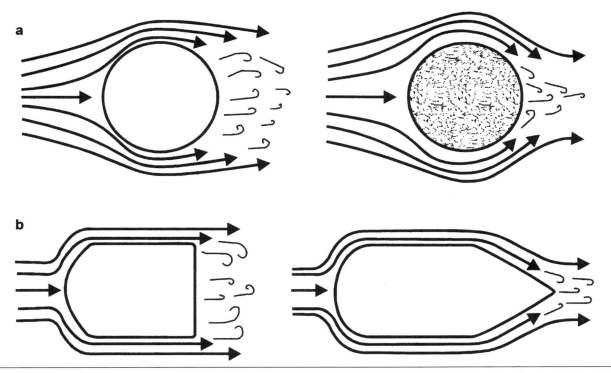

Figure 4.5 Air flow past (*a*) a smooth and a dimpled golf ball and (*b*) a smooth object with and without a fairing. The dimpled ball causes a turbulent boundary layer, which results in the air flow separating from the boundary layer at a point farther back on the ball. Similarly, the fairing results in the air flow separating from the boundary layer at a point further back on the object. The overall result in both cases is a decrease in the drag coefficient.

Reductions in the drag coefficient of bicycles by nearly 50% can be obtained with solid wheels and aerodynamic frames (Pons and Vaughan 1988). Smooth hulls are essential in reducing drag coefficients during rowing, sailing, and windsurfing. Recent bobsled research using wind tunnels has led to a reduction in the drag forces by about 40% (van Valkenburg 1988). Changes in driver and rider positions, helmet configurations, overall shape modifications, and external fairings have led to these substantial reductions of drag forces and improvements of performance. Similar to the dimples in golf balls, fairings delay the separation point of the boundary layer from the air flow, thus reducing the drag (see figure 4.5b). Fairings have also been proposed to reduce drag on luge sleds. Regulations in downhill skiing and ski jumping prevent the use of fairings at the back of ski helmets.

Frontal area and pressure drag are typically influenced more by body position than by athletic equipment. A downhill skier can decrease the drag force by a factor of approximately four by assuming a low tuck versus a standing position (Holden 1988). A speed skater who keeps the upper body horizontal can have a 20% lower drag force than a skater who has an inclined trunk (van Ingen Schenau, de

Boer, and de Groot 1989). Cyclists in a fully crouched position reduce their drag forces by about 30% from an upright posture (Faria and Cavanagh 1978).

There are some instances where reductions in area of athletic equipment have been applied in an attempt to reduce overall drag. Streamlined ski poles and a decrease in the frontal area of ski tips have been used in cross-country and alpine skiing. Streamlining and reducing the projected frontal area of the bobsled and luge have also been attempted.

Mass and Inertia

Mass is defined as the amount of matter in a body that causes it to have weight in a gravitational field and can be thought of as a measure of a body's resistance to translational acceleration. The *moment of inertia* is a quantity that describes the distribution of mass within a body and can be thought of as a measure of a body's resistance to rotational acceleration. The greater the mass or inertia of a piece of equipment, the more energy is required to accelerate the equipment. Conversely, the smaller the mass or inertia, the less energy lost in accelerating the piece of equipment. Any sport requiring equipment to be accelerated or decelerated tends to benefit from a reduction in mass and moment of inertia of the equipment.

Question

Given the same exerted force, compare the kinetic energy of two pieces of equipment with different mass. Assume the following:

General: F = 100 N; t = 2 s.

Equipment A: $m_A = m = 10$ kg.

Equipment B: $m_B = 2m = 20$ kg.

Solution

For equipment A (lighter),

$F = m_A \cdot a.$ Thus,   4.6

$a = 10 \, \text{m/s}^2.$

$v = a \cdot t = 10 \, \text{m/s}^2 \cdot 2 \, \text{s} = 20 \, \text{m/s}.$

$E_k = \dfrac{1}{2} \, mv^2.$

Substituting,

$E_k = \dfrac{1}{2} \cdot 10 \, \text{kg} \cdot (20 \, \text{m/s})^2,$ and

$E_k = 2,000 \, \text{J}.$

For equipment B (heavier),

$F = m_B \cdot a.$ Thus,   4.7

$a = 5 \, \text{m/s}^2.$

$v = a \cdot t = 5 \, \text{m/s}^2 \cdot 2 \, \text{s} = 10 \, \text{m/s}.$

$E_k = \dfrac{1}{2} \, mv^2.$

Substituting,

$E_k = \dfrac{1}{2} \cdot 20 \, \text{kg} \cdot (10 \, \text{m/s})^2,$ and

$E_k = 1,000 \, \text{J}.$

Comments

Under the influence of the same force, the heavier piece of equipment will have half the velocity of the lighter piece of equipment. As a result, the lighter piece of equipment will have twice the kinetic energy of the heavier piece of equipment. Thus, the lighter the equipment the easier it can be accelerated, which leads to higher velocities obtained.

During batting, baseball players who are not strong enough to sufficiently accelerate the bat will "choke up" on the bat. They will move their hands away from the extreme end of the bat and hold the bat farther up the handle. By doing this, they have effectively reduced the moment of inertia of the bat with respect to the axis of rotation, thus making it easier to accelerate. As a result they are able to obtain a higher bat velocity, allowing the possibility of transferring a larger amount of energy to the ball.

Materials such as fiberglass, aluminum, graphite, Kevlar, titanium, and ceramics have had a large influence on decreasing the mass of athletic equipment. Nordic skis, ski poles, bicycles, racquets, golf clubs, bats, boats, paddles, and skates have benefited from the reduced mass of the new materials. The reduction in mass of protective equipment used in contact sports like football and hockey also has a large influence on reducing the energy expenditure of an athlete.

Energy Dissipation in Materials at Impact

When two objects impact one another energy is typically lost to heat. The two objects will have a relative velocity after the impact that is less than their relative velocity before the impact. The change in velocity is dependent on the amount of energy lost and is represented by the coefficient of restitution:

$$e = \frac{v'_A - v'_B}{v_A - v_B} \qquad 4.8$$

where

e = coefficient of restitution,

v_A = velocity of object A before impact,

v_B = velocity of object B before impact,

v'_A = velocity of object A after impact, and

v'_B = velocity of object B after impact.

The contact of a tennis racquet with a tennis ball, a bat with a baseball, a shoe with a soccer ball, or a hand with a volleyball are examples where the coefficient of restitution plays an important role in reducing the energy lost at impact. The coefficient of restitution depends on the structure and materials of both the striking implement and the object being struck. The coefficient of restitution also depends on the impact velocity and temperature of the impacting ball. It has been shown that the coefficient of restitution decreases with increasing impact velocity (Chapman and Zuyderhoff 1986; Snowden and Dowell 1991) and increases with increasing ball temperature (Chapman and Zuyderhoff 1986; Hay 1978).

In the example of a bat and a baseball, the harder the bat, the higher the coefficient of restitution and the higher the amount of energy transferred to the ball (House 1996). The result of the larger amount of energy transferred to the ball is a higher ball velocity. If an athlete uses a softer bat, the athlete would have to swing harder to achieve the same ball velocity. Thus, due to the energy lost between the ball and the bat, the athlete would have to expend more energy to achieve the same results. Bats constructed of metal, graphite, or ceramic materials are generally harder and have a higher coefficient of restitution than their wooden counterparts.

Studies into the rebound heights of tennis balls have determined that the coefficient of restitution increases from the initial value after about 800 impacts and tends to decrease after about 3200 impacts (Rand, Hyer, and Williams 1979). The most likely explanation for this increase is the wearing of the nap on the ball. Additionally, the coefficient of restitution of the tennis balls decreases with age even if the balls are not used.

The coefficient of restitution differs for balls used in different sports (see table 4.4) and is particularly important for sports involving inflatable balls. Energy requirements during volleyball, basketball, soccer, and rugby can be increased substantially if the balls are underinflated and the coefficient of restitution is too low.

Stability

Depending on the activity, athletes may expend large amounts of energy to stabilize their movements. For example, cross-country skiers expend energy to control and stabilize ankle joint movements to appropriately position their skis on the snow. Volleyball and basketball athletes use muscular contraction to stabilize the ankle joint during landing after a jump.

Whenever an athlete uses muscle activation to stabilize a movement, energy is lost. Therefore, if stability can be achieved by other methods, by use of athletic equipment, for example, the energy the athlete would normally expend to stabilize the movement could be conserved and maybe used to enhance performance. Cross-country ski boots use high-cut constructions to increase stability of the ankle joint. It is speculated that stiff high-cut cross-country boots reduce the amount of energy required by the athlete to stabilize the ankle joint. Similarly, basketball shoes are traditionally high-top shoes in order to provide support and stability at the ankle joint, especially during landing movements. Volleyball players commonly use ankle braces to protect their ankle joints during competition. The braces provide external support, which helps reduce the incidence of ankle injuries, but also reduces the stabilizing influence required by the muscles crossing the ankle joint.

Vibrations

The human body is a composition of rigid and nonrigid structures. The rigid structures include primarily bones. The nonrigid structures include muscles, fat, heart, kidney, and others. The nonrigid structures are attached to the rigid structures through connective tissue. Impact forces excite the soft tissues, producing vibrations of the soft tissue relative to the underlying rigid structures. Exposure to long-term or high-energy vibrations often results in subjective discomfort, reduction of performance in the work place, and pathological changes to the nervous and vascular systems (Sakakibara 1994).

External body locations and internal organs have a resonance effect around 4 Hz. The upper extremities show resonance effects at about 10 to 20 Hz (Dupius and Jansen 1981). However, these frequencies (4 to 20 Hz) are right in the frequency range of

Table 4.4 Approximate Coefficient of Restitution of Different Balls for Low-Impact Velocities

Type of ball	Type of surface	Coefficient of restitution	Reference
Golf	Floor	0.83–0.89	Snowden and Dowell 1991
Tennis	Racquet	0.76–0.88	Hatze 1993
Racquet	Floor	0.74–0.88	Snowden and Dowell 1991
Hand	Floor	0.72–0.85	Snowden and Dowell 1991
Rugby	Floor	0.77–0.81	Gallagher and Cooke 1998
Soccer	Floor	0.69–0.80	Snowden and Dowell 1991
Squash	Plywood	0.48–0.60	Chapman and Zuyderhoff 1986

impact forces during landing in sport activities. The musculoskeletal system must activate the muscles accordingly to minimize excessive vibrations at these frequencies. The additional muscle activation, however, costs work. The additional work depends on both the magnitude and frequency of the induced vibrations.

Athletic footwear is one area of equipment that can have an influence on the vibrations transferred to the human body. A recent theoretical model (Nigg and Anton 1995) predicted that viscous materials implemented in a shoe should reduce the work requirements during locomotion. It was speculated that the reduction of work was a result of the damping of the lower extremity vibrations by the viscous material. Results of a recent experimental investigation where oxygen consumption was measured on subjects running in shoes with and without a viscous midsole tend to support the theoretical predictions (Stefanyshyn and Nigg 1998).

Striking implements such as baseball bats, golf clubs, and tennis racquets are prone to vibrations after impact with the ball. The frequency and amplitude of the vibrations depend on two factors. The first is the manner in which the equipment is held; for example, the amplitude of the vibrations depends on the grip pressure (Hatze 1976). The second is the construction of the implement. For example, one way to reduce the vibrations of a tennis racquet (or any striking implement) is to ensure that the ball contacts the center of percussion of the racquet (Elliot, Blanksby, and Ellis 1980). The center of percussion is the point at which a striking implement impacts an object without causing an unbalanced reaction force at the pivot point, which, for most striking implements, is the point where the implement is being grasped by the hands. Athletes will often refer to the center of percussion as the sweet spot. If the ball contacts the implement at a point other than the sweet spot an athlete will feel the vibrations, which may sting the athlete's hands. Additionally, the ultimate performance is compromised because part of the kinetic energy of the implement is lost.

Manufacturers of striking implements have tried to optimize the location of the center of percussion. Conventional tennis racquets did not have the center of percussion in the geometric center of the strings (Brody 1979). By modifying the shape of the racquet head and distributing the weight appropriately to the perimeter of the racquet, manufacturers have been successful in repositioning the center of percussion in the geometric center of the racquet head. Perimeter weighting has also been used suc-

cessfully in golf clubs to enlarge the sweet spot. It has been indicated that the sweet spot of an aluminum bat is larger than that of a wooden bat (Bryant et al. 1977).

Additional Situations With an Unnecessary Loss of Energy

During sprinting, an athlete lands on the ball of the foot. Immediately after impact, the rear of the foot rotates backward toward the track. One may argue that this rotation is counterproductive for the sprinting movement as the athlete loses energy when the leg moves down and has to perform work to lift the heel and the leg up again. Several years ago a spiked shoe was introduced where a wedge was placed under the midfoot to prevent the heel from rotating backward (see figure 4.6). It was speculated that the energy is not lost with this construction and that no additional work needs to be performed. These shoes were used for some exceptional sprinting times. However, it has never been shown experimentally that this wedge was responsible for the excellent sprinting results.

The metatarsophalangeal joint absorbs large amounts of energy as athletes roll onto the ball of the foot during running and jumping movements (Stefanyshyn and Nigg 1997). However, very little energy is produced during takeoff since this joint is only minimally extended. Therefore, energy is dissipated and lost during the bending of the joint in the shoe and foot structures. Improvements in shoe design that reduce the initial bending of the metatarsophalangeal joint may have an influence in reducing the amount of energy the athlete loses at this joint.

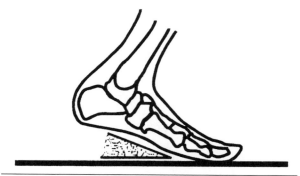

Figure 4.6 Schematic illustration of a theoretical concept to prevent energy loss during sprinting. The wedge placed under the midfoot is to prevent the loss of energy due to the foot rotating backward during ground contact.

Situations Where a Loss of Energy is Advantageous

In some athletic activities, energy loss is a required component of the activity. For example, braking or stopping is essential for control in some sports (e.g., cycling, skating, and skiing). Braking is nothing more than the dissipation of kinetic energy to thermal energy or heat. For most of this chapter, this transfer to thermal energy has been equated to a loss of energy because the energy did not serve a beneficial purpose to the athlete. Thus, although a major goal of athletic equipment is to minimize the loss of energy, there are certain circumstances where athletic equipment is specifically designed to lose energy.

Summary

Over the past 20 to 30 years, sport scientists and sport equipment manufacturers have investigated ways of improving athletic equipment to make sport safer and to enhance performance. The result is equipment that is stronger, lighter, more durable, and more pleasant to use. Consequently, sport performances are faster, higher, longer, and more accurate than they used to be. In fact every world record in sport that was set before 1980 has been broken (Begley and Rogers 1996), an indication of the recent developments in athletic ability and improvements in athletic equipment.

Two of the main principles that have led to the improvements in equipment are

- the increase in the return of conservative energy and
- the reduction of the loss of nonconservative energy.

Energy return refers to the storage and re-use of elastic strain energy. The amount of energy that can be stored by a piece of athletic equipment depends on the stiffness and the deformation of the equipment. Deformation is the more important variable since energy storage increases linearly with increasing stiffness and quadratically with increasing deformation. In general, a system must fulfill several conditions to return energy. It must be able to return the energy at the right time with the right frequency at the right location. Return of energy to improve athletic performance has been studied for different equipment such as trampolines, diving boards, vaulting poles, sport surfaces, and sport shoes.

By reducing the amount of energy lost, an athlete can perform a given task while doing less work. The end result is that the athlete has an additional work capacity to apply toward improving performance. Energy applications in equipment for aspects that are related to an improvement of performance include energy loss due to friction, drag, mass and inertia, and dissipation in materials, stabilization, and vibrations. Reduction of the loss of energy has been studied for equipment such as athletic apparel, skis, balls, skates, and bicycles.

References

Alexander, R.M., and Bennet, M. 1989. How elastic is a running shoe? *New Scientist,* 15 July, 45–46.

Angulo-Kinzler, R.M., Kinzler, S.B., Balius, X., Turro, C., Caubet, J.M., Escoda, J., and Prat, J.A. 1994. Biomechanical analysis of the pole vault event. *Journal of Applied Biomechanics* 10: 147–65.

Baker, J.A., and Wilson, B.D. 1978. The effect of tennis racket stiffness and string tension on ball velocity after impact. *Research Quarterly* 49: 255–59.

Begley, S., and Rogers, A. 1996. How high? How fast? *Newsweek,* 22 July: 22–34.

Bosworth, W. 1981. What? String tighter for more control? *World Tennis,* May, 18.

Braff, T.J., and Dapena, J. 1985. A two-dimensional simulation method for the prediction of movements in pole-vaulting. In *Biomechanics IX.* Vol. B, ed. D.A. Winter et al., 458–63. Champaign, IL: Human Kinetics.

Brody, H. 1979. Physics of the tennis racket. *American Journal of Physics* 47: 482.

Brody, H. 1984. That's how the ball bounces. *Physics Teacher,* November, 494.

Bryant, F.O., Burkett, L.N., Chen, S.S., Krahenbuhl, G.S., and Lu, P. 1977. Dynamic and performance characteristics of baseball bats. *Research Quarterly* 48: 505–9.

Chapman, A.E., and Zuyderhoff, R.N. 1986. Squash ball mechanics and implications for play. *Canadian Journal of Applied Sport Sciences,* March: 47–54.

Davies, C.T. 1980. Effects of wind assistance and resistance on the forward motion of a runner. *Journal of Applied Physiology* 48: 702–9.

de Koning, J.J., de Groot, G., and van Ingen Schenau, G.J. 1992. Ice friction during speed skating. *Journal of Biomechanics* 25: 565–71.

Dupius, H., and Jansen, G. 1981. Immediate effects of vibration transmitted to the hand. In *Man under vibration,* ed. G. Bianchi, K.V. Frolov, and A. Oledzki. Warszawa: Polish Scientific.

Ekevad, M., and Lundberg, B. 1997. Influence of pole length and stiffness on the energy conversion in pole-vaulting. *Journal of Biomechanics* 30: 259–64.

Elliot, B.C. 1982. The influence of tennis racket flexibility and string tension on rebound velocity following a dynamic impact. *Research Quarterly in Exercise and Sport* 53: 277.

Elliot, B.C., Blanksby, B.A., and Ellis, R. 1980. Vibration and rebound velocity characteristics of conventional and oversized tennis rackets. *Research Quarterly in Exercise and Sport* 51: 609.

Ellis, R., Elliot, B., and Blanksby, B. 1978. The effect of string type and tension in jumbo and regular sized tennis racquets. *Sports Coach* 2: 32.

Faria, I., and Cavanagh, P. 1978. *The physiology and biomechanics of cycling*. New York: Wiley.

Frederick, E.C., and Street, G.M. 1988. Nordic ski racing biomedical and technical improvements in cross-country skiing. *Scientific American*, February, T20–22.

Frohlich, C. 1985. Effect of wind and altitude on record performance in foot races, pole vault, and long jump. *American Journal of Physics* 53: 726.

Gallagher, J., and Cooke, C.B. 1998. A comparison of the mechanical properties of five rugby balls with specific reference to place kicking. *Journal of Sport Sciences* 16: 7.

Hatze, H. 1976. Forces and duration of impact and grip tightness during the tennis stroke. *Medicine and Science in Sports and Exercise* 8: 88–95.

Hatze, H. 1993. The relationship between the coefficient of restitution and energy losses in tennis rackets. *Journal of Applied Biomechanics* 9: 124–42.

Hay, J.G. 1978. *The biomechanics of sports techniques*. 2d ed. Englewood Cliffs: Prentice Hall.

Holden, M.S. 1988. The aerodynamics of skiing. *Scientific American*, February, T4–5.

House, G.C. 1996. Baseball and softball bats. In *Sports and fitness equipment design*, ed. E.F. Kreighbaum and M.A. Smith, 117–25. Champaign, IL: Human Kinetics.

Jobse, H., Schuurhof, R., Cserep, F., Schreurs, A., and de Koning, J. 1990. Measurement of push-off force and ice friction during speed skating. *International Journal of Sport Biomechanics* 6: 92–100.

Jones, I.C., Pizzimenti, M.A., and Miller, D.I. 1993. A springboard feedback system: Considerations and implications for coaching. In *US Diving Sport Science 1993 Proceedings*, ed. R. Malina and J.L. Gabriel, 67-79. Indianapolis: U.S. Diving.

Kobayashi, T. 1973. Studies on the properties of ice in speed skating rinks. *Ashrae Journal* 73: 51–56.

Kooi, B.W., and Kuipers, M. 1994. The dynamics of springboards. *Journal of Applied Biomechanics* 10: 335–51.

Leigh, D.C., and Lu, W. 1992. Dynamics of the interactions between ball, strings, and racket in tennis. *International Journal of Sport Biomechanics* 8: 181–206.

Massey, B.S. 1968. *Mechanics of fluids*. London: Van Nostrand.

McMahon, T.A. 1987. The spring in the human foot. *Nature* 325: 108–9.

McMahon, T.A., and Greene, P.R. 1978. Fast running tracks. *Scientific American* 239: 112–21.

McMahon, T.A., and Greene, P.R. 1979. The influence of track compliance on running. *Journal of Biomechanics* 12: 893–904.

Nigg, B.M. 1983. Selected methodology in biomechanics with respect to swimming. In *Biomechanics and medicine in swimming*, ed. A.P. Hollander, P.A. Huijing, and G. de Groot, 72–80. Champaign, IL: Human Kinetics.

Nigg, B.M. 1984. The influence of variation in velocity on the work in cyclic movements. In *Human locomotion III*, edited by A.B. Thornton-Trump. Proceedings of the Third Biannual Conference of the Canadian Society of Biomechanics. 35–36.

Nigg, B.M. 1986. Experimental techniques used in running shoe research. In *Biomechanics of running shoes*, ed. B.M. Nigg, 27–61. Champaign, IL: Human Kinetics.

Nigg, B.M., and Anton, M. 1995. Energy aspects for elastic and viscous shoe soles and playing surfaces. *Medicine and Science in Sports and Exercise* 27: 92–97.

Nigg, B.M., and Segesser, B. 1992. Biomechanical and orthopedic concepts in sport shoe construction. *Medicine and Science in Sports and Exercise* 24: 595–602.

Pons, D.J., and Vaughan, C.L. 1988. Mechanics of cycling. In *Biomechanics of sport*, ed. C.L. Vaughan, 289–315. Boca Raton, FL: CRC Press.

Pugh, L.G. 1970. Oxygen intake in track and treadmill running with observations on the effect of air resistance. *Journal of Physiology* 207: 823–35.

Pugh, L.G. 1971. The influence of wind resistance in running and walking and the mechanical efficiency of work against horizontal or vertical forces. *Journal of Physiology* 213: 255–76.

Rand, K.T., Hyer, M.W., and Williams, M.H. 1979. A dynamic test for comparison of rebound characteristics of three brands of tennis balls. In *Proceedings of the National Symposium of Racquet Sports*, ed. J.L. Groppel, 240. Champaign, IL: University of Illinois.

Sakakibara, H. 1994. Sympathetic responses to hand-arm vibration and symptoms of the foot. *Nagoya Journal of Medical Science* 57: 99–111.

Shorten, M.R. 1993. The energetics of running and running shoes. *Journal of Biomechanics* 26 (Suppl. 1): 41–51.

Snowden, S.R., and Dowell L.J. 1991 The effect of velocity at impact on the coefficient of restitution of balls used in physical education sports classes. *Applied Research in Coaching and Athletics Annual*, March, 107–118.

Stefanyshyn, D.J., and Nigg, B.M. 1997. Mechanical energy contribution of the metatarsophalangeal joint to running and sprinting. *Journal of Biomechanics* 30: 1081–85.

Stefanyshyn, D.J., and Nigg, B.M. 1998. The influence of viscoelastic midsole components on the biomechanics of running. *Abstracts of the Third World Congress of Biomechanics,* 379.

Turnball, A. 1989. The race for a better running shoe. *New Scientist,* 15 July, 42–44.

Valiant, G. 1994. Evaluating outsole traction of footwear. In *Proceedings of the Eighth Biennial Conference of the Canadian Society of Biomechanics,* Calgary, Canada.

Valiant, G.A., McGuirk, T., McMahon, T.A., and Frederick, E.C. 1985. Static friction characteristics of cleated outsole samples on AstroTurf. *Medicine and Science in Sports and Exercise* 17: 222.

van Ingen Schenau, G.J. 1982. The influence of air friction in speed skating. *Journal of Biomechanics* 15: 449–58.

van Ingen Schenau, G.J., de Boer, R.W., and de Groot, G. 1989. Biomechanics of speed skating. In *Biomechanics of sport,* ed. C.L. Vaughan, 121–67. Boca Raton, FL: CRC Press.

van Valkenburg, P. 1988. The aerodynamics of the bobsled. *Scientific American,* February, T10–14.

Vaughan, C.L. 1980. A kinetic analysis of basic trampoline stunts. *Journal of Human Movement Studies* 6: 236–51.

Ward-Smith, A.J. 1985. A mathematical theory of running, based on the first law of thermodynamics and its application to the performance of world class athletes. *Journal of Biomechanics* 18: 337–49.

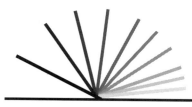

Chapter 5

The Three Modes of Terrestrial Locomotion

A.E. Minetti

"[Q]uadrupedante putrem sonitu quatit ungula campum."
[Hooves shake the dusty ground with the sound of four limbs.]

Virgilio, *Eneide*

When reading this verse with accents on the letters in bold, the sound of a galloping horse emerges (ternary slow canter).

This chapter describes the most diffused terrestrial gaits with the aim to classify them according to the motion dynamics of the body center of mass, both in bipeds and in quadrupeds. Mechanical paradigms, that is, passive objects displaying similar kinetics, are used to illustrate the three energy-saving strategies helping to maintain locomotion at a reasonable metabolic cost over a broad range of speeds. The pendulum and the pogo stick (single and double) show how the different types of mechanical energy (potential, kinetic, and elastic) exchange in the different gaits to reduce the new mechanical work the muscles must supply. For each gait category experimental tracings of mechanical energy and the metabolic cost are discussed, together with some other parameters related to locomotion. In addition, the transition speed between gaits, the effect of gradient on the mechanics and energetics of locomotion, and the current research on moving in different gravitational environments are illustrated.

Mechanics and Energetics of Locomotion

When driving a car it is customary to start with the lowest gear. While accelerating, there comes a time when, often unconsciously, the next gear is chosen, and this process is repeated as many times as required to reach the desired cruising speed. When legged animals move on the ground the situation is very similar: they start with a walk, then switch to a trot (or a run), or even to a gallop to cope with the speed increase. Changing gait or gear is a way to adapt the same actuator (musculoskeletal system or cylinder-piston complex) to different progression speeds. The importance of expanding the foraging territory and the need of escaping predators or hunting prey were probably the greatest evolutionary stimuli for terrestrial species to develop strategies devoted to fast and economic locomotion. The different resulting gaits, which follow similar mechanical paradigms in bipeds and quadrupeds, achieve the body movement by using actuators (muscles), springs (tendons and ligaments), and levers (bones and joints)

...mbinations, with at least one energy-...

...derstand the basic principles ruling loco-motion it is preferable to refer first to the trajectory of the center of mass (COM) of the body, defined as the point (inside or outside the body, depending on the posture) where all of the mass could be considered to concentrate. This simple approach to the mechanical studies of bodies in motion cannot reveal all the determinants of movement, but the overall paradigm is easily detectable. For example, the exchange among different types of energy (potential: vertical position; linear kinetic: linear speed; rotational kinetic: angular speed; see chapter 2) obtainable by the displacement of COM, characterizes many attractions in amusement parks, many toys, and mechanical clocks, where after an initial supply of energy the successive motion passively occurs. It is the case of roller coasters (potential $\leftrightarrow$ kinetic energy) and yo-yos (potential $\leftrightarrow$ rotational/linear kinetic). The other important energy form, not directly detectable by inspecting the COM trajectory, is the elastic energy. This derives from the capability of certain materials to deform (under compressive, tensile, or bending stress) and successively recover the initial shape, storing and releasing mechanical energy in the cycle (see chapter 2). Using a pogo stick, bungee jumping, and bouncing on the trampoline are examples of motion where elastic energy is involved.

These three energy forms (potential, kinetic, elastic) and their interchange are essential to understand gross body movement, as schematically depicted in figure 5.1. At the end of the running phase, the body COM of pole-vaulters possesses mainly kinetic energy, which comes from the running speed. During the pole bending, most of this energy is converted into elastic energy. Subsequently, the pole extension causes the body COM to move upward, increasing the potential energy just enough to permit the body to clear the bar (under optimal circumstances). In many body motions a remarkable amount of mechanical energy is exchanged among the cited forms, reducing the new energy required to be input by muscle contraction. In this chapter the most diffused categories of legged animal gaits are analyzed in this respect, and the mechanical objects (paradigms) more closely resembling the observed body dynamics are suggested. Metabolic aspects of each gait are described at the end of each subsection.

When talking about legged locomotion, many variables are involved:

- The progression speed (m · s^{-1})
- The stride and step length (m)
- The stride frequency (Hz)
- The gradient (in percentage, i.e., the ratio between the vertical and the horizontal distance covered when moving on a slope)
- Duty factor (i.e., the fraction of the stride period at which one foot is in contact with the ground)
- Amount of the vertical (F_y, N) and horizontal (F_x, N) ground reaction force
- Body COM mechanical energy

The body COM mechanical energy includes PE (potential energy), KE_x (kinetic in the horizontal direction), KE_y (kinetic in the vertical direction), and EL (elastic energy). Total energy $TE = PE + KE_y + KE_x + EL$ (J). The mechanical work of locomotion per unit distance is composed of the external work (W_{ext}, needed to move the body COM with respect to the environment, equal to the sum of positive changes in TE) and the internal work (W_{int}, necessary to accelerate the limbs with respect to the body COM (J · m^{-1}); Cavagna and Kaneko 1977). Other variables include the mechanical energy recovery (rec %, reflecting the ability of a system to save energy by exchanging PE and KE), the metabolic cost of locomotion per unit distance (C, ml O$_2$· m^{-1}), and the locomotor efficiency (eff = mechanical work / metabolic work).

In order to compare mechanical and metabolic data, and to allow for different body size, W_{ext} and C are often expressed in J per kg of body mass and per

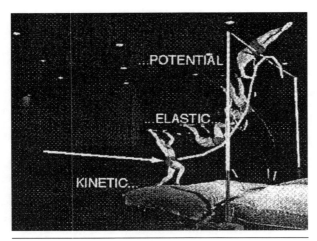

Figure 5.1 Pole-vaulting as an example of mechanical energy exchange.

Adapted, by permission, from T.A. McMahon and J.T. Bonner, 1983, *On size and life.* (New York: Scientific American Library), 158.

meter traveled (1 ml O_2 = 20.1 J). Only a subset of these variables is considered in this chapter. In all the following, humans (mass 75 kg) and horses (mass 538 kg) are proposed as prototypes for bipeds and quadrupeds, respectively, because of the availability of experimental data for these two species.

Slow Speed: Walking and Brachiation

Walking is the gait bipeds and quadrupeds adopt at slow speeds. Walking consists of placing one foot after the other on the ground in a way that at least one of them is always in contact. In bipeds stride phases can exist where both feet are simultaneously in contact (double support phase), while in quadrupeds the limb sequence is set diagonally as [:fore left (FL), hind right (HR), fore right (FR), hind left (HL):], and stride phases where up to three hooves simultaneously touch the ground occur, depending on speed. The absence of a flight phase assures that we are considering a walking gait.

The speed of walking usually ranges from 0.3 to 2.0 m · s⁻¹, both for humans and horses. Figure 5.2 shows the spontaneously chosen speeds at different ages in humans, which were also noticed in adults (Bornstein and Bornstein 1976) to be directly proportional to the population density (slower in villages and faster in big cities).

Mechanics

It is a common experience to find analogies between walking and the movement of a pendulum. In bipeds the upper limbs oscillate forward and backward, while the lower ones behave like a pendulum during the swing and like an inverted pendulum during the stance. Quadrupedal walking is similar to bipedal walking, since it can be viewed as two bipeds, one walking in front of the other, moving their lower limbs with a phase lag of 75% of the cycle between the two. Arboreal primates such as gibbons *(Hylobates lar)* adopt a highly specialized form of suspensory locomotion called brachiation, consisting of swinging the body while hanging from tree branches with one upper limb at a time.

The pendulum and the inverted pendulum are the mechanical paradigms for the gaits described above. In an ideal pendulum (and in a rolling egg, a metaphor also used in the past for walking) the motion can be indefinitely maintained by the exchange of potential and kinetic energy. In fact, when the pendulum is still (KE = 0) the vertical position of the COM is highest (max PE), while at the lowest point of the swing trajectory (min PE) the speed is at a maximum (max KE, see figure 5.3). The constancy of the total mechanical energy of COM, obtained by summing symmetrical (or identical 50% of the cycle out of phase) KE and PE curves, indicates that no extra work (W_{ext} = changes in TE) is needed in an ideal pendulum.

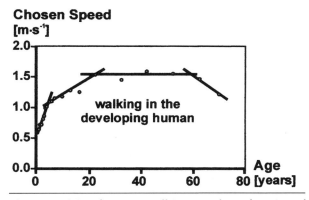

Chosen Speed [m·s⁻¹]

Figure 5.2 Most frequent walking speed as a function of age in humans. Each point corresponds to the average of about 40 measurements. Line segments have been added to the original figure.

Adapted from Rose, J., Gamble, J.G., Lee, J., Lee, R., and Haskell, W.L. 1991. The energy expenditure index: A method to quantitate and compare walking energy expenditure for children and adolescents. *J. Pediatr. Orthop.* 11:571-578; Sutherland, D.H., Olshen, R.A., Biden, E.N., and Wyatt, M.P. 1988. *The development of adult walking.* London: MacKeith Press; Waters, R.L., Hislop, H.J., Perry, J., Thomas, L., and Campbell, J. 1983. Comparative cost of walking in young and old adults. *J. Orthop. Res.* 1:73-76.

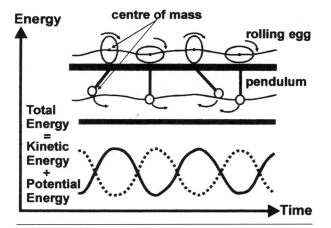

Walking: a pendulum-like gait

Figure 5.3 A rolling egg and a pendulum are shown at different phases of their motion. The vertical position of the center of mass (solid upper curves) determines the pattern of potential energy (lower solid curve), whose sum with the kinetic energy (dotted curve) results in a straight line, the total mechanical energy.

The Speed Limit of Walking

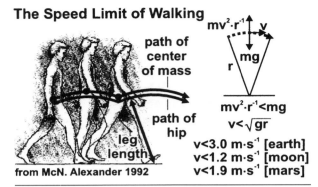

v<3.0 m·s⁻¹ [earth]
v<1.2 m·s⁻¹ [moon]
v<1.9 m·s⁻¹ [mars]

from McN. Alexander 1992

Figure 5.4 The concept of speed limit of walking is illustrated. In the vector diagram on the right-hand side, m is the body mass, g is gravity acceleration, r is the lower limb length, and v is the tangential speed of the inverted pendulum (assumed to correspond to the progression speed).

The drawing on the left is reprinted, by permission, from R.McN. Alexander, 1992. *Exploring biomechanics: Animals in motion.* (New York: Scientific American Library).

An ideal pendulum also has inherent limits. It has already been mentioned that the lower limb moves as an inverted pendulum during stance, thus also the body COM approximately follows a circular trajectory (see figure 5.4). By traveling along the arc two forces act on the COM, downward gravity and upward centrifugal force (radiating from the center of rotation), which depends on the tangential speed and on the radius of the pendulum. There is a speed beyond which the upward force exceeds the downward one, causing the pendulum to leave the ground. As shown in figure 5.4, the limit speed of walking is proportional to the square root of the lower limb length, and to the

acceleration of gravity, resulting in about 3.0 m · s⁻¹ on Earth and about 1.2 and 1.9 m · s⁻¹ on the moon (17% g) and on Mars (40% g), respectively. Beyond these speeds walking is no longer feasible and another gait (running, trotting, or hopping) has to be adopted. However, in a sport discipline called racewalking, athletes can complete the walking race at a speed near to 4 m · s⁻¹. This is done by exaggerating the downward tilt of the contralateral hip, with respect to the stance limb, which results in the lowering of the body COM (this strategy reduces the radius of the circular trajectory). In reality, we do not approach the limit speed before we change gait, in both humans and horses (similarly in cars, we do not force an intermediate gear up to the maximum rpm). The spontaneous transition speed between walking and running, set at about 2.0 m · s⁻¹ for both species (for humans, see Minetti, Ardigò, and Saibene 1994a), seems to be ruled by criteria other than the simple feasibility, such as the minimization of the metabolic energy cost (expressed per unit distance or per stride).

Experimental measurements of COM mechanical energy, obtained via motion analysis (see figure 5.5), show that brachiation and both bipedal and quadrupedal walking occur in agreement with the pendulum paradigm. While it is more difficult to realize it in quadrupeds, the PE and KE curves are out of phase, so that the changes in total energy (TE = PE + KE) are attenuated. This is reflected by a smaller amount of fresh mechanical work (reflected by the smaller oscillations of TE) that has to be input to the system in order to keep it moving with a relatively high total energy. Thus the pendulum-like paradigm can be considered the first energy-saving strategy for terrestrial locomotion.

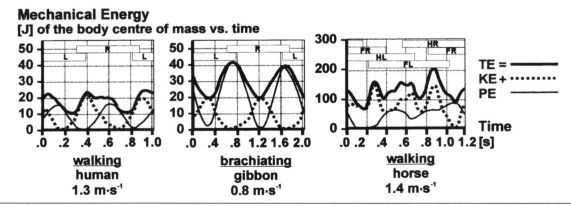

Mechanical Energy
[J] of the body centre of mass vs. time

TE = ———
KE + ·········
PE ———

walking human 1.3 m·s⁻¹

brachiating gibbon 0.8 m·s⁻¹

walking horse 1.4 m·s⁻¹

Figure 5.5 Experimental time course of COM mechanical energies during a typical stride (= 2 steps) in bipeds (left-hand side) and a quadruped (right-hand side). In each graph the energy tracings have been offset in order to allow the shape comparison between them. Data source: human (75 kg; Minetti, Ardigò, and Saibene 1993), gibbon (8 kg; Chang, Bertram, and Ruina 1997), horse (538 kg; Minetti et al. 1999).

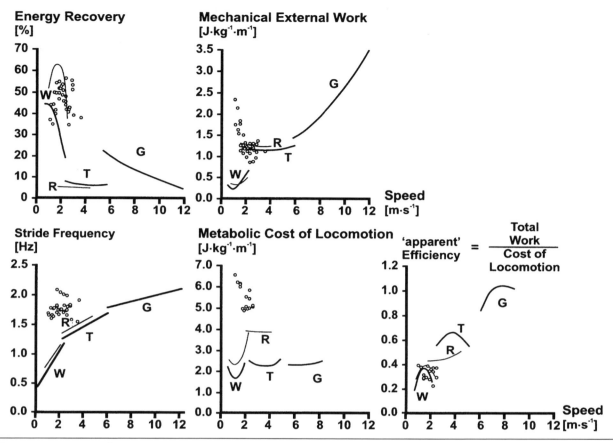

Figure 5.6 A group of gait parameters, shown as a function of speed. Thin and thick curves refer to humans and horses, respectively. Labels reflect the different gait: walking (W), running (R), trotting (T), galloping (G). Open circles represent human skipping.

Adapted, by permission, from A.E. Minetti, 1998. "The biomechanics of skipping gaits: A third locomotion paradigm?" *Proceedings of the Royal Society of London* B265:1227-1235.

Unfortunately, walking does not strictly resemble the ideal pendulum. A parameter called *energy recovery* has been introduced (Cavagna and Kaneko 1977) to account for the amount of energy actually exchanged between PE and KE. Such a parameter, equal to 100% in an ideal pendulum, reaches slightly more than 60% in walking humans (Cavagna and Kaneko 1977) and about 45% in full size horses (Minetti 1998).

Other similarities between bipedal and quadrupedal walking are apparent in figure 5.6. The dependency on speed of the stride frequency and of the mechanical external work (W_{ext}=sum of positive changes in TE curve) are closely related in the two species.

Energetics

To investigate how much metabolic energy is required for locomotion, physiologists collect expired gases. By measuring the amount of oxygen extracted from inspired air in the unit time ($\dot{V}O_2$)

both at rest and during steady state exercise, it is possible to calculate the net metabolic cost of locomotion (C) as (exercise $\dot{V}O_2$ – rest $\dot{V}O_2$) · speed^{-1}. The comparison with mechanical data can be done by converting 1 ml O_2 into 20.1 J. Thus C units are J/m (similar to liters of gasoline per mile traveled in cars) or, by dividing for body mass, J · (kg m)$^{-1}$.

Figure 5.6 shows that the relationship between the cost of locomotion and speed is similar in humans and horses, showing a minimum cost of walking at about 1.1 m · s^{-1}. Thus an optimum walking speed exists at which the metabolic expenditure is minimized for traveling a unit distance. By dividing the mechanical work by the cost of locomotion, we obtain the apparent efficiency of the movement. It is called "apparent" because the accuracy of the numerator is affected by the measuring technique (see later discussion). The numerator of the ratio generating the points plotted in figure 5.6 is the sum of positive W_{ext} and positive W_{int}, the last one accounting for the need to accelerate and decelerate the body segments with respect to the overall COM.

Intermediate Speed: Running, Trotting, and Hopping

Bipeds run at high speed and quadrupeds trot at intermediate speeds (with respect to their maximum). Some quadrupeds, such as the gnu, pass from walking to gallop without using an intermediate gait (Pennycuick 1975). A running biped alternately places the feet on the ground, with a flight phase between each placement. A trotting quadruped can be seen as two running bipeds, one running in front of the other, with a phase lag of 50% of the cycle between the two (a less frequent variation of the running gait in quadrupeds, called the rack, shows 0% phase lag). The speed of running and trotting usually ranges from 2.0 to 5.0 m · s⁻¹, both for humans and horses. The presence of the flight phase assures a bouncing gait (although at slow trotting speeds the flight phase is hardly detectable).

Talking about bouncing and thinking about a ball is one thing. That's why the bouncing ball was classically proposed as the mechanical paradigm for all bouncing gaits. Figure 5.7 shows the energy time course of the ball COM during a bounce. It is noteworthy that both PE and KE are at a minimum during the contact with the ground. This occurs because the position in the trajectory is lowest (minimum PE) and the vertical speed reaches 0 (the vertical component of KE is nil). Thus, if we just sum

the PE and the KE curve, the obtained TE is not a straight line (apparently, the system will not passively move forever, as an ideal ball should do). The missing element in the computation of TE is the elastic energy (EL), stored in the deformation of the ball material during the impact and released when the material restores its original shape (see gray curve in figure 5.7). By adding also this component, the TE curve is constant (thus the system can keep bouncing forever). A more proper paradigm to illustrate the energy exchange occurring in running is the pogo stick. In fact, while a ball (even an ideal one) gains some rotational energy (at the expense of other energy forms) during the bounce, an ideal pogo stick decelerates during the first half of the contact and re-accelerates during the other half with no energy transfer other than the one occurring between (PE + KE) and EL, more similarly to what happens in legged animals. Thus, the main difference with the walking gait is that, during the contact phase, PE and KE do not exchange between each other, but they both exchange with EL. This can be considered the second energy-saving strategy of terrestrial locomotion.

The experimental tracings of COM mechanical energy of running and trotting (see figure 5.8) resemble the paradigm just described, the measured PE and KE qualitatively behaving as in the bouncing ball and the pogo stick. Although it is not possible to obtain experimental elastic energy curves with noninvasive methodology such as motion analysis and force platforms (only direct tendon and ligament elongation could help in that respect), we could expect that it should be similar to what is shown in figure 5.7 (gray curve).

The mechanical principle similarity between bipedal running and quadrupedal trotting is apparent by looking at figure 5.6 (R and T curves). The energy recovery is almost negligible. The dependency of stride frequency on progression speed is less accentuated than in walking, and the mechanical external work shows some speed independence. The differences emerge for the metabolic cost of locomotion, which is constant in human running (i.e., it costs the same amount of fuel to run a meter at a slow and at a very fast speed) and has a minimum at an intermediate trotting speed in horses. The apparent efficiency, that is, the ratio between the total mechanical work and the metabolic cost, is higher in trotting than in running (67% vs. 46% at 4 m · s⁻¹). These values seem remarkably high, considering the limitations expressed in chapter 8. However, since the measured mechanical external work, which is a component of the numerator of the

Running: a bouncing gait

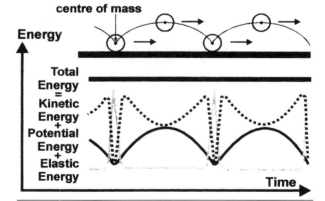

Figure 5.7 A bouncing ball is shown at different phases of its motion. In addition to PE (solid curve) and KE curves (dotted curve), the elastic energy of the ball is represented (gray curve). The sum of the three components results in a straight solid line, reflecting the constant total mechanical energy.

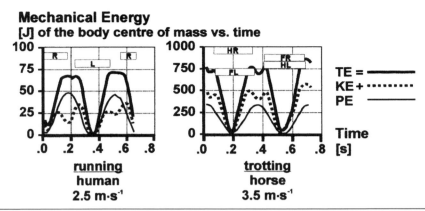

Figure 5.8 Experimental time course of COM mechanical energy during a typical running and trotting stride. In each graph the energy tracings have been offset in order to allow the shape comparison between them. Data sources: human (Minetti, Ardigò, and Saibene 1994b), horse (Minetti et al. 1999).

fraction, does not include the elastic work, as mentioned for the bouncing ball (see figure 5.7), it reflects a work higher than the one actually done by the muscles. Because the mechanical efficiency of muscle ranges from 0 to 35% (Woledge, Curtin, and Homsher 1985), whichever excess with respect to that range can be considered as an indicator of the (minimal) elastic contribution to the observed motion. In the present case the higher apparent efficiency of trotting, and the consequent greater elastic contribution, could be related to the number of limbs involved in each bounce and to the very long tendons attached to short distal limb muscles.

In humans, the Achilles tendon and the ligaments in the arch of the foot have been estimated to store and release 35 and 17 J per running step (Ker et al. 1987; see chapter 3) or 0.6 J · (kg bounce)$^{-1}$. In half-ton horses it has been estimated (Minetti et al. 1999) that during trotting 1.1 J · (kg bounce)$^{-1}$ is stored and released by elastic structures in the limbs. Elastic structures and actuators (muscles) form with the load (ground reaction force to be contrasted) a forced oscillator. Because of some energy dissipation occurring in the musculoskeletal system (for tendons, see chapter 3), it should be better considered as a damped forced oscillator, where great excursions of the COM can be obtained via small excursion of the actuator. Such a device operates optimally at a (resonant) frequency determined by the spring (tendon) and dashpot (dissipation) characteristics. Also the phase of force application within the stride cycle (see also chapter 1) is important to minimize the energy consumption. Recent experiments (Roberts et al. 1997) showed that limb extensor muscles act quasi-isometrically during the contact phase of bouncing gaits, thus supporting the forced oscillator model.

High Speed: Galloping and Skipping

While bipeds use running for the highest speeds, quadrupeds have an additional gear named gallop. The footfall sequence depends on speed and on the gallop variants. The slow gallop, also called canter, is characterised by the sequence [:HL, HR and FL simultaneously, FR, flight:] (this corresponds to the right canter, while the left one is obtained by exchanging L and R). At higher speeds a second flight period is added to the sequence, that is, [:HL, HR, flight, FL, FR, flight:]. This is named right transverse gallop (for the left one, see above). Another variation is the so-called rotatory gallop (with sequence [:HL, HR, flight, FR, FL, flight:]) in the clockwise or counterclockwise variety. It can be noticed that quadrupeds deterministically adopt left or right gallop when they have to turn left or right, respectively, even when the trajectory follows a very shallow curve. Also, when moving uphill, gallop is often preferred to trot. Due to the inherent asymmetry (canter and transverse gallop), the sagittal plane of the quadruped is slightly rotated toward the direction of the gallop (left or right) even during straight locomotion.

In order to understand the mechanics of gallop, and the corresponding paradigm, it is simpler to investigate skipping, a gait humans (but also some birds, jerboas, and lemurs) adopt at the age of about 4.5 years for short bursts of locomotion (Minetti 1998). Skipping consists of placing successively the feet on the ground and having a flight period afterwards. The contact sequence is [:L, R, flight:] in the right unilateral skipping (exchange L and R for the left one) and [:L, R, flight, R, L, flight:] in the bilateral skipping. Two linked bipeds, one in front of the other, performing the same unilateral skipping correspond to a cantering horse if the phase lag between the two is 25% of the stride cycle, or they produce a transverse or rotatory gallop if the phase lag is 50% and they adopt the same or different skipping, respectively (e.g., left-left → left transverse gallop, left-right → counterclockwise rotatory gallop).

Although skipping is not used in everyday life by adults (exception made for training athletes, lateral running, occasionally in cornering and in descending stairs), it has to be considered as a third gait for bipeds because of its peculiarity. In fact, skipping differs from pure walking because it has a significant flight phase and from pure running since a double support period often occurs.

Mechanics

Differently from the procedure used in the previous paragraph, the experimental tracings of skipping and gallop will be analyzed first (see figure 5.9). It is apparent that skipping diverges from the previously considered gaits in many respects. The kinetic energy during the (combined) support phase is exchanged with the potential one, as in walking. Differently from walking, the PE is lowest and KE (here represented as just the horizontal component, KE_x) is highest at midsupport. In addition, as already mentioned, a flight period implies that we are facing a bouncing gait. In the canter or slow gallop (see figure 5.9, right), the potential energy shows a sinusoid-like shape with just one maximum and one minimum per stride (differently from the trot), while KE increases and decreases 3 times. Just by taking into account this threefold difference in shape frequency between PE and KE, it is obvious to

expect some energy exchange between them. In fact, the oscillation of the total mechanical energy curve is attenuated with respect to the sum of oscillations of single PE and KE curves. Another observation is that the KE oscillations indicate that a predominantly accelerative action (increase of KE in the progression direction) occurs during the first half of the combined limb contact, while a substantial braking action is present in the second half (KE decrease), and this can also be noted in skipping (see figure 5.9, left).

A mechanical paradigm for skipping and gallop has been found by rigidly linking two pogo sticks and allowing them to passively land with the first limb to touch the ground in a vertical position (see figure 5.10, right). Differently from the previous graphs, in figure 5.10 the vertical component of KE (KE_y) has been added to PE to represent the vertical mechanical energy of the COM, while the horizontal component (KE_x) has been left alone to represent the horizontal energy of the system (named V and H, respectively, in the lower part of the figure). This helps to classify all the cited gaits according to the mechanical paradigms (energy exchange occurs among V, H, and EL, the elastic energy).

The landing dynamics of the double pogo stick are such that the first limb spring is loaded mainly by storing the potential energy of the body COM (V → EL). Almost simultaneously a portion of V is also transformed into horizontal H (V → H). H further increases in the successive elastic-releasing phase by effect of the first spring recoil (EL → H). At the end of this first part of the landing phase the system total mechanical energy TE is the same as it was before landing, but V is lower and H is higher than before. This excess in H is used in the second part of the landing to load the second limb spring

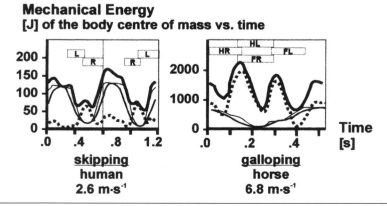

Figure 5.9 Experimental time course of COM mechanical energy during a typical skipping and gallop stride. In each graph the energy tracings have been offset in order to allow the shape comparison between them. Data sources: human (Minetti 1998), horse (Minetti et al. 1999).

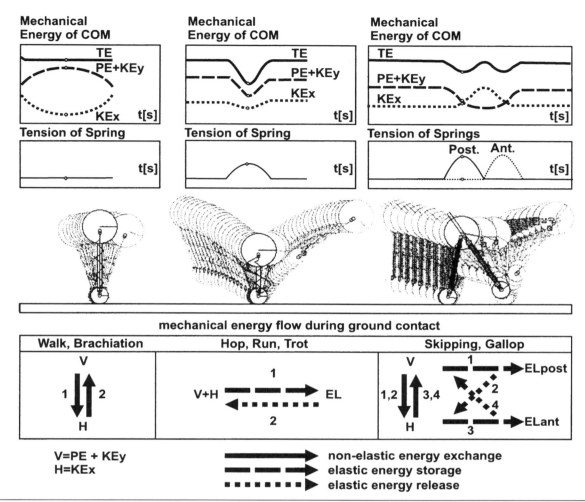

Figure 5.10 Summary of the three illustrated mechanical paradigms of terrestrial locomotion: the (inverted) pendulum (left-hand side), the single pogo stick (central), and the double pogo stick (right-hand side). The upper graphs show the time course of the different forms of mechanical energy. PE and KEy have been added to account for the vertical energy (V) of the system. KEx represents the horizontal energy (H), and the spring tension curve shows when elastic energy (EL) is stored (increase) or released (decrease). The simulations have been obtained by using Interactive Physics (Knowledge Revolution, USA) with different initial conditions applied to a model made of one or two pogo sticks (mass and spring). The diagrams in the lower part of the figure schematically illustrate the mechanical energy exchange during the contact phase in the three paradigms and the associated gaits. The numbers refer to successive subphases in the contact period. ELpost and ELant refer to the mechanical elastic energy that can be stored and released by the posterior and anterior limb springs.

Adapted, by permission, from A.E. Minetti, 1998. "The biomechanics of skipping gaits: A third locomotion paradigm?" *Proceedings of the Royal Society of London* B265:1227-1235.

($H \rightarrow EL$) and, almost simultaneously, to regain V (by raising PE and KE_y needed for the successive flight, $H \rightarrow V$). At the end of the entire contact phase, TE (V + H) has been conserved by applying both strategies, that is, the elastic energy store and release, typical of running, trotting, and hopping, and the interchange of potential and kinetic energy, typical of walking and brachiation.

The illustrated paradigm dynamic works for passive double pogo sticks. What happens in real gaits is that some fresh mechanical work has to be provided by muscle to cope with the inefficiency introduced by the energy dissipation in the locomotor system, and it is likely that the plantarflexion of the first landing limb helps to further raise the KE_x peak observed in real skipping (see figure 5.9, left).

The double pogo stick helps to understand the motion of skipping bipeds (humans, birds, lemurs). Quadruped gallop has been shown to be coherent with the present theory (Minetti 1998). By linking two double pogo sticks, with the phase lag at landing as described above, the energy time course qualitatively resembles the one experimentally obtained in galloping animals.

In conclusion of this section, bipedal skipping and quadrupedal galloping seem to benefit from the same mechanical paradigm, which includes the energy-saving strategies of both walking and running. Other mechanical analogies between skipping and galloping are apparent in figure 5.6 as far as the mechanical external work, the stride frequency, and the energy recovery are concerned.

Energetics

Skipping in humans, the only species where the metabolic cost has been assessed so far, is remarkably expensive (see figure 5.6), resulting in about 50% more cost than running at comparable speed. The associated apparent efficiency is low for a bouncing gait. In contrast, it is evident that in horses the cost of locomotion is similar for trot and gallop (Hoyt and Taylor 1981; Minetti et al. 1999). The capability of using the same amount of fuel per unit distance both at slow and at high speed reinforces the concept of gaits used as gears to cope with increasing speed demand. For each gait there is an optimal speed at which the cost of locomotion is at a minimum. These correspond to the speeds more commonly used by the free-ranging horse (Hoyt and Taylor 1981). The apparent efficiency is highest in galloping, confirming that the deviation from the thermodynamic value for muscular contraction has to be associated with a great deal of elastic energy in the stride that is used to reduce the work that muscles must do.

An additional mechanism capable of storing and releasing elastic energy has been suggested for the gallop by Alexander (1988). The back bending in the quadruped stretches and recoils some ligaments along the spine, allowing exchange of mechanical energy with the swinging limbs. Thus limb tendons (but also the elastic structures within the muscle) and spine ligaments represent the resources available at high speed (gallop) to maintain the apparent mechanical work at a high level at a relatively acceptable metabolic cost. In horses it has been estimated (Minetti et al. 1999) that up to 6.0 J · (kg bounce)$^{-1}$ is stored and released by the elastic structures of the body.

Gait Transition

As already mentioned, humans change from walking to running at the speed of about 2.0 m · s^{-1}. This speed is much less than the limit speed of walking, set at about 3.0 m · s^{-1}. By looking at figure 5.6, it is apparent that the speed at which walking and running require the same metabolic cost is closer to 2.0 m · s^{-1} than to 3.0. This means that at speeds higher than 2.0 m · s^{-1} running is more economical than walking, while the opposite is true at speeds under 2.0 m · s^{-1}. That is why the economy of locomotion has been classically addressed as the variable to be maximized in order to select the proper gait at each speed. A mathematical model of bipedal locomotion (Minetti and Alexander 1997) led to the same conclusions. The same criterion seems to apply in quadrupedal (horse) gait transition from walking to trotting and from trotting to galloping (Hoyt and Taylor, 1981). Speeds of transition between gaits are different if they are crossed while in acceleration or in deceleration (this could be another analogy with the use of gears in a car). The phenomenon, called gait transition hysteresis, is different in bipeds and in quadrupeds.

The determinants of both the transition hysteresis and of the transition speed have been referred to metabolic, mechanical, comfort, and information issues. For example, a force trigger for the trot-gallop transition has been proposed (Farley and Taylor 1991). It has to be mentioned that the speed at which the gait transition occurs is rarely adopted for steady locomotion.

Gradient Locomotion

All of the preceding sections dealt with level locomotion. In that condition, the sum of the increases in the experimental time course of the total mechanical energy of COM equals the sum of the decreases, within one stride. In other words, the COM raising and acceleration (positive work) equals the COM lowering and deceleration (negative work). Thus, in level locomotion the muscles have to do the same amount of positive and negative mechanical work (50–50% partitioning). We have to recall that the metabolic energy that muscles use to do positive work is 5 times greater than that necessary to perform negative work (Abbot, Bigland, and Ritchie 1952).

When moving uphill the positive work tends to prevail due to the increase of PE (which depends on the vertical position of COM), while negative work is predominant in downhill locomotion. The mix of the two work types persists up to certain gradients (–15% and +15%), beyond which the mechanical work of locomotion is entirely negative or positive, respectively. Human walking and running on an incline have been extensively studied by R. Margaria (1938) as far as the metabolic cost is concerned. He found that at each gradient there is an optimal (i.e., less expensive) speed and that, across all gradients, the metabolic cost at the optimal speeds is at a minimum at a steepness of about –10%, both for walking and for running. By using motion analysis of those gaits more recent studies (Minetti, Ardigò, and Saibene 1993; Minetti, Ardigò, and Saibene 1994b) demonstrated that the different partitioning and the different metabolic cost of positive and negative mechanical work explain the observed minimum at –10% gradient. Another result from those studies was that the optimum gradient (i.e., the one minimizing the amount of metabolic energy spent) of a path connecting two sites at different altitudes has to be close to 25%, both for uphill and for downhill walking (Minetti 1995). In case geographic constraints allow only a smaller gradient, the path with the highest possible gradient should be chosen.

While there is some reference in the literature to the effect of gradient on the gait transition speed in humans (Minetti, Ardigò, and Saibene 1994a), nothing appeared about the walk/trot/gallop transition in quadrupeds on incline. It is a common observation that galloping is preferred to trotting on steep paths even at slow speeds.

Different Gravity Conditions

The continuous expansion of the frontiers in space exploration and the realistic opportunity to establish permanent settlements on planets other than Earth present the opportunity to briefly discuss locomotion under different gravity conditions. Remember that gravity changes do not affect our body mass, just our weight. On the moon and Mars our weight is 17% and 40%, respectively, of what it is on Earth, while on Jupiter it could be about 236%. It is difficult to maintain a normal walking pattern when the acceleration of gravity is reduced because of the imbalance between kinetic and potential energy. This is due to the different contribution of the potential energy, which is affected by gravity, to the pendulum-like energy exchange. Margaria and Cavagna (1964) predicted a decrease of the optimal speed of walking to about 1.8 km/h on the moon ($\frac{1}{6}$ of the Earth's gravity). While no measurement of walking speed was taken during the Apollo missions, the collected footage shows a really slowed gait. Recent experiments (Cavagna, Willems, and Heglund 1998) made on airplanes performing parabolic flights capable of simulating different gravita-

tional conditions showed that on Mars (slightly more than $\frac{1}{3}$ of Earth's gravity) the optimal speed could be around 3.4 km/h.

For a gravity acceleration higher than on Earth, pendulum-like mechanics predict higher attainable walking speeds. We could expect, though, that the push made by calf muscles at end-stance could not be strong enough to overcome the weight of the body beyond a given gravitational acceleration.

Among bouncing gaits, skipping and hopping were preferred to running by the astronauts of the Apollo missions on the moon. Such a preference was probably due to the higher vertical ground reaction force during contact determined by the lower duty factor (the fraction of the stride at which one foot is on the ground) of those gaits, with respect to running at the same speed.

Summary and Potential Applications

Figure 5.10 summarizes the main message from this chapter. The three mechanical paradigms, namely the pendulum and the single and double pogo stick, are represented as columns in the picture. The upper graphs illustrate the time course of COM mechanical energy, together with the spring tension, when applicable. The lower diagrams depict the mechanical energy flow occurring in the three paradigms by specifically focusing on the vertical (potential + vertical kinetic), horizontal (horizontal kinetic), and elastic energy. It emerges that walking and brachiation use just the exchange between vertical and horizontal non-elastic energy (pendulum), running, hopping, and trotting benefit from the transformation of (vertical + horizontal) non-elastic energy into elastic energy and back, while skipping and galloping take advantage of the two described energy-saving mechanisms simultaneously.

Considering that each gait is apparently designed around one (or more) energy-saving strategy, it could be anticipated that the transition between gaits would also occur by criteria associated with metabolic cost minimization. In this respect, the different terrestrial gaits behave like the gears in a car. They have been introduced to adapt the same actuator (locomotor and internal combustion systems) to different progression speeds.

Other minor gaits have been observed in legged animals, but we could expect that they follow a combination of the illustrated mechanical paradigms. Future application fields of biomechanical research in locomotion include sport science, prosthetics, and functional electric stimulation. The robot industry already has taken inspiration—from the self-balancing hopping monopod by Raibert (1986), which follows the single pogo stick paradigm—to the Honda hominoid robot model P3 (created in 1997), which walks and ascends stairs just as humans do (also the pendulum-like lateral rocking is present).

References

Abbott, B.C., Bigland, B., and Ritchie, J.M. 1952. The physiological cost of negative work. *J. Physiol. (London)* 117: 380–90.

Alexander, R.McN. 1988. Why mammals gallop. *Am. Zool.* 28: 237–45.

Alexander, R.McN. 1992. *Exploring biomechanics: Animals in motion.* New York: Scientific American Library, W.H. Freeman.

Bornstein, M.N., and Bornstein, H.G. 1976. The pace of life. *Nature* 259: 57.

Cavagna, G.A., and Kaneko, M. 1977. The sources of external work in level walking and running. *J. Physiol. (London)* 268: 467–81.

Cavagna, G.A., Willems, P.A., and Heglund, N.C. 1998. Walking on Mars. *Nature* 393: 636.

Chang, Y.H., Bertram, J.E.A., and Ruina, A. 1997. A dynamic force and moment analysis system for brachiation. *J. Exp. Biol.* 200: 3013–20.

Farley, C.T., and Taylor, C.R. 1991. A mechanical trigger for the trot-gallop transition in horses. *Science* 253: 306–8.

Hoyt, D.F., and Taylor, C.R. 1981. Gait and the energetics of locomotion in horses. *Nature* 292: 239–40.

Ker, R.F., Bennet, M.B., Bibby, S.R., Kester, R.C., and Alexander, R.McN. 1987. The spring in the arch of the human foot. *Nature* 325: 147–49.

Margaria, R. 1938. Sulla fisiologia e specialmente sul consumo energetico della marcia e della corsa a varia velocità ed inclinazione del terreno. *Atti Accad. naz. Lincei Memorie* 7: 299–368.

Margaria, R., and Cavagna, G. 1964. Human locomotion in subgravity. *Aerospace Medicine* 35: 1140–46.

McMahon, T.A., and Bonner, J.T. 1983. *On size and life.* New York: Scientific American Library, W.H. Freeman.

Minetti, A.E. 1995. Optimum gradient of mountain paths. *J. Appl. Physiol.* 79(5): 1698–1703.

Minetti, A.E. 1998. The biomechanics of skipping gaits: A third locomotion paradigm? *Proc. R. Soc. Lond. B* 265: 1227–35.

Minetti, A.E., and Alexander, R.McN. 1997. A theory of metabolic costs for bipedal gait. *J. Theor. Biol.* 186: 467–76.

Minetti, A.E., Ardigò, L.P., Reinach, E., and Saibene, F. 1999. The relationship between the mechanical work and energy expenditure of locomotion in horses. *J. Exp. Biol.* 202: 2329-38.

Minetti, A.E., Ardigò, L.P., and Saibene, F. 1993. Mechanical determinants of gradient walking energetics in man. *J. Physiol. (London)* 472: 725–35.

Minetti, A.E., Ardigò, L.P., and Saibene, F. 1994a. The transition between walking and running in man: Metabolic and mechanical aspects at different grades. *Acta Physiol. Scand.* 150(3): 315–23.

Minetti, A.E., Ardigò, L.P., and Saibene, F. 1994b. Mechanical determinants of the minimum energy cost of gradient running. *J. Exp. Biol.* 195: 211–25.

Pennycuick, C.J. 1975. On the running of the gnu (*Connochaetes taurinus*) and other animals. *J. Exp. Biol.* 63: 775–99.

Raibert, M.H. 1986. *Legged robots that balance.* Boston: MIT Press.

Roberts, T.J., Marsh, E.L., Weyand, P.G., and Taylor, C.R. 1997. Muscular force in running turkeys: The economy of minimizing work. *Science* 275: 1113–15.

Rose, J., Gamble, J.G., Lee, J., Lee, R., and Haskell W.L. 1991. The energy expenditure index: A method to quantify and compare walking energy expenditure for children and adolescents. *J. Pediatr. Orthop.* 11: 571–78.

Sutherland, D.H., Olshen, R.A., Biden E.N., and Wyatt, M.P. 1988. *The development of adult walking.* London: MacKeith Press.

Waters, R.L., Hislop, H.J., Perry, J., Thomas, L., and Campbell, J. 1983. Comparative cost of walking in young and old adults. *J. Orthop. Res.* 1: 73–76.

Woledge, R.C., Curtin, N.A., and Homsher, E. 1985. *Energetics aspects of muscle contraction.* London: Academic Press.

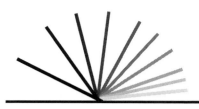

Chapter 6

The Pathways for Oxygen and Substrates

H. Hoppeler and E.R. Weibel

Endurance work is limited in part by the muscles' ability to oxidize substrates in the mitochondria in the process of making ATP by oxidative phosphorylation. This chapter explores the physiological and morphological factors that set the capacity for transfer and metabolism of O_2 in the respiratory system of humans in the perspective of comparative physiology.

The Primary Concepts

Environmental oxygen is transported by the respiratory cascade to the site of oxidation in active tissues. Under conditions of heavy exercise it is ultimately the working skeletal muscle cells that set the aerobic demand, as over 90% of energy is spent in muscle cells. The pathways for oxygen and substrates converge in muscle mitochondria. In mammals, a structural limitation of carbohydrate and lipid transfer from the microvascular system to muscle cells is reached at a moderate work intensity (i.e., at 40–50% of $\dot{V}O_2$max). At higher work rates intracellular substrate stores must be used for oxidation. Because of the importance of these intracellular stores for aerobic work we find larger intramyocellular substrate stores in "athletic" species as well as in endurance-trained athletes. The

transfer limitations for carbohydrates and lipids on the level of the sarcolemma imply that the design of the respiratory cascade from lungs to muscle mitochondria reflects primarily oxygen demand. Comparative studies indicate that the oxidative capacity of skeletal muscle tissue is adjusted by varying mitochondrial content. On the level of microcirculatory oxygen supply it is found that muscle tissue capillarity is adjusted to muscle oxygen demand but that the capillary erythrocyte volume also plays a role. Oxygen delivery by the heart has long been recognized to be a key link in the oxygen transport chain. In humans maximal cardiac output is essentially determined by stroke volume, maximal heart rates being similar in trained and untrained people. The pulmonary gas exchanger offers only a negligible resistance to oxygen flux to the periphery in untrained or moderately trained subjects in normoxia. However, in contrast to all other steps of the respiratory cascade, the lungs appear to have only a minimal phenotypic plasticity, at least in response to endurance exercise training. Because of this lack of malleability, the lungs may ultimately become limiting for $\dot{V}O_2$max when the adaptive processes have maximized O_2 flux in the downstream elements of the respiratory system: heart, microcirculation, and muscle mitochondria.

Historical Highlights

The concept of $\dot{V}O_2$max as a well-defined identifiable plateau of oxygen consumption above which additional power production fails to elicit any further increase in $\dot{V}O_2$ can be traced back to Hill and Lupton (1923). This paper identified the heart or lungs as a possible limiting step of oxygen utilization by muscles. The idea of a single step of the oxygen cascade limiting oxygen flow through the respiratory system has been a very attractive concept and has dominated physiological thinking until recently. Implicitly, most early exercise physiologists assumed the heart to be the most important step of aerobic energy transfer. Christensen (1931) noted that trained subjects had much lower heart rates on any given submaximal work rate than did untrained controls. He also identified stroke volume and hence heart size as critical for aerobic performance, since training observations suggested that trained subjects could reach a given cardiac output at a lower heart rate. When central hemodynamics could be assessed experimentally in exercising humans, a much clearer picture emerged of blood flow distribution and the necessary perfusion pressures during heavy exercise (Bevegard and Shepherd 1967; Holmgren and Astrand 1966). It became established that maximal oxygen uptake was strongly correlated to stroke volume, end diastolic volume, and heart size. These findings were supported by studies using two-dimensional X-ray pictures (Reindell, Klepzig, and Musshoff 1957) and ultrasound (Keul et al. 1981), indicating that atrial and ventricular volumes were enlarged in athletes as a consequence of endurance exercise training. It was realized that the relationship between cardiac oxygen delivery (heart rate · stroke volume · arterial O_2 content) and $\dot{V}O_2$max was close and that exercise training had no effect on maximal heart rate nor on a-v O_2 content difference (Saltin et al. 1968). From this it was concluded that the limitation to $\dot{V}O_2$max had to be the heart, assuming that all other steps in the pathway would be built with excess capacity.

In the 1970s, a wealth of data on the remarkable variability of skeletal muscle oxidative capacity and capillary supply with endurance exercise training became available (Holloszy and Booth 1976; Hoppeler, Lüthi et al. 1973; Saltin and Gollnick 1983), which brought the importance of the periphery for $\dot{V}O_2$max into focus. Why would exercise training induce a major increase in muscle tissue oxidative capacity if the latter was by far in excess of the requirements at $\dot{V}O_2$max? was the salient question remaining unanswered by those purporting cardiac output to be the sole determinant of $\dot{V}O_2$max (Gollnick and Saltin 1982). A new twist to the year-long battle of central versus peripheral limitations of $\dot{V}O_2$max was offered in a paper titled "Metabolic and circulatory limitation to $\dot{V}O_2$max at the whole animal level" by diPrampero (1985). Using a deceptively simple algebraic model, he calculated the contribution of the individual steps of the respiratory cascade viewed as resistors in series. He came to the conclusion that in two-legged exercise in normoxia about 75% of $\dot{V}O_2$max is set by central O_2 transport and the remaining 25% by the periphery. Additionally, he pointed out that the importance of the individual transfer steps was much affected by external conditions such as hypoxia and the size of the working muscle groups. Since then, it has become generally accepted that it is the integrated, interactive effects of all steps in the respiratory cascade that help set $\dot{V}O_2$max (Wagner, Hoppeler, and Saltin 1997): "No single step is *the* limiting one; a change in the capacity of any one step will alter $\dot{V}O_2$max" (p. 2040).

This chapter explores how $\dot{V}O_2$max is influenced by structural capacities and functional regulation at each point in the pathway for oxygen delivery from lungs to skeletal muscle mitochondria. At the level of peripheral supply we additionally analyze how the pathways for oxygen and substrates converge at the mitochondria and how substrate fluxes intervene with setting $\dot{V}O_2$max. This is done considering the phenotypic plasticity of all steps of the respiratory cascade as a consequence of endurance exercise training. In order to study a complex integrated physiological system such as the respiratory cascade, we need to develop an intellectual framework within which we can perform a systems analysis.

Model of the Respiratory System

If an untrained human exercises intensely, his muscles consume about 10 times as much oxygen as at rest. A well-trained athlete doing the same may increase O_2 consumption by over 20 times, matching his higher aerobic performance capacity (see figure 6.1). In either case, this is accompanied by an increase in cardiac output and heart rate as well as by an increase in minute ventilation and respiratory frequency: O_2 uptake in the lung and O_2 transport by the circulation of blood must be matched to O_2 consumption in the mitochondria of the working muscle cells. At the same time substrate supply

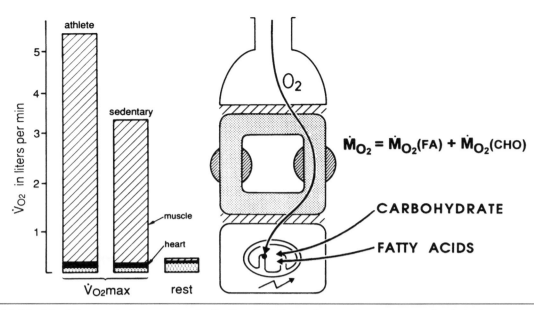

Figure 6.1 Model of the respiratory system indicating that the pathways for oxygen and substrates converge on the mitochondria. The histograms to the left show the distribution of $\dot{V}O_2$ to locomotor muscles, heart, and other tissues under resting conditions and at $\dot{V}O_2$max in sedentary and endurance-trained humans (after Weibel, Taylor, and Hoppeler 1991).

must also be matched to oxidation. These are the basic features of this system.

1. Respiration is an integral function involving the coordinated action of all structures that build the pathway for O_2 from the lung to the respiratory chain enzymes in tissue mitochondria. Under steady state conditions the O_2 flow rate is the same at all levels.

2. Respiration is a regulated process matched to the instantaneous demands of aerobic metabolism: As muscle ATP consumption increases, O_2 demand is increased proportionally, requiring regulation of the various O_2 and substrate transport functions.

3. Respiration on the whole animal level is a limited function. O_2 consumption can rise, up to a limit called $\dot{V}O_2$max, beyond which any additional energy is covered by anaerobic processes. This limit is a characteristic of the individual and is higher in athletes than in untrained subjects; to a certain extent, it is malleable in that it can be elevated by training.

What Sets the Limit for Aerobic Metabolism?

To approach the question of what sets the limit for aerobic metabolism, we must consider the entire pathway for O_2 from the site of uptake in the lung to the sink in the mitochondria (see figure 6.2), as well as the pathways for substrates that fuel oxidation from uptake in the gut to oxidation in the mitochon-

dria. The O_2 pathway is simple: it follows a single path without branches.

The substrate pathways are more complex. Both carbohydrate and lipid pathways branch to form four parallel pathways that converge on the mitochondria (see figure 6.3). Proteins can be ignored because in well-fed subjects their contribution to aerobic metabolism is minimal. During exercise, when rates of substrate oxidation are highest, blood is shunted away from the gut and rates of substrate uptake are lowest. At this point, stores supply most of the fuel for oxidation. Organismic substrate stores outside muscle are in the liver and in the adipose tissue. Inside the muscle cells carbohydrates are stored as glycogen granules and fatty acids as lipid droplets (see figure 6.4). We will have to delineate the importance of the parallel substrate pathways as a function of exercise intensity and duration.

We would like to know whether there is a single step in the oxygen or substrate pathways that sets the upper limit to aerobic metabolism or whether the capacities in each of the steps are approximately matched to each other. We would also like to know whether limiting factors are primarily structural or functional.

Structural Parameters Can Limit $\dot{V}O_2$max

Structural parameters intervene at all levels of the respiratory pathway. For example, the model predicts

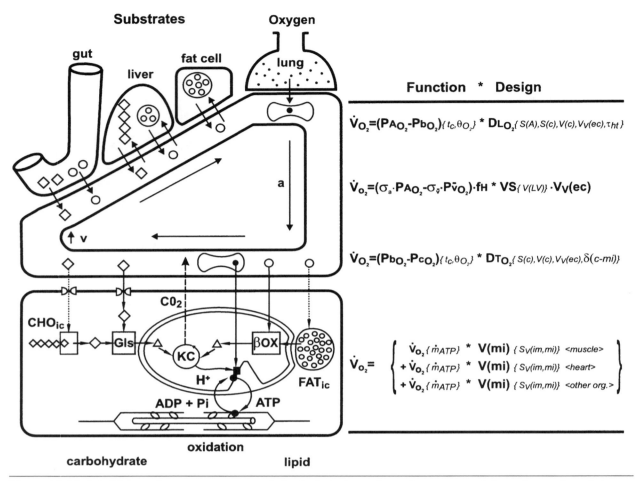

Figure 6.2 Model of the respiratory system expanded to include the pathway for oxygen and the pathways for substrates. The equations to the right express oxygen flow rate ($\dot{V}O_2$) as the product of functional and design parameters shown in boldface; parameters that affect these factors are shown in italics and placed between braces {}. The functional parameters include O_2 partial pressure PO_2; coefficients of hematocrit-specific O_2 capacitance σ, which depend on O_2-hemoglobin dissociation; O_2 binding rate Θ; heart frequency f_H; capillary transit time t_c; and mitochondrial O_2 consumption rate as a function of ATP flux $\dot{V}O_2$(mATP). Design parameters include diffusion conductance D of lung and tissue gas exchangers, which depend on alveolar and capillary exchange surface areas S(A) and S(c); capillary volume V(c); hematocrit Vv(ec); harmonic barrier thickness τ(ht); capillary-mitochondria diffusion distance δ(c-mi); and mitochondrial volume, V(mi); with inner membrane surface density Sv(im,mi).

- that the pulmonary diffusing capacity DLO_2 depends on the surface area available for gas exchange between air and blood,
- that the stroke volume Vs is determined by the size of the heart ventricles,
- that the O_2 transport capacity of the blood is determined by the erythrocyte volume Vv(ec), and
- that the capacity of cells for oxidative phosphorylation is related to the volume of mitochondria V(mi) (see figure 6.2).

In the lung, the actual surface area available for gas exchange can be modified by the perfusion of the capillary bed. However, in general, structural parameters cannot be regulated or changed on a short time scale because this requires morphogenetic processes. Structure can be modified in response to chronically altered demand. This regulation involves transcriptional activation of structure genes (Puntschart et al. 1995). It is well known that in exercise, cardiac output is immediately regulated by increasing heart frequency (f_H) up to a maximum, with stroke volume essentially unchanged; by contrast, as a consequence of exercise training, heart size increases and maximal cardiac output becomes elevated by a larger stroke volume at the same maximal heart rate. From this we conclude that

Convergence of O_2 & Substrate Pathways

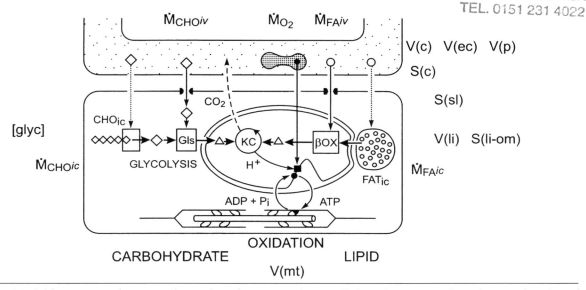

Figure 6.3 Model for structure function relationship of oxygen and intracellular substrate supply to the mitochondria of skeletal muscle cells. Dots indicate oxygen, open circles indicate fatty acids, diamonds indicate glucose, row of diamonds indicates glycogen, and triangles indicate acetyl CoA. The heavy arrows indicate the pathways of intracellular substrate breakdown from the intracellular stores to the terminal oxidase in the mitochondrial membrane (black square). The thin arrows indicate the supply routes of oxygen and substrates from the capillaries, with dotted arrows for the supply route to intracellular stores (temporally split from the phase of oxidation).

Adapted, by permission, from R. Vock, H. Hoppeler, H. Claassen, et al., 1996a. "Design of the oxygen and substrate pathways. VI. Structural basis of intracellular substrate supply to mitochondria in muscle cells," *Journal of Experimental Biology* 199:1689-1697.

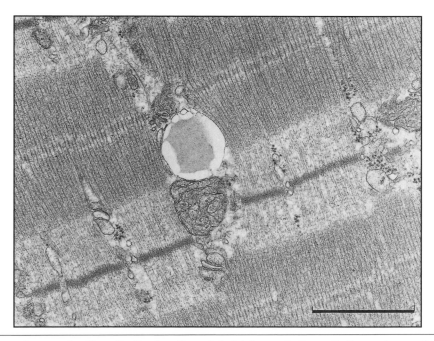

Figure 6.4 Electron micrograph of longitudinal section of skeletal muscle tissue. In the center, on the level of the Z line, we find an interfibrillar mitochondrion with a lipid droplet immediately adjacent. Glycogen granules (small black dots) can be seen preferentially in the region of the A band (marker indicates 1 μm).

structural parameters set the boundaries within which functional parameters regulate oxygen and substrate fluxes according to the instantaneous demands of the organism. Note that the actual functional limitation is heart rate, which has a fixed maximum, as Vs can be increased by training. So it is both functional and structural parameters that set the limit.

The Concept of Symmorphosis

The question, therefore, is whether the structural design of the respiratory system is such that it allows the limits to $\dot{V}O_2$ of the sequential steps to be matched to each other and to the overall limit. Common sense dictates that this should be the case, because it does not make sense for the body to build and maintain, at a high cost, structures of a fundamental and vital functional system that it will never use. To approach this question we have proposed a principle of regulated morphogenesis and of economic design, the hypothesis of *symmorphosis* (Taylor and Weibel 1981). Symmorphosis predicts that the design of all components comprising a system is matched quantitatively to functional demand, "enough but not too much." The principle of symmorphosis is somehow akin to the concept of *homeostasis*. While homeostasis operates in real time, symmorphosis operates on different time scales. It can either be the consequence of genetic selection or it can be brought about by epigenetic structural malleability.

In a pragmatic approach we can test the hypothesis of symmorphosis by taking advantage of the fact that $\dot{V}O_2$max varies among individuals and species. We can study the variation of structural and functional parameters in relation to the variation in $\dot{V}O_2$max. By comparing the changes in structural and functional variables with the variation in $\dot{V}O_2$max we attempt to identify parameters that vary directly in proportion to $\dot{V}O_2$max, whose ratio to $\dot{V}O_2$max is, accordingly, invariant. If a good theoretical model is available (i.e., the Bohr model of oxygen diffusion across the pulmonary gas exchanger), we can test for symmorphosis by using structural data to calculate values for a theoretical functional capacity. We can then experimentally test to what extent this capacity is exploited at $\dot{V}O_2$max.

Study of Limiting Factors Using Comparative Physiology

The limit of aerobic metabolism in humans ranges from less than 40 ml $O_2 \cdot kg^{-1} \cdot min^{-1}$ in untrained subjects to over 80 ml $O_2 \cdot kg^{-1} \cdot min^{-1}$ in top endur-ance-trained athletes. An unknown fraction of this difference is the result of training.

In typical exercise training experiments the gains of $\dot{V}O_2$max are usually much smaller, on the order of up to 1.5-fold (Saltin et al. 1968). We call these changes, which reflect the malleability of the respiratory system, induced variations. Larger variations in $\dot{V}O_2$max are observed among mammalian species as a consequence of two types of selection pressure during evolution: allometric variation and adaptive variation.

Allometric Variation

Metabolic requirements are related to body mass (M_b) in such a way that small animals have higher resting and maximal oxygen consumption rates than large ones. These differences are usually expressed as power functions that describe the change of any functional or structural parameter as body mass changes: $\dot{V}O_2$max is found to be proportional to $M_b^{0.8}$; accordingly, $\dot{V}O_2$max/M_b scales to $M_b^{-0.2}$ (see figure 6.5). Allometric variation accounts for enormous differences in $\dot{V}O_2$max/M_b: a mouse consumes six times more O_2 than a cow per unit body mass (Weibel, Taylor et al. 1981).

Adaptive Variation

In several size classes, nature has selected species for athletic performance (see figure 6.5). Thus we find that nature's athletes such as horses and dogs have a $\dot{V}O_2$max that is more than twofold that of sedentary species of similar size such as cattle and goats (Jones et al. 1989; Taylor et al. 1987). And superathletes such as the pronghorn antelope achieve even higher values (see table 6.1).

For the basic understanding of the design of the mammalian respiratory system and its functional limits, comparative studies using allometric and adaptive variations of $\dot{V}O_2$max have proven to be invaluable. Not only do the studies afford us with large differences in $\dot{V}O_2$max, hence with an excellent signal to noise ratio for our experiments, they also offer the advantage that the possibility of sampling tissue from all steps of the respiratory cascade allows for morphometric quantification of the functionally relevant structural parameters on all levels. For obvious reasons this cannot be done in human experimentation. With human tissue, samples for morphometry have, in general, only been obtained by taking biopsies from single muscles, usually m. vastus lateralis, allowing only a limited analysis of oxygen supply by capillaries and oxygen demand by mitochondria. The analysis is limited in the sense that the exact contribution of the biopsied muscles

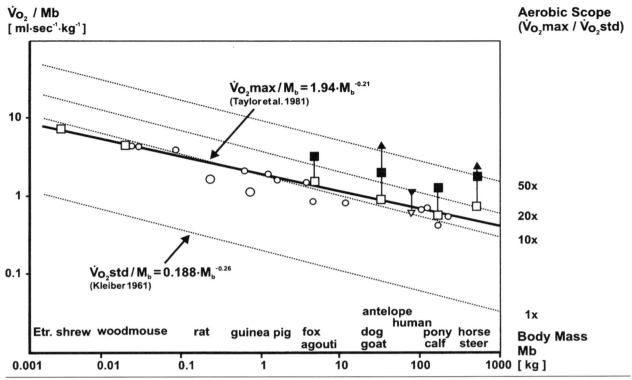

Figure 6.5 Allometric plot of mass-specific O_2 consumption, showing Kleiber's curve (1961) for standard (basal) $\dot{V}O_2/M_b$, isopleths for theoretical aerobic scope (fine lines), and the allometric plot for $\dot{V}O_2 max/M_b$ (open circles, heavy line). Adaptive pairs are shown as open and closed squares for sedentary and athletic species, respectively; closed upright triangles show the highest values measured in these size classes on thoroughbred racehorses and pronghorn antelope. Human $\dot{V}O_2 max$ data are for sedentary people and highly trained athletes (open and closed triangle, respectively).

Adapted, by permission, from E.R. Weibel, C.R. Taylor, and H. Hoppeler, 1992. "Variations in function and design: Testing symmorphosis in the respiratory system," *Respiration Physiology*, 87:325-348.

to $\dot{V}O_2 max$ and their involvement in any particular exercise is not known precisely. We will exploit the comparative data to generate a perspective of the fundamental design principle of the respiratory system in mammals. In addition, at each level of the respiratory system we will try to define the relevant conditions that help to determine aerobic performance in humans.

Muscle Mitochondria Set the Demand for Oxygen

When humans exercise at their maximal rate of oxygen consumption, the mitochondria located in their skeletal muscles consume more than 90% of the oxygen and substrates (see figure 6.2; Mitchell and Blomqvist 1971). We are therefore justified to concentrate on these particular mitochondria in our search for invariant design parameters in the respiratory system. First we need to ask, What is the relevant structural parameter of skeletal muscle

mitochondria to relate to the large variation in $\dot{V}O_2 max/M_b$ with allometric and adaptive variation? The mitochondrion is bounded by a continuous outer membrane and contains a second inner membrane that forms multiple infoldings, the cristae (see figures 6.4 and 6.8). Between the membranes we find the intermembrane space; the innermost compartment is the mitochondrial matrix. In the matrix, substrates are broken down in the Krebs cycle, generating reducing equivalents. The beta-oxidation of fatty acids is also located in the matrix. Thus, the matrix volume is the appropriate structural parameter to characterize substrate catabolism, assuming a constant density of enzymes in the matrix space. In most mammals, matrix volume appears to be proportional to mitochondrial volume and the ratio between mitochondrial volume and matrix volume is also invariant under conditions of endurance exercise training (Davies, Packer, and Brooks 1981). The respiratory enzymes that catalyze oxidative phosphorylation of ATP are built into the inner mitochondrial membrane. It is

Table 6.1 Morphometric and Physiologic Parameters of Three Species Pairs

Differences in morphometric and physiologic parameters of muscle mitochondria and capillaries, and of heart, blood, and lung with adaptive variation of $\dot{V}O_2$max in three species pairs.

Design Function Units	Body $\dot{V}O_2$max/Mb ml·sec⁻¹·kg⁻¹	Mitochondria V(mt)/Mb ml·kg⁻¹	Mitochondria [V(mt)]/{$\dot{V}O_2$max} ml·ml⁻¹·sec	Capillaries V(c)/Mb ml·kg⁻¹	Capillaries [V(c)·Vv(ec)]/{$\dot{V}O_2$max} ml·ml⁻¹·sec	f_H min⁻¹
25–30 kg						
Dog	2.29	40.6	17.7	8.2	1.79	274
Goat	0.95	13.8	14.5	4.5	1.42	268
D/G	2.4*	2.9*	1.2	1.8*	1.26	1.02
150 kg						
Pony	1.48	19.5	13.2	5.1	1.45	215
Calf	0.61	9.2	15.1	3.2	1.63	213
P/C	2.4*	2.13*	0.9	1.6*	0.89	1.02
450 kg						
Horse	2.23	30.0	13.5	8.3	2.05	202
Steer	0.85	11.6	13.7	5.3	2.49	216
H/S	2.6*	2.6*	1.0	1.6*	0.82	0.94
Athletic/sedentary	2.5*	2.5*	1.03	1.7*	0.99	1.0

Design Function Units	Heart Vs/Mb ml·kg⁻¹	Heart [VS·Vv(ec)]/{$\dot{V}O_2$max/fH} ml·ml⁻¹	Blood Vv(ec) Unitless	Lung DLO₂/Mb ml·sec⁻¹·mmHg⁻¹·kg⁻¹	Lung [DLO₂]/{$\dot{V}O_2$max} mmHg
25–30 kg					
Dog	3.17	3.16	0.50	0.118	0.052
Goat	2.07	2.92	0.30	0.080	0.084
D/G	1.53*	1.08	1.68*	1.48*	0.61*
150 kg					
Pony	2.50	2.54	0.42	0.079	0.053
Calf	1.78	3.21	0.31	0.050	0.082
P/C	1.40*	0.79	1.35*	1.57*	0.65*
450 kg					
Horse	3.11	2.58	0.55	0.108	0.048
Steer	1.52	2.58	0.40	0.054	0.064
H/S	2.1*	1.00	1.4*	2.0*	0.76*
Athletic/sedentary	1.7*	0.96	1.5*	1.7*	0.67*

Last lines present overall ratios for athletic/sedentary species. Asterisk denotes ratios significantly different from 1.
Modified from Weibel, Taylor, and Hoppeler 1992.

estimated that some 40% of its surface is made up of the proteins involved in energy transduction (Schwerzmann et al. 1989). Therefore, in principle the surface area of inner mitochondrial membrane appears to be the appropriate structural parameter to relate to $\dot{V}O_2$max. However, several studies have shown that the area of inner mitochondrial membrane per unit volume of mitochondria is invariant at $35\,\mu m^2/\mu m^3$ in most mammals including humans and in muscles varying widely in mitochondrial content (Hoppeler, Mathieu et al. 1981; Schwerzmann et al. 1989).

It follows from these constant relationships that the volume of mitochondria in skeletal muscle is a good measure of matrix enzymes as well as the quantity of respiratory chain present and, therefore, is also an appropriate structural parameter to relate to $\dot{V}O_2$ max. The maximal rate of oxygen consumption by the mitochondria can then be expressed as the product of mitochondrial volume V(mi) and a functional parameter:

$$\dot{V}O_2 max/M_b = V(mi)/M_b \cdot \dot{V}O_2(mi) \qquad 6.1$$

where $\dot{V}O_2$(mi) is the rate at which mitochondria consume oxygen.

How do these parameters change as $\dot{V}O_2$ max/M_b is varied by allometric, adaptive, or induced variation? Because time is the fundamental variable in allometric variation (Lindstedt and Calder 1981; Schmidt-Nielsen 1984), we would anticipate a priori that $\dot{V}O_2$(mi) would vary directly with $\dot{V}O_2$max/M_b, being six times greater in a mouse than a cow, whereas V(mi)/M_b should be invariant. This would seem to be an eminently reasonable design principle because the relative volume of muscle fibers occupied by mitochondria would not increase with demand. This is because relative muscle volume, V(musc)/M_b, follows the general allometric rule of organ size and is invariant with body size. On average, skeletal muscles make up some 40 to 45% of the total body mass (Schmidt-Nielsen 1984). If V(mi)/M_b changed with $\dot{V}O_2$max/M_b the contractile machinery would become progressively more diluted as demand increased; this might pose a serious problem for small animals where mitochondria could occupy a significant fraction of the cell volume.

Contrary to our expectations, small animals have more mitochondria in each gram of muscle than do large animals. Quantitative measurements of mitochondrial volume in animals spanning a range of body masses from less than 20 g (wood mice, Hoppeler, Lindstedt et al. 1984) to over 500 kg (steers and horses, Hoppeler, Jones et al. 1987) re-

veal that the mitochondrial volume of skeletal muscles varies almost directly with $\dot{V}O_2$max, whereas $\dot{V}O_2$ (mi) of the unit mitochondrial volume does not change with size. When compared to the cow, the mouse has six times the volume of mitochondria in its skeletal muscles so that at $\dot{V}O_2$ max each milliliter of mitochondria in both cow and mouse would consume, on the average, 4 to 5 ml $O_2 \cdot ml^{-1} \cdot min^{-1}$ (see figure 6.6). Thus the 5- to 10-fold difference in $\dot{V}O_2$max is matched by corresponding differences in the amount of mitochondria that animals build, whereas the functional parameter, the rate at which each unit of structure consumes oxygen, is the same, irrespective of size.

Why is the widespread fundamental design principle that time constants vary with body size violated with respect to mitochondria? A possible

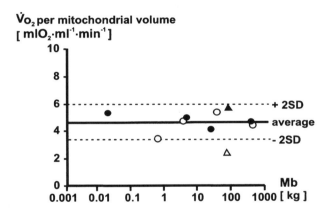

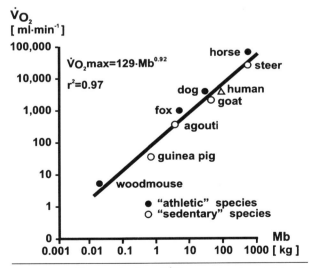

Figure 6.6 Allometric plot of $\dot{V}O_2$ per mitochondrial volume (top) and whole body (bottom) oxygen consumption for species ranging in size from 16 g to 450 kg. Open triangle indicates data for humans considering total muscle mass, and active muscle mass is indicated by closed triangle.

explanation may lie in their origin and their genetics. The enzyme systems of mitochondria and their spatial organization within membranes are very similar to those found in bacteria. So similar, in fact, that it has been suggested that mitochondria evolved from bacteria that were incorporated into the first eukaryotes in an endosymbiotic relationship (Margulis 1981). This is supported by the finding that mitochondria contain their own DNA and ribosomes, which are both similar to those of bacteria. If we consider mitochondria as endosymbiotic bacteria that respond to increased energy demand by reproducing, then their individual size and composition should not change and there is no reason to expect that the rates of their enzymes will change with O_2 demand. A further point may be that the process of oxidative phosphorylation depends on an optimal balance between the elements of the Krebs cycle and the respiratory chain, which is apparently not varied.

If we now consider adaptive variation, we find that mitochondria have adapted to the 2.5-fold differences in aerobic capacity among animals of the same size in the same way that mitochondria adapted to differences with allometry (Hoppeler, Kayar et al. 1987; Mathieu et al. 1981). The higher demand is met by simply building more of the same structure. The 2.5 times greater $\dot{V}O_2max/M_b$ of dogs, ponies, and horses compared to that of goats, calves, and steers is matched by a 2.5 times larger total volume of mitochondria in their muscles, whereas the average rate at which each milliliter of mitochondria consumes oxygen is again the same, namely, 4 to 5 ml O_2 $\cdot$ ml^{-1} $\cdot$ min^{-1} (see table 6.1).

Overall, the comparative approach has revealed that the rate of maximal O_2 consumption by skeletal muscle mitochondria is invariant at

$\dot{V}O_2(mi) = 4$ to 5 ml O_2 $\cdot$ ml^{-1} $\cdot$ min^{-1}, and accordingly,

$V(mi)/\dot{V}O_2 max = 0.2$ ml/ml O_2 $\cdot$ ml^{-1} $\cdot$ min^{-1} = invariant under all circumstances.

Surprisingly, this invariant parameter also appears to apply reasonably well to the mitochondria of the heart. The oxygen consumption of the left heart has been calculated from measurements of work rate for the adaptive pairs while they exercise at $\dot{V}O_2max$. When this value is divided by the mitochondrial volume, one also obtains approximately 4 ml O_2 $\cdot$ ml^{-1} $\cdot$ min^{-1} (Karas et al. 1987b).

The invariant 4 to 5 ml O_2 $\cdot$ ml^{-1} $\cdot$ min^{-1} is not the maximal rate of oxidative phosphorylation of a given mitochondrion; it is an overall measure and assumes that all of the mitochondria of all muscle cells are active and consuming oxygen at the same rate. Although this may be a reasonable assumption for the heart where all of the fibers are active, only 30 to 40% of the muscle fibers are active in the skeletal muscles when animals exercise at $\dot{V}O_2max$ (Armstrong and Taylor 1982). However, it is not as bad as it might seem, since most of the mitochondria are located in these active fibers, because in heavy endurance exercise it is the oxidative, mitochondria-rich fibers that are active. Using a biochemical approach, it has been found that the respiratory capacity of isolated mitochondria is about 5.8 ml O_2 per ml of mitochondria when physiologically reasonable substrates such as succinate are used (Schwerzmann et al. 1989). Using this value it appears that animals are able to exploit 60 to 80% of the in vitro oxidative capacity when they exercise at $\dot{V}O_2max$. This indicates that most of the skeletal muscle mitochondria are utilized at $\dot{V}O_2max$ and that they operate at close to their maximal rates. The invariant estimate of $\dot{V}O_2(mi)$ suggests that the same fraction of skeletal muscle mitochondria is utilized in all quadrupedal mammals.

Humans are different, however. Because of bipedal locomotion, they normally reach $\dot{V}O_2max$ while performing exercises that do not involve all of their muscles, and as a consequence, apparent $\dot{V}O_2(mi)$ is only half of what we find in quadrupedal animals (see figure 6.7). This has led to the notion that, in humans, muscle oxidative capacity is largely exceeding the cardiovascular system's capacity to deliver oxygen to the periphery (Gollnick and Saltin 1982). However, one has to consider that in humans, the cardiovascular pump can only access a fraction of the total peripheral oxidative capacity simultaneously, and that in this active fraction of the total muscle mass, mitochondria likely process oxygen and substrates at the same rate as seen in animals, namely 4 to 5 ml O_2 $\cdot$ ml^{-1} $\cdot$ min^{-1}. This ability of humans to direct blood flow to a subset of muscles greatly increases their repertoire of aerobic tasks. It is of note that untrained humans cannot reach true $\dot{V}O_2max$ when performing an upper body task such as arm-cranking (Rösler et al. 1985). In general, in untrained humans we measure a somewhat smaller peak oxygen consumption with bicycling exercise than with running; this may be related to the fact that during cycling a smaller muscle mass is activated than during running. This discrepancy disappears when well-trained cyclists are tested. They have been reported to reach a higher $\dot{V}O_2max$ when cycling than when running (e.g., Chatterjee and Chakravarti 1986); their specific training may have increased the mitochondrial content

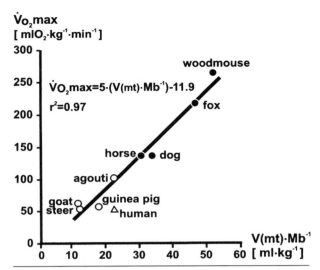

$\dot{V}O_2max$
$[\ mlO_2{\cdot}kg^{-1}{\cdot}min^{-1}\ ]$

$\dot{V}O_2max = 5{\cdot}(V(mt){\cdot}Mb^{-1})-11.9$

$r^2 = 0.97$

woodmouse

fox

horse ● dog

agouti

goat
steer   guinea pig
        △ human

$V(mt){\cdot}Mb^{-1}$
$[\ ml{\cdot}kg^{-1}\ ]$

Figure 6.7 Weight-specific $\dot{V}O_2$max plotted as a function of weight-specific mitochondrial volume V(mi) of skeletal muscle mitochondria for animals differing over five-fold in $\dot{V}O_2$max. Open triangle indicating data calculated for entire musculature of humans. It can be seen that humans have excess mitochondrial volume in comparison to their $\dot{V}O_2$max/M_b. This discrepancy disappears when active muscle mass is considered.

in those muscles activated during cycling to such an extent that their aerobic capacity exceeds that of the larger muscle mass activated when running on a treadmill. In this context it may also be noted that endurance exercise training not only affects trained muscles by greatly increasing their mitochondrial content but also affects muscles that have not been trained (Rösler et al. 1985). Measuring mitochondrial content and fiber size in deltoid muscles during leg training, it was noted that both the volume density of mitochondria and the muscle (fiber) size decreased significantly over a six-week training period.

It should be noted, however, that this type of human experimentation mainly suffers from the fact that only a limited number of muscles can be sampled by small biopsies and that the involvement of these muscles during a particular aerobic exercise regime is not known. Using an animal model of induced variation does not give us a clear-cut answer as to how muscle mitochondrial epigenetic malleability is regulated at the whole animal level (Hoppeler, Altpeter et al. 1995). The changes in $\dot{V}O_2$max that can be induced by exercise training are simply too small to make a meaningful system analysis of the respiratory cascade (even at the level of muscle mitochondria). The available data are compatible with the idea that relatively small shifts of mitochondria occur at the whole animal level, favoring the particular muscle group(s) necessary

to perform a specific aerobic task.

In summary, we can conclude that there is a good match between structure and function in the design of the respiratory system at the level of the mitochondria with allometric and adaptive variations in $\dot{V}O_2$max. In both cases the differences in maximal rates of oxygen consumption are matched by corresponding differences in the amount of mitochondrial structure for both oxygen and substrates, whereas the average rate at which each unit of structure consumes oxygen is invariant. In induced variation the situation is less clear; the current evidence leads us to believe that mitochondrial oxygen consumption remains unchanged and a training-induced increase in demand is met by augmenting mitochondria in those muscles that are exercised while we see a compensatory reduction in other, untrained muscles.

Supply of Substrates From Cellular Stores to Mitochondria

During exercise at $\dot{V}O_2$max, substrates must be transported into the mitochondria at a rate sufficient to fuel oxidative phosphorylation. The oxidation of one mole of glucose consumes 6 moles of O_2, and one mole of fat consumes 23 moles of O_2. It has been clearly established in dogs and goats that circulatory carbohydrates and lipids can only provide a fraction of the total substrate demand of mitochondria in working muscles (Weber et al. 1996a; Weber et al. 1996b). Both carbohydrate and lipid transport from capillaries into muscle cells appears to be maximal at quite low exercise intensities corresponding to approximately 40% of $\dot{V}O_2$max and cannot be up-regulated at higher work intensities. Instead, the mitochondria depend on obtaining their fuels from intracellular substrate stores, whereby at $\dot{V}O_2$max over 80% of the fuel is supplied from glycogen. These results were obtained in a comparative setting of adaptive variation, but they are very similar to the situation observed in exercising humans, where a heavy reliance on intracellular substrate stores at higher work intensities has also been demonstrated (Romijn et al. 1993).

From this it follows that substrates for mitochondrial oxidation at work intensities of around 80% of $\dot{V}O_2$max must primarily be supplied from glycogen granules and lipid droplets inside the muscle cells, with no more than 20 to 30% of the fuel coming from the capillaries. There is indirect evidence that this is due to a limitation of fuel supply from the capillary,

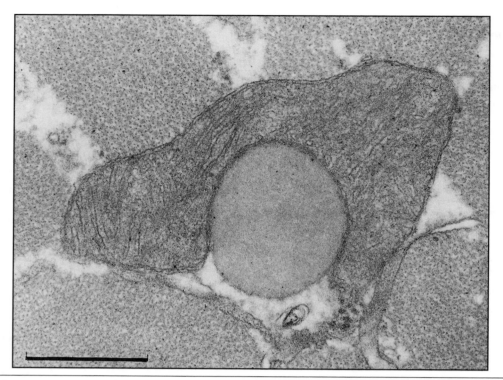

Figure 6.8 Electron micrograph of a cross section of a portion of a muscle fiber. The intimate contact of the mitochondrion outer membrane with the lipid droplet can well be appreciated (marker indicates 0.5 µm).

with the sarcolemma as the main barrier (Vock et al. 1996a).

What are the structures that determine flow through the intracellular substrate pathways? On careful observation, one finds that all of the lipid droplets are in direct contact with the outer mitochondrial membrane (Vock et al. 1996a). They form a tight contact that is often marked by a dent in the mitochondria (see figure 6.8). It seems likely that the fatty acids are transported across the mitochondrial membrane preferentially in these areas of contact, allowing them to circumvent transport problems associated with their low solubility in the cytosol. The maximal flow through the cytosolic pathway is reached at low exercise intensities, and it can supply at best about 20% of the fuel required at $\dot{V}O_2$max. From this it follows that interfibrillar mitochondria likely depend substantially on associated lipid droplets for fatty acid supply.

The importance of intracellular lipid stores for aerobic performance has long been suspected. Already in 1973 (Hoppeler, Lüthi et al. 1973) it was noted that highly endurance trained subjects had considerably larger substrate stores in the form of lipid droplets in their muscles. Later it was found that six weeks of endurance exercise training was sufficient to increase intracellular lipid concentra-

tion by a factor of 2 from 0.5% to 1% of the muscle fiber volume. Highly trained cyclists or endurance runners may have even larger lipid stores in their trained muscles (Hoppeler 1986). Moreover, it was demonstrated in 1976 that the lipid (and glycogen) stores disappeared almost completely after a 100 km run (Kayar et al. 1986; Oberholzer et al. 1976), indicating that the intracellular lipid droplets represent a readily accessible form of substrate reserves in muscle tissue. By contrast, in situations in which the reliance on fatty acid metabolism is reduced, such as in high altitude residents, we find that intracellular substrate stores also are massively reduced (Desplanches et al. 1996).

An additional observation that is likely relevant to the understanding of lipid oxidation in skeletal muscle tissue relates to the distribution of mitochondria within the muscle cell. Whereas a large fraction of mitochondria is found interspersed between myofibrils, there is a smaller population of mitochondria found massed under the sarcolemma close to capillaries (subsarcolemmal mitochondria; see figure 6.9). The location of these mitochondria puts them in close vicinity to the vascular supply of both oxygen and substrates. It has been demonstrated by mathematical modeling that mitochondrial distribution within muscle cells is not likely

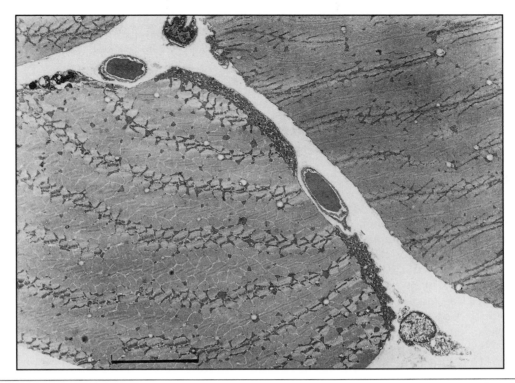

Figure 6.9 Electron micrograph of portions of human skeletal muscle fibers in cross section. Capillaries containing erythrocytes are found between muscle fibers. Subsarcolemmal mitochondria are massed in the periphery of one of the muscle fibers but not immediately between myofibrils nor adjacent to capillaries (marker indicates 10 μm).

to play a major role for oxygen diffusion because the high diffusivity of oxygen in muscle fibers is partly related to the myoglobin content of muscle (Conley and Jones 1996). As a consequence of their location under the sarcolemma, problems of cytosolic transport of substrates is minimal for subsarcolemmal mitochondria. It can be argued that this puts these mitochondria in a favorable position with regard to oxidizing lipids supplied by capillaries (for a review of lipid transfer to mitochondria, see Van der Vusse and Reneman 1996). The notion of a preferential oxidation of vascular lipids by subsarcolemmal mitochondria is further supported by circumstantial evidence indicating a greater relative increase of subsarcolemmal versus interfibrillar mitochondria with endurance exercise training (Hoppeler, Howald et al. 1985). This finding can be interpreted as indicating a general shift toward a larger reliance on fat metabolism, which could have two structural components: an increase of substrate stores favoring lipid metabolism of interfibrillar mitochondria and a large increase in subsarcolemmal mitochondria favoring oxidation of lipids taken from the vasculature.

It is currently not clear what limits triacylglycerol oxidation in exercising muscle. Is it the transfer of fatty acids to the mitochondrial matrix or is it beta-oxidation? In contrast, the rate at which glycogen stores can supply pyruvate to the Krebs cycle is not limited: they can supply it up to the limit of oxidation by the mitochondria and then can achieve a further severalfold increase when fueling anaerobic glycolysis. It therefore appears likely that oxidation in mitochondria is not limited by the supply of fuel but rather by the supply of oxygen from the capillaries or by the amount of mitochondria that can perform oxidative phosphorylation. It is unclear, however, what exactly sets the maximal rate at which mitochondria function in vivo.

Microcirculatory Supply of Oxygen and Substrates

Oxygen diffuses from the capillaries to the mitochondria, and the flow of oxygen at this step can be described as the product of a conductance and a pressure head (see figure 6.2):

$$\dot{V}O_2 max = DTO_2 \cdot (PbO_2 - PcO_2) \qquad 6.2$$

where

DTO_2 is conductance,

PbO_2 is the mean capillary (blood) PO_2, and

PcO_2 is the mean intracellular PO_2.

The conductance extends from the erythrocytes in the capillaries to the mitochondrial oxygen sink. We do not, as yet, have a model or set of measurements that allows us to formulate the dependence of DTO_2 on functional and structural variables. What can we then use as a relevant structural parameter to relate oxygen flow at this step in the respiratory system to the variations in $\dot{V}O_2max$?

In general terms, the maximal conductance must be related to the volume of capillaries in skeletal muscles V(c), a structural parameter that has been measured. Capillary volume V(c) is directly proportional to the capillary surface area available for diffusion of oxygen out of the blood, since the diameter of capillaries does not change with either size or adaptation. Capillary volume also plays an important role in determining the time available for diffusion as blood transits the capillary network.

Using $V(c)/M_b$ as the structural parameter, we can express the maximal flow of oxygen through this step in the respiratory system in the same way we have considered oxygen consumption by the mitochondria, expressed as the product of $V(c)/M_b$ and a functional parameter, $\dot{V}O_2(c)$, the rate at which oxygen diffuses out of each unit volume of capillaries:

$$\dot{V}O_2max/M_b = [V(c)/M_b] \cdot \dot{V}O_2(c). \qquad 6.3$$

We would expect these parameters would change with variations in $\dot{V}O_2max/M_b$ in parallel to the structural and functional parameters of the mitochondria that set the demand. What do we find?

With allometry, small animals have more capillaries in each gram of their muscles than do large animals. When quantitative measurements of capillary volume are made on the same individuals for whom mitochondrial volume measurements were made, one finds that average mass-specific capillary volume, like mitochondrial volume, decreases in direct proportion to $\dot{V}O_2max/M_b$ over the size range of 20 g (mice) to 500 kg (horses), whereas the rate of oxygen delivery per milliliter of capillary, $\dot{V}O_2(c)$, is nearly the same over the entire size range of animals (Taylor et al. 1989). Thus we can conclude that $\dot{V}O_2max/V(c)$ is invariant with size and is equal to 15 ml $O_2 \cdot min^{-1} \cdot ml^{-1}$.

This constant value for $\dot{V}O_2(c)$, like that for $\dot{V}O_2(mi)$, is a minimal value that assumes that at $\dot{V}O_2max$, all of the capillaries are utilized and that oxygen diffuses out of all of them at the same rate. Regardless of whether or not this is the case, it clearly indicates that the capillary surface available for diffusion increases directly with the rate of diffusion of oxygen out of the capillaries over the 5- to 10-fold differences in $\dot{V}O_2max/M_b$ with size. Also, we can see the ratio of V(c)/V(mi) must also be invariant (i.e., ~ 0.3 ml of capillaries for each ml of mitochondria), since both $\dot{V}O_2max/V(mi)$ and $\dot{V}O_2max/V(c)$ are invariant with size (Hoppeler, Mathieu et al. 1981).

With adaptive variation, we find a different pattern of adjustment of capillaries to oxygen demand (Conley et al. 1987). Capillary volume/M_b changes by only 1.7-fold with the 2.5-fold difference in $\dot{V}O_2max/M_b$ between the dog-goat, pony-calf, and horse-steer pairs (see table 6.1).

Using equation 6.3, we can calculate that oxygen diffuses out of each milliliter of capillary (each square centimeter of its surface) at 1.5 times the rate in the more aerobic species. Part of the explanation for this higher rate of diffusion lies in a 1.6-fold higher oxygen concentration of the arterial blood entering the capillary and a corresponding 1.6-fold greater extraction of oxygen as the blood transits the capillary bed. This results from a 1.5-fold higher hemoglobin concentration in the blood of the athletic species, which is due to higher hematocrit or erythrocyte concentration Vv(ec,blood). Multiplying the 1.7-fold greater volume of capillaries by the 1.6-fold greater extraction of oxygen across the capillary provides a 2.7-fold greater oxygen delivery during its transit through the capillary bed.

We can thus conclude that there is a reasonably good match between structure and the oxygen supply function at the level of the capillaries. In the case of allometric variations in $\dot{V}O_2max/M_b$, the higher rates of oxygen consumption are matched by corresponding differences in the amount of capillary structure, whereas the rate at which O_2 diffuses from each square centimeter of capillary surface is invariant in this case. The design of the capillaries at this level is closely matched to that of the mitochondria that consume the oxygen. In adaptive variation, capillaries are incompletely matched, but this is compensated by a higher hemoglobin or erythrocyte concentration. For both allometric and adaptive variation, it therefore appears that the mass of hemoglobin in capillaries is matched to the mass of mitochondria; this is achieved by varying capillary volume and hemoglobin concentration. The invariant ratio of structural parameters, therefore, is expressed as

$$Vv(ec)/V(mi) \approx M(H_b)/V(mi) = \qquad 6.3b$$
$$4 \cdot 10^{-2} g \cdot ml^{-1}$$

and, consequently, we find that the invariant ratio to $\dot{V}O_2$max is expressed as

$$\{V(c) \cdot Vv(ec)\}/\dot{V}O_2max = \qquad 6.3c$$
$$1.8 \ ml \cdot (ml \ O_2 \cdot sec^{-1})^{-1}$$

and thus involves two structural variables, namely, capillary volume and erythrocyte concentration in the blood.

What is the role of capillaries and hematocrit in induced variation? It was noted for the first time in 1977 (Brodal, Ingjer, and Hermansen 1977) that endurance exercise training leads to a significant increase in capillary density in humans (see also Zumstein et al. 1983). In general we note that the increase in capillarity falls somewhat short of the increase in muscle oxidative capacity (Hoppeler, Howald et al. 1985) if both parameters are measured in the same experiment and related to the same reference space (the muscle fiber volume). Typically, exercise training of six weeks in untrained humans leads to an increase in mitochondria of the order of over 40% while the increase in capillary density is less than 30%.

The role of the hematocrit is also of importance for human endurance performance capacity. As demonstrated convincingly in many studies, an increase in hematocrit such as obtained after blood retransfusion (see Ferretti et al. 1992) leads to an immediate and significant increase in whole body $\dot{V}O_2$max. A review of the literature indicates that world class $\dot{V}O_2$maxes can only be reached by athletes who have high hematocrits (Lindstedt et al. 1988). When the appropriate measurements are done it can be demonstrated that the increase in oxygen supply to the periphery with blood doping largely exceeds the gain in $\dot{V}O_2$max obtained with this procedure. The blood doping experiments thus indicate that the periphery can only partly accommodate the excess oxygen, indicating a significant peripheral resistance within the respiratory cascade. The site of this resistance (capillary oxygen release, diffusion of oxygen through interstitial space, and myocyte or mitochondrial oxygen metabolism) is currently debated (Wagner, Hoppeler, and Saltin 1997).

What about the transport of substrates out of the capillaries? The maximum rate appears to be reached at very low exercise intensities, as noted previously. Even at 40% of $\dot{V}O_2$max, only about 20 to 30% of the substrates are supplied from the circulating blood, and the absolute rate does not increase with increasing intensity. Very similar results have been obtained in exercising humans as well as in exercising dogs and goats (Romijn et al. 1993; Vock et al. 1996b).

At work intensities above 40% $\dot{V}O_2$max, intracellular substrate stores, in particular glycogen, are used to fuel aerobic work, as previously noted.

The transport of glucose from capillaries to myocytes has been studied by comparing "athletic" dogs to "sedentary" goats (Vock et al. 1996a). The transfer of glucose from the plasma to the interstitial fluid is assumed to occur through the small pore system at the endothelial intercellular junction. Calculations indicate that neither the pore system nor the interstitial space between the capillary and the myocyte introduces a major resistance to glucose flux across the sarcolemma. Based on morphological evidence it is concluded that the glucose transporters of the sarcolemma are likely saturated at low to moderate exercise intensities and are responsible for the major part of the resistance in the carbohydrate pathway.

The transport and utilization of fatty acids in muscle has recently been reviewed extensively (Van der Vusse and Reneman 1996). Based on data published in many studies these authors come to the conclusion that it is the delivery rather than the oxidative metabolism in myocytes that is the rate-limiting factor in lipid oxidation during exercise. However, many uncertainties remain, and lipid oxidation could be regulated at various steps in the route of transport as well as within the metabolic pathways. The general consensus seems to be that endurance exercise training increases the importance of fatty acids as a source of aerobic energy. Likewise, recent evidence suggests that high fat diets do have a potential for shifting metabolism toward a higher reliance on fatty acid metabolism (Van der Vusse and Reneman 1996). These general observations could be of importance for athletes competing in endurance events, in particular in very long term endurance events where intensities are low and the importance of fat oxidation high. In this context it may be noted that for certain long-term endurance events lasting over days or weeks, such as multistage bicycle races (Tour de France, Giro d'Italia, etc.), the concept of maximal sustained metabolism may be more important than that of $\dot{V}O_2$max (Hammond and Diamond 1997). In these events racers are limited in their performance by the need of having to balance work output to substrate uptake. For humans the limit of energy uptake with nutrition seems to be approximately 7,000 kcal/day. Thus professional bicyclists are capable of elevating metabolism acutely by some 20-fold above basal while average 24-hour metabolism (maximal sustained metabolism) can only be increased 4-fold.

Reviewing the transfer processes for oxygen and substrates from capillaries to myocytes we can conclude that muscle microvasculature is optimally designed for the transfer of oxygen but not for the transfer of substrates. Because of the lack of substantial oxygen stores, oxygen has to be supplied by the circulation so as to match the demand of the contracting muscle cells. By contrast, only a limited quantity of substrates is supplied from the vasculature during exercise. Animals and humans working at intensities above 30% of $\dot{V}O_2$max rely increasingly on intracellular substrate stores, mainly on glycogen at high work intensities. These stores will eventually be exhausted during exercise, leading to muscle fatigue. The intracellular stores can then be replenished during periods of rest at low transfer rates.

Convective Transport of Oxygen and Substrates by the Heart

Oxygen and substrates are transported convectively from the capillaries in the lung to the capillaries in the muscle by the circulation. Delivery of oxygen by the circulation depends on the properties of the heart as a pump and on those of the blood as the carrier, and both can be varied to adjust to differences in demand. Oxygen flow at this step can be described as the product of the maximal cardiac output Q_{max}/M_b times the difference in arteriovenous oxygen concentration $CaO_2 - CvO_2$ (see figure 6.2):

$$\dot{V}O_2max/M_b = (Q_{max}/M_b) \cdot (CaO_2 - CvO_2). \qquad 6.4$$

The relevant structural parameter of the pump is obviously the size of the heart, which determines the amount of blood pumped with each contraction, the stroke volume Vs. The flow of blood at this step is the product of $Vs \cdot M_b$ and a functional parameter f_H, the maximal frequency of contraction of the pump:

$$Q_{max} \cdot M_b^{-1} = f_H \cdot Vs \cdot M_b^{-1}. \qquad 6.5$$

$CaO_2 - CvO_2$ will depend on a second structural parameter, namely, the amount of hemoglobin or erythrocytes contained in the blood.

In allometric variation we would anticipate that time (i.e., f_H) would vary with size, while the structural parameters hemoglobin concentration [Hb] and Vs/M_b would be invariant. What information we have supports this idea. On average, the heart makes up the same fraction of body mass over the size range of mammals from mice to cows, about

0.58% (Prothero 1979); likewise, hemoglobin concentration and oxygen-carrying capacity of the blood do not vary systematically with body size. Mammals spanning a range of body size from bats to horses have, on average, about 13 g of hemoglobin per 100 ml of blood, which can carry 17.5 ml of oxygen (Schmidt-Nielsen 1984). The invariant heart size and O_2 capacity of the blood suggest that body size-dependent differences in circulatory transport are brought about entirely by increases in heart frequency.

The structural and functional variations in circulatory transport of oxygen with adaptive variation are very clear-cut. Here the animals follow the general principles of design: f_H is determined by size, and structures vary with O_2 demand. Maximal heart frequencies of goats and dogs, ponies and calves, and horses and steers are nearly identical pair-wise for animals of the same size, despite a 2.5-fold difference in $\dot{V}O_2max/M_b$ (Jones et al. 1989; Karas et al. 1987b). The structural parameters Vs/M_b and hemoglobin concentration account for all of the 2.5-fold difference in oxygen delivery at this step. These studies indicate that these animals are operating at or close to the upper limit of their structural capacity for convective transport of oxygen in the circulatory system at $\dot{V}O_2max$ (Taylor et al. 1987). Available structures (stroke volume and hemoglobin concentration) for both Q_{max} and $(CaO_2 - CvO_2)$ appear fully exploited by the time an animal has increased its oxygen consumption from resting to maximal levels. Thus at this step of O_2 transport in the respiratory system there appears to be a match between maximal rates of O_2 delivery and the structures involved, a finding that is in accord with the predictions of symmorphosis.

The limited information we have therefore suggests that the structural parameters Vs/M_b and Vv(ec) are invariant in allometric variation but variant in adaptive variation, whereas the functional parameter f_H is invariant in adaptive variation but variant in allometric variation.

The situation in endurance-trained athletes is similar, but not identical, to the situation described for the adaptive pairs. The term *athlete's heart*, denoting a heart increased in size because of athletic activity, was coined over 90 years ago based on careful percussion examinations (Henschen 1899). Until quite recently many cardiologists saw the athlete's heart as a pathological condition (see Keul et al. 1982). This view has been abandoned completely. However, the significance of exercise-induced cardiac hypertrophy is still a controversial issue (Perrault and Turcotte 1994). Whether endurance athletes have larger hearts to start with or whether the trainability of their hearts is better has remained an unresolved issue.

The size of the heart in an untrained human subject is of the order of $10 \, \text{ml} \cdot \text{kg}^{-1} \, M_b$. In comparative terms this clearly indicates that humans belong to the category of "sedentary" mammals. Keul and associates (1982) have reported the heart size of top athletes of many different disciplines. From these data it becomes apparent that heart size is larger in athletes who compete in endurance events. Based on ultrasound and radiographic examinations, heart sizes up to twice those seen in untrained subjects have been documented (Keul et al. 1982). The main variable determining heart size seems to be the total endurance training time.

It is not simply a larger size that determines the better performance capacity of the athlete's heart. Athletes have been shown to be able to increase stroke volume more with increasing heart rates (+ 35%) than sedentary subjects (+ 5%). Moreover, a reduced sympathetic drive, a reduced afterload, and an increase in cardiac compliance all contribute toward a lower cardiac oxygen demand at a given heart rate (Heiss et al. 1976). The stimulus for cardiac growth has been hypothesized to be related to the acceleration of the diastolic filling velocity and the increase in stroke volume; both are likely a consequence of diminished sympathoadrenergic drive (Keul et al. 1982).

As noted above, systemic hematocrit plays an important role in determining cardiac oxygen delivery. The observation that $\dot{V}O_2$max decreases with anemia and can readily be increased with induced erythrocythemia has been one of the main arguments in favor of cardiac limitation of $\dot{V}O_2$max. However, Lindstedt and coworkers (1988) have shown this reasoning to be wrong; and as indicated earlier the prevailing concept today is that limitation is distributed over the entire respiratory cascade with the heart being the main player in humans working with large muscle groups in normoxia. But adjusting the heart is not enough; an extensive review of published data shows that very high $\dot{V}O_2$max values (> 80 ml $\cdot$ min$^{-1} \cdot$ kg^{-1}) can only be observed in subjects that have high hemoglobin concentrations (> 15.5 g $\cdot$ 100 ml^{-1}).

In summary then, while maximal heart rate is important for modulating cardiac output with allometric variation of body mass, in adaptive as well as in induced variation, stroke volume is the main factor determining cardiac output in mammals of a given body size. An important additional component is hematocrit, which is found to be greater in athletic animals. In human athletes a high hematocrit is a prerequisite for achieving a world class aerobic performance level with the caveat that a too high hematocrit increases blood viscosity and may have catastrophic consequences with exercise-induced hemoconcentration.

Oxygen Diffusion in the Lung

In humans, the lung is generally considered to contribute only minimal resistance to oxygen flow to the periphery (di Prampero 1985). In comparative physiology the pulmonary gas exchanger has received considerable interest, and the general design principles that have emerged as a consequence of these analyses are useful for understanding the situation in humans.

The transfer of O_2 from the air to the blood in the lung is achieved by diffusion. The O_2 flow rate is determined by the product of the partial pressure difference as driving force and the conductance of the gas exchanger (Bohr 1909), such that (see figure 6.2)

$$\dot{V}O_2 = DLO_2 \cdot (PAO_2 - PbO_2). \qquad 6.6$$

The partial pressure difference between alveolar air and capillary blood is a functional variable that essentially depends on the ventilation of alveoli through the airways and the perfusion of capillaries by the circulation. In contrast, the diffusion conductance for O_2, the diffusing capacity DLO_2, is largely determined by the following structural parameters (see figure 6.2): the alveolar and capillary surface areas $S(A)$ and $S(c)$ and the harmonic mean barrier thickness of the tissue and of the plasma layer separating erythrocytes from the endothelium and the capillary blood volume $V(c)$.

The morphometric parameters entering the calculation of DLO_2 (see Weibel 1997) are essentially determined by two variables: the lung volume $V(L)$ and the size or density of the "building blocks" of the gas exchange units in lung parenchyma. The ultimate building block of the gas exchanger is the alveolar septum, with morphometric characteristics being the fraction of septum occupied by capillaries, the capillary volume per septal (alveolar) surface $V(c)/S(A)$, the density of erythrocytes or the hematocrit, and the harmonic mean thickness of the tissue barrier (see figure 6.10). These septa are built into the acinus as alveolar walls in the form of a three-dimensional maze; accordingly, the alveolar surface density $Sv(A)$ is a measure of the building block characteristics of lung parenchyma. This hierarchical design provides several options for varying

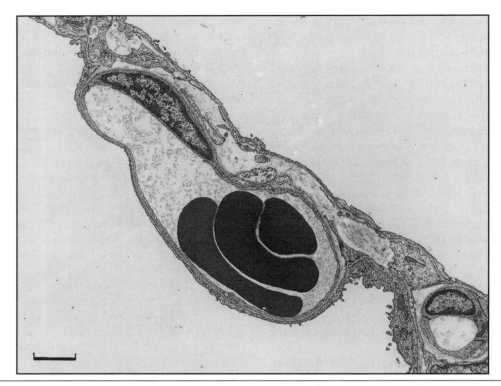

Figure 6.10 Electron micrograph of alveolar septum containing a capillary with three erythrocytes. Note the extremely thin tissue barrier between alveolar air and erythrocytes (marker indicates 2 µm).

diffusing capacity. Thus the total alveolar surface area S(A) is the product of V(L) and the alveolar surface density Sv(A); furthermore, capillary volume is the product of V(c)/S(A) and S(A). To increase DLO_2, the lung would have to increase, for example, the alveolar surface area, and this can be achieved by either increasing lung volume or increasing the alveolar surface density by packing more alveolar septa into the unit volume of lung parenchyma.

Alternatively or additionally, the loading of capillaries onto the septum could be increased or the barrier thickness could be decreased. The first question with respect to design of the gas exchanger is which of these options are used, or whether any of these basic design parameters are invariant with allometric and adaptive variation of $\dot{V}O_2$ max.

Let us first consider lung volume. The general notion is that mass-specific lung volume is invariant with body size. At closer inspection we find, however, that V(L) increases slightly (but significantly) with $M_b^{1.06}$ (Gehr et al. 1981), with the result that $V(L)/M_b$ varies from 35 ml · kg^{-1} in shrews and mice to 60 ml · kg^{-1} in dog, man, and cow and can even reach 100 ml · kg^{-1} in the horse. This last number indicates that, in adaptive variation, relative lung volume is an important variable, being

larger in the athletic species (Constantinopol et al. 1989; Weibel et al. 1987). In a study of the pronghorn antelope, whose $\dot{V}O_2$ max is twice that of the dog, we found the increase in lung volume (to nearly 5 liters for a 20 kg animal!) to account for most of the adaptive lung change (Lindstedt et al. 1991).

Among the building block characteristics, the parameters that characterize septum structure are invariant in adaptive variation (Constantinopol et al. 1989; Weibel et al. 1987), but they show a weak allometric variability (Gehr et al. 1981). The packing of alveolar septa into lung parenchyma, measured by Sv(A) (which is inversely proportional to alveolar diameter), appears to be invariant in the size range of animals between 1 and 100 kg (Weibel, Gehr et al. 1981), assuming values of 400 to 500 cm^{-1} irrespective of whether they are athletic or sedentary. However, it increases drastically in small mammals, up to 1500 cm^{-1} in the shrews, and it falls to 250 cm^{-1} in large mammals, so that over the entire mammalian size range we find Sv(A) to decrease with $M_b^{-0.11}$.

Quite evidently, all of these parameters are subject to a number of constraints. The lung volume is limited by the space available in the chest cavity. The packing of alveolar septa into the air space is limited by the requirements for adequate ventilation as well as by mechanical constraints related to surface tension. The smaller the alveoli, the greater the surface forces; and the larger the alveoli, the more costly the alveolar ventilation. Mammals may indeed have found an optimum range for the size of these building blocks, from which they deviate only in the very small species, and perhaps in the largest but to a lesser extent.

How are these building blocks related to O_2 uptake at $\dot{V}O_2$max? Specifically, is O_2 uptake by the unit capillary volume invariant? We find that it is not. In allometry $\dot{V}O_2$max/V(c) varies with $M_b^{-0.2}$, ranging from 12 ml O_2 min^{-1} · ml^{-1} in large animals to 42 ml O_2 min^{-1} · ml^{-1} in small (500 g) animals, and it may go even higher in mice and shrews (Gehr et al. 1981). In adaptive variation, we find that athletic species load about twice as much O_2 onto their blood per unit time, a rate that appears to be about proportional to their higher hematocrit (Constantinopol et al. 1989; Weibel et al. 1987). Oxygen uptake rate by the pulmonary capillary unit is therefore clearly not invariant, in partial contrast to the muscle capillaries where we have found the discharge rate $\dot{V}O_2$ max/V(c) to be invariant with allometric variation, whereas a similar difference was found between the adaptive pairs. It is noteworthy that the discharge rates are similar in lung and muscle capillaries, at least in the nonathletic species; the observed differences can be explained by different transit times.

When we now consider the total pulmonary diffusing capacity, we must note that it is composed of two main components: the membrane diffusing capacity DMO_2, which is exclusively determined by structural variables, and the blood or erythrocyte diffusing capacity DeO_2, which depends on capillary blood volume and hematocrit, two parameters that are also subject to some functional variation.

The hypothesis of symmorphosis predicts that DMO_2 and DLO_2 should be proportional to $\dot{V}O_2$max. This is not what we find. We note that mass-specific DLO_2 does not change with size, so that the ratio $DLO_2/\dot{V}O_2$ max increases with $M_b^{0.2}$. This means that a 300 kg cow has six times as much diffusing capacity available as a 30 g mouse to accomplish O_2 uptake at $\dot{V}O_2$max. What this means is that the driving force for O_2 uptake in the lung is smaller in the cow than in the mouse, but why this occurs is unknown. Various possibilities have been suggested, such as appreciable differences in capillary transit time (Lindstedt 1984) or differences in the pressure head PAO_2 as a result of the fact that the size of acini varies considerably with body size (Haefeli-Bleuer and Weibel 1988; Rodriguez et al. 1987), and this

could influence alveolar ventilation (Karas et al. 1987a; Weibel, Taylor et al. 1981).

In adaptive variation we find that the athletic species have a larger DLO_2/M_b than the nonathletic animals, but this increase is not proportional to the differences in $\dot{V}O_2$max. Here again, athletic species accomplish their higher O_2 uptake rate by adding to their increased DLO_2 an elevated driving force. In this instance we were able to show that this is partly due to the fact that the athletic species use a greater fraction of their shorter transit time to accomplish equilibration of the capillary blood with alveolar air, namely about 80%, whereas the sedentary species use only 50%, with the remainder appearing as a redundancy (Constantinopol et al. 1989; Karas et al. 1987a).

In conclusion, we find that the ratio $DLO_2/\dot{V}O_2$max is not invariant, neither in allometric nor in adaptive variation. To find an invariant ratio we must consider all structural and functional variables, because only the following equation applies:

$$DLO_2/(\dot{V}O_2max/\Delta PO_2) = \text{invariant.} \qquad 6.6b$$

We must therefore conclude that functional variables are used to a large extent to modulate the rate of O_2 uptake even at $\dot{V}O_2$max. This is possible because the pulmonary gas exchanger maintains an appreciable level of redundancy or excess capacity (Karas et al. 1987a). One is tempted to speculate that maintaining such redundancy in that part of the respiratory system forming the interface with the environment may well be a survival strategy, allowing the organism to cope with adverse environmental factors such as hypoxia. It has indeed been shown that goats can maintain their $\dot{V}O_2$max even at high altitude conditions (Karas et al. 1987a), presumably because they increase cardiac output under hypoxia and their gas exchanger is redundant when judged under sea level conditions.

Humans have a lung structure that is typical for relatively large mammals, and it is unknown whether athletes have a larger pulmonary diffusing capacity. The situation in the exercising trained and untrained human is in some way similar to the situation in athletic and nonathletic species such as horses and steers. Stated in general terms, there is more structural redundancy in untrained than in trained individuals. A major problem with the pulmonary gas exchanger is that it fails to adapt to endurance exercise training, unlike the downstream determinants of oxygen flow of the respiratory system. In humans and animals there is only evidence for lung structural plasticity with chronic hypoxia and following extensive tissue loss after pneumonectomy

(Hsia et al. 1994) but not with whole body endurance type training (Dempsey and Johnson 1992; Dempsey et al. 1977; Hoppeler et al. 1995). The observation that very fit endurance-trained athletes may show arterial hypoxemia (i.e., PaO$_2$ of 75 mm Hg) and reduced saturation (i.e., SaO$_2$ of 85–93%) when working at $\dot{V}O_2$max in normoxia led to the conclusion that in highly endurance trained subjects the pulmonary system can present a significant limitation to oxygen flow to the periphery. One is therefore tempted to speculate that the endurance athlete can improve his or her peripheral capacity for O$_2$ consumption up to a maximum level set by the maximum capacity of the pulmonary system for oxygen uptake.

Conclusions

The factors limiting $\dot{V}O_2$max have been assessed in a systematic analysis of the pathway for oxygen from the lung to skeletal muscle mitochondria taking into account the role played by substrate availability for oxidative metabolism of muscle cells with regard to the pathway of oxygen. Current thinking indicates that the limitation is distributed over all levels of the respiratory system with some steps having more resistance than others under certain conditions. In order to unravel the basic design principles we have taken a comparative approach, comparing structure and function of the respiratory system in animals differing widely in mass-specific $\dot{V}O_2$max and relating our findings to the situation in humans. A detailed analysis of each transfer step of oxygen made use of the concept of symmorphosis, that is, assuming for each level of the respiratory cascade that animals just maintain enough structure to support flux rates at $\dot{V}O_2$max but not more. We found all levels of the cascade to conform with the principle of symmorphosis except for the lungs, which seem to be built with significant, though limited, excess structural capacity. However, this redundancy is variable; it is smaller in athletic than in sedentary species with probably no structural redundancy in the very best of human endurance athletes. The analysis also shows that this system is likely built on the constraint of supplying oxygen rather than substrates to active muscle mitochondria under conditions of maximal aerobic work. Carbohydrate and lipid supply rates are throttled by transport processes at the level of the sarcolemma. In order to ascertain adequate substrate supply at high work loads both lipids and carbohydrates are stored within muscle cells. These substrate stores are replenished at low flux rates during periods of rest to reach a size adequate for high rates of combustion in exercise.

References

Armstrong, R.B., and Taylor, C.R. 1982. Relationship between muscle force and muscle area showing glycogen loss during locomotion. *J. Exp. Biol.* 97: 411–20.

Bevegard, B.S., and Shepherd, J.T. 1967. Regulation of the circulation during exercise in man. *Physiol. Rev.* 47: 178–213.

Bohr, C. 1909. Über die spezifische Tätigkeit der Lungen bei der respiratorischen Gasaufnahme und ihr Verhalten zu der durch die Alveolarwand stattfindenden Gasdiffusion. *Scand. Arch. Physiol.* 22: 221–80.

Brodal, P., Ingjer, F., and Hermansen, L. 1977. Capillary supply of skeletal muscle fibers in untrained and endurance-trained men. *Am. J. Physiol.* 232: H705–12.

Chatterjee, S., and Chakravarti, B. 1986. Comparative study of maximum aerobic capacity by three ergometries in untrained college women. *Jap. J. Physiol.* 36: 151–62.

Christensen, E.H. 1931. Beiträge zur Physiologie schwerer körperlicher Arbeit. IV. Mitteilung: die Pulsfrequenz während und unmittelbar nach schwerer körperlicher Arbeit. *Arbeits-Physiologie* 4: 453–69.

Conley, K.E., and Jones, D. 1996. Myoglobin content and oxygen diffusion: Model analysis of horse and steer muscle. *Am. J. Physiol.* 271: C2027–36.

Conley, K.E., Kayar, S.R., Rösler, K., Hoppeler, H., Weibel, E.R., and Taylor, C.R. 1987. Adaptive variation in the mammalian respiratory system in relation to energetic demand. IV. Capillaries and their relationship to oxidative capacity. *Respir. Physiol.* 69: 47–64.

Constantinopol, M., Jones, J.H., Weibel, E.R., Taylor, C.R., Lindholm, A., and Karas, R.H. 1989. Oxygen transport during exercise in large mammals. II. Oxygen uptake by the pulmonary gas exchanger. *J. Appl. Physiol.* 67: 871–78.

Davies, K.J.A., Packer, L., and Brooks, G.A. 1981. Biochemical adaptation of mitochondria, muscle, and whole animal respiration to endurance training. *Arch. Biochem. Biophys.* 209: 539–54.

Dempsey, J.A., Gledhill, N., Reddan, W.G., Forster, H.V., Hanson, P.G., and Claremont, A.D. 1977. Pulmonary adaptation to exercise: Effects of exercise type and duration, chronic hypoxia, and physical training. *Annals NY Acad. Sci.* 301: 243–61.

Dempsey, J.A., and Johnson, B.D. 1992. Demand vs. capacity in the healthy pulmonary system. *Schweiz. Ztschr. Sportmed.* 40: 55–64.

Desplanches, D., Hoppeler, H., Tüscher, L., Mayet, M.H., Spielvogel, H., Ferretti, G., Kayser, B., Leuenberger, M., Grünenfelder, A., and Favier, R. 1996. Muscle tissue adaptation of high altitude natives to training in chronic hypoxia or acute normoxia. *J. Appl. Physiol.* 81: 1946–51.

di Prampero, P.E. 1985. Metabolic and circulatory limitations to V̇O$_2$ max at the whole animal level. *J. Exp. Biol.* 115: 319–32.

Ferretti, G., Kayser, B., Schena, F., Turner, D.L., and Hoppeler, H. 1992. Regulation of perfusive O$_2$ transport during exercise in humans: Effects of changes in hemoglobin concentration. *J. Physiol. (London)* 455: 679–88.

Gehr, P., Mwangi, D.K., Amman, A., Maloiy, G.M.O., Taylor, R.C., and Weibel, E.R. 1981. Design of the mammalian respiratory system. V. Scaling morphometric pulmonary diffusing capacity to body mass: Wild and domestic animals. *Respir. Physiol.* 44: 61–86.

Gollnick, P.D., and Saltin, B. 1982. Significance of skeletal oxidative enzyme enhancement with endurance training. *Clin. Physiol.* 2: 1–12.

Haefeli-Bleuer, B., and Weibel, E. 1988. Morphometry of the human pulmonary acinus. *Anat. Rec.* 220: 401–14.

Hammond, K.A., and Diamond, J. 1997. Maximal sustained energy budgets in humans and animals. *Nature* 386: 457–62.

Heiss, H.W., Barmeyer, J., Wink, K., Hell, G., Cerny, F.J., Kreul, J., and Reindell, H. 1976. Studies on the regulation of myocardial blood flow in man. I. Training effects on blood flow and metabolism of the healthy heart at rest and during standardized heavy exercise. *Basic Res. Cardiol.* 71: 658.

Henschen, S.W. 1899. Skilauf und Skiwettlauf. Eine Medizinische Sportstudie. Mitteilungen aus der Medizinischen Klinik in Upsala. 15 Jena, Band 2.

Hill, A.V., and Lupton, H. 1923. Muscular exercise, lactic acid, and the supply and utilization of oxygen. *Q. J. Med.* 16: 135–71.

Holloszy, J.O., and Booth, F.W. 1976. Biochemical adaptation to endurance exercise in muscle. *Ann. Rev. Physiol.* 38: 273–91.

Holmgren, A., and Astrand, P.O. 1966. DL and the dimensions and functional capacities of the O$_2$ transport system in humans. *J. Appl. Physiol.* 21: 1463–70.

Hoppeler, H. 1986. Exercise-induced ultrastructural changes in skeletal muscle. *Int. J. Sport Med.* 7: 187–204.

Hoppeler, H., Altpeter, E., Wagner, M., Turner, D.L., Hokanson, J., König, M., Stalder-Navarro, V.P., and Weibel, E.R. 1995. Cold acclimation and endurance training in guinea pigs: Changes in lung, muscle, and brown fat tissue. *Respir. Physiol.* 101: 183–188.

Hoppeler, H., Howald, H., Conley, K.E., Lindstedt, S.L., Claassen, H., Vock, P., and Weibel, E.R. 1985. Endurance training in humans: Aerobic capacity and structure of skeletal muscle. *J. Appl. Physiol.* 59: 320–27.

Hoppeler, H., Jones, J.H., Lindstedt, S.L., Claassen, H., Longworth, K.E., Taylor, C.R., Straub, R., and Lindholm, A. 1987. Relating maximal oxygen consumption to skeletal muscle mitochondria in horses. In *Equine Exercise Physiology II*, ed. J.R. Gillespie and N.E. Robinson, 278–89. Davis, CA: ICEEP.

Hoppeler, H., Kayar, S.R., Claassen, H., Uhlmann, E., and Karas, R.H. 1987. Adaptive variation in the mammalian respiratory system in relation to energetic demand. III. Skeletal muscles: Setting the demand for oxygen. *Respir. Physiol.* 69: 27–46.

Hoppeler, H., Lindstedt, S.L., Uhlmann, E., Niesel, A., Cruz-Orive, L., and Weibel, E.R. 1984. Oxygen consumption and the composition of skeletal muscle tissue after training and inactivation in the European wood mouse *(Apodemus sylvaticus)*. *J. Comp. Physiol. B* 155: 51–61.

Hoppeler, H., Lüthi, P., Claassen, H., Weibel, E.R., and Howald, H. 1973. The ultrastructure of the normal human skeletal muscle: A morphometric analysis on untrained men, women, and well-trained orienteerers. *Pflügers Arch.* 344: 217–32.

Hoppeler, H., Mathieu, O., Krauer, R., Claassen, H., Armstrong, R.B., and Weibel, E.R. 1981. Design of the mammalian respiratory system. VI. Distribution of mitochondria and capillaries in various muscles. *Respir. Physiol.* 44: 87–111.

Hsia, C.C.W., Herazo, L.F., Fryder-Doffey, F., and Weibel, E.R. 1994. Compensatory lung growth occurs in adult dogs after right pneumonectomy. *J. Clin. Invest.* 94: 405–12.

Jones, J.H., Longworth, K.E., Lindholm, A., Conley, K.E., Karas, R.H., Kayar, S.K., and Taylor, C.R. 1989. Oxygen transport during exercise in large mammals. I. Adaptive variation in oxygen demand. *J. Appl. Physiol.* 67: 862–70.

Karas, R.H., Taylor, C.R., Jones, J.H., Lindstedt, S.L., Reeves, R.B., and Weibel, E.R. 1987a. Adaptive variation in the mammalian respiratory system in relation to energetic demand. VII. Flow of oxygen across the pulmonary gas exchanger. *Respir. Physiol.* 69: 101–15.

Karas, R.H., Taylor, C.R., Rösler, K., and Hoppeler, H. 1987b. Adaptive variation in the mammalian respiratory system in relation to energetic demand. V. Limits to oxygen transport by the circulation. *Respir. Physiol.* 69: 65–79.

Kayar, S.R., Hoppeler, H., Howald, H., Claassen, H., and Oberholzer, F. 1986. Acute effects of endurance exercise on mitochondrial distribution and skeletal muscle morphology. *Eur. J. Appl. Physiol.* 54: 578–84.

Keul, J., Dickhuth, H.H., Lehman, H., and Staiger, J. 1982. The athlete's heart: Hemodynamics and structure. *Int. J. Sport Med.* 3:33-43.

Keul, J., Dickhuth, H.H., Simon, G., and Lehmann, M. 1981. Effect of static and dynamic exercise on heart volume, contractility, and left ventricular dimensions. *Circ. Res.* 48: 1162–70.

Kleiber, M. 1961. *The fire of life: An introduction to animal energetics.* New York: Wiley.

Lindstedt, S.L. 1984. Pulmonary transit time and diffusing capacity in mammals. *Am. J. Physiol.* 246: R384–88.

Lindstedt, S.L., and Calder III, W.A. 1981. Body size, physiological time, and longevity of homeothermic animals. *Quart. Rev. Biol.* 56: 1–16.

Lindstedt, S.L., Hokanson, J.F., Wells, D.J., Swain, S.D., Hoppeler, H., and Navarro, V. 1991. Running energetics in the pronghorn antelope. *Nature* 353: 748–50.

Lindstedt, S.L., Wells, D.J., Jones, J.R., Hoppeler, H., and Thronson, H.A. 1988. Limitations to aerobic performance in mammals: Interaction of structure and demand. *Int. J. Sport Med.* 9: 210–17.

Margulis, L. 1981. *Symbiosis in cell evolution.* San Francisco: W.H. Freeman.

Mathieu, O., Krauer, R., Hoppeler, H., Gehr, P., Lindstedt, S.L., Alexander, R.McN., Taylor, C.R., and Weibel, E.R. 1981. Design of the mammalian respiratory system. VII. Scaling mitochondrial volume in skeletal muscle to body mass. *Respir. Physiol.* 44: 113–28.

Mitchell, J.H., and Blomqvist, G. 1971. Maximal oxygen uptake. *New Engl. J. Med.* 284: 1018–22.

Oberholzer, F., Claassen, H., Moesch, H., and Howald, H. 1976. Ultrastrukturelle, biochemische und energetische Analyse einer extremen Dauerleistung (100 Km Lauf). *Schweiz. Zeitschr. Sportmed.* 24: 71–98.

Perrault, H., and Turcotte, R.A. 1994. Exercise-induced cardiac hypertrophy: Fact or fallacy? *Sports Med.* 17: 288–308.

Prothero, J. 1979. Heart weight as a function of body weight in mammals. *Growth* 43: 139–50.

Puntschart, A., Claassen, H., Jostarndt, K., Hoppeler, H., and Billeter, R. 1995. mRNAs of enzymes involved in energy metabolism and mtDNA are increased in endurance-trained athletes. *Am. J. Physiol.* 269: C619–25.

Reindell, H., Klepzig, H., and Musshoff, K. 1957. Neuere Untersuchungsergebnisse über Beziehungen zwischen Grösse und Leistungsbreite des gesunden menschlichen Herzens, insbesondere des Sportherzens. *Deutsche Medizinische Wochenschrift* 82: 613–19.

Rodriguez, M., Bur, S., Favre, A., and Weibel, E.R. 1987. The pulmonary acinus: Geometry and morphometry of the peripheral airway system in rat and rabbit. *Am. J. Anat.* 180: 143–55.

Romijn, J.A., Coyle, E.F., Sidossis, L.S., Gastaldelli, A., Horowitz, J.F., Endert, E., and Wolfe, R.R. 1993. Regulation of endogenous fat and carbohydrate metabolism in relation to exercise intensity and duration. *Am. J. Physiol.* 265: E380–91.

Rösler, K.M., Conley, K.E., Claassen, H., Howald, H., Hoppeler, H., and Gehr, P. 1985. Transfer effects in endurance exercise: Adaptations in trained and untrained muscles. *Eur. J. Appl. Physiol.* 54: 355–62.

Saltin, B., Blomqvist, G., Mitchell, J.H., Johnson Jr., R.L., Wildenthal, K., and Chapmann, C.B. 1968. Response to exercise after bed rest and after training. *Circulation* 38(Suppl. 7): 1–78.

Saltin, B., and Gollnick, P.D. 1983. Skeletal muscle adaptability: Significance for metabolism and performance. In *Handbook of physiology: Skeletal muscle,* ed. L.D. Peachy, R.H. Adrian, and S.R. Geiger, 555–631. Baltimore: Williams & Wilkins.

Schmidt-Nielsen, K. 1984. *Scaling: Why is animal size so important?* Cambridge: Cambridge University Press.

Schwerzmann, K., Hoppeler, H., Kayar, S.R., and Weibel, E.R. 1989. Oxidative capacity of muscle and mitochondria: Correlation of physiological, biochemical, and morphometric characteristics. *Proc. Nat. Acad. Sci.* 86: 1583–87.

Taylor, C.R., Karas, R.H., Weibel, E.R., and Hoppeler, H. 1987. Adaptive variation in the mammalian respiratory system in relation to energetic demand. II. Reaching the limits to oxygen flow. *Respir. Physiol.* 69: 7–26.

Taylor, C.R., and Weibel, E.R. 1981. Design of the mammalian respiratory system. I. Problem and strategy. *Respir. Physiol.* 44: 1–10.

Taylor, C.R., Weibel, E.R., Hoppeler, H., and Karas, R.H. 1989. Matching structure and function in the respiratory system: Allometric and adaptive variations in energy demand. In *Comparative pulmonary physiology: Current concepts,* ed. S.C. Wood, 27–65. New York: Marcel Dekker.

Van der Vusse, G.J., and Reneman, R.S. 1996. Lipid metabolism in muscle. In *Handbook of physiology. Sec. 12. Exercise: Regulation and integration of multiple systems,* ed. L.B. Rowell and J.T. Shepherd, 952–94. New York: American Physiological Society, Oxford Press.

Vock, R., Hoppeler, H., Claassen, H., Wu, D.X.Y., Billeter, R., Weber, J.M., Taylor, C.R., and Weibel, E.R. 1996a. Design of the oxygen and substrate pathways. VI. Structural basis of intracellular substrate supply to mitochondria in muscle cells. *J. Exp. Biol.* 199: 1689–97.

Vock, R., Weibel, E.R., Hoppeler, H., Ordway, G., Weber, J.M., and Taylor, C.R. 1996b. Design of the oxygen and substrate pathways. V. Structural basis of vascular substrate supply to muscle cells. *J. Exp. Biol.* 199: 1675–88.

Wagner, P.D., Hoppeler, H., and Saltin, B. 1997. Determinants of maximal oxygen uptake. In *The lung: Scientific foundations.* 2d ed. Ed. R.G. Crystal, J.B. West, E.R. Weibel, and P.J. Barnes, 2033–41. Philadelphia: Lippincott, Raven.

Weber, J.-M., Brichon, G., Zwingelstein, G., McClelland, G., Saucedo, C., Weibel, E.R., and Taylor, C.R. 1996a. Design of the oxygen and substrate pathways. IV. Partitioning energy provision from fatty acids. *J. Exp. Biol.* 199: 1667–74.

Weber, J.-M., Roberts, T.J., Vock, R., Weibel, E.R., and Taylor, C.R. 1996b. Design of the oxygen and substrate pathways. III. Partitioning energy provision from carbohydrates. *J. Exp. Biol.* 199: 1659–66.

Weibel, E.R. 1997. Design and morphometry of the pulmonary gas exchanger. In *The lung: Scientific foundations.* 2d ed. Ed. R.G. Crystal, J.B. West, E.R. Weibel, and P.J. Barnes, 1147–57. Philadelphia: Lippincott, Raven.

Weibel, E.R., Gehr, P., Cruz-Orive, L.M., Müller, A.E., Mwangi, D.K., and Haussener, V. 1981. Design of the mammalian respiratory system. IV. Morphometric estimation of pulmonary diffusing capacity, critical evaluation of a new sampling method. *Respir. Physiol.* 44: 39–59.

Weibel, E.R., Marques, L.B., Constantinopol, M., Doffey, F., Gehr, P., and Taylor, C.R. 1987. Adaptive variation in the mammalian respiratory system in relation to energetic demand. VI. The pulmonary gas exchanger. *Respir. Physiol.* 69: 81–100.

Weibel, E.R., Taylor, C.R., Gehr, P., Hoppeler, H., Mathieu, O., and Maloiy, G.M.O. 1981. Design of the mammalian respiratory system. IX. Functional and structural limits for oxygen flow. *Respir. Physiol.* 44: 151–64.

Weibel, E.R., Taylor, C.R., and Hoppeler, H. 1991. The concept of symmorphosis: A testable hypothesis of structure-function relationship. *Proc. Nat. Acad. Sci.* 88: 10357–61.

Weibel, E.R., Taylor, C.R., and Hoppeler, H. 1992. Variations in function and design: Testing symmorphosis in the respiratory system. *Respir. Physiol.* 87: 325–48.

Zumstein, A., Mathieu, O., Howald, H., and Hoppeler, H. 1983. Morphometric analysis of the capillary supply in skeletal muscles of trained and untrained subjects: Its limitations in muscle biopsies. *Pflügers Arch.* 397: 277–83.

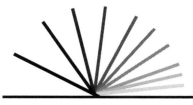

Chapter 7

Energy and Nutrient Intake for Athletic Performance

P.W.R. Lemon

In recent years it has become increasingly apparent that an individual's dietary habits can be a critical determinant of her or his exercise performance. Although less clear, it also appears that chronic (regular) exercise increases the requirements for several nutrients, suggesting that adverse health effects may result if intake of these nutrients is not increased. This exercise-induced increased nutrient need is likely of greatest concern when exacerbated by other preexisting conditions. Specific situations where chronic exercisers must be particularly careful about their diet include the following: when nutrient requirements are already increased above normal such as during periods of rapid growth (children, adolescents, women who are pregnant), when food intake is insufficient due to voluntary restriction (dieters) or extreme energy expenditures (athletes in heavy training), and when overall nutrient intake is inappropriate, that is, when a significant percentage of one's intake consists of foods with low nutrient density (this is particularly prevalent in adolescents, single adults, and the elderly).

Over the past few decades, the various health benefits of chronic exercise have become well understood. As a result, expert committees in both Canada (Canada's Physical Activity Guide to Healthy, Active Living 1998) and the United States (U.S. Department of Health and Human Services 1996) have very recently produced daily physical activity recommendations for all individuals. If chronic exercise increases nutrient needs, it is of interest to consider whether a significant public health concern relative to select nutritional deficiencies might arise if a sizable percentage of the general public heeds these activity recommendations. Fortunately, there appears to be little need for concern because, in general, the increased nutrient needs caused by active living can be quite easily met without any special dietary supplementation, assuming one's food intake comes from a variety of food sources and adequate total energy intake occurs. However, such is not always the case with athletes (especially those with very high energy expenditures), and if the activity-induced increased requirements for select nutrients (e.g., food energy, carbohydrate, protein, fat, water, minerals, and vitamins) are not met, exercise performance and eventually health could be negatively affected (Castaneda et al. 1995; Newsholme and Parry-Billings 1994). Moreover, in terms of enhancing exercise performance, the timing when several of these nutrients

Acknowledgment: The ongoing support of the author's laboratory by The Joe Weider Foundation is gratefully acknowledged.

are ingested relative to training sessions or competitive events is likely critical.

This chapter examines two basic questions. First, does regular physical exercise alter nutritional needs, and second, can minor nutritional manipulations that do not adversely affect overall health enhance exercise performance? The answer to both questions appears to be yes.

In a single chapter, it is impossible to adequately address how regular exercise affects the needs for all nutrients. Neither is it possible to consider the validity of all suggested nutrient intake manipulations relative to their ability to enhance performance. In fact, in both cases, for many nutrients there is little or no scientific data. The nutrients included in this chapter (food energy, carbohydrate (CHO), fat, protein, water, minerals, and vitamins) were selected because, based on sound scientific study, they are the ones whose needs are most likely increased by regular physical exercise or that, with altered quantity or pattern of intake, have been shown to enhance exercise performance. It is likely that there are other nutrient needs that are altered by exercise as well as other nutritional manipulations that can also enhance performance; however, those specifics must be uncovered by future study.

Food Energy

Increases and decreases in body mass depend on the relative balance between energy intake (calorie content of food consumed) and energy expenditure (determined largely by exercise habits). While many people in our primarily sedentary society have difficulty maintaining an adequate expenditure to prevent gains in body fat mass due to excessive energy intake, a substantial number of athletes fail to consume adequate energy to offset the extremely great energy expenditures necessary for training and competition (figure 7.1). As a result, most athletes have much lower body fat content than their more sedentary peers (Lohman 1992), and some may even reduce their fat free (muscle, bone) mass. Assuming that the fat mass does not reach dangerously low values (perhaps 5–7% of body mass in males and 13–15% in females), no adverse effects will result. In fact, typically both health and exercise performance are enhanced by this reduced body fat content, especially in activities where one must carry her or his body mass. In contrast, losses in body fat free mass could adversely affect not only exercise performance but also growth rate and overall health. Based on animal experiments, the fat free mass of females appears to be better protected against insufficient energy intake (Cortright, Rogers, and Lemon 1993; Cortright et al. 1996; Cortright et al. 1997; Pitts 1984; Pitts and Ball 1977). This may explain why, relative to men, women appear to have greater difficulty losing body fat with an exercise program. Perhaps hormonal differences related to reproductive function are involved in this gender difference, but more study is needed to determine the actual mechanism.

Daily food energy requirements for various age groups are known, but these were determined on essentially sedentary subjects (Food and Agricultural Organization 1985; U.S. Food and Nutrition Board 1989; Nutrition Recommendations 1990) and,

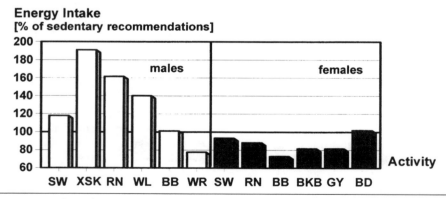

Figure 7.1 Daily energy intakes of young athletes (ages 14–30 yr; males, open bars; females, closed bars) involved in selected physical activities (Di Chen et al. 1989; Horswill, Park, and Roemmich 1990; O'Conner 1994; van Erp-Baart et al. 1989; Wilmore et al. 1992). While males tend to increase energy intake sufficiently with regular physical activity, females frequently do not. Athletes in body mass-restricted sports are the most susceptible to inadequate energy intake regardless of gender. SW, swimming; XSK, cross-country skiing; RN, running; WL, weightlifting; BB, bodybuilding; WR, wrestling; BKB, basketball; GY, gymnastics; BD, ballet dancing.

Reprinted, by permission, from P.W.R. Lemon, 1998, "Effects of exercise on dietary protein requirements," *International Journal of Sport Nutrition* 8(4): 426-447.

due to the energy expended during training and competition, are too low for most athletes. In general, physically active males tend to increase their energy intakes accordingly, but this may not be true for many female athletes (see figure 7.1). Therefore, inadequate intake of a variety of nutrients is possible, especially in the female, and the result could be adverse effects on exercise performance (Frentsos and Baer 1997) or even health (to be discussed).

Although the need for additional food energy is obvious for athletes whose sports have a sizable endurance component, it is also true that strength athletes have increased energy needs (Chen et al. 1989). Unfortunately, this latter fact is frequently overlooked. In contrast to endurance athletes, the increased energy need for strength/power athletes is unlikely caused by high energy expenditures during training (Tarnopolsky et al. 1991; Tesch 1987) because, although strength exercise is very intense, its duration is extremely short. Rather, the elevated energy need is probably caused by an increased resting metabolic rate due to the large muscle mass of these athletes (Pratley et al. 1994). Consequently, many athletes, especially those who have not yet attained adult stature, need to be careful that their diet contains sufficient total energy. A good practice that generally results in a greater total daily energy intake is to regularly consume two or three nutrient-dense snacks in addition to one's three daily meals. If inadequate energy intake remains a problem, intake of supplemental liquid meals is also recommended because these meals frequently enable greater total nutrient intake due to their reduced bulk (Brouns 1993; Brouns et al. 1989). Moreover, liquid food is usually better tolerated than solids not only before and after but, if necessary, during exercise. At greatest risk for inadequate energy intake are the athletes engaged in body mass-restricted activities (e.g., dancers, figure skaters, bodybuilders, gymnasts, wrestlers; see figure 7.1), especially adolescent girls and young women, due to the prevalence of disordered eating habits in this population (Beals and Manore 1994; Sundgot-Borgen 1993a, 1993b; Wilmore 1991). This problem is frequently associated with secondary amenorrhea and osteoporosis (female athlete triad) and is of critical importance, as it can be life threatening. Its causes are unknown but likely multifactorial. As a result, treatment requires significant cooperation from all involved (coach, nutritionist, parents, physician, psychologist, and, of course, the athlete). Hopefully as this serious condition receives more scientific attention a cure or at least optimal treatment strategies will become available.

Carbohydrate

Body carbohydrate (CHO) stores are of extreme importance for the athlete because CHO is the most economic muscle fuel (generates the greatest energy in kJ per liter of oxygen consumed). During intense muscle contraction oxygen delivery to active muscle is limited, and as a result the glycolytic rate accelerates and CHO becomes the major fuel used. Further complicating matters is the fact that total body CHO is so limited (see figure 7.2) that it can be essentially depleted in a single bout of exercise—an hour or two of continuous exercise at moderate to high intensity (Hermansen, Hultman, and Saltin 1967) or as little as 5 to 15 min of intense intermittent exercise (MacDougall, Ward, and Sutton 1977). As CHO availability decreases, other macronutrients (fat and protein) must play a greater role as muscle fuels, and one's intensity of exercise must be reduced. Taken together this means that athletes in a variety of activities, for example, many team games and ball sports, not just endurance athletes, can experience diminished performances due to insufficient stored CHO (Muckle 1973; Simard, Tremblay, and Jobin 1988). Moreover, although CHO depletion is not normally limiting in low intensity or brief competitions, CHO storage is frequently still very important for athletes in many of these activities because it is needed to complete the prolonged training sessions typically used. Chronically low CHO stores would force the athlete to train less

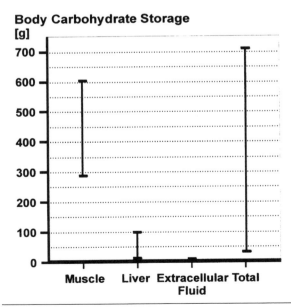

Figure 7.2 Approximate whole body carbohydrate stores for a 70 kg individual (Saltin and Gollnick 1988). The range is due to prior diet and exercise habits.

Table 7.1   Example of a Diet That Contains About 680 g Carbohydrate*

Breakfast:	75 g corn flakes, 500 ml skim milk, 250 ml orange juice, 100 g whole wheat bread (4 slices), 20 g jam, 260 g banana (1 large)
Lunch:	150 g whole wheat bread (6 slices), 110 g turkey breast slices, 110 g lowfat cheese, 300 g apple (2 medium), 500 ml apple juice
Supper:	300 g pasta, 180 g nonmeat sauce, 100 g rolls (2), 30 g margarine, 125 g green beans, 180 g orange (1 medium), 28 g granola bar (1), 500 ml skim milk
Snacks:	150 g apple (1 medium), 180 g orange (1 medium), 28 g granola bar (1)

*~7.5 g · kg body mass^{-1}· d^{-1} for a 90 kg individual.

intensely or over a shorter duration, thereby further adversely affecting competitive performance by minimizing the beneficial effects of her or his training sessions.

The amount of CHO stored in the body is the result of prior diet (filled) and exercise (emptied) routines and can be substantially increased (approximately doubled) with minor manipulations of both. Despite a gain in overall body mass (due to the additional CHO and the water, almost 3 g · g CHO^{-1}, that is stored with it), in events where CHO stores are limiting, athletic performance is significantly enhanced when large quantities of CHO are consumed for a few days prior to the event. However, even if this is understood, it is important to realize that many athletes feel full before intake attains these levels and, therefore, without special effort will not consume sufficient CHO (see table 7.1) or energy to maximize their performance.

The optimal procedure to load CHO appears to be to consume a significant quantity of moderate and high glycemic index foods (e.g., bananas, bread, cake, chocolate, corn, cornflakes, crackers, grapes, honey, oatmeal, oranges, pasta, pastry, potatoes, raisins, rice, sport drinks, sugar, whole wheat cereal), although those with a low glycemic index (e.g., apples, beans, cherries, dairy products, dates, figs, fructose, grapefruit, peaches, peas, plums) are also beneficial. Many believe that ingestion of so much CHO will result in increased body mass, especially fat mass, and, at least in the sedentary individual, excess CHO intake can certainly do so. However, in athletes this is seldom a concern, because it is much more difficult to consume excess CHO due to the high energy expenditures of training and competition. As a result, consumed CHO is used to replenish the CHO stores, which are regularly depleted by exercise as opposed to being converted to fat and stored as adipose tissue. In fact, as alluded to earlier, consuming inadequate energy

occurs much more frequently in athletes than consuming too much. Multiple small meals typically result in a greater overall energy intake than two or three large ones. This means that in addition to their long training sessions, athletes must spend a considerable amount of time each day eating. This requires substantial planning and preparation and can be a problem for younger athletes who frequently do not have good eating habits.

To maximize CHO storage, approximately 0.7 to 0.8 g CHO · kg body mass^{-1} (this would be about 4 slices of bread or 250 ml of a high carbohydrate sport drink for a 70 kg individual; see table 7.2) every two hours is necessary (Blom et al. 1987; Ivy et al. 1988). When this regular eating behavior (grazing) is impossible, the athlete should consume enough CHO in the previous and subsequent meals to make up the discrepancy. In order to consume adequate food energy and to distribute intake more uniformly over each 24-hour period some athletes even keep a snack (frequently a liquid meal) on the nightstand to be consumed during the night should they wake up. Finally, it should be understood that CHO loading has no benefit and can even adversely affect performance (due to the weight gain) in activities where CHO stores are not limiting. Such activities would include short duration (perhaps < 15–30 min) activities unless they are repeated a number of times (MacDougall, Ward, and Sutton 1977; Nicholas et al. 1996).

Before the Event

As mentioned above, in activities where CHO stores are thought to be limiting, CHO loading has been shown to enhance performance. This is best accomplished by consuming large amounts of CHO (approximately 6–10 g CHO · kg body mass^{-1}) for each of three days preceding the event. However, high dietary CHO will have little effect on body stores unless the energy expenditure during training is also cut back. Typically, reducing training loads to

Table 7.2   Moderate to High Glycemic Foods That Provide Approximately 50 g of Carbohydrate

Solid foods

Banana (1 large) 260 g	Oranges 420–600 g
Candy bar 65 g	Pastry 90 g
Cake 90 g	Porridge (oatmeal) 70 g
Corn flakes 60 g	Raisins 80 g
Crackers (plain) 65 g	Rice (white) 170 g
Grapes 315 g	Spaghetti/macaroni 200 g

Drinks

6% sucrose 825 ml	10% corn syrup carbonated 500 ml
7.5% maltodextrin/sugar 660 ml	20% maltodextrin 250 ml

50%, 25%, and 0% five, three, and one day before the competition, respectively, works well (Sherman et al. 1981). Further, loading is maximized if it is preceded by a CHO-depleting exercise bout, perhaps because the enzyme responsible for CHO storage (glycogen synthase) is most active when muscle CHO (glycogen) has been reduced to very low amounts (Bergström and Hultman 1966) or because muscle glucose uptake appears to be maximally stimulated following glycogen-depleting exercise (Ivy et al. 1988). Reductions in training (tapering) during the loading procedure, although often omitted, are essential to allow the surplus dietary CHO to be stored (in both muscle and liver). Moreover, tapering will also enhance performance by lowering the athlete's fatigue level (which should be substantial if her or his training program is sound). Tapering will have no adverse effect on fitness capacity because of the brief duration. Finally, with respect to immediate preevent nutrition, there is considerable evidence that a 200 to 250 g high glycemic index CHO snack that is low in fat and fiber (e.g., eight, 10 cm diameter pancakes, 125 g syrup, 500 ml orange juice, two jelly-filled doughnuts; see table 7.2) about 4 h before a competition will further enhance performance by topping off the CHO stores. However, smaller quantities (60 g) of CHO ingested 45 min before exercise have little effect on fuel use in subjects who have followed the CHO-loading procedures and, as a result, do not seem to affect prolonged endurance performance (Van Zant and Lemon 1997).

During the Event

Carbohydrate intake during the event is also beneficial, especially when the exercise is prolonged. Coyle and associates (1986) observed that trained cyclists could extend the point of exhaustion by 33% (from 3 to 4 h) when consuming 2 g · kg body mass^{-1} of a 50% CHO solution every 20 min versus an artificially sweetened and flavored placebo. Although sporting events typically do not go to the point of actual exhaustion, it is likely that CHO intake during the event would also enable subjects to generate a greater average work rate over a measured distance and kick longer or harder near the finish, resulting in an enhanced performance. Surprisingly, the ergogenic effect of this ingested CHO is not due to a sparing of muscle glycogen use (as this has been shown to be similar under both experimental treatments) (Coyle et al. 1986) but rather to a greater reliance on blood glucose for energy. In addition, it appears (Coyle 1992) that regular intake of small quantities of CHO (30–60 g · h^{-1}) in liquid form (about 500–1000 ml of a sport drink) throughout exercise is the best strategy because it not only provides CHO but also fluid, which is essential (see importance of fluid intake in later discussion). Glucose polymers (maltodextrins), which are chains of 7 to 13 glucose units produced by the hydrolysis of starch and found in many sport drinks, are excellent because they are more readily absorbed (due to less bulk) than many solid CHOs (Brouns 1993; Rehrer 1991). Unfortunately, with the exception of a few activities, for example, cycling, it is difficult to ingest food, or even fluids, while exercise is ongoing. This means that athletes may need to reduce their pace or even stop briefly to ingest CHO and fluid. This is much easier during activities where the exercise is intermittent because periods of low intensity or rest/recovery provide excellent opportunities to take in these nutrients.

Following the Event

There appears to be a short window of opportunity immediately following CHO-depleting exercise (perhaps up to 2 h following) when muscle glycogen storage capacity is enhanced substantially due to a temporary increase in the muscle's rate of glucose uptake (Ivy et al. 1988). Specifically, when CHO (about 2 g · kg body mass[-1]) is consumed after more than 2 h following a glycogen-depleting exercise session, the muscle glycogen resynthesis rate over the next 2 h is considerably slower (by about 45%) than when the CHO is consumed immediately postexercise (see figure 7.3). This information is likely extremely important when several competitions occur on the same or even on successive days (as is the case in many activities, especially youth sport tournaments). Unfortunately, significant CHO intake during this time is difficult because, although athletes are thirsty immediately following strenuous exercise, few are hungry. To maximize CHO resynthesis during this critical period, athletes should consume 50 to 150 g (see table 7.2) of a high CHO sport drink and foods as soon as possible following their contest (preferably before they leave the immediate site). In addition, some evidence suggests that a CHO-protein mixture following exercise may be most advantageous (Zawadski, Yaspelkis, and Ivy 1992). However, when energy intakes are carefully controlled, insulin responses to CHO-protein and CHO alone are similar (Burke et al. 1995). Following strength/power exercise, ingestion of a CHO-protein mixture might be optimal by reducing muscle protein

degradation or enhancing protein synthesis due to the increased availability of amino acids and the associated growth hormone responses (Chandler et al. 1994). Finally, there is some indirect evidence that children and adolescents rely less on CHO for exercise fuel and, as a result, may experience smaller performance benefits from high CHO intakes relative to adults (Bar-Or and Unnithan 1994). Some data even suggest that females benefit less from CHO loading than males, perhaps due to differing hormonal responses (Tarnopolsky et al. 1995). However, despite the fact that these latter two divergent results are interesting, the benefit of high CHO intake for males is well documented, and more study is necessary before it can be determined for certain that children or women respond differently.

Fat

Fat is the major energy reserve in the body (at least 226,000 kJ and typically much more) and in combination with CHO is the primary fuel supply for exercise metabolism. This fat store is so vast that it could fuel moderate intensity exercise continuously for several days. Clearly, fat stores do not limit exercise performance even for the leanest athlete (approximately 6% of body mass as fat).

The vast majority of fat is stored outside the muscle (in adipose tissue) and, therefore, must be transported to contracting muscle before it can be utilized. Lipolysis is influenced by nutritional state (fat provides the majority of energy after an overnight fast even while resting) and exercise patterns (critical factors include both intensity and duration of the exercise bout). Following lipolysis in the adipose cell, which is primarily affected by neurohumoral stimuli and blood flow, free fatty acids and glycerol are released into the blood. The free fatty acids are transported in the blood (bound to albumin) and taken up by muscle, while the glycerol is taken up primarily by the liver for gluconeogenesis. These fatty acids provide the majority of the energy needed for low intensity exercise, especially during the latter stages of prolonged exercise. At exercise intensities around 65% $\dot{V}O_2$max, absolute fat utilization is probably greatest (accounting for perhaps 40–60% of substrate requirements), but the increase over more moderate exercise intensity comes from muscle triacylglycerol (TAG) stores, not from greater use of plasma free fatty acids (see figure 7.4; Romijn et al. 1993). Unfortunately, to produce the same amount of energy, fat oxidation requires more oxygen than

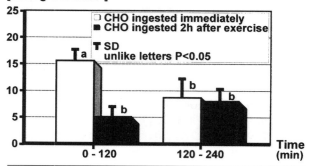

Glycogen Synthesis
[umol·g wet mass⁻¹]

Figure 7.3 Muscle glycogen resynthesis during the first few h after 70 min of glycogen-depleting exercise when CHO (2 g · kg body mass $^{-1}$ of a 25% CHO solution) was consumed immediately (open bars) or at 2 h following (closed bars) the end of the exercise (Ivy et al. 1988). Glycogen resynthesis is much faster when CHO is consumed immediately following exercise.

Whole Body FFA Uptake & Total Fat Oxidation
[umol·kg⁻¹·min⁻¹]

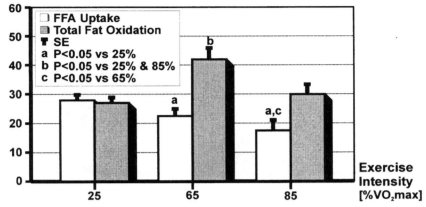

Figure 7.4 Whole body free fatty acid (FFA) uptake from plasma (open bars) and total fat oxidation (closed bars) during 60 min of exercise at 25, 65, and 85% $\dot{V}O_2$max (Romijn et al. 1993). During low intensity exercise (~ 25% $\dot{V}O_2$max) energy requirements are met primarily by FFA uptake. With moderate intensity exercise (~ 65% $\dot{V}O_2$max) total fat oxidation nearly doubles, but this is not due to an increased FFA uptake from plasma, that is, muscle triacylglycerol (TAG) use accounts for most of this increase. At higher intensities (~ 85% $\dot{V}O_2$max), both FFA uptake and TAG use are reduced (but TAG still accounts for ~ 40% of the fat oxidized) as most of energy comes from carbohydrate (CHO) stores (Romijn et al. 1993).

CHO oxidation, and possibly because of the body's limited capacity to supply oxygen (see chapter 6), exercise pace must be reduced when fat provides the majority of the fuel. As mentioned previously, this means that CHO is the preferred fuel during high intensity exercise, but fat may still contribute as much as 25 to 30% at intensities up to 85% $\dot{V}O_2$max (Sadur and Eckel 1982).

One of the most significant adaptations (see figure 7.5) that occurs with endurance exercise training is that a greater percentage of energy can be provided from fat at higher exercise intensities (Bjorntorp 1992; Martin et al. 1993). This effect is possible because of increased muscle TAG use (oxidation of plasma FFA actually decreases) and results in enhanced endurance performance because the athlete's more limited store of CHO is spared until it is needed, for example, for the final kick. This increased muscle TAG use occurs despite the well-known decrease in sympathoadrenal activity caused by endurance training (Hartley et al. 1972a, 1972b). The mechanism of this effect remains to be determined but may relate to an increased β-adrenergic receptor density or to a postreceptor effect induced by training (Martin 1996).

Although some fat intake is necessary in order to obtain several essential nutrients, most sedentary individuals consume more fat than they need. This should be avoided, not only because obesity (and its related complications) is more related to the amount

of dietary fat than to total food energy (Bray 1987; Miller et al. 1990) but also because, at least for sedentary individuals, diets high in fat are associated with cardiovascular disease (Castelli 1984). For most, fat intake should be no greater than 30%* of

Source of Exercise Fuel Utilized
[%]

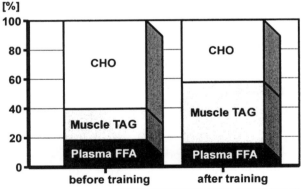

Figure 7.5 Percentage of energy derived from carbohydrate (CHO), plasma free fatty acids (FFA), and muscle triacylglycerol (TAG) during a 90–120 min exercise bout (at 64% of the pretraining $\dot{V}O_2$max) before and after a 12-week endurance training program.

Data from W.H. Martin, G.P. Dalsky, B.F. Hurley, et al., 1993, "Effect of endurance training on plasma free fatty acid turnover and oxidation during exercise," *American Journal of Physiology* 265:E708-E714. W.H. Martin, 1996, "Effects of acute and chronic exercise on fat metabolism," *Exercise and sport science reviews* 203-31.

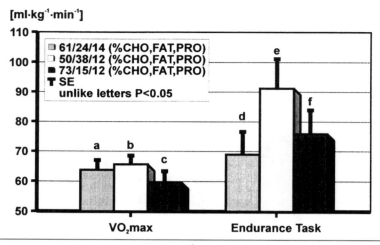

Figure 7.6 V̇O₂max and time to exhaustion on a task (85% V̇O₂max for 30 min, then 75–80% V̇O₂max until exhaustion) following 7 d on a diet containing carbohydrate (CHO), fat, and protein (PRO) percentages of 61/24/14 (moderate CHO, low-moderate fat, adequate PRO), 50/38/12 (moderate CHO, moderate fat, adequate PRO), or 73/15/12 (high CHO, low fat, adequate PRO). The moderate CHO, moderate fat, adequate PRO diet increased time to exhaustion (20–32%) and V̇O₂max (4–11%) relative to the other two diets (Muoio et al. 1994; Leddy et al. 1997).

energy intake. However, the interesting results (see figure 7.6, table 7.3) of two recent studies suggest that dietary fat as high as 38 to 42% of energy intake may enhance endurance performance in active individuals without adversely affecting the blood lipid profile (Leddy et al. 1997; Muoio et al. 1994). These data suggest that, although body fat stores are not limiting from a total fuel standpoint, higher fat intakes may make body fat more available as an exercise fuel, at least in endurance-trained individuals. If dietary CHO is sufficient to avoid significant reductions in CHO stores, this greater fat availability might enhance performance even more than very high CHO diets where fat and food energy intake may be inadequate; that is, assuming adequate total food energy intake, the optimal diet for endurance athletes may be moderate but sufficient CHO (perhaps 6.5 g · kg body mass⁻¹), moderate protein (1.5 g · kg body mass⁻¹), and moderate fat (2.2 g · kg body mass⁻¹). For an individual athlete consuming 220 kJ · kg body mass⁻¹ (~ 53 kcal · kg⁻¹) the energy percentage of such a diet would be 50% CHO, 12% protein, and 38% fat. Results with these kinds of diets need to be replicated but suggest that dietary fat has a different fate in active people, that is, the use of fat as an exercise fuel not only minimizes its accumulation in the body but it may also

prevent the associated adverse health consequences of high fat intakes. This observation is especially important for active young women who tend to undereat because of the current "fat phobia." The resulting low fat/energy intake adversely affects exercise performance and, as mentioned above (section on "Food Energy"), could lead to the life-threatening female athlete triad.

Protein

Protein makes up approximately 15 to 20% of body mass. However, it is not a major exercise fuel because, unlike CHO and fat, protein is not stored in the body as an energy reserve. Rather, each protein molecule has a physiological function, that is, as an enzyme or structural/contractile protein, and so forth. Proteins are comprised of unique mixtures of amino acids. Some of these amino acids (called dispensable or nonessential) can be synthesized in the body from ammonia and another carbon source (fat or CHO). Others (called indispensable or essential) cannot be synthesized by the body and must be obtained in food on a regular basis or the body's ability to synthesize protein is impaired. Together with the ongoing process of protein degradation, this can lead to decreases in muscle mass/strength

---

*A word of caution is necessary here relative to expressing intake of any nutrient as a percentage of energy intake. Such a practice conveys little useful information, as energy intake is frequently unknown. Moreover, a certain intake, that is, 70% CHO, may be inadequate for an athlete on an energy-restricted diet while 50% CHO could be sufficient for an individual with a high energy intake. To avoid this limitation and because nutrient needs are determined largely by body size, it is suggested that recommendations for macronutrient intake be expressed relative to body mass or, if available, lean body mass.

Table 7.3   Effect of Moderate Versus Low-Fat Diets on Cardiovascular Risk in Endurance Runners

Dietary fat content	% of energy intake		
	16	30	42
Adiposity	6	6	6
Body mass	6	6	6
Blood pressure	6	6	6
Serum			
Triacylglycerol	6	6	6
Total cholesterol	6	6	6
LDL cholesterol	6	6	6
HDL cholesterol	6	8	88
Total cholesterol/HDL cholesterol	8	6	6
Apolipoprotein B	6	6	6
Apolipoprotein A1	6	8	8
Apo A1:apo B	6	6	6

6 = no change; 8 = increase; 88 = large increase

Adapted, by permission, from J. Leddy, P. Horvath, J. Rowland, and D. Pendergast, 1997, "Effect of a high or low fat diet on cardiovascular risk factors in male and female runners," *Medicine & Science in Sports & Exercise* 29:17-25.

and, of course, have adverse effects on exercise performance.

Some protein foods traditionally called complete protein foods (e.g., dairy products, eggs, fish, meat) contain all of the indispensable amino acids in adequate amounts and, therefore, are considered to be high quality proteins, that is, they have high protein efficiency ratios (PER). Recently, however, limitations of the PER have become understood (due to differences between humans and rodents) (Food and Agricultural Organization 1989; Food and Drug Administration 1993), and based on the protein digestibility-corrected amino acid score (PDCAAS), it appears that the quality of some proteins differs from traditional understanding, notably isolated soy protein, which compares favorably with the highest quality animal proteins (see table 7.4). Consumption of adequate amounts of these complete protein foods is necessary to maximize protein synthesis (normal growth/repair of body tissue). However, vegetarians, especially growing ones, must be careful to consume appropriate combinations of incomplete protein foods to make complete protein meals (bread plus lentils, corn/rice plus beans, corn plus peas, or any complete protein food with incomplete ones). Problems are most likely if eggs and dairy products are also excluded from the diet, but with some meal planning adequate protein intake is still possible.

The chances of insufficient protein intake or poor combinations of incomplete protein foods are more likely with athletes, because it is clear that exercise has profound effects on protein metabolism. Briefly, during and for some time following exercise

Table 7.4   Protein Digestibility-Corrected Amino Acid Score (PDCAAS) of Selected Food Proteins

Food	PDCAAS
Casein (milk protein)	1.0
Egg white	1.0
Isolated soy protein	1.0
Beef	0.92
Pea flour	0.69
Kidney beans	0.68
Rolled oats	0.57
Lentils	0.52
Peanut meal	0.52
Whole wheat	0.40
Wheat gluten	0.25

Adapted, by permission, from E.C. Henley and J.M. Kuster, 1994, "Protein quality evaluation by protein digestibility-corrected amino acid scoring." *Food Technology* 48:74-77.

Protein Metabolism
[arbitrary units]

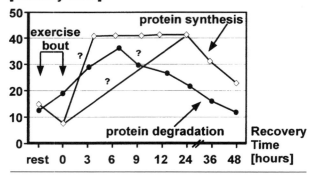

Figure 7.7 Pattern of changes in protein metabolism during and following exercise. The time course and magnitude of the response (synthesis, diamonds; degradation, circles) differs with various types of exercise but the overall pattern is similar.

Reprinted, by permission, from P.W.R. Lemon, 1997, "Dietary protein requirements in athletes," *Journal of Nutritional Biochemistry* 28:52–60.

(depending on the exercise type) protein synthesis is impaired, and protein degradation is enhanced (Biolo et al. 1995; Booth and Watson 1985; Chesley et al. 1992). Then at some point, if sufficient recovery time is allowed, this pattern reverses and protein synthesis exceeds breakdown (see figure 7.7). The result is an increase in the body's protein mass. For endurance athletes the increase is in mitochondrial protein and produces dramatic improvements in prolonged exercise capacity (Faulkner, Green, and White 1994). In contrast, for strength athletes the increase is in contractile protein and manifests itself in enhanced muscle size and strength (Kraemer, Fleck, and Evans 1996).

In view of the changes in protein metabolism induced by exercise it seems reasonable to suggest that the protein requirements of athletes would be greater than sedentary individuals, but the issue is complicated because several factors, including experimental subjects' diet composition, time for adaptation to dietary changes, energy intake, training status, gender, and age as well as exercise intensity/duration and the environment in which it occurs, have not been adequately controlled in many investigations (Lemon 1992). As a result, it is extremely difficult to interpret all of the available data, and there is, as yet, no consensus on how much dietary protein is needed for active individuals (Butterfield 1991; Cathcart 1925; Lemon 1996; Rennie, Smith, and Watt 1994; von Liebig 1842). In fact, the current U.S. recommendations for daily protein intake (0.8 g · kg body mass^{-1}), do not contain an additional allowance for those in-

volved in regular physical activity. It should be understood that these recommendations are based on data derived from essentially sedentary individuals. Furthermore, for reasons that are not clear, the reference list for the current U.S. recommendations does not include a single study on this question published after 1977 (U.S. Food and Nutrition Board 1989). Recently, several well-controlled experiments utilizing the nitrogen balance technique and the metabolic tracer technique have addressed this question, and these data indicate that dietary protein requirements of exercising individuals likely exceed those of the general population (Fern, Bielinski, and Schutz 1991; Lemon et al. 1992; Meredith et al. 1989; Phillips et al. 1993; Tarnoplosky et al. 1992) (see figures 7.8 and 7.9). Based on these data, it is recommended that adult endurance athletes consume about 1.2 to 1.4 g · kg body mass^{-1} · d^{-1} and strength athletes slightly more, perhaps as much as 1.7 to 1.8 g · kg body mass^{-1} · d^{-1} (Lemon 1996). Although this represents 50 to 125% above the current RDA, it is important to realize that protein supplementation will not be necessary because many adult athletes, as a result of their increased energy intake, may already consume this quantity of protein. For example, assuming one's diet contains at least 10% protein (a reasonable assumption; intake is typically around 15% and sometimes much higher, especially with strength athletes), an energy intake of 21,000 kJ (about 5,000 kcal) would contain 125 g of protein, which would be ~ 1.8 g · kg body mass^{-1} · d^{-1} for a 70 kg individual (see table 7.5). As a result, at least in the developed countries, it appears unlikely that there are large numbers of adult athletes who receive insufficient dietary protein. However, the chances are greater that female athletes may be protein deficient due to the tendency of many to undereat relative to their energy expenditure.

Most of the protein studies discussed have used male subjects, and available data suggest that the protein needs of active women, at least those who engage in endurance exercise, may be less than for the male (Phillips et al. 1993; Tapscott, Kasperek, and Dohm 1982; Tarnopolsky et al. 1990; Tarnopolsky et al. 1995). These studies need to be replicated, and investigations with female strength athletes should be carried out to assess if supplemental protein can enhance heavy resistance training for women as it apparently does for men (see figure 7.10; Tarnopolsky et al. 1992). Unfortunately, there is also insufficient data on growing individuals (Lemon 1989), and due to the interaction of the factors mentioned above (poor eating habits and

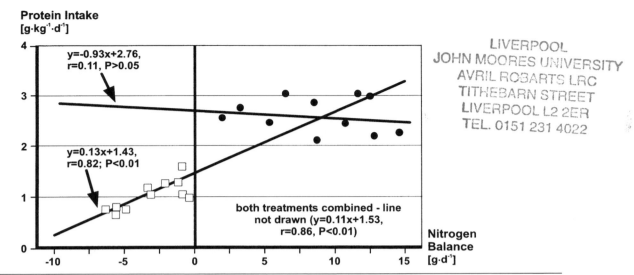

Figure 7.8 Nitrogen balance of a group of novice bodybuilders consuming 2.62 (circles) vs. 0.99 (squares) g of protein · kg body mass^{-1} · d^{-1} (Lemon et al. 1992). Typically, protein requirements are determined as the protein intake where nitrogen intake is equal to nitrogen excretion (y intercept), here 1.43–1.53 g · kg^{-1} · d^{-1}. To account for variability within the population, a safety buffer equal to two standard deviations is added to obtain the recommended protein intake, here 1.63–1.73 g · kg^{-1} · d^{-1}, which is 200–216% of the current recommended dietary allowance.

Adapted, by permission, from P.W.R. Lemon, M.A. Tarnopolsky, J.D. MacDougall, and S.A. Atkinson, 1992, "Protein requirements and muscle mass/strength changes during intensive training in novice bodybuilders," *Journal of Applied Physiology* 73:767-775.

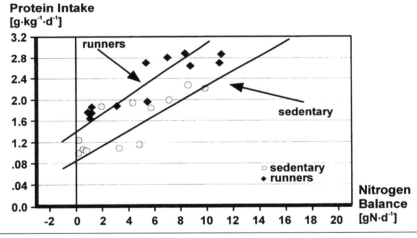

Figure 7.9 Nitrogen balance of endurance runners (diamonds) and sedentary controls (circles) at a range of protein intakes (Tarnopolsky et al. 1988). The observation that the regression line for the runners is upward and to the left relative to the sedentary line indicates that the runners have a greater protein requirement, that is, a greater protein intake is necessary for nitrogen balance (1.37 vs. 0.73 g of protein · kg body mass^{-1} · d^{-1}, respectively).

the pubertal growth spurt), the chances of insufficient protein intake in the young athlete are considerable (Roemmich and Sinning 1996, 1997a, 1997b). Further, some data indicate that protein needs are greater in the elderly (Campbell et al. 1994; Castaneda et al. 1995; Evans 1996), and, as a result, studies need to be completed on senior athletes to determine how regular exercise affects protein needs in this rapidly expanding population. Finally, there is considerable controversy in the area of dietary protein requirements in exercising individuals due in part to changing experimental methodology (Garlick et al. 1994; Rennie, Smith, and Watt 1994; Tessari et al. 1996; Young, Bier, and Pellet 1989), so the debate relative to specific recommendations for dietary protein in the physi-

Table 7.5 Quantity of Protein-Containing Foods Necessary*

Food type	Quantity of food	Quantity of protein
Cheese	50 g	12 g
Egg whites	66 g (2 eggs)	8 g
Fish	85 g	19 g
Meat	175 g	38 g
Milk (skim)	1 L	35 g
Peanuts	36 g	9 g
Beans (kidney)	64 g	4 g

*Actual intake would need to be slightly greater because digestibility varies from about 75% for vegetables to about 95% for milk and egg protein.

Note: Values are based on assumed need to attain an intake of approximately 125 g of protein (1.8 g · kg body mass^{-1} · d^{-1} for a 70 kg individual).

cally active population will likely continue for some time.

Relative to the exercise performance effects of protein supplementation very few data are available (Lemon 1995). Many athletes inappropriately believe supplemental protein or individual amino acids can enhance exercise performance in a variety of activities. One exception may be supplementation with creatine monohydrate (a nitrogen-containing compound found primarily in meat and fish). Recently, this topic has received a considerable amount of attention (Ööpik, Timpmann, and

Medijainen 1995; Volek and Kraemer 1996), and several (Balsom, Söderlund, and Ekblom 1994; Balsom et al. 1995; Birch, Noble, and Greenhaff 1994; Casey et al. 1996; Greenhaff et al. 1993a; Greenhaff et al. 1993b; Greenhaff et al. 1994; Greenhaff 1995; Harris, Söderlund, and Hultman 1992; Kreider et al. 1998; Lemon et al. 1995; Maganaris and Maughan 1998; McNaughton, Dalton, and Tarr 1998; Rossiter, Cannell, and Jakeman 1996; Schneider et al. 1997; Smith et al. 1998; Vandenberghe et al. 1997; Ziegenfuss et al. 1997) but not all (Burke, Pyne, and Telford 1996; Cooke, Grandjean, and Barnes 1995; Mujika et al. 1996; Redondo et al. 1996) studies suggest that a dosage of about 20 g/d for as few as 3 to 5 days can result in gains in body mass (~ 1 kg) and enhanced performance in brief intense exercise tasks (see figure 7.11). Moreover, no adverse side effects have yet been reported. The mechanism appears to be related to enhanced muscle phosphocreatine stores (Harris, Söderlund, and Hultman 1992) and faster rates of phosphocreatine recovery between successive bouts (Greenhaff et al. 1994; Lemon et al. 1995). Apparently a sizable portion of the increased body mass is due to an increase in muscle volume (see figure 7.12a), which appears to be primarily the result of increased intracellular fluid (see figure 7.12b). More study is needed in this exciting area, not only relative to the performance question but also as to whether the increased muscle water can stimulate protein synthesis (Häussinger et al. 1993) and consequently potentiate the anabolic stimulus of strength exercise.

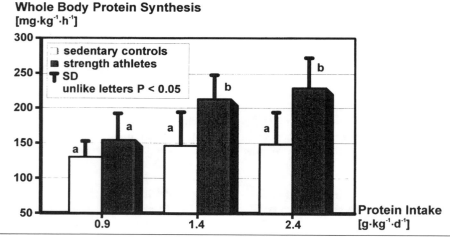

Figure 7.10 In strength athletes (closed bars), the protein synthesis rate was increased when they consumed 1.4 vs. 0.9 g of protein · kg body mass^{-1} · d^{-1}. No such increase was seen in control subjects (open bars). However, increasing the protein intake to 2.4 g · kg^{-1} · d^{-1} had no further anabolic effect (Tarnopolsky et al. 1992).

Adapted, by permission, from M.A. Tarnopolsky, S.A. Atkinson, J.D. MacDougall, et al., 1992, "Evaluation of protein requirements for trained strength athletes," *Journal of Applied Physiology* 73:1986-1995.

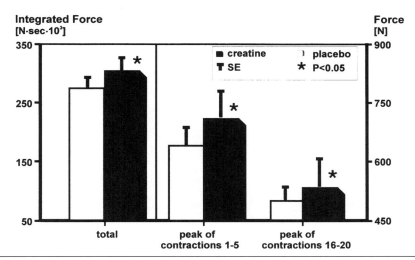

Figure 7.11 In comparison to placebo (open bars), five days of creatine monohydrate ingestion (20 g · d⁻¹, closed bars) increased both total integrated force and peak force of each contraction (first 5 and the last 5 contractions are shown) during 20 maximal, 30 s bouts of plantarflexion exercise, each separated by 16 s recovery periods (Lemon et al. 1995).

Reprinted, by permission, from P.W.R. Lemon, B.R. Newcomer, D.L. Bredle, T.N. Ziegenfuss, M.E. Rogers, and M.D. Boska, 1995, "Creatine intake enhances muscle energetics and force output during repeated maximal isometric plantar flexion in humans," *Medicine & Science in Sports & Exercise* 27(5suppl): S204.

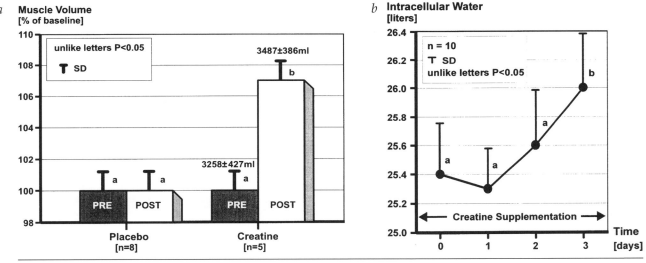

Figure 7.12 Three days of creatine monohydrate ingestion (20 g · d⁻¹) increased thigh muscle volume in elite sprint/power athletes (Ziegenfuss et al. 1997) measured *(a)* by magnetic resonance imaging and also increased whole body intracellular water (Ziegenfuss, Lowery, and Lemon 1998) measured *(b)* by multifrequency bioelectrical impedance.

Water

Water is the largest single component of the body (approximately 50–70% of body mass) and is clearly the single most important nutrient, because it is possible to survive for several weeks without food but only for a few days without water. During strenuous exercise (depending on several factors including training status, exercise duration, and environmental conditions) the vast amount of heat generated by muscle contraction (due to a low me-

chanical efficiency) can produce sweat losses of up to ~ 2 to 3% of body mass (about 2–3 l · h⁻¹). Unfortunately, voluntary fluid intake is rarely sufficient to offset this loss, and the resulting dehydration has been shown to cause as much as a 6 to 7% increase in 5 and 10 km race times (Armstrong, Costill, and Fink 1985). This is critical in terms of performance because the difference between first and second place in competitions is frequently < 2%. More importantly, this dehydration can cause serious medical complications (heat illness). This is especially true with

Table 7.6 Volume · h⁻¹ of Various Carbohydrate (CHO) Drink Concentrations (2–30%) Necessary to Provide Recommended CHO Intake (30–60 g · h⁻¹)

Drink CHO content (g · 100 ml⁻¹)	CHO delivered (g · h⁻¹)			
	30	40	50	60
	Volume of drink necessary (ml · h⁻¹)			
2%	1500	2000	2500	3000
4%	750	1000	1250	1500
6%	500	667	883	1000
8%	375	500	625	750
10%	300	400	300	600
15%	200	267	333	400
20%	150	200	250	230
25%	120	160	200	240
30%	60	80	100	120

Intakes from 625–1250 ml · h⁻¹ are thought to be acceptable by most athletes (underlined). Intakes > 1250 are too large and intakes ≤ 600 ml · h⁻¹ are too small.

Adapted, by permission, from E.F. Coyle, and S.J. Montain, 1992, "Benefits of fluid replacement with carbohydrate during exercise," *Medicine & Science in Sports & Exercise* 24(9suppl): S324-330.

children who, relative to adults, have higher core temperatures at any level of dehydration (Bar-Or 1989; Bar-Or et al. 1980). Therefore, it is essential to do everything possible to minimize the degree of dehydration experienced (American College of Sports Medicine 1996). The most important precautions include consuming fluids before, during, and following exercise, wearing appropriate clothing (porous, light colored), acclimatizing slowly over one to two weeks, and scheduling training sessions so that the hottest times of the day are avoided. Contrary to popular belief, large fluid intakes (up to at least 1,200 ml · h⁻¹) are well tolerated by athletes (Coyle and Montain 1992). All who work with athletes (coaches, trainers, parents) should be absolutely certain that their athletes consume adequate quantities of fluids (see table 7.6).

In recent years, numerous sport drinks have appeared on the market, and athletes and coaches often wonder whether intake of any of these enhances performance relative to plain water. Clearly, the research literature indicates that these drinks can enhance performance by providing additional fluid, CHO, and electrolytes. As one example, Below and coworkers (1994) observed that performance in a final kick following a 50 min cycle ergometer effort (at 80% $\dot{V}O_2$max) was increased by 6% when 1330 ml of water was ingested during the exercise bout (~ 80% of sweat rate) vs. an increase of 12% with 1330 ml of a sport drink was ingested. In general, when substrate is the major limiting factor

(very prolonged efforts when sweat rate is low to moderate, e.g., marathon swimming, prolonged hiking, or marathon running in cool environments) relatively high CHO solutions (150–200 g glucose · l⁻¹) should be consumed because they provide substantial CHO. However, the down side is that fluid availability is reduced due to delayed gastric emptying (Shi et al. 1995). In contrast, lower CHO solutions (< 60 g glucose · l⁻¹) deliver more fluid but less CHO and, therefore, are best when fluid loss, not CHO availability, limits performance (any activity where total sweat loss is high, e.g., > 2–3% of body mass).

The sodium chloride (10–25 mmol · l⁻¹) present in sport drinks is also beneficial because it increases voluntary fluid intake (Maughan 1991; Maughan, Leiper, and Shirreffs 1996b; Wemple, Morocco, and Mack 1997; Wilk and Bar-Or 1996), it helps retain body fluids by reducing urine output (Maughan, Leiper, and Shirreffs 1996a), and it helps prevent hyponatremia, a rare but very serious situation that can occur during prolonged physical exertion (> 4 h), especially if plain water is ingested (Noakes 1992).

Frequent ingestion of small fluid volumes (100–200 ml · 10 min⁻¹) is the best strategy because greater stomach volumes increase fluid delivery to the intestines where it is absorbed. Although drink temperature does not appear to affect fluid absorption (McArther and Feldman 1989), fluid palatability is enhanced when drinks are cold (~ 15° C). Moreover, flavor may be an important consideration, at least for children, as one study indicates that grape drinks are preferred vs. apple, orange, and plain water in 9- to 13-year-old boys and girls (Meyer et al. 1994).

Perhaps the easiest way to assess fluid status is to monitor regularly the body mass of your athletes during both training and competition and be sure the volume of replacement fluid following exercise is greater than these losses. This is necessary in order to cover obligatory urinary losses, which continue even in the dehydrated state. Specifically, within 2 h postexercise rehydration should be about 150 to 200% of sweat losses. This timing is especially important when multiple exercise bouts occur on the same day (Maughan, Leiper, and Shirreffs 1996a). Further, rehydration may be enhanced with a drink with even higher (61 mmol · l^{-1}) sodium chloride contents (at least twice the concentration of most sport drinks) (Shirreffs et al. 1996), although another study indicates that 25 mmol · l^{-1} is optimal (Wemple, Morocco, and Mack 1997). Soft drinks contain adequate CHO (~ 0.6 mol · l^{-1} or 100 g · l^{-1}) but are actually poor rehydration drinks because they are very low in sodium (~ 2–3 mmol · l^{-1}) and some athletes find the carbonation unpleasant.

Postexercise ingestion of solid food is also advantageous because it helps replace fuel stores and may aid rehydration due to the electrolyte content of the food. However, it can reduce associated fluid intake (thereby delaying rehydration) and can cause gastrointestinal problems (Maughan, Leiper, and Shirreffs 1996a). Alcohol-containing beverages are poor replacement drinks because they increase urine output and, therefore, adversely affect rehydration. It is critical to remember that net fluid retention is as important as fuel replacement. Intake of about 500 ml of fluid 30 to 60 min prior to competition is also advisable; however, water cannot be stored for any period of time and ingestion of too much too early will simply be urinated away prior to the event or, worse, force the athlete to take a bathroom break during the event. On the other hand, drinking a large volume of fluid right before starting can be uncomfortable due to stomach or gastrointestinal distension.

Glycerol (1 g · kg^{-1}) plus water (21 ml · kg^{-1}) ingestion prior to endurance exercise followed by additional glycerol (0.1 g · kg^{-1}) plus water (50 ml) every h may be advantageous as its osmotic action can help retain both interstitial and vascular water, thereby improving thermoregulation and cardiovascular function (Lyons et al. 1990; Riedesel et al.

1987). Obviously, fluid intake must be strongly encouraged for all athletes (see table 7.6), but as with all nutritional manipulations, it is important to experiment during training sessions to be certain how each athlete will respond.

Minerals

Minerals are typically classified based on their abundance in the human body: macrominerals (calcium, chloride, phosphorus, potassium, magnesium, sodium, and sulfur) represent more than 0.01% of body mass while the microminerals (arsenic, cobalt, copper, chromium, fluorine, iodine, iron, manganese, molybdenum, nickel, selenium, silicon, tin, vanadium, and zinc) each make up less than 0.001% of body mass. Another and perhaps more practical method to differentiate these two classes of minerals is based on whether the recommended dietary allowance (RDA), or the estimated safe and adequate daily dietary intake (ESADDI) when there is insufficient data to establish a RDA, is greater than 100 mg (macrominerals) or less than 100 mg (microminerals). Although not precise requirements, these guidelines (see table 7.7) are designed to minimize deficiencies. Generally, intakes of ≤ 67 to 75% of these guidelines will lead to problems. In addition, as with most dietary recommendations it is important to remember that both the RDA and ESADDI were determined on individuals with a lifestyle that was basically sedentary and, therefore, could be inadequate for athletes. Mineral status is routinely determined based on assessment of clinical signs/symptoms, and body fluid (urine, blood) or tissue concentrations. Tissue analysis is a better index, but it is also more invasive, and for this reason analysis of body fluids is common. This is especially unfortunate when studying physically active individuals because the often tenuous relationship between blood and tissue mineral concentrations can be further altered by acute or chronic exercise, that is, changes in urine or sweat losses or changes in plasma volume and redistribution between body fluids and tissues.

There are data suggesting that exercise increases the need for some minerals (Brouns 1993; Burke and Deakin 1994; Clarkson 1992ab; Wolinsky and Hickson 1994); however, relative to other areas of exercise nutrition very little is known about mineral requirements. As a result, as discussed with protein, marginal intakes or even those at 100% of current recommendations could lead to problems for the athlete. Clearly our knowledge of the mineral requirements of athletes is rudimentary.

Table 7.7   Current Recommendations (RDA or ESADDI) for Selected Minerals

Age (y)	Calcium (mg)	Iron (mg)	Magnesium (mg)	Zinc (mg)	Copper (mg)	Chromium (μg)	Selenium (μg)
< 0.5	400	6	405	5	0.4–0.6	10–40	10
0.6–1.0	600	10	60	5	0.6–0.7	20–60	15
1–3	800	10	80	10	0.7–1.0	20–80	20
4–6	800	10	120	10	1.0–1.5	30–120	20
7–10	800	10	170	10	1.0–2.0	50–200	30
11–14							
Male	1,200	12	270	12	1.5–2.5	50–200	40
Female	1,200	15	280	15	1.5–2.5	50–200	45
15–18							
Male	1,200	12	400	12	1.5–2.5	50–200	50
Female	1,200	15	300	15	1.5–3.0	50–200	50
19–24							
Male	1,200	10	350	10	1.5–3.0	50–200	70
Female	1,200	15	280	15	1.5–3.0	50–200	55
25–50							
Male	800	10	350	10	1.5–3.0	50–200	70
Female	800	15	280	15	1.5–3.0	50–200	55
> 51							
Male	800	10	350	10	1.5–3.0	50–200	70
Female	800	10	280	10	1.5–3.0	50–200	55

RDA = recommended dietary allowance; ESADDI = estimated safe and adequate daily dietary intake.
Adapted from U.S. Food and Nutrition Board, 1989, *Recommended dietary allowances* (Washington: National Academy Press).

Macrominerals

Calcium is essential for bone growth (99% of the body's calcium is in the skeleton), for blood coagulation, and muscle contraction. High content calcium foods include dairy products, seafood, nuts, and green vegetables. Serum calcium changes with acute exercise are variable (increased, decreased, or no change; Brouns 1993), but chronic exercise (especially if it is weight-bearing) can enhance bone density (Sanborn 1992). However, menstrual irregularities and dangerously low bone densities have been observed in some young female athletes during strenuous athletic training (Drinkwater et al. 1984). This is likely due to overtraining and is definitely a sign that a woman should reduce her training load. It is believed that exercise-induced decreased serum estradiol concentration is the mechanism responsible, because estradiol reduces urinary calcium loss, improves dietary calcium absorption, and reduces bone resorption by increasing calcitonin secretion (Lindsay 1987), but reduced progesterone production may also contribute (Prior et al. 1990). This can be especially devastating for females because, relative to males, females have

reduced bone mineral reserves, typically consume less dietary calcium because it is directly related to energy intake (van Erp-Baart et al. 1989), and experience a much greater loss of bone mass throughout life (especially following menopause) (Highet 1989). This effect is likely exacerbated in athletes due to the additional loss of calcium in sweat (Brouns, Saris, and Schneider 1992; Costill and Miller 1980) and urine, especially if purified protein supplements are consumed without increased phosphate intake, due to the associated increased urinary calcium loss (Allen, Oddoye, and Margen 1979; Flynn 1985). As a result, maintenance of adequate calcium intake is very important. Fortunately, due to a high energy intake, many athletes appear to consume calcium near the RDA; however, individuals in body mass-restricted sports and female athletes frequently do not (Clarkson 1992b). As mentioned above, dairy products are a major source of calcium, for example, an adult woman would approach the RDA for calcium if she consumed one liter of milk · d⁻¹. As a result, one would think that inadequate calcium intake would not pose a problem, but, unfortunately, many females significantly reduce their milk intake as they reach adolescence (presumably due a desire to reduce fat intake, although skim milk has essentially no fat) and, therefore, frequently have difficulty consuming sufficient calcium without supplementation. This could further contribute to the devastating bone health problems discussed above and is likely misguided, not only because, as previously mentioned, the adverse effects of moderate fat intake (25–35% of energy intake) are likely overstated in physically active individuals but also because a variety of low fat dairy products are readily available. It seems prudent that individuals who are working with female athletes make every attempt to be certain that their calcium intake is equal to or above the RDA. It is interesting to speculate that the active lifestyle of the adolescent female athlete coupled with adequate (or surplus) calcium intake might produce sufficient bone density, that is, a greater reserve, to substantially reduce the effects of osteoporosis later in life (Johnston et al. 1992; Wolman et al. 1992). Due to the catastrophic effects of osteoporosis, this area of study also deserves much more attention.

Magnesium is a cofactor in many enzymatic reactions of energy metabolism, it is involved in oxygen extraction by cells, and intracellular losses lead to muscle weakness, neuromuscular dysfunction, and tetany (Brautbar and Carpenter 1984). High magnesium content foods include vegetables, fruits, nuts, grains, and mushrooms. Acute exercise may reduce serum magnesium concentration (Rose et al. 1970; Stendig-Lindberg 1987), and exercise performance is enhanced when exercise-trained subjects with low serum magnesium concentration receive supplementation (von Stock et al. 1979). These data suggest that exercise can affect magnesium status and that reduced serum magnesium concentration may lead to subnormal performances. Low serum magnesium concentration could be caused by magnesium lost in either or both of exercise sweat (Brouns 1993; Costill and Miller 1980) and urine (Deuster et al. 1986). However, the actual significance of these observations must be questioned because many studies indicate that the serum magnesium concentration of athletes is not dramatically different than controls (Clarkson 1992b). Athletes may be protected against increased exercise magnesium losses as a result of their high energy intakes because, as with calcium, magnesium intake is directly related to energy intake (van Erp-Baart et al. 1989). However, individuals involved in body mass-restricted activities should be certain their magnesium intake approaches the upper end of the RDA (350–400 mg · d^{-1} depending on age and gender; see table 7.7). Future studies should focus on this population.

Microminerals

Iron is an important part of hemoglobin, myoglobin, cyctochromes, and a number of enzymes in energy metabolism. Poor iron status is typically indicated by low serum ferritin, increased free erythrocyte protoporphyrin, reduced hemoglobin, or increased transferrin and is most accurately assessed by the iron content of bone marrow. High content iron foods include red meat, organ meats, dark green vegetables, nuts, clams, oysters, and molasses. Although there is some debate, the data collected over the past 10 years indicate that a significant number of athletes have reduced iron stores (Brouns 1993; Clarkson 1992b). Moreover, the resulting performance of many of these athletes is impaired and can be reversed with iron supplementation (Haymes 1987). A number of factors appear to contribute to low iron stores, including diets low in heme-iron (i.e., vegetarian and high fiber diets), increased iron losses (in exercise sweat, in feces due to gastrointestinal bleeding associated with endurance exercise, and in red blood cell damage due to foot strike), and possibly, the effects of exercise on iron absorption. Iron intake is also directly related to energy intake (van Erp-Baart et al. 1989), and, therefore, the highest risk group is once again individuals in body mass-restricted sports, especially females who also have to compensate for monthly iron loss through menstruation. For many of these athletes it is necessary to modify dietary habits to increase iron intake (to at least 15 mg · d^{-1}) and/or to take an iron supplement to avert adverse performance and eventual health impairment.

Zinc is involved in the development of many tissues, including muscle. It is an essential component of numerous enzymes involved in energy metabolism, and it is important in hormonal response, sexual maturation, wound-healing, taste, vision, and immune function. High zinc foods include meat, liver, seafood, whole grains, and wheat germ. Signs of marginal zinc deficiency are slow wound-healing, anorexia, oligospermia, loss of taste and smell, reduced growth, and compromised immune function. Typically, zinc status is assessed based on blood measures; however, redistribution among tissues makes interpretation of this information difficult. A better measure may be the ratio of serum zinc and the sum of serum concentrations of zinc carriers (Couzy, Lafargue, and Guezennec 1990). Zinc intake is also directly related to energy intake (van Erp-Baart et al. 1989), and, therefore, it is not surprising that most dietary surveys of male athletes indicate adequate intake but female athletes (especially those in body mass-restricted sports) fall below the RDA (Clarkson and Haymes 1994). In combination with exercise-induced zinc losses in both urine and sweat (Anderson 1991), these low zinc intakes suggest that many female athletes could benefit from increased zinc intake (at least 15 mg · d^{-1}); however, this must be done with caution because excessive zinc can pose a health risk (Campbell and Anderson 1987; Hackman and Keen 1986). Finally, excess iron intake can affect zinc needs by inhibiting zinc absorption (McDonald and Keen 1988).

Copper is a cofactor in many enzymatic reactions and is involved in erythropoiesis, energy metabolism, iron metabolism, neurotransmitter and brain function, connective tissue formation, and catecholamine regulation. High copper foods include organ meats, seafood, nuts, potatoes, whole grains, and legumes. Copper status is assessed by serum copper, ceruloplasmin (the principal copper-binding protein in plasma), or by the activity of erythrocyte superoxide dismutase (an enzyme that minimizes free radical damage). In North America copper intake is typically marginal and numerous dietary components (zinc, ascorbic acid, iron, calcium, protein, fructose, dietary fiber) are known to reduce copper absorption (Brouns 1993). Although urinary copper loss is minimal (1–2%) and largely unaffected by exercise, exercise-induced sweat and fecal losses can be substantial (Anderson 1991). Exercise effects on blood copper concentration are equivocal (acute exercise has been reported to increase, decrease, or have no effect, and trained athletes usually have normal blood copper concentration), but, as with zinc, these results may be uninterpretable due to redistribution of tissue copper (Clarkson and Haymes 1994). In contrast, several studies have observed decreased ceruloplasmin concentration with endurance training (Campbell and Anderson 1987), and one study reported an increased superoxide dismutase activity in swimmers, perhaps indicating that an adaptation in copper metabolism occurred (Lukaski 1989). It is now well established that exercise increases free radical generation (Jenkins 1993; Kanter 1994). Perhaps copper needs are increased with exercise due to its role in removal of free radicals (superoxide dismutase) (Anderson 1991). More study is needed to determine exactly how chronic exercise affects copper needs.

Chromium potentiates insulin activity and may play a role in maintaining the structural integrity of nucleic acids. High chromium foods include broccoli, asparagus, mushrooms, oysters, organ meats, whole grains, and nuts. Some clinical signs of marginal deficiency are impaired glucose tolerance, elevated insulin concentration, glycosuria, decreased insulin receptor number, elevated blood cholesterol and triacylglycerol, and decreased HDL-cholesterol. Several factors including physical trauma, high simple sugar intake, infection, certain diseases, and exercise exacerbate marginal chromium deficiency. Serum chromium is a poor indicator of chromium status, as it does not reflect body stores. Urine loss may be a more accurate index of utilization because it is the primary route of excretion, and chromium does not appear to be reabsorbed. Endurance exercise significantly increases urinary chromium losses yet endurance-trained subjects appear to have a lower chronic urinary chromium excretion (Anderson 1991). These data may indicate that the chromium stores of trained individuals are substantially lowered or perhaps that some adaptive response to conserve limited stores has occurred. Chromium losses in exercise sweat are unknown. Typically, chromium intake is somewhat low (about 50–60% of recommendations). Although chromium intake of athletes is probably higher, the cumulative effects of high CHO diets and exercise chromium losses may produce adverse effects on exercise performance or health. Preliminary evidence in rodents suggests that chromium supplementation may enhance endurance performance (by increasing glycogen storage or reducing glycogen use) (Anderson 1991). Consequently, based on the available data it appears reasonable to suggest that the chromium intakes of athletes should be toward the upper end of the range of the ESADDI (200 µg · d^{-1}, see table 7.7). Finally, several studies have investigated whether chromium picolinate can enhance gains in muscle mass/strength via its potential effects on amino acid uptake, but there appears to be no beneficial effect (Clarkson and Haymes 1994).

Selenium is an essential component of the enzyme glutathione peroxidase, which is part of the body's antioxidant defense system. Therefore, like copper, the increased free radical generation during exercise may result in increased selenium needs. High selenium foods include meat, fish, kidney, liver, dairy products, and grains. Typically, serum concentration or erythrocyte selenium content has been used to assess selenium status (U.S. Food and Nutrition Board 1989) but platelet glutathione peroxidase activity may be a better assessment tool (Clarkson 1992b). Although little is known of exercise effects on selenium needs, some evidence suggests that selenium supplementation can improve antioxidant status (Dragan et al. 1990; Dragan et al. 1991). However, high intakes (about 200 µg · d^{-1}) appear to be toxic (Levander and Burk 1990), so intakes should not exceed the RDA. More work is needed before selenium supplementation should be recommended for athletes.

Vitamins

Via their action as coenzymes, vitamins are essential for the optimal function of virtually every metabolic process in the body. Vitamin status is usually assessed by food intake analysis or by blood or tissue measures. When intake is inadequate not only will exercise performance be reduced but health will be threatened (van der Beek 1985). Deficiency symptoms can occur quickly (a few weeks for some vitamins), and major debilitating diseases will result if deficiencies become chronic. The thirteen vitamins are divided into two groups: fat soluble (vitamins A, D, E, and K) and water soluble (vitamins B$_1$ [thiamin], B$_2$ [riboflavin], niacin, B$_6$, B$_{12}$, folic acid, biotin, pantothenic acid, and C). Thiamin, riboflavin, vitamin B$_6$, niacin, pantothenic acid, biotin, vitamin C, and vitamin E are of special importance for athletes because of their role in mitochon-drial energy metabolism. In addition, vitamins C and E appear to have antioxidant properties (Jenkins 1993; Kanter 1994). Selected foods with a high content of each vitamin are shown in table 7.8. Actual vitamin intake is determined by both quality and quantity of the food consumed, as well as by the method of preparation. Although individuals who are vitamin deficient have diminished exercise performances that can be improved when the deficiency is reversed, there is no valid evidence that supplementation of individuals with good vitamin status will result in enhanced performance (van der Beek 1992). Unfortunately, the literature is very confusing, largely due to the use of poor methodology (Clarkson 1992a), and many athletes believe that vitamin supplementation is required for elite athletic performance. Some marginal deficiencies have been observed in athletes but this does not seem to be typical, probably because of the high energy intake of many athletes (van Erp-Baart 1989). However, as mentioned previously for several essential nutrients, when food intake or quality is low (body mass-restricted sports), vitamin intake may become inadequate, and health and performance may be adversely affected. Under these conditions vitamin supplements are beneficial. However, there is no good evidence of benefits from megadose

Table 7.8 Selected Food Sources of Vitamins

Vitamin	Food source
Water soluble	
Thiamin (B$_1$)	Unrefined bread/cereals, pulses, pork, potatoes, vegetables, nuts, liver
Riboflavin (B$_2$)	Dairy products, meat, liver, eggs, green leafy vegetables
Niacin	Meat, liver, fish, unrefined bread/cereals, pulses, green leafy vegetables, nuts
Pantothenic acid	Unrefined bread/cereals, potatoes, meat, liver, dairy products
B$_6$	Unrefined bread/cereals, potatoes, vegetables, meat, liver, fish, dairy products, eggs, nuts, bananas
Folic acid	Meat, liver, green leafy vegetables, unrefined bread/cereals, potatoes, fruit
B$_{12}$	Fish, shellfish, meat, liver, dairy products, eggs
Biotin	Liver, eggs, dairy products, fish, nuts
C	Vegetables, fruit, potatoes
Fat soluble	
A	Liver, fish, dairy products, eggs, margarine, butter
Provitamin A	Carrots, dark green leafy vegetables, tomatoes, oranges
E	Liver, unrefined breads/cereals, vegetable/seed oils, margarine, butter, eggs
D	Fish, liver, eggs, dairy products, margarine
K	Liver, green leafy vegetables, cheese, butter

Adapted, by permission, from E.J. van der Beek, 1992, Vitamin supplementation and physical exercise performance. In *Foods, nutrition, and sports performance*, edited by C. Williams and J.T. Devlin (London: E & F.N. Spon), 95-112.

vitamin intake. Moreover, significant problems can result because all fat soluble vitamins and several water soluble vitamins can be toxic if intake is high enough (Alhadeff, Guiltier, Lipton 1984). Finally, as with all nutritional issues, it is important to realize that all individuals involved (physicians, nutritionists, coaches, trainers, and parents) must be sure that their athletes are properly educated.

Summary and Potential Applications

In recent years there has been considerable investigation into the effects of both acute and chronic exercise on nutrient requirements. As a result, it has become increasingly clear that needs for food energy, CHO, protein, water, and several minerals are increased significantly in physically active individuals. Fortunately, this does not result in nutritional deficiencies in the vast majority of athletes because the very high food intakes that typically accompany intense training and competition provide sufficient additional nutrients to cover the increased need. However, some individuals are at increased risk, primarily because, for a variety of reasons, they fail to consume enough food. An athlete who is still growing is especially vulnerable, not only because her or his needs are higher (due to the growth process) but also because of the typically poor eating habits of this population. Those who engage in body mass-restricted sports (due to low energy intake) and especially females (due to the prevalence of disordered eating) are at greatest risk. It is critical that those working with athletes (coaches, nutritionists, physicians, parents, trainers) be alerted to these potential problems because the consequences on health could be very serious, for example, female athlete triad, heat illness, and impaired growth. Finally, it is known that some minor nutritional manipulations, for example, CHO loading and fluid and CHO intake during and following exercise, can enhance performance in activities where CHO availability is limiting, especially when events occur with brief recovery intervals, for example, multiple events on the same or successive days. Recent data suggest that there may also be performance-enhancing effects of other nutrients, that is, creatine, protein, some minerals, and perhaps moderate fat/moderate CHO diets; however, the advantages of these are less clear at present. Much more study is needed to fully understand both the effects of exercise on nutrient needs and the full extent to which selected nutrient supplementation can enhance performance, but it is critical to remember that the increased food intake normally associated with chronic exercise, in most cases, is more than sufficient to cover any increased nutrient requirement. As a result, with only a few exceptions, the heavily promoted and costly practice of food supplementation is rarely necessary. Rather than accepting the various claims that are made or assuming that other athletes are already benefiting from a particular dietary manipulation, the attention of athletes and coaches must shift from hearsay to the scientific literature. Clearly, the best advice is the age-old adage "buyer beware".

As mentioned several times throughout this chapter, there are a variety of topics in the area of exercise nutrition that need to be more fully evaluated. In general, much more information on the nutritional needs of elite athletes is needed because the rigorous training and competition programs utilized make this population different from their sedentary counterparts or even the less serious athlete. In addition, well-controlled performance-oriented studies are desperately needed, because it is widely believed that a variety of nutritional supplements enhance exercise performance despite a dearth of objective evidence supporting most of these claims. Specific topics of current interest include creatine (What are its long-term effects? What are the reasons for the contradictory performance results observed to date?), protein/amino acids (What are the requirements for various activities? Does exercise affect protein needs differently in men vs. women? Are there any benefits to specific amino acid formulations?), sport drinks (Are there better formulations than those currently marketed?), several vitamins and minerals (Does chronic exercise increase needs? If so, can supplementation enhance performance?), fat (Are moderate fat diets the next important ergogenic aid?), and macronutrient formulations (Does the use of liquid meal supplements provide any advantage for the athlete? If so, what is the optimal timing of these to prepare athletes for competition?). These and many other related questions can likely be answered by utilizing our current knowledge base in combination with creative experimental approaches. However, many laboratory measurement techniques are extremely costly, and consequently, substantial additional funding for research must become available. Unfortunately, at the same time government dollars for research have become increasingly more limited. Therefore, it has become essential to look for different sources of support and the success of this search will undoubtedly control the progress of new experimentation in this area. Perhaps by combining resources, industry and academe can provide answers to these important questions. Finally, it is absolutely critical that effective lobbying procedures be put in place to provide the necessary regulations to prevent the sale of any nutritional supplement before it has been shown to be both effective and safe. It seems that the cart has been before the horse for far too long!

References

Alhadeff, L., Guiltier, C.T., and Lipton, M. 1984. Toxic effects of water soluble vitamins. *Nutr. Rev.* 42: 33–40.

Allen, L.H., Oddoye, E.A., and Margen, S. 1979. Protein-induced hypercalciuria: A longer term study. *Am. J. Clin. Nutr.* 32: 741–49.

American College of Sports Medicine. 1996. Position stand on exercise and fluid replacement. *Med. Sci. Sports Exerc.* 28: i–vii.

Anderson, R.A. 1991. New insight on the trace elements, chromium, copper, and zinc, and exercise. In *Advances in nutrition and top sport,* ed. F. Brouns, 38–58. Basel: Karger.

Armstrong, L.E., Costill, D.L., and Fink, W.J. 1985. Influence of diuretic-induced dehydration on competitive running performance. *Med. Sci. Sports Exerc.* 17: 456–61.

Balsom, P.D., Söderlund, K., and Ekblom, B. 1994. Creatine in humans with special reference to creatine supplementation. *Sports Med.* 18: 268–80.

Balsom, P.D., Söderlund, K., Sjodin, B., and Ekblom, B. 1995. Skeletal muscle metabolism during short duration high-intensity exercise: Influence of creatine supplementation. *Acta Physiol. Scand.* 154: 303–10.

Bar-Or, O. 1989. Temperature regulation during exercise in children and adolescents. In *Youth, exercise, and sport,* ed. C.V. Gisolfi and D.R. Lamb, 335–68. Indianapolis: Benchmark.

Bar-Or, O., Dotan, R., Inbar, O., Rotshtein, A., and Zonder, H. 1980. Voluntary hypohydration in 10- to 12-year-old boys. *J. Appl. Physiol.* 48: 104–8.

Bar-Or, O., and Unnithan, V.S. 1994. Nutritional requirements of young soccer players. *J. Sports Sci.* 12: S39–42.

Beals, K.A., and Manore, M.M. 1994. The prevalence and consequences of subclinical eating disorders in female athletes. *Int. J. Sport Nutr.* 4: 175–95.

Below, P.R., Mora-Rodriguez, R. Gonzalez-Alonso, J., and Coyle, E.F. 1994. Fluid and carbohydrate ingestion independently increase performance during one hour of intense exercise. *Med. Sci. Sports Exerc.* 27: 200–10.

Bergström, J., and Hultman, E. 1966. Muscle glycogen synthesis after exercise: An enhancing factor localized to the muscle cells in man. *Nature* 210: 309–10.

Biolo, G., Maggi, S.P., Williams, B.D., Tipton, K.D., and Wolfe, R.R. 1995. Increased rates of muscle protein turnover and amino acid transport after resistance exercise in humans. *Am. J. Physiol.* 268: E514–20.

Birch, R., Noble, D., and Greenhaff, P.L. 1994. The influence of dietary creatine supplementation on performance during repeated bouts of maximal isokinetic cycling in man. *Eur. J. Appl. Physiol.* 69: 268–70.

Bjorntorp, P. 1992. Importance of fat as a support nutrient for energy: Metabolism of athletes. In *Foods, nutrition, and sports performance,* ed. C. Williams and J.T. Devlin, 87–94. London: Spon.

Blom, P.C., Hostmark, A.T., Vaage, O., Kardel, K.R., and Maehlum, S. 1987. Effect of different postexercise sugar diets on the rate of muscle glycogen synthesis. *Med. Sci. Sports Exerc.* 19: 491–96.

Booth, F.W., and Watson, P.A. 1985. Control of adaptations in protein levels in response to exercise. *Fed. Proc.* 44: 2293–300.

Brautbar, N., and Carpenter, C. 1984. Skeletal myopathy and magnesium depletion: Cellular mechanisms. *Magnesium* 3: 57–62.

Bray, G. 1987. Obesity: A disease of nutrient or energy balance. *Nutr. Rev.* 45: 2–33.

Brouns, F. 1993. *Nutritional needs of athletes.* New York: Wiley.

Brouns, F., Saris, W.H.M., Beckers, E., Adlercreutz, H., van der Vusse, G.J., Keizer, H.A., Kuipers, H., Menheere, P., Wagenmakers, A.J., and ten Hoor, F. 1989. Metabolic changes induced by sustained exhaustive cycling and diet manipulation. *Int. J. Sports Med.* 10(Suppl. 1): S49–62.

Brouns, F., Saris, W.H.M., and Schneider, H. 1992. Rationale for upper limits of electrolyte replacement during exercise. *Int. J. Sport Nutr.* 2: 229–38.

Burke, L., and Deakin, V. 1994. *Clinical sports nutrition.* Roseville, NSW, Australia: McGraw-Hill.

Burke, L.M., Collier, G.R., Beasley, S.K., Davis, P.G., Fricker, P.A., Heeley, P., Walder, K., and Hargreaves, M. 1995. Effect of coingestion of fat and protein with carbohydrate feedings on muscle glycogen storage. *J. Appl. Physiol.* 78: 2187–92.

Burke, L.M., Pyne, D.B., and Telford, R.D. 1996. Effect of oral creatine supplementation on single-effort sprint performance in elite swimmers. *Int. J. Sport Nutr.* 6: 222–33.

Butterfield, G.E.. 1991. Amino acids and high protein diets. In *Ergogenics: The enhancement of exercise and sport performance,* ed. D.R. Lamb and M.H. Williams, 87–122. Indianapolis: Benchmark.

Campbell, W.W., and Anderson, R.A. 1987. Effects of aerobic exercise and training on the trace minerals chromium, zinc, and copper. *Sports Med.* 4: 9–18.

Campbell, W.W., Crim, M.C., Dallal, G.E., Young, V.R., and Evans, W.J. 1994. Increased protein requirements in elderly people: New data and retrospective reassessments. *Am. J. Clin. Nutr.* 60: 501–9.

Canada's Physical Activity Guide to Healthy, Active Living. 1998. Ottawa: Health Canada.

Casey, A., Constantin-Teodosiu, D., Howell, S., Hultman, E., and Greenhaff, P.L. 1996. Creatine ingestion favorably affects performance and muscle metabolism during maximal exercise in humans. *Am. J. Physiol.* 271: E31–37.

Castaneda, C., Dolnikowski, G.G., Dallal, G.E., Evans, W.J., and Crim, M.C. 1995. Protein turnover and energy metabolism of elderly women fed a low protein diet. *Am. J. Clin. Nutr.* 62: 40–48.

Castelli, W.P. 1984. Epidemiology of coronary heart disease: The Framingham study. *Am. J. Med.* 76: 4–12.

Cathcart, E.P. 1925. Influence of muscle work on protein metabolism. *Physiol. Rev.* 5: 225–43.

Chandler, R.M., Byrne, H.K., Patterson, J.G., and Ivy, J.L. 1994. Dietary supplements affect the anabolic hormones after weight-training exercise. *J. Appl. Physiol.* 76: 839–45.

Chen, J.D., Wang, J.F., Li, K.J., Zhao, Y.W., Wang, S.W., Jiao, Y., and Hou, X.Y. 1989. Nutritional problems and measures in elite and amateur athletes. *Am. J. Clin. Nutr.* 49: 1084–89.

Chesley, A., MacDougall, J.D., Tarnopolsky, M.A., Atkinson, S.A., and Smith, K. 1992. Changes in human muscle protein synthesis after resistance exercise. *J. Appl. Physiol.* 73: 1383–88.

Clarkson, P.M. 1992a. Vitamins and trace minerals. In *Ergogenics: The enhancement of exercise and sport performance,* ed. D.R. Lamb and M.H. Williams, 123–82. Indianapolis: Benchmark.

Clarkson, P.M. 1992b. Minerals: Exercise performance and supplementation in athletes. In *Foods, nutrition and sports performance,* ed. C. Williams and J.T. Devlin, 113–46. London: Spon.

Clarkson, P.M., and Haymes, E.M. 1994. Trace mineral requirements for athletes. *Int. J. Sport Nutr.* 4: 104–19.

Cooke, W.H., Grandjean, P.W., and Barnes, W.S. 1995. Effect of oral creatine supplementation on power output and fatigue during bicycle ergometry. *J. Appl. Physiol.* 78: 670–73.

Cortright, R.N., Chandler, M.P., Lemon, P.W., and DiCarlo, S.E. 1997. Daily exercise reduces fat, protein, and body mass in male but not female rats. *Physiol. Beh.* 62: 105–11.

Cortright, R.N., Collins, H.L., Chandler, M.P., Lemon, P.W., and DiCarlo, S.E. 1996. Diabetes reduces growth and body composition more in male than female rats. *Physiol. Beh.* 60: 1233–38.

Cortright, R.N., Rogers, M.E., and Lemon, P.W.R. 1993. Does protein intake during endurance exercise affect growth, nitrogen balance, or exercise performance? *Can. J. Appl. Physiol.* 18(4): 403P.

Costill, D.L., and Miller, J.M. 1980. Nutrition for endurance sport: Carbohydrate and fluid balance. *Int. J. Sports Med.* 1: 1–14.

Couzy, F., Lafargue, P., and Guezennec, C.Y. 1990. Zinc metabolism in the athlete: Influence of training nutrition and other factors. *Int. J. Sports Med.* 11: 263–266.

Coyle, E.F. 1992. Timing and method of increased carbohydrate intake to cope with heavy training, competition, and recovery. In *Foods, nutrition and sports performance,* ed. C. Williams and J.T. Devlin, 35–62. London: Spon.

Coyle, E.F., Coggan, A.R., Hemmert, M.K., and Ivy, J.L. 1986. Muscle glycogen utilization during prolonged strenuous exercise when fed carbohydrate. *J. Appl. Physiol.* 61: 165–72.

Coyle, E.F., and Montain, S.J. 1992. Benefits of fluid replacement with carbohydrate during exercise. *Med. Sci. Sports Exerc.* 24(9, Suppl.): S324–30.

Deuster, P.A., Kyle, S.B., Moser, P.B., Vigersky, R.A., Singh, A., and Schoomaker, E.B. 1986. Nutritional survey of highly trained women runners. *Am. J. Clin. Nutr.* 44: 954–62.

Dragan, I., Dinu, V., Cristea, E., Mohora, N., Ploesteanu, E., and Stroescu, V. 1991. Studies regarding the effects of an antioxidant compound in top athletes. *Rev. Roum. Physiol.* 28: 105–8.

Dragan, I., Dinu, V., Mohora, M., Cristea, E., Ploesteanu, E., and Stroescu, V. 1990. Studies regarding the antioxidant effects of selenium on top swimmers. *Rev. Roum. Physiol.* 27: 15–20.

Drinkwater, B.L., Nilson, K., Chestnut, C.H., Bremner, W.J., Shainholtz, S., and Southworth, M.B. 1984. Bone mineral content of amenorrheic and eumenorrheic athletes. *New Engl. J. Med.* 311: 277–81.

Evans, W.J. 1996. Effects of aging and exercise on nutritional needs of the elderly. *Nutr. Rev.* 54(1, Part II): S535–39.

Faulkner, J.A., Green, H.J., and White, T.P. 1994. Response and adaptation of skeletal muscle to changes in physical activity. In *Physical activity, fitness, and health,* ed. C. Bouchard, R.J. Shepherd, and T. Stephens, 343–57. Champaign, IL: Human Kinetics.

Fern, E.B., Bielinski, R.N., and Schutz, Y. 1991. Effects of exaggerated amino acid and protein supply in man. *Experientia* 47: 168–72.

Flynn, A. 1985. Milk proteins in the diets of those of intermediate years. In *Milk proteins '84,* ed. T.E. Galesloot and B.J. Tinbergen, 154–57. Wageningen, Netherlands: Pudoc.

Food and Agricultural Organization, World Health Organization. 1989. Protein quality evaluation. *Food and nutrition paper* No. 51, Rome.

Food and Agricultural Organization, World Health Organization, and United Nations University. 1985. Energy and Protein Requirements. *World Health Organization technical report series* 724, Geneva.

Food and Drug Administration. 1993. Food labeling; general provisions; nutrition labeling; label format; nutrient content claims; health claims; ingredient labeling; state and local requirements and exemptions; final rules. *Fed. Reg.* 58: 2101–6.

Frentsos, J.A., and Baer, J.T. 1997. Increased energy and nutrient intake during training and competition improves elite triathletes' endurance performance. *Int. J. Sport Nutr.* 7: 61–71.

Garlick, P.J., McNurlan, M.A., Essen, P., and Wernerman, J. 1994. Measurement of tissue protein synthesis rates in vivo: Critical analysis of contrasting methods. *Am. J. Physiol.* 266: E287–97.

Greenhaff, P.L. 1995. Creatine and its application as an ergogenic aid. *Int. J. Sports Nutr.* 5(Suppl.): S100–10.

Greenhaff, P.L., Bodin, K., Harris, R.C., Hultman, E., Jones, D.A., McIntire, D.B., Söderlund, K., and Turner, D.L. 1993a. The influence of oral creatine supplementation on muscle phosphocreatine resynthesis following intense contraction in man. *J. Physiol. (London)* 467: 75P.

Greenhaff, P.L., Bodin, K., Söderlund, K., and Hultman, E. 1994. Effect of oral creatine supplementation on skeletal muscle phosphocreatine resynthesis. *Am. J. Physiol.* 266: E725–30.

Greenhaff, P.L., Casey, A., Short, A.H., Harris, R., Söderlund, K., and Hultman, E. 1993b. Influence of oral creatine supplementation of muscle torque during repeated bouts of maximum voluntary exercise in man. *Clin. Sci.* 84: 565–71.

Hackman, R.M., and Keen, C.L. 1986. Changes in serum zinc and copper levels after zinc supplementation in running and nonrunning men. In *Sport, health, and nutrition,* ed. F.I. Katch, 89–99. Champaign, IL: Human Kinetics.

Harris, R.C., Söderlund, K., and Hultman, E. 1992. Evaluation of creatine in resting and exercising muscle of normal subjects by creatine supplementation. *Clin. Sci.* 83: 367–74.

Hartley, L.H., Mason, J.W., Hogan, R.P., Jones, L.G., Kotchen, T.A., Mougey, E.H., Wherry, F.E., Pennington, L.L., and Ricketts, P.T. 1972a. Multiple hormonal responses to graded exercise in relation to physical training. *J. Appl. Physiol.* 33: 602–6.

Hartley, L.H., Mason, J.W., Hogan, R.P., Jones, L.G., Kotchen, T.A., Mougey, E.H., Wherry, F.E., Pennington, L.L., and Ricketts, P.T. 1972b. Multiple hormonal responses to prolonged exercise in relation to physical training. *J. Appl. Physiol.* 33:607–10.

Häussinger, D., Roth, E., Lang, F., and Gerok, W. 1993. Cellular hydration state: An important determinant of protein catabolism in health and disease. *Lancet* 341: 1330–32.

Haymes, E.M. 1987. Nutritional concerns: Need for iron. *Med. Sci. Sports Exerc.* 19: S197–200.

Henley, E.C., and Kuster, J.M. 1994. Protein quality evaluation by protein digestibility-corrected amino acid scoring. *Food Tech.* 48: 74–77.

Hermansen, L., Hultman, E., and Saltin, B. 1967. Muscle glycogen during prolonged severe exercise. *Acta Physiol. Scand.* 71: 129–39.

Highet, R. 1989. Athletic amenorrhea: An update on etiology, complications, and management. *Sports Med.* 7: 82–108.

Horswill, C.A., Park, S.H., and Roemmich, J.N. 1990. Changes in protein nutritional status of adolescent wrestlers. *Med. Sci. Sports Exerc.* 22: 599–604.

Ivy, J.L., Katz, A.L., Culter, C.L., Sherman, W.M., and Coyle, E.F. 1988. Muscle glycogen synthesis after exercise: Effect of time of carbohydrate ingestion. *J. Appl. Physiol.* 64: 1480–85.

Jenkins, R.R. 1993. Exercise, oxidative stress, and antioxidants: A review. *Int. J. Sport Nutr.* 3: 356–75.

Johnston, C.C., Miller, J.Z., Slemenda, C.W., Reister, T.K., Hui, S., Christian, J.C., and Peacock, M. 1992. Calcium supplementation and increases in bone mineral density in children. *New Engl. J. Med.* 327: 82–87.

Kanter, M.M. 1994. Free radicals, exercise, and antioxidant supplementation. *Int. J. Sport Nutr.* 4: 205–20.

Kraemer, W.J., Fleck, S.J., and Evans, W.J. 1996. Strength and power training: Physiological mechanisms of adaptation. In *Exercise and sport science reviews,* ed. J.O. Holloszy, 363–97. Baltimore: Williams & Wilkins.

Kreider, R.B., Ferreira, M., Wilson, M., Grindstaff, P., Plisk, S., Reinardy, J., Cantler, E., and Almada, A.L. 1998. Effects of creatine supplementation on body composition, strength, and sprint performance. *Med. Sci. Sports Exerc.* 30: 73–82.

Leddy, J., Horvath, P., Rowland, J., and Pendergast, D. 1997. Effect of a high or low fat diet on cardiovascular risk factors in male and female runners. *Med. Sci. Sports Exerc.* 29: 17–25.

Lemon, P.W.R. 1989. Nutrition for the muscular development of young athletes. In *Youth, exercise, and sport,* ed. C.V. Gisolfi and D.R. Lamb, 369–400. Indianapolis: Benchmark.

Lemon, P.W.R. 1992. Effects of exercise on protein requirements. In *Foods, nutrition, and sports performance,* ed. C. Williams and J.T. Devlin, 66–86. London: Spon.

Lemon, P.W.R. 1995. Do athletes need more dietary protein and amino acids? *Int. J. Sport Nutr.* 5(Suppl.): S39–61.

Lemon, P.W.R. 1996. Is supplemental dietary protein necessary or beneficial for individuals with a physically active lifestyle? *Nutr. Rev.* 54 (4, II): S169–75.

Lemon, P.W.R. 1997. Dietary protein requirements in athletes. *Nutr. Biochem.* 28: 52–60.

Lemon, P.W.R., Newcomer, B.R., Bredle, D.L., Ziegenfuss, T.N., Rogers, M.E., and Boska, M.D. 1995. Creatine intake enhances muscle energetics and force output during repeated maximal isometric plantar flexion in humans. *Med. Sci. Sports Exerc.* 27(5, Suppl.): S204.

Lemon, P.W.R., Tarnopolsky, M.A., MacDougall, J.D., and Atkinson, S.A. 1992. Protein requirements and muscle mass/strength changes during intensive training in novice bodybuilders. *J. Appl. Physiol.* 73: 767–75.

Levander, O.A., and Burk, R.F. 1990. Selenium. In *Present knowledge in nutrition*, ed. M.L. Brown, 268–73. Washington: International Life Sciences Institute.

Lindsay, R. 1987. Estrogen and osteoporosis. *Physician Sportsmed.* 15: 91–108.

Lohman, T.G. 1992. *Advances in body composition assessment.* Champaign, IL: Human Kinetics.

Lukaski, H.C. 1989. Influence of physical training on human copper nutritional status. *Abtsr. Am. Chem. Soc.* 197: 91.

Lyons, T.P., Riedesel, M.L. Meuli, L.E., and Chick, T.W. 1990. Effects of glycerol-induced hyperhydration prior to exercise in the heat on sweating and core temperature. *Med. Sci. Sports Exerc.* 22: 477–83.

MacDougall, J.D., Ward, G.R., and Sutton, J.R. 1977. Muscle glycogen repletion after high-intensity intermittent exercise. *J. Appl. Physiol.* 42: 129–32.

Maganaris, C.N., and Maughan, R.J. 1998. Creatine supplementation enhances maximum voluntary isometric force and endurance capacity in resistance trained men. *Acta Physiol. Scand.* 163: 279–87.

Martin, W.H. 1996. Effects of acute and chronic exercise on fat metabolism. In *Exercise and sport science reviews*, ed. J.O. Holloszy, 203–31. Baltimore: Williams & Wilkins.

Martin, W.H., Dalsky, G.P., Hurley, B.F., Matthews, D.E., Bier, D.M., Hagberg, J.M., Rogers, M.A., King, D.S., and Holloszy, J.O. 1993. Effect of endurance training on plasma free fatty acid turnover and oxidation during exercise. *Am. J. Physiol.* 265: E708–14.

Maughan, R. 1991. Carbohydrate-electrolyte solutions during prolonged exercise. In *Ergogenics: The enhancement of exercise and sport performance*, ed. D.R. Lamb and M.H. Williams, 35–86. Indianapolis: Benchmark.

Maughan, R.J., Leiper, J.B., and Shirreffs, S.W. 1996a. Rehydration and recovery after exercise. *Sports Sci. Exch.* 9(62): 1–5.

Maughan, R.J., Leiper, J.B., and Shirreffs, S.W. 1996b. Restoration of fluid balance and after exercise-induced dehydration: Effects of food and drink intake. *Eur. J. Appl. Physiol.* 73: 317–25.

McArther, K.E., and Feldman, M. 1989. Gastric acid secretion, gastrin release, and gastric temperature in humans as affected by liquid meal temperature. *Am. J. Clin. Nutr.* 49: 51–54.

McDonald, R., and Keen, C. 1988. Iron, zinc, and magnesium nutrition and athletic performance. *Sports Med.* 5: 171–84.

McNaughton, L.R., Dalton, B., and Tarr, J. 1998. The effects of creatine supplementation on high intensity exercise performance in elite performers. *Eur. J. Appl. Physiol.* 78: 236–40.

Meredith, C.N., Zackin, M.J., Frontera, W.R., and Evans, W.J. 1989. Dietary protein requirements and protein metabolism in endurance-trained men. *J. Appl. Physiol.* 66: 2850–56.

Meyer, F., Bar-Or, O., Salsberg, A., and Passe, D. 1994. Hypohydration during exercise in children: Effect on thirst, drink preferences, and rehydration. *Int. J. Sport. Nutr.* 4: 22–35.

Miller, W.C., Linderman, A.K., Wallace, J., and Niederpruem, M. 1990. Diet composition, energy intake, and exercise in relation to body fat in men and women. *Am. J. Clin. Nutr.* 52: 426–30.

Muckle, D.S. 1973. Glucose syrup ingestion and team performance in soccer. *Br. J. Sports Med.* 7: 340–43.

Mujika, I., Chatard, J.C., Lacoste, L., Barale, F., and Geyssant, A. 1996. Creatine supplementation does not improve sprint performance in competitive swimmers. *Med. Sci. Sports Exerc.* 28: 1435–41.

Muoio, D.M., Leddy, J.J., Horvath, P.J., Awad, A.B., and Pendergast, D.R. 1994. Effect of dietary fat on metabolic adjustments to maximal $\dot{V}O_2$ and endurance in runners. *Med. Sci. Sports Exerc.* 26: 81–88.

Newsholme, E.A., and Parry-Billings, M. 1994. Effects of exercise on the immune system. In *Physical activity, fitness, and health*, ed. C. Bouchard, R.J. Shephard, and T. Stephens, 451–55. Champaign, IL: Human Kinetics.

Nicholas, C.W., Williams, C., Phillips, G., and Nowitz, A. 1996. Influence of ingesting a carbohydrate-electrolyte solution on endurance capacity during intermittent, high intensity shuttle running. *J. Sport Sci.* 13: 283–90.

Noakes, T.D. 1992. The hyponatremia of exercise. *Int. J. Sport Nutr.* 2: 205–28.

Nutrition Recommendations. 1990. *The report of the Scientific Review Committee.* Department of National Health and Welfare, Ottawa.

O'Conner, H. 1994. Special needs: Children and adolescents in sport. In *Clinical sports nutrition*, ed. L. Burke and V. Deakin, 390–448. Roseville, NSW, Australia: McGraw-Hill.

Ööpik, V., Timpmann, S., and Medijainen, L. 1995. The role and application of dietary creatine supplementation in increasing physical performance capacity. *Biol. Sport* 12: 197–212.

Phillips, S.M., Atkinson, S.A., Tarnopolsky, M.A., MacDougall, J.D. 1993. Gender differences in leucine kinetics and nitrogen balance in endurance athletes. *J. Appl. Physiol.* 75: 2134–41.

Pitts, G.C. 1984. Body composition in the rat: Interactions of exercise, age, sex, and diet. *Am. J. Physiol.* 246: R495–501.

Pitts, G.C., and Ball, L.S. 1977. Exercise, dietary obesity, and growth in the rat. *Am. J. Physiol.* 232: R38–44.

Pratley, R., Nicklas, B., Rubin, M., Miller, J., Smith, A., Smith M., Hurley, B., and Goldberg, A. 1994. Strength training increases resting metabolic rate and norepinephrine levels in healthy 50- to 65-year-old men. *J. Appl. Physiol.* 76: 133–37.

Prior, J.C., Vigna, Y.M., Schechter, M.T., Burgess, A.E. 1990. Spinal bone loss and ovulatory disturbances. *New Engl. J. Med.* 323: 121–27.

Redondo, D.R., Dowling, E.A., Graham, B.L., Almada, A.L., and Williams, M.H. 1996. The effect of oral creatine monohydrate supplementation on running velocity. *Int. J. Sport Nutr.* 6: 213–21.

Rehrer, N.J. 1991. Aspects of dehydration and rehydration during exercise. In *Advances in nutrition and top sport,* ed. F. Brouns, 128–46. Basel: Karger.

Rennie, M.J., Smith, K., and Watt, P.W. 1994. Measurement of human protein synthesis: An optimal approach. *Am. J. Physiol.* 266: E298–307.

Riedesel, M.L., Allen, D.Y., Peake, G.T., and Al-Qattan, K. 1987. Hyperhydration with glycerol solutions. *J. Appl. Physiol.* 63: 2262–68.

Roemmich, R.N., and Sinning, W.E. 1996. Sport-seasonal changes in body composition, growth, power, and strength of adolescent wrestlers. *Int. J. Sports Med.* 17: 92–99.

Roemmich, R.N., and Sinning, W.E. 1997a. Weight loss and wrestling training: Effects on nutrition, growth, maturation, body composition, and strength. *J. Appl. Physiol.* 82: 1751–59.

Roemmich, R.N., and Sinning, W.E. 1997b. Weight loss and wrestling training: Effects on growth-related hormones. *J. Appl. Physiol.* 82: 1760–64.

Romijn, J.A., Coyle, E.F., Sidossis, L.S., Gastaldelli, A., Horowitz, J.F., Endert, E., and Wolfe, R.R. 1993. Regulation of endogenous fat and carbohydrate metabolism in relation to exercise intensity and duration. *Am. J. Physiol.* 265: E380–91.

Rose, L.I., Carroll, D.R., Lowe, S.L., Peterson, E.W., and Cooper, K.H. 1970. Serum electrolyte changes after marathon running. *J. Appl. Physiol.* 29: 449–51.

Rossiter, H.B., Cannell, E.R., and Jakeman, P.M. 1996. The effect of oral creatine supplementation on the 1000 m performance of competitive rowers. *J. Sports Sci.* 14: 175–79.

Sadur, C.N., and Eckel, R.H. 1982. Insulin stimulation of adipose tissue lipoprotein lipase activity. *J. Clin. Lab. Invest.* 69: 1119–25.

Saltin, B., and Gollnick, P.D. 1988. Fuel for muscular exercise: Role of carbohydrates. In *Exercise, nutrition, and energy metabolism,* ed. E.S. Horton and R.L. Terjung, 47. New York: MacMillan.

Sanborn, C.F. 1992. Exercise, calcium, and bone density. *Sports Sci. Exch.* 2(24): 1–5.

Schneider, D.A., McDonough, P.J., Fadel, P.J., and Berwick, J.P. 1997. Creatine supplementation and the total work performed during 15 s and 1 min bouts of maximal cycling. *Aust. J. Sci. Med. Sport* 29: 65–68.

Sherman, W.M., Costill, D.L., Fink, W.J., and Miller, J.M. 1981. The effect of exercise-diet manipulation on muscle glycogen and its subsequent utilization during performance. *Int. J. Sports Med.* 2: 114–18.

Shi, X., Summers, R.W., Schedl, H.P., Flanagan, S.W., Chang, R., and Gisolfi, C.V. 1995. Effects of carbohydrate type and concentration and solution osmolality on water absorption. *Med. Sci. Sports Exerc.* 27: 1607–15.

Shirreffs, S.M., Taylor, A.J., Leiper, J.B., and Maughan, R.J. 1996. Postexercise rehydration in man: Effects of volume consumed and drink sodium content. *Med. Sci. Sports Exerc.* 28: 1260–71.

Simard, C., Tremblay, A., and Jobin, M. 1988. Effects of carbohydrate intake before and during an ice hockey match on blood and muscle energy substrates. *Res. Quart. Exerc. Sport* 59: 144–47.

Smith, S.A., Montain, S.J., Matott, R.P., Zientara, G.P., Jolesz, F.A., and Fielding, R.A. 1998. Creatine supplementation and age influence muscle metabolism during exercise. *J. Appl. Physiol.* 85: 1349–56.

Stendig-Lindberg, G., Shapiro, Y., Epstein, Y., Galun, E., Schonberger, E., Graff, E., and Wacker, W.E. 1987. Changes in serum magnesium concentration after strenuous exercise. *J. Am. Coll. Nutr.* 6: 35–40.

Sundgot-Borgen, J. 1993a. Prevalence of eating disorders in elite female athletes. *Int. J. Sport Nutr.* 3: 29–40.

Sundgot-Borgen, J. 1993b. Nutrient intake of female elite athletes suffering from eating disorders. *Int. J. Sport Nutr.* 3: 431–42.

Tapscott, E.B., Kasperek, G.J., and Dohm, G.L. 1982. Effect of training on muscle protein turnover in male and female rats. *Biochem. Med.* 27: 254–59.

Tarnopolsky, L.J., MacDougall, J.D., Atkinson, S.A., Tarnopolsky, M.A., and Sutton, J.R. 1990. Gender differences in substrate for endurance exercise. *J. Appl. Physiol.* 68: 302–8.

Tarnopolsky, M.A., Atkinson, S.A., MacDougall, J.D., Chesley, A., Phillips, S., and Schwarcz, H.P. 1992. Evaluation of protein requirements for trained strength athletes. *J. Appl. Physiol.* 73: 1986–95.

Tarnopolsky, M.A., Atkinson, S.A., MacDougall, J.D., Senor, B.B., Lemon, P.W., and Schwarcz, H. 1991. Whole body leucine metabolism during and after resistance exercise in fed humans. *Med. Sci. Sports Exerc.* 23: 326–33.

Tarnopolsky, M.A., Atkinson, S.A., Phillips, S.M., and MacDougall, J.D. 1995. Carbohydrate loading and metabolism during exercise in men and women. *J. Appl. Physiol.* 78: 1360–68.

Tesch, P.A. 1987. Acute and long-term metabolic changes consequent to heavy resistance exercise. In *Medicine and sports science,* vol. 26, ed. M. Hebbelinck and R.J. Shephard, 67–89. Basel: Karger.

Tessari, P., Barazzoni, R., Zanetti, M., Vettore, M., Normand, S., Bruttomesso, D., and Beaufrere, B. 1996. Protein degradation and synthesis measured with multiple amino acid tracers in vivo. *Am. J. Physiol.* 271: E733–41.

U.S. Department of Health and Human Services. 1996. *Physical activity and health: A report of the Surgeon Gen-*

eral. Atlanta: Center for Disease Control and Prevention, National Center for Chronic Diseases Prevention and Health Promotion.

U.S. Food and Nutrition Board. 1989. *Recommended dietary allowances.* Washington: National Academy Press.

Vandenberghe, K., Goris, M., Van Hecke, P., Van Leemputte, M., Vangerven, L., and Hespel, P. 1997. Long-term creatine intake is beneficial to muscle performance during resistance training. *J. Appl. Physiol.* 83: 2055–63.

van der Beek, E.J. 1985. Vitamins and endurance training: Food for running or faddish claims? *Sports Med.* 2: 175–97.

van der Beek, E.J. 1992. Vitamin supplementation and physical exercise performance. In *Foods, nutrition, and sports performance,* ed. C. Williams and J.T. Devlin, 95–112. London: Spon.

van Erp-Baart, A.M., Saris, W.H., Binkhorst, R.A., Vos, J.A., and Elvers, J.W. 1989. Nation-wide survey on nutritional habits in elite athletes. Part I. Energy, carbohydrate, protein, and fat intake. *Int. J. Sports Med.* 10(Suppl. 1): S3–10.

Van Zant, R.S., and Lemon, P.W.R. 1997. Preexercise sugar feeding doesn't alter prolonged exercise muscle glycogen or protein catabolism. *Can. J. Appl. Physiol.* 22: 268–79.

Volek, J.S., and Kraemer, W.J. 1996. Creatine supplementation: Its effects on human muscular performance and body composition. *J. Str. Cond. Res.* 10: 200–10.

von Liebig, J. 1842. *Animal chemistry or organic chemistry in its application to physiology.* Translated by G. Gregory. London: Taylor & Walton.

von Stock, K., Kron, K.W., Schardt, G., Ertel, H.H., and Schiller, R. 1979. Der einflub orater magnesiumzufuhr auf die leist ungsfahigkeit des menschlichen organismus unter standardisierter ergometrischer bel astung. *Deut. Zeit. Sportsmed.* 1: 22–27.

Wemple, R.D., Morocco, T.S., and Mack, G.W. 1997. Influence of sodium replacement on fluid ingestion following exercise-induced dehydration. *Int. J. Sport Nutr.* 7: 104–16.

Wilk, B., and Bar-Or, O. 1996. Effect of drink flavor and NaCl on voluntary drinking and hydration in boys exercising in the heat. *J. Appl. Physiol.* 80: 1112–17.

Wilmore, J.H. 1991. Eating and weight disorders in the female athlete. *Int. J. Sport Nutr.* 1: 104–17.

Wilmore, J.H., Wambsgans, K.C., Brenner, M., Broeder, C.E., Paijmans, I., Volpe, J.A., and Wilmore, K.M. 1992. Is there energy conservation in amenorrheic compared with eumenorrheic distance runners? *J. Appl. Physiol.* 72: 15–22.

Wolinsky, I., and Hickson Jr., J.F. 1994. *Nutrition in exercise and sport.* Boca Raton, FL: CRC Press.

Wolman, R.L., Clark, P., McNally, E., Harries, M.G., and Reeve, J. 1992. Dietary calcium as a statistical determinant of trabecular bone density in amenorrheic and estrogen-replete athletes. *Bone Min.* 17: 415–23.

Young, V.R., Bier, D.M., and Pellet, P.L. 1989. A theoretical basis for increasing current estimates of the amino acid requirements in adult man with experimental support. *Am. J. Clin. Nutr.* 50: 80–92.

Zawadski, K.M., Yaspelkis, B.B., and Ivy, J.L. 1992. Carbohydrate-protein complex increases the rate of muscle glycogen storage after exercise. *J. Appl. Physiol.* 72: 1854–59.

Ziegenfuss, T.N., Lowery, L.M., and Lemon, P.W.R. 1998. Effect of acute creatine supplementation on body fluid volumes in humans. *J. Exerc. Physiol.* online, 1(3): 1–14. http://www.css.edu/users/tboone2/asep/jan13d.htm.

Ziegenfuss, T.N., Lemon, P.W.R., Rogers, M.E., Ross, R., and Yarasheski, K.E. 1997. Acute creatine ingestion: Effects on muscle volume, anaerobic power, fluid volumes, and protein turnover. *Med. Sci. Sports Exerc.* 29(5, Suppl.): S127.

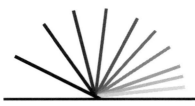

Chapter 8

Intensity of Cycling and Cycle Ergometry: Power Output and Energy Cost

B.R. MacIntosh, R.R. Neptune, and A.J. van den Bogert

It is generally accepted that the intensity of exercise dictates the duration of exercise—the higher the intensity, the shorter the duration the exercise can be sustained. There are several ways in which exercise intensity can be quantified. When using cycle ergometry as the exercise, power output and rate of metabolic energy use are the two principal measures of intensity. In the case of locomotion with a bicycle, velocity can also be used to quantify the intensity of exercise. In this chapter, the determinants and methods of measurement of intensity of exercise are presented, using cycling and cycle ergometry as examples.

Primary Concepts

The themes presented in the following few paragraphs are expanded in the body of this chapter. By the end of this chapter, the reader should have a good understanding of the ways by which the intensity of exercise during cycle ergometry and riding a bicycle can be quantified and should understand the factors that determine the effort necessary for performing such exercise.

It is commonly recognized that the higher the intensity of exercise, the shorter the duration that exercise can be sustained. This fact is represented by the classic intensity-duration relation. The most

appropriate manner of quantifying the intensity of exercise is by the metabolic energy cost of performing the task (energy in). A secondary method of expressing intensity is by mechanical power output (energy out). In cycle ergometry, both of these approaches are used. In cycling, velocity is also measured as an indicator of intensity. Many factors can influence the effort or metabolic energy required to cycle at a given power output or velocity.

The rate of performing mechanical work while cycling can be subdivided into internal and external work (see definitions beginning on page 155). Kinematic and kinetic analyses have been used to quantify these parameters. One advantage of using an assessment of mechanical energy as the criterion for intensity of exercise is that this approach can be used across the full range of possible intensities. A disadvantage is that internal work and external work are not mutually exclusive. Transfer of energy between body segments and between internal and external work can result in overestimation of the total mechanical energy of the system.

All of the metabolic energy for the performance of exercise comes directly from the hydrolysis of adenosine triphosphate (ATP), the common currency of chemical energy in the body. There are three energy systems of the body that work together in an interrelated fashion to replace ATP. In most cases, this occurs as quickly as ATP is hydrolyzed during muscle contraction. The rate of ATP replacement by the long-term energy system, aerobic metabolism, is easily measured and accounts for most of the energy used during exercise when the intensity is less than the capacity for aerobic metabolism. Therefore, oxygen uptake can be measured to quantify exercise intensity. Two important aspects of oxygen uptake are considered in this chapter: steady state oxygen uptake and the slow component of oxygen uptake.

Several factors contribute to the metabolic energy cost of submaximal exercise. In the case of cycle ergometry, cadence and resistance (determinants of external mechanical power output) are of primary importance. To understand the fundamental determinants of the energy cost of exercise, consideration should be given to factors that affect the energy cost of a muscle contraction, including factors that influence efficiency. In addition, the appropriate coordination of muscular effort, motor unit recruitment, and even muscle fiber types are important considerations in determining the metabolic energy cost of exercise.

When cycling exercise is performed on a bicycle rather than a cycle ergometer, the determinants of metabolic energy cost and external mechanical power become air resistance, gravity, inertia, and rolling resistance rather than cadence and flywheel resistance. Velocity is an important consideration in air resistance, and frontal surface area and aerodynamics become important factors that affect the metabolic energy cost of riding at a given speed. Body mass also contributes to the energy cost of cycling. Riding on a hilly course requires a greater energy expenditure than riding on a level course, and the magnitude of the increase in energy cost is related to bicycle and rider mass. There are many factors to consider when discussing the intensity of cycling exercise, and these factors are elaborated below.

Intensity-Duration Relation

The relationship between intensity of exercise and endurance is illustrated in the classic intensity-duration relation (see figure 8.1). In this case, intensity of exercise is presented as both rate of metabolic energy use (joules · s^{-1}) and mechanical power output (watts) with an assumed efficiency of 25%. Efficiency would not only affect the position of the curve, but differences in efficiency at different power outputs would result in differences in the shape of the relationship with power versus that with energy cost. Peak power output is the highest mechanical power output that can be generated in a single

maximal effort. The power output that is associated with maximal oxygen uptake is generally 10 to 25% of peak power output. This intensity of exercise can be sustained for just a few minutes. However, decreases of energy requirement below maximal oxygen uptake for a given exercise can result in progressively longer exercise duration. When the intensity is decreased to anaerobic threshold, which is approximately 60 to 85% of maximal oxygen uptake, the duration can be 30 to 60 min. At 50% of maximal oxygen uptake, the exercise can be sustained for hours. The actual shape of the curve can vary from individual to individual, and factors such as magnitude of maximal oxygen uptake, peak power output, and anaerobic threshold are factors that determine this shape.

In several exercise tasks, the intensity-duration relation has been depicted using velocity as the measure of intensity. This may be useful for some presentations and comparisons (where skill is comparable), but it should be kept in mind that the economy of locomotion, or the rate of metabolic energy use in traveling a given speed, can vary considerably. When comparisons are made between individuals with different skill (or in the case of cycling, different equipment), it should be kept in mind that a common velocity does not necessarily mean a common effort. However, it should also be kept in mind that speed of movement and how long

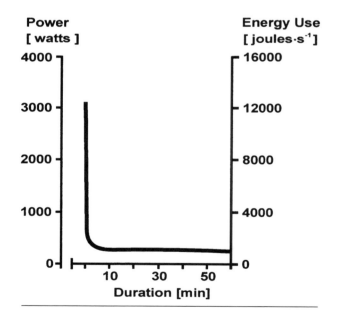

Figure 8.1 The intensity-duration relationship. Intensity is expressed as mechanical power (watts) on the left and rate of metabolic energy use (joules · s^{-1}) on the right, with an assumed efficiency of 25%. Any point on the line can be read on either the power or the rate of energy use scale.

a given speed can be endured are important determinants of success in sports concerned with covering a given distance in the shortest possible time.

It could also be pointed out that two individuals performing cycle ergometer exercise at the same metabolic rate might not perceive the task as requiring the same effort. Exercise physiologists have attempted to normalize such comparisons by expressing the energy cost of the exercise relative to (as a percentage of) maximal oxygen uptake. This normalization results in a better expression of the relative effort required for performance of the task (Brooks and Mercier 1994), but even when the procedure indicates that two performances are of relatively equal effort the exercisers do not both perceive the exercise to be the same intensity.

There are several ways by which the intensity of exercise can be quantified. These methods each have their advantages and disadvantages. In the following section, the use of mechanical power for estimation of exercise intensity in cycling and cycle ergometry is presented. Following this, a discussion of metabolic energy is presented.

External Work and the Rate of Doing Work (Power)

In some activities, the amount of work required to overcome external forces is dependent on the movement pattern. For example, aerodynamic forces in cycling and skating and drag forces in swimming all depend on orientation and velocity of the body segments. The resistive forces for these forms of locomotion can be determined from wind tunnel or water tank experiments and applied to human movement data to obtain an estimate of the associated mechanical work. For these activities, the rate of performing external work (mechanical power) is one indicator of the intensity of exercise, but it gives little indication of the metabolic energy expenditure inside the system; the expenditure is completely external. An example of problems associated with this method is seen in the analysis of walking, which has negligible external work, although metabolic energy is required to perform the task.

In the context of cycling, the external mechanical power or the rate of energy output is a function of the velocity and the force needed to overcome the resistive forces. Recognizing that the force component along the crank does not contribute to the work, the instantaneous external power ($P_{external}$) can be computed from the product of the crank torque (T_{crank}) and angular velocity (ω_{crank}) as

$$P_{external} = T_{crank} \cdot \omega_{crank}. \qquad 8.1$$

Crank torque and angular velocity can be measured experimentally using an instrumented crank or special pedals. Many such systems have been developed, which typically use strain gauge or piezoelectric force transducers and potentiometers or digital encoders to measure crank and pedal displacements (Newmiller, Hull, and Zajac 1988; Wheeler, Gregor, and Broker 1992). Equation 8.1 applies equally to measurement of external power on a cycle ergometer as well as while riding a bicycle. The difference between the two depends on the factors that affect the required torque and angular velocities of the task. Further elaboration of these factors as well as examples of power output across various riding conditions are given in the following sections.

Ergometer Cycling

During stationary ergometer cycling, the external resistive force usually originates from a friction belt or an electromagnetic brake, which is used to model the air drag, rolling resistance, inertial, and gravitational forces normally encountered by a cyclist on the road. The advantage of using an ergometer in a laboratory setting is that it allows very controlled test conditions, where mechanical power output can easily be quantified by measurement of the resistance and angular velocity of the flywheel (or controlling cadence). However, a flywheel is usually present between the crank and the braking mechanism. Therefore, there will be two components of the torque applied to the crank: that necessary to overcome the frictional force induced by the friction belt or the electromagnetic brake and the torque necessary to overcome changes in the flywheel kinetic energy. Hence, the total external power the cyclist will have to overcome becomes

$$P_{external} = \qquad\qquad 8.2$$
$$F_{friction} \cdot V_{friction} + I_{flywheel} \cdot \alpha_{flywheel} \cdot \omega_{flywheel}$$

where

$P_{external}$ = total instantaneous power (watts),

$F_{friction}$ = frictional force applied to the flywheel (N),

$V_{friction}$ = velocity at the point of the $F_{friction}$ application (m · s^{-1}),

$I_{flywheel}$ = moment of inertia of the flywheel (kg · m^{-2}),

$\alpha_{flywheel}$ = angular acceleration of the flywheel (rad · s^{-2}), and

$\omega_{flywheel}$ = angular velocity of the flywheel (rad · s^{-1}).

Accounting for the changes in kinetic energy of the flywheel becomes important when the instantaneous power generated by the cyclist is of interest, since it takes energy to accelerate the flywheel. However, during steady state pedaling conditions, when the average pedaling rate remains nearly constant and the average power over the crank cycle is the only quantity of interest, then the kinetic energy term is negligible since its average goes to zero (average $\alpha_{flywheel} = 0$).

Road Cycling

During road cycling, external power is equal to the rate of doing work to overcome forces from aerodynamic drag, rolling resistance, gravity, and inertia. Each of these quantities is elaborated on below. The following equation can be used to estimate the total external power:

$$P_{external} = \frac{1}{2} C_d \cdot A \cdot \rho \cdot (v - v_{air})^2 \cdot v + M \cdot g \qquad 8.3$$
$$\cdot C_r \cdot \cos\phi \cdot v + M \cdot g \cdot \sin\phi$$
$$\cdot v + M \cdot a \cdot v + I \cdot \alpha \cdot \omega$$

where

C_d = coefficient of drag (air resistance),

C_r = coefficient of drag (rolling resistance),

A = frontal area (m²),

ρ = air density (kg · m⁻³),

v = velocity of forward motion relative to ground (m · s⁻¹),

a = acceleration of forward motion relative to ground (m · s⁻²),

I = moment of inertia of rotating parts (kg · m⁻²),

v_{air} = air velocity (m · s⁻¹)

α = angular acceleration of rotating parts (rad · s⁻²),

ω = angular velocity of rotating parts (rad · s⁻¹),

M = combined bicycle-rider mass (kg),

g = acceleration due to gravity (m · s⁻²), and

ϕ = angle of hill relative to horizontal (degrees).

Aerodynamic Drag

The first term in equation 8.3 quantifies the power required to overcome the aerodynamic drag. Aerodynamic drag increases with the square of velocity, and therefore, power increases with the cube of velocity. If the speed of the cyclist is doubled, the power required to maintain that speed is increased by eight times. A small increase in speed requires a very large increase in mechanical power output. Therefore, aerodynamic drag dominates the power equation at higher speeds and is a primary concern for competitive cyclists.

One factor that can greatly reduce the aerodynamic drag is body position. The area presented to the direction of travel (frontal surface area) is the greatest contributor to drag at any speed, and the magnitude of the effect of changing body position is progressively increased as speed increases (see table 8.1). Typical frontal areas range from 0.50 m² in an upright position to 0.30 m² in a streamlined racing position, which can result in a 30% overall reduction in drag force (Faria and Cavanagh 1978).

The coefficient of drag C_d is also very sensitive to the body position. It, or rather the product of C_d and the frontal area can be obtained from wind tunnel

Table 8.1 External Mechanical Power (Watts) of Cycling

	Speed (km/hr)				
	10	20	30	40	50
Standard setup	13	57	162	358	677
Position (A = 0.30 m²)	11	40	107	228	422
Racing tire (C_r = 0.0022)	10	50	151	344	660
Altitude (2500 m, ρ = 0.957 kg · m⁻³)	12	48	132	288	540
Mass (M = 85 kg)	14	59	165	362	683
Hill (3% grade)	74	179	345	603	983

Note: Comparison of the influence of various factors on the total mechanical power (watts) required to maintain a given speed (equation 8.3). The standard setup was specified with the rider in an upright position (A = 0.50 m²) with coefficient of drag (C_d = 0.78), using a standard road tire (C_r = 0.0039), at sea level (ρ = 1.22 kg · m⁻³) with a bicycle-rider mass (M = 75 kg) on a flat surface with no wind (v_{air} = 0). Each row below the standard setup presents the external power required when only that parameter identified on the left has been changed from the standard setup.

tests (Kyle 1990). Estimates for C_d can also be obtained by measuring the deceleration during a coast-down test (de Groot, Sargeant, and Geysel 1995). But this test protocol is more difficult to control because of influences such as wind and rolling resistance, which is sensitive to surface conditions. Typical values range from 0.78 for a cyclist on a standard road bicycle in a racing position to 0.10 for a bicycle with a full fairing (Kyle 1994). The coefficient of drag C_d is also sensitive to the clothing worn and can be reduced by wearing smooth tight-fitting clothing, which can result in drag reductions of up to 10% (Kyle 1986b).

Aerodynamic drag is also a function of air density, which is proportional to barometric pressure and decreases with altitude. Therefore, at higher altitudes less drag occurs, and less mechanical power output is required to maintain a given speed (see table 8.1). But there is a trade-off. At higher altitudes there is less oxygen available to the rider for aerobic metabolism, so the benefits gained in the reduced air drag are often lost in reduced capacity for aerobic metabolism. For cycling events that are short and primarily anaerobic in nature, such as many track events, cycling at higher altitude has a clear advantage since anaerobic power is not affected by altitude.

Further refinements in aerodynamic drag can be obtained by equipment modifications. These modifications include the following: smaller front wheel, disk wheels, bladed spokes, and elliptical tubing (Dal Monte et al. 1987, Kyle 1988). McCole and coworkers (1990) found that a 7% reduction in oxygen uptake at 40 km/hr could be obtained with an aerodynamic bike in contrast with the subject's own bike. This aerodynamic bike included the following: cow-horn handlebars, down-sloping top tube, 24-inch front wheel, and a rear disk wheel.

The geometry of a racing bicycle is relatively constrained by the rules of the governing bodies. For example, the bicycle seat must be positioned over the crank. This rule prevents the use of recumbent bicycles in conventional racing. Recumbent bicycles, by positioning the seat behind the cranks (body positioned more horizontally), reduce the frontal surface area and therefore air resistance. Fairings are also disallowed by rules of conventional racing. However, the addition of fairings (shell) to a recumbent bicycle creates a vehicle that combines two large advantages in reducing air resistance and therefore increases the speed that can be achieved at a given mechanical power output. These advances in engineering have permitted large improvements in the land speed record for human-powered vehicles (Abbott and Wilson 1995).

The extension of the handlebars forward from the conventional position, along with the provision of elbow rests provides another aerodynamic advantage. These handlebars permit the rider to assume a more aerodynamic position while assuring some level of comfort (reduced effort in supporting the upper body). Clear performance advantages have been obtained with aero handlebars. However, there appears to be a trade-off. Gnehm and coworkers (1997) have shown that the metabolic cost of riding in the aero position at a given mechanical power output is significantly greater than riding more upright with hands on the drop handlebars. The magnitude of the difference is considerably less than the magnitude of advantage obtained by the revised body position, with the net effect still favoring the aero handlebars.

The one-hour cycling record has seen remarkable improvement over the past few years. Much of this improvement can be attributed to modifications in equipment and body position, which have reduced the drag associated with air resistance.

Rolling Resistance

The second term in equation 8.3 quantifies the power required to overcome the rolling resistance. Rolling resistance is proportional to the combined mass of the bicycle and rider and the coefficient of rolling resistance C_r. The coefficient of rolling resistance C_r is bicycle tire- and road surface-specific and is the result of energy dissipated to heat during the interaction between the tire and the road. Similar to C_d, estimates for C_r can be obtained by quantifying the deceleration during coast-down tests (de Groot, Sargeant, and Geysel 1995), but as mentioned previously, the test protocol is difficult to control. A number of factors affect C_r, including surface roughness and tire diameter, material, design, and pressure. Typical values range from 0.0039 for a standard 700 cc wheel to 0.0022 for a high-pressure 700 cc racing wheel (Kyle 1986a). At lower speeds (< 20 km/hr) the rolling resistance dominates equation 8.3; at higher speeds the power to overcome rolling resistance only increases in proportion to v, while the power to overcome the aerodynamic drag increases in proportion to velocity cubed.

Gravity

The third term in equation 8.3 quantifies the power required to perform work against gravity while climbing hills. The magnitude of this effect is illustrated in table 8.1. A slope of 3% relative to horizontal increases the power required to maintain a velocity of 30 km/hr by 113. This power is directly proportional to the mass of the bicycle-rider system. Although an increased mass can penalize the rider on hills, the increase may benefit the rider if the increase is muscle mass and thus allows the rider to produce (or sustain) a higher power output.

Inertia

The last two terms in equation 8.3 represent the mechanical power output required to change the momentum of the bicycle-rider system. The first term quantifies the power required to change the linear momentum of the bicycle-rider mass, while the second term quantifies the power required to change the angular momentum of the rotating parts. The rotating parts include both wheels and the crank arm mechanism. From a practical point of view, the crank arm mechanism contributes very little to the required power output because of its small radius compared to the wheels and can therefore be neglected.

Similar to the ergometer, if the quantity of interest is average power over time during steady state pedaling conditions, then the inertial terms are negligible since the accelerations are small and average to zero. However, if the pedaling conditions are not steady state or if instantaneous power is of interest, then the inertial terms are important and must be included.

Kinematic Methods

A second approach to quantify mechanical energy, called *kinematic methods*, is based on the theory that changes in kinetic and potential energy require mechanical work (the time integral of mechanical power). Therefore, muscles must produce mechanical work to overcome both the external power requirement at the crank and the acceleration and deceleration of the body segments. Computationally, the kinematic approach calculates mechanical work based on changes in kinetic and potential energies of the body center of mass and the individual body segments recorded from external kinematic data. The instantaneous energy of an n-segment body is equal to (Winter 1979)

$$E_B(t) = \sum_{i=1}^{n} PE\,(i,\,t) \; + \; \sum_{i=1}^{n} TKE\,(i,\,t)$$

$$+ \; \sum_{i=1}^{n} RKE\,(i,\,t) \qquad\qquad 8.4$$

where

$E_B(t)$ = instantaneous energy of an n-segment body at time t,

$PE(i,t)$ = potential energy of ith segment at time t,

$TKE(i,t)$ = translational kinetic energy of ith segment at time t, and

$RKE(i,t)$ = rotational kinetic energy of ith segment at time t.

The changes in the energy of the system are then used to estimate the mechanical power and when summed over the entire crank cycle yield the mechanical work. This definition of mechanical work was termed "internal," that is, the apparent energy required to move the segments (Winter 1979):

$$W_{Internal} = \int_{t_0}^{t_f} \left| \frac{dE_B(t_f)}{dt} \right| dt \qquad\qquad 8.5$$

where t_f is the duration of the movement.

When significant external work against the environment exists (e.g., during cycling), the amount of external work is added to the kinematic estimate to determine the total mechanical work for one pedal cycle in the form of (Winter 1979)

$$W_{total} = \int_{t_0}^{t_f} \left| P_{external} \right| dt \; + \; W_{Internal}. \qquad\qquad 8.6$$

This type of analysis has been applied to walking (Wells and Evans 1987), running (Cavagna and Kaneko 1977), and cycling (Widrick, Freedson, and Hamill 1992).

The kinematic method explicitly assumes that energy is freely exchanged among all three forms of energy (PE, TKE, RKE) and among all body segments. Realistically, however, possibilities for such transfers are limited and can only be accounted for by grouping the segments and calculating the contributions of the individual groups separately. Studies have shown that the results of these kinematic methods are very sensitive to assumptions regarding such transfers of energy (Wells 1988; Williams and Cavanagh 1983). Consequently, extensive sensitivity analysis of these assumptions and comparison with metabolic measures (Cavagna and Kaneko 1977) are required.

Kinetic Methods

A third approach, called *kinetic methods*, is based on inverse dynamics analysis and uses the work performed by hypothetical joint torque actuators as a measure of the energetic cost of movement. This approach has recently been applied to cycling (Ingen Schenau et al. 1990; Kautz, Hull, and Neptune 1994) to assess the muscular mechanical energy expenditure MMEE, which represents the sum of the absolute values of the work done by the n intersegmental joint torque actuators as

$$\text{MMEE} = \int_{t_0}^{t_f} \left(\sum_{j=1}^{n} \left| P_j \right| \right) dt. \qquad 8.7$$

Although the kinetic and kinematic methods have been applied extensively throughout the literature, they are not without limitations. A recent study by Neptune and Bogert (1998) using forward dynamic simulations illustrated that these methods can greatly underestimate the MMEE of the muscle fibers. Since the kinetic and kinematic methods rely on net joint moments and segment kinematics, respectively, the amount of muscle cocontractions cannot be uniquely quantified. Cocontractions are inevitable in fast human movements due to the activation dynamics associated with muscle force development (i.e., the delay in muscle force rise and decay) and the need for movement control (e.g., to prevent knee hyperextension).

Several studies using the kinetic method have attempted to identify energy savings by accounting for possible energy transfers between joints via biarticular muscles (Broker and Gregor 1994; Kautz, Hull, and Neptune 1994). This situation can arise when mechanical energy from one joint where energy is being absorbed (negative work) is transferred to an adjacent joint where energy is being generated (positive work) via biarticular muscles. Although these MMEE models have sound theoretical foundations, the results of Neptune and Bogert (1998) indicate that these reductions in MMEE may introduce even greater (up to 40%) errors in the analysis.

These errors indicate the inability of MMEE methods based on external measurements to explicitly quantify the energy transfers between segments and to account for cocontractions of antagonistic muscle groups. These external methods are also limited by their inability to quantify the contribution of elastic energy storage to positive or negative work, the energy cost of isometric contractions, and to explicitly determine how much power is required to accelerate and decelerate the segments. These limitations have led to significant errors in the computation of mechanical efficiencies. Williams and Cavanagh (1983) identified potential errors giving efficiencies ranging from 31 to 197%, which is greater than the estimated maximal efficiency of muscle, which, as outlined below, is approximately 30%. Efficiency of exercise is often considerably less than 30%.

Efficiency of energy transfer from chemical energy within the body to mechanical energy associated with cycling or cycle ergometry is an important link in understanding the relationship between effort necessary to accomplish a task and the apparent work done in completing the task. In the following sections, the metabolic systems that provide (transfer) this chemical energy are described, and factors that determine the energy cost of a muscle contraction are discussed.

Metabolic Energy Systems in the Body

The muscles of the body generate tension and perform work by converting chemical energy into functional contraction, which can result in increased kinetic energy and, in some cases, increased potential energy. These energy systems have been called the immediate, short term, and long term energy systems. The immediate system includes the hydrolysis of ATP to adenosine diphosphate (ADP), the chemical reaction that provides energy for virtually all processes in the body that require energy, followed by replenishment of ATP by transfer of the phosphate group of creatine phosphate (CP) to ADP, resulting in increased cytoplasmic creatine concentration.

The short term system, also called substrate phosphorylation, involves glycolysis, which is the intermediate breakdown of glucose and glycogen to pyruvic acid. Glycolysis is activated by elevated free cytoplasmic Ca^{2+} and increasing ADP concentration. Adrenergic activation of adenyl cyclase will also accelerate glycolysis via increased glycogenolysis. Subsequent conversion of pyruvic acid to lactic acid permits glycolysis to continue in the cytoplasm by providing necessary substrate (NAD^+) for certain key steps in glycolysis (see figure 8.2).

The long term metabolic system is aerobic metabolism; it includes the Krebs cycle and the electron transfer chain, which is coupled with oxidative phosphorylation. The substrates for oxidative metabolism include acetyl CoA from pyruvic acid and β-oxidation of fatty acids, O_2, ADP, and NADH (which is provided for the electron transport chain by either the Krebs cycle within the mitochondria or via glycolysis, which requires transport of reducing equivalents across the mitochondrial membrane). A rising concentration of creatine in the cytoplasm is one of the key features activating oxidative metabolism (Brooks, Fahey, and White 1996). The three energy systems are closely linked, as illustrated in figure 8.2, with all systems serving to maintain ATP concentration as nearly constant as possible.

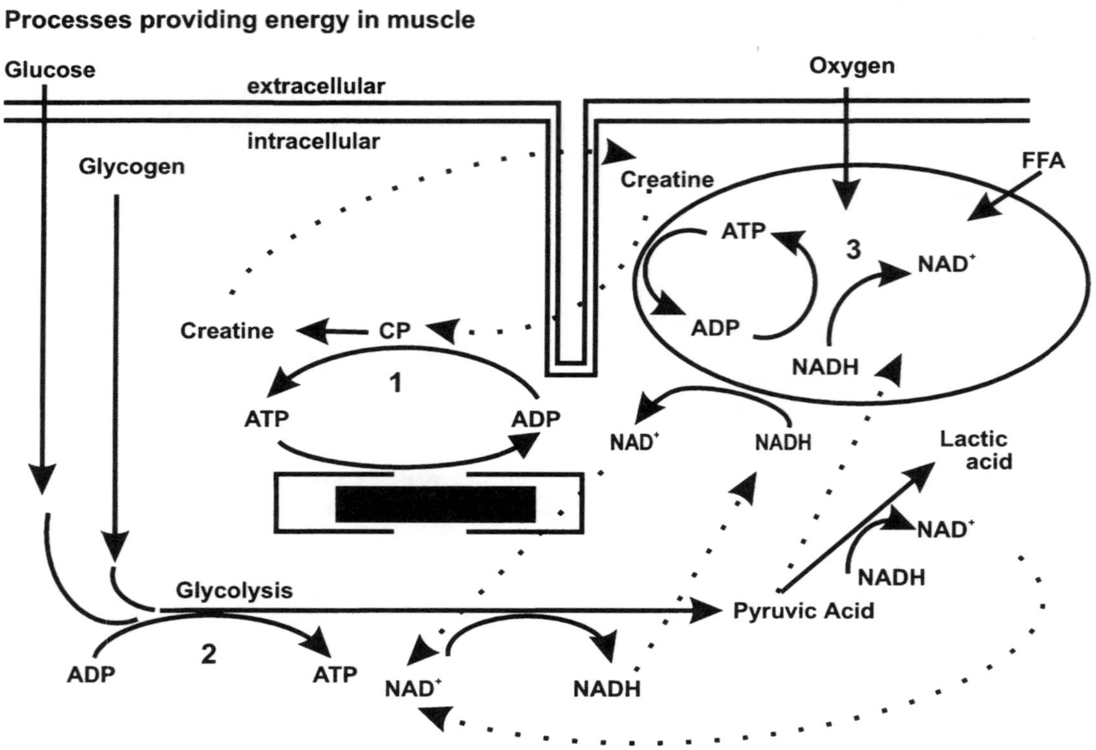

Processes providing energy in muscle

Figure 8.2 Schematic presentation of the metabolic systems of a cell. The double line represents the sarcolemma, including the transverse tubule. Three systems that provide energy are illustrated: (1) immediate (ATP-CP), (2) short term (glycolysis), and (3) long term (aerobic). Dotted lines illustrate interactions among the three systems.

Determinants of Muscle Energetics

The rate at which ATP is hydrolyzed during exercise is primarily determined by what goes on within the muscles that are activated for performance of the task. The primary uses of ATP in contracting muscles include ion pumping and actin-myosin cross-bridge interaction (see figure 8.3). Therefore, the determinants of the energy cost of exercise are dependent on the determinants of ATP splitting by actomyosin ATPase, sarcoplasmic reticulum Ca^{2+} ATPase, and Na^+-K^+ ATPase. Of these three, the Na^+-K^+ ATPase is relatively minor and will not be given further consideration here.

The energy cost of muscle contraction has been determined by heat measurement (Bers, Philipson, and Langer 1981; Hill 1938; Saugen and Vollestad 1996), measurement of oxygen uptake (Stainsby 1970; Stainsby and Barclay 1971; Vollestad, Wesche, and Sejersted 1990; Wilson and Stainsby 1978), or by determination of ATP and CP hydrolysis in muscles poisoned to prevent glycolysis and oxidative replenishment of ATP (Crow and Kushmerick 1982; Foley, Harkema, and Meyer 1991). Collectively these

studies have provided a reasonable picture of the energy cost of muscle contraction, though there is not universal agreement on the importance of the individual determinants.

Activation of a muscle cell is achieved by release of Ca^{2+} into the sarcoplasm, presumably a passive process. Relaxation occurs when cytoplasmic Ca^{2+} concentration is restored to resting values, a process that requires active transport of Ca^{2+} across the membranes of the longitudinal sarcoplasmic reticulum. It is thought that two Ca^{2+} ions are transported for each ATP hydrolyzed (Rüegg 1986), so the total energy required for Ca^{2+} handling depends on the amount of Ca^{2+} released per activation and the total number of activations associated with a contraction.

The rate of use of ATP by actomyosin ATPase is a function of the number of active cross bridges and the rate of turnover of individual cross bridges (Stainsby and Lambert 1979). The number of cross bridges activated is dependent on the mass of muscle that is active, and the intensity of activation within each motor unit, as well as relative muscle length. Intensity of activation is dependent on motor unit recruitment and the force-frequency relation. The

Energy Requiring Processes in Muscle

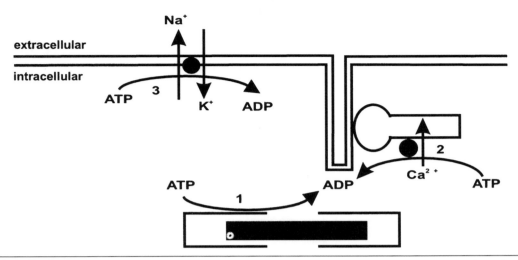

Figure 8.3 Processes in muscle that require energy. The primary energy-requiring process in muscle is the actomyosin ATPase (1); ion pumps, including sarcoplasmic reticulum Ca^{2+} ATPase (2); and sarcolemmal NA^+-K^+ ATPase (3) also require energy.

cross-bridge turnover rate is dependent on myosin isozyme (fiber type) composition and can apparently be modified by shortening, such that shortening is associated with a more rapid turnover of cross bridges. This concept is complicated by the apparent decrease in the number of participating cross bridges when velocity is particularly high (Stainsby and Lambert 1979). The net effect of these two processes (change in length and rate of change in length) is that shortening at slow velocities can result in elevated energy cost, but at rapid velocities there is less energy use per activation.

When a muscle contracts and shortens against a resistance, work is done, but the magnitude of work done is not a direct determinant of the energy cost of a muscle contraction. This is evident from figure 8.4 (top), which illustrates the relationship between load and oxygen uptake, and figure 8.4 (bottom), which shows the corresponding relationship between load and work (per contraction). There is no simple relationship between these two curves. When $\dot{V}O_2$ is plotted against work, it is clear that the oxygen uptake for a given amount of work depends on whether the work was accomplished with a high load (small amount of shortening) or with a low load (large amount of shortening) (see figure 8.5).

Work may not be a direct determinant of the energy cost of muscle contraction, but efficiency (mechanical work/energy in; power/rate of energy use) is a concept that has been given considerable attention in the area of muscle energetics. As indicated previously, velocity of contraction is an

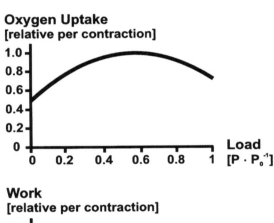

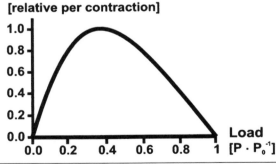

Figure 8.4 The relationship between load and oxygen uptake (top) and between load and work (bottom) are illustrated. The range of values that is observed for oxygen uptake is relatively small in comparison with the range of values for work.

Adapted from W.N. Stainsby and J.K. Barclay, 1972, "Oxygen uptake for brief tetanic contractions of dog skeletal muscle in situ," *American Journal of Physiology* 223:371-375. Similar to W.N. Stainsby, 1982, "Energetic patterns of normally circulated mammalian muscle in situ," *Federation Proceedings* 41:185-188.

Oxygen Uptake
[relative]

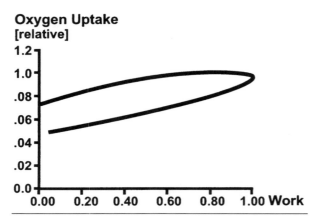

Figure 8.5 The relationship between work and oxygen uptake for mammalian skeletal muscle. The energy required to achieve a given amount of work depends on whether the work is accomplished at a low load or a high load.

Adapted, by permission, from W.N. Stainsby and J.K. Barclay, 1980, "Oxygen uptake for brief tetanic contractions of dog skeletal muscle in situ," *Journal of Applied Physiology* 48(3): 518-522.

important determinant of the energy cost of contraction. Considering the force-velocity properties of muscle, velocity is also an important determinant of the power output that can be achieved with a given contraction. It is known that power output increases as velocity of contraction increases to a peak power output at some optimal velocity and that power output decreases beyond this optimal velocity (see chapter 12). The most work achieved per unit of energy input (highest efficiency) is actually accomplished when the velocity of contraction is at or slightly less than the optimal velocity for power output (Barclay 1996). The optimal velocity for power output (see chapter 12) and therefore the velocity at which efficiency is highest is dependent on the fiber type composition of the muscle (or muscles) under question. Therefore, the efficiency of muscle contraction while shortening at a given velocity is dependent on the fiber type composition of the (active) muscle.

Muscles of the human body use chemical energy to generate tension and accomplish motion. The chemical energy is typically from carbohydrate and lipid sources. Oxidation of these substrates in the Krebs cycle and electron transfer chain results in conversion of ADP to ATP by oxidative phosphorylation. The efficiency of this chemical conversion is approximately 60% (Stainsby and Lambert 1979). Therefore, 40% of the chemical energy available in carbohydrate and lipid is lost as heat, and the rest is conserved as chemical energy in the bond with the terminal phosphate of ATP. This chemical energy is

released in muscle contraction, resulting in accomplishment of work and release of additional heat. This additional heat can account for at least 50% of the free energy of hydrolysis of ATP. Therefore, considering carbohydrate and lipid as the substrates, the highest efficiency that can be achieved by muscles during performance of work is approximately 30%. This estimate includes the energy required for ion pumping, which can account for 10 to 40% of the total energy used in a muscle contraction (Baker et al. 1994; Barclay, Curtin, and Woledge 1993).

The 30% estimate for maximal efficiency is conservatively high. Under conditions of muscle contraction without work accomplished, the efficiency is zero (see figure 8.6). Considering the conditions that affect energy cost of muscle contraction, and the determinants of work per contraction, it can be demonstrated that efficiency is optimal (~ 30%) at conditions near the optimal velocity for power output. Variance from this condition results in more of the energy being dissipated as heat (lower efficiency). Muscle shortening velocity is not constant during cycling, so efficiency cannot be its maximal value. Therefore, it is not surprising that the gross efficiency of cycle ergometry ranges from 15 to 25%, showing a large variation and an upper limit that is less than 30%.

Indirect Calorimetry

Under conditions when the rate of ATP use does not exceed the rate at which it is replenished by oxidative metabolism, measurement of oxygen uptake

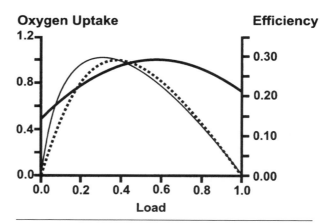

Figure 8.6 Load versus oxygen uptake (thick solid line), work (dashed line), and efficiency (thin solid line). Efficiency was estimated from the ratio of work to $\dot{V}O_2$, normalized to a maximal value of 0.3. The efficiency is dependent on whether or not work is accomplished during a contraction.

can be used to quantify the rate of energy use in muscular exercise. The amount of energy made available for each liter of oxygen used depends on the substrate metabolized. The amount of CO_2 released per liter of O_2 consumed also varies with the substrate metabolized. When it is assumed that carbohydrate and lipid are the only substrates, then the respiratory quotient RQ, which is the ratio of carbon dioxide released to oxygen consumed, can be used to estimate the relative proportions of carbohydrate and lipid that were oxidized to yield ATP for the exercise under question. Measurement of pulmonary exchange of O_2 and CO_2 (respiratory exchange ratio R) provides an estimate of RQ when storage amounts of these gases are not changing. Therefore measurement of oxygen uptake and carbon dioxide production can be used to estimate the metabolic energy cost of exercise when that energy cost does not exceed the rate of replenishment of ATP by oxidative metabolism.

In many cases, oxygen uptake is measured and the energy equivalent is not calculated. Under these circumstances it is assumed that the oxygen uptake is proportional to the energy used during the exercise. In most cases the error associated with this assumption is small. In moderate to high intensity exercise the RQ will vary from 0.85 to 1.0, and this range is associated with 4.862 to 5.047 kcal per liter of O_2 (McArdle, Katch, and Katch 1996). Assuming about 5 kcal per L of O_2 will result in an error in the estimate of energy use ranging up to 2.8%.

Steady State Energetics

When an individual begins to exercise on a cycle ergometer at a cadence and resistance that is compatible with a low intensity, ATP is used in muscle contraction, and it is replaced by hydrolysis of CP and by glycolysis. Increased concentration of creatine, NADH, and pyruvic acid result (see figure 8.2). These products of the immediate and short term energy systems stimulate aerobic metabolism. Therefore, there is a progressive rise in the rate of oxidative metabolism. Oxygen uptake will rise from the resting level within 2 to 4 min to a rate that will sustain replenishment of ATP exclusively by oxidative metabolism.

The magnitude of the rise in oxygen uptake depends on the change in the rate at which ATP is hydrolyzed. This in turn depends on the collective mass of muscles involved in the exercise and the factors presented above concerning the determinants of energy cost of muscle contractions. This includes the number of activations per contraction,

number of motor units activated, and frequency of contractions, as well as the length change and velocity within each contraction and the muscle fiber types involved.

In cycling, these factors are relatively constrained. The geometry of the bike, or cycle ergometer, and crank will dictate the excursion required from individual muscles, and cadence will dictate the pattern of velocity of shortening of a given muscle. The resistance imposed on the flywheel will dictate the number of motor units required (according to their force-velocity properties) to maintain the cadence. Considering that mechanical power output is the product of resistance and velocity of a point on the periphery of the flywheel (proportional to cadence), it is not surprising that oxygen uptake is proportional to power output in cycle ergometry. To increase power output at a given cadence, resistance is increased. This requires recruitment of additional motor units and therefore a higher rate of energy use.

The relatively strong relationship between power output and oxygen uptake has been used by the American College of Sports Medicine (1991) to allow prediction of the oxygen uptake (L/min) when only the power output in watts is known:

$$\dot{V}O_2 = \text{power (watts)} \cdot 12.27 \text{ ml} \cdot \text{watt}^{-1} \qquad 8.8$$
$$+ (3.5 \text{ ml} \cdot \text{kg}^{-1} \cdot \text{kg body wt}).$$

The value of such a prediction is that measurement of power output and heart rate during submaximal exercise can be used to predict maximal oxygen uptake. The fact that such a prediction can be made suggests a constant relationship between oxygen uptake and power output, for cycling. However, there is considerable error associated with this prediction and the equation is only of use for intensities up to that equated with maximal oxygen uptake. The equation has been recently modified by Lang and associates (1992):

$$\dot{V}O_2 = \text{power (watts)} \cdot 11.66 \text{ ml} \cdot \text{watt}^{-1} \qquad 8.9$$
$$+ (3.5 \text{ ml} \cdot \text{kg}^{-1} \cdot \text{min}^{-1} \cdot \text{kg body wt})$$
$$+ 260 \text{ ml} \cdot \text{min}^{-1}.$$

The revised equation greatly reduced the average error of prediction evident with the original equation for the subjects of the latter study. Lang and coworkers (1992) observed that the ACSM equation underestimated the oxygen uptake of a group of 50 to 60 subjects by 0.16 to 0.29 L $\cdot$ min^{-1}. The revised equation provides a better estimate of the average for this group (error = 0 to 0.05 L $\cdot$ min^{-1}), but the standard error of the estimate was still quite large (0.1 to 0.2 L $\cdot$ min^{-1}). It is clear from the relatively low correlation coefficients ($r = .32$ to $.54$) that considerable scatter of the data persists, indicating that several factors appear to influence the oxygen uptake of cycle ergometry over and above the mechanical power output. Some of these factors are reviewed below.

The use of this relationship (equation 8.8 or 8.9) has an underlying assumption that all subjects have the same efficiency. This assumption, of course, has some utility in circumstances where an approximation is suitable, but it should be recognized as a limitation of the method. Efficiency of cycling varies from 19 to 25% among highly trained cyclists (Coyle et al. 1992). The range is greater when less-trained individuals are considered. This equation also assumes that the relationship between $\dot{V}O_2$ and power output is linear, which may not be the case. It has been demonstrated that, at power outputs above the anaerobic threshold, the actual increase in $\dot{V}O_2$ for a given increase in power output is greater than that predicted by extrapolation of the linear (subthreshold) relationship (Zoladz, Rademaker, and Sargeant 1995).

Determinants of Steady State Energy Cost of Cycle Ergometry

As indicated above, the primary determinant of the energy cost of exercise on a cycle ergometer is the mechanical power output that is performed against the cranks and transmitted to the flywheel. There is a relatively linear relationship between power output and energy cost when cadence is kept constant and resistance is increased to require additional power output. However, there are several factors that can affect this relationship. The impact of these factors that influence efficiency are described below.

Cadence and Resistance

Power output is the product of resistance and the linear velocity of a point of application of the resistive force to the flywheel. The distance traveled by this point is a fixed amount per pedal revolution (6 m per pedal revolution on a standard Monark ergometer) so cadence and resistance are the primary determinants of power output in cycle ergometry. Accordingly, there are several combinations of cadence and resistance that can yield a given power output (see figure 8.7). These various combinations result in consistent variations in the energy cost of cycle ergometry.

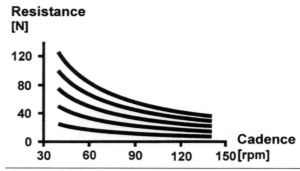

Figure 8.7 Relationship between cadence and resistance showing isopower lines.

Cadence is known to affect the energy cost of cycling (Coast, Cox, and Welch 1986; Londeree et al. 1997; Marsh and Martin 1997; Seabury, Adams, and Ramey 1977). At a given power output, oxygen uptake will have a minimum value at some cadence and will be higher as cadence is changed from this optimal value. That is, there is a unique cadence at which metabolic efficiency is optimized, and when cadence is changed, efficiency decreases. As power output increases, the optimal cadence becomes higher (Seabury, Adams, and Ramey 1977).

The effect of cadence on the energy cost of cycling can be tentatively explained by the determinants of energy cost of muscle contractions. Consider that a cadence of 50 rpm is optimal for a power output of 150 W (watts). Increasing or decreasing cadence while adjusting resistance to maintain this power output will increase oxygen uptake. It is reasonable to assume, therefore, that the velocity of shortening of the active motor units is optimal (highest efficiency) at this cadence. If cadence is increased and resistance decreased to keep power output constant, then the increased velocity of shortening probably results in increased cross-bridge turnover and hence higher demand for ATP.

Assuming a single muscle is called upon to generate a given power output, some fraction of the available motor units would be required. Figure 8.8 (top) illustrates the force-velocity properties that would be expected for progressively increasing recruitment of motor units. Note the assumed size principle, in that the added motor units result in activation of muscle fibers with faster maximal velocity. The corresponding power-velocity relationship is illustrated in figure 8.8 (bottom). As faster motor units are recruited, the optimal velocity shifts to faster velocities. There are two relevant points to obtain from these graphs. In order to achieve a given power output, the appropriate number of motor units must be recruited. This number must

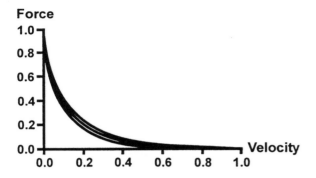

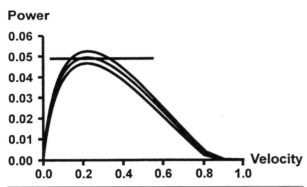

Figure 8.8 Force-velocity and corresponding power-velocity relationships for three submaximal levels of motor unit recruitment. In the top example, all slow twitch motor units have initially been recruited (bottom line), and additional motor units recruited are from the fast twitch pool (upper two lines). Figure 8.7 illustrates that to maintain a given power output at a higher velocity requires lower resistance (and presumably less force). If the decrease in force capability of the active motor units illustrated in the top graph is less than the decrease in the required force, then it becomes necessary to derecruit motor units to maintain the power output at the new (higher) cadence. The horizontal line in the bottom graph illustrates a constant power output ($\approx$0.048). Increasing the velocity from 0.17 to 0.23 can be achieved with fewer motor units recruited (middle line). To increase the velocity further would require recruitment of additional motor units. See chapter 11 for further elaboration.

be adjusted in order to achieve that power output when velocity is changed. When the velocity of contraction corresponds with (or is just less than) optimal velocity for power output, efficiency will be highest. It can be seen that to achieve a higher power output at a given velocity more motor units must be recruited, and when more (faster) motor units are recruited, the velocity associated with the highest efficiency also shifts to higher velocities.

This analogy of a single muscle contracting at various velocities may be an oversimplification of the situation in cycling where several muscles are involved, each with its own (varying) velocity of contraction. However, it has been demonstrated that peak power output has a typical force-velocity relationship (Buttelli et al. 1996; McCartney et al. 1983; Vandewalle et al. 1987), indicating that coordinated activation of several muscles in a multijoint movement follows a force-velocity relationship that is similar to that of a single muscle.

Coordination of Muscular Effort

Considering the factors that determine the energy cost of muscle contractions, it seems logical that inappropriate activation of agonist/antagonist muscles would result in unnecessary metabolic cost of exercise. Exercise on a cycle ergometer requires alternate activation of knee extensors and knee flexors. Any muscle contraction that does not contribute to performance of the task will result in increased energy requirement of the exercise. Similarly, inappropriate activation of muscles will result in greater activation and therefore a higher energy cost than would otherwise be the case. This consideration does not rule out the possible benefit of cocontraction of agonist/antagonist pairs. If knee extensor muscles are activated while flexor muscles are still active, the extensor will be actively stretched prior to the shortening associated with knee extension. It has been demonstrated that eccentric contraction prior to concentric contraction enhances the force that can be generated by a muscle (see chapter 20). If the enhancement results in the requirement of fewer motor units to accomplish the task, then less energy may be required.

Muscle Fiber Types

The efficiency of exercise is affected by the velocity of muscle contraction. This property of exercise is more dependent on the force-velocity properties of muscles than on the energetics of contraction. As was seen in the section on determinants of muscle energetics above, performance of work is a minor determinant of the energy cost of muscle contraction. Therefore, efficiency is largely determined by whether or not the contraction in question results in accomplishment of work. When efficiency of muscle contraction is depicted relative to the force-velocity properties (see figure 8.9), it can be seen that peak efficiency is obtained when the velocity of shortening is just less than the optimal velocity for power output.

Human muscles are composed of two primary fiber types: fast twitch and slow twitch. The subdivisions of the fast twitch fibers do not seriously

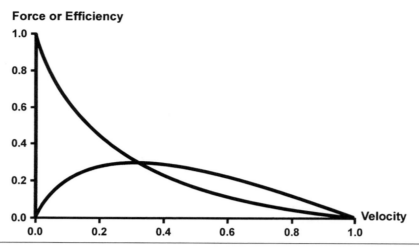

Figure 8.9 Force-velocity and efficiency relationships. Efficiency is a function of velocity, because velocity determines power output. When power is optimized, efficiency is as high as it can be (≈30%).

affect the principles we are concerned with here, so they will not be considered. The force-velocity properties of human fast and slow twitch fibers are considerably different. Therefore, the range of velocities over which efficiency would be expected to be high will be very different for fast and slow twitch fibers. It is not surprising therefore to find that fiber type composition of the principal muscles used in cycling has an effect on the efficiency of cycle ergometry. Most cycle ergometry is done with the subject pedaling at 50 to 90 rpm. Coyle and coworkers (1992) have found that efficiency of cycling at 80 rpm is higher in subjects with more slow twitch muscle fibers in their vastus lateralis muscles. They argue that 80 rpm corresponds with a velocity of shortening that permits a high efficiency of slow twitch fibers and that this cadence would not be efficient for fast twitch fibers.

The above discussion of the metabolic energy cost of cycling has dealt only with steady state conditions. This necessarily limits the application of this information to metabolic rates below anaerobic threshold. What happens when the intensity of exercise is greater than the level that can be satisfied entirely by aerobic energy metabolism?

Slow Component of Oxygen Uptake

When the intensity of cycling exercise dictates that the oxygen uptake exceeds the anaerobic threshold, a steady state is not achieved. Once the rapid kinetic response is complete, the oxygen uptake continues to rise slowly throughout the exercise, even when power output remains constant (see figure 8.10).

This progressive rise in energy requirement is referred to as the slow component of oxygen uptake and is operationally defined as the difference in oxygen uptake from 3 min of exercise to the end (usually 10–20 min). Several factors have been proposed to explain this aspect of the energetics of cycling exercise: lactate metabolism, heat accumulation (Q_{10}), circulating catecholamines, and recruitment of less efficient motor units. Each of these theories is addressed below.

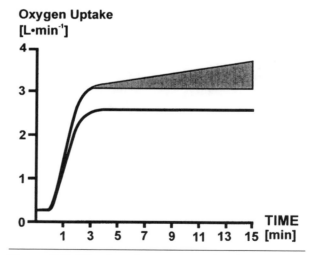

Figure 8.10 Kinetics of oxygen uptake showing two submaximal intensities of exercise: one below anaerobic threshold (lower line) and one above anaerobic threshold. Below the anaerobic threshold, $\dot{V}O_2$ rises to a steady state within 2 to 4 min, and this intensity of exercise can be sustained for a prolonged period of time. Above the anaerobic threshold, $\dot{V}O_2$ does not achieve a steady state. The increment in oxygen uptake beyond 3 min (shaded area) is considered to be the slow component.

Lactate Metabolism

When exercise intensity exceeds the anaerobic threshold, there is a progressive accumulation of lactate in the blood. This increase in blood lactate apparently correlates well with the slow component of oxygen uptake. It has been proposed that the energy cost of disposing of the lactate contributes to the slow component of oxygen uptake. There are two assumptions involved in this theory. The first is that there is a progressive increase in the rate at which lactate is disposed of (at a constant power output), and the second assumption is that disposal of lactate requires energy. However, neither of these assumptions is true.

When the intensity of exercise is below the anaerobic threshold, any net production of lactic acid in some muscles is matched by lactic acid uptake and oxidation in other muscle or nonmuscle tissue. Under these circumstances there may be considerable disposal of lactic acid, and yet oxygen uptake does achieve a steady state, with no apparent slow component of oxygen uptake.

When the intensity of exercise exceeds the anaerobic threshold two factors contribute to the progressive rise in blood lactate at a constant exercise intensity: more motor units with a net production of lactic acid and fewer motor units taking up lactic acid from the blood. Therefore it could be said that the slow component of oxygen uptake occurs when the rate of disposal of lactic acid is diminished relative to what it would have been, at or below the anaerobic threshold.

The second assumption, that removal of lactate requires energy, is derived from the historic concept of a lactacid oxygen debt, which is referred to as the Hill-Meyerhoff theory (Brooks, Fahey, and White 1996). Many years ago, it was proposed that the oxygen uptake during recovery from exercise was used for disposal of lactate that had accumulated during the exercise. It was demonstrated in amphibian muscle that lactate was converted back to glucose or glycogen during recovery. This resynthesis of glucose or glycogen certainly requires energy in the form of ATP, which would be replenished by oxidative metabolism, thereby requiring extra oxygen uptake during a recovery period. Recent research has shown, however, that lactate produced in some muscles during exercise is oxidized in tissues that are not net producers of lactate (Donovan and Brooks 1983). This includes predominantly the heart and inactive or mildly active skeletal muscles. Such metabolism of lactate by oxidative metabolism results in rephosphorylation of ADP to ATP. In other words, the lactate merely becomes a substrate for oxidative metabolism, providing energy for cellular processes that require it.

Infusion of lactate (iso-pH) into a stimulated canine gastrocnemius-plantaris muscle preparation has confirmed that extra oxygen uptake is not required to dispose of the lactate (Poole et al. 1994). No increase in oxygen uptake occurred when muscle lactate was increased by this intraarterial infusion. Therefore, energy use for removal of lactate has not been demonstrated in mammalian muscle and cannot account for the slow component of oxygen uptake.

Heat Accumulation

During cycle ergometry, as in most exercises that are sustained for a period of time, the heat production that results from metabolism and ATP hydrolysis will result in an increase of body temperature. It is known that higher body temperatures are associated with accelerated chemical reactions (Q_{10}) and therefore possibly greater energy requirements. During high intensity exercise, body temperature will rise from 37° C to 40° C. According to most assessments of the Q_{10} effect on metabolic rate, this could contribute to an increase in metabolic rate of 10 to 30% (assuming $Q_{10} = 2.5$). However, it is recognized that temperature can increase during exercise in the absence of increases in metabolic rate (Poole et al. 1991). It should also be noted that as temperature rises during exercise, the rate of rise decreases until a steady temperature is reached. Once a steady temperature is achieved, then further increases in $\dot{V}O_2$ could not be accounted for by a Q_{10} effect. The discrepancies in the kinetics of temperature change and the slow component of oxygen uptake, as well as the absence of a slow component of $\dot{V}O_2$ when temperature rises below the anaerobic threshold, suggest that heat accumulation does not contribute to the slow component of $\dot{V}O_2$. There is no evidence for the role of a Q_{10} effect contributing to the slow component of oxygen uptake.

Circulating Catecholamines

When exercise intensity exceeds the anaerobic threshold, blood catecholamine concentrations are elevated (Kraemer et al. 1991; Poole et al. 1991). It is known that catecholamines can stimulate resting (Chapler, Stainsby, and Gladden 1980) and recovery metabolic rate (Gladden, Stainsby, and MacIntosh 1982), so it is not unreasonable to expect that the catecholamines are directly responsible for the slow component of $\dot{V}O_2$, which is present during exercise when the intensity exceeds the anaerobic threshold. Womack and coworkers (1995) have observed that aerobic training results in a smaller slow component of oxygen uptake as well as a smaller increase in plasma epinephrine concentration, which is consistent with the hypothesis that catecholamines contribute to the slow component. However, when epinephrine was infused during the latter half of a constant power trial, no increase in $\dot{V}O_2$ was seen (Womack et al. 1995). During the infusion, plasma epinephrine concentration reached a level consistent with a large slow component of oxygen uptake, yet $\dot{V}O_2$ was not affected. This observation provides evidence against the hypothesis that the catecholamines affect the slow component of $\dot{V}O_2$. Incidentally, the infusion of epinephrine also resulted in an increased blood lactate concentration, providing further support for the conclusion that disposal of lactate does not contribute to the slow component of oxygen uptake. Gaesser and associates (1994) have presented similar observations.

Recruitment of Less Efficient Motor Units

It has been confirmed that the slow component of $\dot{V}O_2$ is located primarily in the exercising limbs (Poole et al. 1991). Submaximal exercise is thought to require recruitment of only a portion of the available motor units within the active muscles. When exercise intensity exceeds the anaerobic threshold, fatigue is likely to occur in the active motor units. When the force output of a motor unit becomes smaller, additional motor units must be activated in order to sustain a constant power output. The anaerobic threshold probably occurs at an intensity that requires activation of some fast twitch motor units among the primary muscles involved in the task. Therefore recruitment of additional motor units probably will require recruitment of additional fast twitch motor units (Vollestad and Blom 1985). As more fast twitch motor units become involved in the task, optimal velocity for the highest efficiency will increase. If cadence is not increased, then oxygen uptake must increase in order to sustain a constant power output. It should be pointed out that it is not necessary to assume that fast twitch motor units are less efficient, but merely that they are required to work at a velocity that is not particularly efficient for these motor units.

Supramaximal Energy Use, Anaerobic Metabolism

As the energy cost of exercise approaches maximal oxygen uptake, some of the total energy cost is provided by anaerobic biochemical metabolism. When the energy requirement exceeds the demand that can be met with aerobic metabolism, then all additional energy must come from glycolysis. Considering that the peak rate of energy use in the body can exceed maximal aerobic metabolism by a factor of 4 to 10, this represents a considerable range of the potential metabolic cost of exercise. Accurate quantification of energy use other than that provided by aerobic metabolism is not possible.

Measurement of external power output becomes a reasonable alternative for assessment of intensity of exercise when the intensity exceeds maximal oxygen uptake. The duration of exercise in this range of intensities (from maximal oxygen uptake to peak power output) is relatively brief. Highly trained endurance athletes are thought to be able to sustain maximal oxygen uptake for 5 to 7 min. An untrained individual can perhaps continue exercising at maximal oxygen uptake for 1 to 3 min.

The total amount of energy that can be derived from the immediate and short term energy systems has been called the anaerobic capacity. Various methods have been proposed for the measurement of the anaerobic capacity, but there is no universal agreement on the appropriate test (Bangsbo 1996; Bulbulian, Jeong, and Murphy 1998; Green and Dawson 1995). It is sometimes assumed that the anaerobic capacity represents a fixed amount of energy (for a given individual) that can be obtained from nonaerobic sources (Carnevale and Gaesser 1991; Moritani et al. 1981). Various estimates put this total amount of energy at about 20 kJ (of mechanical work) on a cycle ergometer for an average size adult male. Assuming an efficiency of 20%, this would equate to about 100 kJ of chemical energy, or the equivalent of about 4.8 L of accumulated oxygen deficit.

Measurement of the Energy Cost of Cycling

The intensity of exercise while riding a bicycle can be quantified by measurement of oxygen uptake. Two approaches have been used in this endeavor: riding a bicycle on a treadmill, permitting stationary measurement of oxygen uptake, and collecting expired air in a meteorological balloon while a cyclist rides a bike on the open road. This latter approach is more relevant to the energy cost of cycling, while the former approach is not much different from riding a cycle ergometer, with the exception that the rider is permitted to use his or her own bicycle. More recently, telemetered O_2 uptake systems have become available (K2 and K4 Cosmed), which reduces the difficulty of making measurements of oxygen uptake while cycling (Capelli et al. 1998).

McCole and colleagues (1990) developed a system of gas collection that permitted measurement of oxygen uptake while cyclists rode on the open road. The rider breathed through a three-way valve that directed the exhaled air through a long hose for collection of the exhaled air in a meteorological balloon. The meteorological balloon was carried in a pickup truck that followed the cyclist, thereby avoiding interference with the aerodynamics of the cyclist.

Analysis of the results of over 100 trials conducted on level ground permitted McCole and colleagues (1990) to derive an empirical equation for oxygen uptake, which accounted for over 70% of the variability in this measure:

$$\dot{V}O_2 = -4.5 + 0.17\,V_R + 0.052\,V_W + 0.022\,W_R \quad 8.10$$

where

$\dot{V}O_2$ is expressed in liters per minute,

V_R is the velocity of the rider,

V_W is the velocity of the wind (positive for a head wind), and

W_R is the rider weight in kg.

This relationship is illustrated in figure 8.11 to show the importance of velocity and the impact of wind speed and direction on the energy cost of cycling. McCole and associates (1990) are careful to point out that although greater body mass results in greater energy expenditure, the liability of increased energy cost was not as great as the probable increase in maximal $\dot{V}O_2$ associated with a larger muscle mass. Therefore, larger riders have an advantage over smaller riders on a flat course. This advantage is best illustrated by considering an example.

Superimposed on the data of McCole and associates in figure 8.11 are values for oxygen uptake derived from data presented by Capelli and coworkers (1998). These investigators measured oxygen uptake using a telemetered system (K2 Cosmed) in elite Italian cyclists while they rode at constant speed on a velodrome track. The fact that oxygen uptake is lower for these subjects than for the subjects of the McCole study is interesting, but it is difficult to identify the specific reason for this difference. Possibilities include body position, equipment, and economy of locomotion.

Using the equation presented by McCole (equation 8.10), it can be estimated that riding at 35 km · hr^{-1}

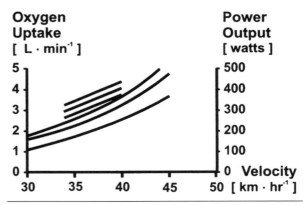

Figure 8.11 The predicted relationship between velocity and oxygen uptake is shown. According to McCole and associates (1990), this relationship can be approximated by a straight line over the range of velocities illustrated (32 -40 km · hr^{-1}). From bottom to top, the three short lines represent (1) control (no wind, 70 kg rider), (2) increased mass (80 kg), and (3) head wind = 10 km · hr^{-1}. The longer upper curved line represents the estimated oxygen uptake for subjects in the study by Capelli and coworkers (1998) who had a bicycle-rider mass of 81 kg. The lowest curved line represents power estimated with equation 8.3, assuming level travel, A = 0.4 m^2, M = 80 kg and other parameters as presented in table 8.1 for the standard condition. The middle longer curved line represents oxygen uptake as estimated for the corresponding power output, using equation 8.8. These oxygen uptakes represent efficiencies ranging from 20 to 22%, which is reasonable for elite cyclists (Coyle et al. 1992). Note that oxygen uptake will underestimate the energy cost of performing exercise when the intensity exceeds the anaerobic threshold.

would require 2.99 L · min^{-1} for a rider weighing 70 kg, and 3.21 L · min^{-1} for a rider weighing 80 kg. Assuming that both riders have a maximal oxygen uptake of 70 ml · kg^{-1} · min^{-1}, this oxygen uptake at 35 km · hr^{-1} represents 61% of $\dot{V}O_2$max for the 70 kg rider and 57% of $\dot{V}O_2$max for the 80 kg rider.

Summary

The intensity-duration relationship is a characteristic property of exercise, which makes measurement of intensity a desirable pursuit. The most appropriate measure of intensity is a measure of the energy input necessary to accomplish the task. This is most easily accomplished by measurement of oxygen uptake, but this approach necessitates that aerobic metabolism account for all of the energy input (i.e., the intensity must be below the anaerobic threshold). When the intensity of exercise exceeds the anaerobic threshold, the intensity of exercise can be estimated by measurement of mechanical power output. In cycle ergometry the power output is determined by the cadence and the resistance applied to the flywheel. If the velocity of the flywheel is not constant, the energy required for acceleration of the flywheel should also be taken into consideration.

In cycling, several factors contribute to the energy required: aerodynamic drag, rolling resistance, work against gravity, and inertia. Of these, aerodynamic drag is the most relevant when cycling at speeds normally used in competition. Aerodynamic drag can be reduced by changes to the bike and to the position of the rider.

Cycling and cycle ergometry serve as useful examples to illustrate the interchange of chemical and mechanical energy that occurs during exercise.

References

Abbott, A.V., and Wilson, D.G. 1995. *Human-powered vehicles*. Champaign, IL: Human Kinetics.

American College of Sports Medicine. 1991. *Guidelines for exercise testing and prescription*. Philadelphia: Lea & Febiger.

Baker, A.J., Brandes, R., Schendel, T.M., Trocha, S.D., Miller, R.G., and Weiner, M.W. 1994. Energy use by contractile and noncontractile processes in skeletal muscle estimated by ^{31}P NMR. *Am. J. Physiol.* 266: C825–31.

Bangsbo, J. 1996. Oxygen deficit: A measure of the anaerobic energy production during intense exercise? *Can. J. Appl. Physiol.* 21: 350–63.

Barclay, C.J. 1996. Mechanical efficiency and fatigue of fast and slow muscles of the mouse. *J. Physiol. (London)* 497: 781–94.

Barclay, C.J., Curtin, N.A., and Woledge, R.C. 1993. Changes in cross-bridge and noncross-bridge energetics during moderate fatigue of frog muscle fibers. *J. Physiol. (London)* 468: 543–56.

Bers, D.M., Philipson, K.D., and Langer, G.A. 1981. Cardiac contractility and sarcolemmal calcium binding in several cardiac muscle preparations. *Am. J. Physiol.* 240: H576–83.

Broker, J.P., and Gregor, R.J. 1994. Mechanical energy management in cycling: Source relations and energy expenditure. *Med. Sci. Sports Exerc.* 26: 64–74.

Brooks, G.A., Fahey, T.D., and White, T.P. 1996. *Exercise physiology: Human bioenergetics and its applications*. Mountain View, CA: Mayfield.

Brooks, G.A., and Mercier, J. 1994. Balance of carbohydrate and lipid utilization during exercise: The "crossover" concept. *J. Appl. Physiol.* 76: 2253–61.

Bulbulian, R., Jeong, J.W., and Murphy, M. 1998. Comparison of anaerobic components of the Wingate and Critical Power tests in males and females. *Med. Sci. Sports Exerc.* 28: 1336–41.

Buttelli, O., Seck, D., Vandewalle, H., Jouanin, J.C., and Monod, H. 1996. Effect of fatigue on maximal velocity and maximal torque during short exhausting cycling. *Eur. J. Appl. Physiol.* 73: 175–79.

Capelli, C., Schena, F., Zamparo, P., Dal Monte, A., Farina, M., and Di Prampero, P.E. 1998. Energetics of best performance in track cycling. *Med. Sci. Sports Exerc.* 30: 614–24.

Carnevale, T.J., and Gaesser, G.A. 1991. Effects of pedaling speed on the power-duration relationship for high intensity exercise. *Med. Sci. Sports Exerc.* 23: 242–46.

Cavagna, G.A., and Kaneko, M. 1977. Mechanical work and efficiency in level walking and running. *J. Physiol. (London)* 268: 467–81.

Chapler, C.K., Stainsby, W.N., and Gladden, L.B. 1980. Effect of changes in blood flow, norepinepherine, and pH on oxygen uptake by resting skeletal muscle. *Can. J. Physiol. Pharmacol.* 58: 93–96.

Coast, J.R., Cox, R.H., and Welch, H.G. 1986. Optimal pedaling rate in prolonged bouts of cycle ergometry. *Med. Sci. Sports Exerc.* 18: 225–30.

Coyle, E.F., Sidossis, L.S., Horowitz, J.F., and Beltz, J.D. 1992. Cycling efficiency is related to the percentage of type I muscle fibers. *Med. Sci. Sports Exerc.* 24: 782–88.

Crow, M.T., and Kushmerick, M.J. 1982. Chemical energetics of slow and fast twitch muscles of the mouse. *J. Gen. Physiol.* 79: 147–66.

Dal Monte, A., Leonardi, L.M., Menchinelli, C., and Marini, C. 1987. A new bicycle design based on biomechanics and advanced technology. *Int. J. Sport Biomech.* 3: 287–92.

de Groot, G., Sargeant, A.J., and Geysel, J. 1995. Air friction and rolling resistance during cycling. *Med. Sci. Sports Exerc.* 27: 1090–5.

Donovan, C.M., and Brooks, G.A. 1983. Endurance training affects lactate clearance, not lactate production. *Am. J. Physiol.* 244: E83–92.

Faria, I.E., and Cavanagh, P.R. 1978. *The physiology and biomechanics of cycling*. New York: Wiley.

Foley, J.M., Harkema, S.J., and Meyer, R.A. 1991. Decreased ATP cost of isometric contractions in ATP-depleted rat fast twitch muscle. *Am. J. Physiol.* 261: C872–81.

Gaesser, G.A., Ward, S.A., Baum, V.C., and Whipp, B.J. 1994. Effects of infused epinephrine on slow phase of O_2 uptake kinetics during heavy exercise in humans. *J. Appl. Physiol.* 77: 2413–19.

Gladden, L.B., Stainsby, W.N., and MacIntosh, B.R. 1982. Norepinephrine increases canine skeletal muscle $\dot{V}O_2$ during recovery. *Med. Sci. Sports Exerc.* 14: 471–76.

Gnehm, P., Reichenbach, S., Altpeter, E., Widmer, H., and Hoppeler, H. 1997. Influence of different racing positions on metabolic cost in elite cyclists. *Med. Sci. Sports Exerc.* 29: 818–23.

Green, S., and Dawson, B.T. 1995. The oxygen uptake-power regression in cyclists and untrained men: Implications for the accumulated oxygen deficit. *Eur. J. Appl. Physiol.* 70: 351–59.

Hill, A.V. 1938. The heat of shortening and the dynamic constants of muscle. *Proc. R. Soc. London B* 126: 136–95.

Ingen Schenau, C.J., van Woensel, W.W., van Boots, P.J., Snackers, R.W., and DeGroot, G. 1990. Determination and interpretation of mechanical power in human movement: Application in ergometer cycling. *Eur. J. Appl. Physiol.* 61: 11–19.

Kautz, S.A., Hull, M.L., and Neptune, R.R. 1994. A comparison of muscular mechanical energy expenditure and internal work in cycling. *J. Biomech.* 27: 1459–67.

Kraemer, W.J., Patton, J.F., Knuttgen, H.G., Hannan, C.J., Kettler, T., Gordon, S.E., Dziados, J.E., Fry, A.C., Frykman, P.N., and Harman, E.A. 1991. Effects of high intensity cycle exercise on sympathoadrenal-medullary response patterns. *J. Appl. Physiol.* 70: 8–14.

Kyle, C.R. 1986a. Mechanical factors affecting the speed of a cycle. In *Science of cycling*, ed. E.R. Burke, 123–36. Champaign, IL: Human Kinetics.

Kyle, C.R. 1986b. Equipment design criteria for the competitive cyclist. In *Science of cycling*, ed. E.R. Burke, 137–44. Champaign, IL: Human Kinetics.

Kyle, C.R. 1988. The mechanics and aerodynamics of cycling. In *Medical and scientific aspects of cycling*, ed. E.R. Burke and M.M. Newson, 235–52. Champaign, IL: Human Kinetics.

Kyle, C.R. 1990. Wind tunnel tests of aero bicycles. *Cycling Science* 3: 57–61.

Kyle, C.R. 1994. Energy and aerodynamics in bicycling. *Clin. Sp. Med.* 13: 39–73.

Lang, P.B., Latin, R.W., Berg, K.E., and Mellion, M.B. 1992. The accuracy of the ACSM cycle ergometry equation. *Med. Sci. Sports Exerc.* 24: 272–76.

Londeree, B.R., Moffitt-Gerstenberger, J., Padfield, J.A., and Lottmann, D. 1997. Oxygen consumption of cycle ergometry is nonlinearly related to work rate and pedal rate. *Med. Sci. Sports Exerc.* 29: 775–80.

Marsh, A.P., and Martin, P.E. 1997. Effect of cycling experience, aerobic power, and power output on preferred and most economical cycling cadences. *Med. Sci. Sports Exerc.* 29: 1225–32.

McArdle, W.D., Katch, F.I., and Katch, V.L. 1996. Exercise physiology: Energy, nutrition, and human performance. Philadelphia: Lea & Febiger.

McCartney, N., Heigenhauser, J.F., Sargeant, A.J., and Jones, N.L. 1983. A constant-velocity cycle ergometer for the study of dynamic muscle function. *J. Appl. Physiol.* 55: 212–17.

McCole, S.D., Claney, K., Conte, J.C., Anderson, R., and Hagberg, J.M. 1990. Energy expenditure during bicycling. *J. Appl. Physiol.* 68: 748–53.

Moritani, T., Nagata, A., DeVries, H.A., and Muro, M. 1981. Critical power as a measure of physical work capacity and anaerobic threshold. *Ergonomics* 24: 339–50.

Neptune, R.R., and Bogert, A.J. 1998. Standard mechanical energy analyses do not correlate with muscle work in cycling. *J. Biomech.* 31: 239–45.

Newmiller, J., Hull, M.L., and Zajac, F.E. 1988. A mechanically decoupled two-force component bicycle pedal dynameter. *J. Biomech.* 21: 375–86.

Poole, D.C., Gladden, L.B., Kurdak, S., and Hogan, M.C. 1994. L-(+)-lactate infusion into working dog gastrocnemius: No evidence lactate per se mediates $\dot{V}O_2$ slow component. *J. Appl. Physiol.* 76: 787–92.

Poole, D.C., Schaffartzik, W., Knight, D.R., Derion, T., Kennedy, B., Guy, H.J., Prediletto, R., and Wagner, P.D. 1991. Contribution of exercising legs to the slow component of oxygen uptake kinetics in humans. *J. Appl. Physiol.* 71: 1245–53.

Rüegg, J.C. 1986. *Calcium in muscle activation: A comparative approach.* Berlin: Springer-Verlag.

Saugen, E., and Vollestad, N.K. 1996. Metabolic heat production during fatigue from voluntary repetitive isometric contractions in humans. *J. Appl. Physiol.* 81: 1323–30.

Seabury, J.J., Adams, W.C., and Ramey, M.R. 1977. Influence of pedaling rate and power output on energy expenditure during bicycle ergometry. *Ergonomics* 20: 491–98.

Stainsby, W.N. 1970. Oxygen uptake for isotonic and isometric twitch contractions of dog skeletal muscle in situ. *Am. J. Physiol.* 219: 435–39.

Stainsby, W.N. 1982. Energetic patterns of normally circulated mammalian muscle in situ. *Fed. Proc.* 41: 185–88.

Stainsby, W.N., and Barclay, J.K. 1971. Relation of load, rest length, work, and shortening to oxygen uptake by in situ dog semitendinosis. *Am. J. Physiol.* 221: 1238–42.

Stainsby, W.N., and Barclay, J.K. 1972. Oxygen uptake for brief tetanic contractions of dog skeletal muscle in situ. *Am. J. Physiol.* 223: 371–75.

Stainsby, W.N., and Lambert, C.R. 1979. Determinants of oxygen uptake in skeletal muscle. In *Exercise and sport sciences reviews.* Vol. 7. Ed. R.S. Hutton and D.I. Miller, 125–52. Santa Barbara, CA: Franklin Institute Press.

Vandewalle, H., Peres, G., Heller, J., Panel, J., and Monod, H. 1987. Force-velocity relationship and maximal power on a cycle ergometer: Correlation with the height of a vertical jump. *Eur. J. Appl. Physiol.* 56: 650–56.

Vollestad, N.K., and Blom, P.C.S. 1985. Effect of varying exercise intensity on glycogen depletion in human muscle. *Acta Physiol. Scand.* 125: 395–405.

Vollestad, N.K., Wesche, J., and Sejersted, O.M. 1990. Gradual increase in leg oxygen uptake during repeated submaximal contractions in humans. *J. Appl. Physiol.* 68: 1150–6.

Wells, R.P. 1988. Mechanical energy costs of human movement: An approach to evaluating the transfer possibilities of two-joint muscles. *J. Biomech.* 21: 955–64.

Wells, R.P., and Evans, N. 1987. Functions and recruitment patterns of one- and two-joint muscles under isometric and walking conditions. *Human Move. Sci.* 6: 349–72.

Wheeler, J.B., Gregor, R.J., and Broker, J.P. 1992. A dual piezoelectric bicycle pedal with multiple shoe-pedal interface compatibility. *Int. J. Sport Biomech.* 8: 251–58.

Widrick, J.J., Freedson, P.S., and Hamill, J. 1992. Effect of internal work in the calculations of optimal pedaling rates. *Med. Sci. Sports Exerc.* 24(3): 376–82.

Williams, K.R., and Cavanagh, P.R. 1983. A model for the calculation of mechanical power during distance running. *J. Biomech.* 16: 115–28.

Wilson, B.A., and Stainsby, W.N. 1978. Relation between oxygen uptake and developed tension in dog skeletal muscle. *J. Appl. Physiol.* 45: 234–37.

Winter, D.A. 1979. A new definition of mechanical work done in human movement. *J. Appl. Physiol.* 46: 79–83.

Womack, C.J., Davis, S.E., Blumer, J.L., Barrett, E., Weltman, A.L., and Gaesser, G.A. 1995. Slow component of O_2 uptake during heavy exercise: Adaptation to endurance training. *J. Appl. Physiol.* 79: 838–45.

Zoladz, J.A., Rademaker, A., and Sargeant, A.J. 1995. Oxygen uptake does not increase linearly with power output at high intensities of exercise in humans. *J. Physiol. (London)* 488: 211–18.

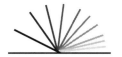

Work and Energy Summary

J. Mester and B.R. MacIntosh

Studying the phenomenon of energy has a very long tradition. It was probably Aristotle who used the term *energy* to express that something is in action. Throughout the progress of physics and physiology, a long list of outstanding scientists improved our knowledge of this phenomenon and developed concepts regarding how movement can be initiated and how this motion can be changed in terms of velocity and direction. It took a long time, however, before Leibniz formulated the principle of conservation of mechanical energy regarding kinetic and potential energy.

Movement, which is accomplished by muscular effort, relies on the transformation of chemical energy to mechanical energy by our muscles. Adenosine triphosphate (ATP) is hydrolyzed by crossbridges in the muscles to accomplish force generation and/or relative sliding of myofilaments which results in movement of the bony levers to which the muscle(s) is/are attached. Replenishment of ATP in the muscles is accomplished by either anaerobic or aerobic means. In the case of aerobic replenishment of ATP, oxygen is required to release the energy from the metabolic substrates (primarily fats and carbohydrates). For this reason, measurement of oxygen uptake can be used to assess the energy cost of exercise. The maximal rate at which oxidative replenishment of ATP can be achieved during whole body exercise is referred to as maximal oxygen uptake.

In part I, the fundamental prerequisites for any kind of movement, from the mechanical as well as from the physiological point of view, have been presented. With the topics of work and energy, a broad approach is offered that starts with the basic principles of physics with respect to mechanical work and energy. It is recognized that optimal performance relies on minimizing energy loss while converting chemical energy to mechanical energy, and that channeling the mechanical energy to the desired task is an important consideration.

Modifications to sports equipment are often used to optimize performance. Examples of equipment design which can influence work and energy include sport shoes, implements like the pole in pole vaulting, and specialized surfaces which allow storage and return of energy to the athlete. In this section, consideration is given to transfer of energy between body segments as well as the role of muscles and/or tendons in storing mechanical energy by deformation and elastic recoil. After these sections dealing mainly with mechanical considerations, the physiological background of energy is presented. The issues covered include the pathways for supply of oxygen and substrates and the role of nutrition in providing not only the substrates, but other nutrients which can influence the capacity to provide or sustain high rates of aerobic chemical energy transduction. Part I concludes with an example of human movement (cycling) which elaborates the determinants of chemical energy use in the muscles and the mechanical factors which dictate the rate at which energy is required for this form of locomotion.

Mechanical aspects of work and energy dictate human locomotion. Chapter 1 by Nigg, Stefanyshyn, and Denoth points out the laws of physics which govern human movement. The total energy of a system includes chemical, mechanical, magnetic, electrical, thermal, and other energy. For all intents and purposes when considering human movement, chemical, mechanical, and thermal energy need to be considered. Chemical energy is added to the system through ingestion of metabolic substrates (food). Mechanical energy can be added or removed through contact with the environment (deformation of the surface, force applied to an implement, etc.). Thermal energy can be lost or gained through convection, conduction and radiation, and thermal energy can also be lost via evaporation of water (sweat) from the skin and airways. Our muscles provide the means by which we interact with our surroundings. They transform chemical energy to mechanical energy and heat. Three strategies commonly applied to improve the work energy-balance during locomotion include:

- storing and returning mechanical energy;
- minimizing energy loss; and
- optimizing muscle functions.

Here one issue of particular scientific interest is the intersegmental transfer of energy, which is important for the understanding of coordination of human movement in the sense of minimizing the loss of energy and optimizing muscle function. This aspect of energy transfer cannot be suitably quantified, making it difficult to determine in a quantitative manner the effectiveness of chemical energy use.

Energy can be stored during human movement in general and during locomotion in particular. An essential consideration is the conservation of energy. This topic is presented from various perspectives in the chapters by Alexander (chapter 2) and Minetti (chapter 5). In these chapters, various aspects of locomotion are considered, and it is demonstrated that in human locomotion, and in other animals, various schemes are used to minimize energy loss from the system. Alexander describes the storage and release of energy in the form of stretching tendons and reveals that some of the energy is lost in the form of heat when a tendon undergoes sequential cycles of stretching and shortening. In addition to tendons acting as springs, storing and releasing energy, ligaments (particularly in the bottom of the foot) can also perform this function. Alexander also points out that during eccentric contraction, greater force can be generated, and therefore a tendon will be stretched to a greater extent. The subsequent potential benefit of this is that during a stretch-shortening sequence, the velocity of shortening can be greater than otherwise expected due to enhanced elastic recoil of the tendon. However, in chapter 3 Hay explains that it is very difficult to demonstrate this enhancement in vivo.

A further application of the "work-energy principle" outlined in chapter 1 can be found in various sports where the characteristics of certain materials (i.e., the surface of sports grounds or shoes) influence the mechanical situation. This is the topic elaborated by Stefanyshyn and Nigg in chapter 4. Simplified, the properties of these materials can be compared to those of a spring. Thus, the storage and return of energy for a linear spring is discussed. For both surfaces of sports grounds as well as shoes it is important that:

- returned energy be substantial enough to make a difference;
- energy be returned at the right time;
- energy be returned with the right frequency; and
- energy be returned to the right location.

In the last few years, sport surfaces have undergone remarkable changes. As a result, many of them fulfill the above mentioned criteria. The energy-return of sport surfaces thus led to a particular improvement of performance in various sports (e.g., track and field and gymnastics). Since sport surfaces have area-elasticity, it is easier than it is for sport shoes to adjust and tune the return of energy. As a matter of fact, the right location and the right frequency of energy-return still remain major impediments for the sport-shoe industry in attempting to develop shoes with appropriate energy return systems.

During the last few years, all manufacturers of sport equipment have developed materials that enhance performance and at the same time try to make sport safer. This results in stronger, lighter, more durable, and more pleasant equipment, which permits better performance, usually by minimizing energy loss or by storing and returning mechanical energy. In terms of energy, this leads to the questions of how athletic equipment can be developed to store and reutilize energy, how energy is lost to the environment, and how this can be minimized. Basically the amount of energy that can be stored in a linear spring is a function of the stiffness of the spring and the degree of deformation. Equipment has very different properties, as far as spring constant and maximal deformation are concerned. The specific requirements by the athlete and his or her physical abilities cover a wide spectrum. Therefore, much scientific effort is concentrated on answering the unsolved question—what kind of material with what mechanical properties is best suited for which athlete.

Equipment enhancements are not restricted to sport surfaces and shoes. Technical development can also minimize the loss of energy to the environment, for example, through friction and drag forces. Depending on the equipment and the sport concerned, it can be the goal to maximize, optimize, or minimize friction forces. In most winter sports (e.g., skating, bobsledding, skiing), it is of utmost importance to minimize friction forces between the implement and the surface, whereas normally in team and other sports, friction forces between the shoes and the surface are imperative for movement control and balance.

In many sports, the velocity of the athlete is the ultimate criterion for the overall performance. In these sports, air and water are the two most common viscous mediums that athletes encounter. Due to the quadratic dependence of drag forces on velocity, minimizing drag is a major goal. Industry research as well as scientific studies focus on this problem of reducing the loss of energy to the environment.

Understanding the mechanical laws and how they restrict and permit movement is critically important for the understanding of human movement. The source of energy for initiating any kind of active and directed human movement, however, is reliant on physiological mechanisms. Human movement is sustained by oxidative metabolism that relies on a series of physiological functions:

- ventilation to introduce high levels of PO_2 to close proximity of the venous blood
- pulmonary diffusion to allow gas exchange between the blood and the atmosphere
- oxygen transport by the blood, primarily bound to hemoglobin
- bulk flow of the blood (cardiac output)
- distribution of the cardiac output
- tissue gas diffusion to provide O_2 to the mitochondria
- mitochondrial respiration

In the chapter written by Hoppeler and Weibel (chapter 6), the energy for endurance exercise and its release in muscles serves as the topic for illustrating how the pathways for oxygen and substrates function. The concept of symmorphosis is introduced to describe a sequence of steps (for the delivery of oxygen from the atmosphere to the mitochondria) which adapt in parallel. The ability to sustain a high rate of aerobic energy transduction relies on each of the above steps being highly developed. The demand for oxygen is set by the muscles in use.

In this section on work and energy, an integrated approach between regulated morphogenesis on the one hand and economic design on the other is considered. Comparative data from various kinds of animals and of the human physiology are reviewed. Both allometric variations (metabolic requirements related to body mass) and adaptive variations, expressed by the selection of certain species by nature for highest aerobic performance (e.g., horses, dogs, pronghorn antelope), are considered in the analyses of these comparative data.

The key element for studies on the provision of energy by means of aerobic metabolism is the mitochondria located in the skeletal muscles, which consume more than 90% of the oxygen and substrates. If there are any invariant parameters of biological design found in the comparative analysis, they should be located at the level of the mitochondria. These data, however, reveal that small animals have more mitochondria per gram skeletal muscle than larger animals. At maximal oxygen uptake, the oxygen uptake normalized to mitochondrial volume does not vary with body size. Therefore differences in $\dot{V}O_2$max per body mass are due to differences in volume of mitochondria per unit of body mass.

An important step in providing oxygen for mitochondrial oxidative phosphorylation is its convective transport by the pumping action of the heart that adjusts the supply according to demand. The structural and functional variability covers the range which is expected by the principle of symmorphosis, maintaining that there is a match between maximal rates of O_2 demand and the structural prerequisites to deliver that oxygen. Chronic high demand for oxygen results in increased heart size, requiring an increased stroke volume. Thus, a higher cardiac output can be achieved. In contrast, lung size does not adapt. The gas-exchanging systems have a high level of redundancy compared to the other steps of the pathway for energy delivery. With training, the increased capacity of the other steps to deliver and use oxygen results in encroachment of the functional capacities on the pulmonary reserve capabilities. Therefore, endurance athletes may be limited in maximal oxygen uptake by the inability of their lungs to maintain a sufficiently high level of ventilation.

The provision of substrates for aerobic energy metabolism is more complex than the process that makes oxygen available. The metabolism of fats is undertaken in the mitochondria; carbohydrates are initially metabolized in the cytoplasm (glycolysis) with subsequent conversion in the mitochondria. The source of substrates can be cellular (lipid droplets and glycogen granules) or extracellular (blood glucose, derived from liver glycogenolysis and gluconeogenesis, and free fatty acids derived from lipolysis in adipose tissue). The rate of uptake of extracellular substrate reaches its maximum at 40% of maximal oxygen uptake. Mitochondria obtain 80% of their fuel from the cellular stores at maximal oxygen uptake, and lipid metabolism is negligible at this intensity.

The section on pathways for oxygen and substrates illustrates the mechanisms and capacities for energy supply during exercise. There is a complex interaction between structure and function in these biological systems, which have a high functional adaptability that can be induced by specific training.

The chemical energy that is released by respiration for the purpose of sustaining muscle contraction has to be obtained by way of nutrition and can be stored in the body in various ways. Recently, the role of nutrition as a constituting factor of health, performance, and well-being has become more and more important and is a focal point of scientific interest in many disciplines. In chapter 7, Lemon discusses the energy and nutrient intake appropriate for maintenance of health and for athletic performance. Two principle concerns of this chapter are:

1. Does regular physical activity alter nutritional needs?
2. Can minor nutritional manipulations enhance performance?

Although both questions can rather easily be answered with "yes," it is difficult to define the exact relation of various elements of certain groups of nutrients (e.g., carbohydrates) or the absolute amount of others (e.g., minerals, vitamins) tuned not only to the kind of exercise (e.g., endurance sport) but also to the individual needs of the athlete. The main reason for this is that nutrition recommendations for daily food requirements are directed to essentially sedentary subjects. In general, these recommendations are too low for athletes in terms of the absolute energy (Calorie) component as well as the quantity of the liquid, minerals, etc.

In terms of energy supply, it is well known that the carbohydrate (CHO) stores are of extreme importance for athletic performance as they deliver the most effective energy for muscle contraction. Therefore, special attention must be paid to a careful and strict diet before, during, and after a sporting event. Recommendations for appropriate strategies are given in this section. With regard to CHO loading, it is useful to consume about 6-10 g CHO · kg body mass^{-1} before the event. In addition to this, the proportion of nutrients with moderate and high glycemic index must be considered. The intake of CHO is imperative in sports with prolonged duration. It is possible to extend the point of exhaustion by 33% by appropriate consumption of CHO. After the event, the consumption of CHO is necessary in order to refill the muscle glycogen stores. The optimal time window for doing this seems to be very small (about 2 h following the event), as the muscle glycogen resynthesis after 2 h is considerably slower. It is also well known that fat is the major energy reserve and together with CHO the most important fuel supply, although the energy supply by fat oxidation requires more oxygen than oxidation of

CHO. The adaptability of biological processes for energy supply by fat oxidation can also be demonstrated in this field. Here one adaptation due to endurance training is that a greater percentage of energy can be provided from fat at a given exercise intensity.

The significant role of protein is quite clear in its physiological function for building up enzymes and structural proteins, for example. The intake of protein for sedentary subjects in the highly developed countries normally covers the requirements completely. The recommendations of 0.8 g · kg body mass^{-1} for daily protein intake applied to mainly sedentary subjects are probably insufficient for elite athletes with a high volume of daily exercise. The difficulty in evaluating this includes various factors, such as time for adaptation to dietary changes, energy intake, training status, gender, and exercise intensity. All these factors may influence the protein turnover significantly. Regarding the high quality of nutrition in the developed countries it is unlikely that a chronic insufficient protein intake occurs in adult male athletes. As for female athletes and younger, growing individuals who engage in chronic high volumes of exercise, a careful assessment of the protein status is recommended. In terms of minerals (macro- and microminerals), there is evidence that regular exercise increases the need for some minerals. The importance and functional role of these minerals is generally accepted. Nevertheless, it is quite difficult to determine a sufficient, specific daily intake.

Chapter 7 stresses the importance of a carefully planned and assessed nutritional behavior, which is basically valid for all individuals. For sedentary subjects, the recommendations are quite clear and accepted. For athletes with rigorous training programs, however, rather little is known. Perhaps it is appropriate to consider recommendations tuned to the requirements of certain sports (e.g., endurance or strength) or even to individual necessities.

Finally, chapter 8 considers an example of locomotion (cycling) and looks at various factors which affect the energetics of performing this task. Both cycle ergometry and cycling are considered. Recognition is given in this chapter to the various factors which determine the energy requirements of a contracting muscle. Factors such as level of activation, load, and muscle length are important considerations. This basic information is taken to the whole body exercise level by consideration of the energetics of cycle ergometry. Power output is a major determinant of the energy cost of cycling, but variation due to cadence can be related to muscle fiber types. As power output increases, the cadence which is most efficient also increases. While the intensity of cycle ergometer exercise is dictated by resistance and cadence, the intensity of effort (metabolic rate) needed to ride a bicycle is dictated by several factors: air resistance, mass of the bicycle rider system, rolling resistance, slope of the terrain, etc. Air resistance can be further broken down to velocity, drag coefficient, air density, and frontal surface area. Improvements in cycling performance can be accomplished by either increasing the metabolic capabilities of the athlete or reducing the energy requirement for travelling a given speed. Recent developments in aerodynamics have permitted tremendous improvements in cycling performance. These developments allow a cyclist to go faster at a given mechanical power output, thus completing the distance with less work, though not necessarily at a different rate of chemical energy use.

Work and Energy Definitions

Many definitions used in this section are based on the definitions outlined in the book *Biomechanics of the Musculo-Skeletal System* (edited by Nigg and Herzog 1994).

Mechanical

compliance: the amount of deformation produced by a given amount of force.

$$\text{compliance} = c = \frac{1}{k} = \frac{\Delta x}{F}$$

where

c = compliance,

k = stiffness,

Δx = deformation, and

F = applied force.

conservation of energy: the sum of the external kinetic and the external and internal potential energy of a system are constant if the external forces and the internal forces acting between the mass points are conservative.

conservative force: a force acting on a system of mass points is called *conservative* if the work needed to move the system from point A to point B is independent of the selected path.

drag: drag is a resistive force imposed on an object, which is passing through a viscous medium.

energy: energy is defined as the ability to do work.

   kinetic energy: energy due to translational or rotational movement.

   potential energy: energy due to the position of a mass.

first law of thermodynamics: the change of the internal energy of a system is equal to the sum of the work and heat added to or removed from the system during a process.

friction: the resistance of two surfaces to slide with respect to each other.

internal energy: a thermodynamical function of state postulated in the first law of thermodynamics.

mass: the amount of matter in an object.

mechanical power: the power is defined as the scalar (dot) product of the force, F, and the velocity, v

$$P = F \bullet v$$

or the power is defined as the rate of change of the kinetic energy of a particle

$$P = \frac{dE_{kin}}{dt}.$$

moment of inertia: the moment of inertia, I, of a mass, m, is the sum of all the axial moments of all its elements, dm.

$$I = \int r^2 \bullet dm$$

where

I = moment of inertia,

dm = infinitesimal mass element in the mass m, and

r = distance of the mass element from the axis for which the moment of inertia has been calculated.

$\bullet$ = sign for the scalar (dot) product of the two vectors.

resultant joint forces/moment: the resultant joint force/moment is the mechanically equipollent force/moment replacing all forces with lines of action crossing the joint.

stiffness: a factor that affects the amount of force required to produce a given amount of deformation

$$k = \frac{F}{\Delta x}$$

where

k = linear stiffness,

F = force acting on a structure, and

Δx = linear deformation of the structure.

scalar: a scalar is a quantity that is determined only by its magnitude.

vector: a vector is a quantity that is determined by magnitude, direction, and sense of direction. Throughout this text, vectors are depicted with **bold** symbols.

work: the work, W, performed by a force vector, **F**, acting on a particle is defined as the line integral

$$W = \int \mathbf{F} \bullet d\mathbf{r}$$

where

W = work performed by the force **F**,

F = force vector acting on the particle,

d**r** = infinitesimal small displacement vector, and

• = sign for the scalar (dot) product of the two vectors.

For the special case where the direction of the force and the displacement are in the same direction:

$$W_{11} = F_{const} \, d_{11}$$

where

W_{11} = work performed by the constant force, F_{const}, along the displacement, which is in the direction of the acting force,

$F_{const} = |\mathbf{F}|$ = constant force acting on the particle, and

$D_{11} = |d\mathbf{r}|$ = displacement parallel to the acting force.

Biological

acute exercise: a single session of physical activity where one's muscles are sufficiently activated to cause an increased utilization of chemical energy. Can involve a variety of exercise modes along a continuum ranging from high intensity, short duration exercise (strength/power activity) to a low intensity, prolonged exercise (endurance activity).

body mass-restricted exercise: activity/sporting event where the rules or subjective aesthetics dictate the maximum body mass/shape that is allowed, e.g., wrestling, gymnastics, ballet, etc.

chronic exercise: regular exercise of sufficient intensity and duration that adaptations (training effects) occur in various tissues and body systems.

estimated safe and adequate daily dietary intake (ESADDI): a provisional recommendation when there are inadequate data to establish a RDA or RNI.

nutrient balance: the situation that exists when intake and excretion of a particular nutrient are equal. Measuring nutrient balance involves quantifying all modes of excretion and intake. Balance is considered positive if intake exceeds excretion (nutrient accumulates in the body tissues) and negative if excretion exceeds intake (nutrient loss from body tissues) (Food and Agricultural Organization 1985; US Food and Nutrition Board 1989; Nutrition Recommendations 1990). Traditionally, this has been the standard experimental technique used to determine nutrient requirements, i.e., intake that elicits balance is thought to be the requirement. Actually, nutrient status is a more precise term for this concept as it avoids the seemingly nonsensical descriptors (positive or negative) for balance; however, the term balance is well entrenched in the literature and its seems unlikely that it will be replaced in the near future. The balance (status) method only assesses net changes in nutrient status and due to the various collections required involves considerable inconvenience for the subjects.

nutrient requirement: amount of a nutrient necessary to elicit balance, i.e., when intake = excretion (Food and Agricultural Organization 1985; US Food and Nutrition Board 1989; Nutrition Recommendations 1990).

protein digestibility-corrected amino acid score (PDCAAS): newer method to assess protein quality of foods which considers digestibility as well as both the indispensable amino acid (amino acids that cannot be synthesized in sufficient quantities by the human body) profile and content of a protein. Recommended for protein quality determinations by the FAO/WHO Joint Expert Consultation on Protein Quality Evaluation (1989) and required since 1993 by the US Food and Drug Administration's Nutrition Labeling Regulations (FDA 1993).

protein efficiency ratio (PER): traditional method used to assess the protein quality of foods. It is based on the ability of a particular protein to support the maximal growth rate in young rodents. PER appears to overestimate the quality of some animal proteins and to underestimate the quality of some vegetable proteins for humans (due to differences in amino acid requirements resulting from the lower maximal growth rates in humans vs. rodents).

recommended dietary allowance (RDA) or recommended nutrient intake (RNI): amount of a nutrient necessary to prevent deficiencies in the vast majority (95-97%) of the population. Determined by adding two standard deviations to the mean intake necessary to elicit balance, i.e., requirement, in a sample that is representative of the population of interest (Food and Agricultural Organization 1985; US Food and Nutrition Board 1989; Nutrition Recommendations 1990). RDA is term used in the United States and RNI in Canada.

respiratory system: cascade of anatomical structures necessary to make environmental oxygen available to the chemical processes responsible for the rephosphorylation of ADP to ATP in mitochondria.

resultant joint forces/moment: the resultant joint force and joint moment with respect to a point, A, are the mechanically equipollent force and moment replacing all forces with lines of action crossing the joint with respect to this point A.

symmorphosis: state of structural design commensurate with functional needs. A functional system is said to be symmorphotic when the quantity of its structure or sequence of structure is matched to what is functionally needed: "enough but not too much."

turnover (flux): the term(s) used to describe the movement of a substance through the body, i.e., the rate at which a substance is metabolized. Measurement of flux/turnover makes it possible to evaluate the various aspects of nutrient metabolism rather than simply net balance. Requires introduction into the body of a traceable nutrient (either radio- or stable-labelled isotopes), as well as the quantification of nutrient intake and excretion, and/or tissue or blood concentration of the nutrient (in some cases one of its metabolites). For example, with time following ingestion (or intra-venous infusion) of a labelled amino acid its concentration increases until equilibration (isotopic steady state) occurs. At this point, input (diet and/or infusion + degradation [in this example, body protein degradation]) into the body's free pools (unbound protein pools) is equal to output (excretion + body protein synthesis) and both are equal to turnover (flux). Turnover (flux) is calculated from the isotope input divided by the steady-state label concentration (specific activity or enrichment in blood, urine, or tissue depending on the nutrient). The time required to reach steady-state depends on several factors but mainly the rate of isotope input (ingestion or infusion). From turnover data, average whole body protein degradation (turnover - protein intake) or protein synthesis (turnover - excretion) can be determined. For a carbon tracer (^{14}C, ^{13}C) excretion is in expired air (CO_2) while for nitrogen (^{15}N) it is in urine (urea and/or ammonia) (Waterlow et al. 1978). In combination with needle biopsy samples from muscle, the turnover technique can be used to assess muscle protein synthesis and degradation (Biolo et al. 1995). Limitations include the cost of the necessary equipment and labelled nutrients

$\dot{V}O_2$max: maximal oxygen uptake or capacity to use oxygen usually given in ml $O_2 \cdot min^{-1} \cdot kg^{-1}$. This value characterizes the global capacity of the respiratory system to transfer and consume oxygen in normoxia when a large muscle mass (> 30% of total muscle mass) is used and is typically measured in an incremental exercise test where a further increase in exercise intensity elicits no further change in oxygen uptake.

$\dot{V}O_2$peak: highest value of O_2 consumption that can be reached under particular conditions. *Comments:* Characterizes limiting conditions for oxygen flux in the respiratory system below $\dot{V}O_2$max (i.e., work with small muscle groups or in hypoxia or when there is uncertainty that $\dot{V}O_2$max has been demonstrated).

$\dot{V}O_2$sust: maximal sustained metabolism. Highest average metabolic rate that can be sustained during days to months with no change in body mass.

Balance and Control
of Movement

Balance and Control of Movement Historical Highlights

Since the gravitation of the earth was the dominant influence on all stages of genesis, from the primitive movements of single-cell organisms to human movement, it is beyond the scope of part II to highlight the entire history. There is good reason to believe that the development of standing upright and walking erect was synchronized with the evolution for controlling all other directed movements under these conditions. Some discoveries indicate that even the evolution of the brain was closely associated with the development of an upright gait.

For the most part, this genesis ceased thousands of years ago, and the evolutionary progress of movement control has slowed to smaller and smaller steps. Today the accuracy and the performance of movement control follow a saturation curve. If we take into consideration that actual movement control and balance are the result of a process that has lasted nearly 40 million years, this optimization is not very remarkable.

About 40 million years ago

Ape-like human predecessors (Propliopithecus and Aegyptopithecus) develop the ability to collect foodstuffs and hunt under gravity with adequate movement control and balance.

About 4.4 million years ago

Australopithecus ramidus, the oldest Hominoid, exists. It is a connecting link between the shape of ape and human.

About 1.5 million years ago

Homo erectus exists. It is a direct ancestor of the actual human form with upright balance and gait. During this era, the perceived advantages of an upright gait (a better view and control of the environment) lead to far-reaching consequences in human movement behavior that we still face today.

1452–1519

Leonardo da Vinci starts dissecting human cadavers to learn more about the action and movement of muscles and improve the realism of his paintings and sculptures.

1475–1564

Michelangelo is one of the early artists/scientists who approaches his work by simultaneously describing the visible world and exploring the functions of human movement. Although no empirical data and no film or video techniques are available to him, earthbound and airborne movements are described and painted in perfectly balanced situations (e.g., *The Creation of Adam*, 1511).

1514–1564

Andreas Vesalius, a Belgian anatomist and physician, lays the foundations of the modern science of anatomy with his dissections of the human body. Vesalius goes on to write an elaborate anatomical work, *De humani corporis fabrica libri septem [On the Structure of the Human Body]* (7 volumes, 1543), based on his own dissections of human cadavers. The volumes are richly and carefully illustrated and are the most accurate and comprehensive anatomical textbooks of the time.

1564–1642

Galilei chronicles revolutionary findings about the overall importance of gravity on earth as well as in the mechanics of the solar system.

1604; 1637

Kepler and Descartes separately study the basic results on the physical fundamentals of human visual perception.

1616

William Harvey, a physician, describes how the heart pumps blood and makes many discoveries regarding muscle contraction.

1687

Isaac Newton publishes *Principia* in which he describes the laws of motion.

1737–1798

Luigi Galvani studies the effects of electricity on animal nerves and muscles. He accidentally discovers that the leg of a frog twitches when touched with an electrically-charged scalpel.

1838–1916

Mach describes pioneering observations (1875) on the physical laws of accelerated movements and the corresponding individual human perception.

1840

German scientists **Hermann von Helmholtz** and **Julius Robert von Mayer** and British physicist **James Prescott Joule** formulate the law of conservation of energy. The conservation of energy law states that the sum of kinetic energy, potential energy, and thermal energy in a closed system remains constant. In classical mechanics, the fundamental laws are the laws of conservation of linear momentum and angular momentum.

1843

Emil Heinrich Du Bois-Reymond finds that a stimulus applied to the electropositive surface of the nerve membrane causes a decrease in electrical potential at that point and that this "point of reduced potential"—the impulse—travels along the nerve as a "wave of relative negativity." He immediately is able to demonstrate that this phenomenon of "negative variation" also occurs in striated muscle and is the primary cause of muscular contraction.

1894

Marey films a cat righting itself during a fall.

1898–1947

In 1898 **Sherrington** observes an exaggerated tonus of the limb extensors in decerebrated cats, investigates reflex responses, and, in 1947, develops a model of the integrative functioning of the neural network and the hierarchy of the human movement control system.

Circa 1900

Brown, Braune, and **Fischer** contribute significantly to the understanding of posture and locomotion.

1924

Gasser and **Hill** determine that the hyperbolic force-velocity relationship is not related to an intervention of the nervous system but is a fundamental characteristic of the muscle fiber itself.

1935

Fenn and **Marsh** mathematically describe the force-velocity relation for muscular contraction using an exponential equation.

1938

While investigating the thermodynamic properties of muscle contraction, **A.V. Hill** develops an equation based on the dynamic constants of muscle that describes the inverse hyperbolic force-velocity equation

$$(P + a) \cdot (V + b) = b \cdot (P_0 + a).$$

1950

Using an isokinetic dynomometer, **D.R. Wilkie** finds that the characteristics of the force-velocity relationship in intact human muscle systems are predicted by the Hill equation.

1957

Eccles publishes prophetic works such as *The Physiology of Nerve Cells* that include reviews of results of intracellular recordings from single cells, leading to a realistic and coherent understanding of the human mind and the neural events leading to movement control.

1974

Nigg identifies the problem of instability for straight somersaults.

1978

Hinrichs speculates that gymnasts make in-flight corrections.

1979

Bosco and **Komi** use multiple joint movements to determine the force-velocity curve and discover a high correlation between muscle fiber distribution and mechanical characteristics.

1987

Vandewalle, Peres, Heller, Panel, and **Monod** use a cycle ergometer with multiple short sprint tests of increasing resistance to plot the torque-velocity curve for cycling.

1994

Playter demonstrates that passive corrections can control instability.

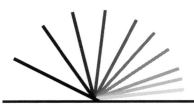

Chapter 9

Basic Concepts of Movement Control

R. Stein, E.P. Zehr, and J. Bobet

Most people move easily, with little effort and little conscious thought, but this ease is deceptive. The muscles and bones, arranged in complex linkages, form a complex mechanical system. The sensory and motor nerves form sophisticated circuits, many of which are not understood. Different structures within the brain also participate in movement, communicating rapidly with each other in intricate patterns to command and monitor it. Given this complexity and sophistication, it is a wonder that human beings can move so well. The sophistication of normal control of movement is most obvious when one tries to artificially control movement. In functional electrical stimulation systems, for example, muscle activity is controlled by artificial stimulation. Even with the best engineering technology, these systems cannot match the control enjoyed by normal subjects.

How does the neuromuscular system turn the nonlinear, time-varying, sluggish properties of muscles and limbs into graceful, intricate movements? In this review, we explain and review the "low level" control of movement, that is, how the muscles, limbs, and reflexes combine to control movement. We first provide a brief tutorial on control systems theory, which is the basis for much of the work on low level motor control. We then review the properties of muscle and those of the reflexes,

two key elements in the control of movement. Finally, we close with a section on the low level control of walking, because of the importance of walking for both normal motion and rehabilitation. Where possible, we compare natural control of movement to artificial control, such as that used to restore movement following spinal cord injury.

Review of Control Theory

There are many approaches to studying the control of movement, from social and behavioral ones to molecular ones. One of the most successful has been the application of engineering control theory. This engineering sense of control is the closest to what we imply by the term movement control in biomechanics, and it is the one we will use here. In control theory, a *controller* is any device, such as a thermostat, that exercises control. The *controlled variable* is the variable being regulated (e.g., temperature) and the *effector* (generally called the *plant*) is the power element (the furnace) that modifies the controlled variable to some desired value or set point (20° C). The simplest type of control, feedforward control, has an input, an output, and some system (the plant) that converts the input to the output. If the operation of the plant is not known, the plant is referred to as a *black box*. Whether or not the plant is known, the aim of control theory is to determine the functional relationship between input and output.

In the area of movement control, muscle is usually the plant. We still don't know the input-output relationship (also known as the transfer function) for muscle, as will be elaborated subsequently (and in chapter 15 by Herzog). Sensors are present in muscle to provide information on length, force, and other variables. When this information is fed back via reflexes, we have feedback or closed loop control. Finally, there are a variety of inputs to the muscle.

Feedforward (Muscle) Control

Figure 9.1 shows an input x(t) and an output y(t) with some function f(t) relating the two. All vary with time, so the output will depend not only on the input at the time of interest (time t) but also on the input at earlier times $t - \tau$. We can write an integral:

$$y(t) = \int x(\tau)\, f(t - \tau)\, d\tau \qquad\qquad 9.1$$

which indicates that the effects at all earlier times τ are summed. This summation is only true for a strictly linear system, a plant in which the output in response to two concurrent inputs equals the sum of the outputs from each input separately. All biological systems are nonlinear, since there are some inputs where linear summation does not apply. Since the control of linear systems is much better understood, biological systems are often treated as if they were linear.

Isometric Contractions

In the simplest example of movement control, the input is a nerve impulse to a motor unit (an α-

motoneuron and the muscle fibers it innervates), and the output is the force measured at a constant position (isometric contraction). The input can be approximated by a unit impulse or Dirac delta function, a very brief rectangular pulse whose area is unity. The output, a "twitch," can be approximated by an equation of the form (Mannard and Stein 1973)

$$f(t) = a\, \{\exp[-b(t - t_0)] - \exp[-c(t - t_0)]\}. \qquad 9.2$$

Equation 9.2 implies that of all the many processes going on in a muscle there are two that are rate limiting, that is, one (b) that determines the rate at which force increases in the muscle and another (c) that determines the rate at which force decreases in the muscle. A system with two exponentials, such as this one, is referred to as a second order system.

We have also included a delay (t_0), which corresponds to the time interval between when the nerve impulse occurs and contraction begins, generally referred to as the excitation-contraction delay. Several processes contribute to this delay (Mannard and Stein 1973), including

1. nerve conduction,

2. synaptic transmission,

3. generation of a muscle action potential,

4. conduction of the potential along the muscle fiber and down into the transverse tubules,

5. release of Ca^{2+} from the sarcoplasmic reticulum,

6. binding of the Ca^{2+} ions to troponin molecules, and

7. a conformational change in the troponin molecule that moves the tropomyosin molecules and allows the binding of myosin heads to actin molecules to begin.

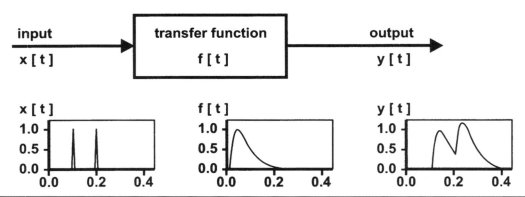

Figure 9.1 Block diagram of a system for transforming an input x[t] to an output y[t] via a transfer function f[t]. In the motor system the input is a series of nerve impulses, each of which produces a twitchlike response.

In biomechanics we often measure the electromyogram (EMG), which is the extracellularly recorded muscle action potential, so the delay does not include the first two steps. The exact contribution of each of the later steps to the total delay, which is approximately 10 ms, is not known.

The increase in force is also due to a number of biochemical processes. Many of these are part of the cross-bridge cycle (Bagshaw 1993; Lymn and Taylor 1971; Warshaw 1996) where

1. a myosin head binds to an actin molecule to form a cross bridge,

2. the head undergoes a conformational change that produces a relative movement, known as the power stroke, between the thick filament containing the myosin and the thin filament containing the actin,

3. as part of the conformational change ADP and inorganic phosphate P_i—formed by splitting the high energy compound ATP—are released,

4. a new molecule of ATP is then bound, and

5. this triggers the release of the myosin head, breaking the cross bridge.

Finally, the myosin head returns to its original position where it is able to bind to another actin molecule and repeat the cycle. The rate of the cross-bridge cycle varies with type of motor unit (fast and slow twitch) and with temperature. It also increases with rate of rise of force (Barany 1967). For these reasons, we relate the constant b in equation 9.2 to the rate of cross-bridge cycling.

The decrease in force at the end of a twitch contraction also involves a number of processes:

1. the detachment of cross bridges following the binding of ATP, as mentioned previously,

2. the release of Ca^{2+} ions from troponin and the associated movements of the tropomyosin molecule,

3. the pumping of Ca^{2+} ions back into the sarcoplasmic reticulum, and

4. viscoelastic movements of the tendon and other structures as the muscle returns to its resting state.

The rate of relaxation of muscle also varies in different types of muscle and at different temperatures. This has been well correlated with the properties of the Ca^{2+} pump in the sarcoplasmic reticulum (Stein, Gordon, and Shriver 1982). More details of the excitation-contraction delay, the cross-bridge

cycle, and the mechanisms for relaxation can be found in the chapter by Herzog (chapter 15). Often, the two rate constants are similar in magnitude, and equation 9.2 then reduces to what is known as a critically damped second order system; that is, one that produces the fastest possible response for a given rate constant without producing oscillation (D'Azzo and Houpis 1966). The form of the equation becomes

$$f(t) = at \exp[-b(t - t_0)] \qquad 9.3$$

where

b is the common rate constant, and

t_0 is the excitation-contraction delay.

If equation 9.2 or 9.3 were always valid for muscle, the feedforward control problem would be solved, and one could predict the outcome for any series of nerve impulses. Figure 9.2 shows the effect of a linearly increasing and decreasing rate of supramaximal stimuli applied to the nerve to soleus muscle of the cat. Rather than increasing linearly, the force saturates at a maximum or tetanic

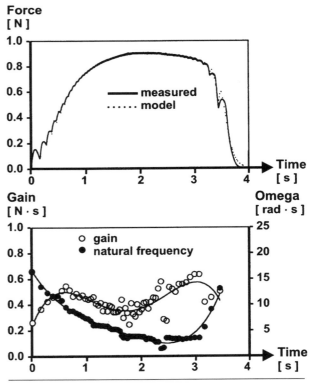

Figure 9.2 The force produced by an increasing and decreasing rate of nerve impulses. A model based on equation 9.3 (dotted line) fits the force (solid line) well, but the parameters (a and b, related to the gain and natural frequency) vary with time.

level where all the available troponin molecules have Ca^{2+} ions bound to them. However, we found that the series of isometric contractions could be well fitted, if the constants a and b in equation 9.3 were allowed to vary from stimulus to stimulus (Bobet, Stein, and Oĝuztöreli 1993). The required variation in gain and natural frequency that is closely related to a and b is also shown in figure 9.2. More recently, the dependence of the parameters on force has been studied, and a sequential model consisting of a linear first order process, a static nonlinearity, and a force-dependent first order process has been developed, which fits the ramp data very well as well as a number of other stimulus patterns (Bobet and Stein 1998).

Force-Length Curve

The force generated by a muscle varies with the length and velocity of the movement. The length dependence comes predominantly from the overlap between thick and thin filaments, as first demonstrated by Gordon, Huxley, and Julian (1966). Above and below the optimal length the possibility of forming cross bridges between actin and myosin is reduced. In the body there will also be geometric factors depending on the angle of the limb with respect to the force-generating axis of the muscle, but the curves can be well fitted by a parabola of the form (Popovic et al. 1999)

$$T_a(\theta) = \begin{cases} c_2\theta^2 + c_1\theta + c_0, & T_a > 0 \\ 0, & T_a \leq 0 \end{cases} \qquad 9.4$$

where

T_a is the active torque at a joint angle θ, and

c_i represent constants, determined by least squares analysis.

T_a must be greater than or equal to zero, so a second line has been added to equation 9.4. Figure 9.3 shows the effect of stimulating the quadriceps muscles at a constant rate (30 Hz) at a variety of knee angles and a variety of relative stimulus intensities. In addition to the active force there is a passive force, which will be due to the ligaments and bone (see chapter 16 by Anderson, Adams, and Hale), and passive tension of the muscles. The passive force has been fitted by a linear relation that determines the stiffness in the midrange of angles and two exponentials that determine the relationship at the two ends of the range, as first suggested by Mansour and Audu (1986):

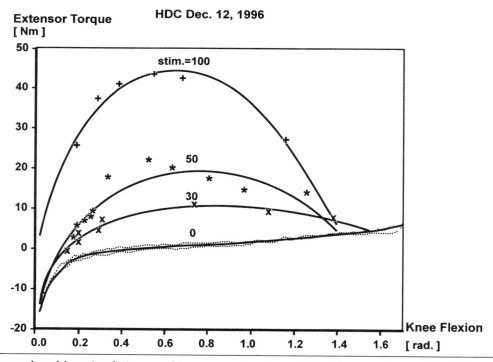

Figure 9.3 Torque produced by stimulating quadriceps muscles at a variety of knee angles and levels between no stimulation (0) and maximal stimulation (100) in a subject with a complete spinal cord injury. Without stimulation the data are well fitted by equation 9.5 in the text, while the extra force produced during stimulation is well fitted by equation 9.4.

$$T_p = d_1(\theta - d_2) + d_3 \exp(d_4\theta) - d_5 \exp(-d_6\theta) \qquad 9.5$$

where

T_p is the passive torque,

d_1 is the stiffness in the midrange,

d_2 is the angle at which the joint is near equilibrium in the absence of gravitational torque,

d_3 and d_4 determine the stiffness near full flexion, and

d_5 and d_6 determine the stiffness near full extension.

With the negative signs as shown, all six parameters should be positive quantities. Figure 9.3 also shows how well this curve fits the passive torque-angle relationship. Further details on measuring the constants experimentally can be found in the paper by Stein and associates (1996).

Force-Velocity Curve

The force generated decreases as the speed of shortening increases. A hyperbolic curve that fits the relationship between force and velocity was suggested 60 years ago (Hill 1938). Huxley (1957) showed how this curve followed from the sliding filament model of muscular contraction. If a muscle is stretched while it is generating active force, the force level initially increases but can then decrease or yield as cross bridges reach their elastic limit and are broken (Hill 1968; Malamud, Godt, and Nichols 1996). Eventually, a steady force is reached that can be considerably more than the isometric level. Typically, the transients are ignored and a modified hyperbola or piecewise linear relation is fitted to the steady state relation (Popović et al. 1999; Winters and Stark 1985). All these processes are working in parallel in a muscle; this is represented in figure 9.4 by three boxes whose outputs are multiplied to give the total force. Viewed in this way, muscle is a three-

input, single-output system. The boxes are not independent and somewhat better fits to the output from electrical stimulation of muscles in cats and spinal cord injured (SCI) humans can be obtained by taking into account interactions among the processes (Chizeck et al. 1991; Shue, Crago, and Chizeck 1995).

Kearney and Hunter (1990), for example, have shown that activation of the muscle increases its apparent stiffness and apparent viscosity dramatically. Forces during a stretch can be fitted to

$$F = kx + bv + ma \qquad 9.6$$

where

F is the extra force required to stretch a muscle with stiffness k, viscosity b, and mass m,

x is change in length,

v is velocity of stretch, and

a is acceleration.

The slope of the force-velocity relation gives the viscosity; however, the stiffness for small displacements can be considerably higher than expected from the length-tension relationship due to the bending of cross bridges. The bending, detachment, and reattachment of the cross bridges gives rise to an apparent viscous drag, which will affect the force-velocity curve, as shown above. The cross-bridge effects are particularly prominent in slow twitch muscle fibers (Malamud, Godt, and Nichols 1996).

The importance of the changing viscoelasticity during activity is shown in figure 9.5, using a simple experiment. The quadriceps muscles of a seated individual were stimulated periodically for 2 s and the knee extended quite smoothly to a stable final position at about 140°. At the end of each stimulation, however, the limb oscillated for several seconds, as expected for a weakly damped pendulum. Clearly, the increased viscosity associated with muscle contraction prevents the oscillation and

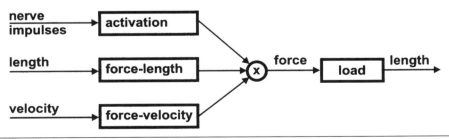

Figure 9.4 The force generated by a muscle depends on activation by nerve impulses as well as the length and velocity of the muscle. To a first approximation the effects of the three factors are multiplicative (×). The load against which the muscle is working affects the force as well as the length changes that are produced.

Knee Angle
[deg]

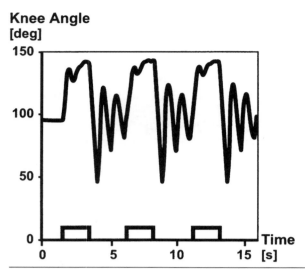

Figure 9.5 The quadriceps muscles of a normal subject were stimulated periodically (boxes at bottom of figure) to produce some knee extension. Note that the movements during stimulation are much more damped (less oscillation) than the movements at the end of stimulation.

permits a smooth movement. Further details on the force-length and force-velocity curves can be found in chapter 15.

There are other factors that influence muscle output, which we only list, although books have been written on each.

Recruitment and Rate Coding

In many of the experiments listed above and in the applications to helping SCI subjects to walk, a muscle(s) is(are) stimulated as a single unit. In normal voluntary contraction, there is a recruitment of motor units (a motoneuron and the muscle fibers it innervates) in order of size from smallest to largest and in order of contractile speed from slowest to fastest. Many other properties are related to the size of a motoneuron, and these properties are known collectively as the size principle (Binder and Mendell 1990; Henneman 1955). Once a motor unit is recruited to fire it will increase its rate of nerve impulses over a certain range as the level of voluntary effort increases (rate coding). These two variables, recruitment and rate coding, are the main neural mechanisms for increasing muscle force.

Fatigue and Training

With repeated or maintained contractions, many of the parameters in the model of figure 9.4 will change due to fatigue within a session or training over a number of sessions. Henneman's size principle, discussed above, will ensure that the slowest motor

units, which are also the most fatigue resistant, are activated first. For larger forces that recruit faster and more fatigable motor units, fatigue becomes more important. Part IV of this book deals with some of the factors involved, so we do not attempt to discuss them in detail. It is important to note, however, that the effects are often particularly profound in disease states where one wishes to use stimulation to replace function after SCI or stroke. For example, we found that with stimulation of the tibialis anterior muscle in SCI subjects at a 50% duty cycle the force declined to about 40% in 3 or 4 min and to near zero after 15 min. With a training program over a period of several months, the force declined to only 80%, and considerable force was maintained over 8 h of stimulation (Stein et al. 1992).

Load

The movement that a muscle can produce depends on the load it is working against. Some of this load arises from the environment: the mass of the object being moved, its friction, and its viscoelasticity. Some also arises from structures within the body: the inertia of a limb or the viscoelasticity of other muscles crossing a joint. The load also dictates what kind of control is appropriate for a muscle. The extraocular muscles, for instance, work against a known and constant load (the eyeball, its socket, and the antagonist muscles). For this reason, the visual system does not need much feedback to control the movement of the eyes. Their control reflects activity of the brain stem nuclei, which control the eye muscles; the nuclei produce a brief pulselike activation to the extraocular motoneurons, followed swiftly by a more prolonged steplike activation. The pulselike activation of the motoneurons moves the eye rapidly to a new position (a movement called a "saccade") and the steplike activation holds it there (Sparks and Mays 1990). Feedback from the visual system helps correct small errors, and information from the vestibular system helps stabilize the eye when the head itself is moving (nystagmus). Other than this, most of the control of the eye muscles is "feedforward" control. This is not true for control of limbs. Limb muscles work against loads that are unpredictable. In walking, for instance, the friction between the foot and the walking surface differs between dirt and ice, sand, or linoleum. In lifting a suitcase, the suitcase's inertia differs if it is full or empty and whether the suitcase contains fluffy sweaters or heavy books. Limb muscles must deal with changes in load arising from the fact that different synergists and antagonists are active in different tasks (chapter 10 by

Herzog). They must also deal with changes in load arising from variation in agonist or synergistic muscles, as previously described. In repetitive movements they must also deal with muscle fatigue, so control has to be continually adjusted. For these reasons, the control of limb movement uses feedback. We now discuss the basics of this feedback.

Feedback (Reflex) Control

A simple feedback controller is shown in figure 9.6a. The α-motoneurons are shown as a summing (+) point, adding a variety of inputs, as first suggested nearly a century ago by Sherrington ([1906] 1961) when he referred to them as the "final common pathway." The forward path contains the muscle and the load, now condensed into a single box that produces a length change. This length change is then fed back to influence the discharge of the motoneuron. Feedback and feedforward control are also referred to as *closed loop* and *open loop* control, respectively, because the sensory information fed back from the output closes the loop back to the motoneuron. Figure 9.6 also shows that by

recording the EMG from the motoneuron during a movement, one cannot tell whether control is open loop or closed loop. Only by adding a perturbation can one determine the extent to which feedback is operating. The thermostat mentioned in the chapter's opening paragraphs is a good example of a feedback system. The temperature has to be measured by a thermometer and compared to the desired temperature to turn the furnace on or off. The thermostat is an example of a finite state controller; that is, the furnace can only be in one of two states (on or off). Other controllers can have an infinite number of states (continuous control). For example, the speed of a car is continuously adjustable depending on the position of the accelerator. In this example, the human is part of the control system and decides the speed and the position of the accelerator to achieve that speed. The process can be automated by the use of cruise control, which can be set to a variety of desired speeds and automatically adjusts the accelerator pedal to maintain that speed despite the terrain. Such a system is often referred to as a *servomechanism*, because the plant is automatically or slavishly (the Latin for

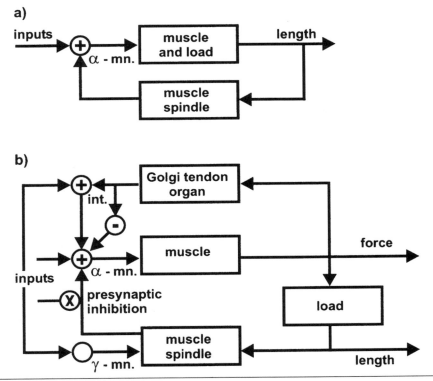

Figure 9.6 *(a)* Feedback system in which muscle and load are combined in the forward loop, and muscle spindles are providing length (and velocity) feedback to the α-motoneurons. *(b)* A more realistic view of feedback control. The Golgi tendon organs provide force feedback that can be switched between inhibitory and excitatory interneurons (Int.) by central inputs. Length feedback through the muscle spindles can be affected by γ-motoneurons as well as presynaptic inhibition, which changes the gain of the feedback to α-motoneurons. Each of the portions of the feedback system receives inputs from a variety of central inputs, as well as other sensory systems.

slave is *servus*) adjusted so as to maintain a desired value.

Stretch Reflexes and Servo-Control

This analogy led Merton (1953) to suggest that muscle spindle sensory feedback could be providing feedback through its direct connection to α-motoneurons (the stretch reflex) as a muscle length servomechanism. A second suggestion was that the desired length was set, not by inputs to α-motoneurons, but to the smaller γ-motoneurons. They innervate small muscle fibers that lie within the muscle spindle itself. The set-point is a common feature of feedback systems (e.g., the desired temperature in the thermostat, the desired speed in the automobile's cruise control). Although Merton's suggestion generated a lot of work, it has now been universally rejected, at least in its original form. There are several reasons for this rejection. First, Merton himself realized that the delays in feedback through the stretch reflex were substantial enough that urgent movements would have to be controlled through direct inputs to α-motoneurons. Secondly, experimental evidence indicates that the stretch reflex is only sufficient to produce a fraction of the maximum voluntary contraction (Matthews 1959; Stein and Kearney 1995), so it couldn't be used to produce the full range of movements. Finally, if the gain of the feedback loop were higher, the system would become unstable (go into oscillation). This is seen in clinical conditions that involve enhanced reflexes such as the spasticity associated with SCI.

To understand the interaction of gain and delay, consider the effect of applying a sinusoidal length change to the system of figure 9.6. This will produce a sinusoidally modulated train of nerve impulses in muscle spindles and α-motoneurons, a sinusoidal modulation of muscle force, and a sinusoidal length change. The gain of the feedback loop in its engineering sense is the ratio of the modulation of output over input. Both should have the same dimensions, such as length, so that the gain is dimensionless. The experiment can be repeated for various sinusoidal frequencies, so loop gain is a function of frequency, not a single number. However, to measure the loop gain, one should open the loop so that the output does not feed back to the input. This can be done in the stick insect, where the stretch receptor can be physically detached from the muscles and joints it controls (Bässler 1986), although this has proven more difficult in mammalian systems. One method is to measure the properties of each box in

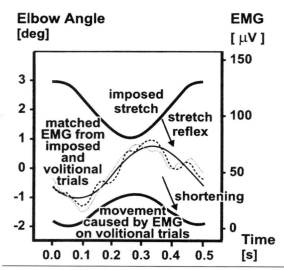

Figure 9.7 A two-part experiment from Bennett, Gorassini, and Prochazka (1994). In the first part the elbow joint was rotated (top trace) with a torque motor that produced a rhythmic EMG. In the second part the subject matched the EMG in a voluntary movement while the torque motor generated a constant torque. The movement that resulted is shown in the bottom trace of the figure.

Reprinted, by permission, from D.J. Bennett, M. Gorassini, and A. Prochazka, 1994, "Catching a ball: Contributions of intrinsic muscle stiffness, reflexes and higher-order responses," *Canadian Journal of Physiology and Pharmacology,* 72:525-534.

figure 9.6 and compute the product to determine the overall loop gain (Stein 1974).

In a clever two-part experiment, Bennett, Gorassini, and Prochazka (1994) measured the loop gain around the human elbow joint (see figure 9.7). In part one, they imposed a sinusoidal stretch of the elbow with a torque motor while the subject was asked to maintain a roughly constant force level and not to react to the movements. The result was a sinusoidally modulated EMG in the elbow muscles. In part two, the subject was asked to match the modulation in EMG while the torque motor was switched to maintain a constant force on the limb. The result was a volitional, approximately sinusoidal movement. The ratio of the output length change from part two with the input length change in part one represents the open loop gain (G_o). In the example shown in figure 9.7, the open loop gain was about 0.5; that is, the response was about half as large as the imposed stretch and basically in the opposite direction.

The stretch reflex, often referred to as a resistance reflex, will resist external displacements. The importance of the loop gain is that it can be used to quantify how much reduction will occur in the external displacement:

$$L_c \cdot L_o^{-1} = 1 \cdot (1 + G_o)^{-1} \qquad\qquad 9.7$$

where

> L_c is the length change that will occur in the presence of feedback (closed loop), and
>
> L_o is the change that will occur in the absence of feedback (open loop).

Thus, for a loop gain of 1 the error will be reduced to a half and for a gain of 9, the error will be reduced to 0.1. As mentioned previously, G_o is a function of frequency and is a vector quantity, since it contains phase as well as gain information. In figure 9.7 there is about a 20° shift in the response, and at some frequency that phase shift will reach 180°. The reflex response in figure 9.7 will then add to the imposed stretch rather than resisting it. In effect, the open loop gain becomes a negative number, and if it equals –1, the ratio in equation 9.7 will be infinite. In other words, the perturbation will cause an oscillation that will build up indefinitely. In practice, the system is nonlinear, as discussed earlier, and the oscillation will build up to a level that is limited by the nonlinearities, that is, until the gain is no longer –1.

Muscle acts sluggishly in that the response to a brief impulse is a longer lasting twitch. In engineering terms it acts as a low-pass filter, responding to high frequency oscillations less well than to low frequency oscillations. As a result, the gain could be higher than 1 at low frequencies and less than 1 at the frequency where the phase shift reaches 180° (typically near the frequency of physiological tremor at about 10 Hz). In fact, Bennett, Gorassini, and Prochazka (1994) found that the gain was near 1 within the normal frequency range of voluntary movements (1–4 Hz). Interestingly, in decerebrate cats Bennett, DeSerres, and Stein (1996) found that the loop gain for the ankle extensor muscles is also near 1 in a postural situation, but it decreased dramatically during locomotion, which can be induced in decerebrate and spinal animals by electrical or chemical means. Similarly, Kearney, Lortie, and Stein, (1999) found that just moving the human ankle joint in the pattern observed during walking decreased the gain of the stretch reflex dramatically (see also Brooke et al. 1996). Further work is needed to determine how reflex gain is modulated during various human movements.

Servo-Assistance and Fusimotor Set

Although, as mentioned earlier, Merton's suggestion of servo-control has been discredited, there is little agreement on exactly how feedback is used. One suggestion that has much more experimental support is the notion of α-γ coactivation. According to this hypothesis α-motoneurons are activated to produce a desired movement, and γ-motoneurons are activated to produce comparable contractions in the small intrafusal muscle fibers within the muscle spindles. The sensory receptors are activated by the contractions of the intrafusal fibers but unloaded by shortening of the parent muscle. In this case, they could serve as a kind of error detector and through the stretch reflex make small corrections between the desired contraction signaled by the γ-motoneuron and the actual contraction produced by the α-motoneuron. This hypothesis has also been referred to as *servo-assistance* (Matthews 1972; Stein 1974) and immediately solves the problems mentioned above. The α-motoneurons get the movement going quickly, eliminating the delay. The stretch reflex only has to correct small errors so it doesn't have to be very powerful, and the gain can be adjusted so that some correction can be produced without compromising stability. The situation is somewhat more complex, in that there are two kinds of muscle spindle sensory fibers (primary and secondary). There are also two kinds of γ-motoneuron (static and dynamic), which have different effects on the sensory receptors, as well as α-motoneurons that innervate both intrafusal and extrafusal muscle fibers (Barker et al. 1977).

To take into account this complexity and the results of experiments in which he continuously recorded from muscle spindle sensory fibers during a variety of behaviors, Prochazka (1989) introduced the concept of fusimotor set, whereby the level and pattern of activity in the two types of fusimotor neurons (another term for γ-motoneurons) would be set by the central nervous system in anticipation of the feedback requirements of the movement before or at the same time as commands to the α-motoneuron that will produce the movement. This idea has intuitive appeal and at least in its broadest description is consistent with much of the experimental data. Servo-assistance could even be considered as one type of fusimotor set.

Inhibitory Modulation

There are other ways in which the central nervous system can and does modify feedback from the periphery. Even the monosynaptic connection from the muscle spindles to α-motoneurons is subject to presynaptic inhibition, and the amount of inhibition is different during standing than it is in walking and in running (Stein and Capaday 1988). Within the step cycle, the amount of inhibition can vary over a few tens of milliseconds. Thus, a stretch of the ankle extensors as the body moves over the foot in the stance phase produces a much larger reflex response to stimulation of muscle spindle sensory fibers than during the swing phase, when the ankle is again flexed by the activation of dorsiflexor muscles (Sinkjær, Jacob, and Birgit 1996). In the stance phase, a stretch reflex can help propel the body forward and upward and assist in push-off, whereas during the swing phase it would lower the toes and could lead to the foot hitting the ground and disturbing the walking. This modulation of the reflexes during the step cycle makes good functional sense in terms of the requirements in the walking cycle. Inhibitory (and excitatory) control can also be applied directly to the motoneuron, but such control is not as specific, since it will change the excitability of the cell to all inputs, not only control the feedback pathway.

Consequently, the diagram of figure 9.6a needs to be made more complex, as shown in figure 9.6b. There are now inputs coming onto α-motoneurons, γ-motoneurons (the complexity of the two types of γ-motoneuron and two types of muscle spindle has been omitted for clarity), and onto presynaptic inhibitory neurons. Since the effect of presynaptic inhibition is to change the gain, this has been indicated by a times (×) sign, rather than the plus (+) sign that was used for the summation done by the α-motoneuron. All three can be modified within and between different behaviors, so we are dealing with a very flexible, adaptive control system.

Golgi Tendon Organs

We have also shown the Golgi tendon organ (GTO) in figure 9.6b and have separated the muscle from the load, so that we can distinguish the force generated by the muscle from the length changes produced by the load. The accepted role of the GTO has gone through a number of stages in the past several years. Originally, GTOs were thought to respond only to large forces and were involved in a protective reflex known as the "clasp knife reflex." This role is still included in many physiology textbooks, although Cleland, Hayward, and Rymer (1990) showed that this reflex came from small, high threshold muscle receptors, not GTOs. Houk and Henneman (1967) showed that, in fact, the GTO was very sensitive to active force generated by the few motor units that impinged on the slip of tendon in which each GTO was found. They suggested that it

would function as a negative feedback loop for force, similar to the way the muscle spindles served as a feedback loop for length. The GTO was also known to have a disynaptic inhibitory pathway; thus, when the force was high, it would reduce the activity of α-motoneurons. Then when the force declined, for example, during fatigue, the inhibition would be reduced (disinhibition), and some compensation could occur. Still, the gain of the pathway was even smaller than for the muscle spindles and in some cases negligible (Houk, Singer, and Goldman 1970; Rymer and Hasan 1980).

More recently, several labs (Gossard et al. 1994; McCrea et al. 1995; Pearson and Collins 1993) have shown that during locomotion the inhibitory postsynaptic potentials (IPSPs) are replaced by excitatory postsynaptic potentials (EPSPs). The negative feedback from GTOs becomes positive feedback and helps to support the weight of the body during the stance phase. The gain of the positive feedback is not enough to produce instability, and several other factors stabilize the reflex loop (Prochazka, Gillard, and Bennett 1997). Essentially, the presence of inhibitory and excitatory interneurons in the feedback pathway from GTOs means that even the sign of the feedback can be changed during walking (see figure 9.6b). Not shown in this figure is that the excitatory path has a longer latency, probably indicating additional interneurons in the pathway and that these interneurons must receive inputs from a variety of sources, including the circuitry that generates the walking rhythm. These recent results further emphasize that the spinal circuitry, rather than being a fixed, simple, even uninteresting level of the motor control system, is indeed a complex, adaptive control system that continues to generate surprises for neuroscientists.

Cutaneous Receptors

Cutaneous and joint receptors, together with some muscle receptors, have traditionally been lumped together under the term flexor reflex afferents, since their stimulation led to a flexor or withdrawal reflex in a decerebrate cat (Lundberg, Malmgren, and Schomburg 1975). The effect of cutaneous receptors has long been known to be more complex, since application of a small piece of acid-soaked paper to the hind limb of a spinal frog may cause the limb to flex or extend. Thus, there is a local sign, depending on the location of the stimulus, and the sign of the response is well correlated with the most effective movement to withdraw the limb from the noxious stimulus (Gordon 1991). Stimulation will also lead to contralateral effects. If one limb is flexed, the opposite limb may have to be extended (crossed extensor reflex) to support the weight of the body. The latency of these cutaneous reflexes is somewhat longer than the earliest reflexes from the muscle spindles and GTOs, suggesting that there are several interneurons involved to compute the appropriate response for a given stimulus.

Only when experiments began with behaving animals and humans did the full complexity of the responses become apparent. Forssberg (1979) investigated the effects of stimulating the dorsal surface of the cat's paw during walking and found that the stimulus during the swing phase elicited a coordinated response at the hip, knee, and ankle. This would be useful to get the paw over the stimulating object and enable locomotion to continue, and he named it the stumbling corrective response. The same stimulus during the stance phase of the step cycle did not elicit this response and might even excite extensors to maintain the support of the body, so this response is also phase dependent. A similar neural and mechanical response has recently been observed in humans using nonnoxious electrical stimulation of the superficial peroneal nerve (figure 9.8), which innervates the dorsal surface of the foot (Zehr, Komiyama, and Stein 1997).

Stimulation of the tibial nerve that innervates the bottom of the foot, again at nonnoxious levels, elicits quite a different pattern. The ankle flexor muscles such as tibialis anterior show a double-burst pattern during walking, the first burst coming early in the swing phase to lift the foot off the ground. The second burst comes late in swing when the foot is about to hit the ground again. Yang and Stein (1990) showed that the effect of stimulating the tibial nerve at the same strength produced an excitation of the TA muscle during the first burst and inhibition of this muscle during the second burst (see figure 9.8), which was later shown to serve a stabilizing function during gait (Zehr, Komiyama, and Stein 1997). DeSerres, Yang, and Patrick (1995) extended the analysis to the level of single motor units and showed again the reflex reversal of the responses. They felt that their results were consistent with the idea of separate excitatory and inhibitory pathways, as shown in figure 9.6b for the GTO. In other words, the cutaneous afferents can activate both excitatory and inhibitory interneurons—which ones are activated depends on the behavior and even the phase of the behavior that is occurring at the time of stimulation. Again, the functional consequences of the reflex reversal are clear. Additional flexion of the ankle, when an object has hit the bottom of the foot or the foot is scraping on the ground, will enable the leg to clear the object and locomotion to continue. In contrast, at the end of swing, when the foot is about to contact the ground, cocontraction of ankle flexors and extensors is needed to stabilize the foot on the ground. If ground contact occurs prematurely, then the response needs to be modified and a suppression of the TA muscle is appropriate to stabilize foot contact on the ground (Zehr, Komiyama, and Stein 1997).

The responses of the cutaneous receptors are qualitatively different than the muscle receptors discussed above. Whereas the muscle receptors

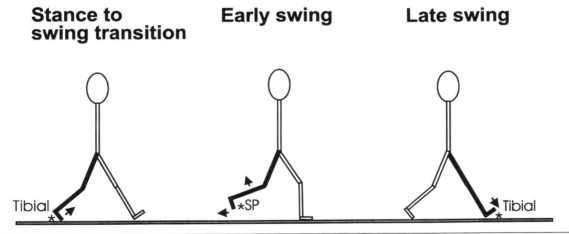

Figure 9.8 Reflex outcome of electrical stimulation of the plantar foot surface (tibial nerve) at the stance to swing transition (left) and late swing (right) and after superficial peroneal (SP) nerve stimulation during early swing in treadmill walking. * indicates area of stimulation and the arrow indicates direction of response.

Reprinted, by permission, from E.P. Zehr, T. Komiyama, and R.B. Stein, 1997, "Cutaneous reflexes during human gait: Electromyographic and kinematic responses to electrical stimulation," *Journal of Neurophysiology* 77:3311-3325.

function to maintain a given state and correct for perturbations, the cutaneous receptors elicit a new prepackaged synergy that may modify the whole behavior so that it can continue (Zehr and Stein 1999). What we have discussed here are the earliest reflex corrections, which all occur within a given step cycle. The presence of distance receptors such as vision may lead to modifications that affect the next or later steps in order to avoid an obstacle. Patla (1991) has carried out a series of elegant experiments to clarify the role of the visual feedback in modifying the locomotor pattern.

Central Control of Locomotion

Although there are various reflexes that modify locomotion, the basic pattern can be generated completely in the central nervous system. The idea of a central pattern generator for animal locomotion is generally attributed to Graham Brown (1911). At that time physiology was under the influence of strong individuals such as Sherrington ([1906] 1961) who had developed the concept of reflexes and favored the idea that locomotion was generated by a closed chain of such reflexes. Similarly, Pavlov was working on conditioning, and he showed that even central behavior could be controlled by pairing peripheral events. The idea of autonomous central generators for a behavior such as locomotion was not fashionable. Only after about 70 years, and the work of another Russian group (Shik and Orlovsky 1976), was it generally accepted. In fact, they rediscovered that the high decerebrate preparation could show walking movements, as Brown himself had demonstrated. In this preparation the higher centers of the brain are removed above the level of the midbrain. Electrical stimulation of a region in the midbrain, which has become known as the *mesencephalic locomotor region* or MLR produces rhythmic alternation of the limbs of a cat placed over a moving treadmill belt (see Whelan 1996 for a review). Because the higher centers have been removed, the cat is not able to balance itself and must be suspended, but the limb movements convincingly show most patterns of normal locomotion. Higher levels of stimulation can produce trotting and even galloping patterns.

Although the midbrain controls locomotion, it is not part of the central pattern generator. This was demonstrated in spinal cats by the application of drugs such as L-dopa and clonidine, which are thought to mimic the action of the transmitters released by descending systems (Grillner 1981;

Pearson and Rossignol 1991). Thus, the pattern generators for locomotion in four-legged animals such as the cat and the rat can be localized to the spinal cord. Indeed, after sectioning the spinal cord between the fore and hind limbs, these drugs will produce alternation of the hind limbs alone. Since the pattern between the two hind limbs can vary from being out of phase (as in walking or trotting) to being in phase (as in galloping), the simplest hypothesis is that there is a central pattern generator controlling each limb and the four generators in a quadruped such as a cat are coordinated through intraspinal connections.

Proof that the circuitry wholly within the spinal cord was sufficient, as well as necessary, was obtained in experiments in which dorsal roots were cut to prevent sensory feedback (reviewed by Grillner 1981; Rossignol 1996). Although the patterns are less precise, the cat still generated walking rhythms. These experiments are not totally convincing, since it is technically difficult to remove absolutely all sensory feedback. In other experiments, however, an animal was paralyzed using the neuromuscular blocking agent curare, so there was no movement and hence no rhythmic feedback from sensors. Despite this, rhythmic bursts of activity in motoneurons were observed when the cord was stimulated electrically from the MLR or from application of drugs. These studies provided solid evidence of the existence of a central pattern generator in experimental animals, but the form of the neural oscillator is still uncertain.

Brown (1911) proposed the idea of mutually inhibitory half-centers, which has dominated thinking ever since (see figure 9.9). In this concept, descending inputs can excite either half-center, but are probably stronger to one center. As a result, one becomes more excited than the other, and the more active center further inhibits the less active one through the mutually inhibitory connections, which in turn removes inhibition from the more active center. Consequently, there is positive feedback around the loop of figure 9.9, since a minus times a minus is a plus. Central to Brown's concept was the notion of fatigue; as the activity in the more active center decreases, at some point it can no longer inhibit the less active center. The two switch roles and this rhythmic alternation continues for some period of time.

Technically, it has been exceedingly difficult to prove or disprove this hypothesis in vertebrate locomotion. Grillner, Wallen, and Brodin (1991) have worked out the circuitry for locomotion in a simple vertebrate, the lamprey, which is illustrated in figure 9.10. Lampreys swim rather than walk so this is

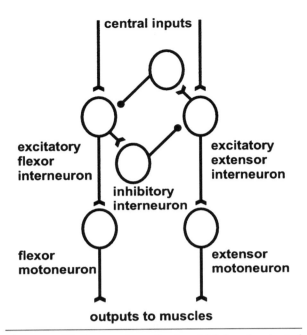

Figure 9.9 Half-center hypothesis, which has dominated thinking on neural oscillators for locomotion, showing mutually inhibitory flexor and extensor centers.

Reprinted, by permission, from R. Tomović, D. Popović, and R.B. Stein, 1995, *Nonanalytical methods for motor control.* (Singapore: World Scientific Publishing).

the form of locomotion illustrated in the figure. In the middle of the diagram are two mutually inhibitory neurons, as expected from the half-center hypothesis. These CCIN neurons have axons that cross over the midline and appear to be involved in alternation of the two sides of the body, which is required for rhythmic swimming. There are other excitatory (EIN) and inhibitory (LIN) interneurons that elaborate and reinforce the basic half-center. Grillner, Wallen, and Brodin (1991) have simulated the patterns using values for synaptic connections and ionic properties that they studied experimentally. The patterns in all the cell types in the simulations agree quite well with the experimental data.

In the lamprey, several mechanisms control the switching between the two sides. The firing rate of the CCIN decreases because of intrinsic mechanisms involving summation of afterpotentials and ionic mechanisms. These are the modern basis for what Brown called fatigue. In addition, the LIN has a high threshold, so its action is delayed. When it eventually fires it will inhibit the CCIN and help to terminate the burst; thus, there are network properties as well as intrinsic mechanisms involved in the switching. Connections from "edge" cells (sensory cells that monitor the bending of the trunk) feed back to the segmental oscillator and reinforce the pattern. Although only the circuit for a single segment is shown, the circuit is reproduced many times and intersegmental connections are present to produce the smooth traveling wave that is necessary for swimming (Cohen 1988). Thus, several mechanisms combine to ensure smooth, coordinated swimming.

Many of the synaptic transmitters used in the lamprey are common to higher vertebrates so there

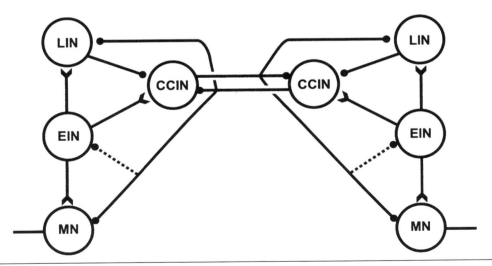

Figure 9.10 Neuronal circuitry for locomotion (swimming) in the lamprey. Note the mutual inhibition between interneurons with crossed connections (CCIN) on the two sides of the body. The bursts of activity of the motoneurons (MN) on each side are reinforced by excitatory interneurons (EIN) and lateral inhibitory interneurons (LIN).

Reprinted, by permission, from S. Grillner, P. Wallen, and L. Brodin, 1991, "Neuronal network generating locomotor behavior in lamprey: Circuitry, transmitters, membrane properties and simulation," *Annual Review of Neuroscience* 14:184. © 1991, by Annual Reviews http://www.AnnualReviews.org.

may well be common principles. Preliminary work on the circuitry for walking in an amphibian suggests that there may be important differences, too (Wheatley, Lawson, and Stein 1994). Cheng and co-workers (1998) found that the flexor and extensor half-centers could generate bursts of activity even when separated from each other. Inhibitory connections are present between the two centers, but they are not necessary for the rhythmicity to occur. Sensory regulation may be much more important in terrestrial locomotion, since the requirements for balance are so much more important. For example, not allowing the hip to extend can prevent the swing phase of locomotion from occurring (Grillner and Rossignol 1978) as can loading at the ankle (Duysens and Pearson 1980). Recent work suggests that the receptors involved in the loading effects are the GTOs, which, as described previously, produce positive feedback to maintain the stance phase (Pearson and Collins 1993).

This trend toward the importance of sensory inputs to maintain balance during forward progression should be even more evident in bipedal animals such as humans. Indeed, there is no definite proof of a spinal pattern generator in humans, although there is some evidence of a half-center organization in paraplegic subjects (Bussel et al. 1988; Calancie et al. 1994; Dietz, Columbo, and Jensen 1994). There have been major reviews recently on the neural basis of locomotion (Rossignol 1996; Whelan 1996), but much work remains to be done. One reason for the concentration of work on locomotion is that it is a rhythmic behavior. Consequently, many cycles can be studied over a period of several minutes so that the details of the behavior can be analysed. We anticipate that similar, exquisitely adapted control strategies will be discovered for other behaviors in the future. As well as biologists and biomechanists learning from engineers who have developed sophisticated artificial control systems, we feel that engineers can learn much in the coming years from the successful control systems that have evolved in natural, biological systems.

References

Bagshaw, C.R. 1993. *Muscle contraction*. London: Chapman & Hall.

Barany, M. 1967. ATPase activity of myosin correlated with speed of muscle shortening. *J. Gen. Physiol.* 50: 197-218.

Barker, D., Emonet-Denand, F., Harker, D.W., Jami, L., and Laporte, Y. 1977. Types of intra- and extrafusal muscle fiber innervated by dynamic skeletofusimotor

axons in cat peroneus brevis and tenuisimus muscles, as determined by the glycogen depletion method. *J. Physiol. (London)* 266: 713-26.

Bässler, U. 1986. Afferent control of walking movements in the stick insect *Cuniculina impigra*. *J. Comp. Physiol.* 158: 351-62.

Bennett, D.J., DeSerres, S.J., and Stein, R.B. 1996. Gain of the triceps surae stretch reflex in decerebrate and spinal cats during postural and locomotor activities. *J. Physiol. (London)* 496: 837-50.

Bennett, D.J., Gorassini, M., and Prochazka, A. 1994. Catching a ball: Contributions of intrinsic muscle stiffness, reflexes, and higher order responses. *Can. J. Physiol. Pharmacol.* 72: 525-34.

Binder, M.D., and Mendell, L.M. 1990. *The segmental motor system*. Oxford: Oxford University Press.

Bobet, J., and Stein, R.B. 1998. A simple model of force generation in isometric skeletal muscle. *I.E.E.E. Trans. Biomed. Eng.* 45: 1010-16.

Bobet, J., Stein, R.B., and Oğuztöreli, M.N. 1993. A linear time-varying model of force generation in skeletal muscle. *I.E.E.E. Trans. Biomed. Eng.* 40: 1000-6.

Brooke, J.D., Cheng, J., Collins, D.F., McIlroy, W.E., Misiaszek, J.E., and Staines, W.R. 1996. Sensori-sensory afferent conditioning with leg movement: Gain control in spinal reflex and ascending paths. *Prog. Neurobiol.* 51: 393-421.

Brown, T.G. 1911. The intrinsic factors in the act of progression in the mammal. *Proc. Roy. Soc.* B84: 308-19.

Bussel, B., Roby-Brami, A., Azouvi, P., Biraben, A., Yakoleff, A., and Held, J.P. 1988. Myoclonus in a patient with spinal cord transection: Possible involvement of the spinal stepping generator. *Brain* 111: 235-45.

Calancie, B., Needham-Shropshire, B., Jacobs, P., Willer, K., Zych, G., and Green, B.A. 1994. Involuntary stepping after chronic spinal cord injury: Evidence for a central rhythm generator for locomotion in man. *Brain* 117: 1143-59.

Cheng, J., Stein, R.B., Jovanović, K., Yoshida, K., Bennett, D.J., and Han, Y. 1998. Identification, localization, and modulation of neural networks for walking in the mud puppy *(Necturus maculatus)* spinal cord. *J. Neurosci.* 18: 4295-304.

Chizeck, H.J., Lan, N., Palmieri, L.S., and Crago, P.E. 1991. Feedback control of electrically stimulated muscle using simultaneous pulse width and stimulus period modulation. *I.E.E.E. Trans. Biomed. Eng.* 38: 1224-34.

Cleland, C.L., Hayward, L., and Rymer, W.Z. 1990. Neural mechanisms underlying the clasp knife reflex in the cat. II. Stretch-sensitive muscular free nerve endings. *J. Neurophysiol.* 64: 1319-30.

Cohen, A.H. 1988. Evolution of the vertebrate central pattern generator for locomotion. In *Neural control of rhythmic movements in vertebrates*, ed. A.H. Cohen, S. Rossignol, and S. Grillner, 129-66. New York: Wiley.

D'Azzo, J.J., and Houpis, C.H. 1966. Feedback control system analysis and synthesis. New York: McGraw-Hill.

De Serres, S.J., Yang, J.F., and Patrick, S.K. 1995. Mechanism for reflex reversal during walking in human tibialis anterior muscle revealed by single motor unit recording. *J. Physiol. (London)* 488: 249–58.

Dietz, V., Colombo, G., and Jensen, L. 1994. Locomotor activity in spinal man. *Lancet* 344: 1260–62.

Duysens, J., and Pearson, K.G. 1980. Inhibition of flexor burst generation by loading ankle extensor muscles in walking cats. *Brain Res.* 187: 321-32.

Forssberg, H. 1979. Stumbling corrective reaction: A phase-dependent compensatory reaction during locomotion. *J. Neurophysiol.* 42: 936-53.

Gordon, A.M., Huxley, A.F., and Julian, F.J. 1966. The variation in isometric tension with sarcomere length in vertebrate muscle fibers. *J. Physiol. (London)* 184: 170-92.

Gordon, J. 1991. Spinal mechanisms of motor coordination. In *Principles of neural science*. 3d ed. Ed. E.R. Kandel, J.H. Schwartz, and T.M. Jessell, 581–95. Amsterdam: Elsevier.

Gossard, J.P., Brownstone, R.M., Barajon, I., and Hultborn, H. 1994. Transmission in a locomotor-related group Ib pathway from hind limb extensor muscles in the cat. *Exp. Brain Res.* 98: 213–28.

Grillner, S. 1981. Control of locomotion in bipeds, tetrapods, and fish. In *Handbook of physiology: Motor control.* Sec. 1, vol. II. Ed. V.B. Brooks, 1179-236. Baltimore: Williams & Wilkens.

Grillner, S., and Rossignol, S. 1978. On the initiation of the swing phase of locomotion in chronic spinal cats. *Brain Res.* 146: 269-77.

Grillner, S., Wallen, P., and Brodin, L. 1991. Neuronal network generating locomotor behavior in lamprey: Circuitry, transmitters, membrane properties, and simulation. *Ann. Rev. Neurosci.* 14: 169-99.

Henneman, E. 1955. Relation between the size of neurons and their susceptibility to discharge. *Science* 126: 1345–46.

Hill, A.V. 1938. The heat of shortening and the dynamic constants of muscle. *Proc. Roy. Soc. Lond.* B126: 136-95.

Hill, D.K. 1968. Tension due to interaction between sliding filaments in resting striated muscle: The effect of stimulation. *J. Physiol. (London)* 199: 637-84.

Houk, J.C., and Henneman, E. 1967. Responses of Golgi tendon organs to active contractions of the soleus muscle of the cat. *J. Neurophysiol.* 30: 466-81.

Houk, J.C., Singer, J.J., and Goldman, M. 1970. An evaluation of length and force feedback in decerebrate cats. *J. Neurophysiol.* 33: 784-811.

Huxley, A.F. 1957. Muscle structure and theories of contraction. *Prog. Biophys. Biophys. Chem.* 7: 255-318.

Kearney, R.E., and Hunter, I.W. 1990. System identification of human joint dynamics. *Crit. Rev. Biomed. Eng.* 18: 55-87.

Kearney, R.E., Lortie, M., and Stein, R.B. 1999. Modulation of stretch reflexes during human walking: Mechanisms and functional implications. *J. Neurophysiol.* 81: 2893-902.

Lundberg, A., Malmgren, K., and Schomburg, E.D. 1975. Convergence from Ib, cutaneous, and joint afferents in reflex pathways to motoneurons. *Brain Res.* 87: 81–84.

Lymn, R.W., and Taylor, E.W. 1971. Mechanism of adenosine triphosphate hydrolysis by actomyosin. *Biochemistry* 10: 4617-24.

Malamud, J.G., Godt, R.E., and Nichols, T.R. 1996. Relationship between short-range stiffness and yielding in type-identified, chemically skinned muscle fibers from the cat triceps surae muscles. *J. Neurophysiol.* 76: 2280–9.

Mannard, A.C., and Stein, R.B. 1973. Determination of the frequency response of isometric soleus muscle in the cat using random nerve stimulation. *J. Physiol. (London)* 229: 275-96.

Mansour, J.M., and Audu, M.L. 1986. The passive elastic moment at the knee and its influence on human gait. *J. Biomech.* 19: 369–73.

Matthews, P.B.C. 1959. The dependence of tension upon extension in the stretch reflex of the soleus muscle of the decerebrate cat. *J. Physiol. (London)* 147: 521-46.

Matthews, P.B.C. 1972. *Mammalian muscle receptors and their central actions.* London: Arnold.

McCrea, D.A., Shefchyk, S.J., Stephens, M.J., and Pearson, K.G. 1995. Disynaptic group I excitation of synergist ankle extensor motoneurons during fictive locomotion in the cat. *J. Physiol. (London)* 487: 527–39.

Merton, P.A. 1953. Speculations on the servo-control of movement. In *The spinal cord: CIBA Foundation symposium,* ed. G.E.W. Wolstenholme, 247-55. London: Churchill.

Patla, A.E. 1991. Visual control of human locomotion. In *Adaptability of human gait,* ed. A.E. Patla, 55–97. Amsterdam: Elsevier.

Pearson, K.G., and Collins, D.F. 1993. Reversal of the influence of group 1b afferents from plantaris on activity in medial gastrocnemius muscle during locomotor activity. *J. Neurophysiol.* 70: 1009–17.

Pearson, K.G., and Rossignol, S. 1991. Fictive motor patterns in chronic spinal cats. *J. Neurophysiol.* 66: 1874–87.

Popović, D., Stein, R.B., Oğuztöreli, M.N., Lebiedowska, M., and Jonić, S. 1999. Optimal control of walking with functional electrical stimulation: A computer simulation study. *I.E.E.E. Trans. Rehab. Eng.* 7: 69-79.

Prochazka, A. 1989. Sensorimotor gain control: A basic strategy of motor systems? *Prog. Neurobiol.* 33: 281-307.

Prochazka, A., Gillard, D., and Bennett, D.J. 1997. Implications of positive feedback in the control of movement. *J. Neurophysiol.* 77: 3237–51.

Rossignol, S. 1996. Neural control of stereotypic limb movements. In *Handbook of physiology: Regulation and integration of multiple systems.* Sec. 12, ed. L.B. Rowell and J.T. Shepherd, 173–216. New York: Oxford University Press.

Rymer, W.Z., and Hasan, Z. 1980. Absence of force-feedback regulation in soleus muscle of the decerebrate cat. *Brain Res.* 184: 203-9.

Sherrington, C.S. [1906] 1961. *The integrative action of the nervous system.* New Haven, CT: Yale University Press.

Shik, M.L., and Orlovsky, G.N. 1976. Neurophysiology of locomotor automatism. *Physiol. Rev.* 56: 465-501.

Shue, G., Crago, P.E., and Chizeck, H.J. 1995. Muscle-joint models incorporating activation dynamics, torque-angle, and torque-velocity properties. *I.E.E.E. Trans. Biomed. Eng.* 42: 212–23.

Sinkjaer, T., Jacob, B.A., and Birgit, L. 1996. Soleus stretch reflex modulation during gait in humans. *J. Neurophysiol.* 76: 1112–20.

Sparks, D.L., and Mays, L.E. 1990. Signal transformation required for the generation of saccadic eye movements. *Ann. Rev. Neurosci.* 13: 309–36.

Stein, R.B. 1974. The peripheral control of movement. *Physiol. Rev.* 54: 215-43.

Stein, R.B., and Capaday, C. 1988. The modulation of human reflexes during functional motor tasks. *Trends Neurosci.* 11: 328-32.

Stein, R.B., Gordon, T., Jefferson, J., Sharfenberger, A., Yang, J., Totosy de Zepetnek, J., and Bélanger, M. 1992. Optimal stimulation of paralyzed muscle in spinal cord injured patients. *J. Appl. Physiol.* 72: 1393–1400.

Stein, R.B., Gordon, T., and Shriver, J. 1982. Temperature dependence of mammalian muscle contractions and ATP activities. *Biophys. J.* 40: 197-207.

Stein, R.B., and Kearney, R.E. 1995. Nonlinear behavior of muscle reflexes at the human ankle joint. *J. Neurophysiol.* 73: 65–72.

Stein, R.B., Zehr, E.P., Lebiedowska, M.K., Popović, D.B., Scheiner, A., and Chizeck, H.J. 1996. Estimating mechanical parameters of leg segments in individuals with and without physical disabilities. *I.E.E.E. Trans. Rehab. Eng.* 4: 201–11.

Tomović, R., Popović, D., and Stein, R.B. 1995. *Nonanalytical methods for motor control.* Singapore: World Scientific.

Warshaw, D.M. 1996. The in vitro motility assay: A window into the myosin molecular motor. *News Physiol. Sci.* 11: 1–7.

Wheatley, M., Lawson, V., and Stein, R.B. 1994. The activity of interneurons during locomotion in the in vitro *Necturus* spinal cord. *J. Neurophysiol.* 71: 2025–32.

Whelan, P.J. 1996. Control of locomotion in the decerebrate cat. *Prog. Neurobiol.* 49: 481–515.

Winters, J.M., and Stark, L. 1985. Analysis of fundamental human movement patterns through the use of in-depth antagonistic muscle models. *I.E.E.E. Trans. Biomed. Eng.* 32: 826–39.

Yang, J.F., and Stein, R.B. 1990. Phase-dependent reflex reversal in human leg muscles during walking. *J. Neurophysiol.* 63: 1109-17.

Zehr, E.P., Komiyama, T., and Stein, R.B. 1997. Cutaneous reflexes during human gait: Electromyographic and kinematic responses to electrical stimulation. *J. Neurophysiol.* 77: 3311–25.

Zehr, E.P., and Stein, R.B. 1999. What functions do reflexes serve during human locomotion? *Prog. Neurobiol.* 58: 185-205.

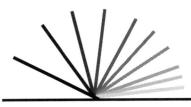

Chapter 10

Muscle Activation and Movement Control

W. Herzog

Voluntary movements are prescribed by the joint moment requirements. The moments about joints, in turn, are caused almost exclusively by muscular forces, although ligaments, bony contact forces, and other soft tissue forces may contribute to the joint moment. Therefore, an understanding of movement control requires knowledge of how a single muscle's force is produced and controlled and how the muscles crossing a given joint interact with each other during a given movement. Other factors also determine movement control. These factors include considerations regarding the control of an entire system rather than just a single joint (after all, we find multijoint muscles virtually everywhere in the human body, therefore muscles crossing one joint may also directly influence the movement at another joint) and considerations regarding the skill with which a movement is performed. Here, we deal almost exclusively with movement control at a single joint for highly skilled movements such as locomotion. Needless to say, the issues surrounding limb and whole body movement control, as well as issues related to skill acquisition, are exciting and cannot be excluded completely, but they will play a minor role in the following chapter. This restriction of the topic is based on the scope of this text and the

available space; it is not based on the importance of the issues involved in movement control.

The aim of this chapter is to cover basic aspects of muscle activation and the control of force in single fibers, motor units, and muscles. Furthermore, issues regarding the synergistic force sharing of muscles crossing a joint and, therefore, moment and movement control at a joint are addressed. This latter topic is covered from an experimental and theoretical perspective. Last, I attempt to summarize basic research findings in the context of applied sport sciences and support my arguments with selected examples. For reasons stated above, this chapter is not comprehensive and should not be viewed as such. Rather, it represents, in the view of the author, a selection of important and interesting aspects of movement control.

Muscle Activation

Muscles are activated by nerve signals. Commands from the central nervous system are transmitted to the muscle via α-motoneurons—neurons that leave the spinal cord through the ventral roots and terminate at specialized sites (the neuromuscular junctions) on muscle fibers. When an action potential

Acknowledgment: The Killam Foundation

of an α-motoneuron reaches the end of the axon (the presynaptic terminal; see figure 10.1), a series of chemical reactions will be triggered that culminate in the release of acetylcholine (ACh) from synaptic vesicles located in the presynaptic terminal. Acetylcholine diffuses across the synaptic cleft, binds to receptor molecules of the postsynaptic membrane, and causes an increase in permeability of the muscle fiber (cell) membrane to sodium (Na^+) ions. Typically, the depolarization of the membrane caused by sodium entry exceeds a critical threshold, and an action potential is generated that will travel along the muscle fiber. Acetylcholinesterase, which is located in the basement membrane of the muscle fiber, terminates the action of ACh by splitting it into acetic acid and choline. The muscle fiber action potential reaches the interior of the fiber at invaginations of the cell membrane called transverse (T) tubules. Depolarization of the membrane within the T-tubules causes a release of calcium ions (Ca^{2+}) from the sarcoplasmic reticulum into the sarcoplasm throughout the myofibrils. Ca^{2+} binds to the troponin-C site on the thin (actin) myofilaments. Troponin C is a subunit of troponin. It contains Ca^{2+}-binding sites to which Ca^{2+} will bind, resulting in removal of the inhibitory influence of the troponin-tropomyosin complex. Cross-bridge formations and cross-bridge cycling now occur, and force production is achieved, presumably by cross-bridge rotation and a sliding of the thin (actin) past the thick (myosin) filaments, provided that the external resistance is small. Cross-bridge interaction requires energy, which is provided by the breakdown of ATP (adenosine triphosphate) into ADP (adenosine diphosphate) plus a phosphate ion. Using energy from ATP, Ca^{2+} is actively pumped back into the sarcoplasmic reticu-

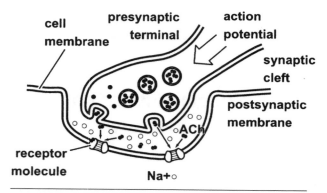

Figure 10.1 Schematic drawing of a neuromuscular junction showing the presynaptic nerve terminal and the motor end plate. Activation crosses the synaptic cleft via a neurotransmitter, acetylcholine, ACh.

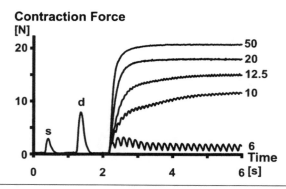

Figure 10.2 Force-time histories of contractions at different frequencies of stimulation in the cat soleus muscle. s = single twitch, d = doublet twitch (i.e., two stimuli separated by 8 ms). Numbers on the right designate frequency (Hz) of stimulation. Note the dramatic increase of the doublet twitch compared to the single twitch and that a stimulation at 20 Hz produces a virtually fused contraction in the slow-twitch fibered soleus.

lum, actomyosin cross-bridge cycling stops, and force production ceases.

The isometric force produced by a muscle fiber of given cross-sectional area is determined by the frequency of stimulation (assuming in vivo, steady state conditions in a nonfatigued fiber). A single motoneuron action potential will elicit a single force response, called a twitch (see figure 10.2). When increasing the frequency of stimulation, force in a fiber (or motor unit or muscle) becomes larger, and force oscillations become smaller. At a critical frequency, force production will become smooth; below this critical frequency, muscular contraction is said to be unfused, above it, the contraction is called fused.

During voluntary contraction, it is not possible to activate a single muscle fiber. A muscle is organized in terms of motor units (MUs); a motor unit is a motor neuron and all the muscle fibers that are innervated by this single neuron. Therefore, activation of a motor neuron will cause force production in all fibers contained in one MU. The smallest unit of force control in a muscle, thus, is the MU and not a single fiber.

Motor units may contain only a few muscle fibers in small muscles in which high precision is required (human eye muscles), or they may contain thousands of fibers in large muscles of the human limbs. It is intuitively obvious that the force in a MU depends on the frequency of stimulation (as for the single fiber) and the size of the MU. Size here refers to the accumulative myofibrillar cross-sectional area or number of fibers contained in a given MU. In this context, it is important to note that the size of MUs in a given muscle may vary considerably.

Finally, activation of the muscle will result in force production. The magnitude of muscle force will depend on the number of motor units that are activated, the frequency of stimulation, and the size of the activated MUs. It is believed that for many forms of contraction, MUs are recruited in a set order. For a graded increase in isometric force, MUs are thought to be recruited according to the size principle (Henneman and Olson 1965; Henneman, Somjen, and Carpenter 1965). The size principle states that small MUs are recruited first when muscle force requirements are low, and with increasing force demands, progressively larger motor units are recruited. Functionally, the small MUs are typically composed of slow twitch fibers, whereas the large MUs are typically composed of fast twitch fibers. A recruitment order according to size is therefore meaningful from a functional point of view, because slow twitch MUs, which are fatigue resistant, will be recruited during low level everyday tasks, which may be performed for long periods of time such as maintaining posture, whereas the fast MUs are only recruited occasionally when the speed or force requirements of the muscle are high, thereby preserving these predominantly fast fatigable MUs whenever possible. Obviously, a muscle is not comprised of two distinct types of fibers with two distinctly different properties. Rather, muscle fiber types and the corresponding properties are continuous. In that sense the above discussion is vastly simplified but applies in general to real muscles with continuous fiber types and fiber properties.

The size principle of MU recruitment within a given muscle is generally accepted as correct for isometric and slow concentric contractions in which forces are built up in a graded (slow) fashion. However, experimental evidence suggests that the MU recruitment order might be changed for high speed and eccentric contractions (Enoka 1996; Nardone, Romanò, and Schieppati 1989). In fact, pilot work suggests that the electromyographic (EMG) activity of the knee extensor muscles during isometric contractions is different when a concentric contraction, compared to an eccentric contraction, is expected to follow the isometric contraction (Grabiner et al. 1995). This change in EMG activity for an isometric contraction in anticipation of a concentric or eccentric contraction suggests that it is not the movement that causes a potential change in MU recruitment order but the plan to perform the movement.

The size principle has been determined for the MU recruitment order in single muscles. It has not been established whether the size principle can also be applied across several muscles that comprise an agonistic group (e.g., the knee extensor muscles). The question of what principles may govern the interaction and coordination of agonistic, antagonistic, and synergistic muscles is the topic of the next section.

Movement Control: Experimental Considerations

The question of movement control can be approached on several levels: whole body movement control, limb movement control, joint movement control, or single muscle or MU control and their contributions to overall joint movements. Here, I would like to focus on single joint movement control or, better, the synergistic force sharing among muscles crossing a single joint. This preference is associated with the fact that multiple muscle force recordings have been restricted to single joints. The primary joint studied is the cat ankle and the corresponding ankle extensor and flexor muscles. Particularly, the cat ankle extensor muscles have been examined and characterized extensively. Selected morphometric data of the primary ankle extensors in the cat (soleus, gastrocnemius, and plantaris) are given in table 10.1.

Walmsley, Hodgson, and Burke (1978) were the first to measure multiple muscle forces from the cat ankle extensor group during locomotor activities. Specifically, they recorded soleus and medial gastrocnemius (MG) forces and EMGs for walking, trotting, galloping, and jumping, spanning a range of speeds from about 0.4 m/s to 3.5 m/s. They found that soleus peak forces and integrated EMGs remained about constant, independent of the speed or mode of locomotion (see figure 10.3). This result was interpreted to show that the soleus was fully activated (EMG) at very low speeds of locomotion, and thus activation could not be increased when ankle extensor requirements increased with increasing speeds of locomotion. Furthermore, the consistency of the soleus peak force magnitudes across speeds was interpreted to mean that force-enhancing changes in the muscle (more and faster speeds of stretch of the soleus at high compared to low speeds of locomotion) were offset by corresponding force-depressing changes (decreased times to develop a fully activated state at fast compared to slow speeds of locomotion). Interestingly enough, results shown by Walmsley, Hodgson, and Burke (1978) for jumping activities showed that their interpretation of the locomotion results could not be correct. Particularly, the proposal that the soleus was fully activated at even the lowest speeds of locomotion must be

Table 10.1 Morphometric Data of the Primary Ankle Extensors in the Cat

		Muscles	
	Soleus	Gastrocnemius	Plantaris
Joints crossed	A	A, K	A, K, M[a]
Fiber lengths (cm)	3.9	2.4	2.1
PCSA (cm^2)	1.0	11.2	4.0
Volume (cm^3)	4.1	27.1	8.5
ST/FT (%)	100/0	21/79	26/74
F_o (N)	36	237	76
Angle of pinnation[b] (degree)	7	19	14

[a]The plantaris has an in-series attachment with the flexor digitorum brevis which crosses the metatarsophalangeal joint.
[b]Angle of pinnation measured in the relaxed state at about optimal length.

The values shown here are averages across 3-10 adult animals with a minimum weight of 35N. A = ankle; K = knee; M = metatarsophalangeal joint; PCSA = physiological cross-sectional area; ST/FT = the percentage of slow- and fast-twitch fibers; F_o = maximal isometric force at optimal muscle length.

questioned in view of their own data showing a large increase in soleus activation (EMG) for jumping compared to all locomotion tasks.

Hodgson (1983) repeated the study performed by Walmsley, Hodgson, and Burke (1978) with similar results. Soleus peak forces and integrated EMGs remained essentially constant for a large range of locomotor activities, whereas MG peak forces and

EMGs increased with increasing speeds of locomotion. In addition, Hodgson (1983) pointed out that during standing, MG may be completely deactivated and carry no force, whereas soleus was activated nearly as much during still standing as during walking and trotting. Hodgson's (1983) interpretation of the results differed from that of Walmsley, Hodgson, and Burke (1978). He argued

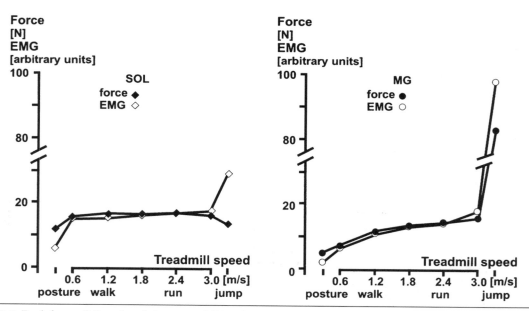

Figure 10.3 Peak forces (N) and peak integrated (50 ms) EMG values (arbitrary units) for the cat soleus (SOL) and medial gastrocnemius (MG) under a variety of movement conditions. Note the steady increase in the peak values of the MG forces and EMGs with increasing ankle extensor requirements and the relatively steady values (at least for all locomotion tests) for the soleus forces and EMGs.

Adapted, by permission, from B. Walmsley, J.A. Hodgson, and R.E. Burke, 1978, "Forces produced by medial gastrocnemius and soleus muscles during locomotion in freely moving cats," *Journal of Neurophysiology* 41:1210.

that increasing the speed of locomotion was associated with an increased excitatory drive to both the soleus and MG. However, with increasing speeds, he argued, there was a selective inhibition of slow MUs through cutaneous and rubrospinal pathways, affecting primarily the slow-twitch fibered soleus (see table 10.1). According to Hodgson (1983), soleus was not fully activated (as claimed by Walmsley, Hodgson, and Burke 1978) even at the fastest speed of locomotion tested.

In a later study, we attempted to resolve the conflicting opinions of Walmsley, Hodgson, and Burke (1978) and Hodgson (1983) regarding soleus activation and force potential. In this study (Herzog and Leonard 1996), we instrumented the cat soleus with indwelling, fine wire EMG electrodes, a force transducer, and a stimulating nerve cuff electrode. With this setup, we could measure forces and activation of the soleus continuously during unrestrained locomotion, and we could stimulate the soleus supramaximally at any instant in time during the step cycle and thus elicit maximal, in vivo

forces (see figure 10.4). We found that for slow walking (0.4 m/s), the soleus was typically not fully activated, that is, a supramaximal stimulus to the soleus nerve (3 × α-motoneuron threshold, 0.1 ms pulse duration, 100 Hz for 100 ms) elicited soleus forces that exceeded the forces observed during voluntary step cycles. Based on these results, we had to reject the idea proposed by Walmsley, Hodgson, and Burke (1978) that soleus is fully activated at even the slowest speeds of walking.

In an effort to gain more insight into the problem of force sharing among synergistic muscles, we developed the techniques to measure more than just two muscle forces at the cat ankle. Typically, we can perform simultaneous force and EMG recordings from soleus (S), gastrocnemius (G), plantaris (P), and the tibialis anterior (TA). When measuring forces across a large range of speeds of locomotion, it appears that P behaves similar to G: its force and activation increases with increasing speeds of locomotion (see figure 10.5). Also, TA forces and EMGs increase with increasing speeds of locomotion.

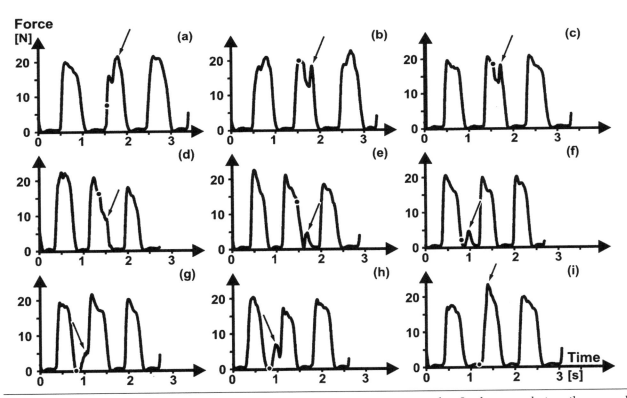

Figure 10.4 Force-time histories of the cat soleus for three consecutive step cycles. In the second step, the normal activation of the soleus was perturbed by superimposing a supramaximal stimulus (100 ms duration, 0.1 ms pulse width, 100 Hz, 3 × α-motoneuron threshold) onto the soleus nerve (indicated by the dot). The arrows indicate corresponding force effects. The dashed arrow indicates a small or no force effect; solid arrows indicate large and identifiable force effects. Note: the nine situations shown are examples of giving the superimposed artificial stimulus at different times during the step cycle.

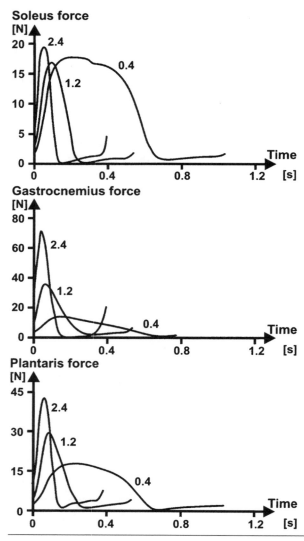

Figure 10.5 Force-time histories of cat soleus, gastrocnemius, and plantaris for walking at speeds of 0.4 m/s and 1.2 m/s and trotting at 2.4 m/s. Note the steady increase in the peak forces of gastrocnemius and plantaris with increasing speeds of locomotion and the virtually unaltered peak forces of the soleus. The traces shown are averaged across 10 consecutive step cycles.

Furthermore, at slow walking speeds S, G, and P forces occur during the stance phase, and TA forces occur during the swing phase of the step cycle. Cocontraction between the TA and any of the ankle extensors is small or even zero. However, at trotting speeds (> 1.2 m/s), TA and ankle extensor forces start to overlap at the transitions from swing to stance and stance to swing.

Cocontraction of antagonistic forces reduces the net joint moment compared to the situation in which the antagonists are silent and further increases instantaneous joint stiffness. It appears pos-

sible that cocontraction of the antagonistic muscles during the transition phase serves two purposes: to build up extensor (flexor) forces for the upcoming task (stance or swing) without producing a dramatic ankle movement, and to stiffen the joint to effectively prevent collapsing of the ankle at paw contact and to accommodate any perturbations that may occur in these transition phases. It appears intuitively appealing that a large preactivation of the ankle extensors prior to paw contact, and a stiffening of the ankle, becomes more important when the speed of locomotion, and therefore the speed and force requirement for the muscles, increases.

The previous results on force sharing among the cat ankle extensors may be summarized as follows: During locomotion, at speeds ranging from still standing (0 m/s) to galloping at about 3.5 m/s, peak forces of S remain about constant and peak forces of G and P increase to satisfy the moment requirements at the ankle with changing speeds and modes of locomotion. Furthermore, during still standing, S will produce virtually the entire ankle moment, whereas G (or more precisely MG) may not be activated at all. In addition, during very fast movements, such as a paw shake or scratching (movements that occur at a frequency of about 8–10 Hz), MG is known to be highly activated, whereas S is almost completely silent (Abraham and Loeb 1985; Smith et al. 1980).

Although the mechanisms underlying the control of cat ankle extensor muscles are not known, the force-sharing behavior among S, G, and P appears a logical consequence of their characteristics and properties (of course, one could argue that the muscles' characteristics and properties have evolved in response to the everyday demands). Soleus is a muscle that contains 95 to 100% slow twitch fibers, which are highly fatigue resistant (see table 10.1). It is ideally suited for tasks that require little force and slow movements and that might be performed for long periods of time. Therefore, it should not be surprising that S produces more active force than G and P during standing and slow locomotion (see figure 10.5) even though its force potential and physiological cross-sectional area are much smaller than those of G and P.

In contrast, G and P contain primarily fast twitch fibers, which are predominantly fast fatigable and have a relatively small capacity for deriving energy for contraction via aerobic pathways compared to the slow twitch fibers of S. Their ability to produce force at high speeds of shortening is better and their force potential is larger than that of S. These properties make G and P good muscles to accommodate increased ankle moments at high speeds of muscle shortening, as might occur when the animal increases its speed of locomotion. At extremely high speeds of movement (paw shake, scratching), S is not activated, presumably because the limited time for activation and deactivation and the high speed of muscle contraction do not allow for effective force production of S even if it was activated maximally together with G and P.

When discussing strategies of force sharing among muscles, one suggestion has come up repeatedly in the literature: muscles are used during movements in such a way as to minimize metabolic cost. As we shall discuss later, this idea has stimulated much theoretical work. However, here we are concerned about exploring this suggestion with the available experimental evidence. Such an undertaking is virtually impossible, because the relation between metabolic cost and force production in skeletal muscle under constantly varying, dynamic situations is not well understood. Nevertheless, some qualitative considerations may still be made. For these considerations to be intuitively appealing, they are made for extreme cases.

Imagine first a high speed movement such as paw shaking or scratching. Let us assume that instead of using G exclusively to perform this movement, S was activated to help in the paw shake response. Since S is considerably smaller than G and contains almost exclusively slow twitch fibers, it would produce little (if any) force during high velocity motion. Therefore, the involvement of S in the paw shake or scratching movement would require a high metabolic cost for little (if any) mechanical work.

Consider second, a change in the natural recruitment of S and G during still standing. When a cat stands still, the ankle extensor moment is primarily produced by S, while G is often silent. If G was activated exclusively for this task and S was silent, the required ankle moment would likely be provided by slow twitch fibers, those of the gastrocnemius. However, G, in contrast to S, is a two joint muscle (see table 10.1) that produces ankle extension and knee flexion. The knee flexor moment produced by G during still standing would have to be overcome by the knee extensor muscles to provide a net knee moment of zero. Therefore, activation of G during still standing may not be more costly or more fatiguing than using S exclusively for the ankle joint but would require more force, and thus more metabolic energy, from the knee extensors to produce stability at the knee.

Movement Control: Theoretical Considerations

The question of how forces are shared among soft tissue structures crossing a joint has become a primary research question in biomechanics. Its importance cannot be sufficiently stressed. Conceptually, forces can be taken up by active (the muscles and attached tendons) or passive structures (bones, ligaments, articular cartilage, menisci, disks). The problem of determining the forces transmitted by soft tissues across a joint has been termed the distribution problem. In the following discussion, some general considerations regarding the distribution problem are made.

The Distribution Problem in Biomechanics

In order to determine theoretically the force magnitudes in structures surrounding a joint, the joint and its associated geometry must be defined. Following that, the intersegmental moments and forces at the target joint can be determined using an inverse systems analysis, and finally, the forces transmitted by soft tissues may be estimated from the intersegmental moment and force using equations describing the mechanical, anatomic, and physiological properties of the joint and its relevant tissues.

Anatomically, a joint is a three-dimensional structure at the interface of two or more bones. Mechanically, a joint is defined by a point or an axis of given direction through a defined point. For example, in order to represent the human ankle, knee, and hip, a transverse axis through the most prominent points of the lateral malleolus, lateral femoral condyle, and greater trochanter have been used. For the mechanical representation of the geometry of a joint, typically only two variables are considered important: the line of action of a force and its moment arm about the defined joint. The line of action of ligaments and tendons crossing a joint can safely be assumed to follow along the longitudinal axis of the tissue, however, decisions about the lines of action of contact forces on the articular cartilage may prove difficult. Similarly, the moment arm of a force transmitted through a ligament or tendon can be readily calculated as the shortest distance from the joint to the line of action of the force of interest, or more generally, by choosing any vector starting from the point defined as the joint to any point on the line of action of the target force.

Once all possible structures that might contribute to the intersegmental moment or force have been described, the general distribution problem may be formulated as follows:

$$\overline{F}^0 = \sum_{i=1}^{m}\left(\overline{f}_i^m\right) + \sum_{j=1}^{l}\left(\overline{f}_j^l\right) + \sum_{k=1}^{c}\left(\overline{f}_k^c\right) + \sum_{t=1}^{o}\left(\overline{f}_t^o\right) \quad 10.1$$

and

$$\overline{M}^0 = \sum_{i=1}^{m}\left(\overline{r}_i^m \times \overline{f}_i^m\right) + \sum_{j=1}^{l}\left(\overline{r}_j^l \times \overline{f}_j^l\right) \quad 10.2$$

$$+ \sum_{k=1}^{c}\left(\overline{r}_k^c \times \overline{f}_k^c\right) + \sum_{t=1}^{o}\left(\overline{r}_t^o \times \overline{f}_t^o\right)$$

where

$\overline{F}$ and $\overline{M}$ are the intersegmental resultant force and moment, respectively; the superscript 0 designates the joint center;

$\overline{f}_i^m$, $\overline{f}_j^l$, $\overline{f}_k^c$, and $\overline{f}_t^o$ are the forces in the ith muscle, jth ligament, kth bony contact, and tth other (so far not considered) force;

$\overline{r}_i^m$, $\overline{r}_j^l$, $\overline{r}_k^c$, and $\overline{r}_t^o$ are the location vectors from the joint center to any point on the line of action of the corresponding forces;

x designates the vector (cross) product; and

m, l, c, o designate the number of muscles/ tendons, ligaments, individual articular contact areas, and other forces, respectively.

The idea behind the distribution approach is to calculate the individual tissue forces on the right-hand sides of equations 10.1 and 10.2 from the known intersegmental resultant force and moment. The intersegmental resultant force and moment are typically determined (in biomechanics) by rigid body, linked system, inverse dynamics methods (Andrews 1974). Equations 10.1 and 10.2 represent two independent vector or six independent scalar equations. Therefore, the system can be solved for two vector or six scalar unknowns. Typically, the lines of action of all forces on the right-hand sides of equations 10.1 and 10.2 are assumed to be known, therefore the system may be solved for six unknown force magnitudes. However, in most cases, the number of unknown soft tissue force magnitudes exceeds six. In these cases (i.e., there are more unknowns than there are system equations), the mathematical system is underdetermined (or redundant), and it has (typically) an infinite number of possible solutions. For example, the equation x + y = 10 where x and y are unknowns, is a simple example of an underdetermined system (i.e., there are two unknowns and one equation). It has an infinite num-

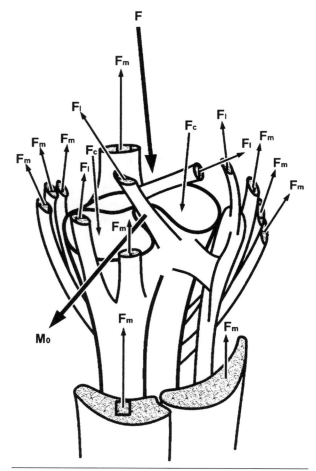

Figure 10.6 Schematic drawing of the human knee (tibial plateau) with the potential force-carrying structures. F and M_o indicate the intersegmental resultant force and moment, respectively. F_m, F_l, and F_c represent the forces transmitted by muscles, ligaments, and bony contacts, respectively.

Adapted, by permission, from R.D. Crowninshield and R.A. Brand, 1981a, "The prediction of forces in joint structures: Distribution of intersegmental resultants," *Exercise Sport & Science Review* 9:159-181. © The Franklin Institute, Philadelphia, PA.

ber of possible solutions, for example, x = 1, y = 9; or x = 4, y = 6.

Similarly, models of human and animal joints typically give an underdetermined set of system equations. For example, consider the human knee (see figure 10.6). It is crossed by many more than six individual muscles; further it contains at least four major ligaments and two distinct contact areas. Therefore, the number of unknown force magnitudes clearly exceeds six, the number of available scalar system equations.

There are a variety of strategies that may be used to solve an underdetermined system of equations. For the distribution problem in biomechanics, mathematical optimization has been used more often than any other approach to solve for the unknown muscle forces across a joint. To solve for ligamentous or bony contact forces, mechanical representations of the ligaments and articular surfaces have been used. Here, we are primarily concerned with the muscular (active) forces, because it is these forces that determine to a large extent the intersegmental resultant moment and, therefore, joint movement. Optimization has been used frequently to determine the individual forces of muscles crossing a joint for two primary reasons: mathematical optimization is a simple and sound way for solving underdetermined mathematical systems, and it has been proposed for a long time (Weber and Weber 1836). Furthermore, it appears intuitively appealing that muscles should be controlled in such a way that movements are performed in accordance with some physiological optimal criterion. One of the basic questions in biomechanics is whether muscles are really controlled according to some physiological optimal criterion and what this criterion might be.

If we assume that the intersegmental resultant moment is primarily caused by the muscular forces, the general distribution problem (equations 10.1 and 10.2) may be approximated as follows:

$$\overline{M}^0 = \sum_{i=1}^{m}\left(\overline{r}_i^m \times \overline{f}_i^m\right).\qquad 10.3$$

Equation 10.3 represents one vector or three independent scalar equations. In order to solve for the unknown muscular forces $\overline{f}_i^m$ using optimization techniques, one needs to define an objective function, the constraint (equality and inequality) functions, and the design variables.

Starting with the design variables, they are the unknown magnitudes of the muscle forces $\overline{f}_i^m$ (we assume that the lines of action of the muscle forces are known from anatomic considerations). The constraint functions in most muscle force-sharing problems are the same. They include the idea that a given joint moment must be satisfied at any time by the sum of the moments produced by all muscles crossing the target joint (i.e., equation 10.3). Further, constraint functions should ensure that the design variables remain within physiological meaningful limits. For example, muscular forces are unidirectional (tensile), therefore, if tensile forces are considered positive, muscular forces cannot be negative ($\overline{f}_i^m \geq 0$).

Many objective functions have been proposed in the literature but only a few have been compared directly with individual muscle forces measured experimentally. It is beyond the scope of this review to summarize all objective functions and discuss their strong and weak sides. However, a few comments should be made. First, there is no objective function that has been shown to predict individual muscle forces accurately for general movements of different speeds and efforts. Second, the idea of minimizing metabolic cost (Hardt 1978), minimizing fatigue (Dul et al. 1984), or maximizing endurance (Crowninshield and Brand 1981a, 1981b) has been repeatedly proposed and, when validated against experimentally recorded muscle forces, has repeatedly failed to give acceptable results (Dul et al. 1984; Herzog and Leonard 1991).

The idea of minimizing the metabolic cost of movement as a physiological criterion for muscle force sharing has great appeal for low level, repetitive, everyday movements such as walking. In the future, the issues surrounding metabolic energy requirements for muscles that contract dynamically should be tackled thoroughly such that this specific objective function could be tested rigorously. Good recordings of multiple muscle forces for a variety of tasks are becoming increasingly more available, providing perfect data to validate proposed theoretical (optimization) models of force sharing among synergistic muscles.

Applications to Sport Science

There are innumerable examples of how the physiology, mechanics, or control of muscles influences sport performances. It is well accepted that people with predominantly slow-twitch fibered muscles are better suited for endurance events (marathon running) than athletes with predominantly fast-twitch fibered muscles, who are more likely to excel in sprint events (100 m sprint). Similarly, it is no surprise that muscular coordination (as revealed by EMG records) might be distinctly different between an expert and a novice athlete, particularly in technically difficult sports (golfing, tennis, pole-vaulting, hurdling, skating, etc.). These well-accepted phenomena will not be discussed here. Rather, I would like to focus on a few selected results obtained in basic research experiments that could have interesting implications for sport performance and training.

Earlier in this chapter, we discussed the force-sharing results of the cat ankle extensor muscles, S, G, and P, for different speeds of locomotion. The result that I found particularly intriguing was that the slow-twitch fibered S did not modulate its peak force when the ankle moment requirements increased, whereas the fast-twitch fibered G and P increased (as expected) their peak force values (albeit, not at the same rate).

The human ankle extensor group also contains the soleus and gastrocnemius; the plantaris is small and unimportant (compared to the cat) and in some people is completely missing. As in cats, human S muscle typically contains a higher percentage of slow twitch fibers than G (Johnson et al. 1973), although the difference in fiber type distribution is not as dramatic as in the cat. Nevertheless, it is perceivable that increasing the speed in a running workout affects primarily the force production in G and not in S. Therefore, if training is specifically aimed at the soleus muscle, a relatively low intensity workout might be sufficient, whereas the largest motor units of G might only be recruited at high intensities.

During cat locomotion, S, G, and P are activated 50 to 120 ms before paw contact (depending on the speed of locomotion). The first part of force production immediately following paw contact is associated with an eccentric contraction in all three muscles. It has been postulated and there is experimental evidence (Enoka 1996; Grabiner et al. 1995; Nardone, Romanò, and Schieppati 1989) that the normal recruitment order of motor units is changed or possibly even reversed during eccentric contractions. If this evidence is correct, eccentric contractions may provide the possibility to train large, fast fatiguing motor units at relatively low joint moments (and thus low muscle forces), because they may be recruited before the small, slow fatiguing motor units. During isometric or concentric contractions in which motor units are presumably recruited according to size, starting with the smallest (slow twitch), the large (fast twitch) motor units might only be recruited and trained at force levels between about 60 to 100% of the muscle's maximum. Approximately 10 years ago, eccentric strength training became popular among athletes for largely unsupported reasons. One effect that eccentric contractions may have, and that has not been recognized to date, is that they might train large, fast twitch motor units effectively and with little total muscular effort.

It is hard, if not impossible, to determine individual muscle forces in sport movements. However, under most circumstances, it is well accepted that the resultant joint moment reflects adequately the moments produced by the muscle forces (equation 10.3). Therefore, an analysis of the joint moments during an activity may reveal how muscles react to changing conditions in training or competi-

tion. For example, Hay and coworkers (1983) studied the knee extensor moments for squat exercises performed using resistances that were equivalent to 40%, 60% and 80% of the four repetition maximum (FRM) for the participating athletes. For a squat exercise, one might assume that the knee extensor moments produced by the athletes would increase with increasing resistance. In fact, if it is assumed that the squat movement remains identical for all three resistance levels, the knee extensor moments can be shown to depend directly on the resistance. However, doubling the resistance from 40 to 80% of FRM had little effect on the knee extensor moments (see figure 10.7). Careful analysis of high speed movement recordings revealed that there were minute changes in the kinematics of the exercise, which resulted in a relative unloading of the knee

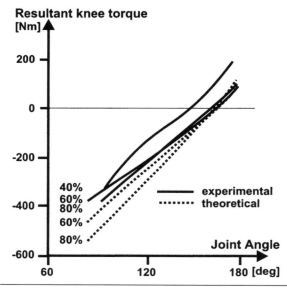

Figure 10.7 Resultant knee extensor moments (negative) vs. knee angles for a barbell squat exercise performed at 40%, 60%, and 80% of the four repetition maximum. The solid lines were experimentally determined. The dashed lines represent the theoretically calculated resultant knee extensor moments. The theoretical values were calculated by taking the movement patterns of the 40% test and increasing the resistance to 60% and 80% of the four repetition maximum. Therefore, there is no difference between the experimental and theoretical curves at 40% of the four repetition maximum, however, the theoretically predicted moments are much larger than the corresponding experimentally determined moments for the 60% and 80% tests.

Adapted, by permission, from J.G. Hay, J.G. Andrews, C.L. Vaughan, and K. Ueya, 1983, Load, speed and equipment effects in strength-training exercises. In *Biomechanics VIII*, edited by H. Matsui and K. Kobayashi. (Champaign, IL: Human Kinetics), 939-950.

(moment) such that the increased resistance had virtually no effect on the knee moment. According to Hay and coworkers (1983), the changes in the kinematics across trials were so subtle that they would likely remain undetected, except possibly for the most skilled and competent observers. This example illustrates that a desired change (increase in the barbell weight to offer increased resistance) might not always produce the desired effect (increased force requirements of the knee extensors during the squat exercise).

As can be seen in table 10.1, the properties of cat soleus, gastrocnemius, and plantaris vary considerably despite the fact that they all are primary ankle extensors. This result indicates either that muscles in an agonistic group have different properties despite their similar functions, or that the different properties reflect different functions of these agonists. For a variety of reasons, the latter explanation is much more appealing. If we presume, therefore, that the mechanical and physiological properties of muscles adapt to the functional demands, one should be able to detect such adaptations quite readily in athletes who use a specific muscle or muscle group differently in their sport than does another group of athletes.

A few years ago, we tested the hypothesis that moment-angle or force-length properties will adapt to the functional requirements imposed on chronically trained muscles of athletes. The rectus femoris of cyclists (n = 3) and runners (n = 4) was used for testing. In cycling, the rectus femoris is used in a chronically shortened position compared to running because of the flexed hip angle adopted in cycling. Therefore, cyclists will train their rectus femoris at relatively short length compared to runners. We speculated that the rectus femoris would be relatively stronger at short lengths (at which it is used in cycling) compared to long lengths in the cyclists and vice versa for the runners (see figure 10.8). In accordance with this expectation, the moment-length relation of all muscles tested in the cyclist had a negative slope, and the muscles of the runners had a positive slope (see figure 10.9). The slopes of the moment-length relationships were significantly different from zero ($\alpha = 0.05$) in all subjects, indicating that the moment-length properties were different in the two groups of athletes. There are a variety of possible explanations for this result:

- The differences in the moment-length properties of the rectus femoris could have existed before the athletes took up competitive sports, and the reason why they excelled in their sport

was because of their peculiar muscle mechanical properties.

- The adaptations could have occurred primarily on a neural level (i.e., the athletes are able to more fully recruit the rectus femoris at lengths used during everyday training).
- The adaptations may have occurred in the muscle itself.

One convenient and appealing adaptive mechanism that could explain the observed results is a change in the number of sarcomeres in series in the muscle, such that optimal length (the length at which the muscle is strongest) coincides with the muscle length used in training. According to this mechanism, the number of sarcomeres in series in

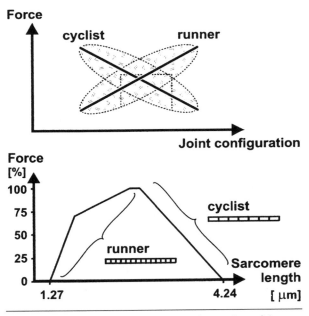

Figure 10.8 Expected force-length relationship of the rectus femoris of elite cyclists and runners (top panel) and the corresponding schematic explanation of how these differences in the force-length relationships between the two groups of athletes might be explained (bottom panel). The idea is that adaptation of the force-length relationship might occur by an increase/decrease in the number of sarcomeres that are aligned in series within a muscle fiber. Runners would be expected to have a large number of sarcomeres in series, so for a given fiber length, the individual sarcomere length would be relatively small. Cyclists would be expected to have relatively few sarcomeres in series, so for a given fiber length, the individual sarcomere lengths would be relatively large. In this way, it could be perceived that the rectus femoris of runners works on the ascending part, and the rectus femoris of cyclists works on the descending part of the force-length relationship within the anatomical range of motion.

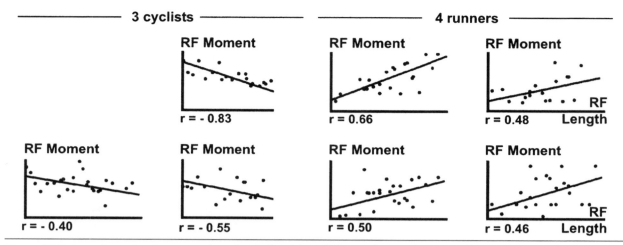

Figure 10.9 Actual results of moment-length relationships of elite cyclists and runners showing that cyclists tended to have a negatively oriented (descending) and runners a positively oriented (ascending) moment-length relationship, as expected from theoretical considerations.

the rectus femoris of a cyclist should be smaller than that found in runners (see figure 10.8). We did not measure the sarcomere numbers in these athletes and therefore are unable to evaluate this particular mechanism of muscle adaptation.

If we assume that the differences in the moment-length relations of the rectus femoris between cyclists and runners are favorable adaptations of the muscle properties to the functional demands, some interesting training theories might be proposed. For example, such a result would suggest that cycling as cross training for running is not recommended in high performance athletes, because the functional demands of the cross-training exercise would change the mechanical properties of the muscles in an undesirable way. Or expressed differently, imagine a long-distance runner who can run a 10 km race in 35 min purely on running training. Now, this athlete starts to compete in duathlons (running, bicycling). In order to prepare properly, the athlete maintains the exact running workouts she or he has always done and adds the cycling. Despite the fact that the athlete does the same running training and adds additional fitness to the body (we assume that there is no overtraining because of the additional cycling), it is predictable that the athlete's 10 km racing time would become worse, because the mechanical properties of the muscles used for running are likely changed by the bicycling and are likely not as optimal for running as they were when the athlete just ran.

In the experimental and theoretical force-sharing research discussed earlier, we have seen that the control of a joint moment is redundant: there are more muscles crossing a joint than there are rota-

tional degrees of freedom in the joint. Therefore, a given resultant joint moment can be achieved with an infinite number of combinations of muscle forces. This statement is only correct for a set of joint moments that may be achieved physiologically. Of course, there cannot be a meaningful physiological solution to the distribution problem if the joint moment is assumed to be so large that even a full activation of all agonistic muscles (those muscles providing a moment contribution in the same direction as the required joint moment), and a deactivation of all antagonistic muscles, cannot produce the required joint moment. Therefore, there are an infinite number of possible solutions when the net joint moment is small, and there is no physiological solution when the net joint moment exceeds a critical threshold. Somewhere between these two general problems, there is a limiting problem that is uniquely determined: the problem in which the net joint moment is exactly of such a magnitude that it can just be satisfied if all agonists are fully activated, and all antagonists are completely deactivated. This specific problem is frequently approached in sprintlike, high intensity, short duration sport events.

For example, a few years ago, we attempted to determine the conditions that gave maximal average power output during a pedal revolution in bicycling by adjusting the geometry of the rider on the bicycle until the total power output of the leg muscles was maximal. The leg model that was used was anatomically correct and displayed the mechanical properties of the corresponding real muscles as well as they were known. In this two-dimensional model, the muscles were always fully activated when they produced positive power and were instantaneously deactivated when they produced negative power. Using this model, we predicted a maximal, average power output per pedal cycle of about 1300 W and an optimal pedaling rate of 150 repetitions per minute (Yoshihuku and Herzog 1990). The power output was well within observed values for subjects comparable to our modeled subject, and the frequency of pedaling that produced the maximal power output was within the range of frequencies observed in sprint cyclists. We concluded from these results that for maximal performances of short duration, the force sharing among synergistic muscles is uniquely determined; all agonists are fully activated, all antagonists are silent. Situations of maximal power output for very short periods of times are encountered in many sport activities; for example, weightlifting, long jump, sprint running, shot put. For these events, the muscle forces involved during maximal power output at joints may be approximated by assuming a full activation of all agonistic muscles (Lutz and Rome 1993).

Final Comments

The aim of this chapter was to cover basic aspects of muscle activation, the control of force in single fibers and muscles, and the control of moments (and thus, movements) at a single joint. The following conceptual results were discussed.

- Force sharing among muscles crossing a joint varies as a function of the movement requirements. Specifically, an increase in the net joint moment does not necessarily mean an increase in all agonistic muscle forces. For an athlete, this result may imply that training at a large range of intensities is important if overall agonistic muscle improvement is to be achieved, and it further implies that one should train under competitive conditions if muscular improvements are required specifically at the intensity of competition.

- Muscular properties appear to adapt physiologically and mechanically to the functional requirements. This observation again suggests specificity of training for an athlete. It also questions the usefulness of cross training in high performance athletes, especially when the cross-training movement is far removed from the movements encountered in the sport of specialization. For example, bicycling appears to be bad cross training for runners, and vice versa.

- We have proposed that during eccentric contractions, large, fast twitch motor units may be recruited at relatively small intensities. If this suggestion is correct, the value of eccentric exercise for training may primarily be in the ability to recruit large, fast twitch motor units at low muscular efforts. During isometric or concentric contractions, these same motor units would presumably only be recruited at near maximal efforts.

References

Abraham, L.D., and Loeb, G.E. 1985. The distal hind limb musculature of the cat (patterns of normal use). *Exp. Brain Res.* 58: 580–93.

Andrews, J.G. 1974. Biomechanical analysis of human motion. *Kinesiology* 4: 32–42.

Crowninshield, R.D., and Brand, R.A. 1981a. The prediction of forces in joint structures: Distribution of intersegmental resultants. *Exerc. Sport & Sci. Rev.* 9: 159–81.

Crowninshield, R.D., and Brand, R.A. 1981b. A physiologically based criterion of muscle force prediction in locomotion. *J. Biomech.* 14: 793–801.

Dul, J., Johnson, G.E., Shiavi, R., and Townsend, M.A. 1984. Muscular synergism. II. A minimum fatigue criterion for load sharing between synergistic muscles. *J. Biomech.* 17: 675–84.

Enoka, R.M. 1996. Eccentric contractions require unique activation strategies by the nervous system. *J. Appl. Physiol.* 81: 2339–46.

Grabiner, M.D., Owings, T.M., George, M.R., and Enoka, R.M. 1995. Eccentric contractions are specified a priori by the CNS. Proceedings of the XV Congress International Society of Biomechanics, 338–39. Jyväskylä, Finland, July 2–6.

Hardt, D.E. 1978. A minimum energy solution for muscle force control during walking. PhD diss., MIT.

Hay J.G., Andrews, J.G., Vaughan, C.L., and Ueya, K. 1983. Load, speed, and equipment effects in strength-training exercises. In *Biomechanics VIII*, ed. H. Matsui and K. Kobayashi, 939–50. Champaign, IL: Human Kinetics.

Henneman, E., and Olson, C.B. 1965. Relations between structure and function in the design of skeletal muscles. *J. Neurophysiol.* 28: 581–98.

Henneman, E., Somjen, G., and Carpenter, D.O. 1965. Functional significance of cell size in spinal motoneuron. *J. Neurophysiol.* 28: 560–80.

Herzog, W., and Leonard, T.R. 1991. Validation of optimization models that estimate the forces exerted by synergistic muscles. *J. Biomech.* 24(Suppl. 1): 31–39.

Herzog, W., and Leonard, T.R. 1996. Soleus forces and soleus force potential during unrestrained cat locomotion. *J. Biomech.* 29: 271–279.

Hodgson, J.A., 1983. The relationship between soleus and gastrocnemius muscle activity in conscious cats: A model for motor unit recruitment? *J. Physiol. (London)* 337: 553–62.

Johnson M.A., Polgar, J., Weightman, D., and Appleton, D. 1973. Data on the distribution of fiber types in 36 human muscles: An autopsy study. *J. Neurolog. Sci.* 18: 111–29.

Lutz, G.J., and Rome, L.C. 1993. Built for jumping: The design of the frog muscular system. *Science* 263: 370–2.

Nardone, A., Romanò, C., and Schieppati, M. 1989. Selective recruitment of high threshold human motor units during voluntary isotonic lengthening of active muscle. *J. Physiol. (London)* 409: 451–71.

Smith, J.L., Betts, B., Edgerton, V.R., and Zernicke, R.F. 1980. Rapid ankle extension during paw shakes: Selective recruitment of fast ankle extensors. *J. Neurophysiol.* 43: 612–20.

Walmsley, B., Hodgson, J.A., and Burke, R.E. 1978. Forces produced by medial gastrocnemius and soleus muscles during locomotion in freely moving cats. *J. Neurophysiol.* 41: 1203–15.

Weber, W., and Weber, E. 1836. *Mechanik der Menschlichen Gehwerkzeuge.* Gottingen: Fischer-Verlag.

Yoshihuku, Y., and Herzog, W. 1990. Optimal design parameters of the bicycle-rider system for maximal muscle power output. *J. Biomech.* 23: 1069–79.

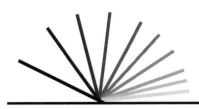

Chapter 11

Power Output and Force-Velocity Properties of Muscle

B.R. MacIntosh and R.J. Holash

It is a general observation that even with maximal effort we can move a heavy object only very slowly. In contrast, a very light object can be moved at a high velocity. Similarly, the acceleration of the human body that can be achieved while sprinting decreases as velocity of motion increases, because the application of force against the ground (in the opposite direction to that of the direction of travel) decreases as the velocity increases (Best and Partridge 1928; Fenn 1930). This observation led to the discovery that skeletal muscle, rather than neural control, is responsible for the characteristic relationship between maximal rate of movement and the ability to apply force. This fundamental property of muscle is known as the force-velocity relationship. For a given muscle or muscle group, and a constant level of activation, the velocity of shortening that can be achieved decreases for contractions against progressively greater loads.

Mechanical power output, the rate of doing mechanical work, is the product of force and velocity. Since maximal force and velocity of muscle shortening are characteristically related, then peak power output is defined and limited by the force-velocity relationship.

In this chapter we briefly present the primary concepts associated with the force-velocity and power-velocity relationships, then we describe each of these primary concepts in detail.

The Primary Concepts

The force-velocity and power-velocity relationships have been studied in detail at various levels of organization: molecular level, single cell level, motor units, whole muscle and multimuscle movements, and single joint and multijoint movements. Each level of organization offers valuable insight into the mechanisms and interactions of force, velocity, and power and contributes to our understanding of the limitations and capabilities of movement resulting from muscular contraction.

The study of the force-velocity relationship requires that one of the two parameters be controlled (independent variable) while the other is measured (dependent variable). Experiments have been conducted with either load (force) or velocity as the independent variable. Load is controlled in isotonic contractions, and velocity is controlled in isokinetic experiments. Regardless of the approach, the characteristic hyperbola (equation 11.1) can be used to describe the relationship between force and velocity of muscle contraction. However, the interpretation of the constants (a and b) in this equation probably requires strict adherence to qualifying criteria (see chapter 10).

Fenn and Marsh (1935) were the first to mathematically describe the relationship between force and velocity of shortening, but Hill (1938) proposed the more commonly used hyperbolic equation to describe the force-velocity relation:

$$(P + a) \cdot (v + b) = b \cdot (P_o + a) = \text{constant} \qquad 11.1$$

where

P is any force,

v is the velocity of contraction at force P,

P_o is maximal isometric force at optimal muscle length, and

a and b are constants (though strictly speaking, $a \cdot P_o^{-1}$ and $b \cdot V_o^{-1}$ are constant).

This equation defines a relationship between force and velocity such that maximal isometric force and maximal (unloaded) shortening velocity (V_o) are identified as intercepts on the respective axes. The constants a and b are determinants of the degree of curvature that is given by the ratio $a \cdot P_o^{-1}$, which is equivalent to $b \cdot V_o^{-1}$. The force-velocity and power-velocity relationships as defined by the Hill equation are presented in figure 11.1. The actual values for the parameters of the force-velocity relationship (isometric force, maximal shortening velocity, and degree of curvature) are affected by various physiological conditions and circumstances. Muscle architecture, including cross-sectional area, muscle length, angle of pennation, and anatomic joint configuration can affect the force-velocity (or torque-angular velocity) properties of a muscle or muscle group. Additional factors that affect these relationships include sarcomere length, muscle fiber type composition, level of activation, and muscle fatigue. The manner in which these factors affect the force-velocity and power-velocity relationships are developed below. Additional factors that are known to affect the force-velocity and power-velocity relationships but that will not be given further consideration in this chapter include training, aging, and disease.

Human movement with maximal effort is dictated by the force-velocity relationships of the specific muscles involved. Multijoint motion, including various forms of locomotion, is similarly limited by this unique property of muscles. In spite of the greater complexity associated with several muscles and multijoint motion, the general relationship between velocity of movement and load, torque, or resistance persists. This is nicely illustrated with an example. Cycle ergometry has been extensively studied with respect to maximal effort and the dependence of velocity (of a point on the periphery of the flywheel) on resistance. This exercise is used to illustrate some of the factors affecting the force-velocity (flywheel resistance-velocity) relationship in human movement including muscle fiber type and motor unit activation.

Shape of the Force- and Power-Velocity Relationships

The force-velocity and power-velocity relationships are fundamental properties of skeletal muscle. An in-depth study of these relationships will permit a better understanding of the limitations and capabilities of human movement.

Concentric Contractions

The general shape of the graphic representation of the relationship between force and velocity of maximal muscle contraction has been described as early as 1924 (Gasser and Hill 1924) and again in 1927 (Levin and Wyman 1927). However, A.V. Hill (1938) is credited with originating the now popular hyperbolic equation that appropriately describes the force-velocity relationship, and which has come to be known as the Hill equation (equation 11.1). This equation has been used to illustrate the force-velocity relationship shown in figure 11.1 with the corresponding power-velocity relationship. Figure 11.2 illustrates the impact of variation in isometric force, maximal shortening velocity, and $a \cdot P_o^{-1}$ on the force-velocity and power-velocity relationships, and table 11.1 presents representative values obtained from the literature for these parameters for a variety of muscles.

Force or Power

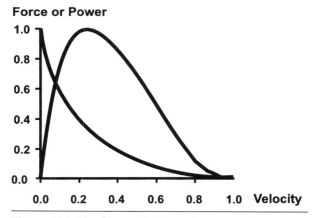

Figure 11.1 The force-velocity and power-velocity relationships are illustrated, with each parameter normalized to a maximal value of 1.0. Equation 11.2 was used to generate values of force for each velocity with $a \cdot P_o^{-1} = 0.27$. Power was calculated as force · velocity at each velocity and normalized for the highest power obtained.

Table 11.1   Reported Values for Parameters of the Force-Velocity Relationship

Muscle[a]	P_o	Units	V_o	$a \cdot P_o^{-1}$	Reference
F tib ant *in vitro*, 2°	0.287	$N \cdot mm^{-2}$	$2.26 \, L \cdot s^{-1}$	0.3	Curtin and Edman 1994
R soleus *in vitro*, 15°	0.13	$N \cdot mm^{-2}$	$0.88 \, L \cdot s^{-1}$	0.06	Bangart, Widrick, and Fitts 1997
R soleus *in vitro*, 20°	0.16	$N \cdot mm^{-2}$	$3.1 \, L \cdot s^{-1}$	0.075	Claflin and Faulkner 1989
R soleus *in situ*, 35°	0.13	$N \cdot mm^{-2}$	$4.5 \, L \cdot s^{-1}$	0.26	Ranatunga and Thomas 1990
R soleus *in vitro*, 35°	0.22	$N \cdot mm^{-2}$	$4.6 \, L \cdot s^{-1}$	0.21	Ranatunga 1982
R EDL *in situ*, 35°	0.13	$N \cdot mm^{-2}$	$13.3 \, L \cdot s^{-1}$	0.31	Ranatunga and Thomas 1990
R EDL *in vitro*, 35°	0.36	$N \cdot mm^{-2}$	$7.7 \, L \cdot s^{-1}$	0.415	Ranatunga 1982
C FDL *in situ*, 37°	5.85	N	$29 \, \mu m \cdot s^{-1}$	0.43	Hatcher and Luff 1987
H quadriceps *in vivo*, 37°	250	Nm	$18 \, rad \cdot s^{-1}$	0.4	Tihanyi, Apor, and Fekete 1982
H elb flexors *in vivo*, 37°	60	Nm	$23 \, rad \cdot s^{-1}$	0.45	Kaneko 1970*
H elb flexors *in vivo*, 37°	60	Nm	$27 \, rad \cdot s^{-1}$	0.4	Wilkie 1950*
H elb flexors *in vivo*, 37°	1550	N	$2.1 \, L \cdot s^{-1}$	0.39	Kojima 1991[b]

*Estimated from published values, with assumed apparent moment arm.
[a]Muscles are from animals: F (frog), R (rat), C (cat), and H (human).
[b]Values for velocity and force were estimated by the authors using biomechanical principles.

Figure 11.2a shows force-velocity curves with variation in maximal isometric force (P_o), while maximal velocity and $a \cdot P_o^{-1}$ are kept constant. In this case, maximal isometric force is normalized relative to the maximal isometric force of the strongest condition. Differences in maximal isometric force like that illustrated in figure 11.2a would be primarily due to differences in myofibrillar cross-sectional area of the muscles represented. Figure 11.2b illustrates the impact of this variation in P_o on peak mechanical power output. With decreases in maximal isometric force, there are proportional decreases in peak power output, but optimal velocity (the velocity that corresponds with peak power output) remains constant.

Figure 11.2c shows force-velocity curves with variation in maximal velocity of shortening, while P_o and $a \cdot P_o^{-1}$ are constant. In this example, maximal velocity of shortening is normalized to the absolute maximal velocity of shortening of the fastest muscle. Variations in maximal velocity are due primarily to differences in fiber types or muscle fiber length. Figure 11.2d illustrates the impact of variations in V_o on power output. With decreases in V_o, there are proportional decreases in peak power output, and optimal velocity also decreases in proportion to the decrease in V_o.

Figure 11.2e shows force-velocity curves with variation in the degree of curvature, while P_o and V_o remain constant. A wide variation in $a \cdot P_o^{-1}$ values has been reported (Baratta et al. 1995a), and this may be a function of the muscle fiber type distribu-

tion (see model below) or conditions of the measurement including temperature, fatigue, and type of contraction. Figure 11.2f illustrates the impact of variations in $a \cdot P_o^{-1}$ on power output. Any decrease in $a \cdot P_o^{-1}$ (greater curvature) results in lower peak power output and a shift in optimal velocity to a slower velocity of contraction.

Eccentric Contractions

An important yet poorly understood aspect of the force-velocity relationship of skeletal muscle deals with the circumstance when the load imposed on the muscle exceeds the maximal isometric force that the muscle is capable of generating. Under these circumstances the muscle is stretched while activated, and this is called an eccentric contraction. The velocity of stretch is very slow when the load just barely exceeds the isometric force but increases with greater load. There is a maximal load with which the muscle is able to limit the velocity of stretch, beyond which the load will uncontrollably accelerate the movement. The eccentric portion of the force-velocity relationship is illustrated in figure 11.3.

When the muscle is forcibly stretched at a given velocity (isokinetic) the muscle generates a braking force, which is greater as the velocity of stretch increases, up to a limit velocity beyond which no further increase in resisting force is observed (Westing and Seger 1989).

The eccentric portion of the force-velocity relationship, which is illustrated in figure 11.3, has

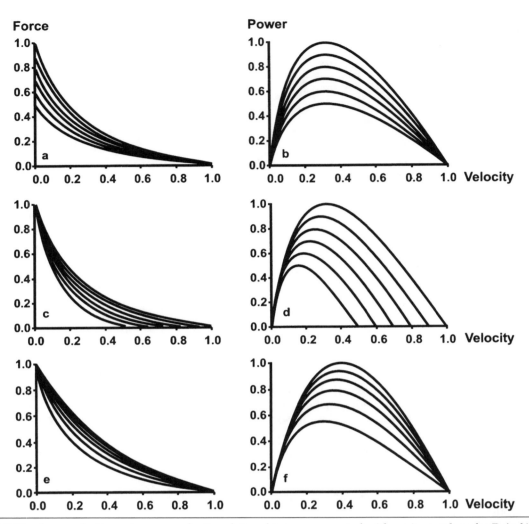

Figure 11.2 The force-velocity and power-velocity relationships are presented with various values for P_o (a, b), V_o (c, d), and $a \cdot P_o^{-1}$ (e, f). In each case, equation 11.2 was used to calculate force for values of velocity from 0 to V_o and power was calculated as force $\cdot$ velocity at each velocity and normalized to the highest power for that series. In (a), values for P_o were 0.5, 0.6, 0.7, 0.8, 0.9, and 1.0. In (c), values for V_o were 0.5, 0.6, 0.7, 0.8, 0.9, and 1.0. In (e) values for $a \cdot P_o^{-1}$ were 0.2, 0.3, 0.4, 0.5, 0.6, and 0.7. Changes in each parameter have specific impact on the magnitude and shape of the power-velocity relationship.

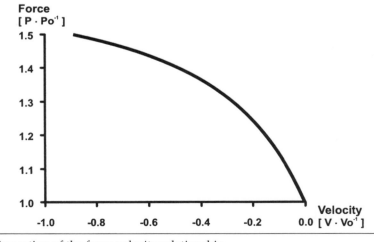

Figure 11.3 The eccentric portion of the force-velocity relationship.

reasonable symmetry between concentric and eccentric contractions and the corresponding concentric properties illustrated in figure 11.1 (page 194). An alternative representation illustrates a yield point—a velocity above which no further braking force is generated, as seen in figure 15.16 (see page 278). A symmetrical relationship appears to be evident in several published eccentric force-velocity curves (Komi 1973; Krylow and Sandercock 1997; Levin and Wyman 1927; Stainsby 1976), but in some cases (Westing, Cresswell, and Thorstensson 1991; Westing and Seger 1989) there is a more evident yield point (no further increase in force or torque with greater velocity or angular velocity). It has been shown that the absence of an increase in torque with increased angular velocity of eccentric contraction is associated with inhibition of muscle activation (Westing, Seger, and Thorstensson 1990).

Muscle Architecture and the Force-Velocity Relationship

The force-velocity relationship presented in figure 11.1 represents the general case, where force and velocity are expressed relative to their respective maximal values. When force and velocity are expressed in absolute values (newtons, $cm \cdot s^{-1}$), various architectural features of muscle structure can influence the parameters of the force-velocity relationship and the corresponding power-velocity relationship. The relevant architectural features include cross-sectional area, pennation, and fiber length (number of sarcomeres in series), and in the case of in vivo study, anatomic joint configuration is an important consideration.

Cross-Sectional Area

The maximal isometric force that can be generated by a muscle is directly proportional to the myofibrillar cross-sectional area, regardless of muscle fiber type (McComas 1996). This architectural feature has a predictable effect on the force-velocity and power-velocity relationships, as illustrated in figure 11.2a. It is common to express force relative to cross-sectional area (see table 11.1) and this is referred to as specific force. In general, cross-sectional area of a parallel-fibered muscle can be measured directly or estimated by volume $\cdot$ length^{-1} (assuming a density of 1.06, volume can be estimated from muscle mass). Caution should be used, however, in situations where the shape of the muscle is not cylindrical (i.e., fusiform or pennate). Volume $\cdot$ length^{-1} underesti-

mates the cross-sectional area in these circumstances. In the case of single-cell preparations, cross-sectional area is often estimated from measurements of cell diameter, assuming the cross section is circular. In most mammalian muscles, the normalization by muscle cross-sectional area gives a value not substantially different from normalization by myofibrillar cross-sectional area. The actual difference between myofibrillar and whole muscle (or cell) cross-sectional area is primarily a function of area occupied by mitochondria and sarcoplasmic reticulum (and extracellular space in the case of whole muscle). In most mammalian muscles this represents a relatively small proportion of the cross section. In some cases, however, the force per fiber cross-sectional area can give a substantial fiber type difference in specific force if the proportion of area occupied by mitochondria is disparate among the fiber types. The proportion of the muscle cross-sectional area occupied by mitochondria is relatively small in human muscle but can reach extremes in flight muscles of hummingbirds and bumblebees (see chapter 6).

Muscle Length

Assuming a given average sarcomere length, muscle fiber length is a function of the number of sarcomeres in series. Since each sarcomere is expected to have an absolute maximal velocity, the maximal shortening velocity of the whole muscle (or muscle fiber) will be dependent on the number of sarcomeres in series. The maximal shortening velocity in $cm \cdot s^{-1}$ will be directly dependent on the fiber length in cm. For a muscle fiber with a maximal velocity of shortening of 2 fiber lengths per second, a 3 cm muscle fiber would have a maximal velocity of shortening of 6 $cm \cdot s^{-1}$. A longer muscle (4 cm) would have a greater maximal velocity (8 $cm \cdot s^{-1}$). Figures 11.2c and 11.2d illustrate the impact of differences in V_o on the force-velocity and power-velocity relationships. A longer muscle can generate a greater power output.

Pennation

The direction of muscle fiber alignment is not always parallel with the principal tendon orientation (line of action of the muscle). In this case the muscle is said to be pennate (feather shaped). Muscles can be unipennate, in which case all fibers have approximately the same angle with respect to the line of action of the muscle or multipennate (see figure 15.8, page 272), in which case different parts of the muscle can have substantially different angles of pennation.

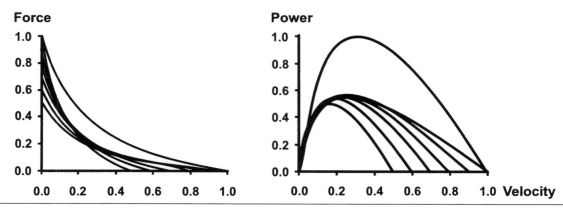

Figure 11.4 When angle of pennation changes for a muscle of fixed volume, muscle fiber length and total cross-sectional area change in opposite directions. This is illustrated here using equation 11.2 to calculate force at each velocity, where P_o increases from 0.5 to 1.0, V_o decreases from 1.0 to 0.5. A value for a $\cdot P_o^{-1} = 0.27$ was used in each case. For comparison, the force-velocity relationship for a parallel-fibered muscle with $P_o = 1$ and $V_o = 1.0$ is shown (non-overlapping line). Power was calculated as force $\cdot$ velocity and normalized to the highest power achieved for any muscle represented. Note that the greatest peak power was generated by the muscle with $P_o = 1$ and $V_o = 1.0$. This muscle would have twice the volume as the pennate muscles.

Pennation has two important physiological effects on the force-velocity relationship. The effective cross-sectional area of a muscle (of a given volume) is increased with pennation. This has a predictable effect on the maximal isometric force. The trade-off for the greater cross-sectional area is that muscle fibers are shorter (fewer sarcomeres in series) than in a parallel-fibered muscle of the same volume. This results in a lower absolute maximal velocity ($cm \cdot s^{-1}$). The impact of muscle fiber pennation on the force-velocity relationship is illustrated in figure 11.4. Pennation angle also directly affects muscle-shortening velocity (Spector et al. 1980). A given fiber length change translates to less muscle shortening for muscles at greater pennation angle. However, this discrepancy can apparently be partly compensated for by an increase in angle of pennation during the muscle contraction (Zuurbier and Huijing 1993).

Anatomic Joint Configuration

The anatomic configuration of a joint and corresponding muscle will have a significant effect on how muscle contraction is translated into movement of the limb. The anatomic configuration of a joint describes not only the joint structure and position in the body, but also how and where the muscles attach on either side of the joint. For movement to occur, the forces and shortening velocities produced by muscular contraction must be translated into torque and angular velocity around the joint, the magnitude of which will vary as much with the location of attachment points of the muscle and the

position of the joint as with the physical properties of the muscle.

Torque is generated when muscular contraction creates a force that acts on a lever arm (bone) some distance from the center (axis) of the joint. The magnitude of the torque that is produced by a muscular contraction is the product of the contraction force and the perpendicular distance from the line of action of the force to the joint center. Torque is calculated through the equation

$$T = f \cdot r \qquad\qquad 11.2$$

where

T = torque,

f = muscular contraction force, and

r = the perpendicular distance from the line of action of force to the joint center.

If we look at the example of the elbow joint we see that for any given joint angle, r will be affected by the angle of the joint and the distance from the joint center that the muscle is attached (see figure 11.5). For this reason even a constant muscular contraction force will produce torque that varies as the joint moves through a range of motion (see table 11.2 and figure 11.6).

The transformation of muscle shortening velocity to joint angular velocity produces a similar although opposite effect to that of the transformation of muscular contraction force to joint torque (see table 11.2). The implication of this is that a constant linear velocity of muscle shortening will produce a

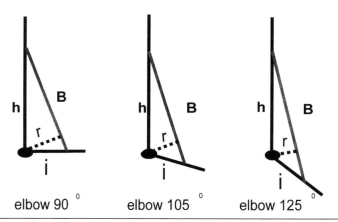

elbow 90 °   elbow 105 °   elbow 125 °

Figure 11.5 Schematic representation of a muscle operating about a joint (B) (i.e., elbow). For a given force of contraction of the muscle, the moment about the joint will change as the joint angle changes. This occurs because the moment arm (r) and the angle of the line of action of the muscle (h) changes with joint angle.

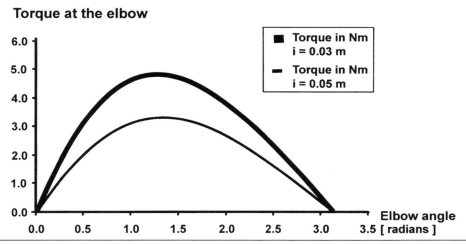

Figure 11.6 Graphic illustration of the impact of joint angle on torque at the elbow for two hypothetical cases: insertion point 0.02 m (thin line) or 0.03 m (thick line). In each case torque is calculated for a fixed muscle face, therefore the length-tension properties of the muscle are ignored.

Table 11.2 Effects of Changing Muscle Insertion Length on Torque and Joint Velocity

Insertion length from joint center (m)	Muscle length (m)	Angular velocity (rads/s)	Joint angle (degrees)	Torque around elbow (Nm)
0.030	0.225	0.055	30	1.670
0.030	0.230	0.042	45	2.308
0.030	0.252	0.034	90	2.979
0.030	0.266	0.041	120	2.439
0.050	0.208	0.033	30	3.002
0.050	0.218	0.024	45	4.063
0.050	0.255	0.020	90	4.903
0.050	0.278	0.026	120	3.889

Calculations are for elbow flexor muscle group undergoing isokinetic contractions at $0.01 \text{ m} \cdot \text{s}^{-1}$ and generating 100 N of force.

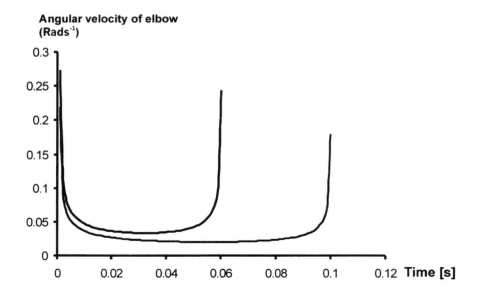

Figure 11.7 Graphic illustration of the impact of varying the distance from the joint center to the point of insertion of the muscle. The estimated length of the muscle is shown during an isokinetic (constant angular velocity) contraction. The length change is greater when the point of insertion is farther from the joint center. Clearly the slope of the line is not constant in either case, indicating a changing velocity of shortening.

changing angular velocity as the joint moves through its range of motion. Once again the magnitude of the change in joint angular velocity will be directly related to the distance from the joint center to the point of insertion of the muscle and the initial joint position (see figure 11.7).

The anatomic configuration of a joint has many functional implications on joint movement and function. For a given force of muscular contraction the torque around the joint will be greatest when the bone (lever arm) is oriented at 90° to the line of action of the force. Increasing the length of the lever arm (insertion distance of muscle on the bone from the joint center) will increase the joint torque throughout the contraction but will also increase the amount of muscle shortening required to move the joint through its range of motion. Conversely, reducing the length of the lever arm (insertion distance of muscle on the bone from the joint center) will increase the angular velocity and displacement of the joint for a given velocity of muscle shortening.

Ways of Studying Force-Velocity Properties

The criteria for appropriate evaluation of the force-velocity relationship include the following: control of one of the two parameters (force or velocity), measurement of the dependent variable (whichever of the two is not controlled), and constant

(usually maximal) activation of the contractile process. The determination of the force-velocity relationship requires that measurements be obtained for several contractions. Each contraction yields one point on the graph of the force-velocity relationship. The levels of organization that can be used in the study of the force-velocity relationship include in vitro, in situ, and in vivo, although strictly speaking, in vivo study should be referred to as torque-angular velocity relationship for any specific joint motion since these are the parameters that are most frequently measured.

In order to determine the relationship between force and velocity of contraction it is necessary to control the load or velocity of contraction and measure the other parameter. In circumstances where a tendon can be attached directly to a measurement device, the load can be controlled electronically with a servomotor or mechanically with an isotonic (constant load) lever.

In Vitro Force-Velocity Relationship

In vitro study of contractile properties entails removal of a muscle (or part of a muscle) from an animal and preservation of contractile function in a laboratory situation. This kind of study has been conducted with single skinned fibers, single intact fibers, and whole muscles. Isolation of the muscle out of the body allows the investigator direct control of the environment (extracellular space in intact cell, intracellular space in skinned fiber) and single fiber work assures homogeneous fiber type representation in the preparation. Limitations include inability to exactly duplicate physiological extracellular (or intracellular) fluid composition, and in vitro work typically is conducted at cold (nonphysiological) temperatures.

In Situ Force-Velocity Relationship

Keeping the muscle in the body of the animal while partly isolating (i.e., connect tendon directly to measuring system) is referred to as in situ. Study of the muscle (or motor unit) while keeping it in the body overcomes some of the disadvantages present in the in vitro preparation. The temperature and composition of the extracellular fluid are more physiological in studies conducted in situ. The major limitation of the in situ approach is the heterogeneous fiber type composition represented in most muscles that can be studied in this way. Differences in rate of fatigue between motor units present special problems affecting interpretation of such experiments. Study of individual motor units in situ overcomes some of these problems but introduces a new consideration. The inactive motor units impose an internal resistance or viscosity that cannot be easily accounted for (Hatcher and Luff 1987).

In Vivo Force-Velocity Relationships

Study of the force-velocity relationship of a muscle while keeping that muscle intact within the body is not a simple task. More often, such studies present a torque-angular velocity relationship rather than the force-velocity relationship, which would require estimation of the moment arm, fiber length, and angle of pennation of the participating muscles. The general principle of the torque-angular velocity relationship is comparable to the force-velocity relationship, but if velocity of shortening of a muscle is not measured, then the true force-velocity relationship is not known. This problem can be overcome to some extent by estimation of muscle shortening using principles of biomechanics, but generally this approach requires several assumptions.

The relationship between the velocity of shortening and the force generated by a muscle has been studied in many different ways. In addition to considering the level of organization (in vitro, in situ, and in vivo), the type of muscle contraction that is used is an important consideration. Type of contraction refers to the form of regulation that is undertaken, and typically scientists have controlled either load (isotonic contractions) or velocity (isokinetic contractions). These two approaches will be consid-

ered below, along with a unique approach, isometric-isotonic contractions.

Isotonic Contractions

Key features of such a measurement system include low mass lever to diminish the impact of inertia, low resistance pivot, afterload stop to permit contraction at the same initial length in repeated trials, and control of a constant load against which the muscle shortens. The constant load has been applied with weights (Gaehtgens, Kreutz, and Albrecht 1979; Hirche, Raff, and Grun 1970), a pneumatic resistance (Ameredes et al. 1992; Fales et al. 1958; Gundersen 1985), or with a motorized lever with feedback to regulate the resistance (Bahler, Fales, and Zierler 1968; Coirault et al. 1993; Fitts et al. 1998; Vandenburgh, Swasdison, and Karlisch 1991).

It is generally regarded that to study the characteristic features of the force-velocity properties of muscle, activation must be maximal, cross-bridge kinetics should be in a steady state, and the parameters (force and velocity) should be measured at a given muscle (sarcomere) length. These criteria are not always adhered to, and failure to follow these may explain why some departures from the classic hyperbolic shape of the force-velocity relationship have been reported. Two approaches have been used to ensure that activation is maximal at the time the measurement is taken: obtain the measurement of force and velocity sufficiently late after the initiation of activation to be assured of this or obtain a steady state of isometric force prior to releasing the muscle to a constant load less than isometric (isometric to isotonic contraction).

Afterloaded Isotonic Contractions

In order to study the velocity of shortening of a muscle while it contracts against a fixed load, scientists have used an afterloaded contraction. In this situation, the muscle is held at the desired initial length while the load is supported by external means. When the muscle is stimulated, isometric tension builds up until the force is sufficient to move the load. If inertia is zero, then the force exerted by the contraction will be equal to the load. Also, while the velocity of contraction is constant, then inertia is inconsequential (Fenn and Marsh 1935), so typically measurements are obtained after the muscle appears to have achieved a constant rate of shortening. Another important consideration is the possible impact of changes in the length of any compliant structures in series with the contractile component of the muscle. If the load on the muscle is constant, then any purely elastic structure cannot be affecting the velocity of shortening. However, it is important to recognize that the contribution of a series elastic component to length of the muscle will change in proportion to the load on the muscle. Therefore, sarcomere length is not the same at a given muscle length with various loads imposed.

Isometric-Isotonic Contractions

To ensure that activation is maximal, some authors have released the muscle to a constant load after maximal isometric force has been permitted to develop. In this situation, the series elastic components of the muscle will have been stretched prior to the isotonic phase of contraction and will contribute to the shortening in the period immediately after the muscle is released to a constant load less than the isometric force. Measurement of the force-velocity properties of muscle should take this into account, and measurement of the velocity must occur after the velocity has slowed to a constant level. Additional care must be exercised to account for the distribution of muscle length between the contractile component and the series elastic component. Once the velocity has slowed to a constant rate of shortening, the length of any elastic structure in series with the contractile component (including those structures attaching the muscle to the measurement system) will be stretched in proportion to the force exerted by the muscle. This means that at any overall muscle length, the length of the contractile component will be shorter when the resisting load is higher. This fact is rarely taken into consideration.

Isokinetic Contractions

A constant velocity of shortening can only be accomplished when the tendon of a muscle is attached to a device that regulates linear translation. Scientists have generally used a motor to achieve this situation (Cecchi, Lombardi, and Menchetti 1984; Ettema, Van Soest, and Huijing 1990). The precautions mentioned above are equally applicable here. If measurements are made at a common muscle length, while length is distributed between contractile component and series elastic component, these measurements will not represent a common sarcomere length.

The term *isokinetic* has frequently been used to represent constant angular velocity. In vivo studies using isokinetic dynamometers (Cybex, Biodex, Kin Com, etc.) have been completed with various portions of the torque-angular velocity relationship represented. There are several pitfalls associated with this approach (Baltzopoulos and Brodie 1989;

Herzog 1988; Winter, Wells, and Orr 1981), and this equipment typically permits estimation of only a limited range of the torque-angular velocity relationship.

Factors Affecting the Force- and Power-Velocity Relationships

The shape of the relationships between force and velocity or power and velocity can be affected by sarcomere length and fiber type of the muscle or fiber. These factors will be discussed below.

Sarcomere Length

Most studies that provide measurements of the force-velocity properties of muscle present measurements at a given muscle length (usually the length at which isometric active force is maximal) or initiate contraction at a given muscle length. Sarcomere length is rarely measured and only occasionally estimated. Considering that a muscle has an elastic component in series with the applied load, measurement at a given muscle length at a variety of isotonic loads assures that the same sarcomere length is not present. Similarly, measuring the force-velocity properties with initial muscle length constant, assures that measurements are unlikely to be obtained at a common sarcomere length.

Maximal velocity of shortening is thought to be independent of muscle length. Julian and Morgan (1981) report that maximal velocity of shortening does not change over the range of sarcomere lengths from 2.2 to 1.75 µm in frog single muscle fibers. Considering the length dependence of maximal isometric force, it seems reasonable to expect the force-velocity properties of a muscle to scale to the length-dependent isometric force, which would yield force-velocity curves similar to that depicted in figure 11.2a. This is similar to that reported by Bahler, Fales, and Zierler (1968).

Scott, Brown, and Loeb (1996) report that in cat soleus muscle, force at any velocity of concentric contraction is decreased in proportion to the decrease in isometric force at lengths shorter than optimal length. This would result in a proportional decrease in power output at short lengths. In contrast, Phillips and Woledge (1992) report that power output is not decreased in spite of the decrease in isometric force at sarcomere lengths less than optimum.

Barata and coworkers (1996) attempted to depict the relationships between load, length, and velocity in a cat muscle preparation in situ. They argue that isokinetic and isometric contractions are somewhat contrived, and that muscle is normally called upon to move a given load. They then offer measurements of shortening velocity with a variety of preloads, which stretch the muscle to an initial length that depends on the passive properties of the muscle. This approach, which has been used in a series of publications (Baratta and Salomonow 1991; Baratta et al. 1995b; Baratta et al. 1996), offers some interesting possibilities but does not represent a realistic force, length, velocity relationship. In the first place, the force generated is neither measured nor calculated (the loads are accelerated during the initial stage of a given contraction). Furthermore, the force needed to move a given load in vivo will change as the joint angle and corresponding moment arm changes (see section on anatomic considerations), and finally, when faced with the task of moving a load, a joint position that permits favorable load, length, velocity condition would likely be assumed. That is, an afterloaded (not preloaded) condition would seem more realistic.

Muscle Fiber Type

Mammalian skeletal muscle can be subdivided into several fiber types, based on myosin isoform composition (Galler, Schmitt, and Pette 1994; Schiaffino et al. 1988; Schuler and Pette 1996), contraction characteristics (Brooks and Faulkner 1991; Burke et al. 1973), or histochemistry (Ariano, Armstrong, and Edgerton 1973; Armstrong and Phelps 1984; Ishihara and Taguchi 1991). There are four distinct myosin isoforms (types I, IIa, IIb, and IIx [also known as IId]) that can be identified by immunohistochemistry (Schiaffino and Reggiani 1994; Staron and Pette 1993). The primary distinguishing contractile characteristic of muscle fibers expressing these myosin isoforms is variation in maximal velocity of shortening and the relative content of myosin light chain 3f, which serves to modulate V_o for each myosin heavy chain isoform, effecting a broad range of values for maximal velocity of shortening (Schiaffino and Reggiani 1996). Table 11.3 presents published values for maximal velocities from skinned fiber experiments, which have compared maximal velocities of muscle fibers for which myosin isoform was determined. The rank order of maximal velocity is I < IIA < IIX <IIB.

Most mammalian muscles are composed of various proportions of these fiber types. Since the fiber types have different contractile characteristics, the force and power-velocity properties of a muscle are determined by the fiber type distribution for a given muscle. Several studies have confirmed that fiber type distribution affects maximal velocity and peak power output (Thorstensson, Grimby, and Karlsson 1976; Tihanyi, Apor, and Fekete 1982).

The most effective way to illustrate how fiber type composition influences the force-velocity properties of muscle is to formulate a mathematical model of a muscle comprised of a number of motor

Table 11.3 Reported Values for Parameters of the Force-Velocity Relationship in Skinned Fiber Preparations

Muscle	Isometric force $(kN \cdot m^{-2})$	Maximal velocity $(FL \cdot s^{-1})$	Temp.	$a \cdot P_o^{-1}$	Source
Rat sol.	130	1.2	15°C		Widrick et al. 1996
Monkey sol.	146	0.7	15°C	0.044	Widrick et al. 1997
Human sol.	145	0.52	15°C	0.037	Widrick et al. 1997
Rat gast.	133	1.34	15°C	0.064	Widrick et al. 1997
Monkey gast.	160	0.69	15°C	0.040	Widrick et al. 1997
Human gast.	136	0.64	15°C	0.034	Widrick et al. 1997
Rat type I	211	0.64	12°C	0.066	Bottinelli, Schiaffino, and Reggiani 1991
Rat type IIA	303	1.4	12°C	0.066	Bottinelli, Schiaffino, and Reggiani 1991
Rat type IIX	439	1.45	12°C	0.132	Bottinelli, Schiaffino, and Reggiani 1991
Rat type IIB	338	1.8	12°C	0.113	Bottinelli, Schiaffino, and Reggiani 1991
Rabbit psoas	246	3.26	20°C	0.10	Wahr and Metzger 1998
Rat sol.	234	1.94	20°C	0.05	Wahr and Metzger 1998

units and use this model to illustrate the force-velocity properties of the muscle. The model described below is an enhanced version of one described previously (MacIntosh et al. 1993).

A Model of Force-Velocity Properties for Mixed Fiber Types

A mammalian skeletal muscle is composed of motor units, each consisting of a motor neuron and a number of muscle cells, which in general express a common muscle fiber type. The force-velocity properties of such a muscle can be expressed in a model by applying the Hill equation to each motor unit and presenting the resulting cumulative force, velocity, and power outputs.

The model muscle that is presented here is composed of 200 motor units of progressively larger size (isometric force). The largest motor unit has an isometric force 50 times greater than the isometric force of the smallest motor unit, and the cumulative isometric force of all motor units = 1.0. Each unit is designated as fast twitch or slow twitch, with all slow twitch motor units having an isometric force smaller than the weakest fast twitch motor unit. Subdivisions of fast twitch motor units (type IIa versus type IIb) have not been considered in this model, but within each fiber type classification, a range of maximal velocities is represented.

The maximal velocities (V_{max}) of individual motor units expressed relative to V_o for the fastest motor unit range from 0.2 to 0.33 for slow twitch motor units and from 0.6 to 1.0 for fast twitch motor units (three times faster than slow twitch units). The frequency distribution of V_{max} within each range of values gives a normal distribution about the mean for slow twitch (0.266) and for fast twitch (0.8) motor units. The designation of a factor of three for the difference in maximal velocity between fast and slow twitch fibers is somewhat arbitrary. However, the apparent difference in maximal velocity of sarcomere shortening between a muscle composed entirely of slow twitch fibers (cat soleus) and a muscle that contains type IIa and IIb fibers (cat medial gastrocnemius muscle) is approximately a factor of three (1,980). Close (1972) has also indicated that within a species, the magnitude of difference in maximal rate of sarcomere shortening between fast and slow muscles is a factor of 2 to 3, which is substantially different from the factors obtained in skinned fiber experiments (see table 11.3).

For each hypothetical muscle (comprised of 30, 40, 50, 60, or 70% fast twitch muscle, based on isometric force), force for each motor unit was calculated at velocities ranging from 0 to V_{max} according to the rearranged Hill (1938) equation

$$Force = [b \cdot (P_o + a) \cdot (v + b)^{-1}] - a \qquad 11.3$$

where variables are defined as for equation 11.1.

There were various additional assumptions made in order to use the Hill equation to calculate the cumulative force-velocity properties of these hypothetical muscles:

- Muscle fiber length is the same for all muscle fiber types.
- The size of a motor unit is represented by its isometric force.
- The ratio of $a \cdot P_o^{-1}$ was arbitrarily set at 0.4 for ST fibers and 0.45 for FT fibers.

The force and power of each muscle was calculated over the full range of velocities (0 to V_{max}) using an Excel spreadsheet. Each column was used to represent an individual motor unit.

Figure 11.8 illustrates the cumulative force-velocity and power-velocity curves for hypothetical muscles of differing fiber type composition. Several distinguishing characteristics are evident from this model:

- The degree of curvature ($a \cdot P_o^{-1}$) decreases as percentage of fast twitch fibers increases.
- At any given velocity (except 0 and 1) force is greater when percentage fast twitch fibers is greater.
- At any given velocity (except 0 and 1) power is greater when percentage fast twitch fibers is greater.
- The velocity at which peak power occurs is greater as percentage of fast twitch fibers increases.

The foregoing discussion of the force-velocity relationship properties of muscle has been presented without concern for the anatomic constraints imposed on the muscle by its confinement within the body. When muscle contracts in vivo, torque is generated and movement typically occurs about some joint axis. Under these circumstances, investigators typically report torque and angular velocity rather than muscle force and velocity of shortening. The shape of the torque-angular velocity relationship is quite similar to the force-velocity relationship, and can be fit by the Hill equation (Tihanyi, Apor, and Fekete 1982; Wilkie 1950).

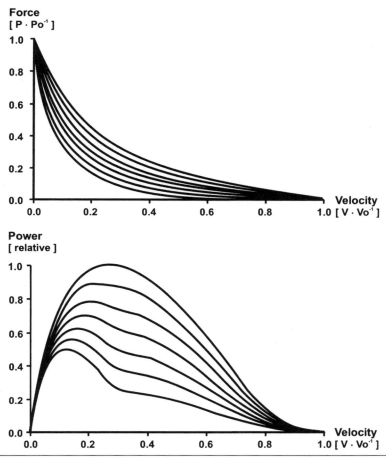

Figure 11.8 The force-velocity and power-velocity relationships are illustrated for hypothetical muscles comprised of 200 motor units each, where, from lowest line to highest line, 20, 30, 40, 50, 60 70, and 80% of isometric force is contributed by fast twitch type fiber. Several observations are relevant. Each muscle has the same maximal velocity. An increase in fast twitch fiber composition results in an increase in apparent $a \cdot P_o^{-1}$ (less curvature), higher peak power output, a broader power curve, and a higher velocity at peak power output.

Conversion of torque and angular velocity to force and velocity of shortening requires an estimate of the moment arm (see figure 11.5, with further elaboration in chapter 15).

Submaximal Activation

The force-velocity properties of a muscle dictate the peak performance (highest force or power at a given velocity) of a muscle. However, sometimes it is desirable to generate less than peak power output at a given velocity. A whole muscle can generate less than peak power output at a given velocity by activating some fraction of the total motor unit pool. An example using the same model as the one described above will serve to illustrate this aspect of muscle performance. It should be noted that motor unit recruitment is not an all-or-nothing response. Regulation of force within a motor unit occurs by frequency modulation. Furthermore, additional

motor units are thought to be recruited before activation (frequency of stimulation) of already recruited motor units is maximal. The example of motor unit recruitment (below) ignores this aspect of frequency modulation and considers only maximal activation of motor units.

Using the model described above, it can be shown that the number of motor units that must be activated to achieve a given submaximal power output is dictated by the velocity (or resistance/load) at which that power output is desired. One additional assumption is required to apply the model to this situation: it is assumed that the size principle dictates the order (smallest to largest) of motor unit activation. The following example illustrates motor unit recruitment needed to achieve 20 to 60% of peak power output for muscles composed of 30, 50, and 70% fast twitch fiber composition. In this case, the values of relative power represent 20 to 60% of peak power of the 70% fast twitch muscle (the muscle with the highest peak power output). In the case of the muscle with 30% fast twitch composition, this would represent 32 to 95% of this muscle's peak power, since this muscle has a peak power output that is just 62.5% of the peak power output of the muscle that is 70% fast twitch.

Figure 11.9 presents the force and power-velocity relationships for the hypothetical muscles composed of 30, 50, and 70% fast twitch fibers, with target power equivalent to 20, 40, and 60% of the peak power output of the muscle composed of 70% FT fiber area. The number of motor units required to be activated to achieve the target power output can be interpreted from the relative isomet-

ric force. Clearly, more motor units must be activated to achieve a given power output when the muscle has a smaller cross-sectional area occupied by FT fibers. Furthermore, it should be recognized that the lines on these graphs represent the minimal motor unit activation to achieve the target power output, and the target power can only be reached within a narrow range of velocities. If it is desirable to achieve a power output equal to 0.4 of the maximal power output of the 70% FT muscle, at the velocity of 0.5 of V_{max}, then the 70% FT muscle can achieve this with 161 motor units. The muscle that is 50% FT will require 184 motor units, and the muscle that is 30% FT will not be able to achieve this target power output at 0.5 of maximum veloc-

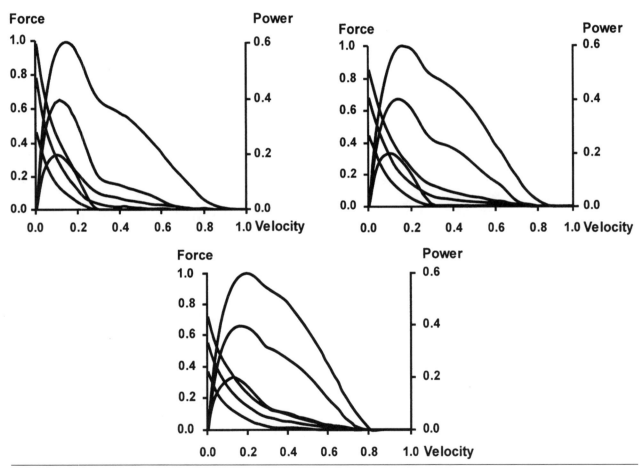

Figure 11.9 Force-velocity and power-velocity curves representing the motor unit recruitment needed to generate target power output equivalent to 40, 60, and 80% of the peak power output of a muscle with 70% of its cross section occupied by fast twitch fibers. In each panel, there are three force-velocity curves and three power-velocity curves. The panels represent muscle that is 30% (top left), 50% (top right), or 70% (bottom) fast twitch composition. The isometric force for each force-velocity curve identifies the relative proportion of that muscle's total cross section that must be activated in order to achieve the target power output. When a muscle is 30% fast twitch, it is necessary to activate 46%, 79%, and 98% of the total cross-sectional area to achieve the target power values. When a muscle is 50% fast twitch, it requires 44%, 68%, and 86% of the total cross section. In the case of a muscle that is 70% fast twitch, it requires activation of 37%, 55%, and 72% of the muscle cross-sectional area.

ity. Several observations can be made from this series of figures:

- The number of motor units that must be activated to achieve a given power output is determined by the target velocity.
- The higher the percentage FT fibers, the lower the number of motor units required to achieve a given power output.
- For each muscle there is a unique velocity at which motor unit recruitment is minimal for a given power output.

Functional Implications of the Force-Velocity Properties of Muscle

Many forms of human movement require power output. A.V. Hill investigated the power generated by subjects as they ran across force platforms, accelerating with maximal effort. It was observed that the force that was exerted by these subjects in the direction opposite to the direction of motion decreased as the velocity of motion increased. Hill subsequently demonstrated that the relationship between force and velocity of motion was similar to the relationship between force and velocity of shortening of a muscle. Many forms of human movement can be similarly represented in an analogous relationship, and these properties of muscle function have particular relevance when power output is an issue.

Cycle ergometry has frequently been used to measure peak power output as the product of resistance and the velocity of a point on the periphery of the flywheel. Since the relationship between resistance applied to the flywheel and the velocity of a point on the perimeter of the flywheel is similar to the relationship between force and velocity of muscle contraction, there is a unique condition (optimum) at which peak power output will occur. This unique condition will be an optimal velocity in isokinetic cycle ergometry or an optimal resistance in ergometry using a friction-loaded cycle ergometer. Both isokinetic cycling (McCartney et al. 1983) and friction-loaded cycle ergometry (Vandewalle et al. 1987) have been used to assess peak power output. The focus here is on the use of the friction-loaded cycle ergometer, since this equipment is more universally available than an isokinetic ergometer.

Early attempts to assess peak power output with cycle ergometry using a friction-loaded ergometer relied on setting the resistance in proportion to body mass (Bar-Or, Dotan, and Inbar 1977). Several studies, however, have noted that setting resistance in proportion to body mass does not give optimal conditions for all subjects. Vandewalle and coworkers (1987) introduced the concept of using the relationship between resistance and velocity of a point on the periphery of the flywheel to individually determine optimal resistance for peak power output. This relationship was determined for each subject by having him or her pedal with maximum effort for a few seconds at each of several resistance loads. The peak velocity achieved, when plotted against the corresponding resistance, gives a linear relationship from which the resistance that is optimal for power output can be determined (MacIntosh and MacEachern 1997; Vandewalle, Peres, and Monad 1987; Vandewalle et al. 1989).

It should be recognized that the relationship between resistance applied to the flywheel and the velocity of a point on the periphery of the flywheel can be expressed as a relationship between torque (resistance · radius) and angular velocity of the flywheel. When the angular velocity is constant, power is equivalent to torque · angular velocity. Recently, it has been recognized that an accurate representation of the power- and torque-angular velocity relationships of cycle ergometry should take into account the torque applied to accelerate the flywheel (Bassett 1989; Lakomy 1986). Continuous measurement of velocity of the flywheel reveals that the velocity not only changes from one pedal stroke to the next, but also oscillates within each pedal stroke. The magnitude of this oscillation at a given power output depends on the resistance imposed on the flywheel (MacIntosh and Oppermann 1994). Calculation of power, therefore, must consider this torque, which contributes to acceleration of the flywheel:

$$\text{Power} = R \cdot v + I \alpha \omega \qquad 11.4$$

where

R = resistance applied to the flywheel,

v = velocity of a point on the periphery of the flywheel,

I = moment of inertia

α = angular acceleration, and

ω = angular velocity.

The torque that is transmitted from the cranks to the flywheel is represented by the following: power · γ^{-1}. The actual torque applied to the cranks would be greater than this, since there is some loss due to friction and acceleration of other moving parts.

Calculation of peak power during a given trial reveals that peak power does not necessarily occur at the instant of maximal velocity. This is evident from figure 11.9. When a subject performs a series of maximal efforts with a variety of resistances (and suitable rest between trials) the resulting torque and angular velocity values obtained at peak power give a fairly linear relationship.

A recent twist on this approach to determining the torque-angular velocity relationship is the use of a single maximal effort, starting from a standstill to obtain several points on the torque-angular velocity curve (Arsac, Belli, and Lacour 1996; Martin, Wagner, and Coyle 1997). A key feature of this technique is that measurements are obtained at a variety of angular velocities as the flywheel is accelerated from zero to an angular velocity corresponding with a cadence in excess of 150 rpm. It has not yet been confirmed that this approach will give the same relationship as the multitrial approach.

The magnitude of peak power output is as much a determinant of the manner in which it is measured as it is of the actual capabilities of the subject. This is because the measurement occurs at a position on the intensity-duration curve where the slope is very steep. The various ways of expressing power include instantaneous (over a fraction of a second), averaged over a complete pedal revolution, and averaged over five seconds. Considering that cycling involves alternating effort by first one leg and then the other, instantaneous power oscillates between high and low values. Quite high power is achieved during the period of acceleration (when the cranks approach the horizontal) and falls to very low levels when the cranks are in a position that does not favor application of torque.

Values for peak instantaneous power output for healthy adult males can vary from 1400 to over 3000 watts. Average power output over a complete pedal revolution (approximately 0.5 s) is generally just over half of peak instantaneous power output.

Assessment of peak power output during cycle ergometry should permit the subject to generate power against an optimal resistance. This resistance can be determined from measurement of the torque-angular velocity relationship.

Concluding Remarks

The force-velocity relationship of muscle represents a characteristic property of muscle that describes the upper limit of performance when power output is a limiting factor. Power, therefore, is a function of muscle size and geometry, sarcomere length, and fiber type. Assessment of peak power output should utilize methods that acknowledge these factors and permit individual conditions that allow maximal power output.

References

Ameredes, B.T., Brechue, W.F., Andrew, G.M., and Stainsby, W.N. 1992. Force-velocity shifts with repetitive isometric and isotonic contractions of canine gastrocnemius in situ. *J. Appl. Physiol.* 73: 2105–11.

Ariano, M.A., Armstrong, R.B., and Edgerton, V.R. 1973. Hind limb muscle fiber populations of five mammals. *J. Histochem. Cytochem.* 21: 51–55.

Armstrong, R.B., and Phelps, R.O. 1984. Muscle fiber type composition of the rat hind limb. *Am. J. Anat.* 171: 259–72.

Arsac, L.M., Belli, A., and Lacour, J.R. 1996. Muscle function during brief maximal exercise: Accurate measurements on a friction-loaded cycle ergometer. *Eur. J. Appl. Physiol.* 74: 100–6.

Bahler, A.S., Fales, J.T., and Zierler, K.L. 1968. The dynamic properties of mammalian skeletal muscle. *J. Gen. Physiol.* 51: 369–84.

Baltzopoulos, V., and Brodie, D.A. 1989. Isokinetic dynamometry applications and limitations. *Sports Med.* 8: 101–16.

Bangart, J.J., Widrick, J.J., and Fitts, R.H. 1997. Effect of intermittent weight bearing on soleus fiber force-velocity-power and force-pCa relationships. *J. Appl. Physiol.* 82: 1905–10.

Baratta, R., and Solomonow, M. 1991. Dynamic performance of a load-moving skeletal muscle. *J. Appl. Physiol.* 71: 749–57.

Baratta, R.V., Solomonow, M., Best, R., Zembo, M., and D'Ambrosia, R. 1995a. Force-velocity relations of nine load-moving skeletal muscles. *Med. & Biol. Eng. & Comput.* 33: 537–44.

Baratta, R.V., Solomonow, M., Best, R., Zembo, M., and D'Ambrosia, R. 1995b. Architecture-based force-velocity models of load-moving skeletal muscles. *Clin. Biomech.* 10: 149–55.

Baratta, R.V., Solomonow, M., Nguyen, G., and D'Ambrosia, R. 1996. Load, length, and velocity of load-moving tibialis anterior muscle of the cat. *J. Appl. Physiol.* 80: 2243–49.

Bar-Or, O., Dotan, R., and Inbar, O. 1977. A 30-second all-out ergometric test: Its reliability and validity for anaerobic capacity. *Israel J. Med. Sci.* 13: 326.

Bassett, D.R., Jr. 1989. Correcting the Wingate test for changes in kinetic energy of the ergometer flywheel. *Int. J. Sports Med.* 10: 446–49.

Best, C.H., and Partridge, R.C. 1928. *Proceedings of the Royal Society (B Series)* 103: 218.

Bottinelli, R., Schiaffina, S., and Reggiani, C. 1991. Force-velocity relations and myosin heavy chain isoform compositions of skinned fibres from rat skeletal muscle. *J. Physiol. (London)* 437: 655–672.

Brooks, S.V., and Faulkner, J.A. 1991. Forces and powers of slow and fast skeletal muscles in mice during repeated contractions. *J. Physiol. (London)* 436: 701–10.

Burke, R.E., Levine, D.N., Tsairis, P., and Zajac, F.E. 1973. Physiological types and histochemical profiles in motor units of the cat gastrocnemius. *J. Physiol. (London)* 234: 723–48.

Cecchi, G., Lombardi, V., and Menchetti, G. 1984. Development of force-velocity relation and rise of isometric tetanic tension measure the time course of different processes. *Pflüg Archiv.* 401: 396–401.

Claflin, D.R., and Faulkner, J.A. 1989. The force-velocity relationship at high shortening velocities in the soleus muscle of the rat. *J. Physiol. (London)* 411: 627–37.

Close, R.I. 1972. Dynamic properties of mammalian skeletal muscles. *Physiol. Rev.* 52: 129–97.

Coirault, C., Chemla, D., Pery, N., Suard, I., and Lecarpentier, Y. 1993. Mechanical determinants of isotonic relaxation in isolated diaphragm muscle. *J. Appl. Physiol.* 75: 2265–72.

Curtin, N.A., and Edman, K.A.P. 1994. Force-velocity relation for frog muscle fibers: Effects of moderate fatigue and of intracellular acidification. *J. Physiol. (London)* 475: 483–94.

Ettema, G.J.C., Van Soest, A.J., and Huijing, P.A. 1990. The role of series elastic structures in prestretch-induced work enhancement during isotonic and isokinetic contractions. *J. Exp. Biol.* 154: 121–36.

Fales, J.T., Lilienthal, J.L., Talbot, S.A., and Zierler, K.L. 1958. A pneumatic isotonic lever system for dog skeletal muscle. *J. Appl. Physiol.* 13: 307–8.

Fenn, W.O. 1930. Work against gravity and work due to velocity changes in running. *Am. J. Physiol.* 93: 433–62.

Fenn, W.O., and Marsh, B.S. 1935. Muscular force at different speeds of shortening. *J. Physiol. (London)* 85: 277–97.

Fitts, R.H., Bodine, S.C., Romatowski, J.G., and Widrick, J.J. 1998. Velocity, force, power, and Ca^{2+} sensitivity of fast and slow monkey skeletal muscle fibers. *J. Appl. Physiol.* 84: 1776–87.

Gaehtgens, P., Kreutz, F., and Albrecht, K.H. 1979. Optimal hematocrit for canine skeletal muscle during rhythmic isotonic exercise. *Eur. J. Appl. Physiol.* 41: 27–39.

Galler, S., Schmitt, T.L., and Pette, D. 1994. Stretch activation, unloaded shortening velocity, and myosin heavy chain isoforms of rat skeletal muscle fibers. *J. Physiol. (London)* 478: 513–21.

Gasser, H.S., and Hill, A.V. 1924. The dynamics of muscular contraction. *Proc. R. Soc. Lond. (Biol.)* 96: 398–437.

Gundersen, K. 1985. Early effects of denervation on isometric and isotonic contractile properties of rat skeletal muscles. *Acta Physiol. Scand.* 124: 549–55.

Hatcher, D.D., and Luff, A.R. 1987. Force-velocity properties of fatigue-resistant units in cat fast twitch muscle after fatigue. *J. Appl. Physiol.* 63: 1511–18.

Herzog, W. 1988. The relation between the resultant moments at a joint and the moments measured by an isokinetic dynamometer. *J. Biomechanics* 21: 5–12.

Hill, A.V. 1938. The heat of shortening and the dynamic constants of muscle. *Proc. R. Soc. Lond. (Biol.)* 126: 136–95.

Hirche, H., Raff, W.K., and Grun, D. 1970. The resistance to blood flow in the gastrocnemius of the dog during sustained and rhythmical isometric and isotonic contractions. *Pflüg Archiv.* 314: 97–112.

Ishihara, A., and Taguchi, S. 1991. Histochemical differentiation of fibers in the rat slow and fast twitch muscles. *Jpn J. Physiol.* 41: 251–58.

Julian, F.J., and Morgan, D.L. 1981. Tension, stiffness, unloaded shortening speed, and potentiation of frog muscle fibers at sarcomere lengths below optimum. *J. Physiol. (London)* 319: 205–17.

Kaneko, M. 1970. The relation between force, velocity, and mechanical power in human muscle. *Res. J. Physical Educ.* 14: 143–47.

Kojima, T. 1991. Force-velocity relationship of human elbow flexors in voluntary isotonic contraction under heavy loads. *Int. J. Sports Med.* 12: 208–13.

Komi, P.V. 1973. Measurement of the force-velocity relationship in human muscle under concentric and eccentric contractions. *Med. Sport* 8: 224–29.

Krylow, A.M., and Sandercock, T.G. 1997. Dynamic force responses of muscle involving eccentric contraction. *J. Biomech.* 30: 27–33.

Lakomy, H.K.A. 1986. Measurement of work and power output using friction-loaded cycle ergometers. *Ergonomics* 29: 509–17.

Levin, A., and Wyman, J. 1927. The viscous elastic properties of muscle. *Proc. Royal Soc. (Biol.)* 101: 218–43.

MacIntosh, B.R., Herzog, W., Suter, E., Wiley, J.P., and Sokolosky, J. 1993. Human skeletal muscle fiber types and force-velocity properties. *Eur. J. Appl. Physiol.* 67: 499–506.

MacIntosh, B.R., and MacEachern, P. 1997. Paced effort and all-out 30 s power tests. *Int. J. Sports Med.* 18: 594–99.

MacIntosh, B.R., and Oppermann, L. 1994. Velocity and power transients in cycle ergometry [Abstract]. Proceedings of the Eighth Biennial Conference and Symposium, Canadian Society for Biomechanics. 62–63.

Martin, J.C., Wagner, B.M., and Coyle, E.F. 1997. Inertial load method determines maximal cycling power in a single exercise bout. *Med. Sci. Sports Exerc.* 29: 1505–12.

McCartney, N., Heigenhauser, J.F., Sargeant, A.J., and Jones, N.L. 1983. A constant-velocity cycle ergometer for the study of dynamic muscle function. *J. Appl. Physiol.* 55: 212–17.

McComas, A.J. 1996. *Skeletal muscle: Form and function.* Champaign, IL: Human Kinetics.

Phillips, S.K., and Woledge, R.C. 1992. A comparison of isometric force, maximum power, and isometric heat rate as a function of sarcomere length in mouse skeletal muscle. *Pflüg Archiv.* 420: 578–83.

Ranatunga, K.W. 1982. Temperature-dependence of shortening velocity and rate of isometric tension development in rat skeletal muscle. *J. Physiol. (London)* 329: 465–83.

Ranatunga, K.W., and Thomas, P.E. 1990. Correlation between shortening velocity, force-velocity relation and histochemical fiber type composition in rat muscles. *J. Muscle Res. Cell. Motil.* 11: 240–50.

Schiaffino, S., Ausoni, S., Gorza, L., Saggin, L., and Gundersen, K. 1988. Myosin heavy chain isoforms and velocity of shortening of type II skeletal muscle fibers. *Acta Physiol. Scand.* 134: 575–76.

Schiaffino, S., and Reggiani, C. 1994. Myosin isoforms in mammalian skeletal muscle. *J. Appl. Physiol.* 77: 493–501.

Schiaffino, S., and Reggiani, C. 1996. Molecular diversity of myofibrillar proteins: Gene regulation and functional significance. *Physiol. Rev.* 76: 371–423.

Schuler, M., and Pette, D. 1996. Fiber transformation and replacement in low-frequency stimulated rabbit fast twitch muscles. *Cell Tissue Res.* 285: 297–303.

Scott, S.H., Brown, I.E., and Loeb, G.E. 1996. Mechanics of feline soleus. I. Effect of fascicle length and velocity on force output. *J. Muscle Res. Cell Motil.* 17:207–19.

Spector, S.A., Gardiner, P.F., Zernicke, R.F., Roy, R.R., and Edgerton, V.R. 1980. Muscle architecture and force-velocity characteristics of cat soleus and medial gastrocnemius: Implications for motor control. *J. Neurophysiol.* 44: 951–60.

Stainsby, W.N. 1976. Oxygen uptake for negative work, stretching contractions by in situ dog skeletal muscle. *Am. J. Physiol.* 230: 1013–17.

Staron, R.S., and Pette, D. 1993. The continuum of pure and hybrid myosin heavy chain-based fiber types in rat skeletal muscle. *Histochemistry* 100: 149–53.

Thorstensson, A., Grimby, G., and Karlsson, J. 1976. Force-velocity relations and fiber composition in human knee extensor muscles. *J. Appl. Physiol.* 40: 12–16.

Tihanyi, J., Apor, P., and Fekete, G. 1982. Force-velocity-power characteristics and fiber composition in human knee extensor muscles. *Eur. J. Appl. Physiol.* 48: 331–43.

Vandenburgh, H.H., Swasdison, S., and Karlisch, P. 1991. Computer-aided mechanogenesis of skeletal muscle organs from single cells in vitro. *FASEB J.* 5: 2860–7.

Vandewalle, H., Peres, G., Heller, J., Panel, J., and Monod, H. 1987. Force-velocity relationship and maximal power on a cycle ergometer: Correlation with the height of a vertical jump. *Eur. J. Appl. Physiol.* 56: 650–6.

Vandewalle, H., Peres, G., and Monod, H. 1987. Standard anaerobic exercise tests. *Sports Med.* 4: 268–89.

Vandewalle, H., Peres, G., Sourabie, B., Stouvenel, O., and Monod, H. 1989. Force-velocity relationship and maximal anaerobic power during cranking exercise in young swimmers. *Int. J. Sports Med.* 10: 439–45.

Wahr, P.A., and Metzger, J.M. 1998. Peak power output is maintained in rabbit psoas and rat soleus single muscle fibers when CTP replaces ATP. *J. Appl. Physiol.* 85: 76–83.

Westing, S.H., Cresswell, A.G., and Thorstensson, A. 1991. Muscle activation during maximal voluntary eccentric and concentric knee extension. *Eur. J. Appl. Physiol.* 62: 104–8.

Westing, S.H., and Seger, J.Y. 1989. Eccentric and concentric torque-velocity characteristics, torque output comparisons, and gravity effect torque corrections for the quadriceps and hamstring muscles in females. *Int. J. Sports Med.* 10: 175–80.

Westing, S.H., Seger, J.Y., and Thorstensson, A. 1990. Effects of electrical stimulation on eccentric and concentric torque-velocity relationships during knee extension in man. *Acta Physiol. Scand.* 140: 17–22.

Widrick, J.J., Bangart, J.J., Karhanek, M., and Fitts, R.H. 1996. Soleus fiber force and maximal shortening velocity after non-weight bearing with intermittent activity. *J. Appl. Physiol.* 80: 981–987.

Widrick, J.J., Romatowski, J.G., Karhanek, M., and Fitts, R.H. 1997. Contractile properties of rat, rhesus monkey, and human type I muscle fibers. *Am. J. Physiol. Regul. Integr. Comp. Physiol.* 272: R34–R42.

Wilkie, D.R. 1950. The relation between force and velocity in human muscle. *J. Physiol. (London)* 110: 249–80.

Winter, D.A., Wells, R.P., and Orr, G.W. 1981. Errors in the use of isokinetic dynamometers. *Eur. J. Appl. Physiol.* 46: 397–408.

Zuurbier, C.J., and Huijing, P.A. 1993. Changes in geometry of actively shortening unipennate rat gastrocnemius muscle. *J. Morphol.* 218: 167–80.

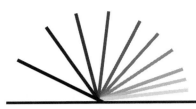

Chapter 12

Stability and Control of Aerial Movements

M.R. Yeadon and E.C. Mikulcik

When a cat falls from a height, it inevitably lands on its feet. For more than a century cats and rabbits have participated in experiments in which they are dropped in an inverted position from which they execute a half twist in midair (Marey 1894). Gymnasts execute similar movements during flight to make a safe landing when a dismount has not gone according to plan. The requirement of such control in aerial movements is not confined to the correction of errors. A gymnast who maintains a fixed configuration in a straight double somersault is unstable and will perform an unwanted half twist in the second somersault. This effect arises from the mechanical characteristics of rotating rigid bodies rather than from an error of the gymnast. Thus, midair control is needed to ensure that twist does not occur. In more complex movements involving multiple twists and somersaults, control is required in order for exactly the correct amounts of somersault and twist rotations to be reached at the time of landing. Strategies for controlling such movements are considered in this chapter.

Statement of the Problem

In aerial sports such as diving, trampolining, and gymnastics, the basic movements comprise somersaults with and without twist. For successful execution of a twisting somersault the athlete must ensure that the required number of twists are completed and for a nontwisting somersault that no twist is seen to occur. Theoretical analyses have shown that a rigid body is stable for rotations about the principal axes corresponding to maximum and minimum moments of inertia but unstable for rotations about the principal axis corresponding to the intermediate moment of inertia (Hinrichs 1978; Marion 1965). This instability arises because rigid body motions fall into two general modes centered on motions about the maximum and minimum principal axes and appear as either wobbling somersaults or twisting somersaults (Yeadon 1993a). Rotations about the intermediate principal axis lie near the boundary separating the two different modes of motion (Yeadon 1993a). For example, a pike (hips flexed) somersault about the lateral axis, corresponding to

Acknowledgment: Much of the material in this chapter has been reprinted from *Journal of Biomechanics* 29, M.R. Yeadon and E.C. Mikulcik, 1341-1348, Copyright 1996, with permission from Elsevier Science Ltd., The Boulevard, Langford Lane, Kidlington OX5 1GB, UK.

maximum moment of inertia, will appear to be a pure somersault since the angular velocity vector will remain close to this principal axis throughout the movement. In the case of a layout (nominally straight) somersault about the lateral axis, corresponding to the intermediate principal moment of inertia, any slight deviation of the angular velocity vector from this principal axis will eventually lead to a substantial buildup of twist sufficient to change backward rotation into forward rotation (Yeadon 1993a). Figure 12.1 shows a computer simulation of a double somersault in which the body maintains a fixed configuration with 1° of asymmetry in the arm abduction angles. In the first somersault there is very little twist while in the second somersault a half twist occurs. This surprising phenomenon can be reproduced practically by trying to throw a video cassette or a teddy bear about its intermediate axis without it twisting. Inevitably the cassette or teddy bear will perform an unwanted half twist that seems to start in the second somersault. The twist rate increases until a quarter twist is reached and then decreases. This means that there is a very slow twist during the first somersault and a rapid twist in the second somersault. This gives rise to the impression that the twist starts in the second somersault. Since gymnasts and trampolinists perform double somersaults in the straight position without twisting, they must have found some means of preventing the buildup of twist. The problem addressed in this chapter is to identify strategies that can be demonstrated to be capable of controlling the instability.

In this study three alternative strategies for controlling nontwisting somersaults using changes of body configuration are investigated. The three strategies comprise (a) arm abduction, (b) arching of the body, and (c) asymmetrical arm adduction/abduction.

Primary Concepts

One possibility of controlling the layout somersault is to adopt a fixed body configuration for which the somersault is stable. This may be accomplished by arching (or flexing) at the hips and spine sufficiently so that the lateral axis corresponds to the largest principal moment of inertia. If a large amount of arching is needed to accomplish this, there may be a problem in that the body may no longer be regarded as being straight and the gymnast's performance will be judged accordingly.

Another possibility is to make in-flight adjustments to prevent the twist from building up. Such corrections could be made with the arms so that when the body starts twisting to the left the arms move in such a way as to produce a twist to the right. If the time delay in the control system is too large it may not be possible for such control to be successful.

Elaboration of Concepts

Various strategies for controlling nontwisting somersaults are investigated and their ability to control twisting somersaults are considered.

Arm Abduction

Nigg (1974) suggested that the arms could be extended laterally in a straight somersault in order to minimize the influence of the instability. While it might be possible that extending the arms could reduce the twist rate since the moment of inertia about the longitudinal axis is greater, the problem of instability still remains and the proposed strategy at best can only delay the inevitable buildup of twist. If the buildup of appreciable twist does not occur until after the completion of two somersaults, then instability would not present a practical problem for a gymnast. If the buildup of twist becomes noticeable after one somersault, then instability would only present a problem for multiple somersaults. If the buildup of twist were noticeable before the completion of one somersault, then instability would pose a problem in all nontwisting straight somersaults. Thus it is of great relevance to determine the amount of somersault at which the twist becomes apparent.

Figure 12.1 A rigid configuration with only 1° of asymmetry in arm abduction angles produces almost a half twist after two somersaults in the straight position.

In order to investigate this strategy for controlling the buildup of twist in straight somersaults, a computer simulation model of aerial movement was used (Yeadon, Atha, and Hales 1990). This model comprised 11 segments representing chest and head, thorax, pelvis, upper arms, lower arms, upper legs, and lower legs. Inertia parameters were calculated from anthropometric measurements of an elite trampolinist using a mathematical inertia model of the human body (Yeadon 1990b). Input to the simulation model comprised (a) initial conditions in the form of the components of angular momentum about the mass center and initial values of three angles defining body orientation, together with (b) time histories of 14 angles defining body configuration. Output of the model comprised the time histories of the angles of somersault, tilt, and twist, which define the orientation of the body in space (Yeadon 1990c).

To determine the effect of arm abduction for somersaults in a straight body configuration, three simulations were first carried out with the arms adducted close to the body. Angular momentum was used, which produced nontwisting double somersaults for symmetrical body configurations. A small perturbation was introduced into each simulation by specifying an initial asymmetry in arm abduction angles. These asymmetries had magnitudes 0.1°, 1°, and 10°. Subsequently three more simulations were carried out in which each arm was approximately perpendicular to the midline of the trunk. Initial arm asymmetries of 0.1°, 1°, and 10° were again used to perturb the motion. A rigid configuration was maintained throughout each simulation.

The introduction of small perturbations into simulations of straight double somersaults resulted in substantial amounts of twist even when the arms were abducted. The twist after one somersault was small when the initial arm asymmetry was 0.1° or 1.0° but was large for an asymmetry angle of 10° (see table 12.1). After two somersaults the twist was large in each of the six simulations. Although separate simulations were carried out for rigid body configurations with arms adducted or abducted, the arms may be considered to have been initially in an adducted position in all cases and to have been instantaneously abducted immediately after takeoff in three of the six simulations. Using the abducted arm configuration did not result in reduced twist. On the contrary, the instability was more evident when the arms were wide. This is because the instability about the intermediate axis is dependent on the three principal moments of inertia. In the case of a body that is symmetrical about the longitudinal axis (e.g., a pencil) the buildup of twist remains slow with a constant twist rate, and there is no sudden half twist. As the two large principal moments of inertia become more different, the twist rate changes more with time so that when the arms are abducted as in figure 12.1 the half twist occurs earlier and faster than when the arms are held close to the body. For an initial arm asymmetry of 1° the movement appeared to be a somersault with little twist followed by a somersault with a half twist (see figure 12.1).

Table 12.1   Twist After One and Two Somersaults as a Function of Arm Symmetry

Somersault	Arm symmetry	Twist Arms adducted	Arms abducted
1	0.1°	0.00	0.01
1	1.0°	0.02	0.07
1	10.0°	0.23	0.35
2	0.1°	0.22	0.41
2	1.0°	0.46	0.49
2	10.0°	0.50	0.52

Somersault and twist values are in revolutions.

Body Arching

Hinrichs (1978) determined the directions of the principal axes during tucked, pike, and layout somersaults from a trampoline. In the tucked and pike somersaults the principal axis corresponding to maximum moment of inertia remained close to the angular momentum vector, while for the layout somersault the intermediate principal axis remained close to the angular momentum vector. This apparent stability about the intermediate axis led Hinrichs to speculate that the trampolinist must have made adjustments during flight to prevent the buildup of twist. On the other hand, it has been shown that small asymmetries do not led to substantial twist in a straight single somersault (see table 12.1) and so in-flight corrections may not have been required. In a straight double somersault, however, even small asymmetries will lead to a half twist. It would be of interest to determine, therefore, the extent to which various amounts of arching of the body can delay the buildup of twist in a double layout somersault. If the strategy of arching is not capable of limiting the effects of instability, the remaining possibility is that in-flight corrections are made to reverse any buildup of twist.

To determine the effect of arching on the stability of layout somersaults, simulations were carried out in which the angle between upper trunk and thighs remained constant. These angles ranged from 120° for an arched configuration to 180° for a straight body. Perturbations were introduced using initial arm asymmetries of 0.1°, 1°, and 10°, and a fixed configuration was maintained throughout each simulation. The results of these simulations established the extent to which arching of the body could limit the buildup of twist during a double somersault.

The adoption of an arched body configuration enabled the effects of instability to be reduced and even eliminated. Adopting an angle of 132° between upper trunk and thighs resulted in stable wobbling somersaults with little twist when the initial arm asymmetry was 0.1° or 1.0°. For an initial arm asymmetry of 10° a body arch of 129° was required in order for the motion to follow the stable wobbling somersault mode. Although in this case the motion was in a technically stable mode, the twist angle oscillated slowly with an amplitude of 0.24 revolutions and reached 0.14 revolutions after two somersaults. From table 12.2 it can be seen that arching provides a progressive improvement in controlling the buildup of twist.

In the simulation shown in figure 12.2 the abduction angle of the left arm is 1° more than that of the right arm so that the principal axes are tilted through a small angle. During the first one and a half somersaults there is a body arch of 145° and a knee angle of 160°. This phase of the motion is unstable but the buildup of twist is slow since the two large principal moments of inertia are approximately equal. In the last half somersault the body moves through the straight position into a stable pike somersault with a body flexion angle of 130°. Thus it is possible to perform an open double back somersault without appreciable buildup of twist providing the body arch is not less than that shown in figure 12.2.

Table 12.2　Twist After Two Somersaults as a Function of Body Arch and Arm Asymmetry

Body arch Arm asymmetry	Twist 0.1°	Twist 1°	Twist 10°	Somersault
120°	0.00 W	0.00 W	0.02 W	Stable
130°	0.00 W	0.01 W	0.16 T	Stable
140°	0.01 T	0.07 T	0.41 T	402°
150°	0.03 T	0.27 T	0.48 T	253°
160°	0.11 T	0.42 T	0.49 T	200°
170°	0.22 T	0.46 T	0.50 T	177°
180°	0.22 T	0.47 T	0.50 T	170°

Twist values are in revolutions. W = wobbling somersault; T = twisting somersault. During the stated somersault angle, the buildup of twist increases by a factor of 10.

Control Using Asymmetrical Arm Movements

In a twisting somersault asymmetrical arm movements may be used to alter the tilt angle to produce twist or to stop or reverse the twist (Yeadon 1993c). It should be possible, therefore, to ensure that the twist angle remains small during an unstable straight somersault by making corrective arm movements. In a backward somersault abduction of the left arm will produce a twist to the right while abduction of the right arm will produce a twist to the left.

To evaluate the capabilities of asymmetrical arm abduction for correcting the buildup of twist, the equations of motion of a model comprising two arms and one body segment were first linearized by assuming that perturbations remained small. This permitted an analytical consideration of the prospective control strategies for arbitrarily small perturbations.

The equations of motion for an asymmetrical arm abduction controller may be derived as follows. Suppose that abduction of one arm is accompanied by adduction of the other so that the sum

Figure 12.2 A double backward somersault with sufficient body arch to limit the buildup of twist.

$\varepsilon_a + \varepsilon_b$ of the abduction angles of the left and right arms from the midline of the trunk remains constant.

Let $\dot{\varepsilon} = -\dot{\varepsilon}_a = \dot{\varepsilon}_b$ be the rate of change of arm abduction angles. As shown in Yeadon (1990c) the equation of motion will be

$$h = I\omega + h_{rel} \qquad\qquad 12.1$$

where

  h is the total angular momentum about the mass center,

  I is the whole body inertia tensor about the mass center,

  ω is the angular velocity of the system, and

  h_{rel} is the angular momentum corresponding to internal movements.

The equation may be written as

$$\begin{bmatrix} h\cos\theta\cos\psi \\ -h\cos\theta\sin\psi \\ h\sin\theta \end{bmatrix} = \qquad\qquad 12.2$$

$$\begin{bmatrix} B & 0 & 0 \\ 0 & A & 0 \\ 0 & 0 & C \end{bmatrix} \cdot \begin{bmatrix} \dot\phi\cos\theta\cos\psi + \dot\theta\sin\psi \\ -\dot\phi\cos\theta\sin\psi + \dot\theta\cos\psi \\ \dot\phi\sin\theta + \dot\psi \end{bmatrix} + \begin{bmatrix} 0 \\ I_a\dot\varepsilon \\ 0 \end{bmatrix}$$

where

  $A > B > C$ are the principal moments of inertia of I,

  $I_a\dot\varepsilon$ is the angular momentum associated with the arm movement, and

  ϕ, θ, and ψ are Cardan angles for somersault, tilt, and twist corresponding to successive rotations about lateral, frontal, and longitudinal principal axes.

Equation 12.2 gives rise to

$$h = B\dot\phi + B\dot\theta \sec\theta\tan\psi \qquad\qquad 12.3$$

$$-h\sin\psi = -A\dot\phi\sin\psi + \qquad\qquad 12.4$$
$$A\dot\theta\sec\theta\cos\psi + I_a\dot\varepsilon\sec\theta$$

$$h\sin\theta = C\dot\phi\sin\theta + C\dot\psi. \qquad\qquad 12.5$$

Eliminating $\dot\phi$ from equations 12.3 and 12.4 gives

$$h(A-B)\sin\psi = AB\dot\theta(\sec\theta\cos\psi + \qquad 12.6$$
$$\sec\theta\sin^2\psi\sec\psi) + I_a B\dot\varepsilon\sec\theta.$$

Eliminating $\dot\phi$ from equations 12.3 and 12.5 gives

$$BC\dot\psi = h(B-C)\sin\theta + BC\dot\theta\tan\theta\tan\psi. \qquad 12.7$$

If the tilt and twist angles θ and ψ are assumed to be small and the controlling angular velocity $\dot\varepsilon$ is small, then equations 12.6 and 12.7 imply that θ and ψ will also be small. The approximations $\sin\theta = \theta$, $\cos\theta = 1$, $\sin\psi = \psi$, $\cos\psi = 1$ are used in equations 12.6 and 12.7 and small quantities of the third order are neglected. Equation 12.6 becomes

$$h(A-B)\psi = AB\dot\theta + I_a B\dot\varepsilon. \qquad\qquad 12.8$$

Equation 12.7 becomes

$$BC\dot\psi = h(B-C)\theta. \qquad\qquad 12.9$$

Differentiating equation 12.9 and using equation 12.8 gives

$$\ddot\psi = k^2\psi - m\dot\varepsilon \qquad\qquad 12.10$$

where $k^2 = h^2(B-C)(A-B)/AB^2C$, and $m = hI_a(B-C)/ABC$.

A value for I_a/A was obtained using a simulation under conditions of zero angular momentum with the model of Yeadon, Atha, and Hales (1990). The arms were moved through a small angle ε and the change in the tilt angle of the longitudinal principal axis was noted. When $h = 0$, equation 12.8 gives $I_a/A = \dot\theta/\dot\varepsilon$ and so θ/ε was used as an approximation to I_a/A.

In equation 12.10 ψ is the output variable, which should remain small if the buildup of twist is to be prevented, and $\dot\varepsilon$ is the control variable. When no control is exercised and a fixed body configuration is maintained $m = 0$ and equation 12.10 has solution

$$\psi = \psi_0 e^{kt}. \qquad\qquad 12.11$$

This demonstrates the exponential buildup of twist for rotations initially close to the intermediate principal axis. If ψ is small the somersault rate will be $\dot\phi = h/B$ and kt may be expressed as $p\dot\phi t = p\phi$ so that equation 12.11 may be written as

$$\psi = \psi_0 e^{p\phi} \qquad\qquad 12.12$$

where $p^2 = (B-C)(A-B)/AC$.

For a given body configuration ψ will increase by a factor 10 when $p\phi$ increases by $\ln(10)$. For arms by the sides of the body this corresponds to 170° of somersault while for arms abducted as in figure 12.1 it requires 163° of somersault. For arched configurations the latent periods for the buildup of twist range from 170° when straight to 402° for a body arch of 140° (see table 12.2). This buildup delay becomes infinite for a body arch of 133°.

In the case where control using asymmetrical arm movement was attempted, proportional plus integral plus derivative (PID) solutions of the form

$$\dot{\varepsilon}\,(t) = K_p\,\psi\,(t-T) + K_i\int\psi\,(t-T) + \qquad 12.13$$
$$K_d\,\dot{\psi}\,(t-T)$$

were sought where K_p, K_i, and K_d are constants and T is the delay in the feedback loop. Thus corrections were based on the state of the system at an earlier time.

Using an analytical solution to the problem of such control for arbitrarily small perturbations, it was found that proportional plus derivative (PD) control was necessary and sufficient to provide stable operation for nonzero time delays (Yeadon and Mikulcik 1996). In other words the best value for the constant K_i was zero. Stable control was possible for a range of values of K_p and K_d providing that the time delay T in the system was not greater than 0.28 somersaults. Figure 12.3 shows that there is a set of controller values that maximizes the time delay T that can be accommodated. Alternatively, it may be stated that for each set of controller values, there is a value of time delay beyond which stable control is not possible. The larger the time delay, the narrower the range of allowable controller parameters, and therefore the greater the difficulty in maintaining stable control.

Control of Finite Perturbations

The above analytical treatment assumes that the perturbations of the system are arbitrarily small. In reality we may have arm asymmetries where the arm abduction angles differ by as much as 10°, and the control system must be able to cope with such disturbances for the duration of a double somersault. On the other hand, stable control may not be necessary since the duration of the movement is limited. Proportional plus derivative control was

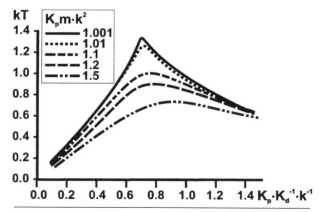

Figure 12.3 The time delay T that can be accommodated for PD parameters K_p and K_d. There are values of K_p and K_d that maximize T. For delays greater than this value of T, stable control is not possible.

incorporated into the 11-segment computer simulation model of Yeadon, Atha, and Hales (1990) by making the arm abduction angles change from ε_a to $\varepsilon_a - \delta\varepsilon$ and from ε_b to $\varepsilon_b + \delta\varepsilon$ over a time interval $\delta T = 0.01\,T_f$, where T_f is the flight time and $\delta\varepsilon = (K_p\psi + K_d\dot{\psi})\,\delta T$. A time delay was also introduced by basing the correction $\delta\varepsilon$ upon earlier values of ψ and $\dot{\psi}$ in a simulation.

Simulations of straight double somersaults with perpendicular arms were run for initial arm asymmetries of 0.1°, 1°, and 10° and values of the proportional and derivative constants were found for which the buildup of twist was controlled. The feedback time delay in the control loop was increased to determine the maximum delay for which control was possible and to obtain the corresponding values for the control parameters. These values were then compared with the corresponding results from the theoretical analysis.

The results of the numerical solutions for control using the computer simulation model were in general agreement with the theoretical findings. Increasing the time delay resulted in smaller ranges of suitable parameter values that gave stable control, and this enabled optimum values to be determined. For a time delay equivalent to 0.02 somersaults and an initial arm asymmetry of 10°, the arms moved rapidly to approximately symmetrical positions that changed little during the movement so that the response was stable (see figure 12.4a). When the delay was increased to 0.12 somersaults the twist was controlled although the amplitude of the arm oscillations did not decrease (see figure 12.4b), and the response could be described as (slightly worse than) neutral control. Increasing the delay to 0.24 somersaults resulted in control of the twist although the difference in arm abduction angles became more than 90° (see figure 12.4c), and the response was unstable.

If the original asymmetry in arm angles is reduced from 10° to 1°, then a delay of 0.24 somersaults produces a response similar to figure 12.4b. A further reduction in arm angle asymmetry to 0.1° with a delay of 0.24 somersaults produces a stable response similar to figure 12.4a. Table 12.3 gives the maximum twist and arm asymmetry in each of these five simulations. Note that in the first simulation the arm asymmetry dropped rapidly from 10° to 2° and then remained at that value.

Since the theoretical analysis of PD control gave a limit on the delay equivalent to 0.28 somersaults, the results of the numerical simulation model may be considered to be comparable. The best values for the proportional and derivative constants K_p and K_d were found by running numerous simulations with

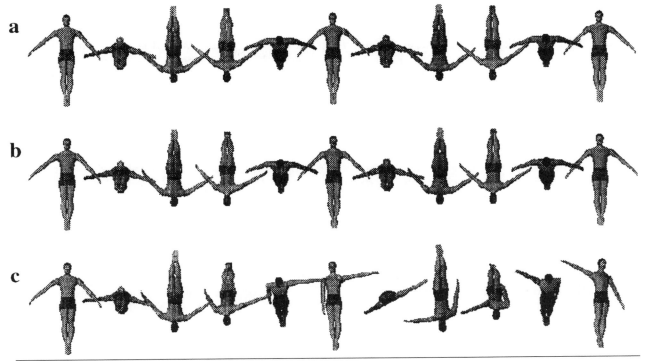

Figure 12.4 Proportional plus derivative control in straight double somersaults with feedback delays of *(a)* 0.02, *(b)* 0.12, and *(c)* 0.24 somersaults.

Table 12.3   Controlling the Twist in Straight Double Somersaults Using Arm Asymmetry

Initial arm asymmetry	Feedback delay (som)	Maximum twist (rev)	Maximum arm asymmetry
10.0°	0.02	0.00	2°
10.0°	0.12	0.01	18°
10.0°	0.24	0.09	143°
1.0°	0.24	0.04	24°
0.1°	0.24	0.00	2°

an initial arm asymmetry of 0.2° and a feedback delay of 0.24 somersaults. The value obtained for $K_p m/k^2$ was 1.01, which agrees well with the analytical result that this parameter should be slightly greater than 1.0 in order to maximize the time delay that can be handled. For a value of 1.01, figure 12.3 indicates a maximum time delay equivalent to 0.26 somersaults for which stable control can be maintained. The value of K_d obtained empirically for the numerical simulations was equal to 0.98 of the optimum value indicated in figure 12.3. It may be concluded that there is good agreement between analytical and numerical results.

Film Analysis

In order to address the question as to whether in-flight corrections are actually used in practice, a straight double somersault performed by an elite trampolinist was filmed using two 16 mm cameras operating at 70 frames per second. In each frame of the flight phase the wrist, elbow, shoulder, hip, knee, and ankle centers were digitized for each camera view and three-dimensional coordinates of these body landmarks were calculated. Orientation and configuration angles were determined (Yeadon 1990a), and segmental inertia parameter values were calculated from anthropometric measurements (Yeadon 1990b).

The orientation of the intermediate principal axis relative to the angular momentum vector was calculated throughout the double somersault in order to determine whether there was sufficient arching or flexing of the body to remove the instability. The film values of the body configuration angles and the initial orientation angles and angular momentum were used as input to the simulation model. This was done in order to establish whether a buildup of twist became noticeable during the simulation. If substantial twist occurred in the simulation this would indicate that the movement was unstable and that the gymnast must have made adjustments during flight, providing that the model was not in error. To demonstrate that such twist occurred as a result of instability rather than due to errors in the model it was necessary to demonstrate that the model could produce a nontwisting double somersault for a body configuration time history within the error limits of the film data. In order to do this, additional simulations were carried out in which the arm abduction angles were modified from the film values using a control strategy with zero delay. It is to be expected that if the simulation model is a close approximation to reality, there exist configuration histories close to those obtained from film for which there is no appreciable buildup of twist in the simulation.

The three-dimensional film analysis of the straight double backward somersault (see figure 12.5a) revealed that the lateral axis through the hips was close to the intermediate principal axis for almost the entire movement. This suggests that control was employed during flight in order to prevent the buildup of twist, although this need not necessarily be so, as shown in table 12.2.

When the film values of the body configuration angles and the initial orientation angles and angular momentum were used as input to the simulation model, the agreement between film and simulation was good during the first somersault (see figure 12.5a, b). During the second somersault the effects due to instability became pronounced and almost one-half twist resulted in the simulated movement. This indicates that without correction the instability would have led to noticeable twist after two somersaults. The discrepancy between simulation and film is to be expected since the estimated error in the arm abduction angles obtained from film was 1.3° and this is sufficient to produce substantial twist in the second somersault (see table 12.1).

The search for modified arm abduction angles that produced a nontwisting double somersault was successful. In a simulation that allowed the arm abduction angles to deviate from the film values by up to a maximum of 1.4° the twist was controlled and the agreement with the film sequence became good throughout the simulation (see figure 12.5a, c). This demonstrated that the twist in figure 12.5b was a result of the instability rather than any inadequacy of the model. Hence the trampolinist must have made corrective movements during flight although what these movements were and what control strategy was used is unknown.

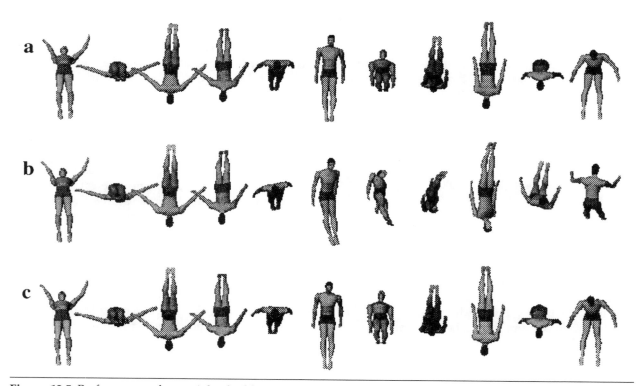

Figure 12.5 Performance of a straight double somersault obtained using (*a*) film analysis, (*b*) simulation, and (*c*) simulation with modified arm abduction angles.

Passive Control

Playter and Raibert (1994) showed that a three-segment model, comprising a body and two arms with torsional springs at the shoulders, automatically makes corrective movements when somersaulting. Providing that the arms are abducted more than about 30° from the midline of the body and that the springs have suitable stiffness, the system is neutrally stable and will perform double and triple somersaults without appreciable twist. The implication of this study of passive stability is that the bodies of gymnasts may automatically and instantaneously make compensatory movements. The majority of straight double somersaults are performed with the arms adducted close to the body. In this configuration passive corrections are insufficient to prevent the buildup of twist, whereas the simulation using PD control based on film data (see figure 12.5c) shows that active corrections can maintain control. This suggests that passive corrections may not be of mechanical importance in the control of twist. On the other hand, the tendency of the limbs to move in the appropriate direction may provide additional feedback information for input to the control system used.

Visual Feedback

An experiment was conducted in order to determine whether visual feedback was necessary for the control strategy used by a trampolinist. The subject performed six straight double somersaults, each from a plain jump. He was then instructed to close his eyes immediately after takeoff and to open them when instructed later in the flight phase.

In the first two attempts at straight double somersaults under these conditions the trampolinist completed one and a half somersaults without twist before being instructed to open his eyes. During the last half somersault a quarter twist occurred on both occasions. The trampolinist reported that he was aware of the instability but was uncertain about trying to correct it with his eyes closed. He was instructed to make adjustments with his eyes closed. In the next four attempts the trampolinist successfully completed a straight double somersault with eyes closed for the first one and a half somersaults. This experiment using visual deprivation of a trampolinist showed that it was not necessary to have visual feedback in order to maintain control in a straight double somersault.

Twisting Somersaults

Since the straight double somersault needs continual correction to prevent twist occurring, it should be relatively easy to produce a straight double somersault with twist. A hypothetical example is shown in the simulation depicted in figure 12.6 in which twist is introduced into a nontwisting somersault by means of asymmetrical arm movements during flight. In the first somersault small arm asymmetries produce a buildup of twist that is accelerated after one somersault by adducting the arms. The tilt out of the vertical somersault plane is apparent after one somersault. In order to obtain a correct landing position the twist must be stopped after one revolution and the tilt angle must be removed. This was

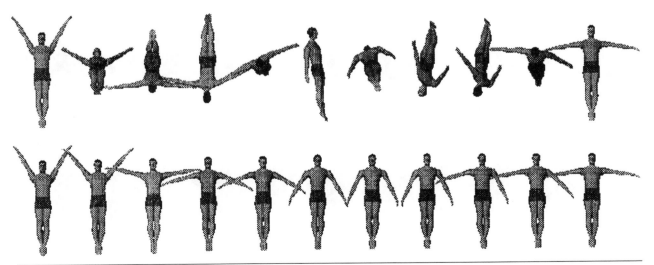

Figure 12.6 Simulation of a back-in full-out straight: a double somersault with one twist in the second somersault. The lower sequence shows the asymmetrical arm movements used in the simulation.

achieved using asymmetrical arm movements toward the end of the second somersault. The right arm is abducted at around the three-quarters twist position in order to slow the twist rotation without increasing the tilt angle. When the twist is almost complete the left arm is abducted in order to remove the tilt. This latter stage of the movement is very similar to the control of the nontwisting straight double somersault, and so it is to be expected that the skills necessary for controlling nontwisting somersaults are also used for controlling twisting somersaults. This may explain why it is rare for instability problems to occur in the learning of straight double somersaults. By this stage the gymnast will have learned how to control unwanted twist both in single straight somersaults without twist and in twisting somersaults.

The other method for preventing twist considered for nontwisting somersaults was to arch or flex sufficiently at the hips in order that the somersault becomes stable. This technique can also be used to change a twisting somersault into a nontwisting somersault. Figure 12.7 depicts a simulation of a hypothetical straight double somersault in which twist is present at takeoff. As the twist reaches one revolution, the body flexes at the hips so that the lateral axis corresponds to the maximum principal moment of inertia. Somersaults about this axis are stable, and so the motion changes from a twisting somersault into a wobbling somersault (Yeadon 1993a, 1993b). Near the end of the second somersault the body extends and the arms are abducted symmetrically. If the timing of these movements is correct it is possible to complete the correct amounts

of somersault and twist rotation without any tilt away from the vertical (see figure 12.7).

Summary and Potential Applications

In this study the abilities of various hypothetical strategies for controlling twist in twisting and nontwisting somersaults have been evaluated using computer simulations. Such theoretical analyses cannot indicate which techniques are actually used by gymnasts but can indicate whether a proposed technique is viable. The results given in this chapter should provide a useful starting point for investigations on the techniques actually employed by competitive athletes.

The strategy of symmetrically abducting the arms during flight, as advocated by Nigg (1974), does not reduce the buildup of twist in straight single and double somersaults (see table 12.1). Arching the body during flight progressively reduces the buildup of twist for small perturbations even when there is insufficient arch to ensure that the motion is in the stable wobbling somersault mode (see table 12.2). In a single somersault it may not be necessary to make corrections (see figure 12.5b). This contradicts the speculation of Hinrichs (1978) that single somersaults about the unstable intermediate axis require in-flight correction. For larger perturbations, arching that is sufficient to ensure the motion is in the stable wobbling somersault mode is not sufficient to ensure that there is no appreciable buildup of twist (see table 12.2). In this case additional arching is required to limit the magnitude of the twist oscilla-

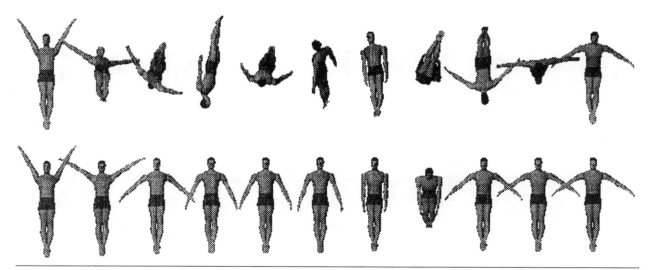

Figure 12.7 Simulation of a full-in back-out pike: a double somersault with one twist in the first somersault. The lower sequence shows the symmetrical arm movements used in the simulation.

tions. Although it is possible to control instability in a double layout somersault using arching, the competitive athlete is likely to lose points due to form breaks (see figure 12.2) since it is an expectation of judges that the body will appear to be straight.

It has been shown that the buildup of twist can be controlled using appropriate arm movements providing that the feedback time delay is not greater than a quarter of a somersault. For double layout somersaults in a gymnastics floor exercise this delay is equivalent to about 150 ms, which cannot be much more than a gymnast's reaction time. In this theoretical study the angular velocity $\dot{\varepsilon}$ of the arm movement was used as the control variable and was a linear function of the twist angle ψ and twist angular velocity $\dot{\psi}$ for PD control. In a practical situation control will be effected using neural stimulation of the appropriate muscle groups, and this input will be related to the joint torques. This suggests that the control variable used by gymnasts will be similar to the angular acceleration $\ddot{\varepsilon}$, which would be a function of $\dot{\psi}$, ψ, and $\ddot{\psi}$ for PID control. Thus the result that PD rather than PID control should be used is equivalent to saying that gymnasts must base their control on the twist angular velocity and acceleration values but not on the twist angle itself. Since the otolith organs of the inner ear can detect angular velocities due to centrifugal effects and the semicircular canals respond to rotational accelerations (Wendt 1951), it is possible that vestibular control is used rather than visual control. This idea is supported by the experiment involving an athlete closing the eyes during a straight double somersault. The main function of the eyes may be to obtain angular information on body orientation in space in order to make in-flight adjustments for correct landing orientation rather than to control instability during flight (Rezette and Amblard 1985).

References

Hinrichs, R.N. 1978. Principal axes and moments of inertia of the human body: An investigation of the stability of rotary motions. Master's thesis. University of Iowa.

Marey, E.-J. 1894. Mecanique animale: Des mouvements que certains animaux executent pour retomber sur leurs pieds lorsqu'ils sont precipites d'un lieu eleve. *La Nature* 10 November: 369–70.

Marion, J.B. 1965. *Classical dynamics of particles and systems.* New York: Academic Press.

Nigg, B.M. 1974. Analysis of twisting and turning movements. In *Biomechanics IV,* ed. R.C. Nelson and C.A. Morehouse, 279–83. London: MacMillan.

Playter, R., and Raibert, M. 1994. Passively stable layout somersaults. In *Canadian Society for Biomechanics: Proceedings of the Eighth Biennial Conference and Symposium.* August 18–20, 1994. University of Calgary. 158–59.

Rezette, D., and Amblard, B. 1985. Orientation versus motion visual cues to control sensorimotor skills in some acrobatic leaps. *Hum. Mvmt. Sci.* 4: 297–306.

Wendt, G.R. 1951. Vestibular functions. In *Handbook of experimental psychology,* ed. S.S. Stevens, 1191–223. New York: Wiley.

Yeadon, M.R. 1990a. The simulation of aerial movement. Part I. The determination of orientation angles from film data. *J. Biomech.* 23: 59–66.

Yeadon, M.R. 1990b. The simulation of aerial movement. Part II. A mathematical inertia model of the human body. *J. Biomech.* 23: 67–74.

Yeadon, M.R. 1990c. The simulation of aerial movement. Part III. The determination of the angular momentum of the human body. *J. Biomech.* 23: 75–83.

Yeadon, M.R. 1993a. The biomechanics of twisting somersaults. Part I. Rigid body motions. *J. Sports Sci.* 11: 187–98.

Yeadon, M.R. 1993b. The biomechanics of twisting somersaults. Part II. Contact twist. *J. Sports Sci.* 11: 199–208.

Yeadon, M.R. 1993c. The biomechanics of twisting somersaults. Part III. Aerial twist. *J. Sports Sci.* 11: 209–18.

Yeadon, M.R., Atha, J. and Hales, F.D. 1990. The simulation of aerial movement. Part IV. A computer simulation model. *J. Biomech.* 23: 85–89.

Yeadon, M.R., and Mikulcik, E.C. 1996. The control of nontwisting somersaults using configurational changes. *J. Biomech.* 29: 1341–48.

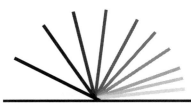

Chapter 13

Movement Control and Balance in Earthbound Movements

J. Mester

During the 40-million-year evolution of human life, movement control and balance had many unique opportunities to adapt to the earth's gravity. There is also good reason to believe that human beings completed this process of adaptation a very long time ago, which may be one reason why human upright movement control and balance have been the focus of scientific interest for so long.

The purpose of this chapter is to discuss, from a general perspective, actual results of research in movement control in earthbound movements with special reference to balance. This means that the basic functioning of muscle itself, such as the interaction of agonists/antagonists or events on the cellular level, are not the focus of this chapter. Instead, the complex framework of movement control and balance are the main elements of discussion.

Since the external forces in various sports can be remarkably high, the requirements for keeping the whole body in balance must be regarded as a function of the time history of a complex regulation phenomenon for which scientific concepts are available only in part. Modern approaches often argue that real balance, in the sense of zero deviation from an equilibrium point, cannot occur in biological or social systems; therefore, the underlying factors regulating this deviation are of major interest in this chapter.

Statement of the Problem

The regulation of balance as a prerequisite for the initiation and continuation of any directed earthbound movement requires high precision for the movement of each segment with a demanding anticipatory processing. The problems for scientific research in this field can be assigned to the number of complex interactions between the central commands from the brain and the various elements of peripheral regulation systems at the spinal level.

One outstanding problem in this regard is that the coordinated output of a directed movement cannot be sufficiently judged by the elements of force-time or distance-time history under the criteria of segment-oriented mechanical optimization. This can be illustrated by the fact that even a simple movement task such as lifting an arm into an upright position requires a regulation process of high complexity, starting with the intention to move and then covering the whole time course of movement until the end. Traditional analyses of movements have not taken into sufficient account the structure of regulation strategies and their effectiveness.

The Primary Concepts

Research findings have given us reason to believe that on a general level a tonic neural pattern generator is responsible for the control of posture. A phasic, rhythm generator then provides for the initiation and continuation of directed movement. It is important to note that the tonic and phasic generators are working simultaneously and in synchrony for the realization of earthbound movements. Since these generators cannot be isolated, either on the central or on the peripheral neuronal level, the networking of these complex neuronal structures must be assumed.

These networks are committed to the central goal of the movement and must include a precise on-line registration of each particular phase. Furthermore, anticipatory control, with respect to the processing of the "future" requirements of ongoing movement, is essential. Since these extensive tasks also need complex network structures, a "network of networks" is assumed to exist that is necessary for movement control and balance in earthbound movements. The functioning of these networks is directly associated with information processing in a network of perceptual elements. As the control of a directed and purposeful movement cannot be isolated from the overall goal of maintaining balance, input from the visual system is, for example, linked functionally with the input from the vestibular system. These holistic considerations complicate the isolated investigation and interpretation of the contributions of the various biological systems. Understanding the complex phenomenon of human motor behavior, however, requires both bottom-up and top-down strategies.

Elaboration of Concepts

The major prerequisite for starting any movement and controlling it is to acquire as much information as possible about the actual environment and any changes in that environment. As gravitational force is exerted on the body, well-known mechanical consequences occur, along with changed torques if the actual position of the segments is altered.

Evolution has made various sensory systems available for detecting gravity and its mechanical consequences (figure 13.1). The continuous perception of the gravitational force (i.e., the somatic graviception) is a combination of the effects of the utricles and the saccules. In addition, there are other sensors that are necessary for an internal representation of balance (in the form of the subjective perception of horizontal body position or the subjective visual vertical axis). It is here that the receptors of the neck muscles and the input from the eye muscles via counterrolling of the eyes play a more important role than has been suggested in previous years (Mittelstaedt 1995).

Consequently, the visual system is significant not only for orientation in space during movement but also for continuous regulation of "simple" tasks such as standing upright. Recent studies have shown that visual control of this regulation during standing is highly associated with regulation velocity. With closed eyes, the regulation velocity of the body sway in the frontal plane increased considerably, an indicator of a lack of input of information (Day et al.

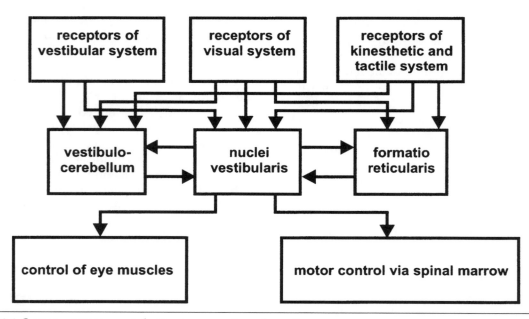

Figure 13.1 Sensory systems contributing to perception of gravity.
Reprinted, by permission, from D.E. Parker, 1980, "The vestibular appartus," *Scientific American*, 243(5):127.

1993). Obviously, an important part of visual perception for maintaining balance is depth information via parallactic elements of the visual field (Warren, Kay, and Yilmaz 1996) and the function of peripheral vision (Assaiante and Amblard 1992).

In spite of the dominance of visual input, tactile and proprioceptive information is also desperately needed, not so much for perception of gravity but for its perception of consequences (e.g., the pressure changes on the feet during standing or on the back during sitting). By their proportional-differential characteristics, most receptors ensure that information about rapid and even small changes in posture are immediately sent to central processing, if needed.

Because testing methods in this area must focus on specific sectors of the vestibular system, different physical stimuli are arranged as a test battery and used to investigate vestibular and balance functions (Honrubia 1995). Posturography concentrates on the vestibulospinal reflex and its influence on balance in the course of vestibular disturbances (Norre 1994), whereas the optokinetic or caloric electronystagmogram investigates the inner ear, brainstem, and cerebellar functions via eye movements. Sinusoidal rotation around the vertical axis is frequently used for measuring horizontal semicircular canal function (Saadat et al. 1995). The function of the brainstem, the cerebellar and oculomotor system can also be evaluated by means of rotation testing (Halmagyi, Colebatch, and Curthoys 1994; Parker 1993).

In recent studies, the interaction between the different sensory systems for graviception became increasingly important. In order to exclude the influence of visual input on balance performance, posturography tests were carried out with closed eyes. The contribution of the foot-ankle proprioception was controlled by having subjects stand on foam rubber. Combinations of these stimuli with various groups of methods have shown that the specific sensory input is integrated in a complex system of processing. If specific input from one system is excluded, it normally leads to destabilization (Norre 1993). If a sensory mismatch is produced, the destabilization and irritation increase.

In our own studies, an attempt was made to identify associated physiological factors in the area of gravity perception and specific vestibular sensory systems. Subjects underwent tests on a rotating chair ($3–5 \ rad \cdot s^{-1}$) with measurements of the nystagmus frequency and standardized gait tests in order to examine the effects of rotation load (Mester 1988). The nystagmus response that occurs during and after a specific rotation load has been frequently used to assess the level of adaptation with respect to the extent of irritation of the vestibular system. The physiological relationship between the generating structures of the nystagmus response and the major vestibular system is determined by the pathways from the nucleus vestibular superior and lateral, to the nucleus paraabducens and abducens, and then to the nucleus oculomotor.

In the course of our tests intra- and postrotational nystagmus was recorded. A comparison between the nystagmus response of highly trained and novice rowers is shown in figure 13.2.

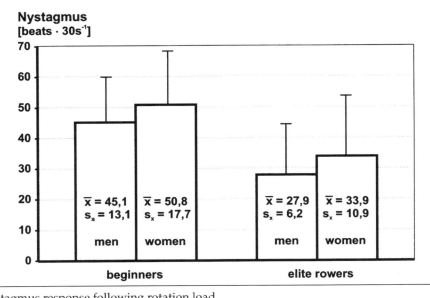

Figure 13.2 Nystagmus response following rotation load.

Reprinted, by permission, from J. Mester, 1988. Diagnostik von Wahrnehmung und Koordination im Sport. Lernen von sportlichen Bewegungen. (Schorndorf: Verlag Hofmann).

The results shown in figure 13.2 illustrate that the greater experience in movement control and rowing-specific balance of the trained rowers is associated with a lower postrotational nystagmus frequency, indicating lower irritation of the vestibular system following the rotation load. This indicates that familiarity with even the relatively low accelerations occurring in rowing may lead to an increased ability to dampen the rotational irritation. This is probably an adaptation in the area of nucleus oculomotor.

Bottom-Up Regulation: Building Up Posture and Balance Against Gravity

Given that the field of gravity on earth is homogenous to a great extent, building up posture against gravity for a human to stand upright could be expected to be an easy task. For a long time, however, it has been known that balance in standing upright is not a matter of fixed, motionless stability. Even if there is no movement of the upper limbs, a lateral and anteroposterior body sway occurs, which is thought to be a regulation process in order to be ready for any perturbation of balance. This permanent regulation implies that no stable equilibrium exists in a mechanical sense. Moreover, the term "static equilibrium" seems to simplify the underlying mechanisms of building up and maintaining balance against gravity. For many years studies have dealt with this topic and a number of parameters for quantification of body sway have been developed.

In earlier studies, body sway was regarded as an overall, complex phenomenon and was mainly investigated by means of the trajectories of the center of force on a force plate and the variation of the foot pressure. In recent studies, more attention has been paid to the problem of regulation strategies or the coordination of simple, quiet standing with the background of musculoskeletal models of the human lower extremities. Kuo (1993), for example, calculated sets of all feasible accelerations of the center of mass with minimal muscle activation as an indicator for the neural effort. He found that the ankle strategy (i.e., controlling the shifts of the center of mass via rotating about the feet and ankles), which is often used in sports for improvement of balance, requires more muscle activation than the hip strategy. Movements of the hips would be most effective in controlling the center of mass with minimum neural effort.

Lekhel and coworkers (1994), who investigated the coordination in unperturbed stance with accelerometric devices at the head, hip, and ankles, also support this idea. Cross-correlation analyses showed that, especially on a soft surface where the greatest body oscillations occurred, the coordination of stance mainly was determined by lateral and forward-backward movements of the hip. The frequency of these movements confirmed the hypothesis from general results on movement control that a tonic descending message from the brain is translated into rhythmic generator motor output at the spinal level. Additional results corroborated that standing upright is a coordinated process with complicated neural network requirements rather than a stable equilibrium. In fact, a comparatively simple task is to lift an arm while standing quietly. As the previous balance is altered with the changing torque of the upper limb movement, a complex order of distinct and precisely coordinated movements is necessary.

Weeks and Wallace (1992) investigated the preparatory and anticipatory events of movement control of a subject standing still as a result of various accelerations of upper limb movements (e.g., lifting of the right arm with different velocities) and on various centers of force recorded on a force plate. Figure 13.3 shows the order in which the various muscles of the lower extremities were enervated and the intensity of the enervation immediately before the onset of the focal movement (lifting of the arm) when the center of force (CoF) was located in different areas of a force plate. If, for example, the CoF was located in the right posterior quadrant (RePos), then the enervation with the sequence that started with the ipsilateral biceps femoris and ended with the contralateral biceps femoris had the highest intensity.

Further results showed that the intended velocity of the upper limb movement had a remarkable influence on the onset and intensity of the premovement enervation. This could be interpreted as the central system for movement control having knowledge of the likely torque as well as the moment of inertia, thus anticipating the required muscular activity for maintaining balance. From this point of view, the muscular activity for regulating the balance must be planned before the physical initiation of movement and must be regarded as a part of the complex movement control. The premovement activity is not simply a stereotypically planned mechanical detail of the response, but a determinant in the accuracy of the overall movement control.

This becomes even clearer when control of movement and balance is needed on a moving/tilting platform. Studies in this area have existed for more

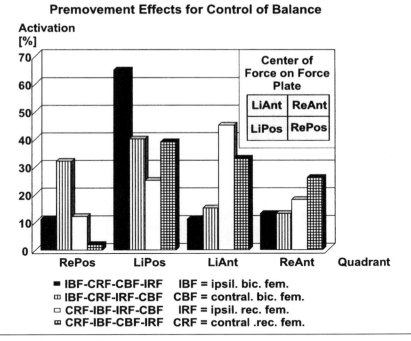

Figure 13.3 Premovement posture effects.

Reprinted, by permission of Elsevier Science, from D.L. Weeks and S.A. Wallace, 1992, "Pre-movement posture and focal movement velocity effects on postural responses accompanying rapid arm movements," *Human Movement Science*, 11:717-734.

than 40 years and have contributed significantly to the understanding of balancing complex biological systems that are based on influences from both a central command and spinal reflexes. One of the classical experiments in this area, by Nashner (1976), showed the influence of the stretch reflex.

In part A, Nashner provoked an unexpected backward movement of the platform, which triggered a long-latency stretch reflex that was easily facilitated with repeated trials, mainly because the function of this reflex was to support balance (see figure 13.4). This was also shown by means of an EMG signal, where the activity started earlier and increased in the course of four trials. In part B of the experiment, the ankle was rotated following upward motion of the platform. Here a long-latency stretch reflex was triggered that adapted (diminished) with repeated trials, because in this case, the reflex served to destabilize posture (Nashner 1976).

The high adaptability for control of seemingly simple earthbound movements such as standing was studied by Dietz and coworkers (1993). In this study, subjects had to maintain their balance on a sinusoidally backward- and forward-moving platform with a frequency of 0.5, 0.3, and 0.25 Hz (amplitude ± 12 cm). Following the mechanical requirements of maintaining balance, the latencies of the compensatory muscle responses to a change

in platform frequency were found to be significantly shorter at the posterior point for the gastrocnemius than for the tibialis anterior, and at the anterior point for the tibialis anterior than for the gastrocnemius. In addition, a similar rapid adaptation within four trials of both the EMG response and biomechanical patterns of the compensating movements was found as with earlier studies by Nashner (1976) and Dietz and associates (1980). Dietz and coworkers (1993) concluded that balance is adjusted in such a way that the forces in respect to the moments acting on the body during platform movements become minimized.

A more demanding task for movement control and balance occurs if the earthbound environment interferes in the coordination of walking with mechanical and other perturbations (Ebersbach, Dimitrijevic, and Poewe 1995). If, for example, a perturbation is induced in walking by a rapid and expected increase of belt speed on a treadmill, the EMG would show a short-latency reflex response (40–60 ms) as a result of spinal neuronal mechanisms involved in the rapid adjustment of gait. Similar results have been confirmed by various large-scale research studies on the research of movement control and balance. Also notable are the results from studies where decerebrated cats were capable of locomotion on a treadmill after recovery

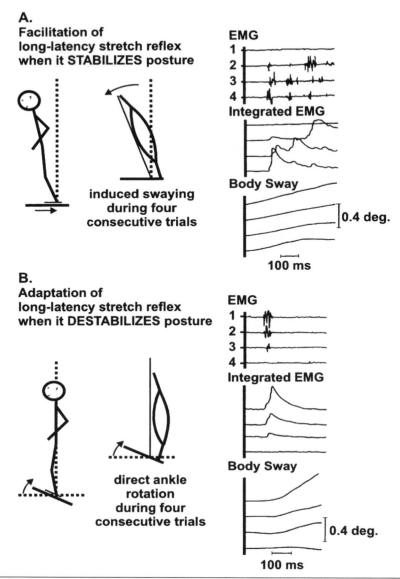

Figure 13.4 *(a)* Facilitation and *(b)* adaptation of the long-latency stretch reflex.

Reprinted, by permission, from L.M. Nashner, 1976, "Adapting reflexes controlling the human posture," *Experimental Brain Research* 26:59-72. © Springer-Verlag.

from surgery and a training period (Iwamoto et al. 1996; Pearson 1976; Wada et al. 1996).

Top-Down Regulation: Central Input and Supervision

Within the framework of this chapter, the term "balance" is defined as the synergy of all processes providing the mechanical and physiological prerequisites for initiating and continuing purposeful movements. Apart from the spinal influences, cen-

tral supervision or central monitoring is the outstanding criterion for human movement control and balance. For many years, the anatomical structures were well known, but only recently were PET techniques and modern high performance computational facilities able to identify the functional circumstances on those levels that offer macroscopical explanations.

These advances in methodology also made possible a renewal of the understanding of the cortical contribution to movement control and balance. Not long ago, studies described the motor cortex as being predominantly composed of centers or specific regions where certain biological areas (e.g., the muscles) were encoded. In the early 1980s, studies began to show that a given movement could be controlled by different motor areas. Soon there was evidence that different motor systems could converge on the same muscles (Hoffmann and Luschei 1980). Finally, a topographically determined idea of cortical functioning was followed by more holistic and functional approaches that described the cooperative networking between completely separate areas of the cortex, as well as between cortical and subcortical structures, depending on the given motor task.

In many recent studies, these complex models of movement control and balance have been successfully investigated and simulated. On the one hand, research has tried to get more deeply into the molecular (even submolecular) levels; on the other, it has attempted to accumulate results and develop complex models for an integrative understanding. McCollum (1994) suggested a set of discrete regions on body position space, defining a mathematical object and unifying discrete with continuous aspects of movement control and balance. Shadmer (1993) elaborated complex postural modules, defined as the synergy of muscles producing a class of torque function that converge at a constant position of balance.

Kuo's (1994) approach to motor control is based on the idea that the coordination of billions of cortical neurons, motor neurons on the spinal level, and more than 600 muscles with 240 degrees of freedom can only be controlled by a highly sophisticated network. This cooperation between central instances (e.g., the motor cortex) and peripheral accountability (e.g., motor units) has not been a common focus of sport science research.

One of Kuo's (1994) most important ideas was the existence of "mapping systems" on the brain level and motor neurons on the spinal level. This idea was important because it explained to a certain degree the fact that every movement, even movements of small body segments such as hand or fingers in standardized situations, has a remarkable amount of variability and that movements are "regulated" in order to achieve a defined goal. The model of a "feature map" and a "motor neuron map" confirmed a high capacity for varied responses. The "zones" of activation in the brain map have not been clearly defined, so slight differences in the starting conditions of the movement, for example, limb positions, must be compensated for and regulated. As the local excitation area varies with the input and the requirements of the movements, the existence of

a flexible semi-topographic organization of the brain can also be supported from this approach. The area of the excitation indicates the direction of the muscular action, whereas the excitation size and the firing rate encode the force. The existence of a similar neuron map, with similar properties, is assumed on the level of spinal motor neurons.

Because it is very unlikely that the starting conditions are exactly the same even for a given and defined movement or motor task, it is not likely useful to activate the same area of cortical neurons and subsequently the same motor neurons. The required homogeneity of activation is provided by the fact that excitatory connections are located in the neighborhood of the neurons, whereas inhibitory connections are established with neurons at a greater distance. This seems to be a very good solution because it provides variability for motor actions with similar tasks or similar starting conditions. Considering the complexity of motor tasks and the highly developed capacity of the motor system to perform these tasks, this motor control system can no longer be regarded as a network, but rather, a network of networks.

On an abstract level, this model can be structured in five steps (see figure 13.5). The CNS control center is faced with the motor task and sends an input to the CNS force distribution center, where a first signal differentiation occurs. Afterward, the signals are transformed into dynamic characteristics and embedded into the dynamic environment of the motor units. The force production of the motor units is then calculated, inducing the body segment dynamics, which can be measured (e.g., in kinematic terms). This model can be applied to single joint movements as well as to whole body movements.

Control and Balance of Directional Earthbound Movements

In a previous section, it was mentioned that there is, in principle, no difference between the control of balance in simple upright standing and the control of directed movements such as grasping, throwing, and hitting. There is a fluent transition between a certain posture, the premovement change of enervation with the anticipatory balance of the expected torques and moments of inertia (see figure 13.3), and the focal movement itself.

When the focal movement is progressing, a dynamic, feedforward control that follows the same principle as the premovement control is necessary.

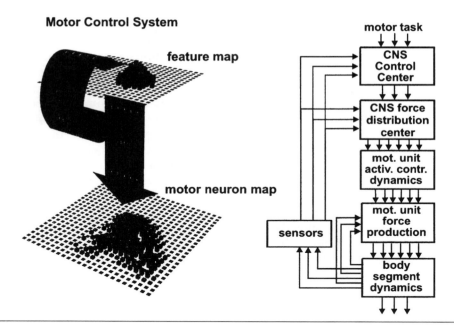

Figure 13.5 Motor control system.

Reprinted, by permission of Elsevier Science, from A.D. Kuo, 1994, "A mechanical analysis of force distribution between redundant, multiple degrees-of-freedom actuators in human: Implications for the central nervous system," *Human Movement Science* 13: 635-663.

This main element can be divided into two parts: the primary setting of stiffness before the intention for a movement or a change of the actual posture is developed and the adaptation of stiffness following the expected requirements for the new balance situation. Georgopoulos (1995) summarized current issues in directional movement control. He pointed out that the direction of the motor output (movement or isometric force) is an important factor for neuronal activity in the motor cortex, at the level of single cells and at the level of greater neuronal populations.

This complex control system, from an isometric starting position to the final position of the limb, is obviously necessary to balance the limb in the process of a directional movement. Around 1965, the *equilibrium point hypothesis* (EPH) emerged from a number of experimental studies and began to penetrate into discussion of a concept of the overall control of directional movements (Asatryan and Feldman 1965). This hypothesis suggested that directional movements may be the result of shifts in the equilibrium state of the motor system associated with the setting of the threshold of the stretch reflex and are, therefore, responsible for establishing balance. Several experimental references were given in the following years on the importance of the stiffness setting for movement control (e.g., Feldman 1986; Houk 1976; Houk 1979; Nichols 1982). One important argument for the EPH is that isometric and isotonic descending commands and, thus, isometric contrac-

tion and auxotonic movement can be explained with one general approach (Latash and Gottlieb 1991).

Over the course of the past 30 years, knowledge of the complex interaction between dominant influences from the spinal level such as the stretch reflex and central input on movement control and balance has grown considerably. Ideas about neural networks and interchange of networks have been developed and improved. The concept that the spectrum of movement control and balance, from simple quiet standing to complex directional motor tasks, forms an integrated unity has become more accepted today.

As was discussed in a previous section of this chapter (see figure 13.3), the postural enervation preceding the focal movement, from a given point in time, anticipates the requirements of the following directional movement and adapts consecutively with the ongoing movement. Since the subsequent biological and mechanical requirements are calculated by the regulating system this cannot simply be obtained by traditional feedback mechanisms; a feedforward controller is needed. From the various concepts dealing with this interaction of movement control and balance in the framework of regulating and communicating networks for directional movements, the approach of Hirayama, Kawato, and Jordan (1993) is worth mentioning. Using an arm movement for the theoretical and empirical approach, the researchers suggested a hybrid cascade neural network model that consisted of a feedforward controller and a postural controller.

The mathematical background for the cascade neural network (see Kawato et al. 1990) was associated with the minimum torque-change criterion together with a minimized number of relaxation interactions. The general symbiosis of the posture controller and feedforward controller is illustrated in figure 13.6.

The dominant role of the given task is emphasized in this model with additional implications from mathematically based information-processing research. Following this concept the feedforward controller obtains information about the initial hand position and the probable duration of the ballistic movement in close interchange with the visual information about the target location. While the feedforward controller, functioning as a phasic rhythm generator, yields the oscillatory muscular enervation in the course of the ballistic movement and the changing of the hand position, the posture controller provides for balance after reaching the target. Here a cooperation between the posture controller and the feedforward controller is respon-

sible for the definition and the onset of the final motor command, which is defined by reducing, but not totally blocking, oscillatory movements and simultaneously increasing balance in a damping process. Hirayama, Kawato, and Jordan (1993) argued that this overall network was comprised of a cascade structure of many elementary units in four layers of neurons and that each of these units reflected the dynamic properties of the controlled subject, that is, the arm. The integration of the various cascade levels provided a forward dynamic model of movement control and balance.

Toward the end of the aimed movement, an active deceleration of the limb movement (Latash and Gottlieb 1991) by means of cocontraction of the antagonist was necessary to prevent the joint from injuries and to establish new balance on a higher level of stiffness. For many years, researchers investigated the problem of if and when the directional and rapid movement could be stopped or altered in terms of kinetic and kinematic parameters. McGarry and Franks (1996) suggested that this point of no return occurred very late and that the movement could be stopped at almost any time. The authors

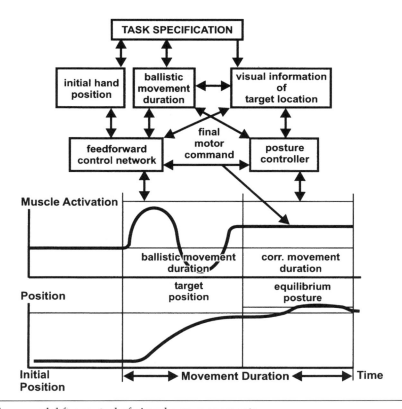

Figure 13.6 Two-phase model for control of aimed arm movements.

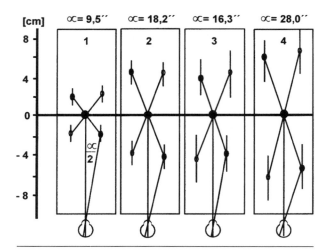

Figure 13.7 Acuity of visual depth perception before and after training. α = angle of stereopsis. Data points represent mean ± SD or ϕ = x ± s_x; 1 = top-class tennis players after training of visual depth perception; 2 = top-class tennis players before training of visual depth perception; 3 = control group (no tennis experience) after training; 4 = control group (no tennis experience) before training.

Reprinted, by permission, from J. Mester, 1988. Diagnostik von Wahrnehmung und Koordination im Sport. Lernen von sportlichen Bewegungen. (Schorndorf: Verlag Hofmann).

also concluded that movement is subject to continuous on-line control.

Similar conclusions concerning on-line control of directional movements were provided by Spijkers and Spellerberg (1995), who pointed out the significance of visual control, especially the actual vision period. Mester (1988) developed a concept with various elements of visual perception with respect to the specific optocontext (e.g., depth perception, peripheral vision, static and dynamic visual acuity). Figure 13.7 shows the acuity of visual depth perception measured by means of the Helmholtz test in various groups of subjects before and after stereopsis training. It can be seen that the group of top-class tennis players had a significantly smaller angle of stereopsis, that is, higher visual depth acuity, than the control group. In both groups, however, it was possible to improve this ability by means of specific visual training.

In the framework of the results just discussed, this permanent visual control is reasonable if a permanent oscillatory motor control drive is exerted that is capable of basic and rapid control of the movement. This hypothesis can be supported by correlation studies between the movement acuity and parameters of visual capability. Figure 13.8 shows that the ability to hit the tennis ball with a given acuity is associated with the static visual acuity and also with the ability for spatial visual perception. These visual factors are responsible for the precise detection and location of the tennis ball in the course of its flight. A higher visual acuity thus enables the player to record the dynamic properties of the flight, whereas a smaller angle of stereopsis yields better spatial information about the actual location of the ball and, more importantly, the change of the spatial parameters of the ball's flight.

As in the actual location of eye fixation due to the distribution of retinal receptors, the area of high visual acuity is very small, and as the amount of reaction time available in tennis is also very small, the spatial and temporal distribution of the eye movements is very important. The intersaccadic suppression of visual perception, which lasts for nearly one second in the course of rapid eye movements, requires a well-planned visual fixation strategy (Mester 1988); yet the flexibility of the biological system can also be demonstrated here.

The accuracy of directional movements generally follows the well-known speed-accuracy trade-off paradigm. The capability of the cascade neural network model expressed by the final position error of the hand can be assessed by means of the iteration number of relaxations for the given movement. In simulations, it has been shown that the error to the desired target position—defined as the Euclidean distance between the target position and the end point of the trajectory after 2,500 iterations or relaxations—was minimized (Hirayama, Kawato, and Jordan 1993). These results indicate the significant influence of planning time on accuracy at a fixed duration of movement. Error was decreased with an increase in the number of iterations, that is, the increased planning time of movement control.

In contrast to this computer simulation, the learning factor has great importance in human neural control events. Additional studies, such as the one by Gorinevsky (1993), would be useful in determining which adaptation effects, for example, with respect to varying load conditions and to various movement speeds, could be analyzed. Gorinevsky (1993) suggested a paradigm called direct motor program learning since the programs for movement control are learned directly without knowing or learning the dynamics of the controlled system. Simulating rapid planar arm motions in the task of point-to-point control, the model must take into account nonlinear arm dynamics, muscle force dynamics, delay in reflex feedback, time dependence of the feedback gains, and coactivation of antagonist muscles.

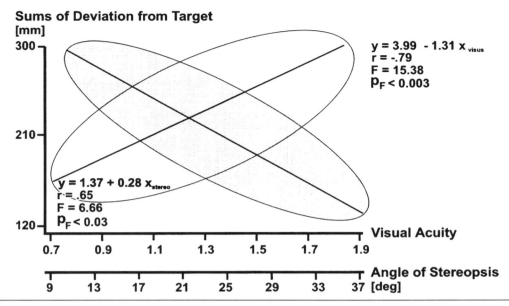

Figure 13.8 Relationship between movement precision, static visual acuity, and angle of stereopsis.

Reprinted, by permission, from J. Mester, 1988. Diagnostik von Wahrnehmung und Koordination im Sport. Lernen von sportlichen Bewegungen. (Schorndorf: Verlag Hofmann).

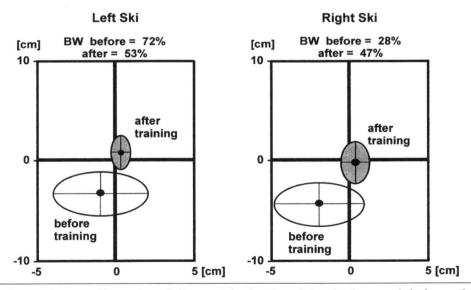

Figure 13.9 Regulation of center of force when gliding straight ahead on alpine skis (case study before and after training). BW = body weight.

Balance in Standing Upright Under High External Forces

Compared to the control of everyday movements or laboratory experiments, some sport requirements place a greater demand on the control system with regard to the time available for the execution of a motor task as well as the load associated with the external forces exerted on the human body. There is reason to believe, however, that there is no differ-

ence, in principle, to these situations of movement control and balance and those that have been discussed up to this point. Therefore, it is not surprising that if we take standing upright and keeping the body in balance on a gliding surface (e.g., skiing on snow) a number of similarities, which have been identified in the laboratory, can be found with the regulation process (see figure 13.9).

Figure 13.9 shows the variability of the center of force (CoF), measured by force plates with strain gauges, for a subject standing on an alpine ski and gliding straight ahead (medium velocity 15 m/s) on a plane. The CoF shifts between the right and the left feet in the anteroposterior and lateral planes, indicating a series of regulation movements in the ankles, knees, hips, upper body, and arms, as represented by an ellipse. The percentage of body weight placed on each foot is also indicated. After each of five runs, feedback about these parameters was given to the subject (i.e., training). Before training, a large variability of the CoF in the lateral plane and the predominant percentage of body weight on the left foot could be noticed. In straightforward gliding this can lead to dangerous situations with respect to the carving of the ski. After training, the CoF of both feet moved to the center of the foot and the diameter of the ellipse was reduced indicating less variation of the CoF in the course of the run.

Kinematically analyzed regulation strategies can be classified mainly as those reported from results in literature (see above). For the sensory input, the mechanoreceptors in the soles of the feet and the lower leg are very important (Spitzenpfeil et al. 1997). Results that have been compiled over the past few years with skiers of various levels of performance have shown that even with a high level of performance, regulation skills can be significantly improved.

Depending on the skiing velocity and the condition of the ski slope, high frequency forces can occur and require demanding reactions to maintain movement control and balance. The high velocity turns of modern ski races with modern equipment can produce frequencies associated with vibrations of the ski that reach 70 Hz (Niessen et al. 1997). The most important parts of the spectrum of the fast Fourier transformation (FFT) of the ground reaction forces significant for movement control and balance in alpine skiing are located between 1 to 3 Hz and between 15 to 20 Hz (see figure 13.10). An area around 1 to 3 Hz is due to the "body sway" necessary for maintaining balance. This movement—which can be explained by genetically fixed motor programs—involves most body segments, and the center of mass is displaced in the anteroposterior and lateral planes. The area around 15 to 20 Hz may have different causes. On one hand, it may be due to vibrations of the ski itself. Some results have suggested, however, that these vibrations are somewhat higher (see above). There is also reason to associate the peak of 15 to 20 Hz with biological properties of muscle stiffness. This is the frequency range of the area of resonant frequency for extended legs (Broch 1994). The closer the frequency of vibration is to this range, the greater is

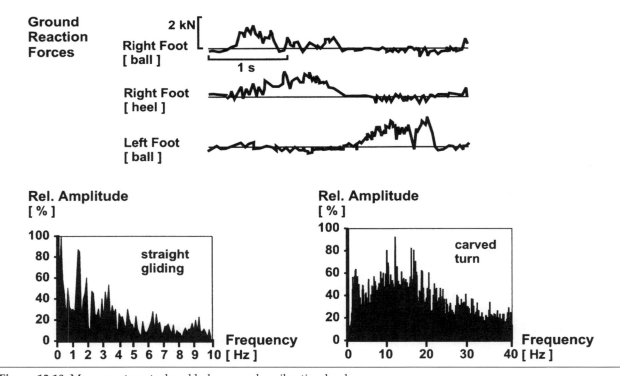

Figure 13.10 Movement control and balance under vibration load.

Reprinted, by permission, from J. Mester, 1997, Movement regulation in alpine skiing. In *Science and skiing*, edited by E. Müller, H. Schwameder, E. Kornexl, and C. Raschner (London: E. & F.N. Spon), 333-348.

the ability of movement control and balance to regulate, compensate, and modulate the ground reaction forces.

Electroencephalographic recordings under vibration load in the laboratory show shifts of activation zones in the area of the motor cortex that support the concept of "mapping zones" (Mester 1997). Additional centers of activation are located in the "integrative area" of the parietal lobe. These findings correspond with other neuroanatomic findings that this cortical region is responsible for the integration of the afferent information for movement control and balance from various sensory systems. These findings also support the hypothesis that the regional interaction of cortical neurons and motor units at the spinal level is not clearly defined for a given motor task (Kuo 1994). The biological system provides potential for variable movement control. This process of regulation can be called one of the most important features of the motor system. It ensures that a given motor task can be fulfilled, even if there are different starting conditions or varying conditions during the movement itself.

Balance in Sitting Under Low External Forces

In many sports, lower external forces occur, which also must be integrated in movement control and balance. A counterexample to the preceding alpine skiing scenario is rowing. In rowing competitions, it is difficult for untrained persons to maintain balance because of the indifferent mechanical equilibrium of the boat. Sitting in a balanced position in a rowboat must also be regarded as a prerequisite for the focal movement, the rowing itself. Keeping one's balance in a rowing boat is a dynamic regulation process, mainly defined by lateral movements of the knees and the sculls. The effects of movement control and balance in rowing were analyzed in the laboratory and on the water (Mester, Grabow, and de Marées 1982; Wagner, Bartmus, and de Marées 1993). Figure 13.11 shows the regression between an anthropometric parameter, the "rowing-specific moment," and the ability to balance the boat. The anthropometric parameter represents the mass and the length of the upper body and has been measured in a sitting position on a special rowing seat by means of a force plate after a horizontal displacement of the subject. The ability to balance the apparatus was measured by means of a potentiometer in the axis of the rowing seat. Results indicated that movement control and balance in rowing tended to be a learned, situation-specific skill rather than one that is routinely acquired or adapted from other motor skills. The reasons for this lie in the general principles of motor learning. In special cases, a transfer of skills from one situation to the other is possible but a similarity between particular motor patterns is needed.

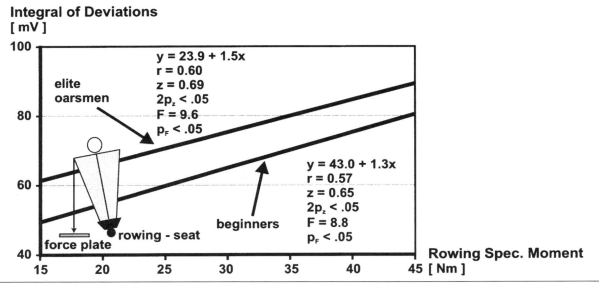

Figure 13.11 Relationship between anthropometric parameter and balance in rowing.

Reprinted, by permission, from J. Mester, 1988. Diagnostik von Wahrnehmung und Koordination im Sport. Lernen von sportlichen Bewegungen. (Schorndorf: Verlag Hofmann).

Movement Control Under Time Pressure

Recent sport applications have been defined as maintaining a certain balance in standing or sitting against gravity plus other external forces (e.g., centrifugal force). Although the focal movements were directly associated or integrated, the dominant prerequisite was maintenance of balance. The general concept of movement control and balance of earthbound movements in preceding sections of this chapter implied a differentiation between those movements that were responsible for building up and regulating balance against gravity and focal movements for fulfilling a given motor task. The same differentiation is also made here. One example of movement control with high requirements for precision under time-pressure is ground strokes and serves in tennis. The tennis serve was chosen because the transition between the integration and interaction of movement control and balance plays an outstanding role.

In spite of all visible differences, successful tennis players have an ability to serve with high velocity, precision, and security. These outcomes depend on ball toss, regulation of the racket movement, and the point of impact on the racket. It has been a major point of research interest to find characteristic differences in the regulation of the strokes of successful tennis players and less-skilled servers (see figure 13.12).

Three-dimensional high speed video (200 fps) was used to record the tennis service actions of internationally ranked players. Ball velocity was measured by radar. Although there was no major intervention of the opponent, and gravity was the most important influencing factor, the ball toss for elite and advanced players showed great variation, even for the flat serve. There were remarkable deviations in the vertical, horizontal, and lateral direction, with the range of the ball toss reaching up to more than 30 cm in the vertical axis (Kleinöder et al. 1995).

This basic variability of the given motor task, that is, tossing up and hitting the ball, can also be illustrated by the point of impact on the racket. The varying locations showed that each player realized a typical pattern with individual variability. The patterns of the internationally ranked players show smaller areas of variation in contrast to less-skilled players, who more frequently produced eccentric points of impact that resulted in lower ball velocities. Compared to less-skilled players, elite players were able to regulate their racket movement much better to meet the variations of their ball tosses, although variability was far from zero.

In both groups, the velocity-time curves of body segments showed the well-known principle of velocity summation from the shoulder to the top of the racket. Analysis of the components of the resultant velocity of the racket revealed decisive characteristics between the expert levels and the advanced players, although there were no significant differences between the two levels regarding the resultant velocities of the racket. The reasons for performance differences are most likely located within the

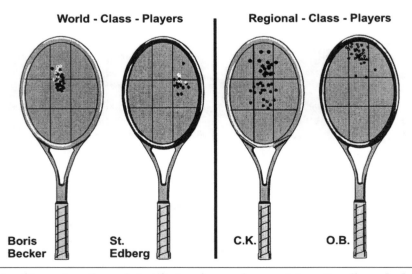

World - Class - Players **Regional - Class - Players**

Boris Becker   St. Edberg   C.K.   O.B.

Figure 13.12 Points of impact on the racket at first and second serve: Internationally ranked players vs. less-skilled players.

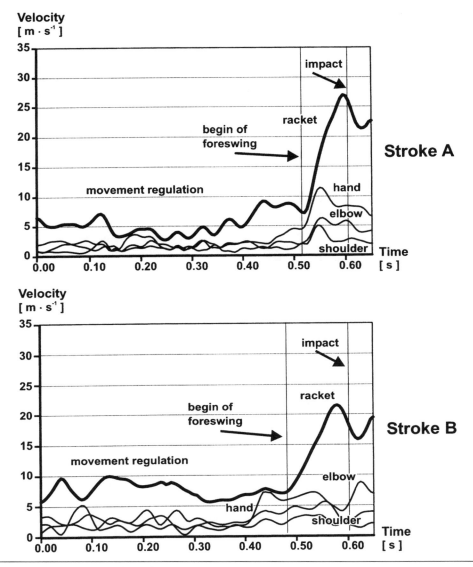

Figure 13.13 Forehand ground strokes of tennis player Pete Sampras.

time structure, the regulation of the acceleration of racket, and the regulation of direction and spin transferred to the ball.

The overall importance of the goal of the movement in the framework of directed movements can be demonstrated by several examples. The results of kinematic analyses of two of Pete Sampras's forehand ground strokes are summarized in figure 13.13. The analysis shows that the biomechanically expected structure, that is, the "staircase effect" with a consecutive order of the maximum segment velocities, can only be proved in one of the strokes (stroke A). This stroke was carried out in a normal situation with no major time pressure. In stroke B the peak velocities occurred in a different sequence. First, the hand is accelerated in order to begin the

swing as soon as possible. Because the other segments are not involved in the mechanically purposeful order, this is accomplished with a compromise to lower acceleration and lower maximum velocity. Nevertheless, the goal of the movement, that is, hitting the ball under high time pressure, was reached.

The high ball velocities in today's international tennis circuit and the well-developed tactical skills in connection with the athletic performance of the players require an adjusted movement control and balance that fits the actual situation but yields the optimum tactical conditions for the following stroke. In this sense the tactical optimum makes use of the high ability of the system of movement control and superimposes the mechanical optimum.

Summary and Potential Applications

The purpose of this chapter was to highlight the importance of the interaction between movement control and balance for human motor behavior in general. As balance of the whole body and the balanced position of body segments in the course of a purposeful movement is part of the evolution of upright standing and upright gait, the underlying processes and limiting factors are difficult to investigate. Modern concepts of movement control no longer differentiate between single motor tasks on the one hand and shifts of body segments in a "static" or "dynamic" framework of balance on the other. On the contrary, they integrate the actual state and resultant change of balance in the performance of every movement, as a part of complex movement control.

During the course of purposeful movement, an interaction of "bottom-up input" from various peripheral receptor systems and spinal processing levels as well as a "top-down" input from the cortical and subcortical structures is responsible for adjusted and variable movements. The processing of information, in direct connection with the motor output and the interaction between the brain and spinal levels, cannot exclusively be assigned to special "centers" but to networks of networks that are able to shift and vary the main neural activity and take into account varying starting conditions and goals of movement. With the ability of "anticipatory regulation" or "feedforward controlling" of the ongoing movement, evolution has provided an outstanding tool for the optimization of motor behavior.

This sophisticated structure yields the necessary prerequisites not only for everyday movements but also for movement control and balance in those sports where remarkable external forces are associated with high requirements for precision of movement. If necessary, even isolated shifts of small body segments are automatically embedded into the anticipatory control framework of the balance of the whole body. Future research should focus on the complexity of the overall movement goals and structure its derivatives before a differentiated movement analysis is carried out. In the course of a complex movement with high external forces or high accelerations of body segments it is likely that a great deal of the segment shifts are necessary, mainly to secure actual balance or meet the forthcoming requirements. The training of isolated or specific movement tasks in sports should likewise be implemented in the framework of the overall goal or context.

References

Asatryan, D.G., and Feldman, A.G. 1965. Functional tuning of the nervous system with control of movement or maintenance of a steady posture. 1. Mechanographic analysis of the work of the joint on execution of a postural task. *Biophysics* 10: 925–35.

Assaiante, C., and Amblard, B. 1992. Peripheral vision and age-related differences in dynamic balance. *Human Move. Sci.* 11: 533–48.

Braune, W., and Fischer, O. 1895. Der Gang des Menschen. Bd. I-VI. In *Abhandlungen der sächsischen Gesellschaft der Wissenschaften*, Leipzig.

Broch, J.T. 1994. Mechanical vibration and shock measurement. In *Application of B&K Equipment*, Naerum.

Day, B.L., Steiger, M.J., Thompson, P.D., and Marsden, C.D. 1993. Effect of vision and stance width on human body motion when standing: Implications for afferent control of lateral sway. *J. Physiol. (London)* 469: 479–99.

Dietz, V., Mauritz, K.H., and Dichgans, J. 1980. Body oscillations in balancing due to segmental stretch reflex activities. *Exp. Brain Res.* 40: 89–95.

Dietz, V., Trippel, M., Ibrahim, I.K., and Berger, W. 1993. Human stance on a sinusoidally translating platform: Balance control by feedforward and feedback mechanisms. *Exp. Brain Res.* 93: 352–62.

Ebersbach, G., Dimitrijevic, M.R., and Poewe, W. 1995. Influence of concurrent tasks on gait: A dual-task approach. *Percept. Mot. Skills* 81: 107–13.

Eccles, J. 1957. *The physiology of nerve cells.* Baltimore: John Hopkins University Press.

Feldman, A.G. 1986. Once more on the equilibrium-point hypothesis (model) for motor control. *J. Mot. Behav.* 18: 17–54.

Georgopoulos, A.P. 1995. Current issues in directional motor control. *Trends Neurosci.* 18: 506–10.

Gorinevsky, D.M. 1993. Modeling of direct motor program learning in fast human arm motions. *Biol. Cybern.* 69: 219–28.

Halmagyi, G.M., Colebatch, J.G., and Curthoys, I.S. 1994. New tests of vestibular function. *Clin. Neurol.* 3: 485–500.

Hirayama, M., Kawato, M., and Jordan, M.I. 1993. The cascade neural network model and a speed-accuracy trade-off of arm movement. *J. Mot. Behav.* 25: 162–74.

Hoffman, D.S., and Luschei, E.S. 1980. Responses of monkey precentral cortical cells during a controlled jaw bite task. *J. Neurophysiol.* 44: 333–48.

Honrubia, V. 1995. Contemporary vestibular function testing: Accomplishments and future perspectives. *Otolaryngol. Head Neck Surg.* 112: 64–77.

Houk, J.C. 1976. An assessment of stretch reflex function. *Prog. Brain Res.* 44: 303–14.

Houk, J.C. 1979. Regulation of stiffness by skeletomotor reflexes. *Ann. Rev. Physiol.* 41: 99–114.

Iwamoto, Y., Perlmutter, S.I., Baker, J.F., and Peterson, B.W. 1996. Spatial coordination by descending vestibular signals. 2. Response properties of medial and lateral vestibulospinal tract neurons in alert and decerebrate cats. *Exp. Brain Res.* 108: 85–100.

Kawato, M., Maeda, M., Uno, Y., and Suzuki, R. 1990. Trajectory formation of arm movement by cascade neural network model based on a minimum torque-change criterion. *Biol. Cybern.* 62: 275–88.

Kleinöder, H., Neumaier, A., Loch, M., and Mester, J. 1995. Cinematographic analysis of the service movement in tennis. In *Tennis: Sportsmedicine and science,* ed. H. Krahl, H.-G. Pieper, W.B. Kibler, P.A. Renström, and W. Rau, 16–21. Düsseldorf: Verlag.

Kuo, A.D. 1993. Human standing posture: Multijoint movement strategies based on biomechanical constraints. *Prog. Brain Res.* 97: 349–58.

Kuo, A.D. 1994. A mechanical analysis of force distribution between redundant, multiple degrees-of-freedom actuators in human: Implications for the central nervous system. *Human Move. Sci.* 13: 635–63.

Latash, M.L., and Gottlieb, G.L. 1991. An equilibrium-point model for fast single joint movement. II. Similarity of single joint isometric and isotonic descending commands. *J. Mot. Behav.* 23: 179–91.

Lekhel, H., Marchand, A.R., Assaiante, C., Cremieux, J., and Amblard, B. 1994. Cross-correlation analysis of the lateral hip strategy in unperturbed stance. *Neuroreport.* 10: 1293–96.

McCollum, G. 1994. Navigating a set of discrete regions in body position space. *J. Theor. Biol.* 167: 263–71.

McGarry, T., and Franks, I.M. 1996. On-line control of speeded motor response. *Percept. Mot. Skills* 82: 636–38.

Mester, J. 1988. *Zur Diagnostik im Bereich der Sinnesorgane.* Schorndorf: Verlag Hofmann.

Mester, J. 1997. Movement regulation in alpine skiing. In *Science and skiing,* ed. E. Müller, H. Schwameder, E. Kornexl, and C. Raschner, 333–48. London: Spon.

Mester, J., Grabow, V., and de Marées, H. 1982. Physiological and anthropometric aspects of vestibular regulation in rowing. *Inter. J. Sports Med.* 3: 174–76.

Mittelstaedt, H. 1995. New diagnostic test for the function of utricles, saccules, and somatic graviceptors. *Acta Otolaryngol. Suppl.* 520: 188–93.

Nashner, L.M. 1976. Adapting reflexes controlling the human posture. *Exp. Brain Res.* 26: 59–72.

Nichols, T.R. 1982. Reflex action in the context of motor control. *Behav. Brain Sci.* 5: 559–60.

Niessen, W., Müller, E., Raschner, C., and Schwameder, H. 1997. Structural dynamic analysis of alpine skis during turns. In *Science and skiing,* ed. E. Müller, H. Schwameder, E. Kornexl, and C. Raschner, 216–25. London: Spon.

Norre, M.E. 1993. Sensory interaction testing in platform posturography. *J. Laryngol. Otol.* 107: 496–501.

Norre, M.E. 1994. Vestibular patients examined by posturography: Sensory interaction testing. *J. Otolaryngol.* 23: 399–405.

Parker, D.E. 1994. Gleichgewichts- und orientierungssinn. In *Physiologie der Sinne,* ed. H.P. Zenner and E. Zrenner, 56–67. Heidelberg: Spektrum Verlag.

Parker, S.W. 1993. Vestibular evaluation: Electronystagmography, rotational testing, and posturography. *Clin. Electroencephalogr.* 24: 151–59.

Pearson, K. 1976. The control of walking. *Sci. Am.* 235: 72–86.

Saadat, D., O'Leary, D.P., Pulec, J.L., and Kitano, H. 1995. Comparison of vestibular autorotation and caloric testing. *Otolaryngol. Head Neck Surg.* 113: 215–22.

Shadmer, R. 1993. Control of equilibrium position and stiffness trough postural modules. *J. Mot. Behav.* 25: 228–41.

Spijkers, W., and Spellerberg, S. 1995. On-line visual control of aiming movements. *Acta Psychol.* 90: 333–48.

Spitzenpfeil, P., Babiel, S., Rieder, M., Hartmann, U., and Mester, J. 1997. The technique of gliding in alpine ski racing: Safety and performance. In *Science and skiing,* ed. E. Müller, H. Schwameder, E. Kornexl, and C. Raschner, 349–55. London: Spon.

Wada, N., Akenaga, K., Takayama, R., and Tokuriki, M. 1996. Descending pathways from the medial longitudinal fasciculus and lateral vestibular nucleus to tail motoneurons in the decerebrate cat. *Arch. Ital. Biol.* 134: 207–15.

Wagner, J., Bartmus, U., and de Marées, H. 1993. Three-axes gyro system quantifying the specific balance in rowing. *Int. J. Sports Med.* 14(Suppl. 1): 535–38.

Warren, W.H., Kay, B.A., and Yilmaz, E.H. 1996. Visual control of posture during walking: Functional specificity. *J. Exp. Psychol. Human Percept. Perform.* 22: 818–38.

Weeks, D.L., and Wallace, S.A. 1992. Premovement posture and focal movement velocity effects on postural responses accompanying rapid arm movements. *Human Move. Sci.* 11: 717–34.

Balance and Control of Movement Summary

J. Mester and B.R. MacIntosh

The evolutionary process of movement control has occurred under the direct influence of gravity. Gravity is the most prevalent influence on human movement, just as it has been on natural earthly environments such as the atmosphere and the climate. The topography of the inner organs as well as the functional capacity of the various physiological systems had to develop under the dominating force of gravity. Upright standing, erect gait, and all acts of movement control are most closely connected to gravity.

It took a million years for motor behavior to change from the quadruped locomotion of the ancient apes to the bipedal gait of *homo erectus*, requiring much more in regard to balance. Modern changes of environment, such as microgravity in space, show considerable effects, e.g., on the precision of grasping tasks, and need quite a long time for adaptation even for highly-developed and high-performance biological motor control capacities. Here the contribution of perception to the effectiveness of balance and motor control becomes quite clear.

Because of the high importance of motor control and balance for mastering daily life, this subject became an early topic of scientific interest. Michelangelo (1475–1564) studied human movement with a combination of artistic and scientific interest. He was followed by Galileo (1564–1642) with his revolutionary findings regarding the overall importance of gravity. Galileo established the foundation for our modern understanding of earthbound movements. The significance of reflex action for motor control and balance under gravity was elucidated by the work of Sherrington (1898). Eccles (1957) revealed the physiology of nerve cells by means of intracellular recordings from single nerve cells.

In chapter 9, Stein, Zehr, and Bobet elaborate the basic concepts of motor control. They describe the complex motor control system, including the classic components of a control system: central controller, effector, and sensors for feedback. Superimposed on this basic control system is a feed-forward con-

trol mechanism that permits a preprogrammed sequence of muscle activations, which can occur with or without relevant feedback from a number of sensory organs (Golgi tendon organs, stretch receptors, vision, cutaneous receptors, and semicircular canals). The various control systems apparently can account for complex interaction of the various factors which influence the contractile response of muscle, including muscle length, force-velocity properties, and prior activity.

Each muscle in the body is organized into motor units. Each motor unit is composed of a single motor neuron and all of the muscle fibers innervated by that neuron. A graded muscle response can be obtained by varying the frequency of activation of individual motor units (rate coding) and by varying the number of involved motor units (recruitment). The motor control system is able to integrate the various sensory feedback information and accomplish smooth coordinated movement by activation of an appropriate number of motor units while simultaneously inhibiting activation of antagonist muscles. This occurs in spite of a complex, history-dependent modulation of the contractile response, which includes prolongation of the response to a single activating stimulus (decreased fusion frequency), short-term enhancement (potentiation) of the contractile response, and a more slowly developing muscle fatigue.

Whereas Stein, Zehr, and Bobet (chapter 9) concentrate on neural regulatory mechanisms, Herzog (chapter 10) elaborates on the contractile response of muscle from motor unit to molecular mechanisms. He pays particular attention to the mechanisms of force production and the various factors that can influence it in muscles: active muscle cross-sectional area, frequency of activation, and muscle length. He discusses the size principle, which dictates that motor units within a given muscle will be recruited in order from smallest to largest. This issue introduces the concept of fiber type, an important consideration with respect to the force-frequency relationship and power output. Coordinated activation of synergistic muscles is investigated, revealing that in general slow-twitch muscles are activated in preference to fast-twitch muscles when movement is relatively slow, but the reverse can occur during rapid movements.

Investigation of the role of a given muscle in contributing to a joint moment is complicated by the infinite number of possible combinations of synergistic muscle involvement across a joint, including cocontraction of antagonist muscle motor units. Most joints of the human body have highly redundant motor capabilities, which means a given moment about the joint can be accomplished by a number of possible motor unit/muscle activation patterns.

Whole body movements can be considered mechanical transmissions of muscle force to limbs and joints. Herzog addresses this important issue as he deals with the mechanical aspects of movement (chapter 10). The force-sharing strategies of agonistic muscles are considered in terms of minimizing metabolic costs with consideration for the interaction of agonist and antagonistic muscles. Additionally the transmission of forces from the active structures (muscles and attached tendons) to the so-called "passive" structures (e.g., bones, ligaments, etc.) is discussed in detail.

There are many applications of motor unit recruitment and the mechanical properties of muscle to sport science. The theoretical approach can be used for a better understanding of the coordination in technically difficult sports, as well as for the further investigation of metabolic phenomena. In order to illustrate the possible application in detail, Herzog offers results of a study that focused on the probable adaptation of moment-angle properties to specific requirements in certain sports, i.e., cycling and running. It was shown that the physiological characteristics of the muscle are appropriate for the activity pattern of the athletes. Specifically it was reported that cyclists who train and compete with relative hip flexion have apparently fewer sarcomeres in series in rectus femoris than runners who train with a more extended hip. If this structural difference represents an adaptation induced by the joint position assumed during regular exercise, then there are serious implications for possible negative influences of practice in an inappropriate body position.

With respect to the mechanical control of a joint moment, Herzog emphasizes that redundancy is an important feature of the biological setup. As there are many more muscles crossing a joint than there are rotational degrees of freedom, a certain joint moment can be produced by an infinite combination of contraction events and properties of the

muscles. Obviously, even from the mechanical point of view, the biological principle of an optimization of adaptability and readiness for unexpected situations keeps its validity.

A special form of motor control is required in certain sports (e.g., diving, trampolining, gymnastics) when the body is under the condition of free fall and aerial adjustments are necessary to avoid instability and realize the goal of the movement. For a long time it has been known that a cat lands on its feet regardless of the initial position from which the fall is initiated. Marey first studied this phenomenon by means of filming techniques in 1894. The association with complex human movements, such as twists and somersaults, is obvious.

The ability to land on one's feet in a safe position cannot simply be explained by aerodynamic influences; active aerial control of movement is needed. In the research presented by Yeadon and Mikulcik (chapter 12), strategies for controlling movements of this kind are discussed. By means of a computer simulation model of 11 segments, three of these strategies are analyzed:

1. symmetrical arm abduction,
2. arching of the body, and
3. asymmetrical arm adduction/abduction.

Using the option of arm abduction, the computer model demonstrates that in contrast to earlier findings, an initial large symmetrical extension of both arms does not reduce the buildup of twist in single or double somersaults. These results can be explained by the fact that instability of the longitudinal axis depends on three principal moments of inertia. The difference between two of them grows with abducted arms so that the twist occurs earlier and faster than with arms held close to the body. Arching the body during flight can reduce the twist for small perturbations. For larger perturbations, the computer simulation shows that the arching must be carried out according to the magnitude of the twist oscillations. An effective way of reducing the buildup of many kinds of twists in aerial movements with arbitrary, small perturbations seems to be the asymmetrical and precisely-controlled movement of the arms, provided that a critical time delay is not exceeded. These findings are presented with the instrument of equations of motions with respect to a model composed of two arms and one body segment.

Following the discussion of Yeadon and Mikulcik, it is important to present practical applications for the understanding of the biological aspects, i.e., sensory input and the effects of the efferent control signals. Considering the normal human reaction time of about 150 ms, a main part of the sensory input with the goal of corrective movements probably arises from the otolith organs and the semicircular canals. Thus the linear and angular acceleration of the body, as opposed to the expected (planned) signals, can be regarded as the required input for corrective movements.

Mester's chapter (chapter 13) focuses on earthbound movements. Here the basic prerequisite for any kind of aimed movement in daily life or in sports is to maintain balance, normally in an upright position. Phenomena of posture, including upright standing and erect gait, involve complex structures of movement control. Therefore, it is important to consider effects of the variability of postural control on the control of directed movements. Phylogenetic development guarantees that both the peripheral reflex control mechanisms at the spine level and the commands from the higher control centers constitute a sophisticated interactive system for maintaining balance in the upright position.

Modern theoretical and empirical approaches do not focus so much on the primary role of the central nervous system but on the circuitry of networks involved. Taking the concept of an interaction of networks as a basis, various phenomena such as premovement posture mechanisms become intelligible. These mechanisms meet the requirements that arise from dynamic changes of posture, such as lifting the arms with varying velocity, when anticipatory adjustments of the tonus of those muscles that are responsible for maintaining balance are provided.

As gravity is the main opponent to balance, a great deal of partly-redundant sensory systems (vestibular, visual, kinesthetic, tactile) continuously monitor the position of the body or rather the changes of the position. Studies with respect to the interaction of the visual, vestibular, and tactile systems have shown that a permanent sensory control of the environment is associated with a permanent oscillatory motor-control drive. These key elements provide not only regulations of balance in the sense of a feedback reaction to certain perturbations, but also anticipatory, feed-forward mechanisms that are highly effective for skilled movements.

Examples from field studies in various sports indicate that even these properties are of a very specific structure concerning the respective requirements. Although center-of-force regulation in straight downhill alpine skiing in the upright position strongly resembles the one in still standing, as proven in neurophysiological lab experiments, there is greater need for specific skill acquisition to permit handling of the high external forces. In particular, studies of movement control under time pressure reveal the high capability of anticipatory adjustments not only of the control of posture, but also of the precise displacements of single segments.

Balance and Control of Movement Definitions

Mechanical

arm abduction: outward rotation of an arm away from the midline of the body.

arm adduction: inward rotation of an arm toward the midline of the body.

axis:

frontal axis: an axis of rotation passing from front to back through the subject's mass center.

lateral axis: an axis of rotation passing from right to left through the subject's mass center.

longitudinal axis: an axis of rotation passing from feet to head through the subject's mass center.

balance: the mechanical (i.e., position of segments) and physiological (e.g., muscle forces/joint stiffness) prerequisites for beginning and continuing a purposeful movement. Often in literature on the basis of mechanical considerations a distinction is made between static balance (e.g., quiet standing upright) and dynamic balance (e.g., running in a curve). As in supposed static situations sophisticated regulation processes for maintaining balance with various displacements of segments are also needed. No distinction between static and dynamic balance is made in chapter 13.

body configuration: the relative orientations of adjacent body segments.

body orientation: the orientation of the body in space described by the angles of somersault, tilt, and twist.

control: producing a desired outcome by means of continual adjustment. In the context of non-twisting somersaults control means the prevention of twisting. For control of twisting somersaults it is necessary to adjust body configuration during flight to achieve the correct amount of somersaults, tilt, and twist at landing.

neutral control: the adjustments needed to maintain control do not increase in size.

stable control: the adjustments needed to maintain control remain small.

unstable control: the adjustments needed to maintain control increase with time and control cannot be maintained indefinitely.

equilibrium: the state of a system (e.g., a body or a position of segments) that is not being accelerated.

equilibrium point hypothesis: the set of active directional movements as the result of shifts of the equilibrium state of the motor system associated with the setting of the threshold of the stretch reflex and thus responsible for establishing balance.

peak power output: the highest power output which can be achieved for a given muscle contraction or movement. Although strictly speaking a peak power output should be an instantaneous event, the term is often used to express the highest average proven output which can be sustained for a fixed brief period of time (i.e., 5 s of maximal effort cycling).

posture: the external visible relative arrangement of body segments in a given upright position representing a discrete event in the time-history of balance.

somersault: rotation about an axis with a fixed direction in space. In the case of a non-twisting somersault the lateral axis will be coincident with this axis.

layout somersault: a somersault in which the body is nearly straight. Typically the body is arched for most of the movement.

pike somersault: a somersault in which the body is flexed at the hips. Typically the angle between the upper trunk and thighs is less than 90°.

stable somersault: a somersault in which the twist remains close to zero. Somersaults about the lateral axis are stable when the body is tucked or pike.

tucked somersault: a somersault in which the body is flexed at the hips and knees. Typically the angles between the trunk and thighs and between the thighs and calves are each less than 90°.

twisting somersault: a somersault in which the twist angle increases. This is one of the two modes of motion of a freely rotating rigid body. In gymnastics terminology a twisting somersault is a somersault with at least half a revolution of twist.

unstable somersault: a somersault in which the twist builds up from near zero to more than a quarter twist. Somersaults about the lateral axis tend to be unstable when the body is straight.

wobbling somersault: a somersault in which the twist angle oscillates. This is the other mode of motion of a freely rotating rigid body. Typically this is the mode of motion for a plain somersault without twist.

tilt: the angle that the longitudinal axis makes with the plane perpendicular to the fixed somersault axis.

twist: rotation about the longitudinal axis which is fixed in the body. Since this axis is fixed within the body, twists are described as either to the left or to the right.

Biological

activation of muscle: the process by which muscle is "turned on" or induced to activity. Muscle response can be graded in terms of motor units activated as well as by frequency modulation of a given cell or motor unit. "Level of activation" refers to the relative magnitude to which the muscle is turned on.

angle of stereopsis: the minimal distance of two fixation objects in depth whose images are located on the Vieth-Müller-horopter circle and so are processed in binocular disparity, thus producing depth perception.

isotonic contraction: a muscle contraction against a constant external load. An isotonic contraction can be preloaded, in which case the muscle is stretched

to the length which can passively support the load before the contraction is initiated, or afterloaded, in which case the initial length is regulated independent of the load. Typically an in vivo contraction cannot be isotonic because a constant resistance or load imposed on a limb will require a changing force as the angle about the axis of rotation changes.

movement control: the synergy of all psycho-neuro-muscular processes which are relevant for realization of the goal of movement.

nystagmus: spasmodic horizontal eye movements associated with rotation loads, e.g., on a rotating chair, indicating irritation of the vestibular system.

pinnation angle: the angle ($>0°$) between the orientation of muscle fibers and the apparent line of action of the muscle.

sensory mismatch: a constellation of the input of the individual sensory systems that differs from the familiar situation. Sensory mismatch occurs for example under microgravity in space when the visual system indicates a vertical upright position of the body by referring to the external visual framework whereas the vestibular system receives (almost) no input via gravitational forces.

static visual acuity: the ability of the visual system (from lens to central processing of visual information in area 17 of the cortex) to distinguish two static outlines (bars, points, etc.) from each other (minimum separable). The static visual acuity is greatest in the region of the fovea centralis. On both the nasal and the temporal side of the fovea the visual acuity drops off sharply. That is why eye movements are needed for scanning the environment.

stereoscopic visual acuity: the ability of the visual system to detect the actual position of a fixation point in depth. Stereoscopic visual acuity is determined by various monocular and binocular visual attributes such as perspective, overlapping of outlines, apparent size of objects, and binocular disparities.

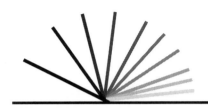

PART III

Load During Physical Activity

Load During Physical Activity Historical Highlights

468–367 B.C.

Among **Hippocrates'** writings is one text on "surgery" in which a detailed study of bandaging is presented.

100 B.C.

Hegator provides the first anatomical definition of a ligament.

A.D. 14–37

Known for his essays on bandages and fractures, **Soranus of Ephesus** lives at the time of Minecrates, physician to Emperor Tiberius Caesar.

A.D. 129

Galen distinguishes ligaments from tendons.

1500s

da Vinci states that tendons are mechanical instruments playing a passive role in carrying out as much work as put on them.

1514

Vesalius details anatomical definitions of joint tissues.

1686

Newton publishes *Philosophiae Naturalis Principia Mathematica [The Mathematical Principles of the Natural Philosophy]* with the three laws that are the cornerstone of all subsequent force discussions.

1691

Havers proposes that bone is composed of organic strings and inorganic plates arranged about tubular cavities.

1739

du Hamel demonstrates the lamellar nature of bone.

1742

Lieutaud suggests that bone is composed of laminae and compact fibers.

1742

Hunter associates articular cartilage with mechanical concepts. He states, "by their [cartilages'] elasticity, the violence of any shock which may happen in running, jumping, et cetera, is broken and gradually spent which must have been extremely pernicious, if the hard surfaces of bones had been immediately contiguous." He also describes fibrous infrastructure radiating perpendicularly from bone surface and deforming upon compression.

1743

Borelli publishes the first part of *De Motu Anumalium [About the Movement of Living Beings]*, discussing the possibilities of quantifying forces in a biological system.

1744

Andry's work influences the development of braces to correct misalignments of the lower extremities.

1830

Schleiden and **Schwann** discover cells and long fibers in dense connective tissues.

1847

Wertheim notes that the stress-strain curve of tendon is non-Hookean.

1855

Kölliker notes that tendons contain "waves" (banding) which disappear when the tissue is tensioned. He attributes these waves to the influence of elastic fibers upon collagen.

1855

Breithaupt describes stress fractures in Prussian military recruits.

1850s

Rudinger and **Hilton** discover nerves in joint tissues and postulate the existence of ligament-muscle feedback systems that mediate movement.

1858

Rankine, a Scottish engineer-physicist, uses a graphic technique for analyzing problems of applied mechanics.

1864

Maxwell shows that reciprocal figures and force polygons are related and may be used in mechanical analyses.

1867

Culman states that there is a similarity between the trabecular arrangement in bone and the structural elements of a crane. In both cases, the principles of highest efficiency and economy are used.

1870

Wolff summarizes various suggestions and statements about bone by stating that there is an interdependence between its form and function. He theorizes that physical laws have strict control over bone growth.

1872

Carlet develops a pneumatic measuring device to quantify the forces between foot and ground.

1875

Culmann, a Swiss scientist, publishes a book dealing with graphical statics and later becomes the leader in the application of graphical methods to enginering problems.

1875

Weisbach, a German engineer, uses diagrams of bodies with forces acting on them (forces indicated by arrows). Some of these diagrams are identical to free body diagrams in the current understanding. Other diagrams include only selected forces.

1881

Gibbs develops the mathematics of vector analysis, which is used to solve the vector equations that result from the free body diagrams.

1883

Roux proposes that bone's trabecular orientation corresponds to the direction of tension and compression stresses and is developed using the principle of maximum economy of material use.

1898

Beiersdorf invents adhesive tape and bandages.

1898

Hultkrantz illustrates surface fiber pattern by pricking cartilage with a sharp tool and noting the direction of "splits" produced.

1910

Smith and **Longley** outline the procedures for handling mechanical problems by using drawings and vectors.

1911

Fick publishes the first biomechanical review of ligaments.

1925

Benninghoff investigates the deeper substance of cartilage. He introduces the detailed "arcade" theory, which confirms Hunter's ideas and describes fibers radiating out from the bone and curving obliquely until parallel to the articular surface.

1928

Reynolds uses the term "free body diagram" and defines it as "the body isolated from all other bodies with arrows representing the forces acting upon it."

1920s and 1930s

Scientists measure apparent elastic properties of articular cartilage (Bar, 1926; Göcke, 1927; Policard, 1936; Muller, 1939) to understand arthritic diseases.

1938

Elftman quantifies ground-reaction forces and the center of force during human locomotion using a force plate (a force-measuring device in the form of a plate).

1942

Maj shows experimentally that cortical tissue of bone becomes weaker with advancing age.

1946

Quigley gives one of the first descriptions of prophylactic use of adhesive ankle taping techniques published as the "figure of eight."

1949

From electron microscopic examination, **Wickoff** observes that there is a helical (spiral) pattern within the collagen fibrils of tendons.

1950

Rollhäuser states that the extensibility of tendon increases with age.

1955

Lissner attempts to interpret the mechanics of the musculo-skeletal system for physiotherapists.

1961

Dempster publishes an article on free body diagrams as an approach to the mechanics of human posture and motion.

1960s and 1970s

Ultrastructure of articular cartilage is described in great detail with the help of the newly developed polarized light microscope and scanning electron microscope (Goodfellow and Bullough 1968; Hunter and Finlay 1973).

1970

Fung formulates a quasi-linear viscoelasticity theory to describe more accurately the mechanical response of soft tissues to load.

1973

Burri first describes functional rehabilitation with a hinged cast system following knee ligament surgery.

1979

Anderson publishes a manuscript describing the first rehabilitational brace.

1980s and 1990s

Mow and coworkers develop multi-phasic theories to describe the mechanical response of articular cartilage.

1980

Noyes provides the first definition of complex structure-function relationships of ligaments.

1981

Woo reports on viscoelastic properties of ligaments.

1983

Nicholas describes the first preventive derotational knee brace, which is promoted as the "Lenox Hill" brace.

1984

The Sports Medicine Committee of the AAOS classifies available types of knee braces for their prophylactic rehabilitative and functional values.

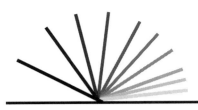

Chapter 14

Forces Acting on and in the Human Body

B.M. Nigg

Forces acting on and in the human locomotor system during physical activities are the reason for locomotion, for deformation of soft and semi-rigid structures of the human body, for growth and development of biological tissue, but also for acute and chronic injuries. This section

- provides the definitions and explanations to apply the force concept appropriately,
- discusses a possible force system analysis determining external forces and moments on a system of interest, and
- provides an overview of the force magnitudes that people experience during physical activities.

Definitions and General Comments

Several terms and expressions are used with specific meanings in this section. These meanings are presented below.

Force

The term *force* is frequently used in the daily language and in scientific terminology. Whoever uses the term has a "good feeling" that he or she under- stands what force is. However, the definition of force is not trivial. Force is—like time, mass, and distance—a quantity that cannot be defined.

Although force is a quantity that cannot be de- fined, effects produced by forces can be described. For the specific biomechanical application, forces can produce biological and mechanical effects. The major *biological effects* of forces include changes in the development of biological tissue and transpor- tation of nutrients through the human body. Muscles may be strengthened, bone may grow more rapidly due to force stimuli, blood moves through the arter- ies and veins. The major *mechanical effects* produced by forces include the ability to *accelerate a mass* (e.g., when kicking a football or throwing a javelin) and the ability to *deform a material* (e.g., to compress a heel pad or to break a bone).

Accelerate a Mass

Newton's second law (for constant mass) describes this ability:

$$\mathbf{F} = m \cdot \mathbf{a} \qquad\qquad 14.1$$

where

$\mathbf{F}$ = resultant force vector acting on the mass m, and

$\mathbf{a}$ = acceleration vector of the mass,

with the unit

$$[F] = [m] \cdot [a] \qquad 14.2$$
$$[F] = kg \cdot m/s^2 = kg \cdot m \cdot s^{-2} = N = newton.$$

The unit of force measurement is 1 N (one newton).

Deform a Material

This ability is described in the simplest case by the equation for the deformation of a linear spring:

$$F = k \cdot \Delta x \qquad 14.3$$

where

F = resultant force vector acting on the spring,

k = linear spring constant (describing the stiffness of the spring), and

Δx = deformation of the spring (vector quantity),

with the units

$[F] = N$,

$[k] = N/m$, and

$[\Delta x] = m$.

Forces are vector quantities for which vector algebra applies. In drawings, force vectors are often depicted as arrows.

Loading Rate

Forces acting on the locomotor system can be studied with respect to their magnitude or their time behavior (frequency). When studying forces one may be interested in the magnitude or in how fast the force is increasing or decreasing. One variable describing this behavior is the loading rate. The *loading rate* of a force-time signal is the time derivative of the force-time curve. The loading rate indicates how fast the force changes in time and can be depicted as the slope of the force-time curve. It is often assumed that the loading rate of a force acting on the locomotor system is associated with the development of movement-related injuries.

Moment

Forces act in a geometric environment. A force acting eccentrically on a disk that is frictionless and mounted on an axis through its center will make this disk start turning. This effect produced by a force is called a *moment*. A moment produced by a force is defined as the turning effect of this force. A moment

occurs when a force acts at a distance d from an axis of rotation A through a point O. The distance d is called the lever or lever arm. In vector notation, the moment is calculated as

$$M_O = r_{PO} \times F_P \qquad 14.4$$

where

M_O = moment (or moment of force) about point O,

r_{PO} = position vector from O to P,

$\times$ = sign for vector product, and

F_P = force acting at point P.

A moment is always determined with respect to a point or an axis (see figure 14.1).

In scalar notation the moment equation can be written as

$$M_O = r_{PO} \cdot \sin \alpha \cdot F_P = d \cdot F_P \qquad 14.5$$

where

M_O = magnitude of the moment (or moment of force) about point O,

r_{PO} = magnitude of the position vector from O to P,

F_P = magnitude of the force acting at point P,

α = angle between the position vector and the force vector, and

$d = r_{PO} \cdot \sin \alpha$ = lever arm.

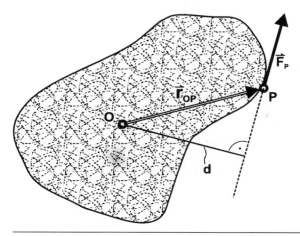

Figure 14.1 Illustration of a force F_P acting at a point P of a disk, which is mounted without friction with an axis A through O. The position vector from O to P is r_{OP}, the distance between the line of action of the force and the fulcrum O is d. The force produces a turning moment M_O, which will cause the disk to turn.

Ground Reaction Force (GRF)

Whenever a part of an athlete's body (e.g., the foot) is in contact with the ground, this part exerts a force on the ground and the ground reacts with an equal and opposite force, the ground reaction force. A *ground reaction force* (GRF) is a force that acts from the ground on the object that is in contact with the ground. Ground reaction forces have been quantified for many different athletic movements (Bates et al. 1982; Cavanagh and Lafortune 1980; Cavanagh et al. 1981; Frederick, Clarke, and Hamill 1984; Hamill et al. 1983; McClay et al. 1994; Nigg 1978). Examples for ground reaction forces for running with heel and with toe landing are illustrated in figure 14.2.

Resultant Forces

Resultant forces are often used to simplify a situation. Typically, resultant forces are a sum of forces of the same kind. The most commonly used resultant forces include body weight and ground reaction force. A *resultant force* is the vector sum of many forces.

$$\mathbf{F}_{res} = \sum_{i=1}^{n} \mathbf{F}_i \qquad\qquad 14.6$$

where

$\mathbf{F}_{res}$ = resultant force vector,

i = index for all forces to be included,

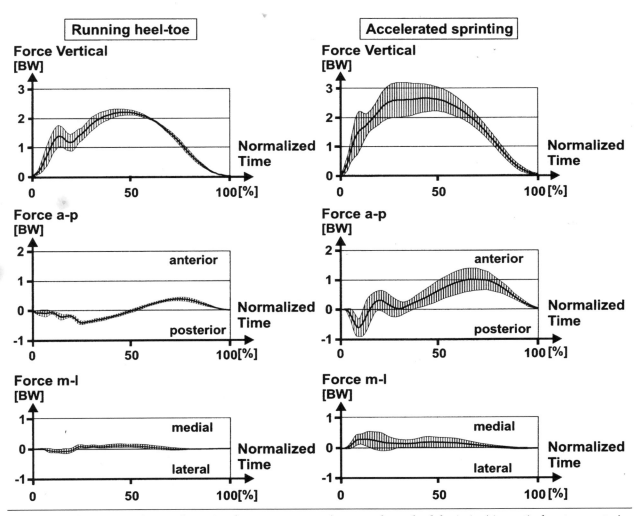

Figure 14.2 Illustration of ground reaction force components (mean and standard deviation) in vertical, anteroposterior (a-p), and mediolateral (m-l) direction for running heel-toe (left) and sprinting with toe contact (right) in units of body weight (BW). Both curves are mean values with standard deviation for five subjects (different for the two tests) with three trials each.

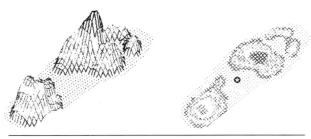

Figure 14.3 Illustration of the normal (perpendicular to the plantar surface of the foot) force distribution between shoe insole and the plantar surface of the foot for a specific time point during midstance in running. On the left side, the normal pressure is depicted with "mountain diagrams," on the right side, the pressure distribution is indicated with different gray scales. The ⊙ symbol indicates the location of the point of application of the (resultant) ground reaction force.

n = number of forces to be included, and

F_i = force vector i to be included in the resultant force.

Body weight is the resultant force for all the segmental weights of a human or animal body. Often, it would be cumbersome to include the various weights of each body segment in a calculation. For instance, one does not want to include the weights of each finger and toe, the weights of feet and hands, and so on in an estimation of jumping height or running speed. Thus, all these segmental weight forces are summarized in one force, the resultant force of body weight, acting at the center of mass of the total body. The final mechanical effect for the movement of the center of mass is the same whether one includes all segmental weights or the resultant body weight in a specific calculation.

The *ground reaction force* is the resultant force of all local forces acting at the foot or the feet of a person. A person contacting the ground with one foot during ground contact in running will experience various forces between the foot and the ground. In midstance, for instance, high local forces act under the heel and under the ball of the foot. However, the forces under the midfoot and under the toes are typically small. These forces can be summarized as one resultant force, the ground reaction force, as illustrated in figure 14.3 for the local forces perpendicular to the plantar surface of the foot. The ground reaction force can replace these individual local forces.

Resultant forces and their point of application are abstract quantities and do not necessarily represent the location where the greatest force is acting. For instance, for a person standing quietly on both feet (20 cm apart) with the weight equally distributed between the two feet the resultant ground reaction force "acts" at the center of force, a point between the feet (see figure 14.4). There is no body part at this point that is in contact with this resultant force. Thus, the resultant force "acts" at a point where no force acts in reality.

Another example for a resultant force that does not represent the actual force acting is the "resultant joint force" a quantity that is often used in biomechanics. This resultant joint force does not represent the actual force that could be measured in a joint, as will be shown later in this chapter.

Impact Forces

Impact forces in human locomotion (Frederick, Hagy, and Mann 1981; Nigg 1978) are typically forces due to the landing of the foot on the ground. Impact

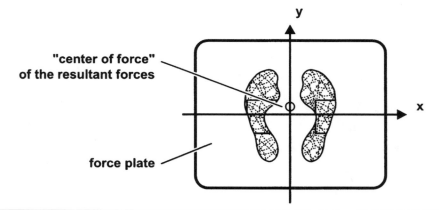

Figure 14.4 Illustration of the location of the center of force for the resultant ground reaction force, the sum of all the local forces acting between the ground and the two feet.

Reprinted, by permission, from B.M. Nigg, 1994, Pressure distribution. In *Biomechanics of the musculoskeletal system*, edited by B.M. Nigg and W. Ed. Herzog. (Sussex, UK: John Wiley & Sons), 227. Copyright John Wiley & Sons Limited.

forces are forces due to the collision of two objects. Impact forces in human locomotion are forces resulting from the collision of the foot with the ground with a force maximum earlier than 50 ms after first contact of the two objects.

Impact forces occur in all three directions, vertically, forward-backward (anterior-posterior, a-p), and mediolaterally (m-l). The most frequently studied impact forces are the vertical impact forces during landing in heel-toe running (Chu et al. 1986; Frederick, Hagy, and Mann 1981; Lafortune and Hennig 1992; Luethi, Nigg, and Bahlsen 1984; Nigg 1978; Nigg et al. 1987; Shorten and Winslow 1992).

Impact forces in sport activities using sport shoes last typically between 20 and 50 ms with peaks typically between 10 and 25 ms. When moving barefoot the impact peaks occur typically between 5 to 10 ms after first contact. The time needed from the moment a signal is registered by the sensory feedback system of the human body to the moment when a force production in a specific muscle is registered as a result of the sensory feedback is typically in the order of magnitude of about 100 ms. Consequently, an athlete cannot adjust the muscular activation during impact to react to the collision. Adjustments have to be made beforehand based on previous foot contacts or based on expectation. Therefore, the collision process during impact is a passive event. Note that these forces were originally called "passive forces" (Nigg 1978).

External Impact Forces

The ground reaction force represents the change in movement of the center of mass CM. It is composed of the sum of all the products of the segmental masses times their accelerations.

$$\mathbf{F}_{ground} = \Sigma \, \mathbf{F}_i = \Sigma \, m_i \cdot \mathbf{a}_i \qquad 14.7$$

where

$\mathbf{F}_{ground}$ = ground reaction force vector,
$\mathbf{F}_i$ = force vector produced by segment i,
m_i = mass of segment i, and
$\mathbf{a}_i$ = acceleration vector of segment i.

The different segments contribute to the total ground reaction force when they are accelerated or decelerated. The magnitude of external impact forces depends on the mass involved in the initial deceleration process. During landing on the forefoot, for instance, only the forefoot and a small part of the leg are involved in the rapid initial deceleration. Thus, the mass involved in the initial deceleration process, the effective mass (Denoth 1986; Nigg 1980), is

rather small (about 1 to 2 kg). Two simple examples illustrate the concept of the effective mass.

Example 1.

We can make the following assumptions:

- All forces act only in vertical direction.
- The peak acceleration during the toe landing process is 200 m/s².

Solution: For an effective mass of 1 kg, the peak force due to the deceleration of the foot (and part of the leg) would be

$$F_{max} = m_{eff} \cdot a_{max} = 1 \text{ kg} \cdot 200 \text{ m/s}^2 = 200 \text{ N}. \qquad 14.8$$

For an effective mass of 2 kg

$$F_{max} = m_{eff} \cdot a_{max} = 2 \text{ kg} \cdot 200 \text{ m/s}^2 = 400 \text{ N}. \qquad 14.9$$

Thus, for a person with a body mass of 80 kg (and a body weight BW of about 800 N) the impact ground reaction force due to the deceleration of the foot and part of the leg during forefoot landing would be between 1/4 and 1/2 of BW.

Example 2.

During landing on the heel of the foot the mass involved in the initial deceleration process, the effective mass, includes the mass of the foot and part of the leg and the thigh. The effective mass for this landing process is about 4 to 10 kg. We can make the following assumptions:

- All forces act only in vertical direction.
- The peak acceleration due to the deceleration of the foot and leg is 200 m/s².

Solution: For an effective mass of 4 kg, the peak force due to the initial deceleration of the foot and part of the leg and thigh would be

$$F_{max} = m_{eff} \cdot a_{max} = 4 \text{ kg} \cdot 200 \text{ m/s}^2 = 800 \text{ N}. \qquad 14.10$$

For an effective mass of 10 kg

$$F_{max} = m_{eff} \cdot a_{max} = 10 \text{ kg} \cdot 200 \text{ m/s}^2 = 2000 \text{ N}. \qquad 14.11$$

Based on these calculations the corresponding external vertical impact force for heel landing would be 1 to 2.5 BW, substantially higher than the corresponding vertical impact forces for toe running. Experimental results confirm these theoretical calculations (see figure 14.3). External vertical impact forces are less than 1 BW for toe running or sprinting, between 1 to 3 BW for heel-toe running, and can reach 5 to 10 BW for extreme landing situations (e.g., landing on the heel in gymnastics or take-off for a long jump).

External impact forces during activities such as running or court games have frequently been associated with the development of overuse injuries (Clarke, Frederick, and Cooper 1983; Falsetti et al. 1983; Lafortune, Hennig, and Lake 1996; Light, McLellan, and Klenerman 1980; Milgrom et al. 1985; Nigg 1978). However, this paradigm is currently under review for normal physical activities such as jogging or playing court games (Nigg 1997). Today, impact force-intensive activities are used to stimulate bone development (Heinonen 1996).

Internal Impact Forces

Internal impact forces in human or animal locomotion are forces that are experienced by biological structures such as bone, cartilage, ligament, and tendon. Internal impact forces can typically not be measured in humans even though isolated attempts have been made (Bergman et al. 1995). Internal impact forces are typically estimated using a modeling approach (Chu et al 1986; Cole et al. 1995; Denoth 1986). Impact forces acting on the shoe sole during landing will be distributed through the shoe sole to the foot and from there to the structures of the lower extremities and the rest of the body. The internal stresses experienced by an individual biological structure (e.g., a bone) depend on the magnitude, the direction, and the point of application of the acting impact force. The importance of the magnitude is obvious. Internal loading will double if the external impact force changes from 1 to 2 BW. However, the direction and the point of application of the external impact force are more important for the internal loading situation. Internal stresses are relatively small when the impact forces act close to the axis of the joint. Internal stresses become excessive when the external impact forces act with a substantial lever to the axis of the joint. The heel of the human foot is round. The levers between the ground reaction force during barefoot landing on the heel and the ankle joint are typically small. Consequently, the internal impact forces for landing on a human heel are typically small.

Active Forces

The literature on forces acting on the locomotor system distinguishes between "impact" and "active" forces. *Active forces* in human locomotion are forces that are completely controlled by muscular activity. Active forces (Nigg 1978) are the primary reason for locomotion. When initiating a movement out of a standing position (e.g., a jump), an athlete activates the muscles and this muscle activation produces the movement. Purely active movements in sports include weightlifting, push-ups, arm swinging, or a javelin throw from a standing position, to mention just a few.

Human locomotion typically has an impact and an active phase (see figure 14.2) with the impact phase at the beginning of ground contact and the active phase during stance and take-off.

Internal and External Forces

The terms *internal* and *external forces* are often used in biomechanical analysis. The notation internal and external depends on the system of interest, which has been arbitrarily defined. Forces are defined as internal/external if they act internally/externally to a system of interest.

The first step when dealing with a force system analysis consists of defining the system of interest. This procedure is described in detail in the section on force system analysis. Once the system of interest is defined all forces acting externally to this system are called external forces. All forces acting within this system are called internal forces.

Example 1: The human body has been defined as the system of interest.

Ground reaction force: external force

Weight: external force

Patella-femoral joint force: internal force

Patellar tendon force: internal force

Quadriceps force: internal force

Example 2: The complete left human leg has been defined as the system of interest.

Ground reaction force: external force

Weight of the leg: external force

Patella-femoral joint force: internal force

Patellar tendon force: internal force

Quadriceps force: internal force

Example 3: The patella has been defined as the system of interest.

Ground reaction force: does not act on this system

Weight of the patella: external force

Patella-femoral joint force: external force

Patellar tendon force: external force

Quadriceps force: external force

Consequently, it is important to identify the system of interest whenever one uses the term *internal* or *external* force.

Frictional Forces

Friction is a property that is determined by the material and surface structures of two contacting surfaces. Friction is important in sport activities. *Frictional forces* are forces due to the resistance of two surfaces to slide with respect to each other.

Friction between the foot and the playing surface has an effect on load, performance, and comfort. High frictional forces generally enhance performance by allowing athletes to apply large forces against the surface without slipping. During cutting and shuffling movements, for example, the magnitude of horizontal forces may reach or exceed the athlete's body weight. Friction, when excessive, may increase the risk of injury (Bonstingel, Morehouse, and Niebel 1975; Nigg and Segesser 1988; Valiant, Cooper, and McGuirk 1986; van Gheluwe, Deporte, and Hebbelinck 1983). High friction between shoe and surface typically results in high forces in the ankle, knee, and hip joint and is discussed as an important factor in the etiology of sport-related injuries, specifically of knee injuries (Torg and Quedenfeld 1971). The discussion on friction is generally divided into two parts, translational and rotational friction.

Translational Friction

The *translational friction* coefficient is typically divided into a static and a dynamic part. Translational friction behavior of two surfaces is characterized by the translational friction coefficient μ_{tr}. The *static translational coefficient* μ_{trsta} is defined as the quotient of the force parallel to the contacting surfaces F_{par} and the normal force F_{norm} immediately before the two surface samples start moving. The *dynamic translational coefficient* is defined as the quotient of the force parallel to the contacting surfaces F_{par} and the normal force F_{norm} for the situation when the two surfaces move with constant velocity with respect to each other.

When the two surfaces do not move with respect to each other the quotient F_{par}/F_{norm} must be smaller than the static translational friction coefficient. Thus, for actual movement conditions with no relative movement between the two surfaces the relationship reads

$$\mu \text{ (movement)} \leq \frac{F_{par}}{F_{norm}} \qquad 14.12$$

indicating that the parallel force (compared to the normal force) is too small to produce a relative movement of the two surfaces (no slipping or sliding!).

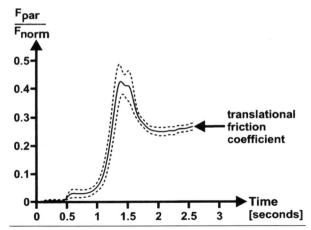

Figure 14.5 Friction coefficient versus time for one specific shoe-surface combination and three independent trials. The initial peak corresponds to the static friction coefficient.

For leisure sport activities the dynamic translational friction coefficient between sport shoe and sport surface is ideally between about 0.6 and 1.0 (Nigg 1990; Ono 1986). High friction coefficients (higher than 1.0) are sometimes demanded from high performance athletes (e.g., in tennis or basketball) to increase the reaction possibilities and the feeling of a "good grip". Low friction coefficients may be the reason for slipping and related injuries.

Typical curves for the translational friction coefficients as a function of time for sport shoes on sport surfaces (see figure 14.5) show an initial maximum, which occurs immediately before the two surfaces begin to slide. This first maximum indicates the static friction coefficient. A subsequent steady state period describes the sliding of the shoe on the surface and corresponds to the dynamic translational friction coefficient.

Ranges for coefficients of static and dynamic friction for specific shoe-surface combinations have been reported in the literature and in technical reports (see table 14.1).

Frictional characteristics, and consequently friction coefficients, may be drastically altered by changes in the surface characteristics. For example, friction coefficients on a dusty floor may be less than 50% of those on a clean floor.

Rotational Friction

Rotational friction is characterized by the moment of rotation or the friction moment M for which the following formula applies (Schläpfer, Unold, and Nigg 1983):

Table 14.1 Ranges of Measured Static and Dynamic Translational Friction Coefficients for Sport Shoe and Sport Surface Combinations

Shoe	Surface	μ_{trsta}	μ_{trdyn}
Jogging	Felt carpet	1.0–1.2	0.9–1.1
	PVC	0.8–0.9	0.8–0.9
	Clay	0.4–0.5	0.3–0.5
	Asphalt	0.7–0.8	0.6–0.8
	Artificial track surface	0.7–1.1	0.7–1.0
Tennis	Artificial surface	0.6–1.7	0.6–1.4
	Artificial surface with granules	0.4–0.6	0.4–0.6
	Artificial grass for tennis	0.7–2.2	0.7–1.8
Football	Artificial turf	1.1–1.6	1.0–1.5
	Sand-filled turf	1.0–1.2	1.0–1.2
Baseball	Artificial turf	1.1–1.5	1.0–1.2
Basketball	Wood	1.0–1.5	1.0–1.3

$$M_{rot} = \mu \cdot \int_{Area} p(r, \theta) \cdot r^2 \, dr d\theta \qquad 14.13$$

where

M_{rot} = moment of rotation or frictional moment,

μ = dynamic translational friction coefficient,

p = normal pressure between the two surfaces, and

r, θ = polar coordinates.

The rotational frictional properties of two surfaces (shoe and surface) can be tested by determining the moment of rotation between the two surfaces during a rotation of one surface with respect to the other one. The quantity of interest, therefore, is M_{max} = maximal moment of rotation.

Sometimes, a *rotational friction coefficient* is used, which is defined as (Stucke, Baudzus, and Baumann 1984)

$$\eta = \frac{M_{rot}}{F_{vert}} \qquad 14.14$$

where

η = quantity describing the resistance to rotation for a vertical loading,

M_{rot} = moment of rotation or frictional moment, and

F_{vert} = vertical force.

For an actual movement with no rotational sliding or slipping the equation changes to

Table 14.2 Maximal Moments of Rotation for Tennis Shoes on Five Different Tennis Surfaces

Shoe	Surface	M_{max} (Nm)
Tennis	Artificial hard court I	16.2–20.1
Tennis	Artificial hard court II	15.8–18.9
Tennis	Artificial hard court III	17.9–24.0
Tennis	Artificial grass	17.1–21.2
Tennis	Artificial with loose granules	9.5–12.3

$$\eta \geq \frac{M_{rot}}{F_{vert}}. \qquad 14.15$$

In analogy to the translational friction coefficient one distinguishes between static and dynamic resistance to rotation.

Ranges for coefficients of maximal moments of rotation measured with subjects during a 180° turn for specific shoe and surface combinations for nine different tennis shoes from three different companies and five different tennis surfaces are listed in table 14.2.

The results illustrated in this table indicate that the differences in rotational resistance between the shoes are smaller than the differences between the surfaces.

Biological Responses

The human body needs force stimuli for the healthy development of its muscles, tendons, ligaments, bones, and cartilage. The biochemical, physiological, and mechanical reactions of biological tissue to force stimuli are defined as *biological responses.*

Humans lying in bed for several weeks loose not only most of their muscle strength but substantial bone mass, cartilage integrity, and ligament and tendon strength. Forces acting on and in the human body, therefore, are something necessary and often have positive effects on tissue development. Forces experienced during common daily activities such as walking are expected to improve the strength of our locomotor system. However, negative effects for the structures of the human body may occur when the local forces, the stresses, are excessive. In extreme situations these negative effects include tears and fractures. Negative effects occur frequently due to repeated overloading. This *wear and tear* is similar to the wear and tear of a nonbalanced wheel of a car. The single force is not excessive. However, the fact that the same location is stressed and strained without sufficient recovery is responsible for an eventual breakdown. The biological responses will be discussed extensively in chapters 15 and 16.

The Force System Analysis

This section concentrates on the description of the mechanical behavior of a system of rigid bodies representing a biological system. One specific approach of analyzing the mechanical behavior of a system of rigid bodies (force system analysis) is presented and discussed. Understanding the definitions presented and the steps involved in the force system analysis will enable one to solve mechanical questions in a systematic way. However, the proposed approach is not the only possible solution for a force system analysis.

The analysis of a force system includes the following components:

- Mechanical system of interest
- Assumptions
- Free body diagram
- Equations of motion
- Mathematical solution

The following paragraphs discuss the listed steps individually. However, it should be kept in mind that every step is required for the successful solution of a mechanical problem.

The Mechanical System of Interest

A mechanical system is a specific collection of particles or rigid bodies representing the body of inter-

est. For example, when analyzing the flight path of a javelin in the air, the system of interest is the javelin. When determining the forces between the foot of a runner and the ground (the ground reaction forces) the system of interest could be the whole runner, the foot of the runner, or the ground. When determining the elbow joint forces during a tennis stroke the system of interest could be the lower arm, the lower arm and the hand, or the upper arm, to list just a few possibilities. In general, to determine forces acting on a body for a specific question, several different systems of interest could be used.

Conceptually, the system of interest must always be a structure on which the force of interest acts as an external force (external to this structure). The system of interest may be determined using the following steps:

1. A sketch of the actual situation is drawn that includes the location for which the force should be determined. For example, when interested in the contact force in the knee joint one may want to draw the total body of an athlete. However, it would also be appropriate to draw just the support leg for which the knee joint force should be determined.

2. The drawn body is subdivided into two parts. The separation is made at the location where the force should be determined. For example, when determining the force between tibia and femur (the knee joint) the human body or the leg is divided at the knee joint into two parts, the lower leg and the rest of the body.

3. The system of interest is part or the whole of one of the two body parts that have been separated from each other. For example, the system of interest in the above example could be the tibia, the lower leg, the femur, or the rest of the body.

For this text the following convention is applied: the system of interest is drawn with solid lines. Other parts of the total body (e.g., human body, ground, shoe, etc.) are not drawn at all.

The selection and definition of the system of interest is an important step in the development of a free body diagram and for the force system analysis. Many errors in the solution of a mechanical problem start right at this point: the system of interest has not been defined properly.

Example

One wants to determine the contact force in the hip joint of the supporting leg during running for a human being. First, a schematic sketch of the runner is made. Second, the drawn sketch of the runner is divided into two parts by cutting at the hip joint. Third, the system of interest is determined. Various segments of the human body may be selected as the system of interest for the same question. For instance, to determine the forces in the hip joint, possible systems of interest include (a) the supporting leg, (b) the supporting femur, (c) the pelvis, and (d) the rest of the body. The selection depends on the knowledge one has about the total system, the mathematical skills available to the scientist, the experimental setup, and the input variables, which can be determined experimentally. The selection of the system of interest often influences the elegance and ease of the mathematical solution.

Assumptions

Once the system of interest is defined one must make assumptions about those aspects that are important and those that are not. Specifically, one must make these assumptions:

- Decide whether the problem is a one-, two- or three-dimensional problem.
- Decide which external forces (e.g., tendon forces, ligament forces) to include/exclude from further consideration.
- Decide the magnitude and direction of possibly known forces.
- Decide about material and structural characteristics of the system of interest.
- Consider other aspects of importance.

The rules that govern the selection of appropriate assumptions are difficult to define. The development of a good set of assumptions depends on the experience and the "feeling" of the researcher. As a simple general rule one may suggest the following: Define what is really important mechanically, and make the assumption that only forces due to those structures are acting.

Free Body Diagram

The development of the free body diagram for the analysis of mechanical and biomechanical problems has taken decades. The most important steps in the development of the free body diagram have been described earlier (Dempster 1961). A free body diagram consists of

- a sketch of the system of interest,
- a representation of all the external forces and moments acting on the system of interest, and
- a reference frame (coordinate system).

Sketch of the System of Interest

The sketch of the system of interest includes only the system of interest and nothing else. If, for instance, the system of interest is the leg, only the leg is drawn. The ground, for instance, or the hip is not drawn. It is important to draw the sketch so that the important geometrical aspects of the system of interest are maintained. For the example where the force in the hip joint is to be determined while standing on one leg, using the leg as the system of interest, the foot should be positioned so that it is about below the center of mass of the whole body. For the example where the force in the ankle joint during heel landing in heel-toe running is to be determined, with the foot as the system of interest, the heel must be drawn lower than the toe. With respect to the shape, it may be perfectly acceptable to simplify the foot as a triangle and the leg as a parallelogram. Furthermore, in this simplified approach it is always assumed that the systems of interest are rigid bodies.

External Forces and Moments

Two types of forces are considered in free body diagrams: contact and remote forces. A *contact force* is a force resulting from physical contact between two objects. An example of a contact force is the force between tibia and femur in the knee joint. A *remote force* is a force not resulting from physical contact, which one object exerts on another. An example of a remote force is the gravitational force.

Contact forces (moments) are forces (moments) caused through the direct contact of the system of interest with other bodies. Examples of contact forces include the following:

- Joint contact forces
- Muscle-tendon forces
- Ligamentous forces
- Forces between the hand and an object
- Forces between the foot and the ground
- Air or water resistance forces

The most important (and typically the only) remote force used in biomechanical force system analyses is the weight of the system of interest. Note that the weight drawn in the free body diagram is the weight of the system of interest and not the weight of the total body! If, for instance, the foot has been selected as the system of interest, the external remote force for this system of interest is the weight of the foot. One does not have to worry about the weight of the rest of the body. It is included implicitly in the forces and moments drawn at the ankle joint and in the ground reaction force.

Often, forces are summarized into *resultant forces*. The weight of the different parts of the hand is typically summarized into the resultant weight of the hand. The various local forces acting between the ground and the foot during walking are typically summarized into the ground reaction force. In general, one resultant joint force and one resultant joint moment can represent any loading situation with respect to a specific point (e.g., a joint center). The resultant joint force is the vector sum of all forces transmitted by muscle-tendon units, ligaments, bones, and soft tissue across the intersegmental joint. The *resultant joint force* is also called *resultant intersegmental joint force*.

Mathematically, the resultant joint force is defined as

$$\mathbf{F} = \sum_{i=1}^{m} \mathbf{F}_{muscle} + \sum_{j=1}^{l} \mathbf{F}_{lig}$$

$$+ \sum_{k=1}^{n} \mathbf{F}_{bone} + \sum_{p=1}^{s} \mathbf{F}_{other} \qquad 14.16$$

where

$\mathbf{F}$ = resultant intersegmental joint force,

$\mathbf{F}_{muscle}$ = forces transmitted by muscles,

$\mathbf{F}_{lig}$ = forces transmitted by ligaments,

$\mathbf{F}_{bone}$ = bone to bone contact forces, and

$\mathbf{F}_{other}$ = forces transmitted by soft tissue, for example.

Note that the resultant joint force is an abstract quantity that is not related to the actual force measured between two bones in a joint (the bone to bone force). The resultant joint moment is the net moment produced by all intersegmental forces with lines of actions that cross the joint. The *resultant joint moments* are also called *resultant intersegmental joint moments*. Resultant joint moments may be subdivided into resultant muscle joint moments, resultant ligament joint moments, resultant bony contact force moments, and other resultant joint moments. The resultant joint moment only reflects the net effect of agonist and antagonist muscles. For that reason, the resultant joint moment is also called *net joint moment*.

In the context of the free body diagrams one can work with actual forces or with resultant forces. These two approaches are briefly discussed.

Actual forces and moments approach.

This approach uses forces and moments that could be actually measured if appropriate transducers were to be used. It starts with the definition of the force-transmitting elements included in the analysis. Subsequently, all actual forces and moments are drawn into the free body diagram. They include the following:

- Remote forces (body weight).
- Contact forces (forces due to adjacent bones, forces due to adjacent ligaments, forces due to adjacent muscle tendon units, reaction forces due to contact with other external bodies, e.g., ground reaction force).
- Frictional joint moments (moments due to friction in the joints). These are usually assumed to be very small and, thus, are typically neglected.

The following is an example of the actual forces and moments approach.

Problem: Draw the FBD that allows determining the force in the Achilles tendon for a person standing on the forefoot of one leg using the actual forces and moments approach.

System of interest: Foot
Assumptions:

- The foot is a rigid structure.
- The foot has one idealized joint, the "ankle joint," which is responsible for plantar- and dorsiflexion between foot and leg.
- The structures responsible for contact forces are the Achilles tendon, the tibia at the ankle joint, and the ground.
- All other contact forces are neglected.
- There is no friction in the ankle joint.
- The weight of the foot can be neglected.
- The problem can be solved two-dimensionally.
- No moments act at the "toe" and the "ankle joint."

The "actual forces and moments approach" is advantageous for simple applications as illustrated in figure 14.6. This approach can be used to determine the order of magnitude of internal forces. It has the advantage that one sees the "actual forces" in the drawing and that one can relate the drawing to the real situation, recognizing forces acting in selected structures. This approach is sufficient for many simple applications and the reader may want

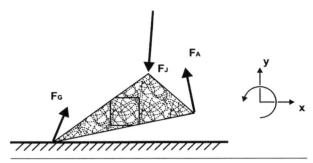

Figure 14.6 Example of a free body diagram estimating the force in the Achilles tendon for a person standing on the forefoot of one leg using the actual forces and moments approach.

to skip the next section if he or she is not interested in complex force system analyses. However, for applications with more complex force distributions, this approach may not be appropriate and can be replaced by another approach, the "resultant forces and moments approach".

Resultant forces and moments approach.

This approach uses equipollent resultant forces and moments (Andrews 1974). The free body diagram includes the following:

- Remote forces (body weight acting at the center of mass)
- Resultant forces and resultant moments

The resultant forces and resultant moments include resultant joint forces and moments and resultant surface forces and moments. For the resultant joint forces and moments, the complicated force distributions at the distal and proximal joints of the segment of interest are replaced by the equipollent resultant joint forces and moments. For the resultant surface forces and moments acting on the system of interest, the sometimes complicated force distributions acting on the segment surface are replaced by an equipollent resultant surface force and moment acting at some arbitrarily but appropriately located point.

The calculated resultant joint forces and moments do not correspond to actual forces and moments and cannot be measured with appropriate transducers as in the previous approach. They are abstract quantities that are often not used as final results, but they are used as input into a second step, the distribution of these forces and moments to the specific structures (ligaments, tendon, bone) that cross a joint.

The following is an example of the resultant forces and moment approach.

Problem: Draw the FBD that allows the determination of the resultant joint force and moment in the ankle joint (which can be used to calculate the force in the Achilles tendon) for a person standing on the forefoot of one leg using the resultant forces and moments approach.

System of interest: Foot
Assumptions:

- The foot is a rigid structure.
- The foot has one idealized joint, the "ankle joint," which is responsible for plantar- and dorsiflexion between foot and leg. The foot does not perform any inversion-eversion or abduction-adduction movements.
- The weight of the foot can be neglected.
- The problem can be solved two-dimensionally.

The resultant forces and moments approach (see figure 14.7) is appropriate for situations in which the complexity of the acting forces prohibits the drawing of a simple diagram, and the subsequent calculations favor a two-step approach (step 1: resultant joint moments and forces, step 2: distribution of these forces and moments to structures crossing the joint).

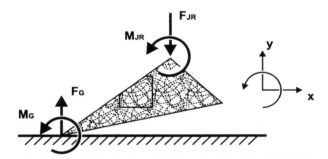

Figure 14.7 Example of a free body diagram estimating the force in the Achilles tendon for a person standing on the forefoot of one leg using the resultant forces and moments approach.

The Reference Frame (Coordinate System)

Each free body diagram must include a reference frame (coordinate system) that defines the direction of the positive axes for translation and rotation. Often the following convention for a Cartesian coordinate system is used:

x = a-p direction.
y = vertical direction.
z = mediolateral direction.

Equations of Motion

The equations of motion are written for translation and rotation. The equations for translation are Newton's second law for mass points or Euler's principle of the motion of the mass center for bodies of finite extent. The equations for rotation use the net moment, the moment of inertia, and the angular acceleration of the body relative to the inertial reference frame R.

The following is a set of equations for a 2-dimensional movement description:

Translation

$$m \cdot a_x = F_{x1} + F_{x2} + \ldots + F_{xn}$$
$$m \cdot a_y = F_{y1} + F_{y2} + \ldots + F_{yn}$$

14.17

Rotation

$$I_z \cdot \alpha_z = M_{z1} + M_{z2} + \ldots + M_{zn}$$

14.18

In the equations

m = mass of the system of interest,

a_x = acceleration of the center of mass of the system of interest in the x direction,

a_y = acceleration of the center of mass of the system of interest in the y direction,

F_{ix} = force component i acting on the system of interest in x direction,

F_{iy} = force component i acting on the system of interest in y direction,

I_z = moment of inertia of the system of interest about an axis perpendicular to the x-y plane through the center of mass of the system of interest,

α_z = angular acceleration of the system of interest with respect to its center of mass, and

M_{zi} = moment produced by the force F_I with respect to the center of mass of the system of interest.

Mathematical Solution

The equations of motion have "n" unknowns. The number of equations "m" can equal, or be smaller or larger than, the number of unknowns. These special cases are

equal, m = n, n determined,

smaller, m < n, n underdetermined,

larger, m > n, n overdetermined.

If the number of unknowns and the number of equations are equal, the system has a unique solution. If not, appropriate mathematical procedures must be applied to find a solution. Extensive descriptions of these approaches can be found elsewhere (Nigg and Herzog 1998).

Order of Magnitude of Forces

Results for external active and impact forces have frequently been reported in the literature. The reported ranges are illustrated in figure 14.8. Maximal external active forces during normal human locomotion are typically below 4 BW. Maximal external impact forces in walking, running, or court games are typically below 4 BW. However, maximal external impact forces can exceed 10 BW in movements such as take-off for a jump, landing in gymnastics, or during triple jump activities.

The external forces produce internal forces that can be high, depending on the structure of interest. The internal forces in the Achilles tendon during take-off in sprinting, for instance, can easily be 8 to 10 BW. The forces in the hip joint during standing on one leg or walking are about 3 to 6 BW. The lever of the external force with respect to the joint about which it acts primarily determines the magnitude of the internal forces.

The forces summarized in figure 14.8 are resultant forces. They are not conclusive indicators of the local stresses in a biological structure responsible for possibly positive or negative effects on that structure. To understand internal forces, the external forces must be combined with the geometry of the given situation to provide the actual internal loading situation in a specific structure of interest (force system analysis). For example, an external force of 2,000 N acting at the heel of the foot during landing produces a force in the ankle joint that is about 4,000 N. The same external force of 2,000 N acting at the ball of the foot during take-off produces a force in the ankle joint that is about 8,000 N (try to calculate these results using the force system approach).

Another aspect to be considered when assessing loading in a biological structure is the local force distribution, the local stress. This aspect has not been discussed in this text. However, for readers interested in more detailed load analysis the determination of local stresses is important. Further theoretical and practical information about this approach can be found elsewhere (Nigg and Herzog 1998). One selected example illustrates the importance of this consideration.

A joint force of 3,000 N acting perpendicular to an idealized tibiofemoral joint with a flat contact area of 30 cm^2 produces an average stress of

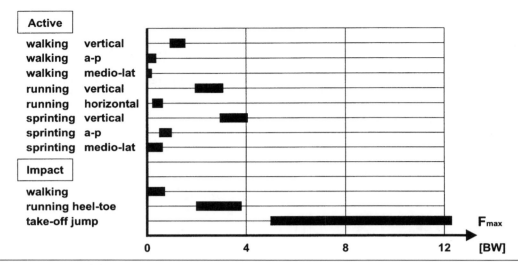

Figure 14.8 Magnitude of the maximal external active and impact forces during walking, running, sprinting, and jumping measured between the foot and the ground with a force plate.

Reprinted, by permission, from B.M. Nigg, 1988, "The assessment of loads acting on the locomotor system in running and other sport activities," *Seminars in Orthopaedics* 3: 197-206.

$$\sigma_{av} = \frac{force}{area} = \frac{3,000 \, N}{30 \, cm^2} = 1 \, MPa. \qquad 14.19$$

The same loading situation for a noncongruent joint with a contact area of only 3 cm² produces an average stress of

$$\sigma_{av} = \frac{force}{area} = \frac{3,000 \, N}{3 \, cm^2} = 10 \, MPa. \qquad 14.20$$

The results of these two examples illustrate that changes in geometry may change forces in joints, tendons, and ligaments by 100 to 1,000%.

Summary

Force is a quantity that cannot be defined. However, effects produced by forces can be described. Forces can produce biological and mechanical effects. The major biological effects of forces include changes in the development of biological tissue and transportation of nutrients through the human body. Muscles may be strengthened, bone may grow due to force stimuli, blood moves through the arteries and veins. The major *mechanical effects* produced by forces include the ability to *accelerate a mass* (e.g., when kicking a football or throwing a javelin) and the ability to *deform a material* (e.g., to compress a heel pad or to break a bone). The literature on forces acting on the locomotor system distinguishes between "impact" and "active" forces. Forces acting on a biological structure can produce translation and rotation. Impact forces are forces due to the collision of two objects. Impact forces in human locomotion are forces resulting from the collision of the foot with the ground with a force maximum earlier than 50 ms after first contact of the two objects. Active forces in human locomotion are forces that are completely controlled by muscular activity. To determine forces acting on a system of interest a procedure called force system analysis can be applied. It includes the determination of the mechanical system of interest, assumptions, a free body diagram, equations of motion, and a mathematical solution.

Forces acting on and within the human body can be quite substantial. Maximal external active forces during normal human locomotion are typically below 4 BW. Maximal external impact forces in walking, running, or court games are typically below 4 BW. However, maximal external impact forces can exceed 10 BW in movements such as take-off for a jump, landing in gymnastics, or during triple jump activities.

References

Andrews, J.G. 1974. Biomechanical analysis of human motion. *Kinesiology IV*, 32–42. Washington, DC: American Association for Health, Physical Education, and Recreation.

Bates, B., James, S.L., Osternig, L.R., Sawhill, J., and Hamill, J. 1982. Effects of running shoes on ground reaction forces. In *Biomechanics VII*, ed. A. Morecki and K. Fidelus. Baltimore: University Park Press.

Bergman, G., Kniggendorf, H., Graichen, F., and Rohlmann, A. 1995. Influence of shoes and heel strike on the loading of the hip joint. *J. Biomech.* 28: 817–27.

Bonstingel, R.W., Morehouse, C.A., and Niebel, B.W. 1975. Torques developed by different types of shoes on various playing surfaces. *Med. Sci. Sports Exerc.* 7: 127–31.

Cavanagh, P.R., and Lafortune, M.A. 1980. Ground reaction forces in distance running. *J. Biomech.* 13: 397–406.

Cavanagh, P.R., Williams, K.R., and Clarke, T.E. 1981. A comparison of ground reaction forces in walking barefoot and in shoes. In *Biomechanics VII,* vol. B. Ed. A. Morecki, K. Fidelus, K. Kedzior, and A. Wit, 151–56. Baltimore: University Park Press.

Chu, M.L., Yasdani-Ardakani, S., Gradisar, I.A., and Askew, M.J. 1986. An in vivo simulation study of impulsive force transmission along the lower skeletal extremity. *J. Biomech.* 19: 979–87.

Clarke, T.E., Frederick, E.C., and Cooper, L.B. 1983. Effects of shoe cushioning upon ground reaction forces in running. *Int. J. Sports Med.* 4: 247–51.

Cole, G.K, Nigg, B.M., Fick, G.H., and Morlock, M.M. 1995. Internal loading of the foot and ankle during impact in running. *J. Applied Biomech.* 11: 25–46.

Dempster, W.T. 1961. Free body diagrams as an approach to the mechanics of human posture and motion. In *Biomechanical studies of the musculoskeletal system,* ed. F.G. Evans, 81–135. Springfield, IL: Thomas.

Denoth, J. 1986. Load on the locomotor system and modeling. In *Biomechanics of running shoes,* ed. B.M. Nigg, 63–116. Champaign, IL: Human Kinetics.

Falsetti, H.L., Burke, E.R., Feld, R.D., Frederick, E.C., and Ratering, C. 1983. Hematological variations after endurance running with hard- and soft-soled shoes. *Phys. Sportsmed.* 11(8): 118–27.

Frederick, E.C., Hagy, J.L., and Mann, R.A. 1981. Prediction of vertical impact force during running [Abstract]. *J. Biomech.* 14: 498.

Frederick, E.C., Clarke, T.E., and Hamill, C.L. 1984. The effect of running shoe design on shock attenuation. In *Sports shoes and playing surfaces: Their biomechanical properties,* ed. E.C. Frederick, 190–98. Champaign, IL: Human Kinetics.

Hamill, J., Bates, B.T., Knutzen, K.M., and Sawhill, J.A. 1983. Variations in ground reaction force parameters at different running speeds. *Human Move. Sci.* 2: 47–56.

Heinonen, A., Kannus, P., Sievänen, H., Oja, P., Pasanen, M., Rinne, M., Uusi-Rasi, K., and Vuori, I. 1996. Randomized controlled trial of effect of high impact exercise on selected risk factors for osteoporotic fractures. *Lancet* 348: 1343–47.

Lafortune, M.A., and Hennig, E.M. 1992. Cushioning properties of footwear during walking: Accelerometer and force platform measurements. *Clin. Biomech.* 7: 181–84.

Lafortune, M.A., Hennig, E.M., and Lake, M.J. 1996. Dominant role of interface over knee angle for cushioning impact loading and regulating initial leg stiffness. *J. Biomech.* 29(12): 1523–29.

Light, L.H., McLellan, G.E., and Klenerman, L. 1980. Skeletal transients on heel strike in normal walking with different footwear. *J. Biomech.* 13: 477–80.

Luethi, S.M., Nigg, B.M., and Bahlsen, H.A. 1984. The influence of varying shoe sole stiffness on impact forces in running. In *Human locomotion III: Proceedings of the Canadian Society of Biomechanics meeting.* 65–66.

McClay, I.S., Robinson, J.R., Andriacchi, T.P., Frederick, E.C., Gross, T., Martin, P., Valiant, G., Williams, K.R., and Cavanagh, P.R. 1994. A profile of ground reaction forces in professional basketball. *J. Appl. Biomech.* 10: 222–36.

Milgrom, C., Giladi, M., Kashtan, H., Simkin, A., Chisin, R., Margulies, J., Steinberg, R., Aharonson, Z., and Stein, M. 1985. A prospective study of the effect of a shock absorbing orthotic device on the incidence of stress fractures in military recruits. *Foot and Ankle* 6: 101–4.

Nigg, B.M. 1978. Die Belastung des menschlichen Bewegungsapparates aus der Sicht des Biomechanikers (Load on the human locomotor system from the perspective of the biomechanist). In *Biomechanische Aspekte zu Sportplatzbelägen,* ed. B.M. Nigg, 11–17. Zürich: Juris Verlag.

Nigg, B.M. 1980. Biomechanische Überlegungen zur Belastung des Bewegungsapparates. (Biomechanical considerations on the loading of the musculoskeletal system). In *Die Belastungstoleranz des Bewegungsapparates,* ed. H. Cotta, H. Krahl, and K. Steinbrück, 44–54. Stuttgart: Thieme Verlag.

Nigg, B.M. 1988. The assessment of loads acting on the locomotor system in running and other sport activities. *Seminars in Orthopaedics* 3: 197–206.

Nigg, B.M. 1990. The validity and relevance of tests used for the assessment of sport surfaces. *Med. Sci. Sports Exerc.* 22: 131–39.

Nigg, B.M. 1997. Impact forces in running. *Curr. Opin. Orthoped.* 8: 43–47.

Nigg, B.M., Bahlsen, H.A. Luethi, S.M., and Stokes, S. 1987. The influence of running velocity and midsole hardness on external impact forces in heel-toe running. *J. Biomech.* 20: 951–59.

Nigg, B.M., and Herzog, W. Ed. 1994. *Biomechanics of the musculoskeletal system.* Sussex, UK: Wiley.

Nigg, B.M., and Herzog, W. Ed. 1998. *Biomechanics of the musculoskeletal system.* 2d ed. Sussex, UK: Wiley.

Nigg, B.M., and Segesser, B. 1988. The influence of playing surfaces on the load of the locomotor system and on injuries for football and tennis. *Sports Med.* 5: 375–85.

Ono, H. 1986. Evaluation method to assess the slipperiness of a sport surface (translation of Japanese title). *Trans. Arch. Ins. Japan* 359: 1–9.

Schläpfer, F., Unold, E., and Nigg, B.M. 1983. The frictional characteristics of tennis shoes. In *Biomechanical aspects of sports shoes and playing surfaces,* ed. B.M. Nigg and B.A. Kerr, 153–60. Calgary: University Printing.

Shorten, M.R., and Winslow D.S. 1992. Spectral analysis of impact shock during running. *Int. J. Sports Biomech.* 8: 288–304.

Stucke, H., Baudzus, W., and Baumann, W. 1984. On friction characteristics of playing surfaces. In *Sport shoes and playing surfaces,* ed. E.C. Frederick, 87–97. Champaign, IL: Human Kinetics.

Torg, J.S., and Quedenfeld, T.C. 1971. Effect of shoe type and cleat length on the incidence and severity of knee injuries among Philadelphia high school football players. *Res. Quart.* 42: 203–11.

Valiant, G.A., Cooper, L.B., and McGuirk, T. 1986. Measurements of the rotational friction of court shoes on an oak hardwood playing surface. *Proceedings of the North American Congress on Biomechanics.* 295–96.

Van Gheluwe, B., Deporte, E., and Hebbelinck, M. 1983. Frictional forces and torques of soccer shoes on artificial turf. In *Biomechanical aspects of sports shoes and playing surfaces,* ed. B.M. Nigg and B.A. Kerr, 161–68. Calgary: University Printing.

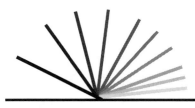

Chapter 15

Force Production in Human Skeletal Muscle

W. Herzog

The force produced by a skeletal muscle depends primarily on the size and structure, the amount of activation, and the contractile conditions (length and speed of contraction) of the muscle. Here, we discuss some aspects of muscular force production, particularly as they relate to voluntary contractions of in vivo human skeletal muscle. In discussing these issues, it is important to understand the process of muscle activation and to have a working knowledge of the molecular mechanism of force production. Therefore, these topics are also presented so that one may read and understand this chapter without having to refer to many external references.

Structure and Morphology of Skeletal Muscle

Skeletal muscle has a consistent structure. In a cross-sectional view (see figure 15.1), this structure is readily revealed. The entire muscle is typically surrounded by a layer of connective tissue called fascia, and further, by a connective tissue sheath known as the epimysium (epi = upon, on top of; mysium = related to muscle). The entire muscle is divided into muscle bundles, called fascicles, by a connective tissue sheath called perimysium (peri = around). Each muscle bundle contains numerous muscle fi-

bers. Muscle fibers are long, thin cells that are enclosed in a thin connective tissue called endomysium (endo = within). Muscle fibers/cells have a plasma membrane, which in combination with the basement membrane, is called sarcolemma (sarco = flesh; lemma = husk, outer shell). Muscle fibers, in turn, contain cylindrical subunits known as myofibrils, and myofibrils are made up of protein subunits called filaments, which are associated with the contractility of the muscle.

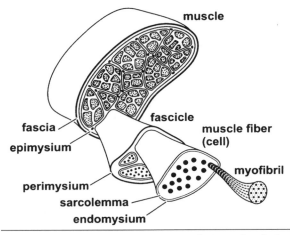

Figure 15.1 Schematic cross-sectional view of a skeletal muscle showing its structural organization.

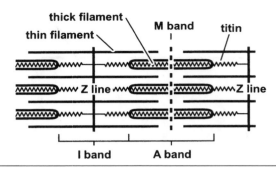

Figure 15.2 Schematic illustration of the longitudinal view of a sarcomere.

In a longitudinal view, myofibrils are arranged in a systematic way such that corresponding filaments within the fibrils line up in an almost perfect fashion. The difference in structure of the filaments gives the myofibrils and the fibers its typical striated pattern, which can be seen under the light microscope. The repeat unit in this pattern is called a sarcomere (see figure 15.2) (sarco = flesh; mere = segment). Sarcomeres are the basic contractile units of skeletal muscle. They are bordered by the so-called Z lines and they contain the thick and thin myofilaments, filaments that are made up primarily of the contractile proteins myosin and actin, respectively.

The Thick Filament

Thick filaments are typically located in the center of the sarcomere, and they cause the dark (anisotropic or A) band of the striation pattern in skeletal muscle (see figure 15.2). A thick filament is made up of myosin molecules. The exact packing of the myosin molecules is uncertain, as is the exact number of molecules that make up the thick filament (estimates range from about 180–300 molecules). Myosin molecules contain a long (156 nm) tail portion that consists primarily of light meromyosin and two identical short globular heads (19 nm) that consist primarily of heavy meromyosin (see figure 15.3). There is a flexible hinge region in the tail portion, which is approximately 43 nm away from the myosin heads. The heads of the myosin molecules extend outward from the thick filament in pairs (see figure 15.4). They contain a binding site for actin and an enzymatic site that catalyses the hydrolysis of adenosine triphosphate (ATP), which releases the energy required for muscular contraction. The heads of the myosin molecules are referred to as cross bridges because of their ability to connect the thick filament (from which they originate) to the thin filament. The cross bridges projecting from the thick filament form a helical array with a periodicity of about 42.9 nm and an axial interval of 14.3 nm (see

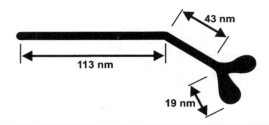

Figure 15.3 A myosin molecule with its two globular heads (19 nm) and its tail, which contains a flexible hinge approximately 113 nm from its end or 43 nm from its heads.

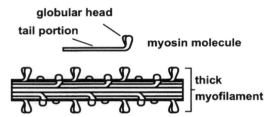

Figure 15.4 Schematic illustration of myosin molecules on a thick myofilament.

Adapted, by permission, from G.H. Pollack, 1990, *Muscles and molecules: Uncovering the principles of biological motion.* (Seattle: Ebner and Sons).

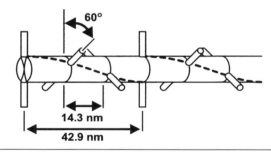

Figure 15.5 Schematic illustration of the longitudinal and rotational arrangement of myosin heads (cross bridges) on a thick myofilament.

Adapted, by permission, from G.H. Pollack, 1990, *Muscles and molecules: Uncovering the principles of biological motion.* (Seattle: Ebner and Sons).

figure 15.5). The myosin molecules in each half of the thick filament are of opposite polarity with their tail ends directed toward the center of the filament. Upon contraction, when the myosin heads attach to the thin filament, the myosin tends to pull the actin toward the center of the sarcomere.

The Thin Filament

The thin filament contains three primary proteins: actin, tropomyosin, and troponin. It appears light in the striation pattern of skeletal muscle (the so-called isotropic or I band) and is bisected by the Z lines of the sarcomere (see figure 15.2). A thin filament

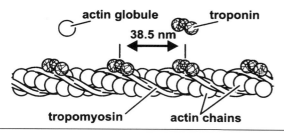

Figure 15.6 Schematic illustration of a thin filament and its primary proteins. Actin globules (G-actin) polymerize to form filamentous strands (F-actin), two of which combine as an α helix to form the major part of the thin filament.

contains about 300 to 350 actin monomers and about 40 to 50 molecules of tropomyosin and troponin (these numbers vary because thin filament length varies among species but is typically about 1.0 μm).

The backbone of the thin filament is two helical strands (F-actin) of polymerized actin monomers (G-actin), which are wound around each other with a pitch of about 73 nm (see figure 15.6). The axial separation of the G-actin monomers is about 5 to 6 nm. The actin filament is a polarized structure with its sense of polarization being opposite on either side of the Z line. Together with the opposite polarization of the thick filament in its two halves, these polarizations cause a force upon actin-myosin interaction, which pulls the thin filaments and Z lines toward the center of the sarcomere.

Tropomyosin is a long fibrous protein that lies in the grooves formed by the actin filaments (see figure 15.6). Together with troponin, tropomyosin performs a control function regulating thick-thin filament interaction. For each tropomyosin molecule there exists a troponin complex. A troponin complex consists of troponin I, troponin C, and troponin T. Troponin is regularly spaced along the thin filament (periodicity of about 38.5 nm), and it contains calcium attachment sites (troponin C), which form an important part in the control of thick-thin filament interaction.

Other Protein Filaments

Aside from the contractile proteins actin and myosin, and the regulating proteins tropomyosin and troponin, there is a variety of other proteins in muscle fibers with structural or passive functional properties. An important protein from a functional point of view is titin.

Titin is a giant protein that spans from Z line to M band in the sarcomere (see figure 15.2). Titin's exact role within the sarcomere remains to be determined; however, it is generally accepted that titin acts as a molecular spring that develops tension when sarcomeres are stretched. Its location has prompted the idea that titin might stabilize the thick filament in the center of the sarcomere. Such a stabilizing function of titin on the thick filament would explain why thick filaments are not pulled to one side of the sarcomere upon contraction when force balance on the two ends of the thick filament is not symmetrical.

Motor Units

In order to understand force production and the control of muscle force, it is necessary to know how voluntary control signals from the central nervous system reach a muscle. The nerve cells that carry action potentials to skeletal muscle fibers are called motoneurons. Motoneurons originate in the central nervous system (spinal cord) and their axons extend to junctions on skeletal muscle fibers. Each motoneuron innervates numerous muscle fibers, therefore a signal in a given motoneuron causes contraction of many fibers, all of the fibers that are innervated by the particular motoneuron. One motoneuron and all the muscle fibers it innervates comprises the smallest unit of force control, and this unit is called a motor unit.

The junction at which a nerve axon meets a muscle fiber is called the neuromuscular junction (see figure 15.7). At the neuromuscular junction, the

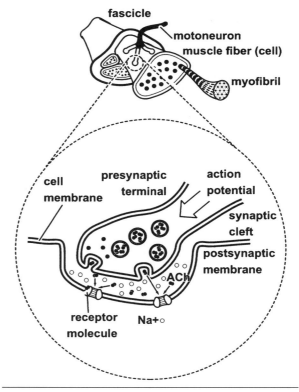

Figure 15.7 Schematic illustration of a neuromuscular junction.

signal (action potential) from the motor axon is transmitted to the muscle fiber via the release of acetylcholine (ACh) from synaptic vesicles located at the terminal end of the motor axon. Once released, acetylcholine diffuses across the gap between axon terminus and muscle fiber (the synaptic cleft), binds to receptor molecules of the muscle fiber membrane, and causes an increase in the permeability of the membrane to sodium (Na^+) ions. When the depolarization of the membrane caused by Na^+ exceeds a critical threshold, an action potential is generated and propagated along the muscle fiber. In order to prevent continuous stimulation of muscle fibers, acetylcholine is rapidly broken down into acetic acid and choline by acetylcholinesterase, which is located in the basement membrane over the end plate region.

Gross Structure of Skeletal Muscle

When considering force production in human skeletal muscle, it is important to realize that motor units may be arranged within a muscle in two basic ways: (1) along the longitudinal axis of the muscle (parallel fibers or fusiform arrangement) or (2) at a distinct angle relative to the longitudinal axis of the muscle (pennate arrangement). These structural arrangements of motor units within a muscle influence the properties of the muscle in a distinct way. Here, I would like to discuss some of these issues in a very schematic but, nevertheless, conceptually correct way. The issues considered deal with the force potential, work potential, and the range of active force production of a muscle. Force potential is defined as the maximal active force a muscle can produce, work potential is defined as the area under the force-length relationship of a muscle, and the range of active force production refers to the difference between maximal and minimal length at which a muscle can just produce active force.

The smallest contractile unit of muscle is a sarcomere. If we assume that sarcomeres in human skeletal muscle all have the same force, work, and active range potential, then it is immediately obvious that the force potential of a muscle depends on the number of sarcomeres that are arranged in parallel to each other, the work potential merely depends on the total number of sarcomeres within the muscle, and the range of active force depends on the number of sarcomeres arranged in series to one another.

As an example, let us consider two muscles of equal volume. Muscle A has parallel fibers; its fibers

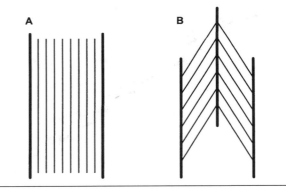

Figure 15.8 Schematic illustration of two muscles of identical volume but different fiber length. Muscle A is a parallel-fibered muscle and muscle B is a bipennate muscle. Muscles A and B have the same maximal work capacity, but muscle A has a larger absolute range of excursion and a lower peak isometric force at optimal length than muscle B.

are assumed to run from one end of the muscle to the other end (see figure 15.8). Muscle B is a bipennate muscle; its fibers run from a central tendon to aponeuroses at the surface of the muscle belly. For the sake of argument, let us assume that the fibers in muscle A are twice as long as those in muscle B. Assuming that a unit length of a fiber in either muscle takes up the same volume, muscle B has twice as many fibers as muscle A. The work potential of muscles A and B shown in figure 15.8 is the same, because they are of equal volume and, therefore, contain about the same number of total sarcomeres. Muscle B is about twice as strong as muscle A, because it has twice the number of fibers (sarcomeres) arranged in parallel compared to muscle A. Finally, muscle A can produce active force over a range twice that of muscle B because it has twice the number of sarcomeres arranged in series in its muscle fibers compared to muscle B.

Needless to say that the above example is just one of many possibilities on how volume, fiber length, and fiber arrangement may differ between two skeletal muscles. However, the above example should be sufficient to illustrate that two muscles that are of the same size (volume or weight) may have completely different contractile properties.

Excitation-Contraction Coupling

Excitation-contraction coupling refers to the events caused by excitation of the muscle fiber through a motoneuron and the subsequent events leading to contraction, that is, force production in the muscle fiber. At rest, a muscle fiber has an electrical charge inside the cell relative to the outside of about –90 mV. When an action potential reaches the nerve terminal, acetylcholine is released, and binding of acetylcholine to the end plate of the neuromuscular junction results in increased permeability of the sarcolemma to Na^+. Inward movement of Na^+ causes depolarization (membrane becomes less polarized), and when the membrane potential reaches a threshold value, additional Na^+ channels are opened, resulting in a rapid change in membrane potential to a peak value of about +40 mV. This reversal in charge, which is followed by a rapid repolarization, is called an action potential and is completed in less than 1 ms in mammalian skeletal muscle at 37° C. The action potential is propagated along the muscle fiber at a speed of about 5 to 10 m/s. This muscle fiber action potential travels along the sarcolemma and enters the "inside" of the cell through specialized invaginations of the sarcolemma, called transverse (T) tubules (see figure 15.9). Once the action potential enters the cell via the T-tubules, calcium ions (Ca^{2+}) are released from specialized storage sites, the sarcoplasmic reticulum (SR), in an (as of yet) unknown way. Ca^{2+} readily binds with the troponin C sites on the thin filaments, which presumably causes an increased affinity of the troponin C for the region of the troponin I that binds to actin, resulting in movement of tropomyosin, exposing the attachment site for the myosin cross bridges on the actin filament (see figure 15.10). Cross bridges now readily attach to actin, and through the breakdown of ATP into ADP plus a phosphate ion, the necessary energy is provided to cause the cross-bridge head to rotate and so tend to pull the thin filament past the thick filament. This action of the cross-bridge head causes shortening (if not resisted) and force production in the muscle fiber. At the end of the cross-bridge cycle, a new ATP molecule attaches to the cross bridge so that the cross bridge can release from its attachment site on actin, return to its original configuration, and be ready for a new cycle

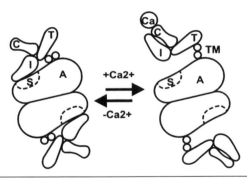

Figure 15.10 Schematic illustration of a cross section through a thin myofilament at the level of a troponin molecule. The arrangement on the left shows the situation in which there is no Ca^{2+} bound to troponin and the cross-bridge attachment site (S) is covered by troponin I (I). The arrangement on the right shows the situation in which Ca^{2+} is bound to troponin C (C) and the troponin I is removed from the cross-bridge attachment site, thereby allowing for cross-bridge attachment. A = actin, TM = tropomyosin, T = troponin T.

Reprinted with the permission of Cambridge University Press from A.M. Gordon, 1992, Regulation of muscle contraction: Dual role of calcium and cross-bridges. In *Muscular contraction*, edited by R.M. Simmons. (Cambridge, UK: Cambridge University Press), 163-179.

of attachment. This actin-myosin interaction through cross-bridge attachments continues as long as Ca^{2+} is bound to troponin C. Ca^{2+} disassociates from troponin C when free Ca^{2+} concentration in the myoplasm falls due to the active transport of Ca^{2+} back into the sarcoplasmic reticulum. The inhibition of cross-bridge attachment to actin is reestablished.

The Cross-Bridge Theory

There are several theories describing the molecular mechanism of force production in skeletal muscle. However, only one of these theories has been accepted as a working paradigm by muscle physiologists and muscle biophysicists: the cross-bridge theory.

The cross-bridge theory of muscular contraction was first formulated by Andrew Huxley (Huxley 1957) four decades ago, and although it has been modified and adapted, the basic principles outlined in Huxley's (1957) treatise are still the foundation of the current thinking. The cross-bridge theory is based on two conceptual ideas: (1) the idea that muscle contraction occurs by a relative sliding of the essentially rigid thick and thin myofilaments and (2) the idea that this sliding motion is produced by a cyclic interaction of sidepieces (cross bridges)

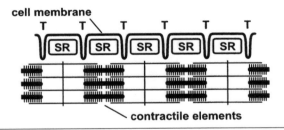

Figure 15.9 Schematic illustration of the arrangement of the transverse (T) tubules in a skeletal muscle fiber. SR = sarcoplasmic reticulum.

extending from one filament (thick) to the other filament (thin), and so creating force and contraction of muscle.

In the initial theory, Huxley (1957) defined two states of the cross bridge, an attached and a detached state. Later, the idea of several attachment states was introduced (Huxley and Simmons 1971) to account better for experimentally observed phenomena, particularly the fast transient force response observed following rapid length changes of muscle fibers. For reasons of simplicity, the 1957 model is discussed here; the interested reader is referred to more extensive reviews of the cross-bridge theory (e.g., Woledge, Curtin, and Homsher 1985).

The essential features of the 1957 cross-bridge theory are illustrated in figure 15.11. The thick (myosin) filament contains sidepieces M that oscillate because of thermal agitation about an equilibrium point that is determined by elastic connections of the M site to the thick filament backbone. These sidepieces can combine with specialized sites A on adjacent (actin) thin filaments. The connections between M sites and A sites are made spontaneously, but the breakage of the bond requires energy, which is provided by the hydrolysis of ATP. Once M-A connections are made, the force in the cross bridge is given by the force in the elastic element that connects the M site to the thick filament. In the original theory, this element was assumed to be linearly elastic, therefore the cross-bridge force f could be calculated as $f = k \cdot x$, where k is the stiffness of the elastic element, and x is the distance from the attachment site A to the equilibrium position of the elastic element.

In order to ensure that cross-bridge attachments produce shortening of the sarcomeres, M-A forma-

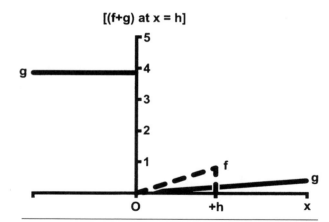

Figure 15.12 Rate function of attachment (f) and detachment (g) as a function of x, where x is the distance from the cross-bridge equilibrium point to the actin attachment site (see figure 15.11).

Adapted, by permission of Elsevier Science, from A.F. Huxley, 1957, "Muscle structure and theories of contraction," *Progress in Biophysics and Biophysical Chemistry* 7: 255-318.

tions had to occur (primarily) on one side of the equilibrium position shown in figure 15.11. This asymmetry of attachment was achieved by formulating rate constants for the attachment and detachment of cross bridges that were an asymmetric function of x (see figure 15.11), the distance from the equilibrium position of M to the next attachment site A on the thin filament. The rate constant for the formation of cross bridges was called f(x), the corresponding detachment constant was called g(x). The constant f(x) was only defined for positive values of x up to a maximal value h, therefore, any newly formed cross bridge could only produce a force causing sarcomere shortening (see figure 15.12).

The detachment function g(x) was defined for all values of x. For $0 < x \le h$, the values of g(x) were smaller than those of f(x), therefore ensuring that more cross bridges are formed than broken within a given period of time. For x < 0 and x > h, cross bridges can only detach, no attachment is possible. Energetically, each cross-bridge cycle (in the classic theory) is associated with the hydrolysis of one molecule of ATP.

For the derivation of the mathematical formulation of the cross-bridge theory, the reader is referred to the original source (Huxley 1957). Some of the assumptions underlying the cross-bridge theory are summarized here for completeness:

- The thick and thin myofilaments are (essentially) rigid.
- Length changes occur because of the relative sliding of thick and thin myofilaments.

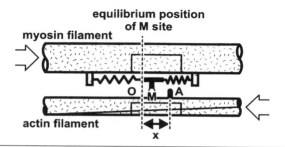

Figure 15.11 Schematic illustration of a mechanical model of cross-bridge interaction as proposed by Huxley (1957). M = cross-bridge head, A = myosin binding site on actin, O = cross-bridge equilibrium position.

Adapted, by permission of Elsevier Science, from A.F. Huxley, 1957, "Muscle structure and theories of contraction," *Progress in Biophysics and Biophysical Chemistry* 7: 255-318.

- Movement and force production are caused by cross-bridge connections between thick and thin myofilaments.

- Each cross-bridge cycle is associated with the hydrolysis of one ATP.

- Each cross bridge is independent of the remaining cross bridges, and its behavior is not dependent on the attachment history.

- Cross-bridge attachment/detachment is governed by functions (see figure 15.12) that depend on x, the distance between a thin filament binding site and the equilibrium position of the M site.

Muscle Force

Having considered the morphology and structural components of skeletal muscle, the events underlying excitation-contraction coupling, and the mechanisms of force production on the molecular level, we can now turn our attention to force production in an entire, in vivo, skeletal muscle. As pointed out earlier, for a given muscle (i.e., given size and structure) the force produced at any instant in time depends primarily on the activation, length, and speed of contraction. In the following, the relationship between these three factors and muscle force will be discussed. I should like to emphasize that these three factors are only the primary factors determining muscular force; they are not the only factors. Several other factors influence muscular force production. These include the time history of the contractile state, the involuntary inhibition of in vivo force production associated (presumably) with reflex pathways, and the interaction of the three primary factors mentioned above in dynamic, voluntary contractions, to mention just a few of the secondary factors that affect muscular force. For reasons of space limitation, not all of these secondary factors are considered here; however, aspects of history dependence of force production as well as the interaction of the three primary factors and their influence on force production are addressed briefly.

Force as a Function of Activation

Activation, or the active state of a muscle, has been defined in many ways. Investigators working on the muscle fiber level typically associate activation with the concentration of free calcium in the sarcoplasm or the calcium bound to the troponin C sites of the thin filament. Here, I would like to adopt a definition of activation that relates in a direct way to force production in a whole muscle. Therefore, activation is defined in this chapter by the number of active motor units and their corresponding firing frequency.

For illustration, let us assume we have a muscle consisting of 1,000 motor units. As discussed, these motor units are not uniform in size and properties. Small motor units (i.e., those with few muscle fibers) are typically of the slow twitch type. These motor units have a high oxidative capacity and are extremely fatigue resistant. Large motor units are stronger than the small motor units (because of the larger number of muscle fibers), and, in a mixed fiber muscle (most human skeletal muscles are of mixed fibers) they are primarily of the fast twitch type.

According to the size principle of motor unit recruitment (Henneman and Olson 1965), the small, slow motor units are recruited first in a graded muscular contraction. Then, with increasing force demands two things happen; an increasing number of motor units is recruited and the already recruited motor units increase their firing rates. This example illustrates that muscular force can be regulated in two ways: (1) by increasing the number of active motor units and (2) by increasing the firing frequency of the motor units. In the large muscles of the human arms and legs, it is typically assumed that both of these mechanisms are at work for isometric forces of up to about 60% of the maximal force. At about 60% of the maximal isometric force, it is assumed that most (if not all) motor units are recruited; therefore, the remaining 40% of force production required to reach the maximal isometric force is accomplished almost exclusively by increasing the firing rate of the active motor units. Figure 15.13 illustrates the effect of increasing stimulation rates of all motor units in the cat soleus muscle

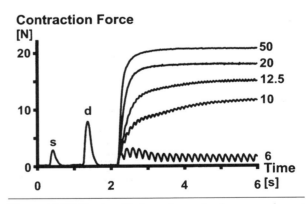

Figure 15.13 Contractile force in a cat soleus muscle as a function of the rate of stimulation of the soleus nerve. s = single twitch stimulation; d = doublet stimulation (i.e., two stimulation pulses separated by 8 ms); 6 Hz, 10 Hz, 12.5 Hz, 20 Hz, and 50 Hz stimulation of the soleus nerve.

(95–100% slow twitch fibers) on the corresponding force. As discussed above, an increase in the firing rate within the physiological range (i.e., about 5–50 Hz for the cat soleus) is associated with an increase in force.

The most common way to capture qualitatively the activation (as defined here) of a skeletal muscle in vivo is by measuring the electromyographic signal (Basmajian and de Luca 1985). Each action potential produced by a stimulating pulse results in a distinct electrical signal that can be recorded on the surface of the muscle. The more action potentials per muscle fiber and the more fibers that are active at a given time, the larger the electromyographic signal. Therefore, at a first glance it appears obvious that one can and should relate the electromyographic signal directly to the muscle force, particularly when realizing that the individual fiber action potentials add algebraically (i.e., linearly in time and space) to the total electromyographic signal. However, relating electromyographic signals and force during dynamic contractions has not been successful to date for a variety of reasons:

- The electromyographic signal of one fiber may look different in magnitude and shape than the signal from a different fiber in the same muscle because of the location of the recording electrodes relative to the fibers.

- The electromyographic signal and the corresponding force are distorted in time in a variable, nonlinear way.

- Because of the straight algebraic summation of fiber action potentials, there is a large amount of (statistically random) cancellation of positive with negative signals.

Despite these difficulties, electromyographic signals must be considered a good qualitative indicator of muscle activation, and, as a bonus, electromyographic signals can be recorded easily and noninvasively.

Force as a Function of Length

The fact that the force potential of a muscle depends on its length has been known for at least a century (Blix 1894). In 1966, Gordon, Huxley, and Julian performed the classic study in which they recorded the force of frog striated muscle fibers as a function of sarcomere length. These investigators showed that specific sarcomere lengths could be related to specific changes in the force potential of muscle fibers. The dependence of muscle force on muscle

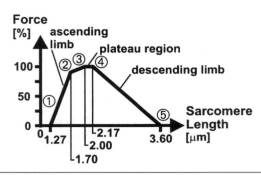

Figure 15.14 Sarcomere force-length relationship of frog skeletal muscle derived using sarcomere length control.

Adapted, by permission, from A.M. Gordon, A.F. Huxley, and F.J. Julian, 1966, "The variation in isometric tension with sarcomere length in vertebrate muscle fibres," *Journal of Physiology (London)* 184: 170-192.

length has been described by the force-length relationship.

Force-length relationships of whole muscles are obtained under precisely controlled experimental conditions; the muscle is held isometrically at a series of different lengths, the activation is maximal for voluntary and supramaximal for artificially activated contractions, and the relationship is obtained in the nonfatigued state. Figure 15.14 shows the sarcomere force-length relationship obtained by Gordon, Huxley, and Julian (1966) with selected, schematic sarcomere lengths corresponding to crucial events on the force-length trace. Starting with the longest sarcomere length (3.60 μm), it is observed that there is no active force production. The thick-thin filament arrangement is such that the two filaments do not overlap at this length. According to the cross-bridge theory (Huxley 1957), no cross-bridge formations between thick and thin filaments can occur at this length; therefore, there is no active force.

At a sarcomere length of 2.17 μm, force production is maximal, because thick and thin myofilament overlap is such that a maximal number of cross-bridge attachments can be made between the two filaments. At 2.0 μm, the overlap between thick and thin filaments is complete; however, no additional force is produced (as one might suspect), because there is no change in the number of cross bridges that can be engaged.

At progressively shorter lengths, from 2.0 to 1.7 μm, force decreases. The exact reason for this decrease in force is not known, but the following scenarios have been proposed: the thin filaments start overlapping and so interfere with one another; the thin filaments may make cross-bridge attachments on the opposite side of the half sarcomere, and therefore are being pushed by these cross-bridges rather than pulled; the lateral distance between thick and thin filaments may increase because of volume preservation, and so cause unfavorable cross-bridge kinetics compared to a more stretched state; and the free calcium concentration in the sarcoplasm may be reduced at short compared to long sarcomere lengths. Although researchers may not agree on the mechanism underlying the force-length relationship of skeletal muscle, there is unanimous acceptance of the fact that maximal isometric force of a muscle depends (often to a large degree) on its length.

Determining force-length relationships of individual human skeletal muscles in vivo is a difficult task for a variety of reasons: force (typically) cannot be measured directly, maximal effort voluntary contractions are hard to repeat in a consistent manner, and the length of the contractile elements of muscle (the muscle fibers or sarcomeres) is hard to determine. Nevertheless, several attempts have been made to obtain the force-length relationships of individual human skeletal muscles with some surprising results. These results are given here for the purpose of illustration and completeness; however, the reader is referred to the original sources for the methodological details and the precise interpretation of the results.

• Force-length relationships of in vivo human skeletal muscles show nonzero forces over a much larger range than one would expect based on the fiber length and the sarcomere geometry (i.e., the thick and thin filament lengths in human skeletal muscle) (Herzog and ter Keurs 1988).

• Only part of the force-length relationship of some muscles is used to cover the entire anatomic range of joint movement. For example, the human gastrocnemius appears to operate predominantly on the ascending part and the plateau region of the force-length relationship. The descending part is not utilized even at extreme joint configurations (Herzog, Read, and ter Keurs 1991). Interestingly, this result agrees conceptually with observations on the force-length relationship of the gastrocnemius in other mammals (e.g., the cat, Herzog et al. 1992; the striped skunk, Goslow and van de Graaf 1982).

• The force-length relationship appears to be plastic, that is, it may be changed by chronic, long-term training as it occurs in high performance athletes. For example, it has been shown that the force-length relationship of human rectus femoris is different between high performance runners and cyclists. In the runners, the force-length relationship has a positive slope (i.e., force increases with increasing muscle length), whereas in the cyclists, it has a negative slope (i.e., force decreases with increasing muscle length) (Herzog, Hasler, and Abrahamse 1991).

In summary, much of the isometric force potential in skeletal muscle can be explained by the force-length properties of the sarcomeres and the cross-bridge theory of muscular contraction. However, the force-length properties of entire muscle differ sufficiently from those of sarcomeres (and possibly those of isolated fibers); therefore, care must be taken when interpreting force-length properties across structural levels.

Force as a Function of the Rate of Change in Length

When a muscle shortens or is lengthened, the force of a maximally activated muscle will depend on the rate of change in muscle (or better, contractile element) length. The rate of change in muscle length is typically referred to as "velocity," and the relationship that exists between force and the rate of change in muscle length is known as the "force-velocity" relationship. This terminology is not quite correct because "force" and "velocity" are vector quantities, whereas the "force-velocity" relationship of muscles is a scalar property. For convenience, and for historical reasons, I will refer to the relation between force and the rate of change in muscle length as the force-velocity relationship; however, no vector relationship should be implied.

The fact that the maximal force a muscle can exert depends on the velocity of contraction has been known for at least 70 years (Gasser and Hill 1924). Probably the first formal description of the force-velocity relationship was given by Fenn and Marsh (1935), who described the relationship in terms of an exponential function.

Hill (1938) arrived at a description of the force-velocity relationship of frog striated muscle while performing experiments on the heat of shortening. Hill (1938) found that during shortening of a fully activated muscle, extra heat (heat exceeding that of a corresponding isometric contraction) was liberated. This extra heat was found to be proportional to the amount of shortening, that is, extra heat = ax, where x is the amount of shortening and "a" is a constant describing the heat produced (in excess of the isometric heat) per unit distance of shortening. Since the work done by an isotonically (i.e., constant force) contracting muscle is Px, where P is the force, the total energy produced by a muscle in excess of that observed isometrically is total extra energy = ax + Px = (P+a)x. The rate of extra energy liberation becomes (P+a) · dx/dt or (P+a) · v, where v is the velocity of shortening. Hill (1938) found that the rate of extra energy liberation was inversely and linearly related to the force of the muscle, therefore

$$(P + a) v = b (P_o - P) \qquad 15.1$$

where b is a constant defining the absolute rate of energy liberation and P_o is the isometric force.

Equation 15.1 may be solved for the muscle force P as a function of the velocity of shortening

$$P = \frac{P_o b - av}{b + v}. \qquad 15.2$$

The force-velocity relationship described by equation 15.2 represents a rectangular hyperbola with asymptotes P = −a and v = −b (see figure 15.15). When measuring the force and speed of shortening of a muscle simultaneously, equations 15.1 and 15.2 can be verified without heat measurements.

Hill's (1938) force-velocity relationship is only correct for maximally activated muscle, for isotonic shortening contractions, and at optimal length of the muscle. Such contractions likely do not exist in our daily movement repertoire. Nevertheless, Hill's equation is probably used more often than any other equation to describe the force of a muscle as a function of its contractile speed, even for submaximal contractions, under nonisotonic conditions, and far away from the optimal length. Under these various conditions, a rectangular hyperbola (see equation 15.2) may not adequately describe the relationship between force and speed of contraction, and the constants should not be interpreted with respect to energetics of muscle contraction.

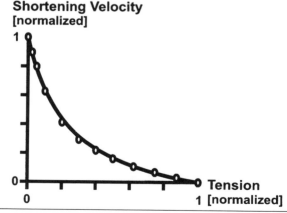

Shortening Velocity
[normalized]

Figure 15.15 Schematic force-velocity relationship of a skeletal muscle for shortening contractions.

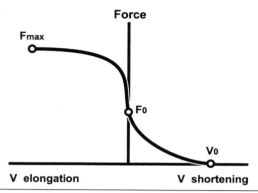

Figure 15.16 Schematic force-velocity relationship of a skeletal muscle for shortening and lengthening contractions.

For lengthening (eccentric) contractions, the force a muscle can exert exceeds the isometric force by a substantial amount (see figure 15.16). Eccentric forces increase rapidly as a function of the speed of lengthening, much faster than the forces decrease as a function of the speed of shortening. Therefore, the slope of the force-velocity relationship around the isometric point is said to be discontinuous. Whether this statement is correct is not known; however, muscle physiologists are in general agreement that the slope of the force-velocity relationship must change quickly (but not necessarily discontinuously) around the isometric point.

Maximal force values for in situ, eccentric contractions range from about 1.5 to 2.0 times the maximal isometric force. Interestingly, there is strong evidence that such high force values cannot be achieved during voluntary contractions. For example, Westing, Seger, and Thorstensson (1990) report that the maximal eccentric forces in the knee extensor muscles are about the same as the corresponding isometric forces. When superimposing a maximal electrical stimulation on the femoral nerve during knee extensor contractions, Westing, Seger, and Thorstensson (1990) found that the additional force produced by electrical stimulation beyond that obtained during voluntary contractions was much higher for the eccentric than the isometric contractions, indicating that there are neural pathways that are inhibited and do not allow for full activation of the knee extensors during eccentric contractions. Therefore, despite the fact that muscle tissue is inherently capable of producing much larger forces during eccentric than isometric contractions, the central nervous system appears to inhibit access to the full eccentric force potential during voluntary contractions, at least in human knee extensor muscles. Such an inhibition may be caused via reflex pathways to avoid injury to muscles and tendons.

Force as a Function of Activation, Length, and Speed

In the previous sections, we dealt with activation, length, and speed of muscular contraction independently. In real life, dynamic contractions occur with variable activation, length, and speed. Considering these three factors simultaneously is difficult, and probably the most important message regarding this topic is that there is no constitutive law for muscle that has been able to describe muscle force as a function of the above three parameters for all conceivable contractile scenarios.

The scope of this chapter does not allow me to go into the details on how one might combine the effects of activation, length, and speed to obtain muscle force. However, I would like to give a few examples to illustrate some of the difficulties and non-linearities associated with combining these factors.

Activation and Length

Rack and Westbury (1969) investigated the effects of activation and length on muscle force. Activation was produced by stimulation of five independent ventral root filaments, while length and force were recorded from the cat soleus. Force-length curves were obtained for rates of stimulation between 3 to 60 Hz, providing a large range of activation levels. Rack and Westbury (1969) found that the shape of the "force-length" relationship changed in a highly nonlinear way for different levels of activation (see figure 15.17), thereby pointing out the difficulty of

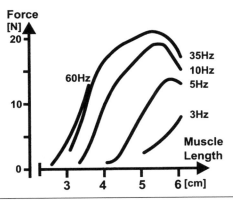

Figure 15.17 Force-length relationship of cat soleus muscle when stimulated supramaximally at different frequencies. The frequencies of stimulation are indicated for each curve in Hz.

Adapted, by permission, from P.M.H. Rack and D.R. Westbury, 1969, "The effects of length and stimulus rate on tension in the isometric cat soleus muscle," *Journal of Physiology (London)* 204: 443-460.

extrapolating force-length properties obtained under maximal activation to the corresponding submaximal properties.

Length and Velocity

Hill's (1938) characteristic equation (equations 15.1 and 15.2) for the force-velocity relationship of shortening muscle was derived for muscles operating at optimal length. Hill (1938) argued that the characteristic equation would always be correct if P_o (equation 15.2), the maximal isometric muscle force at optimal length, was replaced by the maximal isometric muscle force at the length of interest, $P_o(l)$. Therefore, equation 15.2 might be rewritten in a general form that should be correct for shortening contractions at any length:

$$P = \frac{P_o(l)\, b - av}{b + v}. \qquad 15.3$$

Calculating the maximal speed of shortening v_o from equation 15.3 by replacing v by v_o and setting the left hand side of equation 15.3 equal to zero (i.e., there is no external muscle force at the maximal speed of shortening), we obtain

$$v_o = \frac{P_o(l)\, b}{a}. \qquad 15.4$$

Since "a" and "b" are assumed to be constants for a given muscle, temperature, and contractile condition, it is obvious from equation 15.4 that the maximal speed of shortening v_o is linearly proportional to $P_o(l)$, the maximal isometric force of the muscle as a function of its length. Hill's (1938) idea of the generalizability of his equation to all muscle lengths was supported by others (Abbott and Wilkie 1953). However, it has been shown to be not correct. Edman (1979) demonstrated that the maximal speed of fiber shortening was independent of sarcomere length over a large range (i.e., a range sufficient to produce large changes in the values $P_o(l)$, and, therefore, by implication, the values for v_o), thereby disqualifying Hill's (1938) idea of the relationship between force, speed, and length. More likely, the relationship between force, speed, and length follows an equation of the form

$$v_o = \frac{P_o\, b - av}{b + v} \cdot f \qquad 15.5$$

where f represents the force of a muscle as a function of its length normalized to the absolute maximum value.

History Dependence of Force Production

For a given activation, length, and speed of contraction, muscle force also depends on the history of contraction. This history dependence of force production has been known and accepted for nearly half a century (Abbott and Aubert 1952). Nevertheless, history dependence of force production is not only not considered in the cross-bridge theory, it was actively excluded by Huxley (1957) by requiring that the probability of attachment and the amount of force production of cross bridges did not depend on the cross-bridge's contractile history.

The most frequently studied phenomenon of history dependence is the depression of force following shortening contractions. Force depression here refers to the fact that a muscle's isometric force at a given length is smaller following a shortening than during an isometric contraction. Although there has been much argument about the mechanism causing force depression following shortening, the issue is not resolved (Herzog and Leonard 1997). Recent experimental evidence from cat soleus suggests that the force depression following shortening is directly related to the amount of shortening (see figure 15.18) and to the mean force during the shortening phase (see figure 15.19). Based on these results and others, Herzog and Leonard (1997) sug-

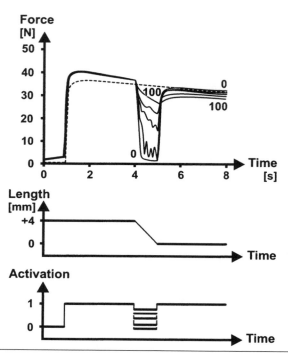

Figure 15.19 Force depression in cat soleus as a function of the amount of force (activation) during the shortening phase.

gested a new mechanism for force depression following muscle shortening, which (quantitatively) describes the known experimental observations in this area of research. The proposed mechanism is associated with a stress-related inhibition of cross-bridge attachments in the overlap zone between thick and thin myofilaments that is newly formed during shortening. The inhibition of cross-bridge attachments is postulated to be caused by the deformation of the thin myofilaments during contraction. The molecular details of this mechanism must be tested in the future.

Final Comments

To date, there are no constitutive laws that accurately describe force production of skeletal muscle for all contractile conditions. In a recent symposium on skeletal muscle, I heard a keynote lecturer state that such constitutive laws can never be derived, arguing that if one wants to know the force of a muscle for given contractile conditions, one must measure it experimentally. I am not quite as pessimistic as that particular lecturer is. I believe that forces in skeletal muscles may be calculated accurately in the near future based on the knowledge of just a few basic properties (maybe the force-length relationship and the maximal speed of shortening)

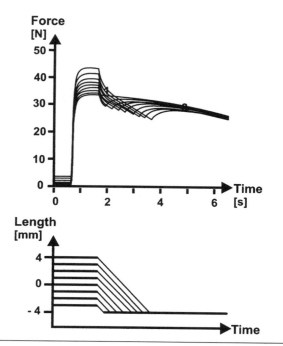

Figure 15.18 Force depression in cat soleus as a function of the magnitude of muscle shortening.

and an understanding of the detailed mechanisms of force production in entire skeletal muscle. It should be the goal of experimentalists to elucidate these mechanisms and the goal of theoreticians to implement these mechanisms into constitutive equations of muscle contraction.

References

Abbott, B.C., and Aubert, X.M. 1952. The force exerted by active striated muscle during and after change of length. *J. Physiol. (London)* 117: 77–86.

Abbott, B.C., and Wilkie, D.R. 1953. The relation between velocity of shortening and the tension-length curve of skeletal muscle. *J. Physiol. (London)* 120: 214–23.

Basmajian, J.V., and de Luca, C.J. 1985. *Muscles alive: Their functions revealed by electromyography.* Los Angeles: Williams & Wilkins.

Blix, M. 1894. Die Laenge und die Spannung des Muskels. *Skand. Arch. Physiol.* 5: 149–206.

Edman, K.A.P. 1979. The velocity of unloaded shortening and its relation to sarcomere length and isometric force in vertebrate muscle fibers. *J. Physiol. (London)* 291: 143–59.

Fenn, W.O., and Marsh, B.O. 1935. Muscular force at different speeds of shortening. *J. Physiol. (London)* 85: 277–97.

Gasser, H.S., and Hill, A.V. 1924. The dynamics of muscular contraction. *Proc. Royal Soc.* B96: 398–437.

Gordon, A.M. 1992. Regulation of muscle contraction: Dual role of calcium and cross bridges. In *Muscular contraction,* ed. R.M. Simmons, 163–79. Cambridge, UK: Cambridge University Press.

Gordon, A.M., Huxley, A.F., and Julian, F.J. 1966. The variation in isometric tension with sarcomere length in vertebrate muscle fibres. *J. Physiol. (London)* 184: 170–92.

Goslow, G.E., Jr., and van de Graaf, K.M. 1982. Hind limb joint angle changes and action of the primary ankle extensor muscles during posture and locomotion in the striped skunk (*Mephitis mephitis*). *J. Zool. (London)* 197: 405–19.

Henneman, E., and Olson, C.B. 1965. Relations between structure and function in the design of skeletal muscles. *J. Neurophysiol.* 28: 581–98.

Herzog, W., Hasler, E., and Abrahamse, S.K. 1991. A comparison of knee extensor strength curves obtained theoretically and experimentally. *Med. Sci. Sports Exerc.* 23: 108–14.

Herzog, W., and Leonard, T.R. 1997. Depression of cat soleus forces following isokinetic shortening. *J. Biomech.* 30(9): 865–72.

Herzog, W., Leonard, T.R., Renaud, J.M., Wallace, J., Chaki, G., and Bornemisza, S. 1992. Force-length properties and functional demands of cat gastrocnemius, soleus, and plantaris muscles. *J. Biomech.* 25(11): 1329–35.

Herzog, W., Read, L.J., and ter Keurs, H.E.D.J. 1991. Experimental determination of force-length relations of intact human gastrocnemius muscles. *Clin. Biomech.* 6: 230–8.

Herzog, W., and ter Keurs, H.E.D.J. 1988. A method for the determination of the force-length relation of selected in vivo human skeletal muscles. *Pflügers Archiv.* 411: 637–41.

Hill, A.V. 1938. The heat of shortening and the dynamic constants of muscle. *Proc. of Roy. Soc. Lon.* 126: 136–95.

Huxley, A.F. 1957. Muscle structure and theories of contraction. *Prog. Biophys. Biophys. Chem.* 7: 255–318.

Huxley, A.F., and Simmons, R.M. 1971. Proposed mechanism of force generation in striated muscle. *Nature* 233: 533–38.

Pollack, G.H. 1990. *Muscles and molecules: Uncovering the principles of biological motion.* Seattle: Ebner.

Rack, P.M.H., and Westbury, D.R. 1969. The effects of length and stimulus rate on tension in the isometric cat soleus muscle. *J. Physiol. (London)* 204: 443–60.

Westing, S.H., Seger, J.Y., and Thorstensson, A. 1990. Effects of electrical stimulation on eccentric and concentric torque-velocity relationships during knee extension in man. *Acta Physiol. Scand.* 140: 17–22.

Woledge, R.C., Curtin, N.A., and Homsher, E. 1985. *Energetic aspects of muscle contraction.* London: Academic Press.

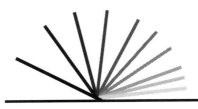

Chapter 16

Mechanical Effects of Forces Acting on Bone, Cartilage, Ligaments, and Tendons

D.D. Anderson, D.J. Adams, and J.E. Hale

Viewed within the context of human motion, the structural building blocks of the musculoskeletal system must perform at a certain functional level to allow and support movement without incurring injury. This is often considered the primary role of the musculoskeletal tissues. In this chapter we discuss the structural and material responses of bone, articular cartilage, ligaments, and tendons to external forces, both within the normal physiological range and at levels associated with traumatic tissue failure. In light of its active role in initiating movements of the body, skeletal muscle is not considered in this section (see chapter 15).

The Primary Concepts

Among the musculoskeletal tissues, the range of material and structural behavior is logically separated into hard and soft tissue response. Bone, by virtue of its predominantly crystalline mineral nature, exhibits fundamentally different behavior than soft tissues such as articular cartilage, ligament, and tendon. In the case of soft tissues, water and collagen content play a more central role in carrying and transmitting load, albeit to varying degrees among the different tissues.

Physical activity involves overcoming external forces by generating internal forces of muscle contraction. These forces are transmitted through individual musculoskeletal tissue structures in different ways according to their mechanical function and biological composition (i.e., form follows function). Bone, articular cartilage, ligament, and tendon exemplify the unique differentiation among musculoskeletal tissues to perform specific biomechanical functions in transmitting force and generating motion. Mineral crystals within bone tissue provide structural rigidity to transmit large forces efficiently without significant dissipation of energy. Suitably, the deformation of bone tissue is relatively small and primarily elastic. The composition and tensile stiffness of tendons reflect their role in generating motion by efficiently transmitting muscle contraction forces across joints. On the other hand, ligaments and articular cartilage experience relatively large deformations, dissipating energy through viscoelastic and poroelastic processes. This reflects the mechanical function of ligaments to restrict excessive motion, and of articular cartilage to distribute large forces across moving joints. A general understanding of the compositional differences between the musculoskeletal tissues elucidates the unique structure-function-property relationships of bone, articular cartilage, ligament, and tendon in generating and controlling movement.

Our discussion of the mechanical effects of forces acting on musculoskeletal tissues begins with a basic introduction to the characterization of mechanical properties. A special emphasis is placed upon distinguishing between *structural* and *material* properties, as this point is of great importance in understanding the fundamental relationship between structure and function of the musculoskeletal connective tissues. It is in this context that the composite, hierarchical structure that is the hallmark of biological materials is likewise introduced. The body works with relatively few building blocks to construct the structures that it needs to support movement. The economy and efficiency with which these tissues perform their respective duties is indeed remarkable.

Four specific musculoskeletal tissues are treated individually in the present chapter: bone, articular cartilage, ligament, and tendon. Bone is considered first, partly because of its unique hard tissue characteristics and partly because it has been the most extensively studied. The soft tissues are discussed subsequently, beginning with articular cartilage. Much progress has been made in elucidating the mechanical properties of articular cartilage in the past 15 to 20 years, attributed mostly to a greater understanding of the mechanical contribution of water to load support. Finally, the primarily tensile load-bearing soft tissues, ligament and tendon, are addressed. These two tissues share many common features at both the molecular and structural level, reflecting similarity in their contributions to support movement.

Characterizing Mechanical Properties

Mechanical properties provide a measure of a material's ability to resist deformation when subjected to externally applied forces. These properties can be determined experimentally by applying a known compressive, tensile, or shearing force or displacement to a material specimen at a constant rate and monitoring the resulting displacement or force, respectively. By plotting applied load as a function of the displacement, a representative load-displacement curve can be generated. The relationship between applied load and observed displacement is dependent upon not only the material being tested, but also the size and shape of the test specimen. For example, a metacarpal and a femur will exhibit dramatically different behavior when subjected to the same load, despite the fact that both are composed of bone. Characteristics derived from the load-displacement data are referred to as structural properties.

In discussing the tensile properties of ligaments and tendons, it is especially important to distinguish between the structural properties of the tissue-bone complex and the material properties of the tissue. The structural properties (e.g., stiffness, energy absorbed, ultimate load, and ultimate elongation) are measurements of the tensile behavior of the tendon or ligament as a functional unit: a composite bone-tissue complex of connective tissue substance and its insertion into bone. Direct experimental measurement of load (using a load cell) and deformation (based on test machine crosshead or clamp-to-clamp displacement) is relatively straightforward. Results of such tensile testing are frequently reported as load-displacement data. These structural properties are dependent on a number of parameters, including the material properties of the tissue substance, the geometry of the tissue (cross-sectional area, length, and shape), and the properties of the bone-substance and muscle-substance junctions.

By accounting for the geometry of a suitably chosen specimen, the material properties of the tissue itself can be determined from load-displacement data. This involves converting load and displacement data into stress and strain, respectively. For a uniformly distributed load, stress within a material is equal to the magnitude of the applied load divided by the cross-sectional area upon which it acts, expressed in units of Newtons per square meter (N/m^2) or Pascals (Pa). (For nonuniformly distributed loads, the spatial distribution of the load relative to discrete portions of the cross-sectional area must be known to map the stress distribution.) Consistent with the three types of loading described above, three types of stress are commonly considered: compressive, tensile, and shear. A combination of two or more types of stress may act on a material, such as in bending, when one surface is loaded in tension and the other in compression. Strain is defined as the change in length of a given segment divided by that segment's unloaded length (the gauge length) and thus has dimensions of length per length (mm/mm). In addition to strain in the direction of loading, smaller strains occur in orthogonal directions, which are opposite in sign (e.g., tensile loading produces narrowing in the transverse direction). These changes are easily observed by stretching a rubber band and noting the decrease in its width (and thickness). The extent to which a material exhibits such behavior is characterized by the ratio of transverse (lateral) strain to longitudinal (axial) strain, as first formulated by S.D. Poisson in

1828. Poisson's ratio for isotropic materials ranges between 0.0 and 0.5.

Changes in the geometry of the material specimen that occur during loading (e.g., narrowing, or necking, of a tensile specimen) affect calculations of stress and strain. Stress and strain values based on the unloaded cross-sectional area and a preset gauge length, respectively, are referred to as nominal or engineering stress and strain. True stress and true strain account for highly localized changes in geometry. True stress is obtained by normalizing the applied load to the instantaneous cross-sectional area over which it is distributed. True strain is more difficult to define, since basic strain measurements will always depend upon a chosen gauge length. To remove gauge length from the determination, true strain can be defined as a ratio of the local change in cross-sectional area to the original cross-sectional area. The magnitudes of true stress/strain are greater than that of the corresponding nominal stress/strain. Because true stress/strain are considerably more difficult to determine in practice, nominal stress/strain are usually reported.

The relationship between stress and strain is characterized by the shape of the stress-strain curve (see figure 16.1) and is described mathematically by a constitutive equation. Depending on the shape of the curve, separate constitutive equations may be required to describe accurately the material's behavior across specific ranges of load magnitudes. Most materials exhibit a linearly elastic response over at least some part of their load range, with stress simply related to strain by a proportionality constant. (The earliest description of this relationship, known as Hooke's law, was for metals.) For compressive or tensile loading, this constant is referred to as the elastic modulus, or Young's modulus, and it is equal to the slope of the stress-strain

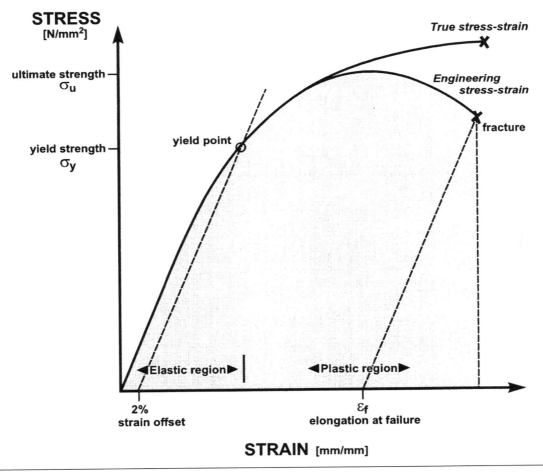

Figure 16.1 Stress-strain curve plotted by converting measures of force and deformation appropriately to represent characteristic material behavior.

Reprinted, by permission, from D.D. Anderson, 1999, Orthopaedic biomechanics. In *Orthopaedic surgery: The essentials*, edited by M.E. Baratz, A.D. Watson, and J.E. Imbriglia (New York: Thieme Publishers), 9.

curve over the linear range. A shear modulus is defined similarly for torsion or shear loading. The magnitude of the modulus indicates the ability of a material to resist deformation; for a given stress, a material with a high elastic modulus deforms less than a material with a low modulus. Within the elastic region, unloading of the material follows the same curve as loading and results in complete recovery of any deformation.

Many materials exhibit a nonlinear response at extreme (high or low) loads, requiring a higher order constitutive equation to represent the entire stress-strain relationship. Because some degree of permanent or plastic deformation of the material persists following removal of higher loads, this latter portion of the stress-strain curve is referred to as the plastic region. Beginning at the point of highest applied stress, unloading typically follows a path parallel to the elastic portion of the loading curve and intersects the strain axis (zero stress) at some nonzero value. The area bounded by the loading and unloading curves (hysteresis loop) represents the energy expended in permanently deforming a material.

The transition point between the elastic and plastic regions of the stress-strain curve is referred to as the proportionality limit or yield stress/strain. This point may be determined by the intersection of the stress-strain curve and a line parallel to the linear portion of the curve with an arbitrarily specified strain offset (often 0.2% for metals). The ultimate stress or strength of a material is the highest stress that particular material is able to withstand prior to failure. Depending on the definition of failure, ultimate stress can be defined as either the absolute maximum value attained or as some submaximum level at which failure initiates (typically indicated by a slight decrease or discontinuity in load). The corresponding strain value is referred to as the ultimate strain.

The energy expended in deforming a material is equal to the combined area under the elastic and plastic portions of the stress-strain curve (see figure 16.2). The toughness of a material is a measure of the

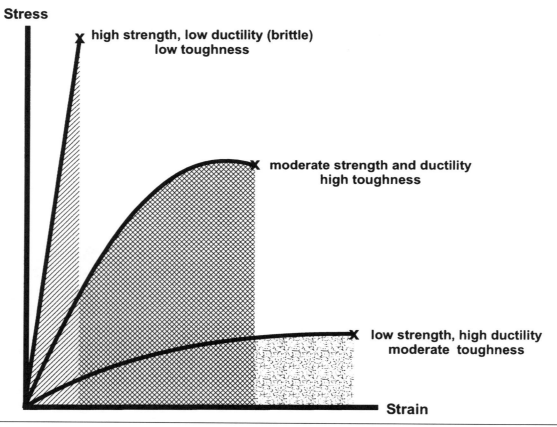

Figure 16.2 Stress-strain curves showing a range of brittle to ductile behavior. The crosshatched area under a curve reflects a material's toughness.

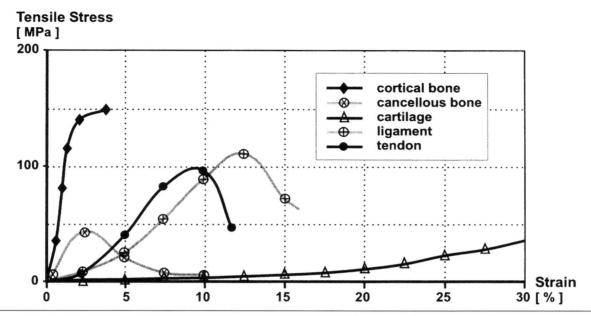

Figure 16.3 Stress-strain curves representing the range of material behavior exhibited in the musculoskeletal tissues considered subjected to tensile loading.

total energy needed to produce failure. Toughness has the dimensions of work or energy per unit volume (newton meter per cubic meter = joule per cubic meter) and is therefore also referred to as strain energy density. A material that fails at a relatively low stress but undergoes very large strain is said to be ductile. A material that is able to withstand higher stresses but fails at relatively low strain is referred to as brittle. Although ductile and brittle materials exhibit dramatically different behavior, the areas under their respective stress-strain curves can be quite similar.

Failure may be due to a single catastrophic loading event, as considered above, or may occur as a result of repeated or cyclic loading and unloading. Under the latter condition, it is possible for a material to fail at stress levels below the ultimate stress as a result of accumulated microdamage. As the number of cycles increases the stress required to produce failure decreases, approaching asymptotically the fatigue or endurance limit, which represents the stress value below which the material could theoretically undergo an infinite number of loading cycles without experiencing failure. Given the cyclic demands associated with many activities of daily living, the fatigue properties of a material are crucial to its long-term survival.

The range of material and structural behavior considered in the present chapter can logically be separated into hard and soft tissue response. The basic constituents among the musculoskeletal tissues are very similar, although present in different relative quantities (see figure 16.3). Bone, a hard tissue with fully 75% inorganic mineral phase by wet weight, exhibits fundamentally different behavior than soft tissues. Collagen (primarily types I and II) and proteoglycans are the complex structural macromolecules that account for the majority of dry weight in soft tissues. Water content plays a fundamental role in carrying load, albeit to varying levels in the different soft tissues. Basic differences in the mechanical properties of ligament and tendon are dictated by structure and relative content of the primary solid constituents.

Soft tissue mechanical behavior may be generally characterized as viscoelastic, possessing both load rate and load history dependence. Another characteristic feature of soft tissues is two-phase loading behavior whereby low force/stress can produce substantial deformation of the structure/material over a region termed the toe region. At higher force/stress, the structures can exhibit significantly more stiff and nearly linear stress-strain response. As a result of internal energy dissipation, the loading and unloading curves of soft connective tissues do not follow the same path but instead form a hysteresis loop. Other important viscoelastic characteristics of articular cartilage, ligaments, and tendons are creep, an increase in deformation over time under a constant load, and stress relaxation, a decline in stress over time under a constant deformation. While bone also exhibits some of these features, they occur over a much longer time scale and involve much smaller deformations. The viscoelastic response of articular cartilage depends on two fundamentally different physical mechanisms: (1) the intrinsic viscoelastic properties of the organic solid matrix components and (2) the frictional drag arising from the flow of interstitial fluid through the porous-permeable solid matrix (poroelastic).

Hierarchy: A Composite Structural Theme in Musculoskeletal Tissues

The concept of structural hierarchy, or multiple levels of organization, is inherent to the musculoskeletal tissues. Basic molecular units combine to form highly complex and ordered structures with multiple levels of hierarchy, delivering unique properties to the different tissues. In defining material properties as opposed to structural properties in tissues, a distinction is usually made at some intermediate level in the hierarchy. While the specific complexity in molecular structure of different tissues (e.g., bone, articular cartilage, ligament, and tendon) is beyond the scope of this text, the hierarchical composition of tendons is presented below as an example of this general concept.

Tendon is a highly complex material tuned finely to its specific functional demands. In tendon, the basic structural molecule is a specific form (type I) of the protein collagen. The primary structure of collagen consists of amino acid groups arranged in a chain. A secondary structure involves the arrangement of these chains in a left-handed helical structure. Its tertiary structure combines three of these collagen chains in a right-handed triple helix held together by hydrogen and covalent bonds. The quarternary structure involves the organization of groups of five of these collagen molecules into ordered units of microfibrils, microfibrils into subfibrils, and subfibrils into fibrils (see figure 16.4). Fibrils are in turn arranged in closely packed, parallel bundles that are oriented in a distinct longitudinal pattern. At this level, proteoglycans and glycoproteins in association with water are incorporated in a matrix, binding the fibrils together to form fascicles. The fascicles are bound together by loose connective tissue, which permits longitudinal movement of collagen fascicles and supports blood vessels, lymphatics, and nerves. Similar levels of mi-

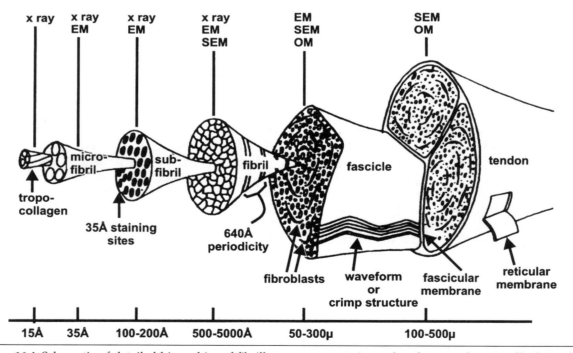

Figure 16.4 Schematic of detailed hierarchies of fibrillar arrangements in tendon down to the microfibril size. EM = electron microscope; SEM = scanning EM; OM = optical microscope.

Reprinted, by permission, from J. Kastelic, A. Galeski, and E. Baer, 1978, "The multi-composite structure of tendon," *Connective Tissue Research* 6:11-23. Used with permission of Gordon and Breach, Science Publishers Ltd. Copyright © 1996 -2000 OPA (Overseas Publishers Association) N.V.

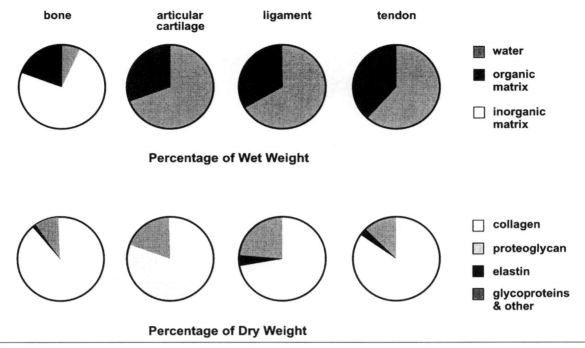

Figure 16.5 Representation of the composition of selected musculoskeletal tissues by wet and dry weight.

crostructural hierarchy impart unique properties to each of the musculoskeletal tissues.

Bone

The skeleton provides a rigid framework of support and protection for muscles, soft connective tissues, and organs. Forces of muscle contraction are transferred from bone to bone across joints to effect movement. Bones are living organs, housing the production of blood cells (hematopoiesis) and serving as storehouses of calcium and other minerals. The 206 individual bones in the human body are commonly classified according to shape (long, short, or flat) and according to location (axial, peripheral, cranial, facial, vertebral, auditory). Sesamoid bones, such as the patella, serve to increase the mechanical leverage of tendons at joints.

Bones are composed of two macroscopically distinct forms of hard connective tissue: cortical (or compact) bone and cancellous (or trabecular) bone. Cortical bone is dense, whereas cancellous bone has macroscopically large porosity. Cortical bone comprises the entire diaphyses of long bones and provides a solid outer shell around regions containing cancellous bone.

Bone tissue consists of viable cells and extracellular matrix. The matrix is composed of an organic (osteoid) phase and an inorganic (mineral) phase, each comprising approximately 50% of the bone volume (see figure 16.5). Organic osteoid consti-

tutes 25% of bone tissue by weight. Ninety percent of the organic phase is fibrous type I collagen, while the remaining 10% is amorphous ground substance (glycoproteins and glycosaminoglycans). Inorganic mineral constitutes 75% of bone tissue by weight, primarily in the form of calcium phosphate crystals similar to hydroxyapatite ($[Ca_{10}(PO_4)_6](OH)_2$). The collagen fibers contribute to the tensile strength of bone tissue, whereas the mineral phase gives bone a relatively higher compressive strength.

The fundamental differences between types of cortical bone have been classified according to microscopic appearance or function. Woven bone and lamellar bone (plexiform or osteonal) constitute the broadest categorization of bone tissue types. Woven bone is relatively porous and hypercellular and contains loosely packed and randomly oriented collagen fibers. It forms rapidly in response to damage or overload, such as in fracture callus, to provide structural strength. Although the mechanical properties of woven bone are inferior to dense and organized bone tissue, the chemical and molecular constituents generally differ only in relative proportion.

The microstructure of lamellar bone is more organized than that of woven bone. Primary lamellar bone is arranged in dense layers, forming circumferential rings approximately 5 microns thick in long bones. Cancellous bone also is composed of primary lamellae, having a porous structure of thin

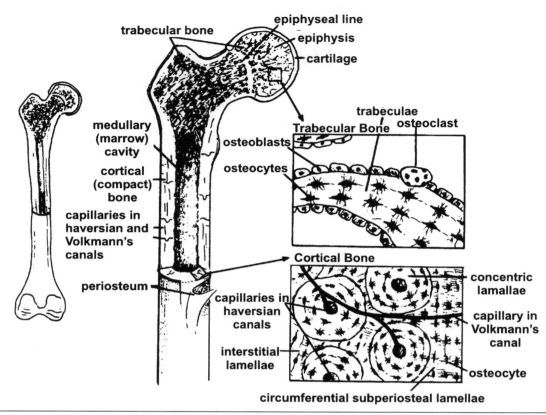

Figure 16.6 Schematic diagram of cortical and trabecular bone.

Reprinted, by permission, from W.C. Hayes and M.L. Bouxsein, 1997, Biomechanics of cortical and trabecular bone: Implications for assessment of fracture risk. In *Basic orthopaedic biomechanics* (2nd ed.), edited by V.C. Mow and W.C. Hayes. (Philadelphia: Lippincott-Raven Publishers) 69-111.

plates and trabecular spicules. The relatively large surface area of cancellous bone exposed to marrow may account for an estimated eightfold higher metabolic rate over that of cortical bone (Martin, Dannucci, and Hod 1987). Moreover, it may provide an explanation as to why the predominance of early signs of osteoporosis occur in regions of cancellous bone, which retain hematopoietic (red) marrow into adulthood (e.g., vertebrae, iliac crest, distal radius, and proximal femur). That is, skeletal calcium is most efficiently mobilized into the vascular system at those sites rich in red bone marrow.

Plexiform bone (one type of lamellar bone) is formed when bone tissue must be deposited relatively quickly, as in rapidly growing children or animals (Currey 1984; Martin and Burr 1989). Named for its appearance of intertwining vascular plexuses, plexiform bone formation initiates by subperiosteal or subendosteal budding, perpendicular to the bone surface. The buds subsequently bridge and fill with lamellar plates, analogous in form to dendritic crystallization of cooling liquids. The resulting tissue is composed of blocks of plexiform bone

approximately 100 to 150 μm on a side. The increased surface area, vascularity, and structural rigidity of plexiform bone (as compared to woven bone) reflect the physiological demand imposed by a rapidly developing skeleton (Currey 1960).

In humans and some animals, concentric lamellae of cortical bone form around blood vessels to create osteons, or haversian systems, approximately 200 to 300 microns in diameter and up to 10 mm long (see figure 16.6). Osteonal bone can be found primarily in the diaphyses of long bones, with individual osteons aligned predominantly in the longitudinal direction. *Primary* osteons develop within organized primary lamellar bone as modified vascular channels, with concentric layers of lamellar bone filling the channel. As hypothesized for plexiform bone, formation of primary osteons may be related to the rate of skeletal growth (Enlow 1962; Rajaram and Ramanathan 1982). *Secondary* osteons later replace primary osteonal or lamellar bone in a process of tunneling and refilling (Frost 1963a, 1963b; Tomes and De Morgan 1853). The principal morphological distinction of primary osteons from secondary osteons is the absence of "cement" or "reversal" lines formed during the remodeling process along the circumferential border of secondary osteons. Because complete mineralization of the osteoid (new bone tissue) within secondary osteons requires weeks or months, secondary osteons are also weaker and more compliant than primary osteons (Ascenzi and Bonucci 1967; Currey 1959; Vincentelli and Grigorov 1985).

Cancellous bone consists of lamellar bone organized into trabecular struts, plates, and spicules. This structure is observed readily, often demonstrating preferred alignment along predominant directions of load transfer to reinforce the bone structure. Characterizing the mechanical properties of cancellous bone as a porous material requires that specimens be machined to a size large enough to include a representative volumetric sample of this structure. A sufficient sample allows cancellous bone to be regarded as a material continuum rather than a structure. Elastic properties of the bone tissue comprising individual trabecula have been measured directly using micromechanical test methods (Ashman and Rho 1988; Kuhn et al. 1989; Rho, Ashman, and Turner 1993; Runkle and Pugh 1975; Ryan and Williams 1989; Townsend, Rose, and Radin 1975), as well as indirectly by relating continuum level results to specimen-specific computer models of trabecular structure (Mente and Lewis 1989; Williams and Lewis 1982). These studies reported an elastic modulus for trabecular bone tissue ranging from 1 to 14 GPa, suggesting that trabecular tissue is not as stiff as cortical tissue, which has an elastic modulus of 15 to 20 GPa (Reilly, Burstein, and Frankel 1974).

The stress-strain behavior of bone demonstrates a linearly elastic response from the onset of loading. Substantial plastic deformation occurs when bone is loaded in either tension or compression. More plastic deformation occurs in osteonal bone than in primary bone, attributed to debonding of and partial pull-out of osteons at their cement lines. Rate of loading, or strain rate, also affects the quantitative response of bone to loading. This rate dependence has been addressed in many experimental studies, with the data from McElhaney (1966) providing a time-honored example of strain rate effects (see figure 16.7). Strength and elastic modulus of bone tissue are higher when loaded at high strain rates, with less energy absorbed than at lower strain rates.

Anisotropy in material strength of human cortical bone can be appreciated by examining the properties measured by Reilly and Burstein (1975) (see

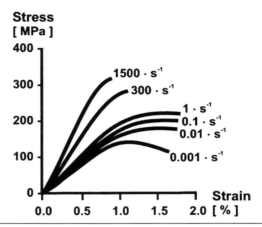

Figure 16.7 Strain rate dependence of cortical bone material behavior. Both modulus and strength increase for increased strain rates.

Reprinted, by permission, from J.H. McElhaney, 1966, "Dynamic response of bone and muscle tissue," *Journal of Applied Physiology* 21: 1233.

table 16.1). Compact bone is approximately 50% stronger in compression than in tension and greatly stronger along its longitudinal axis than in transverse directions (radial and circumferential). Cortical bone is strongest in longitudinal compression (approximately 200 MPa), while torsional shear strength is only a third of that (approximately 70 MPa). These directional and loading mode dependencies reflect the predominant longitudinal orientation of osteons. For purposes of comparison, the tensile and compressive strengths of wood (parallel to the grain) range between 60 to 100 MPa and 35 to 60 MPa, respectively (U.S. Department of Agriculture 1987).

The material properties of bone are affected by many inherent factors: bone type, porosity, mineralization, and age being most identifiable and measurable. Bone type, or structural organization, as well as its orientation within test specimens, can be examined histologically. Porosity is simply the

Table 16.1 Mechanical Strengths of Human Femoral Cortical Bone

	Tension	Compression	Shear*
Longitudinal	133 MPa	193 MPa	—
Transverse	51 MPa	133 MPa	68 MPa

*Torsional shear strength, cylinder axis = longitudinal axis of bone.

From D.T. Reilly and A.H. Burstein, 1975, "The elastic and ultimate properties of compact bone tissue," *Journal of Biomechanics* 8: 393-405.

relative void volume within bone and may be attributed to microscopic voids, lacunae and canaliculi, and vascular spaces in compact bone, as well as macroscopic voids within cancellous bone. The degree of mineralization usually is quantified on the basis of volume, excluding porosity, as a specific density of mineral. Mineralization may also be examined noninvasively through the use of computed tomography and photon absorptiometry. Although the effects of aging on mechanical properties of bone are interesting and worthy of examination, the associated changes in porosity, mineralization, osteon density, remodeling rates, and collagen integrity are the keys to understanding relationships between the biology and mechanical function of bone tissue. A specific region of remodeled bone tissue may not reflect the chronological age of the individual. A primary concern of aging effects on bone is that of senile osteoporosis. Intracortical porosity can nearly double between the ages of 40 and 70 (Martin, Pickett, and Zinaich 1980) as a result of increased numbers of haversian canals, resorption spaces, and partially remodeled osteons. Such increases in porosity correspond to dramatic weakening of bone tissue (see figure 16.8).

Although bone demonstrates plasticity at large loads, physiological loading remains well within the elastic region of the stress-strain curve. Detailed investigation of bone material properties and elasticity requires assessment of inherent anisotropy. Directional dependencies in material behavior correspond to the structural orientation of bone tissue. For example, in the case of osteonal bone, strength and stiffness are highest in the longitudinal direction of haversian systems. Likewise, the concentric lamellar rings of primary bone are analogous to the cylindrical symmetry in trees, with principal material directions corresponding to longitudinal, radial, and circumferential orientation of lamellae. The constitutive behavior of bone tissue usually exhibits distinct differences in these three perpendicular directions, and it therefore can be regarded as an orthotropic material.

Discussion of the constitutive elasticity of bone tissue requires a mathematical understanding of a generalized Hooke's law, relating each of the six stress components to all six strain components:

$$\sigma_1 = C_{11}\varepsilon_1 + C_{12}\varepsilon_2 + C_{13}\varepsilon_3 + C_{14}\varepsilon_4 + C_{15}\varepsilon_5 + C_{16}\varepsilon_6. \quad 16.1$$
$$\sigma_2 = C_{21}\varepsilon_1 + C_{22}\varepsilon_2 + C_{23}\varepsilon_3 + C_{24}\varepsilon_4 + C_{25}\varepsilon_5 + C_{26}\varepsilon_6.$$
$$\sigma_3 = C_{31}\varepsilon_1 + C_{32}\varepsilon_2 + C_{33}\varepsilon_3 + C_{34}\varepsilon_4 + C_{35}\varepsilon_5 + C_{36}\varepsilon_6.$$
$$\sigma_4 = C_{41}\varepsilon_1 + C_{42}\varepsilon_2 + C_{43}\varepsilon_3 + C_{44}\varepsilon_4 + C_{45}\varepsilon_5 + C_{46}\varepsilon_6.$$
$$\sigma_5 = C_{51}\varepsilon_1 + C_{52}\varepsilon_2 + C_{53}\varepsilon_3 + C_{54}\varepsilon_4 + C_{55}\varepsilon_5 + C_{56}\varepsilon_6.$$
$$\sigma_6 = C_{61}\varepsilon_1 + C_{62}\varepsilon_2 + C_{63}\varepsilon_3 + C_{64}\varepsilon_4 + C_{65}\varepsilon_5 + C_{66}\varepsilon_6.$$

In tensor or matrix notation, this can be written as $\sigma_i = C_{ij}\varepsilon_j$, with summation inferred by the subscripts. The 36 elastic constants C_{ij} are analogous to elastic modulus (E) for a uniaxial test ($\sigma = E\varepsilon$). Because $C_{ij} = C_{ji}$, the matrix of elastic constants is symmetric and reduces mathematically to just 21 independent constants for completely *anisotropic* materials. Symmetry in material structure further reduces the number of constants. Materials with different properties in three orthogonal directions are said to be *orthotropic*, with their elasticity characterized by nine independent elastic constants. If the properties are different in only two directions, such as longitudinal and transverse, the material is *transversely isotropic* and can be described by five elastic constants. Finally, the elastic behavior of *isotropic* materials is described by only two independent elastic constants, which can be expressed in terms of elastic modulus (E), shear modulus (G), and Poisson's ratio (υ). Further explanation of elasticity theory and anisotropy can be found in texts on composite materials (e.g., Agarwal and Broutman 1990; Jones 1975).

Establishing reliable constitutive data for bone is problematic. Individual elastic constants must be evaluated based on numerous directional measurements, allowing only partial constitutive measurements by mechanical testing of individual speci-

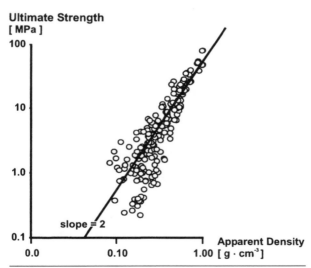

Ultimate Strength [MPa]

slope = 2

Apparent Density [g · cm⁻³]

Figure 16.8 Ultimate compressive strength as a function of apparent density for trabecular bone. In general, compressive strength varies as a power law function of density with an exponent of approximately two.

Reprinted, by permission, from T.M. Keaveny and W.C. Hayes, 1993, Mechanical properties of cortical and trabecular bone. In 7, *Bone: Bone Growth*. Edited by B.K. Hall (Boca Raton, Fla.: CRC Press), 328.

Table 16.2   Elastic Properties of Cortical and Cancellous Bone, Measured by the Ultrasound Wave Propagation Technique

Cortical bone (human femur)[a]		Cancellous bone (human tibia)[b]	Average ± standard deviation	Range
E_{radial} (GPa)	12.0	E_1 (MPa)	347 ± 218	110 – 1,230
$E_{circumferential}$ (GPa)	13.4	E_2 (MPa)	457 ± 282	140 – 1,750
$E_{longitudinal}$ (GPa)	20.0	E_3 (MPa)	1,107 ± 634	340 – 3,350
G_{rc} (GPa)	4.5	G_{12} (MPa)	98 ± 66	30 – 380
G_{rl} (GPa)	5.6	G_{13} (MPa)	133 ± 78	35 – 410
G_{cl} (GPa)	6.2	G_{23} (MPa)	165 ± 94	45 – 460
υ_{rc}	0.38			
υ_{rl}	0.22	Orientation:	1 / anterior-posterior	
υ_{cl}	0.24		2 / medial-lateral	
υ_{cr}	0.42		3 / superior-inferior	
υ_{lr}	0.37			
υ_{lc}	0.35			

[a]Ashman et al. 1984.
[b]Ashman et al. 1989.

mens. An alternative to mechanical measurement has been to assess the elastic properties using a method of ultrasonic wave propagation through tiny, multifaced specimens of bone (Ashman and Rho 1988; Ashman et al. 1984). The velocity of wave propagation is related to material density and elastic stiffness, thus allowing assessment of material elasticity in multiple directions. The method has been used to measure the elastic properties of both cortical and cancellous bone (see table 16.2), providing constitutive data very similar to those obtained via mechanical testing (Ashman, Corin, and Turner 1987).

Efforts to explain the mechanical behavior of cancellous bone have been to describe its anisotropy and density as a structural material, much as plastic foam has been. The mechanical properties of open- or closed-cell materials obey power law relationships, which prompted researchers to examine power law behavior of cancellous bone. Mechanical testing of cancellous specimens has shown that both strength and modulus are proportional to approximately the square of the apparent density (Carter and Hayes 1977; Rice, Cowin, and Bowman 1988). Moreover, a strong correlation has been shown between compressive strength and modulus in all specimen orientations (Rice, Cowin, and Bowman 1988; Williams and Lewis 1982), further suggesting

that cancellous bone strength and modulus are related to densitometric or morphological features in a similar manner. Specific mechanical properties of cancellous bone can vary tremendously between sites. Carter and Hayes (1977) reported compressive strengths ranging from below 1 MPa up to 15 MPa, while compressive modulus ranged between 20 MPa and 200 MPa. The presence of marrow within the trabecular bone specimens was seen to influence compressive strength and modulus only at high dynamic strain rates beyond 10 Hz. Pugh, Rose, and Radin (1973) reported a purely elastic response at low strains and strain rates. No hydraulic strengthening or viscous component of practical magnitude was seen up to 3000 Hz. Pooling various experimental data, Rice, Cowin, and Bowman (1988) concluded that for cancellous bone, elastic modulus and strength are proportional to strain rate raised to the 0.06 power. That is, each order of magnitude increase in strain rate strengthens and stiffens cancellous bone by just 15%. In these and subsequent studies, it has been concluded that trabecular bone may be treated as an elastic material, with the elasticity controlled by the structure, orientation, and density of the cancellous bone.

The elastic and failure properties of bone tissue provide estimates of bulk material response to prescribed stress states. Fractures of bones involve a single traumatic overload or the accumulation and propagation of cracks. Thus, both the fracture and fatigue behaviors of bone are relevant toward understanding its mechanical integrity. Conceptually, fracture properties are measured in terms of the energy expended to propagate a crack by some finite distance. Various experimental techniques and measurements have been used to quantify the fracture mechanics of bone (Bonfield 1987). The experiments are often tedious and problematic but provide a unique measure of resistance to crack propagation. The integrity of a material in terms of limiting crack propagation, or fracture toughness, provides a measure of failure at the microstructural level of organization. In cortical bone specimens, fracture toughness varies with bone density and the orientation of the crack relative to that of the constituent osteons (Behiri and Bonfield 1989). Fracture toughness is two times higher in the transverse direction (perpendicular to osteons) than in the longitudinal direction (parallel to osteons). The reason for this is clear. Cracks in cortical bone propagate between osteons and along cement lines, which are oriented longitudinally. Fatigue involves a progressive loss of mechanical integrity due to damage induced by cyclic loading. Unlike many materials, bone does not exhibit a substantial endurance limit of stress magnitude, below which failure would not occur for any number of cycles. Normal physiological loading causes microcracks to form in bone, but under "normal" conditions the damage is repaired and does not accumulate beyond the fracture threshold. Inhibition of the repair process, excessive loads, or increased repetition can upset this balance and lead to fatigue fractures.

Articular Cartilage

Hyaline articular cartilage is a major component of all synovial joints, covering the articulating surfaces of apposing bones. These joints serve the mechanical function of allowing motion between body segments under dynamic loads. This function involves distributing the load over articulating surfaces and allowing nearly frictionless sliding of the surfaces over one another. Under normal conditions, hyaline articular cartilage can perform these essential biomechanical functions with little damage over seven or eight decades of life. However, damage associated with trauma or chronic and progressive degenerative processes may compromise the functional life of the tissue.

Hyaline articular cartilage consists of an extensive extracellular matrix with a relatively small number of highly specialized cells (chondrocytes)

distributed throughout the tissue. The primary components of the matrix are proteoglycans, collagens, and water, with other proteins and glycoproteins present in lower amounts (see figure 16.5, page 289). These constituents combine to provide the tissue with its complex structure and mechanical properties. Normal hyaline articular cartilage has a water content ranging from 65 to 80% of its total wet weight. The water is nonhomogeneously distributed, decreasing in concentration from roughly 80% at the articular surface to 65% near the subchondral bone (Armstrong and Mow 1982). Over 50% of the dry weight of hyaline articular cartilage consists of collagen, of which 90 to 95% is type II. Collagen fibers are intertwined within the extracellular matrix to form a cross-linked network.

A distinction should be made from the start between hyaline articular cartilage and fibrocartilage. Fibrocartilage is a transitional cartilage found at the margins of some joint cavities, in the joint capsules, and at the insertions of ligaments and tendons into bone. It also forms the menisci interposed between the hyaline articular cartilage of some joints (most notably the knee) and composes the annulus fibrosus of the intervertebral disk. Though closely related histologically to hyaline articular cartilage, fibrocartilage has different biochemical and mechanical properties. Roughly 55% of fibrocartilage wet weight is water. Over 75% (dry weight) of the extracellular matrix is collagen, primarily type I (55–65% dry weight). Proteoglycans and glycoproteins are present in quantities of only about 10% of that in hyaline articular cartilage. In compression, fibrocartilage exhibits similar material properties as hyaline articular cartilage. In tension, fibrocartilage is nearly 10 times stiffer than hyaline articular cartilage, as one might expect based on its higher collagen content. For the remainder of this section, the term articular cartilage will be used to refer specifically to hyaline articular cartilage.

Articular cartilage is anisotropic; its material properties differ for tissue specimens obtained from different orientations relative to the joint surface. This anisotropy is thought to be related to varying collagen fibril arrangements within planes parallel to the articular surface. Adult articular cartilage is highly nonhomogeneous, with a pronounced variation of tensile stiffness and strength with depth below the articular surface. This correlates with observed ultrastructural variation in collagen fiber orientation as well as variation in collagen content through the depth of articular cartilage (see figure 16.9). The uppermost 10 to 20% of the cartilage layer thickness constitutes a region known as the

Zones

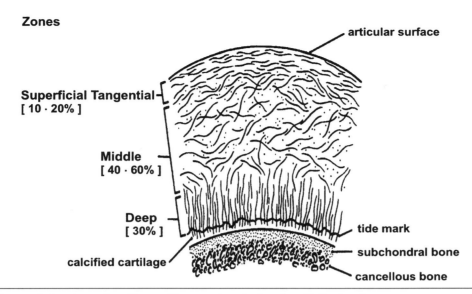

Superficial Tangential [10 · 20%]

Middle [40 · 60%]

Deep [30%]

calcified cartilage

articular surface

tide mark

subchondral bone

cancellous bone

Figure 16.9 Layered structure of cartilage collagen network showing distinct regions.
Reprinted with permission from V.C. Mow, C.S. Proctor, and M.A. Kelly, 1989, Biomechanics of articular cartilage. In *Basic biomechanics of the musculoskeletal system*, 2nd ed., edited by M. Nordin and V.H. Frankel (Philadelphia: Lea & Febiger), 34.

superficial tangential zone, where collagen fibers are oriented predominantly parallel to the articular surface. In the next 40 to 60% (the middle zone) of the cartilage thickness, collagen fiber orientation is fairly random. In the deepest layer (final 30%, the deep zone) collagen fibers are oriented more perpendicular to the articular surface, inserting into the underlying calcified cartilage. Studies have shown that cartilage specimens from the superficial zone have much higher stiffness and failure stresses than those from the deeper zones (Akizuki et al. 1986), correlating with the observation that collagen content is greatest near the articular surface.

At the articular surface, anisotropy is usually described with respect to the direction of "split lines." Split lines are elongated fissures produced by piercing the articular surface with a small needle, and their pattern over the articular surface is generally believed to correspond to collagen fibril direction. Studies have repeatedly shown that cartilage specimens are stiffer and stronger in tension along a direction parallel to the split lines than in a direction perpendicular to them.

The tensile properties of articular cartilage have been studied extensively. Articular cartilage exhibits viscoelastic behavior in tension, appearing stiffer with increasing strain (Woo et al. 1980). For this reason, it is difficult to attribute a single tensile stiffness value to cartilage. Tangent stiffness values are often used to reflect behavior at a given strain level. The tangent modulus is defined by the tan-

gent to the stress-strain curve. Data from simple uniaxial tests show that the tensile modulus of articular cartilage may vary from 5 to 50 MPa, depending on the location, depth, and orientation of the test specimen, as well as presence of surface fibrillation and compositional changes (Kempson 1979; Roth and Mow 1980; Woo, Akeson, and Jemmott 1976; Woo et al. 1979). Under normal conditions and at physiological (less than 15%) strain levels the tangent modulus ranges from 5 to 10 MPa (Akizuki et al. 1986). In all reported cases, the tensile modulus and strength of normal articular cartilage are far less than those of tendons and ligaments. This can be attributed to the significantly lower collagen content in articular cartilage (see figure 16.5, page 289).

Early mechanical studies characterizing the compressive properties of cartilage used indentation experiments to measure an "apparent" elastic modulus. Studies were performed by Bär (1926), Coletti, Akeson, and Woo (1972), Gocke (1927), Hayes and associates (1972), Hirsch (1944), Hori and Mockros (1976), Kempson, Freeman, and Swanson (1971), and Sokoloff (1966). A constant step load was applied to test specimens through either a plane-ended cylinder or a spherical tip, and deformation was measured over time. All studies noted an instantaneous elastic indentation followed by creep to an asymptotic deformation. Researchers thus reported two moduli, one at or shortly after load application and the other at equilibrium following compressive creep. Hirsch (1944) reported an imperfect elastic recovery of cartilage, leaving questions about the nature of cartilage compaction. This matter was later resolved by Elmore and coworkers (1963), who showed that persistent deformation could be recovered over time in a saline bath; they concluded that the creep response of articular cartilage was largely due to the flow of interstitial fluid from the tissue. Sokoloff (1966) considered articular cartilage to be comparable mechanically to a medium hardness rubber. He calculated an instantaneous Young's modulus of 2.28 MPa at 0.8 s after loading and an equilibrium modulus value of 0.69 MPa at 1 h after loading. The mathematical analysis for spherical indentation of a layer of linearly elastic material attached to a rigid foundation (an idealization of articular cartilage attached to bone) was carried out by Hayes and associates (1972) and later advanced by Hori and Mockros (1976). None of these indentation studies explicitly considered fluid flow through cartilage.

Other investigators performed uniaxial compression tests on cylindrical-shaped cartilage specimens harvested from articular surfaces. McCutchen (1962) used porous platens to measure an instantaneous and a 30 min equilibrium Young's modulus of 11.1 MPa and 0.32 MPa, respectively. A porous platen was used to allow fluid flow from the cartilage specimen to the platen. Johnson, Dowson, and Wright (1975) performed similar tests using a sinusoidally varying compressive force at a frequency of 1 Hz to more closely simulate cyclic physiological loading. Their results showed that the elastic modulus in uniaxial compression increased from 12 to 45 MPa with increasing strain.

The basic assumption in the previously mentioned studies is that cartilage behaves elastically when subjected to sufficiently fast load application. In an ideal sense this is only true immediately after load application, before fluid in the cartilage has time to flow, or at equilibrium, when movement of interstitial fluid ceases. The importance of fluid flow in the determination of stresses that occur during normal gait cycles or sudden load application as in athletic participation is not clear. Paul (1967) showed that during the normal gait cycle, the entire loading and unloading sequence of a hip joint occurs within one second, with peak loads reached within an average 0.5 seconds. In impact studies, the load times are considerably less, on the order of milliseconds. Therefore, for consideration of cartilage deformation under most functional loads, such

as walking and running, and for impact loads, it would seem reasonable to ignore the time dependency and approximate cartilage response using short time properties.

Under other loading conditions with a longer time scale, the mechanical behavior of articular cartilage has been shown to depend upon its multiphasic (mixed solid and fluid phases) structure and its fluid-bathed environment. While most early investigators recognized the intrinsic link between fluid flow and cartilage deformation, no theoretical attempts were made to model these effects. The root for understanding cartilage deformation in relation to fluid flow is found in soil consolidation theory. Biot (1941) formulated a general theory of three-dimensional consolidation, in which he considered three phases in soil settling. The first phase was due largely to water flowing out of soil directly beneath an applied load. The second phase involved water flowing from loaded to unloaded regions in the soil. The final phase was linked to the restraining effect of the unloaded region upon the loaded region.

McCutchen (1962) hypothesized that the fluid phase carries most of the load in cartilage. Zarek and Edwards (1963) considered cartilage as a biphasic (two-phase) structure consisting of a porous elastic solid matrix made up of collagen fibers, ground substance, and cells saturated by a pore fluid. Load applied to the tissue was thought to be carried partly by pressure in the fluid and partly by compressive stresses in the solid matrix. In 1967, Edwards (1967) loaded cartilage in confined compression using a porous piston. These experiments allowed negligible flow parallel to the loaded surface while allowing unrestrained flow in the normal direction. They showed that the solid matrix could not be modeled accurately as a linearly elastic material (i.e., the solid matrix is viscoelastic).

Mow and colleagues (1980, 1984) developed a general biphasic mixture theory for articular cartilage. They suggested that a mechanical model for cartilage should also include the influence of mobile electrolytes and consider the solid phase as a fiber-reinforced composite porous matrix.* Mow's biphasic theory models articular cartilage as a soft porous elastic solid permeated by water. The theory accounts for time-dependent aspects of cartilage load-deformation behavior in terms of resistance to interstitial fluid flow. Limited permeability of the solid matrix is responsible for flow resistance asso-

---

* A triphasic theory, integrating effects of mobile electrolytes, has been reported by Lai, Hou, and Mow (1991) and recently extended by Gu, Lai, and Mow (1998).

ciated with both direct mechanical drag and with ionic and electrostatic forces. The theory depends on three fundamental material coefficients: permeability, equilibrium or aggregate modulus, and solid phase Poisson's ratio. The linear biphasic theory assumes that the solid phase is isotropic and linearly elastic and the fluid phase is a linearly viscous fluid. Experimentally, Mow and coworkers (1980) studied the relationship between permeability and applied compressive load, as well as the intrinsic moduli of the solid matrix phase. They reported an intrinsic elastic aggregate modulus for human articular cartilage of 0.70 ± 0.09 MPa.

Given the current understanding of the poroelastic behavior of articular cartilage, one can correlate physical mechanisms with observed mechanical behavior. The creep response of articular cartilage involves several phases. Initially, fluid flow occurs rapidly, evidenced by a rapid rate of increased deformation, and diminishes gradually as permeability decreases with increasing strain until flow cessation. During creep, load applied at the surface is balanced by the compressive stress developed within the solid matrix and by frictional drag generated by interstitial fluid flow. Creep ceases when the compressive stress within the solid matrix balances the applied stress. For relatively thick human cartilage measuring up to 4 mm, it takes 4 to 16 h to reach creep equilibrium. The time to reach creep

equilibrium varies inversely with the square of the tissue thickness. At equilibrium, when fluid flow ceases, the deformation can be used to determine the intrinsic compressive aggregate modulus (H_A) of the solid matrix. H_A ranges from 0.4 to 1.5 MPa for human articular cartilage.

Stress relaxation in articular cartilage involves a slightly different mechanism (see figure 16.10). During compression (time points A and B), stress rises due to fluid pressurization, since exudation is constrained by matrix permeability. Subsequent stress relaxation (time points C and D) is associated with fluid redistribution within the solid matrix. This stress relaxation process ceases when the compressive stress developed equals the equilibrium stress level determined by the intrinsic compressive modulus of the solid matrix (time point E). Under physiological loading characterized by relatively low strain rates, high stress levels are difficult to maintain in cartilage because stress relaxation rapidly occurs. This necessarily leads to rapid spreading of the contact area in the joint during articulation.

In general, a biphasic theory has been shown to be sufficiently accurate to explain much of the observed mechanical behavior of articular cartilage in compression. It does not accurately predict behavior at short time periods (< 1 s) following abrupt load application. Many physical activities are characterized by relatively high strain rates. Under these

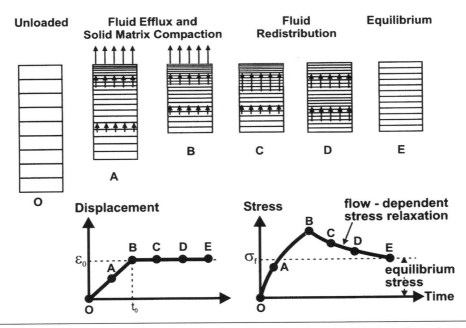

Figure 16.10 Schematic representation of fluid exudation and redistribution within cartilage during a rate-controlled compression stress-relaxation experiment..

Reprinted, by permission, from V.C. Mow, C.S. Proctor, and M.A. Kelly, 1989, Biomechanics of articular cartilage. In *Basic biomechanics of the musculoskeletal system*, 2nd ed., edited by M. Nordin and V.H. Frankel (Philadelphia: Lea & Febiger), 41.

loading conditions, the fluid within the cartilage does not have time to flow through the solid matrix. Therefore, articular cartilage may be reasonably assumed to behave as a single phase, linear elastic solid. Using experimental data, Eberhardt and co-workers (1990) presented arguments supporting the use of linear elastic models to study cartilage failure. In using a simplified linear elastic approximation, the ability to differentiate load-carrying mechanisms between solid and fluid phases of cartilage is sacrificed.

Animal studies and in vitro testing have documented articular damage and resultant articular wear associated with impact loading. With normal compressive physiological loading, articular cartilage undergoes surface compaction with lubricating fluid being exuded through this compacted region near the surface. Because the subchondral plate underlying the cartilage layer is nearly impermeable, fluid flow to deeper layers is precluded. Fluid redistribution within the articular cartilage relieves the stress in this compacted region quite quickly; the stress may decrease by 63% within 2 to 5 s (Mow, Holmes, and Lai 1984). Typical mean compressive stress levels of 3 to 7 MPa may be seen in cartilage during daily activities, with higher stresses experienced in localized regions. If compressive loads are supplied quickly with insufficient time for internal fluid redistribution to relieve the compacted region, high stresses within the solid matrix may result in damage. Repo and Finlay (1977) have shown that at peak compressive stress levels of 25 MPa (nominally 25% strain, strain rates of 500%/s and 1000%/s), chondrocyte death and cartilage fissuring occur. Zimmerman and coworkers (1988) found that repetitive compressive stresses of 6.9 MPa (nominal strain rate of 250%/s) were sufficient to produce accelerated fissuring in plugs of articular cartilage and subchondral bone.

Ligament

Ligaments are passive soft connective tissue structures that connect bones across joint articulations. Ligaments guide and restrict joint motion to maintain the position and orientation of bones within a normal physiological range. Histological examination of ligament tissue, as well as reflex activation via experimental stimulation of nerves, suggests the presence of neural elements that provide proprioceptive feedback, further aiding to maintain joint position.

Ligaments are composed primarily of water (60–70% wet weight) and a hierarchical structure of fibrous collagen, quite similar to tendon (see figures 16.4 and 16.5, pages 288 and 289). Tropocollagen

molecules form microfibrils, which in turn comprise subfibrils, fibrils, and fibers. The fibers demonstrate a periodic wave, or crimp, along their length. In some ligaments, bundles of fibers form distinct fascicles, which may serve a functional role in distributing forces differentially throughout a range of motion (Takai et al. 1993). Ligament substance also contains smaller amounts of elastin, actin, and fibronectin proteins, as well as glycosaminoglycans. Elongated fibroblasts reside along and between the collagen fibers, functioning to produce these matrix molecules. The hierarchical aggregation of fibrous type I collagen serves to resist and transmit tensile forces through the midsubstance and insertions to bone.

The structure of ligament insertions reflects their critical function in transmitting force to restrict joint motion. Tissue morphology within the insertion exhibits distinct zones of transition from ligament substance to fibrocartilage, from fibrocartilage to calcified cartilage, and from calcified cartilage to bone. Rounded, chondrocyte-like cells reside within the cartilaginous zones, producing the molecular constituents for aggregation of fibrocartilage matrix components. As in articular cartilage, histological staining reveals a distinct irregular border or "tidemark" between the zones of fibrocartilage and calcified cartilage, with clusters of mineral crystals evident within the calcified zone. Collagen fibers extend throughout all four zones; connecting and bridging collagen fibers extend between zones to provide a strong, "direct" insertion of ligament into bone. In some cases, the ligament insertion is "indirect," attaching to and within the periosteum, rather than directly into cortical bone tissue.

Experimental measurement of the force and deformation within a ligament pulled in tension demonstrates a force-elongation curve characterized by a small, nonlinear "toe" region, a relatively large linear region, and often a second nonlinear region that may plateau or include discontinuities (see figure 16.11). The nonlinear toe region corresponds to low forces associated with everyday physiological activity. The length, diameter, and crimp in collagen fibers are distributed heterogeneously throughout the ligament substance, such that when a ligament structure is pulled in tension, individual fibers are recruited and elongated as the force is increased. The linear region of the tensile force-elongation curve corresponds to full recruitment and uncrimping of individual fibers, prior to failure. With the crimp extended, all of the collagen fibers resist deformation, exemplified by a larger, linear stiffness. As the ligament extends further,

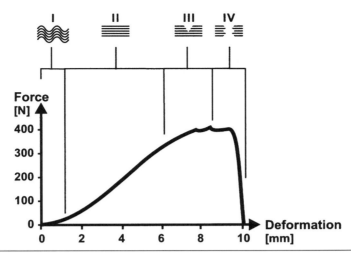

Figure 16.11 A typical force-deformation curve for ligament for monotonic forcing. I = toe region; II = linear region; III = region of microfailure; IV = failure region. At top are schematic representations of fibers going from crimped (I), through recruitment (II), to progressive failure (III and IV).

Reprinted, by permission, from C.D. Frank and N.G. Shrive, 1994, Ligaments. In *Biomechanics of the musculoskeletal system*, edited by B.M. Nigg and W. Ed. Herzog. (Sussex: John Wiley & Sons), 116. Copyright John Wiley & Sons Limited.

some fibers begin to break, such that the effective structural stiffness drops. Subsequent failure of large groups of fibers may produce discontinuities in the force-elongation curve, until the substance is torn apart in its entirety. Because the size and compliance of ligaments (and tendons) renders them difficult to hold for tensile testing to failure, the tests are usually performed with the ligament substance and insertions intact, as a bone-ligament-bone construct, and often with the involved joint intact as well. This aids to preserve the anatomic orientation of the insertions at a particular joint flexion angle. Thus the force-elongation data represent the *structural* behavior of the ligament-bone construct, including the ligament substance and its insertions.

Many studies of the mechanical integrity of ligaments have involved the cruciate and collateral ligaments of the knee. Sport-related injuries of the knee commonly include a tear of one or more of these ligaments. While the collateral ligaments heal with minimal or no surgical invasion, complete tears of the cruciate ligaments do not heal at all. This discrepancy in healing ability is primarily due to the extraarticular location of the medial and lateral collateral ligaments versus the intraarticular location of the anterior and posterior cruciate ligaments. The cruciate ligaments reside within the synovial fluid of the joint cavity, without direct apposition to vascular nutrition. Because injury and surgical reconstruction of the anterior cruciate ligament (ACL) is commonplace, its properties have been studied more extensively than many other ligaments. A

pronounced reduction in failure load and structural stiffness of the ACL-bone construct occurs over the life of humans (Woo et al. 1991). The force required to break the ACL reduces by nearly 60% between the age of 20 and 60 (see figure 16.12). Interestingly, the properties of patellar ligament-bone constructs, which are used to replace the torn ACL, do not exhibit this reduction in mechanical integrity with aging (Flahiff et al. 1995).

In order to reduce the force-elongation data to a stress-strain curve, cross-sectional area and changes in length of the ligament substance must be measured. Cross-sectional area usually is measured with a custom-made micrometer by compressing the region of interest into a rectangular shape. Alternately, ligament or tendon cross sections can be reconstructed computationally, using a laser system to scan the profile of the ligament at discrete increments of rotation about the ligament longitudinal axis (Woo et al. 1990). Similar computational reconstruction methods using optical principles also have been devised (Iaconis, Steindler, and Marnozzi 1987). Likewise, accurate measurement of axial strain can be measured directly by painting (with histological stain) a pair of transverse parallel lines on the ligament surface. By videotaping the tensile test and replaying the analog video through dimensional analysis hardware, the separation distance between the parallel lines can be measured accurately. Similar techniques using small dots sprayed on the surface allow determination of the local surface strain field. An obvious advantage to this

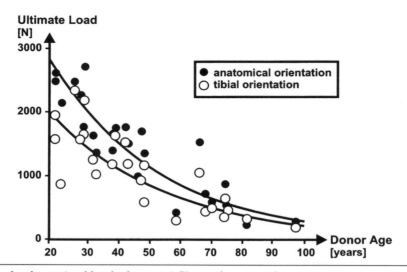

Figure 16.12 Ultimate load sustained by the human ACL as a function of age.

Reprinted, by permission, from S.L-Y. Woo, J.M. Hollis, D.J. Adams, R.M. Lyon, and S. Takai, 1991, "Tensile properties of the human femur-anterior cruciate ligament-tibia complex: The effect of specimen age and orientation," *American Journal of Sports Medicine* 19(3): 222.

method is the archived record of the test for later analysis. Measurement of strain using displacement recorded by the test machine generally is avoided, as it includes deformation within the ligament substance, insertions, bone, and fixtures used to hold the bone-ligament-bone specimen. Analysis of the tensile stress-strain curve provides measures of ligament *material* properties.

The *structural* and *material* properties are both important, as each elucidates the mechanical integrity of ligament function. Changes in structural properties between two test groups (e.g., control versus immobilized joints) that are not reproduced in the material properties (i.e., ligament substance) suggest that the mechanical behavior of the ligament insertions was changed. In fact, this phenomenon occurs when joints are immobilized for a period of weeks (as may occur when casting a fracture). A reduction in mechanical integrity of both ligament substance and insertions occurs with immobilization and disuse. Upon remobilizing the joint, the ligament substance regains its mechanical integrity before the insertions, rendering the insertions vulnerable to damage if challenged beyond relatively low physiological levels of activity.

Tendon

Tendons are the dense fibrous attachments that connect muscle to bone. Tendons possess a complex composite structure of collagen fibrils embedded in a matrix of proteoglycans, glycoproteins, and water and contain relatively few living cells. Fibrils are bound together in closely packed, highly ordered

parallel bundles to form fascicles (see figure 16.4, page 288). The major constituent of tendon after water is type I collagen (86% of dry weight). Tendons differ from ligaments in two important ways. First, tendons contain a higher percentage of collagen and a lower percentage of ground substance than ligaments. Second, the collagen in tendons is more highly oriented than in ligaments, with fibers nearly all oriented along the long axis of the tendon. Tendon possesses one of the highest tensile strengths of any soft tissue in the body, attributed both to the fact that collagen is the strongest of fibrous proteins and that the collagen fibers are principally aligned along the direction of tensile loading.

Morphology can vary over the length of a given tendon and between different tendons, and it can be correlated with the predominant types of loading that the tendon experiences. For instance, in certain situations tendon must not only carry a tensile load but also direct that load over the surface of a geometrically complex bone. In so doing, it must glide smoothly over the bony surface in order to allow smooth motion while subjected to substantial transverse compressive forces. A tendon is enclosed in a sheath that acts as a pulley and, with the assistance of synovial fluid, facilitates this gliding motion. Interestingly, Vogel and Koob (1989) have observed that tendon structure differs near sites in contact with bone. One would presume that these adaptive morphological differences are related to the specific function of a given tendon or a given segment of tendon (see chapter 17). Tendons generally have large, parallel fibers that insert uniformly into bone. Near the bone, the structural and material properties of the tendon change. This is consistent with a complex transition of cell type and biochemical constitution from tendon to bone.

The primary role of tendon is to efficiently transmit the force of an associated skeletal muscle to bone. Tendons directly influence how muscle contraction imparts motion of the bones, by controlling where and in what direction muscle forces act. They deliver mechanical advantage to muscles by maximizing the moment arm and reducing the amount of force required to rotate a given joint. Tendons also couple different actions motored by a single muscle and can cross one or more joints in so doing. In the hand, tendons make it possible for extrinsic muscles, located remotely in the forearm, to move the fingers. Supporting structures (such as retinaculum and tendon sheath) exert control over how tendons are routed along their course and across joints, preventing bowstringing (tightening of the tendon along a straight line) from the muscle attachment to its bony insertion (see figure 16.13).

The material behavior of tendon is similar to that of other collagenous soft tissues such as ligament and skin. The stress-strain curve begins with a toe region in which the tendon stretches (strains) without much force. This behavior has been attributed to the straightening of crimped collagen fibrils and the reorienting of fibers along the direction of loading. The toe region is rather small in tendon because the collagen fibers are nearly parallel with the long axis of the tendon, and less fiber alignment is required prior to frank fiber loading than in ligaments. The size of the toe region manifest by tendons decreases with age because the amount of crimp has been shown to decrease with age (Kastelic and Baer 1980).

As strains are increased, the toe region is followed by a fairly linear region, the slope of which is

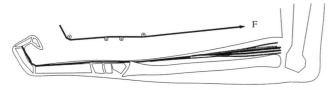

Figure 16.13 Tendons crossing the wrist, traveling along the metacarpals, and inserting on the middle phalanx allow extrinsic muscle of the forearm to effect flexion of the proximal interphalangeal joint. Tendon gliding is facilitated over its course by retinacular and ligamentous pulley structures.

Reprinted, by permission, from J.E. Hale, 1999, Biomaterials. In *Orthopaedic surgery: The essentials*, edited by M.E. Baratz, A.D. Watson, and J.E. Imbriglia (New York: Thieme Publishers), 16.

often used to represent the elastic modulus of the tendon. Following the linear region, at large strains, the stress-strain curve can end abruptly or curve downward as a result of irreversible fiber failure or permanent stretching in the tendon. Thus, to fully describe the stress-strain curve, the slope of the linear region (the elastic modulus), the maximum stress (ultimate tensile strength) and strain (ultimate strain), and the area under the curve (the strain energy density to failure) are required. Typically, the toe region lies within 0 to 3% strain. The ultimate or failure strain of tendon is approximately 8 to 10%. The region of reversible or linear strain extends to roughly 4%, beyond which some amount of permanent deformation occurs. During physiological activities, the strains experienced by tendon range from 2 to 5% (Fung 1981).

The material properties of tendon are dependent on many factors and can vary widely. In humans, the elastic modulus ranges from 1.2 to 1.8 GPa, the ultimate tensile strength ranges from 50 to 105 MPa, and the ultimate strain ranges from 9 to 35%. The elastic strain energy recovered when a tendon is unloaded is 90 to 96% per cycle at physiologically relevant strain rates, indicating that tendons efficiently transmit force without dissipating much energy during activity. When muscle is activated and the muscle-tendon unit is stretched, as on impact during locomotion, the energy associated with the deformation of the tendon is almost completely recovered when the deforming force is removed.

Most in vitro investigations of the stress-strain properties of tendons consider the entire stress-strain curve from zero load to failure. Others have confined their scope simply to the point of ultimate tensile strength. The primary reason for studying the properties of an isolated tendon over the entire stress-strain range is to describe the tissue throughout its physiological and supraphysiological (failure) capacity. However, tendons function in vivo as parts of a functional unit of bone-tendon-muscle-tendon-bone. The healthy tendon is seldom the weakest link in this structural unit, provided that the tensile stress is applied along the line of main fiber direction. The musculotendinous junction is generally believed to be the weak link in this unit, and failure most often occurs there. Traumatic tendon ruptures in humans likely are attributable to uneven distribution of forces on the fiber bundles, since the offending external force in most cases is applied as a combination of loading types.

Like those of ligaments and other collagenous tissues, the mechanical properties of tendon are strain-rate dependent. An increased elongation speed (i.e., a higher strain rate) results in increased stiffness values. However, for tendons, this strain-rate dependency is not as pronounced as it is for other soft tissues. The viscoelastic characteristics are thought to be in part attributable to viscous properties of the interfibrillar matrix (ground substance). Tendon, which contains more collagen and less ground substance than other soft tissues, demonstrates diminished viscoelastic behavior and is more purely elastic than other soft tissues.

The most common tendon injuries are lacerations, ruptures, or tendinitis. Lacerations and ruptures may be considered acute traumatic injuries, although a tendon may be predisposed to rupture by chronic tendinitis or repeated injury. Tendinitis (tendon inflammation) is a more common tendon injury and can be disabling when chronic. Generally, tendinitis is described as an overuse syndrome in which the tendon is loaded repeatedly to a point where the normal repair process cannot keep pace with the rate at which microdamage accumulates in the tissue. When this happens, a typical healing response is initiated but the inflammatory stage persists. Chronic inflammation may ultimately lead to permanent scarring of the tendon.

Elliot and Crawford (1965) found a strong correlation between the maximum force a muscle could produce during isometric tetanic stimulation and the size of a tendon. They also found that tendons of fusiform muscles had smaller cross-sectional areas relative to muscle strength than those of penniform muscles. Harkness (1968) estimated from these data that the tensile strength of a tendon is four times the maximum force produced by its muscle. Hirsch (1974) came to the same conclusion after measuring the maximum force of a muscle in vivo and performing a tensile failure test of its tendon in vitro. The structural properties of human flexor tendons, in particular, have been studied extensively. Average breaking forces are approximately 1000 N. Failure strains are in the range of 11 to 15%. Given that normal tendon function occurs within the toe region of the stress-strain curve, reaching only occasionally into the linear region, the safety margin is considerable.

Summary and Potential Applications

The utility of understanding the mechanical function and properties of the musculoskeletal connective tissues serves many scientific endeavors. Measurements of structural and material properties provide an assay of mechanical and functional integrity. The effects of injury and disease alter the mechanical response of connective tissues to force and deformation, diminishing function and increasing susceptibility to further tissue damage. For example, the effects of osteoporosis are quantified by testing the mechanical and structural deficiencies of bone samples. Arthritic processes are studied by relating the biological changes in cartilage to altered mechanical integrity. Healing of repaired tendons and ligaments is assessed in animal models by mechanical testing of the tissue. Likewise, the suitability of biological or synthetic substitutes for damaged tissues includes an assessment of how well the mechanical behavior is matched.

The need for dynamic loading in maintaining bone tissue homeostasis has long been recognized, with notable treatises on functional adaptation dating to the nineteenth century (Roux 1881; Wolff [1892] 1986). Immobilization, bed rest, and exposure to microgravity provoke atrophy of bone tissue, yet even modest levels of activity and mobility maintain healthy bones. Beyond a broad range of moderate activity, exercise regimens and activities that transmit relatively high dynamic forces through bones promote hypertrophic responses of bone cells. A discussion of functional adaptation of bone and soft tissues follows in the next chapter.

Joint loading and motion are required to maintain normal adult articular cartilage composition, structure, and mechanical properties. The type, intensity, and frequency of loading necessary to maintain normal articular cartilage vary over a broad range. When the intensity or frequency of loading exceeds or falls below these necessary levels, the balance between synthesis and degradation processes will be altered, and changes in the composition and microstructure of cartilage may ensue.

Reduced joint loading, in the form of rigid immobilization or casting, leads to atrophy or degeneration of the cartilage. Increased joint loading, either through excessive use, increased magnitudes of loading, or impact, will also affect articular cartilage. Catabolic effects can be induced by a single impact or repetitive trauma and may serve as the initiating factor for progressive degenerative changes. Disruption of the intraarticular structures, such as menisci or ligaments, will alter the forces acting on the articular surface in both the magnitude and areas of loading. The resulting joint instability is associated with profound and progressive changes in the biochemical composition and mechanical properties of articular cartilage. In experimental animal models, responses to transection of the anterior cruciate ligament or meniscectomy, consistent with clinically observed degenerative changes in humans, have included fibrillation of the cartilage surface, increased hydration, changes in the proteoglycan content, reduced number and size of proteoglycan aggregates, joint capsule thickening, and osteophyte formation. Altered mechanical properties also have been observed together with these histological and biochemical composition changes.

Tendons and ligaments exhibit stress relaxation, creep, and hysteresis as a result of viscoelasticity. When performing a cyclic test by repeated loading and unloading of a tendon or ligament, the stress-strain curve shifts to the right (becomes less stiff). This is important physiologically, because the loading of tendons encountered in activities of daily living usually is cyclic. The viscoelastic response will regulate not only the tension, but also the elongation and, thus, the characteristics of muscle contraction. For example, in an isometric contraction, the length of the muscle-tendon unit remains constant; however, because of creep, the tendon elongates, allowing the muscle to shorten. Physiologically, the change in length of the muscle decreases the rate of muscle fatigue. Thus, tendon creep increases muscle performance in an isometric contraction.

The viscoelastic behavior of tendons and ligaments has other important clinical significance. During walking or jogging, the range of applied strain magnitudes and strain rates are fairly consistent. Cyclic stress relaxation will effectively soften tissue substance with continuous decreases in peak stress as cycling proceeds. This phenomenon may help to prevent fatigue failure of ligaments and tendons. Similarly, deformation increases slightly during cycles to a constant load, demonstrating creep behavior of tendons and ligaments. These changes have been noted clinically, with temporary softening of all of these tissues and thus increases of test excursion (laxity) in exercised joints. This phenomenon may in part explain the benefit of beginning exercise with a brief warm-up period. After a short recovery period following exercise, there is a return to normal joint stiffness and apparent length.

Indirect tendon injury depends heavily upon anatomic location, vascularity, and skeletal maturity, as well as the magnitude of the applied forces. When the force (stress) in the muscle-tendon-bone complex exceeds the tolerance of this structure, injury (failure) occurs in the weakest link. Most tendons can withstand tensile forces larger than can be exerted by their muscles or sustained by the bones. As a result, avulsion fractures (in which the tendon pulls off of its bony attachment) and tendon ruptures at the musculotendinous junction occur much more commonly than midsubstance tendon ruptures. In the flexor tendons of the hand, for example, avulsion of the bony insertion of the tendon occurs mainly in young adults during athletic participation.

Exercise has a positive long-term effect on the structural and mechanical properties of tendons. The stiffness, ultimate tensile strength, and weight of tendons increase as a result of long-term exercise training. The crimp angle and crimp length have been found to be influenced significantly by exercise. Exercise may also enhance collagen synthesis; this possibility is confirmed by results of biochemical studies of collagen metabolism after physical stress (Michna and Hartmann 1989). Furthermore, tendons subjected to exercise have a relatively high percentage of larger diameter collagen fibrils. The thick fibrils can be expected to withstand greater tensile forces than thin fibrils because they contain a higher number of intrafibrillar covalent cross-links.

It is within the context of human motion that the structural building blocks (tissues) of the musculoskeletal system must function to allow and support movement without incurring injury. In this chapter we have discussed the structural and material responses of bone, articular cartilage, ligaments, and tendons to external forces, both within the normal physiological range and at levels associated with traumatic tissue failure. Forces are transmitted through individual musculoskeletal tissue structures in different ways according to their mechanical function and biological composition (i.e., form follows function). A general understanding of the compositional differences between the musculoskeletal tissues elucidates the unique structure-function-property relationships of bone, articular cartilage, ligament, and tendon in generating and controlling movement. We have touched only briefly upon the nature of how the musculoskeletal tissues adapt to their environment as it changes according to a person's age, activity level, or disease processes. The details of this unique biological feature (adaptation) and its implications in musculoskeletal function supporting movement are the topic of the next chapter.

References

Agarwal, B.D., and Broutman, L.J. 1990. *Analysis and performance of fiber composites.* 2d ed. New York: Wiley.

Akizuki, S., Mow, V.C., Muller, F., Pita, J.C., Howell, D.S., and Manicourt, D.H. 1986. Tensile properties of human knee joint cartilage. I. Influence of ionic concentrations, weight bearing, and fibrillation on the tensile modulus. *J. Orthop. Res.* 4: 379–92.

Armstrong, C.G., and Mow, V.C. 1982. Variations in the intrinsic mechanical properties of human articular cartilage with age, degeneration, and water content. *J. Bone Joint Surg.* 64A: 88–94.

Ascenzi, A., and Bonucci, E. 1967. The tensile properties of single osteons. *The Anatomical Record* 158: 375–86.

Ashman, R.B., Corin, J.D., and Turner, C.H. 1987. Elastic properties of cancellous bone: Measurement by an ultrasonic technique. *J. Biomech.* 20: 979–86.

Ashman, R.B., Cowin, S.D., Van Buskirk, W.C., and Rice, J.C. 1984. A continuous wave technique for the measurement of the elastic properties of cortical bone. *J. Biomech.* 17: 349–61.

Ashman, R.B., and Rho, J.Y. 1988. Elastic moduli of trabecular bone material. *J. Biomech.* 21: 177–81.

Ashman, R.B., Rho, J.Y., and Turner, C.H. 1989. Anatomic variation of orthotropic elastic moduli of the proximal human tibia. *J. Biomech.* 22: 895–900.

Bär, E. 1926. Elasticiatssprufunger der gelenkknorpel. *Arch. Entwickklungsmech Organ.* 108: 739–60.

Behiri, J.C., and Bonfield, W. 1989. Orientation dependence of the fracture mechanics of cortical bone. *J. Biomech.* 22: 863–72.

Biot, M.A. 1941. General theory of three-dimensional consolidation. *J. Appl. Physiol.* 12: 155–64.

Bonfield, W. 1987. Advances in the fracture mechanics of cortical bone. *J. Biomech.* 20: 1071–81.

Carter, D.R., and Hayes, W.C. 1977. The compressive behavior of bone as a two-phase porous structure. *J. Bone Joint Surg.* 59A: 954–62.

Coletti, J.M., Akeson, W.H., and Woo, S.L-Y. 1972. A comparison of the physical behavior of normal articular cartilage and the arthroplasty surface. *J. Bone Joint Surg.* 54A(1): 147–60.

Currey, J.D. 1959. Differences in the tensile strength of bone of different histological types. *J. Anat.* 98: 87–95.

Currey, J.D. 1960. Differences in the blood supply of bone of different histological types. *Quart. J. Micro. Sci.* 101: 351–70.

Currey, J.D. 1984. *The mechanical adaptations of bones.* 262–63. Princeton: Princeton University Press.

Eberhardt, A.W., Keer, L.M., Lewis, J.L., and Vithoontien, V. 1990. An analytical model of joint contact. *J. Biomech. Eng.* 112: 407–12.

Edwards, J. 1967. Physical characteristics of articular cartilage. *Proc. Inst. Mech. Eng.* 181: 16–24.

Elliot, D.H., and Crawford, G.N.C. 1965. The thickness and collagen content of tendon relative to the strength and cross-sectional area of muscle. *Proc. R. Soc. Lond.* 162B: 137–46.

Elmore, S.M., Sokoloff, L., Norris, G., and Carmeci, P. 1963. Nature of imperfect elasticity of articular cartilage. *J. Appl. Physiol.* 18: 393–96.

Enlow, D.H. 1962. Functions of the haversian system. *Am. J. Anat.* 110: 269–82.

Flahiff, C.M., Brooks, A.T., Hollis, J.M, Vander Schilden, J.L., and Nicholas, R.W. 1995. Biomechanical analysis of patellar tendon allografts as a function of donor age. *Am. J. Sports Med.* 23: 354–58.

Frost, H.M. 1963a. *Bone remodeling dynamics.* Springfield, IL: Charles C. Thomas.

Frost, H.M. 1963b. Dynamics of bone remodeling. In *Bone biodynamics,* edited by H.M. Frost. Springfield, Ill.: Charles C. Thomas.

Fung, Y.C. 1981. *Biomechanics: Mechanical properties of living tissues.* 222. New York: Springer-Verlag.

Gocke, E. 1927. Elastiziatats studien am junger und alten gelenkknorpel. *Verh. Deutsch Orthop. Ges.* 22: 130–47.

Gu, W.Y., Lai, W.M., and Mow, V.C. 1998. A mixture theory for charged hydrated soft tissues containing multielectrolytes: Passive transport and swelling behaviors. *J. Biomech. Eng.* 120: 169–80.

Harkness, R.D. 1968. Mechanical properties of collagenous tissues. In vol. 2.of *Treatise on collagen: Biology of collagen,* ed. B.S. Gold, 247–310. London: Academic Press.

Hayes, W.C., and Bouxsein, M.L. 1997. Biomechanics of cortical and trabecular bone: Implications for assessment of fracture risk. In *Basic orthopaedic biomechanics.* 2d ed. Ed. V.C. Mow and W.C. Hayes, 69–111. Philadelphia: Lippincott-Raven.

Hayes, W.C., Keer, L.M., Herrmann, G., and Mockros, L.F. 1972. A mathematical analysis for the indentation tests of articular cartilage. *J. Biomech.* 11: 407–19.

Hirsch, C. 1944. A contribution to the pathogenesis of chondromalacia of the patella. *Acta Chir. Scand.* 83(Suppl.): 1–106.

Hirsch, G. 1974. Tensile properties during tendon healing. *Acta Orthop. Scand.* 153(Suppl.): 1–145.

Hori, R.Y., and Mockros, L.F. 1976. Indentation tests of human articular cartilage. *J. Biomech.* 9: 259–68.

Iaconis, F., Steindler, R., and Marnozzi, G. 1987. Measurements of cross-sectional area of collagen structures (knee ligaments) by means of an optical method. *J. Biomech.* 20: 1003–10.

Johnson, G.R., Dowson, D., and Wright, V. 1975. A new approach to the determination of the elastic modulus of articular cartilage. *Ann. Rheum. Dis.* 34(Suppl.): 116–17.

Jones, R.M. 1975. *Mechanics of composite materials.* Washington, DC: Scripta Book.

Kastelic, J., and Baer, E. 1980. Reformation in tendon collagen: The mechanical properties of biological materials. Proceedings of the XXXIV Symposium of the Society for Experimental Biology.

Kastelic, J., Galeski, A., and Baer, E. 1978. The multicomposite structure of tendon. *Connect. Tissue Res.* 6: 11–23.

Keaveny, T.M., and Hayes, W.C. 1993. Mechanical properties of cortical and trabecular bone. In vol. 7 of *Bone: Bone growth*. Ed. B.K. Hall, 285–344. Boca Raton, FL: CRC Press.

Kempson, G.E. 1979. Mechanical properties of articular cartilage. In *Adult articular cartilage*. 2d ed. Ed. M.A.R. Freeman, 333–414. Tunbridge Wells, UK: Pitman Medical.

Kempson, G.E., Freeman, M.A.R., and Swanson, S.A.V. 1971. The determination of a creep modulus for articular cartilage from indentation tests on the human femoral head. *J. Biomech.* 4: 239–50.

Kuhn, J.L., Goldstein, S.A., Choi, K.W., London, M., Feldkamp, L.A., and Matthews, L.S. 1989. Comparison of the trabecular and cortical tissue moduli from human iliac crests. *J. Orthop. Res.* 7: 876–84.

Lai, W.M., Hou, J.S., and Mow, V.C. 1991. A triphasic theory for the swelling and deformation behaviors of articular cartilage. *J. Biomech. Eng.* 113(3): 245–58.

Martin, R.B., and Burr, D.B. 1989. *Structure, function, and adaptation of compact bone.* New York: Raven Press.

Martin, R.B., Dannucci, G., and Hod, S. 1987. Bone apposition rate differences in osteonal and trabecular bone. *Trans. Orthop. Res. Soc.* 12: 178.

Martin, R.B., Pickett, J.C., and Zinaich, S. 1980. Studies of skeletal remodeling in aging men. *Clin. Orthop. Rel. Res.* 149: 268–82.

McCutchen, C.W. 1962. The frictional properties of animal joints. *Wear* 5: 1–17.

McElhaney, J.H. 1966. Dynamic response of bone and muscle tissue. *J. Appl. Physiol.* 21: 1231–36.

Mente, P.L., and Lewis, J.L. 1989. Experimental method for the measurement of the elastic modulus of trabecular bone tissue. *J. Orthop. Res.* 7: 456–61.

Michna, H., and Hartmann, G. 1989. Adaptation of tendon collagen to exercise. *Int. Orthop.* 13: 161–65.

Mow, V.C., Holmes, M.H., and Lai, W.M. 1984. Fluid transport and mechanical properties of articular cartilage: A review. *J. Biomech.* 17(5): 377–94.

Mow, V.C., Kuei, S.C., Lai, W.M., and Armstrong, C.G. 1980. Biphasic creep and stress relaxation of articular compression: Theory and experiments. *J Biomech. Eng.* 102: 73–84.

Paul, J.P. 1967. Forces transmitted by joints in the human body. *Proc. Inst. Mech. Eng.: Lubrication and wear in living and artificial joints* 181: 8–15.

Pugh, J.W., Rose, R.M., and Radin, E.L. 1973. Elastic and viscoelastic properties of trabecular bone: Dependence on structure. *J. Biomech.* 6: 475–85.

Rajaram, A., and Ramanathan, N. 1982. Tensile properties of antler bone. *Calcified Tissue Int.* 34: 301–5.

Reilly, D.T., and Burstein, A.H. 1975. The elastic and ultimate properties of compact bone tissue. *J. Biomech.* 8: 393–405.

Reilly, D.T., Burstein, A.H., and Frankel, V.H. 1974. The elastic modulus of bone. *J. Biomech.* 7: 271–75.

Repo, R.U., and Finlay, J.B. 1977. Survival of articular cartilage after controlled impact. *J. Bone Joint Surg.* 59A(8): 1068–76.

Rho, J.Y., Ashman, R.B., and Turner, C.H. 1993. Young's modulus of trabecular and cortical bone material: Ultrasonic and microtensile measurements. *J. Biomech.* 26: 111–19.

Rice, J.C., Cowin, S.C., and Bowman, J.A. 1988. On the dependence of the elasticity and strength of cancellous bone on apparent density. *J. Biomech.* 21(2): 155–68.

Roth, V., and Mow, V.C. 1980. The intrinsic tensile behavior of the matrix of bovine articular cartilage and its variation with age. *J. Bone Joint Surg.* 62A: 1102.

Roux, W. 1881. *Der zuchtende Kampf der Teile, oder die "Teilauslese im Organismus (Theorie der 'funktionellen Anpassung')"* Leipzig: Wilhelm Engelmann.

Runkle, J.C., and Pugh, J. 1975. The micromechanics of cancellous bone. II. Determination of the elastic modulus of individual trabeculae by a buckling analysis. *Bulletin of the Hospital for Joint Diseases* 36(1): 2–10.

Ryan, J.C., and Williams, J.L. 1989. Tensile testing of rodlike trabeculae excised from bovine femoral bone. *J. Biomech.* 22: 351–55.

Sokoloff, L. 1966. Elasticity of aging cartilage. *Proc. Fed. Am. Soc. Exp. Biol.* 25: 1089–95.

Takai, S., Woo, S.L-Y., Livesay, G.A., Adams, D.J., and Fu, F.H. 1993. Determination of the in situ loads on the human anterior cruciate ligament. *J. Orthop. Res.* 11: 686–95.

Tomes, J., and De Morgan, C. 1853. Observations on the structure and development of bone. *Phil. Trans. Roy. Soc. Lond.* 143: 109–39.

Townsend, P.R., Rose, R.M., and Radin, E.L. 1975. Buckling studies of single human trabeculae. *J. Biomech.* 8: 199–201.

U.S. Department of Agriculture. Forest Service. 1987. Wood as an engineering material. In *Wood handbook: Agriculture handbook 72*. Washington, DC: Forest Products Laboratory.

Vincentelli, R., and Grigorov, M. 1985. The effect of haversian remodeling on the tensile properties of human cortical bone. *J. Biomech.* 18: 201–7.

Vogel, K.G., and Koob, T.J. 1989. Structural specialization in tendons under compression. *Int. Rev. Cytol.* 115: 267–93.

Williams, J.L., and Lewis, J.L. 1982. Properties and an anisotropic model of cancellous bone from the proximal tibial epiphysis. *J. Biomech. Eng.* 104: 50–56.

Wolff, J. [1892.] 1986. The law of bone remodeling (Das gesetz der Transformation der Knochen). Translated by P. Maquet and R. Furlong. New York: Springer-Verlag.

Woo, S.L-Y., Akeson, W.H., and Jemmott, G.F. 1976. Measurements of nonhomogeneous directional mechanical properties of articular cartilage in tension. *J. Biomech.* 9: 785–91.

Woo, S.L-Y., Danto, M.I., Ohland, K.J., Lee, T.Q., and Newton, P.O. 1990. The use of a laser micrometer system to determine the cross-sectional shape and area of ligaments: A comparative study with two existing methods. *J. Biomech. Eng.* 112: 426–31.

Woo, S.L-Y., Hollis, J.M., Adams, D.J., Lyon, R.M., and Takai, S. 1991. Tensile properties of the human femur-anterior cruciate ligament-tibia complex: The effect of specimen age and orientation. *Am. J. Sports Med.* 19: 217–25.

Woo, S.L-Y., Lubock, P., Gomez, M.A., Jemmott, G.F., Kuei, S.C., and Akeson, W.H. 1979. Large deformation nonhomogeneous and directional properties of articular cartilage in uniaxial tension. *J. Biomech.* 12: 437–46.

Woo, S.L-Y., Simon, B.R., Kuei, S.C., and Akeson, W.H. 1980. Quasi-linear viscoelastic properties of normal articular cartilage. *J. Biomech. Eng.* 102: 85–90.

Zarek, J.M., and Edwards, J. 1963. The stress-structure relationship in articular cartilage. *Med. Electron. Biol. Eng.* 1: 497–507.

Zimmerman, N.B., Smith, D.G., Pottenger, L.A., and Cooperman, D.R. 1988. Mechanical disruption of human patellar cartilage by repetitive loading in vitro. *Clin. Orthop. Rel. Res.* 229: 302–7.

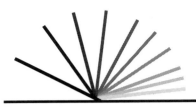

Chapter 17

Biological Response to Forces Acting in the Locomotor System

K.J. Fischer

Of all the passive connective tissues, bone is probably the most recognized for its ability to adapt to varied mechanical stimuli. It has an amazing capacity to heal, atrophies when mechanical stimulation is reduced, and clearly hypertrophies under increased mechanical stimulation. The adaptation of bone to mechanical stresses has been postulated since the early nineteenth century. In 1938, Bourgery theorized that bone adapts according to a minimum mass (or alternately a "maximum" stress) principle (Roesler 1987). In the middle of the nineteenth century, von Meyer and Culmann combined evidence from anatomy and engineering stress analysis to postulate that bone adapts to the applied principal stresses. Soon after, Julius Wolff popularized the idea that bone adapts according to strict mathematical laws and aligns itself with the principal stresses in the bone (Roesler 1987). In the late nineteenth century, Roux introduced the ideas of functional adaptation, a quantitative self-regulating mechanism, and a cellular control process. Since that time, many people have made observations of bone adaptation and have sought to determine the underlying mechanisms. Modern stress analysis techniques, biochemistry, molecular biology, and gene transfer have brought us closer to a full understanding of the bone adaptation process, but the underlying transduction pathway (likely multiple pathways) from

mechanical stimulus to adaptation remains a mystery.

With regard to adaptation of soft tissues, the history is less clear. Certainly, by the early twentieth century, there was an awareness that mechanical stimulation also affects soft tissue mechanical properties and can modulate tissue phenotype. In recent years, modern analysis methods have allowed investigators to begin documenting and correlating changes in cartilage, tendon, and ligament tissue structure and properties with changes in mechanical stimulation. As with bone adaptation, the underlying transduction pathways have been elusive.

Statement of the Problem

Exercise can take the form of professional athletics, amateur athletics, general maintenance of physical fitness, or rehabilitation following an injury or disease. In all of these cases, it is useful and important to know how the exercise regimen will affect the normal, injured, or diseased musculoskeletal tissues. In general, we know that mechanical stimulation is good for tissues, and they tend to become more stiff and strong with regular stimulation. This chapter examines the extent of adaptation that can be expected for a given tissue, its stimulus, and the time over which the stimulus is applied. The research considered provides insight into which activities and training regimens are most likely to result in injury from overuse, and indicates warning signs, such as tissue degradation, so that such injuries might be avoided. Finally, this critical investigation of tissue adaptation considers potential exceptions to the benefits for these tissues of mechanical stimulus from exercise.

The Primary Concepts

The main body of this chapter is broken down into four different sections, each addressing a different passive connective tissue and its characteristics with regard to adaptation to mechanical stimulus. The end of each section includes a summary and discussion of the data available for each tissue. A close examination of each tissue type should provide some important considerations for those working in the exercise, rehabilitation, and sports medicine fields.

Adaptation of Bone to Mechanical Stimulus

Compared to the other passive connective tissues, bone adaptation to mechanical stimulus has been the most widely studied. This section first describes the data from clinical and animal studies that demonstrate the phenomenon of bone adaptation and the extent of adaptation possible. Then, the evidence of the potential for bone adaptation to exercise from cross-sectional studies and prospective exercise intervention studies is presented. In addition, evidence of bone adaptation to a reduction in mechanical stimulus from bed rest or paralysis, for example, is considered. Finally, the various theories of the transduction of the mechanical stimulus from the applied load to the cellular response are briefly discussed.

Adaptation of Articular Cartilage to Mechanical Stimulus

Adaptation of cartilage, especially as it relates to the causes and cures of osteoarthritis, has also been extensively studied. Most studies consider changes in chemical composition but do not address the question of mechanical properties. This section reviews studies of articular cartilage adaptation from both a chemical composition perspective and a mechanical properties perspective. Data on mechanical properties, while rather scant, are consistent with biochemical changes. There is substantial information available on what appear to be the

beneficial and detrimental stimuli for cartilage. Also, there is limited data that provide some insight into the role of exercise as a risk factor for osteoarthritis and its role with regard to patients with existing osteoarthritis. The latter part of this section provides a very practical insight into the risk associated with osteoarthritis (OA) from exercise. Recent data, which indicate that low impact or moderate exercise may be beneficial to arthritic joints and still provide benefits to the cardiovascular and pulmonary systems, is also presented.

Adaptation of Tendon to Mechanical Stimulus

The evidence that mechanical stimulus can induce changes in the microstructural composition and the mechanical properties of tendons is described in this section. Tendons are commonly affected by overuse injuries, and many tendon studies have focused on such injuries. This section examines data indicating the probable reasons for the development of such injuries.

Adaptation of Ligament to Mechanical Stimulus

Ligament adaptation to increased or decreased mechanical stimulus has been studied less than all the musculoskeletal tissues previously discussed. Changes in ligament properties with increased loading are not as well documented as for the other passive connective musculoskeletal tissues. This section examines the available evidence and considers a possible explanation for the sometimes limited ligament adaptation that has been observed.

Bone Adaptation

Research has demonstrated that mechanical forces encountered during normal activity can play a significant role in determining the external morphology and internal structure of a bone. A 1969 study that compared the throwing arm of professional pitchers to the contralateral arm found hypertrophy in the shaft of the humerus and bone ends of the elbow in the pitching arm (King, Brelsford, Tullos 1969). Relative densification and hypertrophy of the proximal radius and ulna were also evident from radiographs. Because all hormonal levels and systemic biochemical processes should be equivalent in both arms, this adaptation can be attributed, at least in a general sense, to loading conditions. In a study similar to that for professional pitchers, the effect of increased loading was also shown in pro-

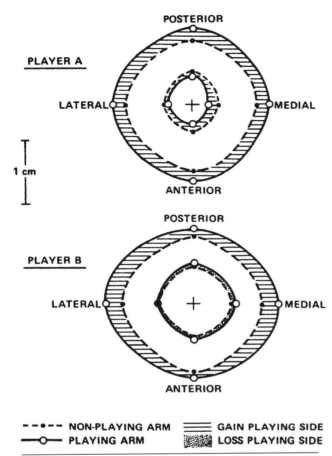

PLAYER A

POSTERIOR

LATERAL — MEDIAL

ANTERIOR

1 cm

PLAYER B

POSTERIOR

LATERAL — MEDIAL

ANTERIOR

- - • - - NON-PLAYING ARM   ═══ GAIN PLAYING SIDE
—○— PLAYING ARM   ▓▓ LOSS PLAYING SIDE

Figure 17.1 Comparison of humerus midshaft periosteal and endosteal diameters from the dominant and nondominant arms in tennis players, determined from orthogonal roentgenograms. The lines joining the measurement points were inferred. Both subjects were international tennis players with 18 years' experience.

Reprinted, by permission from H.H. Jones, J.D. Priest, W.C. Hayes, et al, 1977, "Humeral hypertrophy in response to exercise," *Journal of Bone and Joint Surgery* 59A: 207.

fessional female and male tennis players (Jones et al. 1977). These female and male athletes had 28.4% and 34.9%, respectively, greater humeral cortical thickness in the dominant arm than in the nondominant arm (see figure 17.1).

A study of femoral density in three groups with different levels of activity found that athletes had significantly more dense femurs than nonathletes (Nilsson and Westlin 1971). Also, the leg of preference in the athletes had significantly higher femoral bone density than the contralateral leg. Even non-elite athletes who exercised regularly had significantly higher femoral density than those who did not exercise (Nilsson and Westlin 1971). While this was a cross-sectional study and the groups were self-selected, the consistent trend of increasing den-

sity with increasing activity strongly supports the notion of bone adaptation to loading conditions. Dalén and Olsson (1974) performed a study comparing bone density at several anatomic sites for a group of 15 male cross country runners, 24 sedentary male controls, and 19 male office employees who entered a short-term training (walking/running) program. They found higher bone mineral content in cross country runners than in sedentary control subjects (Dalén and Olsson 1974). The short-term prospective study with the training group did not show a significant increase in bone mineral content at the end of the three-month training period, but this may be due to insufficient time or inadequate levels of exercise. CT evaluation of lumbar spine density found significantly higher values for elite women distance runners than for sedentary women of similar age, but radius density was not different (Marcus et al. 1985). Also, a CT study of middle-aged runners and paired controls matched for age, sex, years of education, occupation, and (for women) years postmenopausal found that the running groups averaged 40% more bone density in the first lumbar vertebra than the matched pairs (Lane et al. 1986).

A few prospective studies have been performed on the role of physical activity in maintaining or increasing bone mass. A year-long study of healthy postmenopausal women examined the effect of conditioning and aerobic exercises (1 h, three times each week) on total body calcium and forearm density (measured by single photon absorptiometry). A significant rise in total body calcium (781 ± 95 g to 801 ± 118 g) was observed in the exercise group, while each subject in the control group exhibited a decrease in total body calcium (Aloia et al. 1978). Though the differences in total body calcium changes were highly significant for the groups, no changes were noted in forearm mineral density in either group. The lack of forearm changes may be due to the choice of a measurement site not highly loaded during exercise or to inadequate precision of the measurements. A study of 20 men who were in training for a marathon competition showed a 3% increase in calcaneal bone density for 7 runners who ran more than 16 km consistently in weekly training sessions (Williams et al. 1984). The 13 runners who did not run consistently long distances (less than 16 km) showed only a marginally significant 1% increase in calcaneal bone density (about twice the increase of the control group). The correlation of the distance run with percentage change in bone density was significant for the consistent runners (Williams et al. 1984). Weightlifters may show dramatically higher bone density than matched controls. Men who have lifted weights for over a year were found using dual-photon absorptiometry to have higher lumbar spine (11%), femoral neck (16%), and greater trochanter (15%) bone mineral density (BMD) than age-matched controls (Colletti et al. 1989). Density in the midradius was not different between weightlifters and controls.

Several prospective studies have been conducted in postmenopausal women, because they are at risk for osteoporotic bone loss. The results from some of these studies are not as consistent or promising. One study actually noted a greater decrease in bone density of the radius in the exercise group than in the control group (Smith et al. 1984). Such results are not well understood but have been attributed to redistribution of bone mass from nonweight-bearing bones to the more stimulated weight-bearing bones (Smith et al. 1984). Also, differences in physiological age of participants may affect physical and mental function in these studies. Some of the problems of existing studies include questionable choice of bone density measurement sites (forearm bone for primarily lower body activity). Also, the age of the participants may limit the loading to insufficient levels (Marcus and Carter 1988). But some studies do indicate that exercise can prevent bone loss. One study of "moderate-for-age" exercise in women 50 to 73 found that lumbar spine and distal forearm density were unchanged in those who exercised, while the control group had bone density decreases of 2.7% in the lumbar spine and 1.8% in the distal forearm (Krølner et al. 1983).

While an increase in mechanical loading through exercise appears to increase bone density, the converse also appears to be true. A reduction in loading can result in a significant decrease in bone density. This bone loss has been documented in a study of prolonged bed rest (for 17 weeks) (Leblanc et al. 1990). Even after six months of reambulation only the calcaneus returned to near normal density levels, though several areas showed positive increases in density over the reambulation period. Evidence of reduced bone density also comes from studies of fetal neuromuscular disease (Rális et al. 1976; Rodriguez et al. 1988a, 1988b). With this disease, there is decreased bone diameter, decreased cortical thickness, and decreased mineralization. These changes reduce both the strength and stiffness of the tissue and the overall structural properties of the bone. While factors other than mechanics may be implicated in these changes, bones of these children do respond to increased loading after birth and do form callus at fracture sites (Rodriguez et al. 1988a).

Loss in bone density (and muscle atrophy) is also noted in astronauts, especially those subjected to prolonged periods in a weightless environment (Baldwin et al. 1996; Cavanagh, Davis, and Miller 1992; Tipton and Hargens 1995). Substantial effort has been expended to develop appropriate and effective countermeasures to the loss of gravity loading during spaceflight. To date, no exercise program in this weightless environment has been fully successful in preventing muscle and bone atrophy.

There is also evidence of human bone adaptation in response to the altered mechanical environment induced by an orthopedic implant. Fracture plates used to reconstruct complex or severe fractures can shield the fracture site and surrounding bone from mechanical stresses and can result in a loss in bone density (Cordey et al. 1985). Femoral surface replacements, used at one time to treat severe arthritis, produce both localized stress shielding and increased stress that result in poor short-term outcomes due to component loosening (Murray and Van Meter 1982). Total hip replacements, more commonly used to treat severe arthritis, cause stress shielding that results in substantial bone loss in the proximal-medial femoral cortex (Engh and Bobyn 1988; Engh, Bobyn, and Glassman 1987). Such adverse remodeling, if continued for several years, will compromise the mechanical support for the implant (see figure 17.2), and the implant may loosen or fail. While early failures of such implants are relatively rare, the failure rate for cemented Charnley implants was found to increase dramatically after about 10 years of implantation (Schulte et al. 1993).

While human cross-sectional and prospective studies are confounded by problems of self-selected groups and subject compliance, animal studies have provided perhaps even more convincing evidence of bone adaptation to mechanical stimulus from exercise. For example, the axial skeletal calcium and bone volume in the rat was found to increase in exercised (running) rats compared to nonexercised controls (Saville and Whyte 1969). The ratio of muscle mass to bone was not significantly different between the groups, indicating that each increased proportionally to the exercise loads. Also, rats exposed to chronic 3-G centrifugation developed bones with higher bending strength than 1-G controls (Wunder 1977). Several surgical intervention studies performed with rabbits reported positive bone apposition in response to intermittent mechanical loading (Hert, Liskova and Landa 1971a; Hert, Pribylova, and Liskova 1972; Hert, Sklenska, and Liskova 1971; Liskova and Hert 1971). Canine fore-

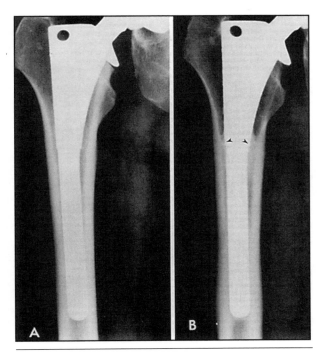

Figure 17.2 Postoperative (A) and one year (B) anterior-posterior roentgenograms of a one-third porous-coated AML hip stem. Note the extensive loss of bone density in the proximal medial femoral neck and also laterally. Also note extensive new bone formation near the junction of porous and smooth implant surfaces (arrows) and cortical hypertrophy near the stem tip.

Reprinted, by permission, from C.A. Engh and J.D. Bobyn, 1988, "The influence of stem size and extent of porous coating on femoral bone resorption after primary cementless hip arthroplasty," *Clinical Orthopaedics and Related Research* 231: 10.

limbs were found to hypertrophy under constant, sustained, extreme loads, and the ulnar hypertrophy was noted after resection of a portion of the radius (Chamay and Tschantz 1972). Tibial bone mass was found to increase in dogs running with lead jackets (Martin et al. 1981). Similarly, exercised pigs were observed to have increased cortical thickness and cross-sectional area, and ulnar resection resulted in hypertrophy of the radius (Woo et al. 1981a). Still, quantifying the loads in most animal studies remains problematic. Strain gauges on the bones in question can provide very useful information about loads and adaptation. Principal strains in the pig radius were more than doubled immediately after ulnar resection, indicating a doubling of the loads (Goodship, Lanyon, and McFie 1979). However, the strains returned to normal values after three months, indicating that bone adapts such that it achieves a specific strain level. One unique test method that noninvasively applies known in vivo loads to the rat tibia using a four-point bending

fixture may help improve understanding the quantitative relationship between mechanical stimulus and bone adaptation (Forwood and Turner 1994; Forwood et al. 1996; Turner et al. 1994). But even these experiments are somewhat confounded by the loading to the tibia from normal activities of the rat.

Animal studies of unloading have also demonstrated that bones adapt to reduced mechanical stimulus. Cartilaginous limb buds from the chick and rat were shown to develop to bones even when transplanted to a host with little or no mechanical stimulus (Felts 1959; Murray and Huxley 1925; Willis 1936). The bones, however, were shorter and narrower than normal bones at the same stage in development. Several animal studies of denervation have also reported reduced bone mass, reduced cortical area, and reduced mechanical properties in response to reduced muscular loading (Biewener and Bertram 1994; Gillespie 1954; Hall and Herring 1990; Wong et al. 1993). Cast immobilization, denervation, and excision of the humerus all caused a reduction in bone mass in growing and adult canine bones (rarefaction of trabecular bone and increased porosity in cortical bone) due to the unloading from each treatment (Allison and Brooks 1921). In addition, fracture strength of those unloaded canine bones was only one-third to one-half that of the contralateral controls. Other immobilization studies in dogs have also shown consistent and dramatic decreases in cortical thickness, increases in cortical porosity, and endosteal resorption in the immobilized limb (Jaworski, Liskova-Kiar, and Uhthoff 1980; Uhthoff and Jaworski 1978; Uhthoff, Sékaly, and Jaworski 1985). The cortical bone thickness, however, was shown to return to (at least near) normal upon reloading of the immobilized limb. The return of normal bone density was also observed in the rabbit calcaneus after removal of the immobilization cast, though bone apposition was slower than the initial bone resorption (Geiser and Trueta 1958). Similarly, the periosteal bone growth in rats is reported to essentially stop during the microgravity of space flight, but the diaphysis soon returns to normal thickness once the rats return to normal gravity (Morey and Baylink 1978). A recent study of bone resorption, with a disuse model in the turkey, reported that bone loss was primarily (84%) from uniform endosteal expansion, but loss from increased cortical porosity was preferentially located in the thickest areas of the cortex, those regions with perhaps previously higher levels of load (Gross and Rubin 1995).

Computational studies have illustrated that realistic normal femoral morphology (characterized by the apparent density distribution) could be predicted using finite element analysis of the mechanical stimulus (in the form of strain energy density) coupled with a mathematical theory of bone adaptation to the strain energy density levels (Beaupré, Orr, and Carter 1990; Huiskes, Weinans, and Dalstra 1989). Also, the results of strain energy-based computer simulations of bone adaptation around implants followed the trends of long term density changes around experimentally analyzed canine implants (Weinans et al. 1993). Similarly, realistic patterns of endochondral ossification were observed using a computational simulation with a mechanics-based ossification criterion (Wong and Carter 1990). In addition, results from simulations of human cross-sectional long bone growth based primarily on mechanics quantitatively corresponded to actual growth data (van der Meulen, Beaupré, and Carter 1993). These studies have illustrated the link between mechanics and bone adaptation but have not provided information about the specific form of the mechanical signal (or signals) or the mechanism by which the mechanical stimulus is transduced by the cell.

The most commonly held theory is that adaptation is driven directly by local strain levels (Cowin and van Buskirk 1978; Frost 1964; Hart, Davy, and Heiple 1984; Lanyon 1987; Rubin and Lanyon 1984b). In addition to the study by Goodship, Lanyon, and McFie (1979) that has already been discussed, this view is supported in part by evidence that dynamic strains are similar in a wide range of animal species of various sizes (Biewener 1991; Rubin and Lanyon 1984a). The dynamic strain similarity, however, could be caused by another measure of mechanical stimulus. Discussions about stress versus strain stimulus are rather pointless, since they are directly related by constitutive relationships. In addition, mechanical stimulus may be indirectly mediated to the cells. Some proposals for such mediation include mechanoelectrical transduction (Chakkalakal 1989), fatigue microdamage (Burr et al. 1985; Carter 1982; Prendergast and Taylor 1992), solubility changes for hydroxyapatite or extracellular fluid pressure changes (Treharne 1981), and fluid flow around the cells (Weinbaum, Cowin, and Zeng 1994). The specific form of the mechanical stimulus and its transduction pathway may someday be better understood through the use of modern molecular biology and biochemistry in conjunction with quantifiable experiments of mechanical stimuli. Regardless of the specific form of the stimulus or the transduction pathways for bone or any other tissue, it is very apparent that exercise, especially weight-bearing exercise, can increase bone density and structural properties.

In summary, bone adaptation to mechanical stimulus is a well-documented phenomenon. The response to increased mechanical stimulus is not as rapid or pronounced as the response to decreased mechanical stimulus. The return of bone density after a period of decreased stimulus is also much slower than the loss of bone density. Thus, periods of prolonged inactivity, such as bed rest, should be avoided if possible. The detrimental effects of decreased stimulus may be even more severe in the elderly, who are also at risk for osteoporosis.

From the perspective of one caring for the elderly, who is concerned with prevention and treatment of osteoporosis, exercise has been shown to be generally beneficial. Weight-bearing exercise especially may slow the process or prevent osteoporosis, and it has other health benefits as well. Certainly, the level of exercise must be tailored to the capabilities of each subject. For those following patients or studying bone loss, it appears that the femur and the lumbar spine are the most appropriate locations to perform radiological studies of bone density. Several papers have suggested, and the combined evidence strongly indicates, that the radius (or forearm) is not an appropriate site for monitoring bone density changes (Aloia et al. 1978; Colletti et al. 1989; Marcus et al. 1985; Smith et al. 1984). While total body calcium or density in weight-bearing bones is increasing, the density in the radius may remain unchanged or may decrease. Thus, the site should be avoided in future studies.

From the athlete's point of view, bone adaptation is generally beneficial, providing a stronger frame to carry the extreme loads during training and competition. Stress fractures are one problem that may affect the elite athlete. At the beginning phases of bone adaptation, there is an activation of osteoclasts, which remove existing extracellular matrix. If the bone-forming osteoblasts cannot replace this bone quickly enough, microdamage from the increased loading may begin to accumulate and could lead to a stress fracture (Ovara, Puranen, and Ala-Ketola 1978). While not catastrophic if treated and allowed to heal, a stress fracture could certainly prevent continuation of training and participation in competition. Only a period of reduced loading (i.e., rest) will allow the stress fracture to heal. In cases of extreme physical exertion, there is indication that bone mass can actually be lost, even though the stimulus is very high (Pouilles et al. 1989). Thus, while bone clearly adapts positively to changes in mechanical stimulus, training levels should be increased moderately, steadily, and continuously to allow time for positive bone adaptation to occur.

The principle of slow increases in activity should also be (and generally is) observed for those who may be undergoing rehabilitation after severe injury requiring prolonged bed rest.

Articular Cartilage Adaptation

Articular cartilage is a skeletal connective tissue that is not noted for its ability to adapt to mechanical stimulus. This avascular tissue is normally quiescent, with minimal cellular activity in the adult. Articular cartilage is considered an excellent load-bearing surface for the articular joints, due to its viscoelastic properties and low coefficient of friction. But articular cartilage is also considered somewhat vulnerable to excess loading and injury, with a poor ability to restore normal structure and function of the tissue.

Does cartilage have the ability to positively adapt to increased mechanical stimulus from exercise or to repair itself after injury? Recent studies indicate that it has the ability for both. In one study, young female beagle dogs (15 weeks old at the initiation of the study) ran on a treadmill inclined at 15° (Kiviranta et al. 1988). After a 10-week period for the dogs to become accustomed to the treadmill, they ran for an hour a day, five days a week, at 4 km/h, for 15 weeks. When the dogs reached the age of 40 weeks, histological tissue samples were collected from 11 different anatomic sites of the right knee joint. The thickness of the uncalcified cartilage was 19 to 23% greater on the lateral condyle and patellar surface of the femur of the trained dogs (as compared to unexercised controls). Enhancement in cartilage thickness was less in other parts of the articular surfaces. Total glycosaminoglycans were increased by 28% in the summits on the femoral condyles (more on the medial condyle than on the lateral condyle). The increase in glycosaminoglycans appeared to be primarily due to additional condroitin sulfates localized in the intermediate, deep, and calcified zones of the cartilage. The superficial zone did not show any changes. The results of this study indicate that running exercise locally alters the biological properties of young articular cartilage in regions that bear the highest loading. These changes appear to be a positive adaptation that tends to offset the effects of increased loading. Because of the young age of these growing animals, however, it is difficult to extrapolate these results to adult humans. Such adaptation may only be possible during the growth phase of life.

In a subsequent study of animals, animals of the same age as in the previous study were trained more vigorously, reaching a maximum running level of 40 km/day (Lammi et al. 1993). This study considered cartilage adaptation in the hip joint. No macroscopic or microscopic signs of articular degeneration or injury were visible in any of the animals. Running increased the amount of proteoglycans resistant to extraction, particularly in the less weight-bearing tissue. The proteoglycan increase could indicate the presence of a stiffer collagen network and suggests a tendency for positive adaptation. The authors concluded that the cartilage of the femoral head appeared to have a greater capacity to adapt to the increased mechanical loading than the cartilage of the femoral condyles. They ascribed this difference to the relative congruency of the hip joint. Certainly, many other differences, such as joint range of motion and ligamentous and bony constraints, could also have played a role in the results.

Immobilization of the knee joint of young beagle dogs has been shown to reduce the concentration of articular cartilage glycosaminoglycans (GAGs) by 13 to 47%, indicating a weakening of the tissue (Kiviranta et al. 1987). After immobilization for 11 weeks and remobilization for 15 weeks, the GAG concentration was restored in the patellofemoral region and the tibial condyles (Kiviranta et al. 1994). On the femoral condyles (especially at the periphery), GAG concentration remained 8 to 26% less than the control values. On the load-bearing areas of the femoral condyles, the thickness of the uncalcified cartilage was as much as 15% less than in the age-matched controls. The changes induced by unloading were somewhat reversible, but recovery appears to be very slow and some of the alterations may be permanent. These canine studies indicate an increase in thickness and tissue structural content of the cartilage with increased loading and atrophy of the cartilage tissue with unloading. Thus, the cartilage does appear to continually adapt to its mechanical stimulus, but the rate of adaptation appears to be relatively slow. Again, one should note that all these canine studies were performed in growing animals. Adaptation of the cartilage may be substantially enhanced by the growth factors present in these young animals.

A study of the knee meniscus in a rat model showed that menisci from rats, exercised on a treadmill at 1.8 km/h at 12% grade for 1 h per day, had increased concentrations of collagen, proteoglycan, and calcium in the posterior region of the lateral menisci (Vailas et al. 1986). However, no significant changes were found in the anterior region of the lateral meniscus. The authors point out that these region-specific changes in meniscal concentrations of calcium and matrix macromolecules are consistent with the distinctly different functional roles of the anterior and posterior regions of the rat knee meniscus. Because the rat knee remains flexed during normal gait, the posterior region is believed to carry greater loads throughout the step cycle. Thus, even the meniscus, with its low metabolic activity and relatively poor blood supply, demonstrates an ability to adapt to increased mechanical stimulus from exercise.

Biochemical investigations have suggested a link between mechanical stimulus and the synthesis of the structural proteoglycans biglycan and decorin. An in vitro study of cultured immature bovine articular cartilage found that synthesis of decorin, but not biglycan, was significantly higher in cartilage subjected to 0.2 MPa (at 2 Hz) loading for a week (Visser et al. 1994b). Decorin may aid in adaptation of articular cartilage through its interaction with collagen type II. A similar study of mature bovine articular cartilage found the same increase in decorin synthesis with no increase in biglycan synthesis (Visser et al. 1994a). In the mature tissue, there was a decrease in biglycan synthesis from the initial levels, but the initial levels of biglycan synthesis were maintained in the immature cartilage. This difference indicates the possible reduction in adaptation of mature cartilage.

Another study of cartilage explants showed that short-term (14 h) static compression (see figure 17.3) reduces the biosynthetic rates of cartilage, as measured by sulfate and proline incorporation (Boustany et al. 1995). Those markers are used to determine synthesis of proteoglycan, a major structural component of cartilage extracellular matrix. Surprisingly, when the duration of the compression was increased to 60 hours, there was a modest increase in biosynthetic activity. These results demonstrate the phenomenon of biosynthetic changes with mechanical stimulus, but they also indicate that the relationship between specific loading conditions and cartilage adaptation is quite complex. Further work is required to better characterize the response of cartilage to specific pressure states.

Similar work has shown that dynamic, cyclic pressure has a greater potential to stimulate cartilage adaptation (Suh et al. 1996). A study of cyclic hydrostatic pressure on bovine cartilage explants indicates that there is a frequency dependency of the response, and more recent work suggests that it may be the subambient pressures (which are induced within the implant during the release phase of a positive cyclic compressive waveform) that are most stimulatory to proteoglycan synthesis. Recent

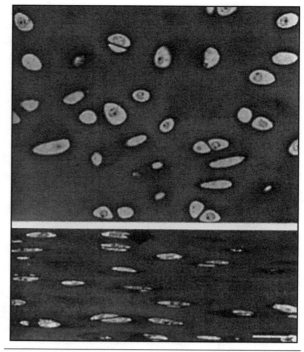

Figure 17.3 Histological sections of a 2 mm diameter cartilage disk that was fixed (top) under free-swelling conditions and (bottom) while compressed to 50% of the cut thickness. With this extreme compression, there is clear deformation of the chondrocytes in the cartilage.

Reprinted, by permission, from N.N. Boustany, M.L. Gray, A.C. Black, and E.B. Hunziker, 1995, "Time-dependent changes in the response of cartilage to static compression suggest interstitial pH is not the only signaling mechanism," *Journal of Orthopaedic Research* 13:740-754.

work by the same group appears to confirm that the relative negative pressure in the tissue (induced by dynamic unloading) is the most important component of the mechanical stimulus (Suh et al. 1997). Again, further work is needed to fully understand the specific stimuli and the corresponding responses.

With respect to mechanism of transduction from the stimulus to the adaptation response, there is some evidence that compression of the cartilage provides a direct mechanical signal to the cell (Buschmann et al. 1996). With compression of a cartilage explant, inhomogeneity was noted in the aggrecan synthesis, which was not present prior to compression. In addition, there were reductions in cell and nucleus volume and radii in the direction of compression (see figure 17.3). These observations lend support to the hypothesis that the cellular signals may be primarily mechanical or mechanically mediated chemical pathways.

One experimental study examined the relationship between articular cartilage compressive modulus and average joint stress (Yao and Seedhom 1993). Indentation testing of cartilage in the human ankle and knee revealed that compressive modulus was significantly correlated with the compressive mechanical stress. While this study demonstrated a relationship between mechanical stress and cartilage compressive modulus, it does not prove that mechanical load alters cartilage modulus. This study does, however, confirm observations and conclusions of the animal studies discussed above. The study was limited, in that compressive stresses were measured in only one position for each joint. Certainly the stresses could change substantially with joint position.

The studies discussed previously consider some of the effects of mechanical stimulus on cartilage, but the athlete would like to know how exercise affects the cartilage in the human knee. Specifically, is exercise beneficial or detrimental to the health of one's cartilage? The results of the animal studies discussed above indicate that exercise would modify the mechanical properties and thickness of the articular cartilage in a beneficial way. But exercise has been implicated in the development of osteoarthritis (OA) and is generally prohibited for patients diagnosed with severe OA. Is prohibition of exercise warranted, especially in light of the fact that

regular exercise improves general health and may increase longevity whether it is recreational or competitive (Lane 1996; Paffenbarger et al. 1986)?

Cross-sectional studies of jogging report conflicting results about its impact on development of OA. One study found that female runners have more radiographic evidence of knee subchondral sclerosis and osteophytes than controls but found no differences between male runners and controls (Lane et al. 1986). Another study found that age, pace per mile, and weight were predictors of the radiographic knee OA score (Lane et al. 1993). Long-term follow-up found no differences in radiographic scores for knee OA between those still running, those who stopped running, and those who had never run (Lane 1996). One convincing set of data comes from a 10-year canine study of dogs that ran at 3 km/h for 75 min, 5 days a week for 527 weeks, carrying weighted jackets with 30% of their body weight (Newton et al. 1997). The cartilage of the knee joints of 11 experimental (exercised) animals and 10 control animals was inspected for evidence of joint injury or degeneration. No joints showed evidence of ligament or meniscal injuries, cartilage erosion, or osteophytes. Using light microscopy, the investigators found no evidence of cartilage fibrillation (see figure 17.4) or

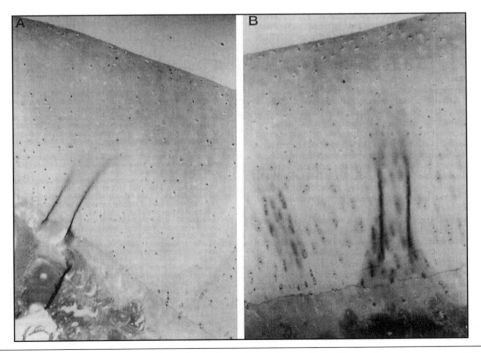

Figure 17.4 Hematoxylin and eosin stained histological sections of a control (A) specimen from an unexercised dog and of an experimental (B) specimen from a trained dog. Note that both specimens have a smooth and intact cartilage surface.
Reprinted, by permission, from P.M. Newton, V.C. Mow, T.R. Gardner, et al, 1997, "The effect of lifelong exercise on canine articular cartilage," *American Journal of Sports Medicine* 25(3): 285.

difference in safranin O staining of the tibial articular cartilage between the two groups. Tibial articular cartilage thickness and mechanical properties did not differ between the two groups (Newton et al. 1997). These data strongly indicate that a lifetime of weight-bearing exercise should not lead to degeneration of normal joints.

For high intensity exercise, the data more strongly indicate that such exercise may increase the risk of OA, though some evidence is conflicting. A study of Finnish elite runners, mean age 55 years and an average of 21 years of running, found no increased radiographic or clinical OA compared to similar-aged controls (Puranen et al. 1975). A study of former college varsity cross country runners reported that their activity did not predispose them to knee or hip OA (Sohn and Micheli 1985). However, former Swiss national long-distance runners were found to have more radiographic evidence of hip OA than bobsled competitors and controls (Marti et al. 1989). Age and mileage run were found to be independent, significant, positive predictors of radiological signs of degenerative hip disease in that study. Also a retrospective study of OA of the hip, knee, and ankle reported that athletes from all types of competitive sports had a small increase in risk for OA requiring hospitalization (Kujala, Kaprio, and Sarna 1994). Athletes participating in sports requiring high power maneuvers (such as soccer, ice hockey, basketball, wrestling, and weightlifting) were at higher risk for painful OA at a younger age than endurance athletes were (Kujala et al. 1995). A study of lifetime loading histories in a group of elite athletes showed that 3% of shooters, 29% of soccer players, 31% of weightlifters, and 14% of runners had radiographic evidence of knee OA. The authors suggested that increased risk in soccer players may be due to injury and increased risk in weightlifters may be due to high body mass (Kujala et al. 1995). The injury theory is supported by further research, which showed that prevalence of knee OA was 1.6% in controls, 4.2% in non-elite soccer players, and 15.5% in elite soccer players (Roos et al. 1994). When players with documented injuries were analyzed separately, 13% of non-elite players had knee OA compared to only 3% of uninjured non-elite players. However, 11% of uninjured elite soccer players were still noted to have knee OA (Roos et al. 1994). This finding indicates that substantially overloading a normal joint may increase the risk of OA.

Certainly in the case of abnormal joints, due to anatomic variance or injury, exercise is highly correlated with OA. One study of long-distance runners found that six of 20 had radiographic and clinical evidence of OA, and each of the six with OA had either anatomic variances or a record of past injury (McDermott and Freyne 1983). A study of soccer players (20 or more years after knee joint injuries and partial meniscectomies) reported that, of those with an intact anterior cruciate ligament (ACL), 24% had knee OA (Neyret et al. 1993). Of those who had ruptured an ACL, 77% had radiographic OA. These data indicate that injury to the meniscus or the ACL appears to accelerate the development of OA (Lane 1996).

An interesting twist comes from recent data indicating that low impact to moderate exercise may be beneficial to arthritic joints as well as normal joints. While the exercise may not reverse and repair the effects of OA, it does appear to reduce pain and improve function of the joint and certainly has benefits to the cardiovascular and pulmonary systems (Minor and Lane 1996). Short-term prospective studies have found that moderate supervised exercise can benefit patients with OA. In one such study, patients were able to tolerate weight-bearing exercises and experienced a decrease in arthritis pain (and a reduction in medication use) with fitness walking (Kovar et al. 1992). A study of home-based and hospital-based exercise programs to strengthen and increase endurance of the knee extensors in patients with OA of the knee found that patients had decreased pain, increased function, increased maximal weight lifted, and increased endurance after four weeks of therapy (Chamberlain, Care, and Harfield 1982). A control group of nonexercisers, however, was not included in the study. Suggested goals for such exercise programs in patients with OA included increasing or maintaining joint motion, increasing strength and endurance of periarticular muscles, increasing aerobic capacity, assisting in weight loss, and improving functional capacity in activities of daily living (Minor and Lane 1996). The importance of the aerobic component was illustrated in a study that compared aerobic walking, aerobic aquatics, and nonaerobic controls (Minor et al. 1989). The aerobic groups showed significant improvements in aerobic capacity, walking speeds, depression, anxiety, and physical activity after 12 weeks of exercise. Such programs must be carefully planned and patients should be educated and monitored, because overuse of an actively inflamed joint is likely to aggravate the inflammatory process and increase joint deterioration (Minor and Lane 1996). Current knowledge suggests that vigorous activity is ill advised in the presence of active inflammation.

In summary, data from animal studies, in vitro studies, and cadaveric studies all confirm a link between mechanical stimulus and the mechanical integrity of cartilage tissue. While the animal study data demonstrate that cartilage can adapt to increased mechanical loading in growing animals, there is a lack of data from studies of mature animals. Cadaveric studies confirm a correlation between adult cartilage thickness and mechanical stress, but this could be a result of developmental processes as well. In vitro studies indicate that the chondrocytes can and do respond, to increase the mechanical integrity of the extracellular matrix. The question remaining is at what rate and to what extent can this adaptation take place? Further studies will be required to better characterize the adaptation of articular cartilage to mechanical stimulus.

This section has also reviewed data that provide a framework for beginning to assess the risks of developing OA as a result of participation in moderate exercise or athletics. There are several limitations that must be noted with regard to this clinical data. The radiographic methods and criteria may not be consistently applied in all studies. Most of the studies are cross-sectional in nature and only provide information about single points in time. Certainly, the onset of OA is time dependent. Cross-sectional studies that indicate no difference between groups cannot provide conclusive data, since subjects must be followed longitudinally to determine whether disease may yet develop. Cross-sectional studies that indicate a difference between groups must be treated with caution, since those more athletic individuals may have a body type that predisposes them to OA. In addition, many of the studies rely on retrospective data that may, in some instances, be unreliable or inadequate. Clearly, more thorough longitudinal studies would be beneficial.

Still, the data do provide some guidelines that can be helpful with regard to prevention and treatment of OA. Moderate exercise should pose little or no threat of OA for the normal joint. Elite competition appears to predispose even a normal joint to a higher incidence of OA. And even moderate exercise appears to increase the risk of OA in abnormal or previously injured joints. Also, low impact exercise (even weight-bearing exercise) appears to be beneficial for patients with OA, as long as the joint is not actively inflamed. The fact that data from some studies conflict, however, indicates that firm conclusions may not yet be drawn. Further, more controlled and more thorough studies may help sort out the risk factors for OA.

Tendon Adaptation

Tendon adaptation has primarily been described with morphometry, histology, molecular biology, and ultrasonography. Relatively few studies have addressed the issue of the mechanical strength of the tissue, except by inference from other measures of tissue structure. Detailed information on normal tendon structure is available in book chapters and review articles (Viidik 1990).

Several investigations have focused on the recognition and treatment of tendon overuse injuries (Archambault, Wiley, and Bray 1995; Hess et al. 1989; Kvist 1994; Scioli 1994; Williams 1986, 1993). These have identified external factors related to such injuries, such as the specifics of the training program, increases in intensity, duration, and frequency of physical activity, and a change in the way an activity is performed (Williams 1986). In addition to such training errors, factors such as the external temperature, the condition of the surfaces during training or competition, the shock absorption of the playing surface, and improper equipment are also factors in tendon overuse injuries (Hess et al. 1989; Kvist 1994). Intrinsic factors such as joint malalignment, limb length discrepancies, muscular imbalance, and muscular insufficiency have also been identified as risk factors for tendon overuse injury (Hess et al. 1989; Kvist 1994). Diseases such as gout or systemic corticosteroid treatment are also risk factors for tendinitis (Scioli 1994). Still, it is not clear exactly what magnitude, duration, or frequency of loading (or what type of loading) is most likely to cause tendon overuse injuries. Even the normal loading of tendons in activities of daily living has not been fully quantified. There are data on tendon loads measured in humans (Archambault et al. 1995), but this is limited both with respect to the number of tendons measured and the number of activities considered.

Exercise has been shown to induce specific structural and biochemical changes in tendons. Changes in tendon cells, collagen fibrils, and ultrastructure have been documented in various animal models in response to excessive loading. An overload model of the rat plantaris tendon (produced by ablation of the synergists) showed that there was a notable transformation of quiescent fibrocytes into active fibroblasts without the presence of inflammatory cells (Zamora and Marini 1988). Collagen bundles in the tendon were disrupted, and the investigators noted empty longitudinally oriented spaces. The authors proposed that as the tendon begins to hypertrophy it undergoes a transient period of mechanical weakness, as indicated by the ultrastructural changes. Because the level of loading and how the load relates to human activities is unknown, it is somewhat unclear whether this report of transient structural weakness is a factor in human tendon overuse injuries.

Other investigators have used specific cellular markers to examine the effect of high intensity exercise on immature tendons (Curwin, Vailas, and Wood 1988). For white leghorn roosters in a running group, they found that collagen turnover was increased and mature pyridinoline cross-links decreased in immature tendons. Collagen deposition was increased, but the number of cells (DNA content) and total collagen content did not change. This short-term study did not determine if the tendons returned to normal with continued exercise or with maturation of the animals.

Other studies that indicate a transient structural weakness during adaptation to increased mechanical stimulus examined the effect of exercise on the flexor digitorum longus tendon in young mice (Michna 1984; Michna and Hartmann 1989). At one week after initiation of the treadmill exercise program, increases in mean diameter (30%), number of fibrils (15%), and cross-sectional area of tendon fibrils (15%) were noted (Michna 1984). During the time between three to seven weeks, there was a fall in the level of the controls of mean fibril diameter (26%), number of fibrils (26%), and fibril cross-sectional area (17%) (Michna 1984). In the long term (at 10 weeks), the mean fibril diameter of the exercise group was either not significantly different from controls or only marginally smaller. The collagen fibrils found at 10 weeks in the exercisers did not appear substantially different from those in tendons not subjected to training, though the density of fibrils was somewhat lower (see figure 17.5). The reduced fibril diameters in the midportion of the training program also indicate that the mechanical strength of the tendon is reduced for a period of time. The tendons of mice that underwent long-term exercise had larger fibril diameters than those of mice that did not undergo training, but they also had a broader distribution of fiber size, including very thin fibrils. These findings indicate that if the exercise level is not so high as to cause an overuse injury the tendon will eventually adapt to a higher level of mechanical strength. Because of the young age of the mice in these studies, it not clear whether there is a growth or age-related constraint on the adaptation, or whether increased fibril diameters would be more pronounced in a model with mature animals. More obvious long-term adaptation was

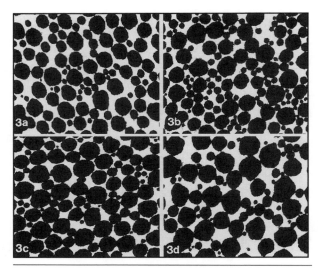

Figure 17.5 Distribution of collagen fibrils in a control mouse tendon (*3a*), a tendon from an exercised mouse after one week (*3b*), a control tendon after 10 weeks (*3c*), and a tendon from an exercised animal after 10 weeks (*3d*). Note larger number of fibrils and higher fibril density after one week. There is a lower fibril density after 10 weeks, but there is an increase in the number of small collagen fibrils.

Reprinted, by permission, from H. Michna and G. Hartmann, 1989, "Adaptation of tendon collagen to exercise," *International Orthopaedics.* 13: 162. © Springer-Verlag.

found in the rat calcaneal tendon with endurance treadmill running, where an increase in mean fibril cross-sectional area (32%) was present after 10 weeks of training (Enwemeka, Maxwell, and Fernandes 1992). Food restrictions were found to prevent any significant increase in fibril diameter in the rat calcaneal tendon.

It has been shown that changes in the mechanical stimulus from treadmill running exercise can produce changes in the levels of insulin-like growth factor-1 (IGF-1) in the fibroblasts and the paratenon of a rat tendon (Hansson 1988). Because IGF-1 stimulates collagen synthesis and cell replication, the IGF-1 level is likely an important indicator of tendon adaptation. Another study of rat tendon adaptation showed that hind-limb suspension reduced the collagen and proteoglycan concentrations in the patellar tendons, compared with nonsuspended controls (Vailas et al. 1988). These data suggest that tendons adapt to reduced loading (in a manner opposite to that for increased loading) by decreasing tendon material properties.

Given all the evidence for tendon adaptation presented above, it is clear that exercise will induce an adaptation response that will, in the long term, result in improved ultrastructure of the tendon. It also seems likely that, if the stimulus is too great or unrelenting, the initiation of remodeling may weaken the tendon to a point at which it becomes susceptible to overuse injury. Still, the studies of tendon adaptation discussed thus far only consider the ultrastructure and chemical environment of the tendon. Are these findings substantiated by investigations of the mechanical integrity of the tendon with strenuous exercise?

In the studies presented earlier, several authors suggested that the mechanical integrity of the tendon is reduced in the early stages of tendon remodeling. The mechanical properties of tendons exposed to exercise training have actually been evaluated in only a few cases. The tendons of rabbits trained on a running machine for 40 weeks did not show significant increases in stiffness, energy absorbed to failure, or maximum load (Viidik 1969). The authors did, however, note a trend toward a slight increase in those parameters. Digital extensor tendons in swine, however, did produce an increase in the collagen concentration, the tendon cross-sectional area, and the maximal load (also the ultimate strain was decreased) after 12 months of treadmill exercise (Woo et al. 1980). This long-term result is consistent with the ultrastructural studies on flexor digitorum longus tendons in mice (Michna 1984; Michna and Hartmann 1989). The results were somewhat different for the flexor tendons of the swine. While there was a significant increase in the stiffness and failure load of exercised tendons, there was no significant increase in the collagen content (Woo et al. 1981b). The differences could be due to variations in the loading of these tendons during locomotion.

Tendon adaptation has also been studied in rats subjected to programs of speed, endurance, and combined speed and endurance running exercise (Sommer 1987). The Achilles tendon cross-sectional area and ultimate strength were measured at time periods of 4, 8, 12, and 16 weeks after the initiation of the exercise program. After 16 weeks of running exercise (in three different categories), the cross-sectional area of tendons increased, but the ultimate tensile strength decreased, when compared to the tendons of control animals. These results suggest that the ultrastructure of the tendons was not normal and further substantiate the claims of transient structural weakness. Perhaps the affects of maturation in these rats also affected adaptation of the tendons to increased mechanical stimulus. Also, the failure load of the tendons from exercised animals was not reported, but a close look at the reported averages suggests that failure loads (and thus the overall structural properties) of control

and exercised animals were likely not statistically different.

The tensile properties and fibril alignment in rat tendons were also studied using a treadmill running exercise model (Vilarta and Vidal 1989). The average tensile strength of tendons was increased in animals in the exercise group, and birefringence measurements showed that the fibrils were better aligned in tendons of exercised animals. These results indicate that the overall strength of the tendon is increased by increasing cross-sectional area and improving the ultrastructure of the tendon during adaptation to increased mechanical stimulus.

In addition to the studies of adaptation to exercise, investigators have also considered the effect of compressive stresses, caused when a tendon wraps around a bone, on cellular and tissue changes in the tendon. The compressive zone corresponded to a major fibrocartilagenous region (see figure 17.6) in the flexor digitorum profundus tendons in the canine forefoot (Okuda et al. 1987). The rabbit flexor digitorum profundus tendon normally has a region of fibrocartilage where it wraps around the talus. This region is characterized by rounded chondrocyte-like cells, as opposed to elongated fibroblasts in the bulk of the tendon. In addition, the tissue is identified as cartilaginous within the tendon by the presence of type II collagen in the extracellular matrix (Giori, Beaupré, and Carter 1993). When the tendon was removed from the groove in the talus, the transverse compression on the tendon was eliminated,

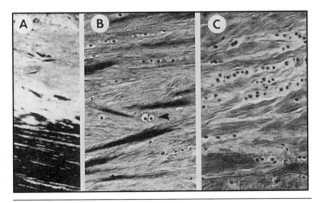

Figure 17.6 Low power view of the fibrocartilage region in a tendon (left), a high power view of the fibrocartilage region using Movat stain (middle), and a high power view of the fibrocartilage region using H and E stain (right). The arrow in the middle section indicates chondrocyte-like cells arranged in columns.

Reprinted, by permission from Y. Okuda, J.P. Gorski, K.N. An, and P.C. Amadio, 1987, "Biochemical, histological and biomechanical analyses of canine tendon," *Journal of Orthopaedic Research* 5: 63.

and the investigators noted a rapid modulation of the fibrocartilagenous region toward decreased GAG levels similar to those of tension-transmitting tendon (Gillard et al. 1979). Upon returning the tendon to the groove, the GAG levels returned to those of the original fibrocartilagenous tissue, confirming the modulation of the cell behavior by the mechanical environment. Finite element modeling of the tendon wrapping has demonstrated that the fibrocartilagenous regions correspond to a region of high compressive hydrostatic stresses in the tissue (Giori, Beaupré, and Carter 1993). Thus, tendons not only respond to the overall loading levels, but also adapt to the local mechanical environment.

While these studies provide some insight into adaptation response of tendon to mechanical stimulus from exercise, there is some evidence to suggest an indirect mechanism in adaptation and overuse injuries. Using a laser Doppler flowmeter, blood flow was recently measured in the human Achilles tendon, and it was found that flow generally decreased with contraction of the triceps surae muscles (Åström and Westlin 1994b). Interestingly, a slight increase in blood flow was noted during recovery following the contraction. A similar study found that flow in chronically symptomatic human Achilles tendons was increased compared to that in control tendons, but flow was still reduced by muscle contraction (Åström and Westlin 1994a). Animal studies have also demonstrated significant increases in blood flow in the Achilles tendon of the running dog (Bülow and Tøndevold 1982). Thus, tendon blood flow is necessary for normal maintenance and adaptation of the tendon and appears to be upregulated by exercise. In addition, it is interesting that areas of high compression, as a tendon wraps around a bone, will tend to inhibit vascularity, and these regions are where fibrocartilage is typically formed, as discussed previously. Thus, there may be interplay among mechanical stresses, vascularity, and tissue adaptation.

Based on the studies discussed above, it is clear that tendons do adapt at the microstructural level to mechanical stimulus from exercise and that in the long term these changes result in increased tissue organization and increased mechanical properties. Tendons first respond to increased loading by increasing the tendon cross-sectional area, but increases in tensile strength or overall failure strength of the tendon are slow. Increased blood flow in the tendon appears to be related to both positive adaptation and overuse injuries. The root cause of overuse injuries is still unidentified, but it appears that it may be related to a transient period of mechanical

weakness in the early stages of tendon remodeling. Some biochemical and histological evidence indicates that the tendon is weakened in the early stages of adaptation. Mechanical testing of tendons subjected to increased loading confirms this loss in strength. This finding may have great implications for the origin of overuse injuries such as tendinitis. Tendons display an even more clear reduction in tissue quality and mechanical strength during periods of reduced mechanical stimulus. Evidence of the modulation of tendon tissue and cells toward a cartilage phenotype imposed by local stresses demonstrates the adaptability of the passive connective musculoskeletal tissues. We can learn much more about tendon adaptation through further studies that are more thorough and complete with regard to quantifying the increased level of loading, measuring ultrastructural and mechanical properties, and taking measurements at several time points throughout the testing period.

For the athlete or the therapist, as was suggested with regard to bone, a slow and gradual increase in training level, activity, or rehabilitation exercises appears to be the primary recommendation with regard to tendons. A sudden increase in mechanical stimulus will trigger a remodeling response, and if the remodeling and repair does not exceed the rate of damage from overload, an overuse injury is likely to occur. Because a change in activities may shift the primary exertion to different muscle groups (and associated tendons) that are not well trained, even the elite athlete should begin gradually when attempting to learn a new athletic activity.

Ligament Adaptation

Of all the passive connective musculoskeletal tissues, the least amount of research has been performed with regard to ligament adaptation. Still, from the limited studies that have been performed, we find evidence that mechanical stimulus can induce changes in the microstructural composition and the mechanical properties of ligaments. While the adaptation response noted was not always proportional to changes in mechanical loading, there appear to be both hypertrophy in response to increased loading and atrophy in response to decreased loading.

A study of several groups of rats with different activity levels showed that trained (running) animals and trained-detrained animals both had significantly higher separation force of the medial collateral ligament (MCL) than nontrained controls (Tipton, Schild, and Tomanek 1967). In addition, the same study found that cast immobilization significantly reduced the separation force of the MCL. Because failures occurred at the tibial attachment, the authors postulated that the adaptation occurred at the ligament insertion site (Tipton, Schild, and Tomanek 1967). Thus, there is a question as to whether the ligament adapted or whether the bone at the insertion site adapted to the changes in stimulus. The question was partially answered by a study that considered both failure load and ligament elasticity (Viidik 1968). Anterior cruciate ligaments (ACLs) from rabbits that were trained by running exhibited a tendency for higher energy to failure and maximum load, but again, the failure site was usually the tibial insertion. They did note differences in load-relaxation behavior between trained and untrained groups and attributed these to changes in the ligamentous tissue of trained rabbits (Viidik 1968). It is true, however, that the insertion site properties may have some effect on stiffness and relaxation behavior of the entire complex. Many of these early studies on ligament adaptation were confounded by avulsions and insertion site failures, making their data somewhat difficult to interpret (Laros, Tipton, and Cooper 1971; Tipton et al. 1970). These studies did, however, show an increase in structural properties of ligaments with increased physical activity.

At least part of the reduction in strength of the ligament complex is probably due to underlying bone resorption in the insertion site region, which increases the tendency toward avulsions. Ash content analysis of the talus and calcaneus indicates bone loss in the immobilized limb, and histology of the ACL insertion site shows increased porosity through enlarged haversian canals (Noyes 1977). Histological analysis has also shown that fibrous tissue replaces the resorbed bone, deep to the canine MCL and LCL insertions, apparently extending the insertions deeper into the remaining bone (Laros, Tipton, and Cooper 1971). This phenomenon was not observed in other knee ligament insertions.

Several studies have provided data that indicate the ligament itself also adapts to increased or decreased loading. Rats that ran on a treadmill as endurance exercise were found to have significantly stronger and stiffer ACLs than nonexercised controls (Cabaud et al. 1980). In this study, nearly all ligaments failed in the midsubstance, so the strength parameters (see figure 17.7) can be associated with the ligament tissue itself. The likely reason that these investigators were able to achieve consistent midsubstance failures was their use of a high strain rate of 95%/s. Note that the increases in mechanical and structural properties were only on the order of 10%. Surprisingly, they found significantly higher maximum linear ACL load in the groups exercised 30 min/day than in those exercised 60 min/day, suggesting that short periods of exercise may be more beneficial. While those rats exercised 6 days/

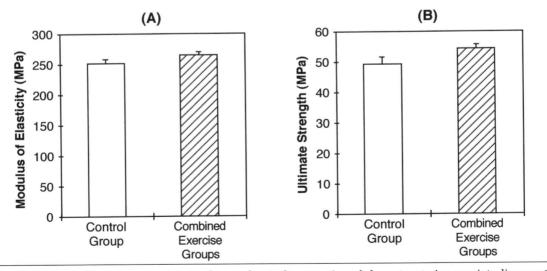

Figure 17.7 The effect of running exercise on the mechanical properties of the rat anterior cruciate ligament. The comparisons shown here are between the control (unexercised) group and the combined data from four groups with different exercise regimens. The differences for both comparisons are statistically significant ($p < .05$). Data compiled from Cabaud et al. 1980.

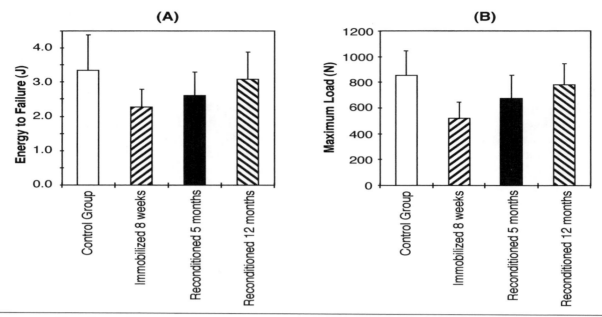

Figure 17.8 The effects of immobilization and retraining on the structural properties of the anterior cruciate ligaments of rhesus monkeys. Immobilization caused a highly significant ($p < .001$) drop (39%) in maximum failure load and similar changes in energy absorbed to failure. Retraining improved ligament properties, and though the average properties are still below the control after 12 months of reconditioning, those differences are not statistically significant. Data compiled from Noyes 1977.

week showed a trend toward higher average maximum load than those exercised only 3 days/week, the differences were not significant.

Exercise and immobilization both appear to regulate ligamentous tissue mechanical properties. The effects of limb immobilization and reconditioning on ligament properties have been investigated in a primate animal model (Noyes et al. 1974). With 53 male rhesus monkeys in four different groups, they observed a decrease in failure load (39%) and energy absorbed to failure (32%) and found an increase in compliance of the ACL after 8 weeks of lower limb immobilization. After 20 weeks of resumed activity, there was only partial recovery in ligament strength, although compliance had nearly returned to normal. In this study, the most common type of failure was a midsubstance failure of the ligament itself, but there was an increased tendency toward femoral avulsion fractures in immobilized specimens (Noyes et al. 1974). The distribution of failures returned to normal after retraining for 12 months (Noyes 1977), and the average strength properties (see figure 17.8) were within 10% of the controls (not statistically different). This study provided strong evidence that the ligament itself, not just the insertion site, adapts to the change in mechanical stimulus.

A similar study of 50 rats examined five different groups to determine the effects of immobilization

and retraining on the strength and stiffness of the anterior and posterior cruciate ligaments (PCL) (Larsen, Forwood, and Parker 1987). The ACL from the right leg and the PCL from the left leg of rats from each treatment group were tested in tension while still attached to the femur and tibia. After only four weeks of immobilization, the failure load of the anterior ligament declined by 25% and the stiffness of both ligaments declined 25% and 33%, respectively. No significant reduction in PCL failure load was noted. After an additional six-week retraining program of swimming, the failure load and stiffness values returned to control levels. In this model, part of the failure load component included bone strength, as failures included both avulsions (in the immobilization group) and insertion site failures. Still, the stiffness changes are indicative of a reduction of mechanical properties of the ligament with immobilization and return of those properties with retraining.

Ligament adaptation to immobilization was also demonstrated in a rabbit model (Woo et al. 1982). The researchers noted significant decreases in ultimate load and stiffness of MCLs from immobilized limbs as compared to contralateral controls. Stiffness, ultimate load, and energy absorbed to failure of the immobilized MCLs were approximately one-third of those from the contralateral, nonimmobilized control. Because of the substantial drop in structural properties and (assumed) minimal change in cross-sectional area of the ligament, the authors deduced that the mechanical properties of the ligament were reduced. Similar reductions in ultimate load and stiffness were found in isolated lateral collateral ligament (LCL) tissue of immobilized rabbits (Amiel et al. 1982).

The response of a healing MCL to exercise has also been investigated in a rat model (Vailas et al. 1981). Compared to immobilization, progressive exercise markedly enhanced the healing process by stimulating a more rapid recovery of the ligament separation force to within "normal" limits. Not only was the separation force of repaired ligaments in exercised animals greater than that of continuously immobilized and nonexercised (repaired) ligaments, but the exercise group had higher (though not significantly) separation force than the uninjured control ligaments. In addition, biochemical analysis revealed that DNA concentration (cellularity) was closest to normal levels in the exercised group, and those ligaments contained the greatest amount of total collagen of any from the experimental groups.

Biochemical and histological studies also provide evidence of ligament midsubstance adaptation to mechanical loading in normal ligaments, regardless of changes in the structural or mechanical properties. Dogs that underwent training had higher MCL collagen content than untrained and immobilized dogs, and histology revealed that the trained group had large fiber bundle diameters (Tipton et al. 1970). The periarticular connective tissues in immobilized rabbit knees (including tendon, ligament, capsule, synovium, and fascia) exhibited reduced water content and significant reductions in glycosaminoglycan content and in chondroitin sulfates (Akeson et al. 1973). Because the tissue was mixed, however, it is difficult to draw specific conclusions about ligament adaptation. Another study found a higher turnover of collagen in MCLs of immobilized rabbits, increasing the concentration of reducible collagen cross-links and decreasing the mechanical properties of the ligaments (as shown in the LCLs of the same animals) (Amiel et al. 1982). After eight weeks of immobilization, the MCLs of rabbit knees were grossly reduced in width and decreased in weight by 20% (Gamble, Edwards, and Max 1984). The ligament fascicles were noted to be more widely spaced and irregular. Perhaps more importantly, enzymatic changes indicated that the cells switched from an anabolic synthetic state to a catabolic degradative state during immobility (Gamble, Edwards, and Max 1984).

Because connective tissue cell morphology is indicative of activity and phenotype, the varying morphology of the ligament insertion site also provides evidence of tissue adaptation in ligaments. The cells in the midsubstance of ligaments are typically elongated. But near the ligament insertion sites, the cells become rounded and the tissue appears fibrocartilagenous. The gradient in cell shape from the midsubstance to the insertions has been hypothesized to reflect a gradient in the mechanical stresses (Matyas et al. 1995). To test the hypothesis, morphometric measurements were correlated with mechanical stresses and strains as predicted by finite element analysis. The areas with the roundest cells corresponded to the areas with the highest principal compressive stresses, and the areas with the flattest cells corresponded to the areas with the lowest compressive stresses in the model. Thus, the ligament cells appear to adapt to the local mechanical environment and change their phenotype accordingly. This study also suggests that ligaments have the ability to adapt both to the overall mechanical stimulus to the ligament and the specific type of local tissue stimulus.

Further evidence of ligament adaptation comes from the natural history of autograft or allograft replacements for the injured ACL. When one considers a ligament autograft, the tissue is not normal ligament when implanted. Patellar tendon is a common autograft for the ACL. These autografts were assessed grossly, histologically, and biochemically with respect to time in a rabbit model (Amiel et al. 1986). The histological observations showed that the autografts became acellular in the central region but had cells in the periphery after two weeks. After three weeks, the grafts exhibited a proliferation of cells in the central region, and after four weeks postoperation the cellularity of the grafts was homogeneous. Necrosis of the graft followed by cellular proliferation indicates that extrinsic cells repopulate the graft, and the native fibroblasts do not survive. The autografts showed a gradual conversion to the properties of the intact ACL. By 30 weeks posttransplantation, the tissue characteristics were ligamentous in appearance. The concentration of type III collagen increased from 0 to 10% (similar to that in the normal ACL) in 30 weeks. This study indicates that even nonligamentous grafts can adapt to the mechanical environment and change the tissue phenotype to that of a normal ligament. The insertion/attachment site changes were not reported in this study, but these results would also be of great interest, since tissue and cell phenotype in the ligament change near the insertion.

Certainly, the evidence overwhelmingly supports active adaptation of ligaments to the mechanical environment. It is also clear that changes in ligament structural and mechanical properties, morphology, and biochemistry are much more pronounced in the immobilization models than in exercise models. One might tend to conclude that the ligament has limited ability to increase its mechanical properties, but an alternate view also seems plausible. Because the ligaments provide a guiding and restraining function in the joint, increases in activity will not necessarily greatly increase ligament loads. The muscles, the active component of the musculoskeletal system, will certainly have increased loads, but their action may tend to protect the ligaments from large increases in loading. Thus, only small increases in ligament properties, on the order of 10%, are observed in experimental studies. Of course, this hypothesis can only be confirmed by direct measurement of ligament forces, but it is consistent with data from exercise and immobilization studies. Also, this theory is consistent with the theory of homeostatic maintenance of tissue stimulus throughout the musculoskeletal system. And certainly, the morphological and biochemical findings that ligaments are more metabolically active than tendons indicates a greater potential for adaptation (Amiel et al. 1984).

With regard to application of data for ligament adaptation to athletic training, sports medicine, and rehabilitation, the issues to consider are not as clear as for other tissues. Certainly, immobilization will weaken the ligaments, and therefore it is not benign with regard to any of the musculoskeletal tissues. Thus, for rehabilitation, early joint motion should improve healing of an injured ligament. Any training regimen that does not endanger the bones, cartilage, or tendons should not pose a threat to the ligaments. However, muscle fatigue could certainly play a role in ligament injury, since the muscle action appears to normally protect (at least partially) the ligaments from the increased loads due to exercise. Certainly, caution should be used before allowing return to strenuous exercise. If the muscle control is reduced, reaction time is increased, or if poor proprioception causes a fall, the ligaments will be at increased risk of injury.

Summary and Potential Applications

The data presented in this chapter demonstrate that the passive musculoskeletal tissues do adapt to mechanical loading. The adaptation of each tissue generally follows the mechanical stimulus. Increased mechanical stimulus results in increased mechani-

cal properties. This indicates that each tissue adapts to maintain a constant ratio of mechanical stimulus to tissue mechanical properties. Bone demonstrates this adaptation most obviously, but the soft tissues also change their ultrastructure and chemical composition to adjust their mechanical properties according to the applied mechanical stimulus.

Prolonged bed rest, immobilization, or other means of reducing the mechanical stimulus will have long-lasting effects on the musculoskeletal tissues. Bone mass will be lost, and the return to normal density levels (and thus bone strength) can take much longer than the time over which the density decreased. Cartilage thickness will be reduced, and tendons and ligaments will have impaired mechanical properties. Thus, if prolonged reduction in mechanical stimulus is required, the retraining program should be gradual and steady, rather than intense and aggressive.

With regard to increased loading, all of the passive musculoskeletal tissues appear to be able to adapt to increases in loading, though this adaptation is slower than that for reduced loads. Thus, athletic training programs should also be designed with a relatively gradual increase in intensity and duration of activities. This will protect all tissues from the possibility of overuse injury. For the bones, a sudden rise in intensity or duration may initiate a stress fracture, requiring the need for a long pause in the training schedule. Similarly, tendinitis could be brought on by sudden and extreme levels of exertion. Muscle fatigue is implicated in both stress fractures and ligament injuries, indicating that training sessions should be discontinued prior to extreme levels of muscle fatigue.

In addition, muscle fatigue can play a role in increasing impact loads in the joints and thus may endanger the cartilage as well. Muscle fatigue can also result in falls or momentary balance loss that can result in musculoskeletal injury. Certainly, any action that can be taken to protect the joint from injury will reduce the chances of direct cartilage damage and long-term degeneration due to secondary osteoarthritis. In addition, athletes should be properly trained in the early stages of learning a new activity, to prevent abnormal joint motion during the activity that may tend to overload the cartilage and predispose a joint to osteoarthritis.

References

Akeson, W.H., Woo, S.L-Y., Amiel, D., Coutts, R.D., and Daniel, D. 1973. The connective tissue response to immobility: Biochemical changes in periarticular connective tissue of the immobilized rabbit knee. *Clin. Orthop. Rel. Res.* 93: 356–62.

Allison, N., and Brooks, B. 1921. Bone atrophy. *Surg. Gynec. Obstet.* 33: 250–60.

Aloia, J.F., Cohn, S.H., Ostuni, J.A., Cane, R., and Ellis, K. 1978. Prevention of involutional bone loss by exercise. *Ann. Intern. Med.* 89: 356–58.

Amiel, D., Frank, C., Harwood, F., Fronek, J., and Akeson, W. 1984. Tendons and ligaments: A morphological and biochemical comparison. *J. Orthop. Res.* 1: 257–65.

Amiel, D., Ing, D., Kleiner, J.B., and Akeson, W.H. 1986. The natural history of the anterior cruciate ligament autograft of patellar tendon origin. *Am. J. Sports Med.* 14: 449–62.

Amiel, D., Woo, S.L-Y., Harwood, F.L., and Akeson, W.H. 1982. The effect of immobilization on collagen turnover in connective tissue: A biochemical-biomechanical correlation. *Acta Orthop. Scand.* 53: 325–32.

Archambault, J.M., Wiley, J.P., and Bray, R.C. 1995. Exercise loading of tendons and the development of overuse injuries. *Sports Med.* 20: 77–89.

Åström, M., and Westlin, N. 1994a. Blood flow in chronic Achilles tendinopathy. *Clin. Orthop. Rel. Res.* 308: 166–72.

Åström, M., and Westlin, N. 1994b. Blood flow in the human Achilles tendon assessed by laser Doppler flowmetry. *J. Orthop. Res.* 12: 246–52.

Baldwin, K.M., White, T.P., Arnaud, S.B., Edgerton, V.R., Kraemer, W.J., Kram, R., Raab-Cullen, D., and Snow, C.M. 1996. Musculoskeletal adaptations to weightlessness and development of effective countermeasures. *Med. Sci. Sports Exerc.* 10: 1247–53.

Beaupré, G.S., Orr, T.E., and Carter, D.R. 1990. An approach for time-dependent bone modeling and remodeling application: A preliminary remodeling simulation. *J. Orthop. Res.* 8: 662–70.

Biewener, A.A. 1991. Musculoskeletal design in relation to body size. *J. Biomech.* 24(Suppl. 1): 19–29.

Biewener, A.A., and Bertram, J.E.A. 1994. Structural response of growing bone to exercise and disuse. *J. Appl. Physiol.* 76: 946–55.

Boustany, N.N., Gray, M.L., Black, A.C., and Hunziker, E.B. 1995. Time-dependent changes in the response of cartilage to static compression suggest interstitial pH is not the only signaling mechanism. *J. Orthop. Res.* 13: 740–50.

Bülow, J., and Tøndevold, E. 1982. Blood flow in different adipose tissue depots during prolonged exercise in dogs. *Pflügers Arch.* 392: 235–38.

Burr, D.B., Martin, R.B., Schaffler, M.B., and Radin, E.L. 1985. Bone remodeling in response to in vivo fatigue microdamage. *J. Biomech.* 18: 189–200.

Buschmann, M.D., Hunziker, E.B., Kim, Y.-J., and Grodzinsky, A.J. 1996. Altered aggrecan synthesis correlates with cell and nucleus structure in statically compressed cartilage. *J. Cell. Sci.* 109: 499–508.

Cabaud, H.E., Chatty, A., Gildengorin, V., and Feltman, R.J. 1980. Exercise effects on the strength of the rat anterior cruciate ligament. *Am. J. Sports Med.* 8: 79–86.

Carter, D.R. 1982. The relationship between in vivo strains and cortical bone remodeling. *CRC Critical Reviews in Biomedical Engineering* 8: 1–28.

Cavanagh, P.R., Davis, B.L., and Miller, T.A. 1992. A biomechanical perspective on exercise countermeasures for long-term spaceflight. *Aviat. Space Environ. Med.* 63: 482–85.

Chakkalakal, D.A. 1989. Mechanoelectric transduction in bone. *J. Mater. Res.* 4: 1034–46.

Chamay, A., and Tschantz, P. 1972. Mechanical influences in bone remodeling: Experimental research on Wolff's law. *J. Biomech.* 5: 173–80.

Chamberlain, M.A., Care, G., and Harfield, B. 1982. Physiotherapy in osteoarthritis of the knees: A controlled trial of hospital versus home exercises. *Int. Rehab. Med.* 4: 101–6.

Colletti, L.A., Edwards, J., Gordon, L., Shary, J., and Bell, N.H. 1989. The effects of muscle-building exercise on bone mineral density of the radius, spine, and hip in young men. *Calcif. Tiss. Int.* 45: 12–14.

Cordey, J., Schwyzer, H.K., Brun, S., Matter, P., and Perren, S.M. 1985. Bone loss following plate fixation of fractures? Quantitative determination in human tibiae using computed tomography. *Helv. Chir. Acta* 52: 181–84.

Cowin, S.C., and van Buskirk, W.C. 1978. Internal bone remodeling induced by a medullary pin. *J. Biomech.* 11: 269–75.

Curwin, S.L., Vailas, A.C., and Wood, J. 1988. Immature tendon adaptation to strenuous exercise. *J. Appl. Physiol.* 65: 2297–301.

Dalén, N., and Olsson, K.E. 1974. Bone mineral content and physical activity. *Acta Orthop. Scand.* 45: 170–74.

Engh, C.A., and Bobyn, J.D. 1988. The influence of stem size and extent of porous coating on femoral bone resorption after primary cementless hip arthroplasty. *Clin. Orthop.* 231: 7–28.

Engh, C.A., Bobyn, J.D., and Glassman, A.H. 1987. Porous-coated hip replacement: The factors governing bone ingrowth, stress shielding, and clinical results. *J. Bone Joint Surg.* 69B: 45–55.

Enwemeka, C.S., Maxwell, L.C., and Fernandes, G. 1992. Ultrastructural morphometry of matrical changes induced by exercise and food restrictions in the rat calcaneal tendon. *Tiss. Cell* 24: 499–510.

Felts, W.J.L. 1959. Transplantation studies of factors in skeletal organogenesis. *Am. J. Phys. Anthrop.* 17: 201–15.

Forwood, M.R., Owan, I., Takano, Y., and Turner, C.H. 1996. Increased bone formation in rat tibiae after a single short period of dynamic loading in vivo. *Am. J. Physiol.* 270: E419–23.

Forwood, M.R., and Turner, C.H. 1994. The response of rat tibiae to incremental bouts of mechanical loading: A quantum concept for bone formation. *Bone* 15: 603–9.

Frost, H.M. 1964. *The laws of bone structure.* Springfield, IL: Charles C. Thomas.

Gamble, J.G., Edwards, C.C., and Max, S.R. 1984. Enzymatic adaptation in ligaments during immobilization. *Am. J. Sports Med.* 12: 221–28.

Geiser, M., and Trueta, J. 1958. Muscle action, bone rarefaction, and bone formation. *J. Bone Joint Surg.* 40B: 282–311.

Gillard, G.C., Reilly, H.C., Bell-Booth, P.G., and Flint, M.H. 1979. The influence of mechanical forces on glycosaminoglycan content of the rabbit flexor digitorum profundus tendon. *Conn. Tiss. Res.* 7: 37–46.

Gillespie, J.A. 1954. The nature of the bone changes associated with nerve injuries and disuse. *J. Bone Joint Surg.* 36B: 464–73.

Giori, N.J., Beaupré, G.S., and Carter, D.R. 1993. Cellular shape and pressure may mediate mechanical control of tissue composition in tendons. *J. Orthop. Res.* 11: 581–91.

Goodship, A.E., Lanyon, L.E., and McFie, H. 1979. Functional adaptation of bone to increased stress: An experimental study. *J. Bone Joint Surg.* 61A: 539–46.

Gross, T.S., and Rubin, C.T. 1995. Uniformity of resorptive bone loss induced by disuse. *J. Orthop. Res.* 13: 708–14.

Hall, B.K., and Herring, S.W. 1990. Paralysis and growth of the musculoskeletal system in the embryonic chick. *J. Morphol.* 206: 45–56.

Hansson, H.-A. 1988. Somatomedin C immunoreactivity in the Achilles tendon varies in a dynamic manner with the mechanical load. *Acta Physiol. Scand.* 134: 199–208.

Hart, R.T., Davy, D.T., and Heiple, K.G. 1984. A computational method for stress analysis of adaptive elastic materials with a view toward applications in strain-induced bone remodeling. *J. Biomech. Eng.* 106: 342–350.

Hert, J., Liskova, M., and Landa, J. 1971a. Reaction of bone to mechanical stimuli. I. Continuous and intermittent loading of tibia in rabbit. *Folia Morphol.* 19: 290–300.

Hert, J., Pribylova, E., and Liskova, M. 1972. Reaction of bone to mechanical stimuli. III: Microstructure of compact bone of rabbit tibia after intermittent loading. *Acta Anat.* 82: 218–30.

Hert, J., Sklenska, A., and Liskova, M. 1971b. Reaction of bone to mechanical stimuli. V. Effect of intermittent stress on the rabbit tibia after resection of the peripheral nerves. *Folia Morphol.* 19: 378–87.

Hess, G.P., Cappiello, W., Poole, R.M., and Hunter, S.C. 1989. Prevention and treatment of overuse tendon injuries. *Sports Med.* 8: 371–84.

Huiskes, R., Weinans, H., and Dalstra, M. 1989. Adaptive bone remodeling and biomechanical design considerations for noncemented total hip arthroplasty. *Orthopedics* 12: 1255–67.

Jaworski, Z.F., Liskova-Kiar, M., and Uhthoff, H.K. 1980. Effect of long-term immobilization on the pattern of bone loss in older dogs. *J. Bone Joint Surg.* 62B: 104–10.

Jones, H.H., Priest, J.D., Hayes, W.C., Tichenor, C.C., and Nagel, D.A. 1977. Humeral hypertrophy in response to exercise. *J. Bone Joint Surg.* 59A: 204–8.

King, J., Brelsford, H.J., and Tullos, H.S. 1969. Analysis of the pitching arm of the professional baseball pitcher. *Clin. Orthop.* 67: 116–23.

Kiviranta, I., Jurvelin, J., Tammi, M., Säämänen, A.-M., and Helminen, H.J. 1987. Weight bearing controls glycosaminoglycan concentration and articular cartilage thickness in the knee joints of young beagle dogs. *Arthritis Rheum.* 30: 801–9.

Kiviranta, I., Tammi, M., Jurvelin, J., Arokoski, J., Säämänen, A.-M., and Helminen, H.J. 1994. Articular cartilage thickness and glycosaminoglycan distribution in the young canine knee joint after remobilization of the immobilized limb. *J. Orthop. Res.* 12: 161–67.

Kiviranta, I., Tammi, M., Jurvelin, J., Säämänen, A., and Helminen, H.J. 1988. Moderate running exercise augments glycosaminoglycans and thickness of articular cartilage in the knee joint of young beagle dogs. *J. Orthop. Res.* 6: 188–95.

Kovar, P.A., Allegrante, J.P., MacKenzie, C.R., Peterson, M.G.E., Gutin, B., and Charlson, M.E. 1992. Supervised fitness walking in patients with osteoarthritis of the knee. *Ann. Intern. Med.* 116: 529–34.

Krølner, B., Toft, B., Nielsen, S.P., and Tøndevold, E. 1983. Physical exercise as prophylaxis against involutional vertebral bone loss: A controlled trial. *Clin. Sci.* 64:541-546.

Kujala, U.M., Kaprio, J., and Sarna, S. 1994. Osteoarthritis of weight-bearing joints of lower limbs in former elite male athletes. *Br. Med. J.* 308: 231–34.

Kujala, U.M., Kettunen, Y., Paananen, H., Aalto, T., Battié, M.C., Impivaara, O., Videman, T., and Sarna, S. 1995. Knee osteoarthritis in former runners, soccer players, weightlifters, and shooters. *Arthritis Rheum.* 38: 539–46.

Kvist, M. 1994. Achilles tendon injuries in athletes. *Sports Med.* 18: 173–201.

Lammi, M.J., Häkkinen, T.P., Parkkinen, J.J., Hyttinen, M.M., Jortikka, M., Helminen, H.J., and Tammi, M.I. 1993. Adaptation of canine femoral head articular cartilage to long-distance running exercise in young beagles. *Ann. Rheum. Dis.* 52: 369–77.

Lane, N.E. 1996. Physical activity at leisure and risk of osteoarthritis. *Ann. Rheum. Dis.* 55: 682–84.

Lane, N.E., Bloch, D.A., Jones, H.H., Marshall, W.H., Wood, P.D., and Fries, J.F. 1986. Long-distance running, bone density, and osteoarthritis. *JAMA* 255:1147-1151.

Lane, N.E., Michel, B., Bjorkengren, A., Oehlert, J., Shi, H., Bloch, D.A., and Fries, J.F. 1993. The risk of osteoarthritis with running and aging: A five-year longitudinal study. *J. Rheum.* 20: 461–68.

Lanyon, L.E. 1987. Functional strain in bone tissue as an objective and controlling stimulus for adaptive bone remodeling. *J. Biomech.* 20: 1083–93.

Laros, G.S., Tipton, C.M., and Cooper, R.R. 1971. Influence of physical activity on ligament insertions in the knees of dogs. *J. Bone Joint Surg.* 53A: 275–86.

Larsen, N.P., Forwood, M.R., and Parker, A.W. 1987. Immobilization and retraining of cruciate ligaments in the rat. *Acta Orthop. Scand.* 58: 260–4.

Leblanc, A.D., Schneider, V.S., Evans, H.J., Engelbretson, D.A., and Krebs, J.M. 1990. Bone mineral loss and recovery after 17 weeks of bed rest. *J. Bone Miner. Res.* 5: 843–50.

Liskova, M., and Hert, J. 1971. Reaction of bone to mechanical stimuli. II. Periosteal and endosteal reaction of tibial diaphysis in rabbit to intermittent loading. *Folia Morphol.* 19: 301–17.

Marcus, R., Cann, C., Madvig, P., Minkoff, J., Goddard, M., Bayer, M., Martin, M., Gaudiani, L., Haskell, W., and Genant, H. 1985. Menstrual function and bone mass in elite women distance runners. *Ann. Intern. Med.* 102: 158–63.

Marcus, R., and Carter, D.R. 1988. The role of physical activity in bone mass regulation. *Adv. Sport Med. Fitness* 1:63-82.

Marti, B., Knobloch, M., Tschopp, A., Jucker, A., and Howald, H. 1989. Is excessive running predictive of degenerative hip disease? Controlled study of former elite athletes. *Br. Med. J.* 299: 91–93.

Martin, R.K., Albright, J.P., Clarke, W.R., and Niffenegger, J.A. 1981. Load-carrying effects on the adult beagle tibia. *Med. Sci. Sports Exerc.* 13: 343–49.

Matyas, J.R., Anton, M.G., Shrive, N.G., and Frank, C.B. 1995. Stress governs tissue phenotype at the femoral insertion of the rabbit MCL. *J. Biomech.* 28: 147–57.

McDermott, M., and Freyne, P. 1983. Osteoarthritis in runners with knee pain. *Brit. J. Sports Med.* 17: 84–87.

Michna, H. 1984. Morphometric analysis of loading-induced changes in collagen fibril populations in young tendons. *Cell Tiss. Res.* 236: 465–70.

Michna, H., and Hartmann, G. 1989. Adaptation of tendon collagen to exercise. *Int. Orthop.* 13: 161–65.

Minor, M.A., Hewett, J.E., Webel, R.R., Anderson, S.K., and Kay, D.R. 1989. Efficacy of physical conditioning exercise in patients with rheumatoid arthritis and osteoarthritis. *Arthritis Rheum.* 32: 1396–405.

Minor, M.A., and Lane, N.E. 1996. Recreational exercise in arthritis. *Rheum. Dis. Clin. No. Am.* 22: 563–77.

Morey, E.R., and Baylink, D.J. 1978. Inhibition of bone formation during spaceflight. *Science* 201: 1138–41.

Murray, P.D.F., and Huxley, J.S. 1925. Self-differentiation in the grafted limb bud of the chick. *J. Anat.* 59: 379–84.

Murray, W.R., and Van Meter, J.W. 1982. Surface replacement hip arthroplasty: Results of the first 74 consecutive cases at the University of California, San Francisco. In *The hip: Proceedings of the Tenth Open Scientific Meeting of the Hip Society,* ed. J.P. Nelson, 156–66. St. Louis: Mosby.

Newton, P.M., Mow, V.C., Gardner, T.R., Buckwalter, J.A., and Albright, J.P. 1997. The effect of lifelong exercise on canine articular cartilage. *Am. J. Sports Med.* 25: 282–87.

Neyret, P., Donell, S.T., Dejour, D., and Dejour, H. 1993. Partial meniscectomy and anterior cruciate ligament rupture in soccer players: A study with a minimum 20-year follow-up. *Am. J. Sports Med.* 21: 455–60.

Nilsson, B.E., and Westlin, N.E. 1971. Bone density in athletes. *Clin. Orthop.* 77: 179–82.

Noyes, F.R. 1977. Functional properties of knee ligaments. *Clin. Orthop. Rel. Res.* 123: 210–42.

Noyes, F.R., Torvik, P.J., Hyde, W.B., and DeLucas, J.L. 1974. Biomechanics of ligament failure. *J. Bone Joint Surg.* 56A: 1406–18.

Okuda, Y., Gorski, J.P., An, K.-N., and Amadio, P.C. 1987. Biochemical, histological, and biomechanical analyses of canine tendon. *J. Orthop. Res.* 5: 60–68.

Ovara, S., Puranen, J., and Ala-Ketola, L. 1978. Stress fractures caused by physical exercise. *Acta Orthop. Scand.* 49: 19–27.

Paffenbarger, R.S., Hyde, R.T., Wing, A.L., and Hsieh, C.-C. 1986. Physical activity, all-cause mortality, and longevity of college alumni. *New Eng. J. Med.* 314: 605–13.

Pouilles, J.M., Bernard, J., Tremollieres, F., Louvet, J.P., and Ribot, C. 1989. Femoral bone density in young male adults with stress fractures. *Bone* 10: 105–8.

Prendergast, P., and Taylor, D. 1992. Design of intramedullary prostheses to prevent bone loss: Predictions based on damage-stimulated remodeling. *J. Biomed. Eng.* 14: 499–506.

Puranen, J., Ala-Ketola, L., Peltokallio, P., and Saarela, J. 1975. Running and primary osteoarthritis of the hip. *Br. Med. J.* 2: 424–25.

Rális, Z.A., Rális, H.M., Randall, M., Watkins, G., and Blake, P.D. 1976. Changes in shape, ossification, and quality of bones in children with spina bifida. *Dev. Med. Child Neurol. Suppl.* 18: 29–41.

Rodriguez, J.I., Garcia-Alix, A., Palacios, J., and Paniagua, R. 1988a. Changes in the long bones due to fetal immobility caused by neuromuscular disease: A radiographic and histological study. *J. Bone Joint Surg.* 70: 1052–60.

Rodriguez, J.I., Palacios, J., Garcia-Alix, A., Pastor, I., and Paniagua, R. 1988b. Effects of immobilization on fetal bone development: A morphometric study in newborns with congenital neuromuscular diseases with intrauterine onset. *Calcif. Tiss. Int.* 43: 335–39.

Roesler, H. 1987. The history of some fundamental concepts in bone biomechanics. *J. Biomech.* 20: 1025–34.

Roos, H., Lindberg, H., Gärdsell, P., Lohmander, L.S., and Wingstrand, H. 1994. The prevalence of gonarthrosis and its relation to meniscectomy in former soccer players. *Am. J. Sports Med.* 22: 219–22.

Rubin, C.T., and Lanyon, L.E. 1984a. Dynamic strain similarity in vertebrates: An alternative to allometric limb bone scaling. *J. Theor. Biol.* 107: 321–27.

Rubin, C.T., and Lanyon, L.E. 1984b. Regulation of bone formation by applied dynamic loads. *J. Bone Joint Surg.* 66A: 397–402.

Saville, P.D., and Whyte, M.P. 1969. Muscle and bone hypertrophy: Positive effect of running exercise in the rat. *Clin. Orthop.* 65: 81–88.

Schulte, K.R., Callaghan, J.J., Kelley, S.S., and Johnston, R.C. 1993. The outcome of Charnley total hip arthroplasty with cement after a minimum twenty-year follow-up: The results of one surgeon. *J. Bone Joint Surg.* 75A: 961–75.

Scioli, M.W. 1994. Achilles tendinitis. *Orthop. Clin. N. Am.* 25: 177–82.

Smith, E.L., Smith, P.E., Ensign, C.J., and Shea, M.M. 1984. Bone involution decrease in exercising middle-aged women. *Calcif. Tiss. Int.* 36: S129–38.

Sohn, R.S., and Micheli, L.J. 1985. The effect of running on the pathogenesis of osteoarthritis of the hips and knees. *Clin. Orthop.* Sep(198): 106–9.

Sommer, H.-M. 1987. The biomechanical and metabolic effects of a running regime on the Achilles tendon in the rat. *Int. Orthop.* 11: 71–75.

Suh, J.-K., Baek, G.H., Malin, C., Rizzo, C.F., and Westerhausen-Larson, A. 1997. Effect of intermittent negative pressure on the biosynthesis of articular cartilage. *1997 ASME Summer Bioengineering Conference.* Sun River, OR. 481–82.

Suh, J.-K., Marui, T., Malin, C., and Bai, S. 1996. Dynamic hydrostatic pressure and cartilage biosynthesis: Theoretical and experimental analyses. *Transactions of the 42d Annual Meeting of the Orthopaedic Research Society.* February 19–22, Atlanta. 329.

Tipton, C.A., and Hargens, A. 1995. Physiological adaptations and countermeasures associated with long-duration spaceflights. *Med. Sci. Sports Exerc.* 28: 974–76.

Tipton, C.M., James, S.L., Mergner, W., and Tcheng, T.-K. 1970. Influence of exercise on strength of medial collateral knee ligaments of dogs. *Am. J. Physiol.* 218: 894–902.

Tipton, C.M., Schild, R.J., and Tomanek, R.J. 1967. Influence of physical activity on the strength of knee ligaments in rats. *Am. J. Physiol.* 212: 783–87.

Treharne, R.W. 1981. Review of Wolff's law and its proposed means of operation. *Orthop. Rev.* 10: 35–47.

Turner, C.H., Forwood, M.R., Rho, J.Y., and Yoshikawa, T. 1994. Mechanical loading thresholds for lamellar and woven bone formation. *J. Bone Miner. Res.* 9: 87–97.

Uhthoff, H.K., and Jaworski, Z.F. 1978. Bone loss in response to long-term immobilization. *J. Bone Joint Surg.* 60B: 420–9.

Uhthoff, H.K., Sékaly, G., and Jaworski, Z.F. 1985. Effect of long-term nontraumatic immobilization on metaphyseal spongiosa in young adult and old beagle dogs. *Clin. Orthop.* 192: 278–83.

Vailas, A.C., Deluna, D.M., Lewis, L.L., Curwin, S.L., Roy, R.R., and Alford, E.K. 1988. Adaptation of bone and tendon to prolonged hind limb suspension in rats. *J. Appl. Physiol.* 65: 373–76.

Vailas, A.C., Tipton, C.M., Matthes, R.D., and Gart, M. 1981. Physical activity and its influence on the repair process of medial collateral ligaments. *Conn. Tiss. Res.* 9: 25–31.

Vailas, A.C., Zernicke, R.F., Matsuda, J., Curwin, S., and Durivage, J. 1986. Adaptation of rat knee meniscus to prolonged exercise. *J. Appl. Physiol.* 60: 1031–34.

van der Meulen, M.C.M., Beaupré, G.S., and Carter, D.R. 1993. Mechanobiological influences in long bone cross-sectional growth. *Bone* 14: 635–42.

Viidik, A. 1968. Elasticity and tensile strength of the anterior cruciate ligament in rabbits as influenced by training. *Acta Physiol. Scand.* 74: 372–80.

Viidik, A. 1969. Tensile strength properties of Achilles tendon systems in trained and untrained rabbits. *Acta Orthop. Scand.* 40: 261–72.

Viidik, A. 1990. Structure and function of normal and healing tendons and ligaments. In *Biomechanics of diarthrodial joints,* ed. V.C. Mow, A. Ratcliffe, and S.L-Y. Woo, 3–38. New York: Springer-Verlag.

Vilarta, R., and Vidal, B.D.C. 1989. Anisotropic and biomechanical properties of tendons modified by exercise and denervation: Aggregation and macromolecular order in collagen bundles. *Matrix* 9: 55–61.

Visser, N.A., Vankampen, G.P.J., Dekoning, M.H.M.T., and Vanderkorst, J.K. 1994a. The effects of loading on the synthesis of biglycan and decorin in intact mature articular cartilage in vitro. *Conn. Tiss. Res.* 30: 241–50.

Visser, N.A., Vankampen, G.P.J., Dekoning, M.H.M.T., and Vanderkorst, J.K. 1994b. Mechanical loading affects the synthesis of decorin and biglycan in intact immature articular cartilage in vitro. *Int. J. Tiss. Reac.* XVI: 195–203.

Weinans, H., Huiskes, R., van Rietbergen, B., Sumner, D.R., Turner, T.M., and Galante, J.O. 1993. Adaptive bone remodeling around bonded noncemented total hip arthroplasty: A comparison between animal experiments and computer simulation. *J. Orthop. Res.* 11: 500–13.

Weinbaum, S., Cowin, S.C., and Zeng, Y. 1994. A model for the excitation of osteocytes by mechanical loading-induced bone fluid shear stresses. *J. Biomech.* 27: 339–60.

Williams, J.A., Wagner, J., Wasnich, R., and Heilbrun, L. 1984. The effect of long-distance running upon appendicular bone mineral content. *Med. Sci. Sports Exerc.* 16: 223–27.

Williams, J.G.P. 1986. Achilles tendon lesions in sport. *Sports Med.* 3: 114–35.

Williams, J.G.P. 1993. Achilles tendon lesions in sport. *Sports Med.* 16: 216–20.

Willis, R.A. 1936. The growth of embryo bones transplanted whole in the rat's brain. *Proc. Roy. Soc. Lond. (Ser. B)* 120: 496–98.

Wong, M., and Carter, D.R. 1990. A theoretical model of endochondral ossification and bone architectural construction in long bone ontogeny. *Anat. Embryol.* 181: 523–32.

Wong, M., Germiller, J., Bonadio, J., and Goldstein, S.A. 1993. Neuromuscular atrophy alters collagen gene expression, pattern formation, and mechanical integrity of the chick embryo long bone. *Prog. Clin. Biol. Res.* 383B: 587–97.

Woo, S.L., Kuei, S.C., Amiel, D., Gomez, M.A., Hayes, W.C., White, F.C., and Akeson, W.H. 1981a. The effect of prolonged physical training on the properties of long bone: A study of Wolff's law. *J. Bone Joint Surg.* 63A: 780–7.

Woo, S.L-Y., Gomez, M.A., Amiel, D., Ritter, M.A., Gelberman, R.H., and Akeson, W.H. 1981b. The effects of exercise on the biomechanical and biochemical properties of swine digital flexor tendons. *J. Biomech. Eng.* 103: 51–56.

Woo, S.L-Y., Gomez, M.A., Woo, Y.-K., and Akeson, W.H. 1982. Mechanical properties of tendons and ligaments. II. The relationships of immobilization and exercise on tissue remodeling. *Biorheol.* 19: 397–408.

Woo, S.L-Y., Ritter, M.A., Amiel, D., Sanders, T.M., Gomez, M.A., Kuei, S.C., Garfin, S.R., and Akeson, W.H. 1980. The biomechanical and biochemical properties of swine tendons: Long-term effects of exercise on the digital extensors. *Conn. Tiss. Res.* 7: 177–83.

Wunder, C.C. 1977. Femur-bending properties as influenced by gravity. III. Sex-related weakness after 4-G mouse growth. *Aviat. Space Environ. Med.* 48: 339–46.

Yao, J.Q., and Seedhom, B.B. 1993. Mechanical conditioning of articular cartilage to prevalent stresses. *Brit. J. Rheum.* 32: 956–65.

Zamora, A.J., and Marini, J.F. 1988. Tendon and myotendinous junction in an overloaded skeletal muscle of the rat. *Anat. and Embryol.* 179: 89–96.

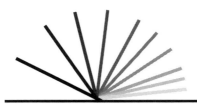

Chapter 18

Prevention of Excessive Forces With Braces and Orthotics

A. Gollhofer, W. Alt, and H. Lohrer

The following chapter is focused on the passive and active mechanisms for an effective joint stabilization. Differences in the anatomic and biomechanical properties of the ankle joint and the knee joint need different approaches to evaluate and validate the various orthotic systems. Basic knowledge is provided in order to understand the physiological and mechanical aspects of joint stabilization. Historical and epidemiological studies are presented in order to verify the many different approaches in the current literature.

The Primary Concepts

The joint complexes of the lower extremities are more susceptible to sport injuries than other joints. Ankle sprains or lesions of the knee joint ligaments are common, particularly during ball games and other competitive sports. Especially for the ankle joint system, several studies have demonstrated the efficacy of tapes or other orthotic devices in reducing the probability of injury. As a result, many of these devices are employed as preventive measures.

The Ankle Joint

In an effort to develop valid and quantitative methods in the diagnosis of ankle injuries, some investi-

gators have examined the mechanical behavior of the joint complex using in vitro studies after sectioning the ligaments (Cass, Morrey, and, Chao 1984; Johnson and Markolf 1983; McCullough and Burge 1980; Rasmussen 1985; Siegler, Chen, and Schneck 1990). Contradictory results have been found with many of these studies, often due to incompatible methodological approaches. In addition, the complexity of the joint mechanics of the talocrural and subtalar joint makes it difficult to identify single causal determinants. Research conducted on cadaveric specimens needs to be transferred to living subjects in order to improve the value of available diagnostic techniques.

Both clinical and biomechanical research has investigated the various aspects of joint stabilization. Two principal mechanisms—passive and active—appear to influence the stabilization of the joint complexes in the lower extremity. Physiologically, the "passive" mechanical fitting of the bony structures of the joint architecture ensures a secure framework that is effective in situations with axial loading. Within the spatial constraints provided by the ligaments, tendons, and joint capsules, these passive factors represent the individual predisposition for stabilization of the joint complex. Muscles acting around the joints of the lower extremity contribute "actively." By contracting agonist and antagonist muscles, joints can be stiffened. This can occur by voluntary contraction or reflex-induced activation. These mechanisms seem to provide active joint stabilization during daily activities or sport.

The determination of dynamic stability of the joint complex has been the aim of several investigations. Proprioceptive exercises (Fiore and Leard 1980), orthotic devices (Gross et al. 1987; Scheuffelen, Gollhofer, and Lohrer 1993; Scheuffelen et al. 1993; Stover 1980; Tropp, Ekstrand, and Gillquist 1985), functional footwear (Johnson, Dowson, and Wright 1976; Lohrer, Alt, and Gollhofer 1996), and the application of adhesive tapes (Alt et al. 1996) or wraps have been used to externally stiffen the joint as a way to limit its range of motion and reduce the internal loading of the joint complex.

Taping is most frequently employed by athletes, because of its low volume and anatomic fit. There is still some debate concerning its stabilizing effects over time and its influence on motor performance. The effectiveness of orthoses and functional footwear has been presented in numerous studies (Greene and Wright 1990; Gross et al. 1987; Scheuffelen et al. 1993; Stover 1980), but athletes' opinions are still extremely diversified.

Injury-Related Epidemiology

According to a statistical survey of 15,212 injuries from 13,296 athletes, more than 60% were related to injuries of the lower extremities (Steinbrück 1997) (see figure 18.1). Of these injuries, 20% were related to the ankle joint, which was most frequently injured in ball games.

Many researchers have investigated the differential support functions provided by ankle taping or

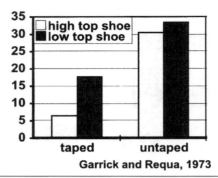

Ankle Sprains [1000 player games]

Garrick and Requa, 1973

Figure 18.2 Block diagram showing the probability rates of ankle sprains in ball games with and without external support. Modified according to the data published by Garrick and Requa (1973).

orthoses. The most valid criterion for determining the effectiveness of a joint protection system is its ability to reduce the risk of injury in the relevant anatomic area.

One of the first publications in this area (Quigley, Cox, and Murphy 1946) reported a more than 50% reduction in ankle injuries when the athletes taped their ankles before competition or practice. Since Garrick and Requa's (1973) classic study, it has been documented that injury reduction is influenced by essentially two factors: the tape application and the individual's ankle sprain history. In a study on injury rate in basketball, Garrick and Requa investigated 2,562 player games. They observed a clear reduction in ankle sprains when the players were taped prophylactically (see figure 18.2), particularly in players with histories of recurrent ankle sprains. For preinjured ankles, the researchers found that taping reduced the probability of sprains from 27.7 to 16.4 per 1,000 player games.

The influence of previous injury has also been studied (Surve et al. 1994; Tropp, Ekstrand, and Gillquist 1985). Soccer players with a history of recurrent sprains showed a significant increase in sprain probability compared to players with stable ankles. In the Tropp study a group with sprain history, protected by a semi-rigid orthotic system and trained with proprioceptive exercises, was investigated. This group showed similar low injury rates when compared with stable controls. The study also demonstrated the importance of proprioceptive function in ankle stabilization.

In conclusion, there exists a general consensus about the value of tape or orthoses to reduce ankle joint injury. The functional mechanisms that explain

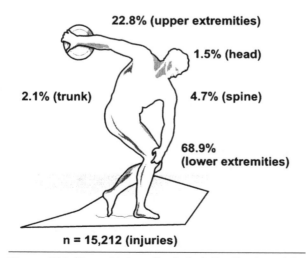

Figure 18.1 Sites of injury. Statistical distribution of the body areas that are subjected to injury.

the reduced incidence of such injuries, however, are still under debate. Controlled longitudinal studies with a large number of subjects are necessary to investigate the complex structure of ankle joint injury. Currently, there is little information available about the pretraumatic status of joint stability and its relationship to the incidence of unexpected inversion trauma.

Evaluation Criteria Assessment

Athletes, coaches, and health professionals have investigated numerous devices that were designed to enhance the external stability of the knee and ankle joints. Two aspects have guided the evaluation of acceptability:

1. The extent to which the orthoses interfere with the normal joint mechanics
2. The value of the preventive outcome with respect to its effects on motor performance

Influence of Ankle Orthoses on Load to the Joint Complex

Load and strain in the ligaments of the ankle joint complex have been rarely studied, mainly because of the size of these tiny anatomic structures. Sauer, Jungfer, and Jungbluth (1978) performed an in vitro investigation of the physical characteristics of the medial and lateral ligaments of the ankle joint, including the anterior and posterior tibiofibular ligament. Using a loading rate of 8 mm · min⁻¹, they determined the limits to failure of the bone-ligament-bone complex.

Tensile strength was found to be greater in males. Remarkably higher forces were found in the posterior (1,167 N) and anterior (931 N) tibiofibular syndesmosis and in the ligamentum deltoideum (628 N) for males. The lowest values were found in the anterior talofibular ligament for both females (147 N) and males (186 N). In accordance with the high injury rates, these findings supported the idea that inversion stress to the ankle joint is primarily harmful to talofibular and calcaneofibular ligaments.

For anterior-posterior and eversion-inversion loading, Hollis, Blasier, and Flahiff (1995) found that both ankle and subtalar joint laxity are affected by flexion angle and ligamentous stability. Hollis and coworkers also reported that the subtalar joint has its highest instability in plantarflexion position, a potential risk for ligament injury.

Ligamentous loading patterns have also been found to change under axial load conditions. An in vitro study by Cawley, France, and Paulos (1991) demonstrated significant differences in the strain behavior of ligaments compared to axially load-free test conditions. Axial loading tended to accentuate the strain patterns identified in both the anterior talofibular and the calcaneofibular ligaments.

Influence of Ankle Orthoses on Joint Function

The mechanical effects of tape or ankle orthoses on the movement excursion of the ankle joint complex have been well documented. Using electrogoniometers (McCorkle 1963), torsiometers (Rarick et al. 1962), or similar types of apparatus, for example, the Inman ankle-testing machine (Fumich et al. 1981), studies examining the relationship between structural tension and angular excursion have provided some insight into the mechanical aspects of the limitation of the joint. These ROM tests are performed before and after mechanical loading, where the joint complex is passively or actively examined. In one of the first studies of this kind Rarick and coworkers (1962) determined the mechanical support of various methods of taping before and after a 10 min exercise of running, jumping, pivoting, quick starts, and quick stops. The authors reported, even for exercisers with 10 min duration, a 40% loss of support strength as early as 10 min following the exercise routine. Follow-up studies have focused on the absolute restriction of motion. Fumich and associates (1981) tested the combined foot and ankle motions before and after 3 h of football practice with taped ankles. The researchers reported a more than 50% loss in stability compared to preexercise values. Similarly, Greene and Hillman (1990) demonstrated a differential effect when subjects were treated with adhesive tape or a semi-rigid orthosis. Following the identical test exercises, the group with the taped ankles showed a significantly decreased stability, whereas the group with the semi-rigid orthoses retained their preexercise stability.

One recent study (Alt et al. 1996) demonstrated that the loss of stability in taped ankles was determined both by the material and the technique. This particular study utilized a very tight tape material (Beiersdorf, Hamburg, Germany). After 10 and 20 min of vigorous exercise involving running and hopping the preexercise stability was reduced, depending on the technique and material of the tape, but was never reduced below 15% loss of stability (see figures 18.3 and 18.4).

To summarize, protection of the joint with tapes or orthoses is effective in mechanically stabilizing or stiffening the joint. With adhesive tapes, however, the gain in stability is often reduced after even

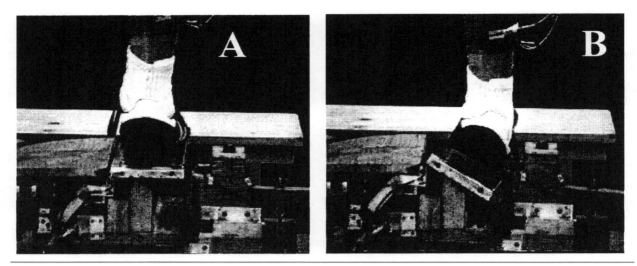

Figure 18.3 *(a)* Vertical loading of the ankle with full body weight. *(b)* Mechanical stimulation of the ankle by sudden inversion of 30° combined with plantarflexion of 15°. Angular excursion is registered by electrogoniometers.

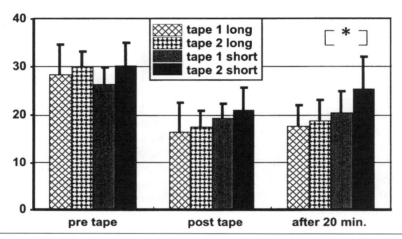

Figure 18.4 Angular excursion following mechanical stimulation (inversion 30°, plantarflexion 15°; n = 10). The means and SDs denote the maximum inversion amplitudes before and after taping and after 20 min of vigorous exercising.

short periods of exercise. This instability is not only dependent on the technique but also on the material used for taping. Orthoses do not seem to lose strength or stability.

In most studies investigating the mechanisms of joint instability the experiments have been conducted with normal, stable ankles. It is questionable, however, whether or not the effects and characteristics seen in stable ankles are comparable to patients with chronic ankle instability. Very few epidemiological surveys have made a distinction between previously sprained and not previously sprained ankles (e.g., Garrick and Requa 1973). Based on laboratory reports, it can be hypothesized that the value of external support operates differently in these populations, partly because of the lower injury rate observed in stable ankles. More-

over, it is reasonable to expect that the active and passive mechanisms are also different in previously sprained ankles.

In one X-ray study, Larsen (1984) found a strong relationship between the reduction of talar tilt displacement immediately after taping and clinical ankle instability. After 20 min of exercise (running), however, these stabilizing effects were reduced. Gollhofer, Scheuffelen, and coworkers (1991) investigated the complex joint stabilization by means of a tilt platform, similar to the device originally presented by Nawoczenski and associates (1985). Relative changes of inversion kinematics were compared in subjects with stable ankles and those with chronic ankle instability. The results indicated (see figure 18.5) that chronic instability is not necessarily accompanied by an increase in angular excursion.

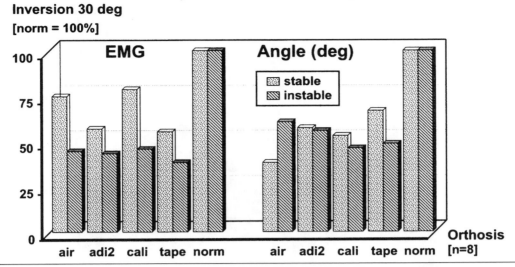

Figure 18.5 Inversion amplitudes and mean EMG amplitude of m. peroneus in eight subjects with stable and chronic instable ankle joints. The applied amplitude of inversion was 30°. Mean EMG amplitudes are normalized with respect to the value obtained in the norm condition.

Further mechanisms must be taken into consideration in the evaluation of joint complex support. More recent studies (Lohrer et al. 1994; Scheuffelen, Gollhofer, and Lohrer 1993) have shown that even in unstable ankles the talar tilt—as a measurement of ligament stress—is not primarily related to the dimensions of the induced displacement. Within a 30° calcaneal inversion displacement limit, the radiologically observed talar tilt is largely suppressed when the ankle joint is loaded. Results have indicated that axial loading is an essential feature in the determination of joint stability. Even in subjects with chronic ankle instability, only minimal talar tilts in inversion displacements of more than 30° are demonstrated when the joint complex is loaded (see figure 18.6).

Tension measurements of the lateral ligaments of the ankle joint complex have shown that the motion of the foot in relation to the ankle joint stresses the three ligaments to a different extent (Attarian et al. 1985; Hintermann and Nigg 1995; Nigg et al. 1990; Renström et al. 1988; Wirth, Küsswetter, and Jäger 1978). Tensile strength measurements conducted by Rasmussen and Kromann-Andersen (1983) presented evidence that the incidence of ankle sprain injury is related to the complex motion of plantarflexion plus internal rotation and inversion of the foot (Renström and Theis 1993; Stüssi, Stacoff, and Segesser 1992). With respect to the congruency of the bony structures and the stability functions of the ligaments, it can be assumed that axial loading of the ankle joint increases stability (Cawley and France 1991; Hollis, Blasier, and Flahiff 1995;

McCullogh and Burge 1980; Michelson and Hutchins 1995; Stormont et al. 1985).

Ankle orthoses have a considerable impact on the joint mechanics. Depending on the type and material of the orthotic device, mechanical support may be reduced after only short periods of mechanical usage. However, axial loading is highly effective in stabilizing the joint complex passively. It appears that passive stabilization serves as a self-protecting system to prevent high stress on ankle joint ligaments.

Influence of Ankle Orthoses on Motor Performance

Abdenour and coworkers (1979) investigated the influence of taping on contraction characteristics (maximum force) and on ROM. They found no significant differences between taped and untaped ankles, except for the ROM of inversion measured in high dynamic contractions. The authors concluded that taping does not influence strength and power but may provide some mechanical restriction that would be expected to prevent inversion ankle sprains.

Greene and Wright (1990) reported a relative impairment of running time in a four-corner base run when subjects wore an air-stirrup orthosis. No other orthosis that was tested in the study had any effect on running ability. Pienkowski and coworkers (1995) compared three variants of in-shoe orthoses and found no significant differences with respect to complex motor skills.

As a means of a direct comparison, Greene and Hillman (1990) investigated the mechanical restric-

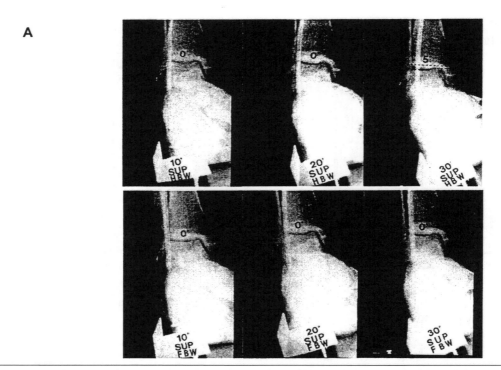

Figure 18.6a X-ray graphics (a-p direction) of one subject. The ankle joint is tilted in 10°, 20°, and 30° position. In the upper graphic the ankle joint is loaded with one-half body weight. In the lower graphic the ankle is loaded with full body weight.

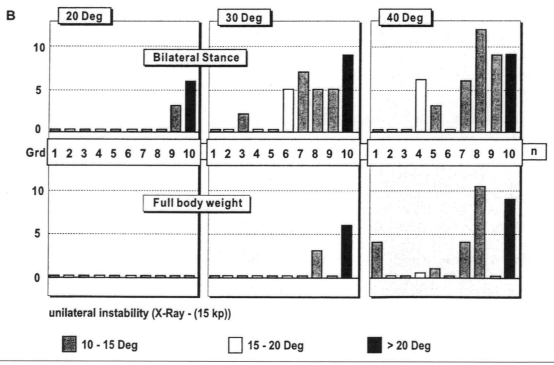

Figure 18.6b Mean talar tilts in 10 subjects with chronic ankle instability when loaded with half (upper) and full body weight (lower). The bars denote the individual talar tilts in static platform position of 20°, 30°, and 40° inversion. Notice that only in the 40° displacement condition is a considerable talar tilt measurable.

tion of taping versus semi-rigid orthoses in volleyball. After a 3 h volleyball practice, ankles that were taped showed a significant loss of stability; motor performance (measured as vertical jumping ability), however, was not affected by either support system. The authors concluded that prevention of initial inversion sprains might be more effective with semi-rigid orthoses than with tape. Other findings contradict those of Greene and Hillman. Juvenal (1972), Mayhew (1972), and Burks and associates (1991) found significantly reduced motor performance in vertical jumping ability and running with the use of semi-rigid ankle orthoses.

The influence of external ankle protection on the control of motor skill or even motor performance has not been systematically investigated. Laboratory measures are most often "validated" with tests that address motor function specific to a particular sport. Additional research is necessary to investigate the relationship between the biomechanical properties of various orthotic devices, their influence on laboratory measures, and their consequences on motor performance and skill. A clear distinction between performance and skill needs to be assessed.

Active Mechanisms: Proprioceptive Input

Several large clinical and prospective studies (Firer 1990; Garrick 1977; Rovere et al. 1988; Sitler et al. 1994; Surve et al. 1994) have found that the probability of injury to the ankle joint complex can be significantly reduced by wearing tapes or other functional orthoses. Additional findings that some of these protection systems lose 50% of their supportive function after only 10 min of exercise suggest, however, that mechanisms other than mechanical ones must be considered. Furthermore, Tropp, Ekstrand, and Gillquist (1985) showed proprioceptive training to be as effective as taping in the prevention of ankle reinjury.

In general, three factors emerge that may have some relevance to the incidence of chronic ankle joint instability:

1. Elongation, partial rupture, or rupture of the ligaments
2. Dysfunction of the afferent setup of the arthron, commonly described as proprioception (Freeman, Dean, and Hanham 1965)
3. Functional weakness of the peroneal muscles (Bosien, Staples, and Russel 1955)

Investigations into the neuromuscular aspects of

joint stabilization are rare and results have been contradictory. In 1967, Freeman and Wyke reported that excitation of cutaneous and articular mechanoreceptors in cats facilitates or inhibits gamma-motoneurons, directly influencing articular reflexes. For the knee joint it has been shown that the lateral ligaments are receptor-bearing structures (De Avilla et al. 1989). Cooper and associates (1995) have found similar structures of mechanoreceptive sensors in the ligaments of the ankle joint.

Most research studies investigating proprioceptive properties have been focused on the determination of reflex latency elicited by mechanical stimulation. Löfvenberg and coworkers (1995), who defined differences in proprioception as differences in reflex latencies, found prolonged reaction times in subjects with chronic ankle instability. Karlsson and Andreasson (1992) reported significantly longer latencies in the peroneal muscles of subjects with unilateral instability on the affected side, following a 30° inversion. Johnson and Johnson (1993), in accordance with Isakov and associates (1986), reported no differences in the reflex latencies across groups that (1) received conservative treatment, (2) received surgical treatment, or (3) demonstrated chronic instability. Mechanical stimulation in the three groups was performed by externally induced inversions of 35° amplitude.

The question of which neuromuscular parameter explains the dysfunction that causes recurrent sprain is still a matter of debate. The determination of reflex latencies following an inversion stimulus does provide some information about alterations in the pathways of the reflex contribution. From a functional point of view, however, prolonged reflex latencies explain only that the reflex responses induced by the inversion arrive too late at the joint stabilizing muscles. The strength of the response is not considered. When quantifying the effectiveness of reflex contributions, this information is extremely important; therefore, the evaluation of the strength of the afferent contribution needs to be considered.

Differences in reflex latency likely reflect alterations in spinal transduction or processing, although they may also be influenced by the types or numbers of response-generating receptors. With respect to the mean latency reported in the literature (i.e., 60–90 ms for m. peroneus), it may be assumed that the stretch-evoked responses are polysynaptic (Johnson and Johnson 1993; Konradsen, Ravn, and Sorensen 1993; Scheuffelen et al. 1993). Similar responses are observed in posture. Functionally, they are necessary for maintenance and balance of the equilibrium of the center of gravity; mechanically, they are important for the compensation of externally induced perturbations (Dietz et al. 1990; Horstmann, Gollhofer, and Dietz 1988).

From a physiological point of view, the prolongation in reflex latencies may only be influenced by the fastest responding structures. These structures may not necessarily have an important function in muscle activation, meaning that pure latency registration does not explain the entire mechanism.

In situations with large inversion amplitudes it is functionally most important to have early access to activate the peroneal muscles. The muscle must also be supplied with an adequate amount of efferent input in order to resist mechanical perturbation and stiffen the joint complex. In these cases, a measure of integrated EMG would be valuable.

Research on the proprioceptive function of the ankle joint complex has been conducted by means of a three-dimensional force platform. Konradsen, Ravn, and Sorensen (1993) and Tropp (1986) analyzed the posture of healthy subjects and those with chronic ankle instability. Other approaches have investigated the sensory angular reproduction of different joint dynamics under either active or passive conditions (Freeman, Dean, and Hanham 1965; Glick, Gordon, and Nishimoto 1976; Löfvenberg et al. 1995; Tropp 1986). These findings demonstrated a proprioceptive deficit in reproducing distinct angular dynamics in cases of chronic ankle instability.

Although enhancement of proprioceptively generated muscle activation has been speculated from experiments on the knee joint (Perlau, Frank, and Fick 1995) and ankle (Jerosch and Bischof 1994), the literature lacks data from controlled studies to support this hypothesis. In experiments with regional anesthesia of the ankle and foot, Konradsen, Ravn, and Sorensen (1993) concluded that intact afferent input from the ligaments is a necessary prerequisite for precise ankle positioning. After local anesthesia of the LTFA and LFC, however, Feuerbach and coworkers (1994) found no significant changes in the proprioception of the ankle.

Despite the fact that selective anesthesia of a single ligament is technically difficult, the authors assumed that influences from skin, muscle, tendon, and joint receptors were extremely relevant. They also observed significantly improved sensitivity of joint positioning with orthoses. From microgravity experiments (Roll et al. 1993) it is known that postural control is influenced not only by visual or vestibular input but also by proprioceptive afferents from the leg muscle receptors and other mechanoreceptors.

Taken together, these results are not sufficient to adequately explain the precise role of proprioception. At least two questions remain:

1. Can the data from posture and position experiments explain the incidence of chronic instability?

2. How can the proprioceptive stimulating effects of orthoses or tape be evaluated with greater efficacy?

One approach to investigating proprioceptive activation in relation to different orthoses was presented by Scheuffelen and coworkers (1993). Subjects were tilted from an upright uni- or bilateral stance (see figure 18.3, page 334) with a constant amplitude of 30° inversion combined with 15° plantarflexion. The tilt series was carried out under conditions that involved tapes, stabilizing shoes, various ankle orthoses, and no orthosis. The latter served as a control condition for individual comparison. Each condition was repeated five times to minimize response fluctuations. Average patterns were calculated using the superficially recorded EMGs from mm. gastrocnemius, tibialis ant., m. peroneus, and vastus med. and the traces from a twin axis electrogoniometer (Penny and Giles, Blackwood, UK) during angular excursion of the ankle joint complex in the dorsal/plantar and inversion/eversion planes. The averaging process was based on a trigger signal elicited at the start of the tilt movement.

Figure 18.7 illustrates EMG profiles of the m. peroneus and its respective goniometer signal from the inversion movement. These patterns demonstrate that the angular velocity is twice as high in the unprotected leg as in the protected leg. Despite the largely reduced angular velocity, however, the EMG profiles are similar in size. This contradicts the findings of Gollhofer and Rapp (1993), who showed a close relationship between stretch velocity and EMG response size. One explanation for the disparity between stretch velocity and registered EMG quantity may be the differences in proprioceptive stimulation of the ankle orthosis. Differential fitting of tape or ankle orthosis could either alter the sensitivity (gain) of the receptor itself or modulate the spinal transmission of the afferent signals. Both of these alternatives might explain why reflex latencies are primarily not affected by tape or orthosis. Furthermore, studies that have reported improved sensory feedback in positioning could also be explained by this hypothesis.

It is important to recognize that functionally active mechanisms are beneficial only in the second phase of an inversion injury. In concordance with

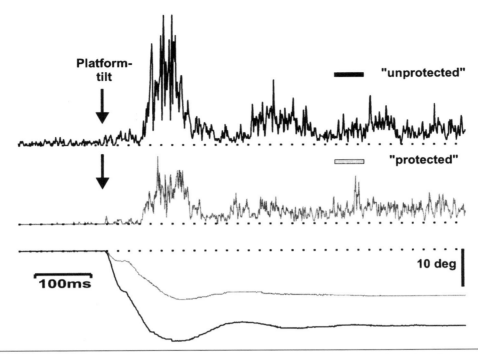

Figure 18.7 Comparison of the averaged and rectified EMG patterns of m. peroneus (n = 5) and traces of the angular movements (inversion) registered from electrogoniometers. In the situation without brace, the velocity and ROM are approximately twice the observable values in the braced condition.

the literature (Johnson and Johnson 1993; Karlson and Andreasson 1992; Konradsen, Ravn, and Sorensen 1993; Lynch et al. 1996), the reflex-induced activity appears after approximately 70 ms (refer to figure 18.7). From this it can be assumed that after an electromechanical coupling time of approximately 50 ms (Cavanagh and Komi 1979), effective muscular action cannot be expected before 120 ms following the initiation of the inversion displacement. Assuming a tilt velocity of $400° \cdot s^{-1}$, an unprotected ankle joint would already be in a 40° inversion position within this time period. The consequence of this (as shown in figure 18.6) is a substantial talar tilt movement, perhaps forceful enough to sprain the joint ligaments.

The assessment of mechanical constraint caused by tape or orthoses is necessary to evaluate effectiveness of such treatment, because only by reducing the angular velocity of inversion movement do the active neuromuscular systems have a chance to effectively stabilize the joint complex. The height of the afferent response, modulated to a large extent by the quality of the fit of ankle orthosis, is the second measure necessary to describe the active mechanisms for an effective joint stabilization.

In summary, active and passive joint stabilizing systems interact and are functionally coupled. One approach that assesses both criteria simultaneously was proposed by Alt, Lohrer, and Gollhofer (1999). Here, the ratio between afferent reflex integral and angular displacement was calculated, to determine the "proprioceptive gain factor." This measurement provides high scores if the neuromuscular activity is effectively combined with reduced angular movements.

With regard to reflex activities, it can be concluded that appropriate joint stabilization by active neuromuscular stiffening of the joint complex is highly dependent on the proprioceptive function of the sensory systems (Gollhofer, Strass, et al. 1991). Unfortunately, there is still a lack of evidence that elaborates the precise role of the sensory structures and their influence on the proprioception itself. From a functional point of view, an effective muscular action cannot be expected before 120 ms after initiation of the inversion. Consequently, positioning of the joint and sense of motion may be more important in the earlier phases of an inversion stimulus, whereas active joint stabilization has significant compensatory functions in the later phases.

The Knee Joint

The evolution from quadrupedal to bipedal gait is associated with an increased loading of the joints of the lower extremity. The knee joint in particular, interposed between the two long levers of the femur and the tibia, presents a higher risk for injury and degenerative disease. In response to this, knee joint stabilizers have been developed, ranging from virtually immobilizing devices to flexible high-tech systems.

There was little improvement in knee orthotics before 1980. In ball sports and sport activities involving direct body contact, the incidence of knee injuries has increased rapidly. Because ligamentous knee lesions were often diagnosed, industrial promotion has been focused on leisure activities. The increasing number of knee injuries with degenerative disease has inspired biomechanical researchers to seek an understanding of the inherent factor that stabilizes the joint complex. On the basis of anatomic and biomechanical knowledge new surgical concepts for injured knee joint ligaments have been developed. In the 1960s, large postsurgical plaster casts were applied for about six weeks. Following this period of immobilization, rehabilitation programs that achieved a return to full function were rare.

In his extensive monograph, Werner Müller (1982) provided an excellent report on the current knowledge of knee anatomy, biomechanics, and surgical treatment; however, he did not refer to knee orthoses nor to the functional stabilizing aspects after treatment. In professional sports, athletes with specific deficits on the cruciates demanded functional braces to reestablish sport abilities following a knee injury. In the late 1960s and early 1970s, a "derotational" brace was designed for preventive reasons. At the same time, the first rehabilitational brace was introduced for the functionally unstable knee joint.

More recently, surgical techniques followed by primarily functional treatment have received attention. With the introduction of rehabilitative knee orthoses, the range of motion of the knee joint could be restricted to prevent overload of the healing tissues. As a consequence, medical companies began the development and production of various orthoses.

Scientific research on braces was demanded following the 1984 symposium of the Sports Medical Committee of the American Academy of Orthopedic Surgeons on knee bracing (see Branch and Hunter 1990). The symposium concluded that despite the growing number of knee orthoses, there were anecdotal reports but very few scientific papers on this issue. As a result the 1990s saw an increase in the publication of studies concerned with different aspects of knee stabilization.

Injury-Related Epidemiology

Tears of the ligamentous structures of the knee commonly occur in high-risk sports, such as football. In the prevention of ligamentous disruptions, controversy exists about the efficacy of prophylactic bracing and the application of functional rehabilitational braces following injury or surgery. In an effort to minimize the potential risk for further injury and to optimize tissue healing many amateur and professional teams currently utilize prophylactic braces on uninjured players and functional braces on players who have sustained previous injuries. The results of epidemiological studies, however, are contradictory. In contrast to the uniform reports obtained from studies regarding the ankle joint, three sets of results can be found for the knee joint:

1. Only minor protective effect regarding the use of prophylactic bracing
2. No effect of prophylactic bracing
3. An increased injury rate induced by the use of prophylactic bracing

In 1990, Sitler and coworkers published a prospective, randomized study with 1,396 football players. After a two-year follow-up they reported that the use of a preventive brace significantly reduced the frequency of knee injuries in both the number of injured athletes and the number of medial collateral ligament injuries. The severity of injuries was unaffected. A comparison of offensive and defensive players found significantly fewer cases of injury among defensive players who were braced. This finding contradicted a review of epidemiological studies carried out by Garrick and Requa (1988) who did not confirm any injury protection effect of braces.

Hewson, Mendini, and Wang (1986) reported no differences between type and severity of knee injuries between braced and unbraced college football players; however, they determined a twofold increase in injury rate for high-risk players with and without a prophylactic knee brace.

In 1987, Rovere, Haupt, and Yates published a two-year control study on football players. They determined a higher rate of knee injury, an increase in episodes of muscle cramping in the triceps surae muscle, and a greater number of ACL tears when subjects were equipped with braces. Among injured players, the nature and type of injury were not altered by the use of prophylactic braces. Teitz and associates (1987) obtained similar results from a retrospective two-year study. A protective effect to the MCL or ACL was not found. They reported a significant increase in injuries to the knee with prophylactic braces. The severity of injuries was not affected by wearing a brace.

Grace, Skipper, and Newberry (1987) observed an increase in the rate of knee injuries with the use of a single upright knee brace and an increase in injury to the ipsilateral ankle and foot associated with either type of prophylactic knee brace (see figure 18.8). The braces, however, were not randomly assigned to the study participants, so voluntary selection of the braces could have contributed to the results. It is important to note, however, that the severity of knee injury doesn't appear to be affected by prophylactic bracing.

Compliance, bias, brace fit, actual injury mechanisms, and psychological factors all play an important role in evaluating the effectiveness of prophylactic braces. As a result, findings range from "effective in giving minimal protection" to "serious harmful consequences." This is reflected in the large discrepancy between industry-promoted assertions on the "healing and preventive characteristics" of newly designed knee orthoses and the scarcity of scientific knowledge and proof of benefits from these products. In fact, the use of knee orthoses has

increased immensely over the past 15 years, especially as a substitute for casting.

Evaluation Criteria Assessment

Evaluation of the mechanical effects of bracing, bandaging, and taping is based on the relationship of form and function of the investigated structure. Menschik (1974) describes the femorotibial joint with its connecting cruciates and collateral ligaments as a four-bar linkage system that combines rolling and gliding motions. Because of the convex form of the femoral condyles in the sagittal plane and the corresponding flat tibial plateau, the menisci are necessary to reduce load on the joint cartilage and form a functional congruency (Müller 1996). Motion in the knee joint centers around six degrees of freedom. There are three rotations—extension-flexion, varus-valgus, and internal-external—and three translations—anterior-posterior, medial-lateral, and distraction-compression.

Data obtained directly from in vitro measurements (Berns, Hull, and Patterson 1992) showed the load dependence of the collateral and cruciate ligaments from the flexion angle. The angle-dependent effect of different fiber bundle loading was shown in the ACL. At full flexion only the most anterior fiber bundle of the ACL was tightened. Müller (1996) called this effect a "progressive fiber recruitment with gaining strength corresponding to the mechanical demand" (p. 105).

In contrast to the ACL, the PCL is most prone to injury in full flexion. Flexion-dependent changes in strain were shown in a recent study by Hull and coworkers (1996). In fully extended position the posterior superior site of MCL was significantly more loaded than was found in 30° of flexion. Combined loads such as valgus and external moments increased strain in MCL more than only unidirectional load. A detailed explanation of ligament forces in the cruciates can be found in Mommersteeg and coworkers (1997). A complex biomechanical approach, including in vitro studies with X-ray stereophotogrammetric analysis and inverse dynamic modeling, revealed that ACL force decreased from 130 N in extension to 0 N in 90° flexion when 150 N compressive force was applied. In addition, the stabilizing effects of the ligaments seemed to be supported by axial loading with body weight.

Knee Sprains
[1000 player games]

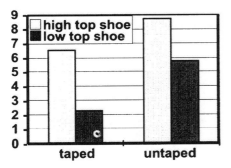

Ankle Sprains
[1000 player games]

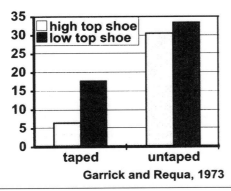

Garrick and Requa, 1973

Figure 18.8 Knee and ankle sprains per 1,000 player games in both taped and untaped situations. The incidence of injury is largely dependent on the type of tape.
From Garrick and Requa, 1973.

Influence of Knee Orthoses on Load to the Joint Complex

Valid assessment of the stabilizing effects of orthoses have been described in many studies. Different biomechanical and clinical approaches have also been developed to elucidate their functional properties, with special emphasis to the following areas: (1) stress transferred to the ligaments in healthy versus chronic unstable subjects' proprioceptive motor performance, and (2) the active part in the stabilization of the knee joint complex.

In 1992, Beynnon and coworkers investigated seven knee orthoses for their in vivo strain shielding effects. In low load conditions (i.e., < 20 N) none of the tested devices showed a significant reduction in strain in ACL. Four of the seven tested devices, however, showed a significantly reduced strain under load conditions of 100 N to 180 N. Wearing a brace was found to reduce internal rotation torque by 1.3%. Systematic differences in conditions with or without orthotic protection were not established for isometric contraction force or active ROM measurements. The Beynnon group's study produced evidence for the preventive function of orthoses in reducing the load on knee ligaments, especially the ACL. A similar approach was published by Jonsson and Kärrholm (1990). They found no differences between custom-made and off-the-shelf braces. They argued that there might be an interaction between the technique of attachment and the hinge design as well as the interface with the muscles of the lower limb.

Joint architecture and forces and moments acting on the knee joint have been areas of specific focus and interest for Regalbuto, Rovick, and Walker (1989). The results of this study supported the importance of optimal brace fit, especially the optimization of the brace movement axis with respect to the anatomic constraints of the knee joint. For example, although only slight differences were found in forces between various brace designs, a considerable difference was found when the rotation axis of the braces was displaced (12 mm) from the optimized joint position. In clinical tests Rupp, Lanta, and Schulz (1995) demonstrated in 20 subjects with chronic instability that the anterior tibial translation was significantly reduced when subjects were protected with orthotic devices. This was shown for the manual drawing maneuver and standardized pulling tests (KT 1000).

Differences between prophylactic and functional braces in the ability to reduce load to the knee ligaments have been found by Baker and associates (1989). Transducers implanted to the MCL and ACL of cadaver knees were used to register forces, while electrogoniometers measured the angular excursion. With a fully extended knee, prophylactic braces had a limited capacity to reduce direct lateral stress to the MCL; however, this effect was greatly reduced at 20° of flexion. Brown, van Hoeck, and Brand (1990) tested eight braces with a mechanical surrogate model under dynamic valgus load conditions. All brace types tested reduced MCL strain and increased failure load. In a two-stage model, France and colleagues (France et al. 1987; France and Paulos 1990) investigated the in vitro effectiveness of bracing. Under static test conditions, the braced specimen tended to preload the MCL. The authors described a center axis shift of forces away from the MCL to the cruciates and concluded that for low rate dynamic valgus none of the braces provided significant protection against MCL injury. In the second part of their study, France and coworkers (1987) evaluated the clinical relevance of preloading. Using a combination of in vivo EMG and a mechanical surrogate knee model, they found that the impact responses in the braced knee (across six brace types) showed no clinical significance because MCL preloading was compensated by axial or dynamic joint compression. In order to compare the braced versus unbraced conditions, the authors created an impact safety factor (ISF) calculated from peak MCL tension and impact momentum. Using this factor, only one brace provided a minimum standard of 30% reduction in MCL load or an overall ligament protection of 50%.

In summary, exact fitting of an orthosis seems most important for it to have a protective effect. Slight aberrations from optimal positioning can lead to significant external loads to the knee ligaments. Positive orthotic effects have been proven for low level external loads, although most of the experimental data are without functional relevance because of the static or in vitro measurement conditions.

Influence of Knee Orthoses on Motor Performance

There is a basic uncertainty whether or not the employment of preventive orthoses in sport activities influences motor performance. Braces have been found effective immediately after application (Tegner et al. 1988; Veldhuizen et al. 1991). Although the performance parameters were reduced by approximately 10% over the first few days, they regained initial values after a four-week period. No change in atrophy was noted. In running tests, $\dot{V}O_2$max, heart rate, and blood lactate level were reduced by 8% to 10%. Similar results were presented by Zetterlund, Serfass, and Hunter (1986). In subjects running on a treadmill with a brace (n = 10), they observed an increase in submaximal $\dot{V}O_2$ and heart rate of about 5%. Kinematics was not altered in braced running when compared to free running.

A comparison of two different orthoses in 21 subjects by Knutzen and coworkers (1987) found no group-specific differences in ground reaction forces or knee kinetics in running. A general brace-induced reduction in knee angular excursion found during the stance and swing phase led to the assumption that knee braces enhanced joint stiffness during movement. From kinematic data and ground reaction forces from healthy subjects, DeVita and coworkers (1996) calculated joint torque and power patterns with one functional knee brace. One of the major findings was an increase in extensor torque of hip and ankle joint. Compared to walking without a functional brace, more work was produced in the hip and less work in the knee joint. These findings corresponded with EMG data from Branch and Hunter (1990) and support the conclusion that adaptation of the hip extensors reduces stress on the ACL.

Active Mechanisms

Brace-induced influences were analyzed in 12 subjects with ACL deficits (Acierno et al. 1995). In subjects with a "high activity history" strength quality during extension remained unaltered even in braced conditions, whereas in subjects with a "low activity history" an increase in extensor quadriceps EMG was associated with a decrease in hamstring activation.

Measurements of the intramuscular tension (Styf, Nakhostine, and Gershuni 1992) revealed that an increased compartment pressure exists in several orthoses, even in a relaxed situation. Based on these findings, it can be argued that compressive effects may also influence muscular function when distinct knee braces are worn.

Classification of Systems

Currently, there is no classification system available that is generally accepted to evaluate knee orthoses on the basis of their purpose. Knee braces are classified by the American Academy of Orthopedic Surgeons (AAOS 1984, cited in Cawley 1990) by type and function (see table 18.1).

Most current orthotic research was carried out using this classification system, but it is easy to see the difficulty in categorizing an individual brace to this system (Baker 1990; Baker et al. 1987, 1989; Brown, van Hoeck, and Brand 1990; Cawley 1990). In clinical practice, individual braces are used for prophylactic as well as functional and rehabilitational purposes. In this sense, all braces are functional.

Table 18.1 Knee Brace Classification System

Type	Function
Prophylactic	Prevention of knee injury in uninjured subjects
Functional	Knee stabilizing while performing sport activities
Rehabilitational	Protection of the healing ligaments after knee surgery

Cited in Branch and Hunter 1990.

By applying table 18.1 definitions, it is evident that most orthoses can function for each of the three purposes; in other words, their functions overlap. According to Jonsson and Kärrholm (1990) identical braces have been used for functional and rehabilitational purposes.

Other classification systems are based on brace design and constructional criteria:

- Double hinge—single hinge (Walker, Rovick, and Robertson 1988)
- Hinge post strap—hinge post shell (Cawley, France, and Paulos 1991)
- Polycentric hinge—monocentric hinge (Styf, Nakhostine, and Gershuni 1992)
- Off-the-shelf—custom-made (Beynnon et al. 1992)
- Elastic strapping—rigid strapping (France, Cawley, and Paulos 1990)

Based on current literature, these discriminations are not useful because in most cases the biomechanical and clinical properties are too similar (Brown, van Hoeck, and Brand 1990; Styf, Nakhostine, and Gershuni 1992; Wojtys et al. 1990). For simplicity, we propose a classification system similar to that for ankle orthoses that includes sleeves, bandages, taping, and braces.

Knee sleeves are circular knitted elastic systems, delivered off the shelf. Several sizes are produced to ensure a smooth fit. They are inelastic and reach from the distal aspect of the quadriceps muscle to the proximal aspect of the shank. To prevent longitudinal sleeve contraction, they are equipped with a flexible, nonhinged collateral bar at each side.

Knee bandages are similar to knee sleeves, but with certain functional elements added (Gutenbrunner et al. 1997). Medial and lateral collateral stabilizers without hinges support knee ligaments (e.g., neoprene bandage) or even correct patellar motion patterns (Isakov et al. 1986; Schaff et al. 1995).

Knee taping was the first form of functional orthotic treatment (Roser, Miller, and Clawson 1971); however, there are only a few studies that have investigated its protective purpose (Cushnaghan, McCarthy, and Dieppe 1994; Reese and Burruss 1986). No systematic studies are available demonstrating the efficacy of patellar taping against chondropathic, osteoarthritic, or even subluxating patella (Worrell, Ingersoll, and Farr 1994).

Knee braces are built on lateral or medial hinged stiff bars. Fitting to the knee and leg is achieved by specific shell and pad systems, which are adapted by more or less rigid plastic materials with extended skin surface contact areas at the shank and thigh (Cook, Tibone, and Redfern 1989). Fixation to the limb is highly variable by elastic or rigid strapping, so migration is the biggest potential problem of the braces because they can produce undesired shear forces to the knee (Blauth, Ulrich, and Hahne 1990). According to France, Cawley, and Paulos (1990), a functional brace is "the least important component in the treatment chain" (p. 745). Cawley (1990) predicted that advanced surgical technology "will undoubtedly reduce the need for those orthoses in the coming years" (p. 768) and this will also hold true for the other brace systems.

Further Bracing Applications

Historically, a large variety of orthoses were developed for patients with neuromuscular diseases and for patients with pseudarthroses. Stabilization of mechanically unstable joint systems or correction of deformities to enable improved function was the basic rationale. Today many custom-made or off-the-shelf orthoses exist with the potential for conservative and postoperative treatment. These orthoses are rather stiff and usually embrace more than one joint.

Hip Joint

In a classical orthotic treatment, newborn children with hip dysplasia are equipped with "spreading-out orthoses" to correct the laterally acting forces. The efficacy of these corrections corresponds to the early diagnosis of the deformity. An early postnatal ultrasound hip screening within the first week of birth can minimize the risk of hip osteotomies. This example demonstrates how external forces (produced by the orthosis) can positively modify bone growth, as stated by Pauwel's law.

Upper Extremities

In the upper extremities orthotic treatment is not as frequent as in the lower extremities. Sometimes orthoses are applied to stabilize the finger joints and wrist in degenerative diseases and following ligament lesions. In situations when the elbow is prone to overload injuries ("tennis arm" or "golfer's elbow") orthotic treatment is helpful. Here, the basic principle is to exert local compressive forces to the painful tendon insertions. Ligamentous lesions or degenerative joint diseases may be further indications for orthotic therapy. The shoulder joint has the highest physiological mobility of all human joints. Functionally, the acromioclavicular and sternoclavicular joints must be considered as one entity. Thus, after shoulder injuries or surgical treatments orthoses can either act as temporary immobilizing agents or they may limit external rotation and abduction.

Spine

The cervical region can be stabilized with circular orthoses made of foam or with hard plastic materials. Applications depend not only on the grade but also on the time course of injury. Most often instabilities and degenerative and disk diseases are indications for orthotic treatment. At the thoracic spine, kyphotic deformities, especially those emerging in adolescence, are treated with three-point stabilizing orthoses. Stabilization of the lumbar spine is achieved with circular elastic orthoses equipped with stiff elements directed at the lumbosacral region. This type of orthosis can effectively treat most lumbar syndromes, disk injuries, and instabilities.

In summary, with the exception of spreading hip orthoses, orthotic treatment must be regarded as only one possible therapy. In most cases active physiotherapy and, depending on the individual type and stage of the injury or disease, functional electrotherapy or even medication may be required.

Summary and Potential Application

Despite the fact that all epidemiological surveys demonstrate that injuries of the joint complexes of the lower limb are the most frequently injured body areas, there is still no consensus about the nature and mechanisms of the chronic instability of these joints. Controlled longitudinal studies must be performed in order to elaborate the relationship between pretraumatic status and probability for incidence of injury.

Several investigations report a drastic reduction in injury rate when the joints are protected with external support mechanisms. From taping studies, it is known that the primary stabilization function may be lost after only a short period of mechanical stress. The underlying mechanisms, however, that explain the reduced injury rates are still restricted to the "proprioceptive function." Neurophysiological and neuromuscular properties of the various supportive devices are the more recent focus of research projects.

Whereas past investigations have concentrated on the determination of the reflex latencies, current projects are more oriented toward the amount of proprioceptively generated neuromuscular activity. Additional research is needed to elaborate the cause and effect of the afferent gain in assessment of the proprioceptive functions of various orthotic devices.

Improved knowledge about mechanical and neurophysiological aspects of joint stabilization will help to develop an effective support system for recovery from injury. In addition, a deeper understanding of the joint stabilizing aspects will improve the training programs necessary for injury rehabilitation. Specific emphasis needs to be given to injury prevention. External support of the joint reduces the incidence of traumatic injury; however, there are still important questions concerning the adaptive mechanisms of chronically worn external joint support over time.

Conclusions

Despite the large body of literature that has been produced in the last 20 years, there is still no agreement on the basic mechanisms of joint stabilization, although there is general agreement about passive and active mechanisms.

The passive systems stabilizing a joint complex are highly specific for the ankle and knee joints, so it is difficult to generalize the load-reducing function of braces and orthoses. The anatomic architecture of the ankle joint complex requires different functions of an external support system than for the knee joint.

Active systems are most probably influenced by afferent feedback from various types of sensory receptors in the joint complex and in the adjacent muscles. Specific emphasis is given to the receptor-bearing structures found in the ligaments of the knee and in the ankle joint. Results presented in the literature provide some suggestions for explaining stable and unstable joint systems.

Epidemiological surveys are consistent with respect to the injury-reducing function of braces, tapes, and other orthoses for the ankle joint. For the knee joint, however, the reports are contradictory and largely dependent on the methodological approach.

In the future, controlled studies following the long-term aspects of external joint support on passive and active structures could help to fill the gap between our epidemiological findings and our physiological knowledge. Comprehensively designed, complex studies are needed to examine the multiple influences on the nature of chronic joint instability.

References

Abdenour, T.E., Saville, W.A., White, R.C., and Abdenour, M.A. 1979. The effect of ankle taping upon torque and range of motion. *Ath. Training* 14: 227–28.

Acierno, S.P., D'Ambrosia, C., Solomonow, M., Baratta, R.V., and D'Ambrosia, R.D. 1995. Electromyography and biomechanics of a dynamic knee brace for anterior cruciate ligament deficiency. *Orthopedics* 18: 1101–7.

Alt, W., Lohrer, H., and Gollhofer, A. 1999. Functional properties of adhesive ankle taping. Neuromuscular and mechanical effects before and after exercise. *Foot & Ankle* 20: 238–45.

Alt, W., Lohrer, H., Gollhofer, A., and Scheuffelen, C. 1996. Preventive ankle taping: Evaluation of mechanical, neuromuscular, and thermal effects before and after exercise. In *Proceedings XIV Symposium on Biomechanics in Sports*, ed. J.M.C.S. Abrantes. Lisboa Codex, Portugal: Edicoes FMH. pp. 537–40.

Attarian, D.E., McCrackin, H.J., De Vito, D.P., McElhaney, J.H., and Garrett, W.E. 1985. Biomechanical characteristics of human ankle ligaments. *Foot & Ankle* 6: 54–58.

Baker, B. 1990. The effect of bracing on the collateral ligaments of the knee. *Clin. Sports Med.* 9: 843–51.

Baker, B.E., van Hanswyk, E., Bogosian, S.P., Werner, F.W., and Murphy, D. 1987. A biomechanical study on the static stabilizing effect of knee braces on medial stability. *Am. J. Sports Med.* 15: 566–70.

Baker, B.E., van Hanswyk, E., Bogosian, S.P., Werner, F.W., and Murphy, D. 1989. The effect of knee braces on lateral impact loading of the knee. *Am. J. Sports Med.* 17: 182–86.

Berns, G.S., Hull, M.L., and Patterson, H.A. 1992. Strain in the anteromedial bundle of the anterior cruciate ligament under combination loading. *J. Orthop. Res.* 10: 167–76.

Beynnon, B.D., Pope, M.H., Wertheimer, C.M., Johnson, R.J., Fleming, B.C., Nichols, C.E., and Howe, J.G. 1992. The effect of functional knee braces on strain on the anterior cruciate ligament in vivo. *J. Bone Joint Surg.* 74A: 1298–312.

Blauth, W., Ulrich, H.W., and Hahne, H.J. 1990. Sinn und Unsinn von Knieorthesen. [Sense and nonsense of knee orthosis.] *Unfallchirurg* 93: 221–27.

Bosien, W.R., Staples, O.S., and Russel, S.W. 1955. Residual disability following acute ankle sprains. *J. Bone Joint Surg.* 37A: 1237–43.

Branch, T.H.P., and Hunter, R.E. 1990. Functional analysis of anterior cruciate ligament braces. *Clin. Sports Med.* 9: 771–97.

Brown, T.D., van Hoeck, J.E., and Brand, R.A. 1990. Laboratory evaluation of prophylactic knee brace performance under dynamic valgus loading using a surrogate leg model. *Clin. Sports Med.* 9: 751–63.

Burks, R.T., Bean, B.G., Marcus, R., and Barker, H.B. 1991. Analysis of athletics performance with prophylactic ankle devices. *Am. J. Sports Med.* 19: 104–6.

Cass, J.R., Morrey, B.F., and Chao, E.Y.S. 1984. Three-dimensional kinematics of ankle instability following serial sectioning of lateral collateral ligaments. *Foot & Ankle* 5: 142–49.

Cavanagh, P.R., and Komi, P.V. 1979. Electromechanical delay in human skeletal muscle under concentric and eccentric contractions. *Eur. J. Appl. Physiol.* 42: 159–63.

Cawley, P.W. 1990. Postoperative knee bracing. *Clin. Sports Med.* 9: 763–70.

Cawley, P.W., and France, E.P. 1991. Biomechanics of the lateral ligaments of the ankle: An evaluation of the effects of axial load and single plane motions on strain patterns. *Foot & Ankle* 12: 92–99.

Cawley, P.W., France, E.P., and Paulos, L.A. 1991. The current state of functional knee bracing research. *Am. J. Sports Med.* 19: 226–33.

Cook, F.F., Tibone, J.E., and Redfern, F.C. 1989. A dynamic analysis of a functional brace for anterior cruciate ligament insufficiency. *Am. J. Sports Med.* 17: 519–24.

Cooper, P.S., McKeon, B., Gossling, H.R., and Ciesielski, T.E. 1995. Neural anatomy of the lateral ankle ligaments. Paper presented at the first combined meeting of the American, British, and European surgeons, Dublin.

Cushnaghan, J., McCarthy, C., and Dieppe, P. 1994. Taping the patella medially: A new treatment for osteoarthritis of the knee joint? *BMJ* 308: 753–55.

De Avilla, G.A., O'Connor, B.L., Visco, D.M., and Sisk, T.D. 1989. The mechanoreceptor innervation of the human fibular collateral ligament. *J. Anat. Lond.* 162: 1–7.

DeVita, P., Torry, M., Glover, K.L., and Speroni, D.L. 1996. A functional knee brace alters joint torque and power patterns during walking and running. *J. Biomech.* 29: 583–88.

Dietz, V., Gollhofer, A., Horstmann, G.A., and Trippel, M. 1990. Postural adjustments and gravity. *J. Physiol. (London)* 420: P21.

Feuerbach, J.W., Grabiner, M.D., Koh, T.J., and Weiker, G.G. 1994. Effect of an ankle orthosis and ankle ligament anesthesia on ankle joint proprioception. *Am. J. Sports Med.* 22: 223–29.

Fiore, R.D., and Leard, J.S. 1980. A functional approach in the rehabilitation of the ankle and rear foot. *Ath. Training* 14: 231–35.

Firer, P. 1990. Effectiveness of taping for the prevention of ankle ligament sprains. *Br. J. Sports Med.* 24: 47–50.

France, E.P., Cawley, P.W., and Paulos, L.E. 1990. Choosing functional knee braces. *Clin. Sports Med.* 9: 743–51.

France, E.P., and Paulos, L.E. 1990. In vitro assessment of prophylactic knee brace function. *Clin. Sports Med.* 9: 823–41.

France, E.P., Paulos, L.E., Jayaraman, G., and Rosenberg, T.D. 1987. The biomechanics of lateral knee bracing. II. Impact response of the braced knee. *Am. J. Sports Med.* 15: 430–8.

Freeman, M., Dean, M., and Hanham, I. 1965. The etiology and prevention of functional instability of the foot. *J. Bone Joint Surg.* 47B: 678–85.

Freeman, M., and Wyke, B. 1967. Articular reflexes at the ankle joint: An electromyographic study of normal and abnormal influences of ankle joint mechanoreceptors upon reflex activity in the leg muscles. *Br. J. Surg.* 54: 990–1001.

Fumich, R.M., Ellison, A.E., Guerin, G.J., and Grace, P.D. 1981. The measured effect of taping on combined foot and ankle motion before and after exercise. *Am. J. Sports Med.* 9: 165–70.

Garrick, J.G. 1977. The frequency of injury, mechanism of injury, and epidemiology of ankle sprains. *Am. J. Sports Med.* 5: 241–42.

Garrick, J.G., and Requa, R.K. 1973. Role of external support in the prevention of ankle sprains. *Med. Sci. Sports Exerc.* 5: 200–3.

Garrick, J.G., and Requa, R.K. 1988. Prophylactic knee bracing. *Am. J. Sports Med.* 16: 118–23.

Glick, J.M., Gordon, R.B., and Nishimoto, D. 1976. The prevention and treatment of ankle injuries. *Am. J. Sports Med.* 4: 136–41.

Gollhofer, A., and Rapp, W. 1993. Recovery of stretch reflex responses following mechanical stimulation. *Eur. J. Appl. Physiol.* 66: 415-20.

Gollhofer, A., Scheuffelen, C., and Lohrer, H. 1993. Neuromuskuläre Stabilisation im oberen Sprunggelenk nach Immobilisation. *Sportverl Sportschad.* 7(Sonderheft 1): 23–28.

Gollhofer, A., Scheuffelen, C., Lohrer, H., and Terreri, S. 1991. Functional significance of braces and special shoes in subjects with and without chronic instability of the ankle joint. In *Book of abstracts*, ed. R.N. Marshall,

G.A. Wood, B.C. Elliott, T.R. Ackland, and P.J. McNair, 173–75. Hamilton Hill, Perth, Australia: PK Commercial & Instant Print.

Gollhofer, A., Strass, D., Kyroelaeinen, H., Dietz, V., and Trippel, M. 1991. Neuromuscular control mechanisms as a function of variable load conditions. *Int. J. Sports Med.* 12: 92–98.

Grace, T.G., Skipper, B.J., and Newberry, J.C. 1987. Prophylactic knee braces and injury to the lower extremity. *J. Bone Joint Surg.* 70A: 422–27.

Greene, T.A., and Hillman, S.K. 1990. Comparison of support provided by a semi-rigid orthosis and adhesive ankle taping before, during, and after exercise. *Am. J. Sports Med.* 18: 498–506.

Greene, T.A., and Wright, C.R. 1990. A comparative support evaluation of three ankle orthoses before, during, and after exercise. *J. Orthop. Sports Phys. Ther.* 11: 453–66.

Gross, M.T., Bradshaw, M.K., Ventry, L.C., and Weller, K.H. 1987. Comparison of support provided by ankle taping and semi-rigid orthosis. *J. Orthop. Sports Phys. Ther.* 9: 33–39.

Gutenbrunner, C., Hildebrand, H.D., Schaff, P., and Gehrke, A. 1997. Untersuchung über Wirkung und Wirksamkeit funktioneller Kniebandagen bei Chondropathia patellae und Gonarthrosen. *Orthopädische Praxis* 33: 52–58.

Hewson, G.F., Jr., Mendini, R.A., and Wang, J.B. 1986. Prophylactic knee bracing in college football. *Am. J. Sports Med.* 14: 262–66.

Hintermann, B., and Nigg, B.M. 1995. *In vitro* kinematics of the axially loaded ankle complex in response to dorsiflexion and plantarflexion. *Foot & Ankle* 16: 514–18.

Hollis, J.M., Blasier, R.D., and Flahiff, C.M. 1995. Simulated lateral ankle ligamentous injury. *Am. J. Sports Med.* 23: 672–77.

Horstmann, G.A., Gollhofer, A., and Dietz, V. 1988. Reproducibility and adaptation of the EMG responses of the lower leg following perturbations of upright stance. *Electroenceph. Clin. Neurophysiol.* 70: 447–52.

Hull, M.L., Berns, G.S., Varma, H., and Patterson, H.A. 1996. Strain in the medial collateral ligament of the human knee under single and combined loads. *J. Biomech.* 29: 199–206.

Isakov, E., Mizrahi, J., Solzi, P., Susak, Z., and Lotem, M. 1986. Response of the peroneal muscles to sudden inversion of the ankle during standing. *Int. J. Sports Biomech.* 2: 100–9.

Jerosch, J., and Bischof, M. 1994. Der Einfluß der Propriozeptivität auf die funktionelle Stabilität des oberen Sprunggelenkes unter besonderer Berücksichtigung von Stabilisierungshilfen. *Sportverl. Sportschad.* 8: 111–21.

Johnson, E.E., and Markolf, K.L. 1983. The contribution of the anterior talofibular ligament to ankle laxity. *J. Bone and Joint Surg.* 62B: 81–89.

Johnson, G.R., Dowson, D., and Wright, V. 1976. A biomechanical approach to the design of football boots. *J. Biomech.* 9: 581–84.

Johnson, M.B., and Johnson, C.L. 1993. Electromyographic response of peroneal muscles in surgical and nonsurgical injured ankles during sudden inversion. *J. Orthop. Sports Phys. Ther.* 18: 497–501.

Jonsson, H., and Kärrholm, J. 1990. Brace effects on the unstable knee in 21 cases: A comparison of three designs. *Acta Orthop. Scand.* 61: 313–18.

Juvenal, J.P. 1972. The effects of ankle taping. *Ath. Training* 10: 146–49.

Karlsson, J., and Andreasson, G.O. 1992. The effect of external ankle support in chronic lateral ankle joint stability: An electromyographic study. *Am. J. Sports Med.* 20: 257–61.

Knutzen, K.M., Bates, B.T., Schot, P., and Hamill, J. 1987. A biomechanical analysis of two functional knee braces. *Med. Sci. Sports Exerc.* 19: 303–9.

Konradsen, L., Ravn, J.B., and Sorensen, A.I. 1993. Proprioception at the ankle: The effect of anesthetic blockade of ligament receptors. *J. Bone Joint Surg.* 75B: 433–36.

Larsen, E. 1984. Taping the ankle for chronic instability. *Acta Orthop. Scand.* 55: 551–53.

Löfvenberg, R., Kärrholm, J., Sundelin, G., and Ahlgren, O. 1995. Prolonged reaction time in patients with chronic lateral instability of the ankle. *Am. J. Sports Med.* 23: 414–17.

Lohrer, H., Alt, W., and Gollhofer, A. 1996. Orthesen. In *GOTS-Manual Sporttraumatologie*, edited by M. Engelhardt, B. Hintermann, and B. Segesser. Bern: Huber Verlag. pp. 473–82.

Lohrer, H., Scheuffelen, C., Gollhofer, A., and Alt, W. 1994. Der Bewegungsablauf und die Beanspruchung des Sprunggelenkes unter dynamischer Belastung, edited by A. Verdonk and M. Wiek Biokinetika. Tagungsband. 111–20.

Lynch, S.A., Eklund, U., Gottlieb, D., Renström, P.A., and Beynnon, B. 1996. Electromyographic latency changes in the ankle musculature during inversion moments. *Am. J. Sports Med.* 24: 362–69.

Mayhew, J.L. 1972. Effects of ankle taping on motor performance. *Ath. Training* 7: 10–11.

McCorkle, R.B. 1963. A study of the effect of adhesive strapping techniques on ankle action. Master's thesis, Springfield College, Springfield, MA.

McCullogh, C.J., and Burge, P.D. 1980. Rotatory stability of the load-bearing ankle: An experimental study. *J. Bone Joint Surg.* 62B: 460–4.

Menschik, A. 1974. Mechanik des Kniegelenkes. *Z. Orthop.* 112: 481–95.

Michelson, J.D., and Hutchins, C. 1995. Mechanoreceptors in human ankle ligaments. *J. Bone Joint Surg.* 77B: 219–24.

Mommersteeg, T.J.A., Huiskes, R., Blankevoort, L., Kooloos, J.G.M., and Kauer, J.M.G. 1997. An inverse dynamics modeling approach to determine the retraining function of the human knee ligament bundles. *J. Biomech.* 30: 139–46.

Müller, W. 1982. *Das Knie.* Berlin: Springer-Verlag.

Müller, W. 1996. Form and function of the knee. *Am. J. Sports Med.* 24: 104–6.

Nawoczenski, D.A., Owen, M.G., Ecker, M.L., Altman, B., and Epler, M. 1985. Objective evaluation of peroneal response to sudden inversion stress. *J. Orthop. Sports Phys. Ther.* 7: 107–9.

Nigg, B.M., Skarvan, G., Frank, C.B., and Yeadon, M.R. 1990. Elongation and forces of ankle ligaments in a physiological range of motion. *Foot Ankle Int.* 11: 30–40.

Perlau, R., Frank, C., and Fick, G. 1995. The effect of elastic bandages on human knee proprioception in the uninjured population. *Am. J. Sports Med.* 23: 251–55.

Pienkowski, D., McMorrow, M., Shapiro, R., Caborn, D.N., and Stayto, J. 1995. The effect of ankle stabilizers on athletic performance. *Am. J. Sports Med.* 23: 757–62.

Quigley, T.B., Cox, J., and Murphy, J. 1946. Protective device for the ankle. *J. Am. Med. Assoc.* 123: 924–26.

Rarick, G.L., Bigley, G., Karst, R., and Malina, R.M. 1962. The measurable support of the ankle joint by conventional methods of taping. *J. Bone Joint Surg.* 44A: 1183–90.

Rasmussen, O. 1985. Stability of the ankle joint. *Acta Orthop. Scand.* 56(Suppl.): 1–75.

Rasmussen, O., and Kromann-Andersen, C. 1983. Experimental ankle injuries. *Acta Orthop. Scand.* 54: 356–62.

Reese, R.C. Jr., and Burruss, T.P. 1986. Athletic training techniques and protective equipment. In *The lower extremity and spine in sports medicine,* ed. J.A. Nicholas and E.B. Hershman, 245–68. St. Louis: Mosby.

Regalbuto, M.A., Rovick, J.S., and Walker, P.S. 1989. The forces in a knee brace as a function of hinge design and placement. *Am. J. Sports Med.* 17: 535–43.

Renström, P., and Theis, M. 1993. Biomechanik der Verletzung der Sprunggelenkbänder. *Sportverl. Sportschad.* 7(Sonderheft 1): 29–35.

Renström, P., Wertz, M., Incavo, S., Pope, M., Ostgaard, H.C., Arms, S., and Haugh, L. 1988. Strain in the lateral ligaments of the ankle. *Foot & Ankle* 9: 59–63.

Roll, J.P., Popov, K., Gurfinkel, V., Lipshits, M., Andre-Deshays, C., Gilhodes J.G., and Cuoniam, C. 1993. Sensorimotor and perceptual function of muscle proprioception in microgravity. *J. Vestibular Res.* 3: 259–73.

Roser, L.A., Miller, S.J., and Clawson, D.K. 1971. Effects of taping and bracing on the unstable knee. *Northwest Med.* 70: 544–46.

Rovere, G.D., Clarke, T.J., Yates, C.S., and Burley, K. 1988. Retrospective comparison of taping and ankle stabilizers in preventing ankle injuries. *Am. J. Sports Med.* 16: 228–33.

Rovere, G.D., Haupt, H.A., and Yates, C.S. 1987. Prophylactic knee bracing in college football. *Am. J. Sports Med.* 15: 111–16.

Rupp, S., Lanta, P., and Schulz, H. 1995. Reduction of the anterior drawer of the knee joint by rehabilitation orthoses: Comparison of the MVP orthosis vs. the Donjoy-Gold Point orthosis. *Unfallchirurg.* 98: 474–77.

Sauer, H.-D., Jungfer, E., and Jungbluth, K.H. 1978. Experimentelle Untersuchungen zur Reißfestigkeit des Bandapparates am menschlichen Sprunggelenk. *Hefte Unfallheilkunde* 131: 37–45.

Schaff, P.S., Luber, M., Mößmer, Ch., and Rosemeyer, B. 1995. Der Effekt infrapatellarer Sehnenbandagen auf das EMG-Muster. *Sportorthop. Sporttraumat.* 11: 2–7.

Scheuffelen, C., Gollhofer, A., and Lohrer, H. 1993. Neuartige funktionelle Untersuchungen zum Stabilisierungsverhalten von Sprunggelenksorthesen. *Sportverl Sportschad.* 7: 30–36.

Scheuffelen, C., Rapp, W., Gollhofer, A., and Lohrer, H. 1993. Orthotic devices in functional treatment of ankle sprain. *Int. J. Sports Med.* 14: 1–9.

Siegler, S., Chen, J., and Schneck, C.D. 1990. The effect of damage to the lateral collateral ligaments on the human characteristics of the ankle joint: An in vitro study. *J. Biomech. Eng.* 112: 129–37.

Sitler, M., Ryan, J., Hopkinson, W., Wheeler, J., Santomier, J., Kolb, R., and Polley, D. 1990. The efficacy of a prophylactic knee brace to reduce knee injuries in football: A prospective, randomized study at West Point. *Am. J. Sports Med.* 18: 310–15.

Sitler, M., Ryan, J., Wheeler, B., McBride, J., Arciero, R., Anderson, J., and Horodyski, M. 1994. The efficacy of a semi-rigid ankle stabilizer to reduce acute ankle injuries in basketball. *Am. J. Sports Med.* 22: 454–61.

Steinbrück, K. 1997. Epidemiologie. In *GOTS-Manual Sporttraumatologie,* edited by M. Engelhardt, B. Hinterman, and B. Segesser. Bern: Huber Verlag. pp. 19–29.

Stormont, D.M., Morrey, B.F., Kai-Nan A., and Cass, J.R. 1985. Stability of the loaded ankle: Relation between articular restraint and primary and secondary restraints. *Am. J. Sports Med.* 13: 295–300.

Stover, C.N. 1980. Air-stirrup management of ankle injuries in the athlete. *Am. J. Sports Med.* 8: 360–5.

Stüssi, E., Stacoff, A., and Segesser, B. 1992. Biomechanische Überlegungen zur Belastung der Sprunggelenke. *Orthopäde.* 21: 88–95.

Styf, J.R., Nakhostine, M., and Gershuni, D.D. 1992. Functional knee braces increase pressures in the anterior compartment of the leg. *Am. J. Sports Med.* 20: 46–49.

Surve, I., Schwellnus, M.P., Noakes, T., and Lombard, C. 1994. A fivefold reduction in the incidence of recurrent ankle sprains in soccer players using the sport-stirrup orthosis. *Am. J. Sports Med.* 22: 601–5.

Tegner, Y., Pettersson, G., Lysholm, J., and Gillquist, J. 1988. The effect of knee rotation brace on knee motion. *Acta Orthop. Scand.* 59: 284–87.

Teitz, C.C., Hermanson, B.K., Kronmal, R.A., and Diehr, P.H. 1987. Evaluation of the use of braces to prevent injury to the knee in collegiate football players. *J. Bone Joint Surg.* 69: 2–9.

Tropp, H. 1986. Pronator muscle weakness in functional instability of the ankle joint. *Int. J. Sports Med.* 7: 291–94.

Tropp, H., Ekstrand, J., and Gillquist J. 1985. Prevention of ankle sprains. *Am. J. Sports Med.* 13: 259–62.

Veldhuizen, J.W., Koene, F.M., Oostvogel, H.J., von Thiel, T.P., and Verstappen, F.T. 1991. The effects of supportive knee brace on leg performance in healthy subjects. *Int. J. Sports Med.* 12: 577–80.

Walker, P.S., Rovick, J.S., and Robertson D.D. 1988. The effects of knee brace hinge design and placement on joint mechanics. *J. Biomech.* 21: 965–74.

Wirth, C.J., Küsswetter, W., and Jäger, M. 1978. Biomechanik und Pathomechanik des oberen Sprunggelenks. *Hefte Unfallheilkunde* 131: 10–22.

Wojtys, E.M., Loubert, P.V., Samson, S.Y, and Viviano, D.M. 1990. Use of a knee-brace for control of tibial translation and rotation. *J. Bone Joint Surg.* 72A: 1323–29.

Worrell, T.W., Ingersoll, C.D., and Farr, J. 1994. Effect of patellar taping and bracing on patellar position: An MRI case study. *J. Sport Rehab.* 3: 146–53.

Zetterlund, A.E., Serfass, R.C., and Hunter, R.E. 1986. The effect of wearing the complete Lenox Hill derotation brace on energy expenditure during horizontal treadmill running at 161 meters per minute. *Am. J. Phys. Med.* 14: 73–76.

Load During Physical Activity Summary

J. Mester and B.R. MacIntosh

Human movement is inseparably linked to a certain mechanical load, which when exceeded may induce ruptures of tissues, fractures of bones, and many other threats to health. Therefore, mechanical load and the ways to prevent threats to health have been subjected to scientific interest. Long before Christ, bandages for fractures were used. During the 15th and 16th centuries, famous scientists/artists created important concepts of the anatomical structures that are subjected to physical load. There has been much discussion on the properties of the passive structures, such as ligaments, tendons, and cartilage, and on the recovery of bone structures after intensive load and/or after adaptation following a chronic load. These adaptation processes represent behavior patterns that can be observed in most biological structures.

The acting principle behind a certain load is force. Those forces that act on and in the human body are discussed by Nigg in chapter 14. After defining the major terms, Nigg highlights the basics of ground reaction forces as indicators of external load. Impact forces can reach remarkably high values and thus have been suspected as being detrimental to health. They can be characterized as internal and external impact.

Nigg offers a survey of the magnitude of forces as an additional, important topic for further discussion on the biopositive and/or bionegative effects of mechanical load on the biological structures involved. External active forces may range from simple body weight (BW) to fourfold BW in sprinting (vertical component). Impact forces can reach up to twelve times BW in the takeoff phase of jumping events.

Apart from those external forces that can be simply derived from the mass and acceleration involved, the contraction of muscles and the properties of the so-called "passive" elements are responsible for the generation and transmission of force. The chapter written by Herzog (chapter 15) focuses on structure and morphology of the skeletal muscle. Herzog introduces the most important facts in the area of molecular interaction with respect to the molecules involved in force generation and the interaction between thick/thin filaments as well as neural connections. These facts enable the reader to observe and understand force production at this level.

A breakthrough in the understanding of muscular contraction and force production was made by Huxley (1957), who introduced the sliding filament and cross-bridge theories. It is accepted that muscle contraction is made possible by the relative sliding of the thick and the thin filaments. The idea that the motion is produced by a cyclic interaction of side-pieces (cross-bridges) that protrude between the thick and the thin filaments is accepted as the basis for muscle contraction and force production.

In vivo, the force generated by a given muscle depends on size and structure and is influenced by the level of activation, muscle length, and the speed of contraction, which partly is a function of level of activation but is limited by the external load applied. In real contractions, e.g., in the dynamic situations of sports, these parameters of muscle contraction demonstrate nonlinear properties, making it difficult to predict specific values for these parameters at any instant in time. Nevertheless, it is important to note that these properties of muscle basically constitute the force production capabilities associated with movement.

Taking into account the entire structure of the muscle, down to the filaments and the molecules involved in the process of contraction, it is rather easy to understand that even nowadays the problem of calculating muscle forces for all contractile elements and conditions has not been solved to a satisfactory extent. An optimistic view holds that such a holistic concept can be developed only with some basic properties, such as the force-length relationship and the force-velocity properties of muscle. However, such attempts must account for the impact of prior activity (duration, magnitude, and length changes) on subsequent force-generating capabilities.

After discussing the function of the physiological "active" structures that provide force production, it is important to face the mechanical effects of these

351

forces acting on the "passive" systems, such as bone, cartilage, ligaments, and tendons. It is well known that these musculoskeletal tissues first of all serve as force transmitters according to their specific mechanical functions and biological structures and thus express their unique structural hierarchies. Moreover, ligaments and articular cartilage can show distinct deformation properties. They are able to dissipate energy through viscoelastic/poroelastic processes. Mechanical topics like these are the center of attention in the chapter by Anderson, Adams, and Hale (chapter 16).

The main principles of mechanical effects on the musculoskeletal tissues, i.e., the basic characteristics (strain) of material behavior that is exposed to a certain mechanical load (stress), are presented at the beginning of this chapter. Simplified, the range of mechanical reaction of these tissues can be grouped into hard and soft responders. So hard tissues, such as bone, with approximately 75% inorganic mineral content, react totally differently than soft tissues, such as collagen, consisting of macromolecules and water, which itself plays an important role as a mechanical load carrier. These soft tissues can exhibit extremely interesting behavior in terms of mechanical response. As a result a two-phase response can often be analyzed. The first phase is characterized by substantial deformation of the material due to low loads. In the second phase a higher stiffness and nearly linear stress-strain response can occur. A further important reaction to load in these soft tissues can be a creep, i.e., an increase of deformation over time, even under a constant load, that can lead to injuries.

In chapter 16, Anderson, Adams, and Hale discuss the particular properties of bones, articular cartilage, ligaments, and tendons. In general, the viscoelastic behavior of these structures has important clinical significance. The findings in this chapter can help us better understand and prevent injuries or failure due to inappropriate sudden or chronic mechanical load. For a long time, it was assumed that these passive tissues more or less were not able to adapt to chronic exercise. Now we know that there are indeed long-term effects of exercise on passive structures. In the case of the tendons, this includes an increase in stiffness, ultimate tensile strength, and weight. In addition to the fact that physical stress can enhance collagen metabolism, exercise can also improve collagen synthesis.

By an adjustment of synthesis and degradation, articular cartilage and the corresponding mechanical properties in general are subject to adaptation induced by mechanical loading associated with ex-

ercise. In contrast, a lack of mechanical stimulation can induce atrophy and severe forms of degeneration. Especially in older individuals, it is important to carefully plan not only the intensity of exercise with respect to cardiac stress, but also the mechanical load of the passive tissues. In order to quantify these preventive considerations and develop concepts of performance improvement at the level of elite sport, it is imperative to intensify scientific efforts in this field.

The concepts of mechanical load discussed by Anderson, Adams, and Hale are closely connected to the chapter written by Fischer (chapter 17) on biological response to forces acting in the locomotor system. Athletes with chronic high mechanical load have significantly more dense femurs than nonathletes. Many studies have supported the notion of bone adaptation to loading conditions. The converse is also true; there is negative adaptation, such as bone loss, under reduced chronic load. This still remains a major problem for astronauts staying in a weightless environment for an extended period. This problem is of major importance not only for future space programs but also for clinical settings and the elderly because the biological response and negative adaptation to decreased mechanical stimuli is more rapid than the positive response to increased mechanical stimuli in terms of exercise. For individuals engaged in high exercise and training loads, bone adaptation is generally beneficial as it provides a stronger frame to carry the extreme loads increased in training and competition.

In the study of articular cartilage, for a long time there have been no results stating that this tissue is adaptive to mechanical stimuli, due to its avascular structure and minimal cell activity. Recent animal studies, however, indicate that articular cartilage does have the ability to positively adapt to mechanical stimuli and also to repair itself after injury. This ability is associated with the concentration of glycosaminoglycans and the biosynthetic rate of the structural proteoglycans biglycan and decorin, which seem to react following the application of mechanical load, e.g., compression. Some evidence suggests that articular cartilage requires a certain physiological mechanical load to maintain its function.

For the tendons and ligaments, scientific findings support the ability of these passive tissues to adapt to increased physical activity respiratory chronic mechanical load. Thus, for the tendons it is quite clear that long-term exercise will result in an improvement of the ultrastructure of the tendon. Although there are not as many findings on the adaptive ability of ligaments, Fischer notes the results of his own studies that mechanical stimuli can induce changes in the microstructural composition and the mechanical properties of ligaments. In the animal model, a ligament adaptation to immobilization also was demonstrated in the form of a decrease in ultimate load and stiffness.

The results of Fischer and those of Anderson et al. note that the so-called passive tissues, such as bones, articular cartilage, tendons, and ligaments, also show the basic biological ability of adaptation in response to a chronic mechanical stimulus. With an increase in mechanical load, all passive skeletal tissues obviously are able to adapt, although this adaptation occurs more slowly than the negative adaptation following reduced loads. This slow adaptation should be taken into consideration in dictating the rate of increase of intensity or duration of training load for programs intended to increase strength.

These considerations of prevention are also dealt with in chapter 18, which focuses on the prevention of excessive forces by means of mechanical devices such as braces and orthotics. As these devices can remarkably reduce the probability of injuries for athletes in ball games remarkably, it is important to discuss criteria in terms of functional properties. Gollhofer, Alt, and Lohrer present evaluation criteria for examining these devices to enhance the external stability of the knees and the ankle joints. These criteria mainly consider the possible interference of the joint function imposed by ankle and knee orthosis, such as sleeves, bandages, taping, and braces, and the validity claims for injury prevention by these devices. An important area of research here is the interaction of the mechanical, i.e., stabilizing, effect of the various devices and the effects on proprioceptive functions.

Generally speaking, most of the protective devices can lead to a reduction of the injury rate, although the basic mechanisms of joint stabilization are still controversial. Further research is needed to explore the specific mechanisms of injury and possible injury prevention strategies. These strategies need to consider not only braces and orthotics, but also training programs targeted at injury prevention.

Load During Physical Activity Definitions

Mechanical

adduction/abduction: angular movement where the angle between the moving segment and the adjacent segment is decreased/increased. Rotation of the foot around the inferior-superior axis.

anisotropic: a material is anisotropic if the magnitude of a particular property is not the same for all orientations of a test sample.

anisotropic (A) band: dark band in the middle of a sarcomere caused by the thick myofilaments.

axes of the foot:

> **anterior-posterior axis (a-p):** the axis defined from the center of calcaneus to the cross between second and third metatarsal phalanges.
>
> **inferior-superior axis (i-s):** the axis from the plantar to the dorsal surface of the foot perpendicular to the plantar surface of the foot.
>
> **medio-lateral axis (m-l):** the axis from the medial to the lateral side of the foot, parallel to the plantar surface of the foot.

body weight: the resultant force for all the segmental weights of a human or animal body.

constraint functions: functions that limit the solution space in an optimization procedure.

coordinate system: mathematical system of axes that establishes a reference frame for defining material behavior. A Cartesian system of three orthogonal axes (x,y,z) commonly is used, although a cylindrical system (r,θ,z) often describes the natural orientation of the material (e.g., bone and wood).

creep: time-dependent increase in strain from an initial baseline level following an instantaneous application of a constant force to a test sample.

deformation: change in size of a structure in response to external forces acting on it.

design variables: variables that are adjustable in an optimization approach and are contained in the objective function.

distribution problem: a problem in biomechanics in which the resultant intersegmental (joint) force and moment are assigned to individual structures crossing the joint of interest.

dorsi-/plantarflexion: dorsi-/plantarflexion occurs around the medio-lateral axis defined by the malleoli decreasing/increasing the ankle angle in the sagittal plane.

elastic: a material behaves elastically if its deformation follows along the same load-deformation curve during both the loading and unloading portion of a load-unload cycle, returning to zero deformation at zero force. The definition can alternately be applied to stress and strain relationship, substituting for load and deformation, respectively.

elastic modulus (E): a measure of a material's resistance to deformation in a prescribed direction. It is calculated as the ratio of stress to strain ($E = \sigma/\varepsilon$) in a uniaxial test. Elastic, or Young's, modulus typically refers to a material or range of strain for which the deformation is recovered immediately upon release of force. Units of elastic modulus are the same as for stress (e.g., N/mm^2 = mega-Pascal or MPa).

extension: angular movement, where the angle between the moving and adjacent segments is increased.

flexion: angular movement, where the angle between the moving and adjacent segments is decreased.

force: force cannot be defined. However, effects produced by forces can be described. Mechanical effects of forces are:

> A force can accelerate a mass: $\mathbf{F} = m \cdot \mathbf{a}$
>
> A force can deform a spring: $\mathbf{F} = k \cdot \Delta \mathbf{x}$
>
> **active force:** force generated by activity.
>
> **external/internal force:** force acting externally/internally to the system of interest.
>
> **ground reaction force (GRF):** a force which acts from the ground on the object which is in contact with the ground.

impact force: force that results from a collision of two objects, reaching its maximum earlier than 50 milliseconds after the first contact of the two objects. The most often used application of impact force is the landing of the heel of the foot on the ground.

normal force: force perpendicular to the surface of interest.

shear force: force parallel to the surface of interest.

fracture toughness: a material property specifying the amount of energy required to propagate a crack or tear some finite distance. Units = work/crack area (e.g., Joules/m², equivalent to N/m).

frequency: number of oscillations per time unit.

natural frequency: a frequency of vibration that corresponds to the elasticity and mass of a system under the influence of its internal forces and damping. The lowest natural frequency of a system is known as the fundamental frequency. Higher natural frequencies are harmonics of the fundamental value. Some systems exhibit several coupled or uncoupled modes of vibration.

resonance frequency: frequency of a forced vibration input on a system that corresponds to a natural frequency of the system.

friction:

dynamic translational friction: the dynamic translational coefficient, μ_{trdyn}, is defined as the quotient of the force parallel to the contacting surfaces, F_{par}, and the normal force, F_{norm}, for the situation when the two surfaces move with constant velocity with respect to each other.

$$\mu_{trdyn} = \frac{F_{par}}{F_{norm}} = \text{dynamic translational coefficient}$$

static translational friction: the static translational coefficient, μ_{trsta}, is defined as the quotient of the force parallel to the contacting surfaces, F_{par}, and the normal force, F_{norm}, immediately before the two surface samples start moving.

$$\mu_{trsta} = \frac{F_{par}}{F_{norm}} = \text{static translational coefficient}$$

functional instability:

ankle joint: a tendency for the foot to give way after an ankle sprain (Freeman 1965). Possible causes of functional instability are:

- presence of proprioceptive deficit mainly caused by scarring of capsular and ligament structures;
- functional weakness of the muscles acting around the joint (i.e., peroneal muscles);
- anatomical joint laxity due to accused sprain of the calcaneofibular and anterior talofibular ligaments.

knee joint: a tendency of the knee to give way most often caused by ligamentous lesions or mechanical incarcerations, chondral lesions, or atrophic muscles.

heterogeneity/homogeneity: heterogeneity is the nonuniform distribution of a property or composition throughout a test sample. Conversely, homogeneity implies the same properties or composition occur at all points in the sample.

inversion/eversion: rotation of the foot around an anterior-posterior axis through the foot (Nigg and Segesser 1992). Inversion is characterized by inward rotation of the foot elevating the medial border of the foot. Eversion is characterized by outward rotation of the foot elevating the lateral border of the foot.

isotropic: a material is isotropic if the magnitude of a particular property is the same in all directions of a test sample. Materials can be isotropic in terms of mechanical, thermal, and electrical behavior.

joint proprioception: sensory ability to perceive joint movement and position in space. Proprioceptive function is mediated by afferent pathways originating from muscle spindles. This multi-sensory input is organized and processed in the spinal and transcortical nervous system.

line of action: a line that extends the line of a force vector to infinity.

loading rate: a time derivative of the force-time curve.

material property: a mechanical property linked intrinsically to the tissue or material, irrespective of size and shape. Stress, strain, elastic modulus, and strain energy density are material properties.

mathematical system: equations

mega-Pascal (MPa): an SI unit of pressure or stress; 10^6 N/m²

moment: turning effect of a force applied at point P, F_P, about a base point O with:

$$M_O = r_{PO} \times F_P$$

where

M_O = moment (or moment of force) about point O

r_{PO} = position vector from O to P

$\times$ = sign for vector product

F_P = force acting at point P

objective function: a mathematical function based on design variables that describes the value of the desired objective in an optimization approach.

optimization: branch of mathematics.

orthotropic: a material is orthotropic if its property is different in all three orthogonal directions of the prescribed mathematical coordinate system.

pronation/supination: supination and pronation are a combination of inversion-eversion plus plantar/dorsiflexion plus adduction/abduction (Nigg and Segesser 1992).

range of movement (ROM): angular excursion experienced by rotations around distinct joint axes induced either by external forces (i.e., passive ROM; PROM) or by internal muscular actions (i.e., active ROM; AROM).

resultant forces: the vector sum of many forces.

stiffness: force required per unit length to deform a sample. Slope of a force-deformation curve. Analogous to a spring constant; units = N/m or N/mm.

strain (ϵ): relative change in size in a prescribed direction ($\epsilon = \Delta L/L_0$). Units of strain = length/length (e.g., m/m); thus, strain is dimensionless. For purposes of discussion only, not in mathematical calculations, strain is often expressed as a percentage ($\times$ 100).

stress (σ): applied force per unit area. Units = MPa = N/mm^2.

stress-relaxation: time-dependent decrease in stress from an initial peak level following instantaneous application of a constant deformation to a test sample. After a rapid initial decrease, stress asymptotically approaches an equilibrium level.

structural property: a mechanical property which depends specifically upon the size and shape of a test sample. Force, deformation, stiffness, and absorbed energy are structural properties. These properties are useful in evaluating the integrity of whole tissue structures such as ligaments and tendons.

system of interest: sum of mass points or segments which are considered as a basic unit to discuss and solve a problem. In biomechanical analysis the system of interest is typically considered as being rigid. All forces acting from the outside on this system are called external forces for this system. Vice versa, all forces acting within the system of interest are called internal forces for this system.

torque: result of a force couple.

ultimate strength: a material property which reflects the maximum singular application of stress (σ_{ult}) a given material or tissue can withstand before failing. The loading mode typically is specified (e.g., tension, compression, shear) with units of stress.

underdetermined: a mathematical system that contains more unknowns than system.

vector: a two-dimensional quantity defined by its magnitude and direction.

vector cross product: operation in vector algebra similar to a multiplan with ordinary numbers.

viscoelasticity: characteristic of a material the deformation of which depends upon loading rate and time history of loading. Creep and stress-relaxation behavior are commonly studied to assess viscoelastic properties.

Biological

α-motoneuron: neurons which carry signals from the central nervous system (spinal cord) to extrafusal skeletal muscle fibers.

acetylcholine: substance that transmits nerve signals to the muscle fiber across the synaptic cleft.

action potential: depolarization/repolarization wave that travels along the muscle fiber and triggers contraction.

adenosinediphosphate (ADP): chemical product of ATP hydrolysis: ATP $\rightarrow$ ADP + P$_i$

adenosinetriphosphate (ATP): high energy phosphate whose breakdown provides the energy for muscular contraction. Adenosine with three phosphates attached. ATP is considered the common currency for energy in cells. Hydrolysis of ATP yields energy for biological and mechanical work.

angle of pinnation: angle between the main (average) fiber direction and the line of action of the entire muscle.

anterior cruciate ligament (ACL): major ligament of the knee. Primary ligamentous restraint to anterior displacement of the tibia with respect to the femur.

bone mineral density (BMD): an estimated average bone density measure determined from a dual-photon absorptiometry measurement of bone mineral content and the estimated volume of the examined bone.

contraction: mechanical response to activation of a muscle, resulting in generation of tension and/or shortening of the muscle.

> **concentric contraction:** contraction in which muscle (fiber, sarcomere) length decreases.
>
> **eccentric contraction:** contraction in which muscle (fiber, sarcomere) length increases.
>
> **fused contraction:** a muscle contraction resulting from repeated stimulations which does not contain oscillations in force.
>
> **isometric contraction:** contraction in which muscle (fiber, sarcomere) length remains constant.
>
> **isotonic contraction:** contraction of a muscle against a constant resistance.
>
> **unfused contraction:** a muscle contraction resulting from repeated stimulations that is associated with oscillations in force.

cross-bridge: presumed link that forms between thick and thin myofilaments and produces force and movement between the thick and thin myofilaments.

electromyography: measurement of the electrical signal that occurs during muscular activation.

endomysium: thin connective tissue sheath that surround a muscle fiber.

epimysium: connective tissue sheath that surrounds muscle.

firing frequency: frequency of action potentials of a motor neuon and the corresponding muscle fibers.

force-length relationship: the relationship in skeletal muscle that exists between the maximal, isometric force and the muscle or sarcomere length.

force-velocity relationship: the relationship in skeletal muscle that exists between the force and the speed of contraction of the shortening/lengthening.

glycosaminoglycan (GAG): a polysaccharide chain in proteoglycans. Hyaluronate, chondroitin sulfate, keratan sulfate, heparan sulfate, and heparin are the major GAGs. In proteoglycan from cartilage, keratan sulfate and chondroitin sulfate chains are covalently attached to a polypetide backbone called the core protein.

isotropic (I) band: light band at the end of a sarcomere caused by the thin myofilaments.

lateral collateral ligament (LCL): major ligament of the knee. Primary ligamentous restraint to varus rotation of the tibia with respect to the femur.

M-band: center band in the sarcomere.

medial collateral ligament (MCL): major ligament of the knee. Primary ligamentous restraint to valgus rotation of the tibia with respect to the femur. Commonly injured in sports due to a strong blow to the lateral knee with the foot firmly planted.

motor unit: a motor neuron and all the muscle fibers it innervates.

muscle:

> **agonistic (antagonistic) muscle:** a muscle that produces a moment about a joint that is in the same (opposite) direction as the resultant joint moment.
>
> **fusiform muscle:** muscle with fiber alignment approximately parallel to the line of action of the muscle.
>
> **pennate muscle:** muscle with fiber alignment at a distinct angle (or angles) relative to the line of action of the muscle.
>
> **synergistic muscle:** a muscle that participates in the production of a moment with other muscles.

myofibrils: bundles of contractile proteins which are systematically arranged in muscle fibers.

myofilament:

> **thick myofilament:** contractile filament which contains primarily myosin proteins.
>
> **thin myofilament:** contractile filament which contains primarily the proteins actin, troponin, and tropomyosin.

myosin, actin: contractile proteins in skeletal muscle.

neuromuscular junction: site where an α-motoneuron meets a muscle fiber and permits transduction of a nerve signal (action potential) to cause activation of the muscle fiber via neuromuscular transmission.

optimal muscle length: length at which a muscle can produce its maximal isometric force.

osteoarthritis (OA): an inflammation of a joint associated with bone and cartilage degeneration. OA is normally associated with aging or may occur secondary to an injury to the joint. Inflammation levels may be so low as to go unnoticed until the cartilage degeneration becomes severe.

perimysium: connective tissue sheath that surrounds muscle fiber bundles (i.e., fascicles).

physiological cross-sectional area: theoretical area of the muscle that is obtained by summing the cross-sectional areas of all muscle fibers arranged in parallel to each other.

posterior cruciate ligament (PCL): major ligament of the knee. Primary ligamentous restraint to posterior displacement of the tibia with respect to the femur.

postsynaptic membrane: muscle membrane at the neuromuscular junction.

sarcolemma: plasma membrane and basement membrane around a muscle fiber (cell).

sarcomere: basic contractile unit of skeletal muscle which extends from Z-line to Z-line.

sarcoplasmic reticulum: organelle in muscle fiber that serves as a storage container for calcium and participates in activation of a muscle fiber by controlled release and reuptake of calcium.

size principle: principle that states that motor units are recruited in accordance with their size; small units first, large units last. Size can refer to soma, volume, axon diameter and number of associated muscle fibers.

slow (fast) twitch fibers: muscle fibers whose twitch response characteristics and maximal velocity of shortening are slow (fast).

synaptic cleft: gap between the end of an α-motoneuron and a muscle fiber.

titin: giant protein that spans from the Z-line to the M-band in the sarcomere of striated muscle.

transverse (T-) tubules: invaginations of the muscle fiber membrane which extend into the interior of the fiber, thereby permitting action potentials to reach the interior of fibers easily.

troponin, tropomyosin: regulatory proteins on the thin myofilament.

twitch: force response of a muscle caused by a single stimulus.

ventral root: site where outgoing (efferent) signals leave the spinal cord.

PART IV

Fatigue and Exercise

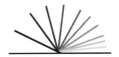

Fatigue and Exercise Historical Highlights

1665

Grimaldi "explains" that the audible sounds heard while clenching the fists after stopping the ears with the thumbs are due to continuous hurried motion of animal spirits (cited in Wollaston 1810).

1810

Wollaston gives an empirical evaluation of the frequency content of the muscle sound during voluntary contraction. Comparing the tone of the muscle sound during voluntary contraction and the noise of the wheels of a carriage on a cobbled street, he first estimates the main frequency component around 25 Hz.

1849

Using a galvanometer, DuBois-Reymonds detects muscle electrical activity during muscle voluntary contraction.

1890

Mosso designs an "ergographerz" that is able to follow the reduction of the displacement of a weight during fatiguing repetitive flexion of the third digit. The graphic output is a "fatigue curve" typical for each subject and specific individual condition.

1923

Cobb and Forbes report an increase in the amplitude of the electromyogram and a shift toward the low end of the frequency components.

1948

Gordon and Holbourn define the sounds recorded at the muscle surface as the mechanical counterparts of the muscle "action currents."

1956

R.G. Edwards and O.C.J. Lippold perform one of the first time-domain analyses of EMG during fatiguing exercise. They report an increase of integrated EMG at fatigue.

1963

Eberstein and Sandow present evidence that fatigue results from failure of excitation-contraction coupling. Twitch amplitude decreases while membrane action potential changes very little. Potassium contracture remains at the same amplitude as prior to the fatigue, indicating that if sufficient activation is achieved, the force of contraction is maximal. These authors propose the principle mechanism of fatigue as either reduced calcium release or reduced effectiveness of calcium.

1968

Kadefors presents one of the first modern spectral analyses of EMG at fatigue. The shift toward the lower frequencies with fatigue is confirmed.

1985

Barry first reports changes in acoustic myogram due to fatigue.

1989

Keidel and Keidel apply spectral analysis to "vibromyogram" during fatiguing exercise.

1985–1997

Cooke, Pate, White, Franks, Lucianai, and Myburgh publish a series of papers on skinned skeletal muscle fibers, demonstrating changes in calcium sensitivity under a variety of conditions which mimic fatigue.

1989

Allen, Lee, and Westerblad use calcium measurements in single frog muscle fibers to demonstrate that fatigue results from decreased amplitude calcium transients.

1997

Westerbald, Bruton, and Lannergren show that the effect of acidosis on calcium sensitivity is greatly attenuated at physiological temperatures, suggesting that fatigue may not be due to a direct effect of acidosis on reducing calcium sensitivity.

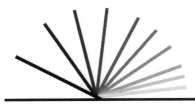

Chapter 19

Contractile Changes and Mechanisms of Muscle Fatigue

B.R. MacIntosh and D.G. Allen

Exercise typically involves repeated activation of the skeletal muscles in a coordinated fashion. The intensity of the exercise, which is determined by muscle activation parameters including repeat interval, duration of contraction, frequency of activation, and proportion of the total motor unit pool activated (for a given muscle), will determine the length of time that the exercise can be sustained.

Persistent exercise at a given intensity ultimately results in failure of the ability to sustain that intensity. This is true whether the exercise involves a single muscle, a muscle group, or a large fraction of the total body muscle mass. We do not fully understand all of the factors that limit endurance of exercise, but considerable advance has been made in understanding individual factors that may contribute to muscle fatigue.

It has been determined that in some cases the cause of the failure to continue exercise is central, and in some cases the cause of the failure is peripheral. When the cause is central, activation of the motor neural pathway fails. Under these circumstances, the individual is apparently unable or unwilling to achieve activation of sufficient numbers of motor units to permit continuing the exercise at the specified or desired intensity. Even when evidence of central fatigue is present, there is concurrently deterioration in the contractile response of the muscle, which contributes to the inability to continue with the exercise task. In many cases, evidence of central fatigue is absent, and the cause of the inability to continue the exercise can be localized to changes within the muscles involved.

Primary Concepts

This chapter presents three principal aspects of the study of muscle fatigue: the ways in which fatigue is studied, the consequences of repeated activation of the skeletal muscles, and the basic mechanisms of fatigue. Fatigue can be studied at three organizational levels: in vitro, in situ, or in vivo. Each of these approaches offers advantages and disadvantages, and each provides important information in gaining an understanding of the consequences and mechanisms of muscle fatigue. There are several characteristic changes in contractile response that can be considered symptoms of fatigue. These changes include reduced force output, slowing of relaxation, reduced economy or efficiency, and reduced maximal velocity. In considering the basic mechanisms of fatigue, metabolic aspects are briefly

Acknowledgments: Research in the laboratory of DGA was supported by the National Health and Medical Research Council of Australia.

addressed, then the following potential mechanisms are considered in greater detail: failure or impairment of T-tubular conduction, voltage sensor/release channel inactivation, reduced Ca^{2+} sensitivity, impaired cross-bridge dynamics, and changes in sarcoplasmic reticulum Ca^{2+} pump and uptake. The emphasis in discussing such mechanisms is on changes in a single muscle or muscle fiber.

Consideration is not given in this chapter to central fatigue or failure of motor neuron activation and conduction of action potentials. For information on central fatigue, the reader is directed to alternative sources (Gandevia, Allen, and McKenzie 1995; Liepert et al. 1996; Newsholme and Blomstrand 1995; Taylor et al. 1996).

This chapter is not intended to be comprehensive in covering all aspects of muscle fatigue or surveying all literature dealing with muscle fatigue. In many cases there are several studies that could have been chosen to illustrate a given point, and our selection is somewhat arbitrary. There have been several reviews written over the past few years (Allen, Lännergren, and Westerblad 1995a; Edman 1995; Enoka and Stuart 1992; Fitts 1994; Roed 1991; Westerblad et al. 1991), and the reader is directed to these for additional information on skeletal muscle fatigue.

Ways of Studying Fatigue

The purpose of research in fatigue is to gain an understanding of the circumstances, consequences, and mechanisms of attenuation of the contractile response that can be attributed to acute repetitive activation. Several approaches are used in such research, and the following section describes these ways of studying fatigue. The various approaches can be divided into three categories: in vitro, in situ, and in vivo. Each of these, along with its respective advantages and disadvantages, is presented below.

In Vitro

Research is considered to be in vitro when investigation occurs outside the natural setting. This could entail study of subcellular components, intact cells, or tissue, usually in a test tube or isolated tissue bath. An example of in vitro study would be removal of a single intact muscle cell from an animal to study in a superfusion system (Lännergren and Westerblad 1987; Lee, Westerblad, and Allen 1991). There are several advantages to using the in vitro approach to study muscle fatigue. By removing the muscle (or part of it) from the body, control over the environment in which the muscle is located can be exerted more rigorously. This includes control of

ion composition of a superfusion solution, temperature, and substrate. This control permits evaluation of change in extracellular fluid composition with precise control. Control of temperature is an important issue in that historically skeletal muscle has been studied in vitro at room temperature or colder. It is not always appropriate to extrapolate such findings to conditions at physiological temperatures.

Additional advantage can be gained in single cell preparations since measurement of sarcomere length and membrane potential are possible. In the case of membrane disrupted or subcellular preparations, control can extend to the apparent intracellular environment (Cooke and Pate 1985; Cooke et al. 1988). With skinned fiber preparations, the objective is often to measure the force response to varying concentrations of Ca^{2+} (Fabiato and Fabiato 1978) and the impact of variations in the intracellular environment on the force-Ca^{2+} relation (Cooke and Pate 1985; Cooke et al. 1988; Myburgh and Cooke 1997).

The major disadvantage of the in vitro preparation is that it is likely that systemic factors such as hormones and the influence of the autonomic nervous system are eliminated. Furthermore, with a single cell superfused preparation, exchange between the intracellular and extracellular environments can proceed without altering the composition of the extracellular space.

There is another disadvantage that is evident when multicellular preparations are removed from a host and studied in vitro. Removal from the host eliminates the functions served by the circulatory system. Delivery of oxygen and substrates for energy metabolism, as well as removal of products of metabolism, rely on diffusion across the thickness of the isolated preparation. It has been demonstrated that in an isolated whole muscle, the rate of disappearance of glycogen from the central core is accelerated (Segal and Faulkner 1985). Hypoxia of the central core in particular may alter the contractile consequences associated with fatigue, and the possible mechanisms of fatigue of such preparations. An exception to this disadvantage is that the hypoxic conditions mimic the situation of a single sustained contraction in vivo, during which blood flow is arrested by intramuscular pressure.

In Situ

When isolation of the tissue under study is partial, such an approach is referred to as in situ. In this situation, there may be partial removal of the tissue from the body, for example, for attachment of a tendon to a force transducer or lever for direct measurement of force of contraction (MacIntosh and Gardiner 1987; Stainsby and Barclay 1971) or control of load and shortening (Ameredes et al. 1992; Stainsby and Barclay 1971). The advantage gained by keeping the tissue in the body is that many of the disadvantages associated with the in vitro preparation are overcome. An intact circulation assures that diffusion distances are kept to realistic values. The temperature at which the preparation is kept is usually more physiological. In most cases the motor nerve is maintained intact, permitting the study of a nerve-muscle preparation, and, in some cases, individual motor units can be studied (Bevan et al. 1992; Hatcher and Luff 1987).

The disadvantages of an in situ preparation include loss of many of the advantages present in the in vitro preparation. The investigator generally has no control over the ion and hormonal composition of the blood perfusing the muscle under study. Furthermore, changes in the contractile response of a muscle cannot be assumed to result from a uniform change in all cells of the preparation. There is therefore some loss of certainty of mechanisms responsible for certain changes in biochemical as well as contractile response. In the case of fiber type differences within a single muscle under study, this disadvantage can at least be partly overcome by study of single motor units (Bakels and Kernell 1995; Robinson, Enoka, and Stuart 1991; Seburn and Gardiner 1995), a technique that requires laminectomy and dissection of single motor axons. A further disadvantage of the in situ preparation is that generally supramaximal indirect stimulation of the muscle is accomplished with synchronous activation of the entire motor unit pool. This form of stimulation does not emulate the asynchronous activation of select motor units within the muscle.

In Vivo

Research involving the study of an intact animal or person, making measurements with minimal invasiveness, is considered in vivo. Examples of this would be measurement of the ability of a subject to sustain effort against a measuring system like a cycle ergometer or an isokinetic dynamometer, while measuring muscle activity electromyographically (Stokes and Dalton 1991; Vaz et al. 1996). The obvious advantage of this system is that the outcome represents the true physiological consequence of such exercise. Recent advances in the use of magnetic resonance spectroscopy have made this approach useful for noninvasive measure of certain aspects of the energetics of muscle (Kent-Braun, McCully, and Chance 1990; Kreis and Boesch 1996; McCully et al. 1991). It would seem that the goals of

muscle research, and in particular research in muscle fatigue, are to understand how and why muscle force response becomes attenuated during exercise. For this reason, in vivo research is essential in linking the mechanistic studies that are done in situ and in vitro with a sense of reality. It is one thing to be able to demonstrate or prove that a given mechanism is responsible for effecting change in the force response in vitro. It cannot simply be assumed that the same mechanism is effective in vivo. This must be demonstrated for any theory to be supported.

The disadvantages of in vivo research include the loss of resolution of mechanisms, similar to the problems with in situ measurement. A further disadvantage is the indirect manner in which force is usually measured, although this can be overcome in animal models with implanted force transducers (Herzog, Leonard, and Stano 1995; Walmsley et al. 1978). A similar approach has been used with limited scope in human subjects (Gregor et al. 1991).

It would appear that to understand the mechanisms of fatigue, it is important to consider evidence from a cross section of methodological approaches. There is a complementary nature to the various preparations. Each has advantages and disadvantages, yet the disadvantages of one approach can be overcome with the use of another preparation.

Fatigue Protocol

The pattern of activation that is used in the study of fatigue is probably an important determinant of the contractile consequences of repeated activation and the specific mechanism of fatigue (Edwards 1983). In this regard, it is important to recognize some of the more common fatigue protocols that are in use. Activation of muscle in studies of muscle fatigue is either voluntary or stimulated. When activation is voluntary, the pattern of activation (frequency and duration) is not well controlled, but this is an important approach, since it is the ultimate goal of such research to understand what is happening during human exercise. Various approaches have been used with voluntary activation, including sustaining a constant (Bigland-Ritchie et al. 1983) or submaximal force, or intermittent contractions that may be maximal (Beelen et al. 1995; Bigland-Ritchie et al. 1983), or contractions with specific submaximal target forces (Saugen and Vollestad 1996) or work (Fulco et al. 1996).

When stimulation of the muscle(s) under study is the approach used, the stimulation parameters can be precisely controlled. Under these conditions, the stimulation can be continuous or intermittent. The frequency of stimulation can be low (≤ 10 Hz), intermediate (10–50 Hz), or high (≥ 50 Hz). The frequency of contractions and duration of stimulation within each contraction can also be regulated. The most common approach is with brief intermittent contractions at intermediate frequencies (10–50 Hz), with the classic Burke protocol (Burke et al. 1973) and slight variations on this basic pattern used more often than any other stimulation pattern. The Burke protocol involves intermittent stimulation (once per second), with pulses at 40 Hz sustained for 330 ms. This pattern of stimulation is typically sustained for 2 to 6 min. The original purpose of the Burke protocol was to permit classification of individual motor units as either fatigue-resistant or fatigable motor units (Burke et al. 1973). This purpose necessarily imposes a limitation of the use of the Burke protocol in the study of fatigue in fatigue-resistant muscle or motor units. By design, the protocol minimizes fatigue in such motor units.

More recently, considerable study of isolated single muscle cells has been conducted using a fatigue protocol introduced by Westerblad and Lännergren (1991) (Allen, Lännergren, Westerblad 1997; Kawta 1983; Westerblad and Allen 1992a), which involves intermittent contractions (70–100 Hz for 350 ms) beginning with one contraction each 4 s, and decreasing the interval every 2 min. A modification of this protocol has been used with an in situ preparation (Chin et al. 1995). The advantage of such an approach is that fatigue-resistant fibers, which may not otherwise express depression of force if the conditions are not severe enough, are subjected to conditions that will result in fatigue. Progressive increase in the duty cycle will ultimately result in failure regardless of the fiber type under study. This represents a major difference from the approach by Burke and coworkers (1973), who tried to avoid fatigue in certain motor units to facilitate classification of fiber types.

Contractile Consequences Associated With Fatigue

Several aspects of the contractile response are modified in fatigued muscle, including the following: the force-frequency relation, force generating capacity, rate of relaxation, force-velocity relation, and efficiency or economy of contraction. These modified contractile properties can be used to characterize the extent and type of fatigue. Each of these factors is discussed below.

Force-Frequency Relation

Low and high frequency fatigue are defined in terms of the consequences of force depression relative to the frequency of muscle activation used for evaluation. This is illustrated in figure 19.1, which shows three possible consequences with respect to changes in the force-frequency relation: (1) depression at low but not high frequencies, (2) depression at high but not low frequencies, and (3) uniform depression at all frequencies evaluated.

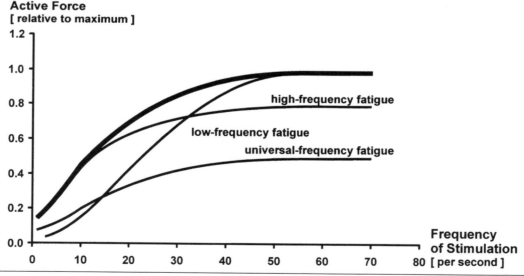

Figure 19.1 Force-frequency relationship for human muscle (control adapted from McComas 1996). From top to bottom on the right-hand side of the graph, conditions represent the control (thicker line), low frequency fatigue, high frequency fatigue, and universal frequency fatigue.

Only by assessing the force-frequency relation can high frequency and low frequency fatigue be identified.

Edwards and coworkers (1977) first characterized low frequency fatigue. They demonstrated a prolonged depression of force in response to low frequency stimulation (20 Hz) relative to high frequency stimulation (50 Hz) in human subjects. Low frequency fatigue occurred in subjects following intermittent tetanic stimulation, intermittent maximal voluntary contractions, and following a stepping exercise or cycle ergometer exercise (Edwards et al. 1977). There is some confusion in the literature in that some authors have used the terms low frequency or high frequency fatigue to refer to the frequency of stimulation used to induce the fatigue (Cairns and Dulhunty 1995; Fratacci et al. 1996; Vandenboom and Houston 1996). Although it appears to be true that only high frequency stimulation will result in high frequency fatigue, there are reports of low frequency fatigue occurring as a consequence of repetitive high frequency stimulation. Golgeli, Ozesmi, and Ozesmi (1995) observed a rightward shift in the force-frequency relation after fatigue of rat diaphragm strips due to low frequency (5 Hz), high frequency (50 Hz), or intermittent (25 Hz for 160 ms each s) stimulation, confirming that low frequency fatigue can be a consequence of all three of these stimulation paradigms. Others have also studied low frequency fatigue following high frequency stimulation (Chin and Allen 1996b; Westerblad, Duty, and Allen 1993).

The twitch response in the experiments of Edwards and coworkers (1977) was depressed to the same extent as the force during low frequency stimulation, but this observation was not consistent. The authors suggest that result was due to the large variability often seen with twitch responses with human subjects and recommend that the twitch not be used to assess fatigue. However, when the motor nerve is placed directly on the stimulating electrode (rather than be activated by percutaneous stimulation), as with in situ experiments with animal muscles, the twitch contraction is less variable and appears to be a good indication of low frequency fatigue (MacIntosh, Stainsby, and Gladden 1983; MacIntosh et al. 1994).

Fitts and coworkers (1982) have reported depression of force at both ends of the force-frequency relation (twitch and tetanic force) following a seven hour swim to exhaustion in rats. The depression in force was greater for fast twitch edl (74%) than for slow twitch soleus (25%). The twitch-tetanus ratio

did not change from the control values for either muscle.

When frog sartorius muscle is stimulated with 1.5 ms pulses at 30 per min for 15 min in the absence of O_2, there is a depression of both twitch and tetanic force, with an apparently different time course of recovery. However, after 20 min of aerobic recovery, the twitch-tetanus ratio is the same as before the period of stimulation, with both twitch and tetanic force still just 80% of the prefatigue value (Fitts and Holloszy 1978), indicating incomplete recovery.

The importance of considering a period of recovery in the frequency-dependent nature of muscle fatigue is emphasized by a study from Houston's lab (Vandenboom and Houston 1996). Intermittent stimulation of mouse skeletal muscle (150 Hz for 400 ms, 1 per s for 2 min) resulted in greater depression of the high frequency response than of the twitch response. Recovery was not monitored in this series. It should be noted that the amplitude of twitch contractions was evaluated between tetanic contractions, and twitch potentiation was evident by 15 s in this sequence of stimulation. The likely continued presence of twitch potentiation during the period of intermittent stimulation may have prevented a diagnosis of low frequency fatigue. Low frequency fatigue was observed in a subsequent series of experiments when a recovery period (20 min) was permitted following the intermittent contractions (150 Hz for 400 ms, 0.75 trains per s for 10 min) reported in the same paper.

Decreased Force Capacity

Depression of the force response in skeletal muscle can be a consequence of decreased activation, decreased sensitivity to calcium, or decreased force capacity. This is illustrated in figure 19.2, which shows the force-calcium relation and the three potential mechanisms of fatigue. The capacity of a muscle to produce force can only be assessed when maximal activation is achieved. This would be analogous to testing the maximal force that can be exerted by a skinned muscle fiber. To evaluate this, it would be necessary to test with supramaximal free calcium concentration and to be sure the force was not increased by an increase in calcium concentration. Maximal activation is often assumed to be obtained with tetanic stimulation, or with maximal voluntary contraction, when a superimposed twitch (Merton 1954) does not elicit an increment in tension. However, if fatigue has resulted in impaired Ca^{2+} release, then

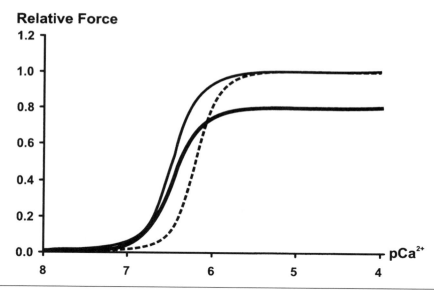

Figure 19.2 Force-calcium relationship. The thin solid line represents the steady state relationship between negative log of Ca^{2+} concentration, $p[Ca^{2+}]$, and active isometric force. Hill coefficient (n) = 2.6, $p[Ca^{2+}]_{50}$ (k) = –6.5. The thick solid line represents fatigue resulting from depressed maximum force (P_o = 0.65), n = 2.6, k = –6.5. The dashed line represents fatigue resulting from decreased Ca^{2+} sensitivity; n = 3.0, k = –6.22.

tetanic stimulation is not assured to give maximal activation.

Addition of caffeine in millimolar concentrations has been used to evaluate maximal contraction force. In circumstances when tetanic stimulation elicits decreased amplitude of contraction, addition of caffeine results in enhanced Ca^{2+} release and possibly greater force (Kawta 1983). If the caffeine-supplemented contraction is still depressed, then it can be concluded that maximal force is depressed.

Slowing of Relaxation

Relaxation from isometric contraction in skeletal muscle is a complex process (Gillis 1985) that potentially relies on a series of events: uptake of Ca^{2+} by cellular buffers and longitudinal SR, dissociation of Ca^{2+} from troponin, and cross-bridge kinetics (Westerblad and Allen 1996). Measurement of the time course of relaxation has been done in a variety of ways, including the following: half-relaxation time measured from the peak of contraction or from some point on the apparently exponential decline in tension, or peak rate of relaxation. These various ways of measuring relaxation may reflect different processes, and the time course of change in these parameters during fatigue and recovery from fatigue is different (MacIntosh 1991).

There is a general perception that slowing of relaxation is a given in fatigued skeletal muscle.

Certainly it has been demonstrated that repetitive activation leads to slowing of relaxation, and this seems to correspond with fatigue (Fitts 1994; Gordon et al. 1990; Westerblad and Allen 1994a; Westerblad and Lännergren 1991). However, slowing of relaxation actually occurs while activity-dependent potentiation still permits developed tension to be equal to or greater than the initial developed tension (Pate et al. 1994), so its relationship with fatigue is not simple. Slowing of relaxation tends to reduce the fusion frequency and shifts the force-frequency relation to the left. There is a marked slowing of relaxation during twitch contractions in fatigued frog muscles (Thompson et al. 1992), and considering also the prolonged contraction time can explain the apparent smaller effect of fatigue on the twitch of these muscles when the tetanus is severely reduced (Thompson et al. 1992). However, fatigue can depress the low frequency response such that the fusion frequency appears to be shifted to the right, even when slowing of relaxation is evident (Golgeli, Ozesmi, and Ozesmi 1995). When fatigue is present, a short period of recovery results in restoration of the rate of relaxation, in spite of persistent low frequency fatigue (MacIntosh et al. 1994).

One factor seems to accentuate the slowing of relaxation. When fatigue develops under hypoxic conditions the decline in developed tension is rapid, and the magnitude of the slowing of relaxation is quite pronounced. Under these conditions, recovery of oxygenation (blood flow) results in rapid restoration both of developed tension and rate of relaxation (Bergström and Hultman 1991). In contrast, during 10 Hz or 5 Hz stimulation in dog gastrocnemius plantaris muscle preparation with spontaneous blood flow, there is virtually an absence of slowing of relaxation yet developed tension decreases considerably (MacIntosh, Stainsby, and Gladden 1983). Recent observations in human muscle in vivo (Vollestad, Sejersted, and Saugen 1997) suggest that relaxation may even be accelerated during repetitive activation consistent with fatigue.

Impaired Force-Velocity Properties

The majority of studies dealing with fatigue are concerned only with isometric contractions. However, there have been a number of studies that have considered dynamic contractions and the impact of fatigue on the ability of muscle to generate tension while shortening.

The changes in the characteristic properties of the force-velocity relation due to fatigue can be quantified by consideration of isometric force (P_o), maximal velocity (V_o) and a/P_o, or the curvature of the force-velocity relation. Curtin and Edman (1994) observed in fatigued frog single fibers a reduction in V_o that was half as great as the reduction in P_o, and

an increase in a/P_o, which indicates less curvature in the force-velocity relation. This decrease in curvature would minimize the impact of decreases in P_o and V_o on the force developed at intermediate velocities.

Hatcher and Luff (1988) observed a sustained depression of V_o in fast twitch muscle of the cat (in situ), whereas isometric force recovered from a depression of similar magnitude within 45 min. There was also a transient increase in a/P_o, but this returned to control levels after 15 min of recovery.

Curtin and Edman (1994) reported that maximal velocity, measured by the slack test or by curve-fitting the Hill equation $[(P + a)(v + b) = (P_o + a)/b)]$, decreases in frog single fiber subjected to fatiguing stimulation. An increase in a/P_o (less curvature) permitted some compensation (less reduction in peak power). Their results are presented in figure 19.3. These observations are consistent with previous results (Edman and Mattiazzi 1981).

In contrast with the results of Edman and colleagues (Curtin and Edman 1994; Edman and Mattiazzi 1981), Hatcher and Luff (1987) observed no change in a/P_o in cat flexor digitorum longus muscle motor units. Unique in the study by Hatcher and Luff (1987) was that they imposed partial chronic denervation of this muscle and allowed some recovery, which resulted in enlarged motor units due to

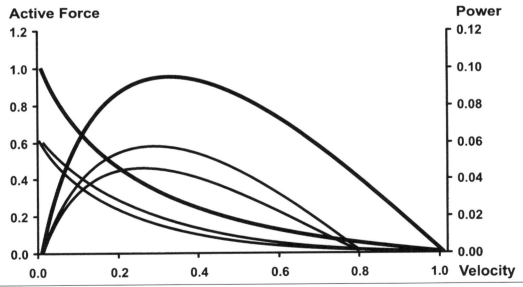

Figure 19.3 Force-velocity (concave lines) and power-velocity (upward curve) relationships. The control condition is illustrated with the thicker line. The upper thin line represents fatigue, corresponding to a 40% decrease in P_o and a 20% decrease in V_{max}. The Hill equation was used to draw each force-velocity curve, with the following conditions: control ($a/P_o = 0.25$, $P_o = 1$, $V_{max} = 16$, $b = 0.25$); fatigue ($a/P_o = 0.4$, $P_o = 0.6$, $V_{max} = 0.8$, $b = 0.32$). The lower thin line illustrates the consequence if fatigue did not change a/P_o. The same (fatigue) values were used for P_o (0.6) and V_{max} (0.8), but $a/P_o = 0.25$.

sprouting of surviving motor neurons. The authors claim that without undertaking this precaution, internal resistance would have been substantial, and this affects measurement of maximal velocity and hence a/P_o. Without this treatment, small motor units (< 15% whole muscle P_o) seemed to have very high a/P_o ratio (less curvature) without imposing fatigue.

Barclay (1996), studying bundles of mouse soleus and extensor digitorum longus fibers at 25 °C, observed reduced curvature (increased a/P_o) of the force-velocity relation and a decrease in maximal velocity. These changes resulted in a decrease in peak absolute power with fatigue, but when power was estimated with force and velocity expressed relative to the new (fatigued) maximal values (P_o and V_{max}), peak power was relatively enhanced in the fatigued muscles. The effects of fatigue on the force-velocity properties were greater in edl than in soleus in spite of greater contractile demand on soleus (1 s at 75 Hz for soleus and 0.2 s at 125 Hz for edl, both at s intervals for 30 contractions).

Training apparently depresses the magnitude of change in force-velocity properties with fatigue. Fitts and Holloszy (1977) reported that in untrained rats, indirect stimulation of the soleus muscle in situ (37.5 °C) at 100 Hz for 250 ms, 110 trains per min for 30 min resulted in a depression of P_o by 32% and V_{max} by 12%. In the trained animals, the same stimulation protocol resulted in only an 8% decrease in P_o and no change in V_{max}. Training in these rats consisted of 15 weeks of treadmill running, at the end of which rats were running for 2 hours at 32 m · min⁻¹ up a 15% grade, 5 times per week.

De Haan, Jones, and Sargeant (1989) note that depression of V_{max} results in a shift in the optimal velocity for peak power output. If power is measured at a common velocity before and after fatigue, then the depression in power output will be overestimated. Although this observation was made with a rat medial gastrocnemius muscle preparation in situ, the results are possibly quite important for human studies, where optimal conditions are not always considered.

Efficiency and Economy During Muscle Fatigue

Changes in the energy cost of muscle contraction have been reported in studies of muscle fatigue. The rate of ATP turnover is apparently decreased in sustained muscle contraction (Crow and Kushmerick 1982), yet the rates of heat production and oxygen uptake are increased during repeated submaximal voluntary isometric contractions (Saugen and Vollestad 1996).

Vollestad and colleagues have done extensive work in studying the metabolic consequences of submaximal voluntary isometric contraction (Saugen and Vollestad 1996; Vollestad 1995; Vollestad and Verburg 1996; Vollestad, Wesche, and Sejersted 1990). When subjects sustained 30 or 50% of maximal voluntary force for 6 s with 4 s rest, the metabolic cost, measured by rate of temperature rise or by oxygen uptake, increased throughout the exercise, which was sustained to exhaustion (Saugen and Vollestad 1996). After 3 min of recovery, the economy of contraction was still low. The mechanism for this change in energy cost is not clear, but several theories have been entertained, including the following: progressive uncoupling of oxidative phosphorylation, nonspecific Q_{10}, progressive recruitment of fast twitch motor units. Of these mechanisms, the motor unit recruitment hypothesis seems most reasonable. It is known that the economy of isometric contraction is less in fast twitch muscle fibers than in slow twitch muscle fibers (Crow and Kushmerick 1982). It is also known that 30% of maximal isometric force can be achieved with recruitment of a portion of the motor unit pool, which for most individuals does not include fast twitch motor units. In order to sustain the target force, recruitment of less economical fast twitch motor units must eventually occur. A similar theory has been used to explain the slow component of oxygen uptake during cycle ergometer exercise (Barstow et al. 1996; Gaesser and Poole 1996) (and see chapter 8). It should be kept in mind that this proposed mechanism is hypothetical. Further research will need to be done in order to confirm this theory.

Cellular Mechanisms of Muscle Fatigue

Much of modern research on fatigue attempts to investigate selected parts of the complex process of muscle activation and mechanical response and understand how these change during repeated activity. In the present section we apply this process to the sequence of events that is generally considered to be the most important contributor to fatigue (Bigland-Ritchie, Furbush, and Woods 1986; Merton 1954).

For simplicity we will consider muscle function to require the sequential participation of the following processes and events:

- The action potential generation and propagation, including conduction down the T-tubule
- Activation of T-tubular voltage sensor and sarcoplasmic reticulum Ca^{2+} release channel
- Ca^{2+} binding to troponin
- Cross-bridge function
- Uptake of calcium by the sarcoplasmic reticulum Ca^{2+} pump

For each process we consider briefly how repeated activity might change function and how such changes in function might contribute to muscle fatigue. With such a large series of topics the coverage must inevitably be superficial and selective and we try to overcome this by reference to detailed reviews in each area.

A general question, which applies to all these sections, is the nature of the cellular changes that drive the alterations in function. Traditionally, the metabolic changes associated with intense muscle activity have been regarded as the primary cause of functional change. It is well recognized that maximal muscle activity can consume energy in the form of ATP at a rate greatly in excess of the ability of the muscle to resynthesize ATP. When this occurs various immediate and short-term energy sources are accessed; initially phosphocreatine (PCr) and subsequently glycogen, and associated with these changes are increases in the end products of these reactions, creatine, inorganic phosphate, and lactic acid. When these short-term stores are consumed with continuing high rates of energy use, ATP itself will be broken down with increases in its metabolic products, ADP, AMP, and inorganic phosphate (P_i), and various metabolic products of these. These metabolic changes have been considered in many reviews (e.g., Fitts 1994; Sahlin 1991; Spriet 1992; Vollestad and Sejersted 1988) and will not be further considered here. However, there are many other changes associated with activity that might potentially contribute to the changing muscle properties. For instance, there are both intracellular and extracellular changes in ion concentrations, particularly of Na^+ and K^+, which could well affect the action potential and hence Ca^{2+} release. Likewise, when a muscle is tetanized the intracellular calcium is elevated 10- to 20-fold during each tetanus so that the time-averaged $[Ca^{2+}]_i$ is greatly elevated by repeated tetani. Such long-term rise in $[Ca^{2+}]_i$ may activate proteases, proto-oncogenes, or other cell regulatory pathways, leading to protein catabolism and subsequent impaired function. Another possible factor contributing to attenuation of function is that repeated activation of the muscle at high levels of force or especially when the muscle is stretched while contracting can lead to mechanical damage of various sorts.

The Action Potential and Conduction Through the T-Tubular System

The muscle action potential must travel along the surface of the muscle and then enter the T-tubular membrane where the close contact between the T-tubular voltage sensor and the SR Ca^{2+} release channel occurs. It is established that Ca^{2+} release and contraction depend on both the magnitude and the duration of the depolarization (Ashley and Ridgway 1970), and it follows that changes in the action potential may affect force production. For these reasons there has been great interest in the way the action potential changes during repeated activity and in possible ionic or metabolic changes that could affect the action potential.

During each action potential Na^+ ions enter the muscle and K^+ ions leave. The resulting ionic changes are generally quite small, for instance Hodgkin and Horowicz (1959) calculated that $[Na^+]_i$ would rise by ~8 μmol per action potential while $[K^+]_i$ would decrease by ~5 μmol per action potential. The extracellular changes depend on the volume of the relevant extracellular space and how quickly this equilibrates with the plasma. Furthermore, the action of the sodium/potassium pump will reverse these changes over a period of minutes (Cairns et al. 1997). However the extracellular space within the T-tubules represents a special case because of their small volume and relatively slow diffusion equilibrium with the extracellular space (~1 s) (Nakajima, Nakajima, and Peachey 1973). Within the T-tubular space it has been estimated that K^+ rises by 0.28 mmol while Na^+ falls by 0.5 mmol per action potential (Adrian and Peachey 1973). This leads to the possibility that when muscles are stimulated at high frequencies and for long periods there will be substantial increases in T-tubular K^+ and decreases in Na^+. Alternatively, if muscles are ischemic during tetani, which happens spontaneously because contractions that are above about 50% maximal cause collapse of the blood vessels, then such changes may occur to the surface membrane as well as the T-tubules. If extracellular K^+ accumulated, this would tend to cause depolarization of the resting membrane potential, which would cause failure of reactivation of the Na$^+$ channels and cause a reduced overshoot. The reduction in extracellular Na$^+$ would also cause a reduced overshoot of the action potential.

Do such changes in the action potential occur and do they contribute to muscle fatigue? Unfortunately it is not easy to give a clear-cut answer to these questions. Taking the T-tubular case first, it is not currently possible to measure the action potential within the T-tubular network, and recordings with conventional microelectrodes are dominated by the surface membrane action potential. Likewise it is not possible to measure the extent of ionic changes within the T-tubular space. What has been done is to use either the contraction of central myofibrils (Edman and Lou 1992; Garcia et al. 1991) or the central [Ca^{2+}]$_i$ (Duty and Allen 1994; Westerblad et al. 1990) as an indicator of the inward spread of activation. All these methods are in agreement that during long, high frequency tetani there is a failure of central activation that is likely to be caused by ionic changes in the T-tubules and that this process makes at least some contribution to the reduction of force.

Another possible mechanism that could lead to changes in the form of the action potential is the activation of ATP-sensitive K$^+$ channels. These channels are known to exist in skeletal muscle and are activated by a reduction in ATP, with isolated channels showing half maximal activation (K$_{0.5}$) at 0.5 mmol (Spruce, Standen, Stanfield 1985). It is also known that acidosis contributes to the opening of these channels (Davies, Standen, and Stanfield 1992) so this provides another possible mechanism that might contribute to channel opening in muscle fatigue. If the ATP or pH fell low enough to activate a significant fraction of these channels, then one would predict a more rapid repolarization and a shorter action potential, which would reduce Ca^{2+} release and consequently force production. In extreme metabolic depletion this appears to occur (Lüttgau 1965), but the evidence that this mechanism operates during physiological fatigue is not strong. One approach to evaluate this has been to block the channel with inhibitors such as glibenclamide, but this has not consistently produced an increase in [Ca^{2+}]$_i$ or force (Duty and Allen 1995; Renaud et al. 1996). Probably the absence of an effect during fatigue is because the ATP does not fall sufficiently low to activate enough channels. ATP is variously reported to fall to around 4 to 6 mmol during fatigue (Fitts 1994; Westerblad et al. 1991). With a K$_{0.5}$ of 0.5 mmol

only a very small fraction of the channels would be opened under these conditions.

The T-Tubular Voltage Sensor and Sarcoplasmic Reticulum Ca^{2+} Release Channel

If a muscle is fatigued by repeated tetani it is known that the tetanic [Ca^{2+}]$_i$ becomes smaller, suggesting that failure of Ca^{2+} release occurs. This failure would contribute to the decline of force (Allen, Lännergren, Westerblad 1989; Baker et al. 1993; Westerblad and Allen 1991). Experiments with ion probe analysis show that the sarcoplasmic reticulum (SR) of fatigued muscle contains the same amount of Ca^{2+} as normal (Gonzalez-Serratos et al. 1978), while measurements of the charge movement associated with the voltage sensor suggest that the voltage sensor operates normally during fatigue (Györke 1993). These results suggest that the critical failure in fatigue is either the coupling of the voltage sensor to the SR Ca^{2+} channel or the opening of the SR Ca^{2+} channels. Thus the question becomes which of the possible metabolic or other changes that occur during prolonged intense activity of muscle could affect the coupling or opening of the SR Ca^{2+} channel.

One popular hypothesis is that acidosis causes the failure of Ca^{2+} release. This issue has been discussed at length elsewhere (Allen, Westerblad, Lännergren 1995), and we believe the evidence against this hypothesis is strong. The main reasons are that failure of Ca^{2+} release occurs in situations where there is little acidosis (Westerblad and Allen 1992b), and when acidosis does occur the failure of Ca^{2+} release appears to be very similar (Chin and Allen 1996a). Furthermore, when muscles are made acidotic the Ca^{2+} transient becomes larger, though this does not necessarily mean that SR Ca^{2+} release is enhanced but could instead result from the reduction in SR Ca^{2+} pumping rates in acidotic conditions (Westerblad and Allen 1993a, 1993b), combined with possible changes in Ca^{2+} buffering.

Another interesting possibility is that either Mg^{2+} or ATP affects SR Ca^{2+} release. ATP has a high affinity for Mg^{2+} and normally occurs in cells in the form of MgATP. When there is net breakdown of MgATP there will be a rise in [Mg^{2+}]$_i$ and a fall in MgATP. It is well established that increasing [Mg^{2+}] closes the SR Ca^{2+} release channels (Lamb and Stephenson 1991; Smith, Coronado, and Meissner 1986), so the rising [Mg^{2+}]$_i$ during fatigue could cause or contribute to the failure of Ca^{2+} release. We have shown that [Mg^{2+}]$_i$ rises during fatigue (from 0.8 mmol to 1.6 mmol) but only during the late phase of fatigue, which is when failure of Ca^{2+} release occurs (Westerblad and Allen 1992b). To test whether this rise in $[Mg^{2+}]_i$ could be the cause of failure of Ca^{2+} release, $MgCl_2$ was injected into unfatigued fibers. These fibers showed substantial force development despite very high levels of $[Mg^{2+}]_i$ and to reduce force to 50% required a $[Mg^{2+}]_i$ of 2.9 mmol. The interpretation of these results was that elevated $[Mg^{2+}]_i$ played only a small role in the failure of Ca^{2+} release. However, if the fact that the myofibrils have both reduced sensitivity to $[Ca^{2+}]_i$ and a reduced maximum force during fatigue (Westerblad and Allen 1993a) is taken into account, then it may be that the effects of Mg^{2+} can explain a substantial fraction of the failure of Ca^{2+} release during fatigue. Further experiments will be needed to resolve this issue.

It is also possible that the fall in ATP affects SR Ca^{2+} release or coupling. Metabolic studies suggest that ATP falls from 6.0 to 4.5 mmol during intermittent, repeated tetani, while calculations from the change in $[Mg^{2+}]_i$ suggest that ATP could fall as low as 1.75 mmol (Westerblad and Allen 1992b). How could the change in ATP affect excitation-contraction coupling? There are many possibilities, which have been discussed at length elsewhere (Allen, Lännergren, and Westerblad 1997; Korge and Campbell 1995).

One possibility is that the reduced ATP has a direct effect on the SR Ca^{2+} release channel (Owen, Lamb, and Stephenson 1996; Smith, Coronado, and Meissner 1985). Such studies have shown that ATP is required for SR channel opening and that the $K_{0.5}$ is around 0.7 mmol. Even the larger fall in ATP calculated for single fibers would have only minor effects on this basis. Another factor that reduces the likely effect of ATP is that the breakdown products in muscle (ADP and P_i) also seem able to cause some channel opening, though at a lower sensitivity (Kermode, Sitsapesan, and Williams 1995; Meissner 1994). Another possibility is that the reduced ATP might cause opening of the ATP-sensitive K^+ channel in the surface membrane (discussed previously).

In summary, it is clear that Ca^{2+} release from the SR fails in fatiguing muscles, but the mechanism is unclear. It is also of interest that Ca^{2+} release shows a small but persistent reduction during recovery from fatigue (for review, see Allen, Lännergren, and Westerblad 1995b). Metabolites have recovered back to normal at this time and recent experiments suggest that this kind of failure of Ca^{2+} release may be a consequence of sustained elevation of resting $[Ca^{2+}]_i$ (Lamb, Junankar, Stephenson 1995; Chin and Allen

1996b). Thus there are probably multiple mechanisms involved in the failure of Ca^{2+} release associated with fatigue.

Ca^{2+} Sensitivity of the Contractile Proteins

Contraction occurs when Ca^{2+} binds to troponin and activates cross-bridge cycling. Much is known of the intervening details (for review, see Ashley, Mulligan, and Lea 1991) but for the present purpose the multiple interactions can be characterized by the steady state relation between $[Ca^{2+}]_i$ and force (Fabiato and Fabiato 1978; Godt and Nosek 1989; Pate et al. 1995), as shown in figure 19.2. The relationship is S-shaped and to a good approximation the relationship can be defined by a Hill curve (see Westerblad and Allen 1993b), which is characterized by the following parameters of the curve: $K_{0.5}$ ($[Ca^{2+}]_i$ which causes half maximal contraction); the n value of the Hill equation, which represents the steepness of the curve; and the P_{max} (force at saturating $[Ca^{2+}]_i$). When a muscle is activated at its maximum stimulation frequency the resulting $[Ca^{2+}]_i$ is close to that which causes saturation of the contractile proteins. Consequently under these conditions small increases or decreases in the level of $[Ca^{2+}]_i$ have little effect on the resulting tension. Equally small changes in the sensitivity of the contractile proteins to Ca^{2+} have little effect on the force. However, these are not the normal conditions under which muscles are activated in the body. Normally, muscle fibers are only partially activated and consequently the $[Ca^{2+}]_i$ lies on the steep part of the $[Ca^{2+}]_i$/force curve, and small changes in the $[Ca^{2+}]_i$ or the Ca^{2+} sensitivity lead to large changes in force. We have already noted that changes in Ca^{2+} release occur in fatigued muscles: we now show that changes in Ca^{2+} sensitivity are important in fatiguing muscles.

Traditionally this subject has been studied in "skinned" muscles in which the surface membrane has been removed either chemically or mechanically. This allows the experimenter control over the solution in which the myofilaments are bathed. Naturally, attention was first concentrated on changes that are known to occur in the intracellular environment as a consequence of prolonged or intense muscular activity. Lactic acid is often produced in muscles and will accumulate when the mechanisms for removing it are saturated. Consequently acidosis is one of the most studied consequences of prolonged, intense muscular activity.

Fabiato and Fabiato (1978) showed that acidosis reduced the Ca^{2+} sensitivity of the contractile proteins and proposed that this was an important mechanism of force reduction in fatigue. While this mechanism has been widely invoked (Fitts 1994), there are also powerful arguments against this hypothesis. One clear issue is that in many types of fatigue there is little or no acidosis (Vollestad et al. 1988; Westerblad and Allen 1992a). A more fundamental observation is that the reduction of Ca^{2+} sensitivity observed in skinned fibers at room temperature is much reduced at more physiological temperatures (Pate et al. 1995; Westerblad, Bruton, and Lännergren 1997). The experiments of Westerblad and coworkers were on intact fibers so that acidosis could have had a range of effects on Ca^{2+} release, metabolic pathways, as well as Ca^{2+} sensitivity and P_{max}. Nevertheless, at near physiological temperatures acidosis had much less effect than at room temperature.

It is also possible to measure Ca^{2+} sensitivity in intact fibers (Westerblad and Allen 1993b) and such studies have shown that Ca^{2+} sensitivity falls in fatigued fibers, at least at room temperature. This fall in sensitivity occurs in fibers that do not show acidosis (Westerblad and Allen 1992a) so some other change in the intracellular environment seems to be involved. A likely contender is inorganic phosphate, which rises sharply during fatigue (Cady et al. 1989), and which has been shown to reduce Ca^{2+} sensitivity of the myofilaments (Godt and Nosek 1989). Whether this effect persists at near physiological temperatures is not known. Thus, for the moment, it seems likely that increase in P_i is the cause of the reduced Ca^{2+} sensitivity observed during fatigue. The reduced Ca^{2+} release and the reduced Ca^{2+} sensitivity both make important contributions to the reduced force observed in fatigued muscles. The further reduction in force, or lack of recovery of force during a subsequent period of quiescence in spite of recovery of metabolic status, indicates that metabolic parameters are not directly involved in the impaired Ca^{2+} release, but that some long-term change in the processes leading to Ca^{2+} release is impaired.

Cross-Bridge Function

The force produced by a muscle depends on the number of available cross bridges (XBs), the fraction that are attached, and the force produced by each XB. The number of available cross bridges depends on overlap of the myofilaments and the degree of activation by Ca^{2+}, but both of these can be regarded as constant when considering the maxi-

mum force (P_{max}) at a given length. The factor that therefore potentially changes P_{max} during muscle activity is the fraction of attached bridges, which in turn depends on the attachment and detachment rate (f and g) and on the force/XB (P_{XB}). For simplicity in this account we will only consider the changes in P_{max} during muscle activity and the metabolic or other factors that cause them, but we will not attempt to determine whether f, g, or P_{XB} are involved (for more detail on this aspect see Edman and Lou 1992).

To determine whether P_{max} changes during muscle fatigue it is necessary to maximally activate the muscle in control and fatigued state. In mammalian muscle this can be done by application of caffeine, which causes a substantial increase in Ca^{2+} release and, in muscles stimulated at a high frequency, causes $[Ca^{2+}]_i$ to rise sufficiently high to maximally activate the contractile proteins. Such experiments generally show that a moderate (10–30%) reduction in P_{max} is an early feature of muscle activity (Edman and Lou 1992; Lännergren and Westerblad 1991; Westerblad and Allen 1991).

What causes this reduction in P_{max}? The main possibility is the effects of metabolite changes on the contractile proteins. This topic has been extensively reviewed (Allen, Lännergren, Westerblad 1995a; Fitts 1994; Westerblad et al. 1991), and the present account will be very brief. As with Ca^{2+} sensitivity, the use of skinned muscle fibers is the main approach to this topic. Acidosis causes a rather small fall in P_{max} (Fabiato and Fabiato 1978) and is probably not the major cause as discussed above. Accumulation of P_i causes a relatively large fall in P_{max}. For instance, in humans P_i can increase from 2 mmol to 20 mmol (Cady et al. 1989). Studies on mammalian skinned fibers suggest that such a change in P_i would cause P_{max} to fall by about 45% (Millar and Homsher 1990). These workers also showed that the fall of force was approximately a logarithmic function of the P_i. Consequently, the starting P_i is more important than the final P_i and this makes quantitation difficult because the P_i under control conditions is often difficult to measure.

Given the changes in pH and P_i that occur in muscle, do these adequately explain the fall in P_{max}? Westerblad and Lännergren (1994) have considered this issue and conclude that the predicted fall in force caused by these two metabolites is greater than the observed fall in force. This has led them to the proposal that some other metabolite, possibly ADP, may enhance the force, and they have considerable evidence for this possibility (Westerblad and Lännergren 1995). However there are a range of alternative possibilities that could contribute to this discrepancy. One possibility is the possible confounding effects of temperature, which have not been examined for P_i. Another is the possibility that structural disorder in the sarcomere pattern develops during isometric contractions. Another is the possibility that oxygen radicals produced during repeated activity have a direct effect on P_{max} (Wilson et al. 1991).

Sarcoplasmic Reticulum Ca²⁺ Pump

The sarcoplasmic reticulum Ca^{2+} pump is an ATP-driven pump that returns Ca^{2+} from the myoplasm to the SR and therefore plays an essential role in relaxation and in the normal cycling of Ca^{2+}. The biochemistry of the pump has been extensively studied as an example of ATP-driven chemical pumps (MacLennan and Holland 1975), and a number of reviews of the role of the SR pump in relaxation are available (Gillis 1985; Lüttgau and Stephenson 1986). In the present context we consider the evidence that the pump slows during muscle fatigue and the possible causes of this slowing. We will also briefly consider the effects on muscle performance of reduced SR Ca^{2+} pump function. This topic has been reviewed in more detail elsewhere (Allen, Lännergren, and Westerblad 1995b).

Two main approaches have been used to assess SR Ca^{2+} pump function during fatigue. One approach is to measure the rate of decline of $[Ca^{2+}]_i$ after a tetanus and deduce the SR pump function from this information (Klein et al. 1991; Westerblad and Allen 1993b). This approach has shown that the apparent pump rate seems to fall to about 15% of the initial rate during a period of repeated tetani and then shows a partial recovery to about 50% of initial rate after 1 hour of rest (Westerblad, Duty, and Allen 1993; Westerblad and Allen 1993a). This method requires a number of assumptions, can only be performed on isolated muscle fibers, and gives no direct information about the causes of the changes in function. On the other hand, it does have the advantage of measuring the overall changes in pump function, whatever their cause. An alternative method that has become popular in recent years is to take samples from human or animal muscles at various stages during activity. These samples are then processed to form vesicles of predominately SR membrane origin, which are

suspended in an appropriate solution. The rate of uptake of Ca^{2+} can then be measured following addition of ATP. This method suffers from the difficulty that the functional integrity of the SR may be affected by the isolation procedure (Chin et al. 1994) and that metabolic changes that might affect the pump function will not be apparent, since the samples are suspended in a standard solution. Of course it is possible to measure the effects of specific metabolic changes using this preparation and it has long been known that the pump rate has an optimal pH and falls in acid or alkaline solutions (Shigekawa et al. 1978). Methods of this sort have shown a reduction in pump function to about 50% in various types of human muscle fatigue and that this partially recovers over a period of about one hour (Byrd, Bode, and Klug 1989; Gollnick et al. 1991).

The reductions in pump rate seen in isolated SR cannot have a direct metabolic origin and suggest that some structural change in the SR occurs during fatigue. Suggested mechanisms are the increase in temperature caused by repeated activity, the increased concentration of oxygen radicals, or the depletion of glycogen, which may affect the pump rate even after isolation (Booth et al. 1997).

In the studies in intact fibers the reductions of pump rate were greater, probably because both structural factors and metabolic factors inhibited the pump activity. As noted above it is well established that acidosis can slow the pump but this is not a factor in the reported experiments because the mouse mammalian fiber shows little acidosis under the conditions of these experiments (Westerblad and Allen 1992a). By default it seems possible that the rise in P_i slows the pump, as has been observed in isolated SR (Yamamoto 1998). This would explain why recovery of the rate of relaxation occurs over the course of a few minutes after fatiguing stimulation (MacIntosh et al. 1994).

The functional consequences of the observed slowing of the pump are far from clear. It is often proposed that slowing of the pump would inhibit recycling of Ca^{2+}, producing less Ca^{2+} release and a smaller tetanic $[Ca^{2+}]_i$ signal. However, in skeletal muscle Ca^{2+} that is not pumped back into the SR remains in the myoplasm, principally bound to buffers such as parvalbumin. Consequently, even though Ca^{2+} release may be smaller, the tetanic $[Ca^{2+}]_i$ resulting can be larger because the Ca^{2+} buffers are nearer to saturation. Thus the use of drugs that inhibit the SR Ca^{2+} pump in skeletal muscle usually cause larger tetanic $[Ca^{2+}]_i$ signals and increased force (Bakker, Lamb, and Stephenson 1996; Westerblad and Allen 1994c). This suggests that the fall in tetanic $[Ca^{2+}]_i$ observed during muscle fatigue occurs in spite of slowing of the Ca^{2+} pump rather than because of it.

Relaxation usually slows during fatigue and is dependent on both the rate of uptake of Ca^{2+} by the SR and the rate of detachment of cross bridges. Thus some studies have shown a correlation between SR Ca^{2+} pump rate and relaxation and have implied that slower pump rate causes the slowing of relaxation (Gollnick et al.1991). However other studies have found that relaxation showed no slowing despite a small reduction in SR Ca^{2+} pump rate (Booth et al. 1997). This topic has also been investigated in single fibers, where it was possible to estimate both the contribution of the SR Ca^{2+} pump and slowed XB detachment to the slowing of relaxation. In mouse muscle the slowing of relaxation appeared to arise mainly through the XB (Westerblad and Allen 1994b), whereas in frog muscle both processes contributed to the slowing of relaxation (Westerblad, Lännergren, and Allen 1997). It seems that with these two processes in series the rate-limiting step differs in different animals and muscles.

Summary

A wide range of processes within muscles have their function altered by activity. Muscles fulfill many different functional roles in an animal and it is not surprising that they have specialized in different ways to meet these needs. Muscles such as the heart have generous blood supply, dense capillaries, and a large aerobic capacity and can meet the needs for ATP production without drastic changes in key metabolites and show no obvious fatigue under normal circumstances. In contrast, many fast muscles have limited oxidative capacity and are designed for brief periods of activity when they use mainly stored energy. If these muscles are used for prolonged periods the resulting changes in metabolites alter many aspects of function and some combination of these and other factors contribute to the observed fatigue. As the investigation of muscles becomes more cellular and molecular, the challenge is to identify specific mechanisms that are important in particular types of activity. Armed with such knowledge one can begin to understand fatigue and identify training or other regimes that will optimize muscle function under particular circumstances.

References

Adrian, R.H., and Peachey, L.D. 1973. Reconstruction of the action potential of frog sartorius muscle. *J. Physiol. (London)* 235: 103–31.

Allen, D.G., Lännergren, J., and Westerblad, H. 1995a. Muscle cell function during prolonged activity: Cellular mechanisms of fatigue. *Exp. Physiol.* 80: 497–527.

Allen, D.G., Lännergren, J., and Westerblad, H., 1995b. Role of ATP in the regulation of Ca^{2+} release in normal and fatigued mouse skeletal muscle [Abstract]. *Proceedings of the Australian Physiological and Pharmacological Society* 26: 123P.

Allen, D.G., Lännergren, J., and Westerblad, H. 1997. The role of ATP in the regulation of intracellular Ca^{2+} release in single fibers of mouse skeletal muscle. *J. Physiol. (London)* 498: 587–600.

Allen, D.G., Lee, J.A., and Westerblad, H. 1989. Intracellular calcium and tension during fatigue in isolated single muscle fibers from *xenopus laevis. J. Physiol. (London)* 415: 433–58.

Allen, D.G., Westerblad, H., and Lännergren, J. 1995c. The role of intracellular acidosis in muscle fatigue. *Adv. Exp. Med. Biol.* 384: 57–68.

Ameredes, B.T., Brechue, W.F., Andrew, G.M., and Stainsby, W.N. 1992. Force-velocity shifts with repetitive isometric and isotonic contractions of canine gastrocnemius in situ. *J. Appl. Physiol.* 73: 2105–11.

Ashley, C.C., Mulligan, I.P., and Lea, T.J. 1991. Ca^{2+} and activation mechanisms in skeletal muscle. *Q. Rev. Biophys.* 24: 1–73.

Ashley, C.C., and Ridgway, E.B. 1970. On the relationships between membrane potential, calcium transient, and tension in single barnacle muscle fibers. *J. Physiol. (London)* 209: 105–30.

Bakels, R., and Kernell, D. 1995. Measures of "fastness": Force profiles of twitches and partly fused contractions in rat medial gastrocnemius and tibialis anterior muscle units. *Pflüg. Archiv.* 431: 230–6.

Baker, A.J., Longuemare, M.C., Brandes, R., and Weiner, M.W. 1993. Intracellular tetanic calcium signals are reduced in fatigue of whole skeletal muscle. *Am. J. Physiol.* 264: C577–82.

Bakker, A.J., Lamb, G.D., and Stephenson, D.G. 1996. The effect of 2,5-di-(tert-butyl)-1,4-hydroquinone on force responses and the contractile apparatus in mechanically skinned muscle fibers of the rat and toad. *J. Muscle Res. Cell. Mot.* 17: 55–67.

Barclay, C.J. 1996. Mechanical efficiency and fatigue of fast and slow muscles of the mouse. *J. Physiol. (London)* 497: 781–94.

Barstow, T.J., Jones, A.M., Nguyen, P.H., and Casaburi, R. 1996. Influence of muscle fiber type and pedal frequency on oxygen uptake kinetics of heavy exercise. *J. Appl. Physiol.* 81: 1642–50.

Beelen, A., Sargeant, A.J., Jones, D.A., and De Ruiter, C.J. 1995. Fatigue and recovery of voluntary and electrically elicited dynamic force in humans. *J. Physiol. (London)* 484: 227–35.

Bergström, M., and Hultman, E. 1991. Relaxation and force during fatigue and recovery of the human quadriceps muscle: Relations to metabolite changes. *Pflüg. Archiv.* 418: 153–60.

Bevan, L., Laouris, Y., Reinking, R.M., and Stuart, D.G. 1992. The effect of the stimulation pattern on the fatigue of single motor units in adult cats. *J. Physiol. (London)* 449: 85–108.

Bigland-Ritchie, B., Furbush, F., and Woods, J.J. 1986. Fatigue of intermittent submaximal voluntary contractions: central and peripheral factors. *J. Appl. Physiol.* 61: 421–29.

Bigland-Ritchie, B., Johansson, R., Lippold, O.C.J., and Woods, J.J. 1983. Contractile speed and EMG changes during fatigue of sustained maximal voluntary contractions. *J. Neurophysiol.* 50: 313–24.

Booth, J., McKenna, M.J., Ruell, P.A., Gwinn, T.H., Davis, G.M., Thompson, M.W., Harmer, A.R., Hunter, S.K., and Sutton, J.R. 1997. Impaired calcium pump function does not slow relaxation in human skeletal muscle after prolonged exercise. *J. Appl. Physiol.* 66: 1383–89.

Burke, R.E., Levine, D.N., Tsairis, P., and Zajac, F.E. 1973. Physiological types and histochemical profiles in motor units of the cat gastrocnemius. *J. Physiol. (London)* 234: 723–48.

Byrd, S.K., Bode, A.K., and Klug, G.A. 1989. Effects of exercise of varying duration on sarcoplasmic reticulum function. *J. Appl. Physiol.* 66: 1383–89.

Cady, E.B., Jones, D.A., Lynn, J., and Newham, D.J. 1989. Changes in force and intracellular metabolites during fatigue of human skeletal muscle. *J. Physiol. (London)* 418: 311–25.

Cairns, S.P., and Dulhunty, A.F. 1995. High frequency fatigue in rat skeletal muscle: Role of extracellular ion concentrations. *Muscle & Nerve* 18: 890–8.

Cairns, S.P., Hing, W.A., Slack, J.R., Mills, R.G., and Loiselle, D.S. 1997. Different effects of raised $[K^+]_o$ on membrane potential and contraction in mouse fast and slow twitch muscle. *Am. J. Physiol.* 273: C598–611.

Chin, E.R., and Allen, D.G. 1996a. Fatigue in mouse fibers at different work intensities [Abstract]. *Proceedings of the Australian Physiological and Pharmacological Society.* 27: 112P.

Chin, E.R., and Allen, D.G. 1996b. The role of elevations in intracellular $[Ca^{2+}]$ in the development of low frequency fatigue in mouse single muscle fibers. *J. Physiol. (London)* 491: 813–24.

Chin, E.R., Green, H.J., Grange, F., Dossett-Mercer, J., and O'Brien, P.J. 1995. Effects of prolonged low frequency stimulation on skeletal muscle sarcoplasmic reticulum. *Can. J. Physiol. Pharmacol.* 73: 1154–64.

Chin, E.R., Green, H.J., Grange, F., Mercer, J.D., and O'Brien, P.J. 1994. Technical considerations for assessing alterations in skeletal muscle sarcoplasmic reticulum Ca^{++} sequestration function in vitro. *Mol. Cell. Biochem.* 139: 41–52.

Cooke, R., Franks, K., Luciani, G.B., and Pate, E. 1988. The inhibition of rabbit skeletal muscle contraction by hydrogen ions and phosphate. *J. Physiol. (London)* 395: 77–97.

Cooke, R., and Pate, E. 1985. The effects of ADP and phosphate on the contraction of muscle fibers. *Biophys. J.* 48: 789–98.

Crow, M.T., and Kushmerick, M.J. 1982. Chemical energetics of slow and fast twitch muscles of the mouse. *J. Gen. Physiol.* 79: 147–66.

Curtin, N.A., and Edman, K.A.P. 1994. Force-velocity relation for frog muscle fibers: Effects of moderate fatigue and of intracellular acidification. *J. Physiol. (London)* 475: 483–94.

Davies, N.W., Standen, N.B., and Stanfield, P.R. 1992. The effect of intracellular pH on ATP-dependent potassium channels of frog skeletal muscle. *J. Physiol. (London)* 445: 549–68.

De Haan, A., Jones, D.A., and Sargeant, A.J. 1989. Changes in velocity of shortening, power output, and relaxation rate during fatigue of rat medial gastrocnemius muscle. *Pflüg. Archiv.* 413: 422–28.

Duty, S., and Allen, D.G. 1994. The distribution of intracellular calcium concentration in isolated single fibers of mouse skeletal muscle during fatiguing stimulation. *Pflüg. Archiv.* 427: 102–9.

Duty, S., and Allen, D.G. 1995. The effects of glibenclamide on tetanic force and intracellular calcium in normal and fatigued mouse skeletal muscle. *Exp. Physiol.* 80: 529–41.

Edman, K.A.P. 1995. Myofibrillar fatigue versus failure of activation. *Adv. Exp. Med. Biol.* 384: 29–43.

Edman, K.A.P., and Lou, F. 1992. Myofibrillar fatigue versus failure of activation during repetitive stimulation of frog muscle fibers. *J. Physiol. (London)* 457: 655–73.

Edman, K.A.P., and Mattiazzi, A.R. 1981. Effects of fatigue and altered pH on isometric force and velocity of shortening at zero load in frog muscle fibers. *J. Muscle Res. Cell Mot.* 2: 321–34.

Edwards, R.H.T. 1983. Biochemical bases of fatigue in exercise performance: Catastrophe theory of muscular fatigue. In *Biochemistry of exercise,* ed. H.G. Knuttgen, J.A. Vogel, and J. Poortmans, 3–28. Champaign, IL: Human Kinetics.

Edwards, R.H.T., Hill, D.K., Jones, D.A., and Merton, P.A. 1977. Fatigue of long duration in human skeletal muscle after exercise. *J. Physiol. (London)* 272: 769–78.

Enoka, R.M., and Stuart, D.G. 1992. Neurobiology of muscle fatigue. *J. Appl. Physiol.* 72: 1631–48.

Fabiato, A., and Fabiato, F. 1978. Effects of pH on the myofilaments and the sarcoplasmic reticulum of skinned cells from cardiac and skeletal muscles. *J. Physiol. (London)* 276: 233–55.

Fitts, R.H. 1994. Cellular mechanisms of muscle fatigue. *Physiol. Rev.* 74: 49–94.

Fitts, R.H., Courtright, J.B., Kim, D.H., and Witzmann, F.A. 1982. Muscle fatigue with prolonged exercise: Contractile and biochemical alterations. *Am. J. Physiol.* 242: C65–73.

Fitts, R.H., and Holloszy, J.O. 1977. Contractile properties of rat soleus muscle: Effects of training and fatigue. *Am. J. Physiol.* 233: C86–91.

Fitts, R.H., and Holloszy, J.O. 1978. Effects of fatigue and recovery on contractile properties of frog muscle. *J. Appl. Physiol.* 45: 899–902.

Fratacci, M.D., Levame, M., Rauss, A., and Atlan, G. 1996. Rat diaphragm during postnatal development. II. Resistance to low and high frequency fatigue. *Repro. Fertil. Develop.* 8: 399–407.

Fulco, C.S., Lewis, S.F., Frykman, P.N., Boushel, R., Smith, S., Harman, E.A., Cymerman, A., and Pandolf, K.B. 1996. Muscle fatigue and exhaustion during dynamic leg exercise in normoxia and hypobaric hypoxia. *J. Appl. Physiol.* 81: 1891–1900.

Gaesser, G.A., and Poole, D.C. 1996. The slow component of oxygen uptake kinetics in humans. In *Exercise and sport sciences reviews.* Vol. 24. Ed. J.O. Holloszy, 35–70. Baltimore: Williams & Wilkins.

Gandevia, S.C., Allen, G.M., and McKenzie, D.K. 1995. Central fatigue: Critical issues, quantification, and practical implications. *Adv. Exp. Med. Biol.* 384: 281–94.

Garcia, M.D.C., Gonzalez-Serratos, H., Morgan, J.P., Perreault, C.L., and Rozycka, M. 1991. Differential activation of myofibrils during fatigue in phasic skeletal muscle cells. *J. Muscle Res. Cell. Mot.* 12: 412–24.

Gillis, J.M. 1985. Relaxation of vertebrate skeletal muscle: A synthesis of the biochemical and physiological approaches. *Biochim. Biophys. Acta* 811: 97–145.

Godt, R.E., and Nosek, T.M. 1989. Changes of intracellular milieu with fatigue or hypoxia depress contraction of skinned rabbit skeletal and cardiac muscle. *J. Physiol. (London)* 412: 155–80.

Golgeli, A., Ozesmi, C., and Ozesmi, M. 1995. Dependence of fatigue properties on the pattern of stimulation in the rat diaphragm muscle. *Indian J. Physiol. Pharmacol.* 39: 315–22.

Gollnick, P.D., Körge, P., Karpakka, J., and Saltin, B. 1991. Elongation of skeletal muscle relaxation during exercise is linked to reduced calcium uptake by the sarcoplasmic reticulum in man. *Acta Physiol. Scand.* 142: 135–36.

Gonzalez-Serratos, H., Somlyo, A.V., McClellan, G., Shuman, H., Borrero, L.M., and Somlyo, A.P. 1978. Composition of vacuoles and sarcoplasmic reticulum in fatigued muscle: Electron probe analysis. *Proc. Nat. Acad. Sci. USA* 75: 1329–33.

Gordon, D.A., Enoka, R.M., Karst, G.M., and Stuart, D.G. 1990. Force development and relaxation in single motor units of adult cats during a standard fatigue test. *J. Physiol. (London)* 421: 583–94.

Gregor, R.J., Komi, P.V., Browning, R.C., and Järvinen, M. 1991. A comparison of the triceps surae and residual muscle moments at the ankle during cycling. *J. Biomech.* 24: 287–97.

Györke, S. 1993. Effects of repeated tetanic stimulation on excitation-contraction coupling in cut muscle fibers of the frog. *J. Physiol. (London)* 464: 699–710.

Hatcher, D.D., and Luff, A.R. 1987. Force-velocity properties of fatigue-resistant units in cat fast twitch muscle after fatigue. *J. Appl. Physiol.* 63: 1511–18.

Hatcher, D.D., and Luff, A.R. 1988. Contractile properties of cat skeletal muscle after repetitive stimulation. *J. Appl. Physiol.* 64: 502–10.

Herzog, W., Leonard, T.R., and Stano, A. 1995. A system for studying the mechanical properties of muscles and the sensorimotor control of muscle forces during unrestrained locomotion in the cat. *J. Biomech.* 28: 211–18.

Hodgkin, A.L., and Horowicz, P. 1959. Movements of Na and K in single muscle fibers. *J. Physiol. (London)* 145: 405–32.

Kawta, H. 1983. Effects of twitch train on the tetanic contractility of the frog skeletal muscle. *Japan. J. Physiol.* 33: 429–48.

Kent-Braun, J.A., McCully, K.K., and Chance, B. 1990. Metabolic effects of training in humans: A ^{31}P-MRS study. *J. Appl. Physiol.* 69: 1165–70.

Kermode, H., Sitsapesan, R., and Williams, A.J. 1995. ADP and inorganic phosphate activate the sheep cardiac sarcoplasmic reticulum Ca^{2+} pump properties in frog skeletal muscle [Abstract]. *J. Physiol. (London)* 487: 171P.

Klein, M.G., Kovacs, L., Simon, B.J., and Schneider, M.F. 1991. Decline of myoplasmic Ca^{2+}, recovery of calcium release and sarcoplasmic Ca^{2+} pump properties in frog skeletal muscle. *J. Physiol. (London)* 441: 639–71.

Korge, P., and Campbell, K.B. 1995. The importance of ATPase microenvironment in muscle fatigue: A hypothesis. *Int. J. Sports Med.* 16: 172–79.

Kreis, R., and Boesch, C. 1996. Spatially localized, one- and two-dimensional NMR spectroscopy and in vivo application to human muscle. *J. Magn. Reson. Series B* 113: 103–18.

Lamb, G.D., Junankar, P.R., and Stephenson, D.G. 1995. Raised intracellular $[Ca^{2+}]$ abolishes excitation-contraction coupling in skeletal muscle fibers of rat and toad. *J. Physiol. (London)* 489: 349–62.

Lamb, G.D., and Stephenson, D.G. 1991. Effect of Mg^{2+} on the control of Ca^{2+} release in skeletal muscle fibers of the toad. *J. Physiol. (London)* 434: 507–28.

Lännergren, J., and Westerblad, H. 1987. The temperature dependence of isometric contractions of single, intact fibers dissected from a mouse foot muscle. *J. Physiol. (London)* 390: 285–93.

Lännergren, J., and Westerblad, H. 1991. Force decline due to fatigue and intracellular acidification in isolated fibers from mouse skeletal muscle. *J. Physiol. (London)* 434: 307–22.

Lee, J.A., Westerblad, H., and Allen, D.G. 1991. Changes in tetanic and resting [Ca^{2+}]$_i$ during fatigue and recovery of single muscle fibers from *Xenopus laevis*. *J. Physiol. (London)* 433: 307–26.

Liepert, J., Kotterba, S., Tegenthoff, M., and Malin, J.P. 1996. Central fatigue assessed by transcranial magnetic stimulation. *Muscle & Nerve* 19: 1429–34.

Lüttgau, H.C. 1965. The effect of metabolic inhibitors on the fatigue of the action potential in single muscle fibers. *J. Physiol. (London)* 178: 45–67.

Lüttgau, H.C., and Stephenson, D.G. 1986. Ion movements in skeletal muscle in relation to activation of contraction. In *Physiology of membrane disorders*, ed. T.E. Andreoli, J.F. Hoffman, B.B. Farmer, and S.G. Schultz, 449–68. New York: Plenum.

MacIntosh, B.R. 1991. Relaxation of skeletal muscle: Staircase in fatigue and with dantrolene [Abstract]. In *Muscle fatigue: Biochemical and physiological aspects*, ed. G. Atlan, L. Beliveau, and P. Bouissou, 67. Paris: Masson.

MacIntosh, B.R., and Gardiner, P.F. 1987. Posttetanic potentiation and skeletal muscle fatigue: Interactions with caffeine. *Can. J. Physiol. Pharmacol.* 65: 260–8.

MacIntosh, B.R., Grange, R.W., Cory, C.R., and Houston, M.E. 1994. Contractile properties of rat gastrocnemius muscle during staircase, fatigue, and recovery. *Exp. Physiol.* 79: 59–70.

MacIntosh, B.R., Stainsby, W.N., and Gladden, L.B. 1983. Fatigue from incompletely fused tetanic contractions in skeletal muscle in situ. *J. Appl. Physiol.* 55: 976–82.

MacLennan, D.H., and Holland, P.C. 1975. Calcium transport in the sarcoplasmic reticulum. *Ann. Rev. Biophy. Bioeng.* 4: 377–403.

McComas, A.J. 1996. *Skeletal muscle: Form and function*. Champaign, IL: Human Kinetics.

McCully, K.K., Kakihira, H., Vandenborne, K., and Kent-Braun, J. 1991. Noninvasive measurements of activity-induced changes in muscle metabolism. *J. Biomech.* 24(Suppl. 1): 153–61.

Meissner, G. 1994. Ryanodine receptor/Ca^{2+} release channels and their regulation by endogenous effectors. *Ann. Rev. Physiol.* 56: 485–508.

Merton, P.A. 1954. Voluntary strength and fatigue. *J. Physiol. (London)* 123: 553–64.

Millar, N.C., and Homsher, E. 1990. The effect of phosphate and calcium on force generation in glycerinated rabbit skeletal muscle fibers: A steady state and transient kinetic study. *J. Biol. Chem.* 265: 20234–40.

Myburgh, K.H., and Cooke, R. 1997. Response of compressed skinned skeletal muscle fibers to conditions that simulate fatigue. *J. Appl. Physiol.* 82: 1297–304.

Nakajima, S., Nakajima, Y., and Peachey, L.D. 1973. Speed of repolarization and morphology of glycerol-treated frog muscle fibers. *J. Physiol. (London)* 234: 465–80.

Newsholme, E.A., and Blomstrand, E. 1995. Tryptophan, 5-hydroxytryptamine, and a possible explanation for central fatigue. *Adv. Exp. Med. Biol.* 384: 315–20.

Owen, V.J., Lamb, G.D., and Stephenson, D.G. 1996. Effect of low [ATP] on depolarization-induced Ca^{2+} release in skeletal muscle fibers of the toad. *J. Physiol. (London)* 493: 309–15.

Pate, E., Bhimani, M., Franks-Skiba, K., and Cooke, R. 1995. Reduced effect of pH on skinned rabbit psoas muscle mechanics at high temperatures: Implications for fatigue. *J. Physiol. (London)* 486: 689–94.

Pate, E., Wilson, G.J., Bhimani, M., and Cooke, R. 1994. Temperature dependence of the inhibitory effects of orthovanadate on shortening velocity in fast skeletal muscle. *Biophys. J.* 66: 1554–62.

Renaud, J.M., Gramolini, A., Light, P., and Comtois, A. 1996. Modulation of muscle contractility during fatigue and recovery by ATP-sensitive potassium channel. *Acta Physiol. Scand.* 156:203-212.

Robinson, G.A., Enoka, R.M., and Stuart, D.G. 1991. Immobilization-induced changes in motor unit force and fatigability in the cat. *Muscle & Nerve* 14: 563–73.

Roed, A. 1991. Selective potentiation of subtetanic and tetanic contractions by the calcium-channel antagonist nifedipine in the rat diaphragm preparation. *Gen. Pharmacol.* 2: 313–18.

Sahlin, K. 1991. Control of energetic processes in contracting human skeletal muscle. *Biochem. Soc. Trans.* 19: 353–58.

Saugen, E., and Vollestad, N.K. 1996. Metabolic heat production during fatigue from voluntary repetitive isometric contractions in humans. *J. Appl. Physiol.* 81: 1323–30.

Seburn, K.L., and Gardiner, P. 1995. Adaptations of rat lateral gastrocnemius motor units in response to voluntary running. *J. Appl. Physiol.* 78: 1673–78.

Segal, S.S., and Faulkner, J.A. 1985. Temperature-dependent physiological stability of rat skeletal muscle in vitro. *Am. J. Physiol.* 248: C265–70.

Shigekawa, M., Dougherty, J.P., and Katz, A.M. 1978. Reaction mechanism of Ca^{2+}-dependent ATP hydrolysis by skeletal muscle sarcoplasmic reticulum in the absence of added alkali metal salts. I. Characterization of steady state ATP hydrolysis and comparison with that in the presence of KCl. *J. Biol. Chem.* 253: 1442–50.

Smith, J.S., Coronado, R., and Meissner, G. 1985. Sarcoplasmic reticulum contains adenine nucleotide-activated calcium channels. *Recent Adv. Physiol.* 316: 446–49.

Smith, J.S., Coronado, R., and Meissner, G. 1986. Single channel measurements of the calcium release channel from skeletal muscle sarcoplasmic reticulum. *J. Gen. Physiol.* 88: 573–88.

Spriet, L.L. 1992. Anaerobic metabolism in human skeletal muscle during short-term, intense activity. *Can. J. Physiol. Pharmacol.* 70: 157–65.

Spruce, A.E., Standen, N.B., and Stanfield, P.R. 1985. Voltage-dependent ATP-sensitive potassium channels of skeletal muscle membrane. *Nature* 316: 736–38.

Stainsby, W.N., and Barclay, J.K. 1971. Relation of load, rest length, work, and shortening to oxygen uptake by in situ dog semitendinosis. *Am. J. Physiol.* 221: 1238–42.

Stokes, M.J., and Dalton, P.A. 1991. Acoustic myography for investigating human skeletal muscle fatigue. *J. Appl. Physiol.* 71: 1422–26.

Taylor, J.L., Butler, J.E., Allen, G.M., and Gandevia, S.C. 1996. Changes in motor cortical excitability during human muscle fatigue. *J. Physiol.* 490: 519–28.

Thompson, L.V., Balog, E.M., Riley, D.A., and Fitts, R.H. 1992. Muscle fatigue in frog semitendinosus: Alterations in contractile function. *Am. J. Physiol.* 262: C1500–6.

Vandenboom, R., and Houston, M.E. 1996. Phosphorylation of myosin and twitch potentiation in fatigued skeletal muscle. *Can. J. Physiol. Pharmacol.* 74: 1315–21.

Vaz, M.A., Zhang, Y., Herzog, W., Guimaraes, A.C.S., and MacIntosh, B.R. 1996. The behavior of rectus femoris and vastus lateralis during fatigue and recovery: An electromyographic and vibromyographic study. *Electromyogr. Clin. Neurophysiol.* 36: 221–30.

Vollestad, N.K. 1995. Metabolic correlates of fatigue from different types of exercise in man. *Adv. Exp. Med. Biol.* 384: 185–94.

Vollestad, N.K., and Sejersted, O.M. 1988. Biochemical correlates of fatigue. *Eur. J. Appl. Physiol.* 57: 336–47.

Vollestad, N.K., Sejersted, O.M., Bahr, R., Woods, J.J., and Bigland-Ritchie, B. 1988. Motor drive and metabolic responses during repeated submaximal contractions in humans. *J. Appl. Physiol.* 64: 1421–27.

Vollestad, N.K., Sejersted, I., and Saugen, E. 1997. Mechanical behavior of skeletal muscle during intermittent voluntary isometric contractions in humans. *J. Appl. Physiol.* 38: 1557–65.

Vollestad, N.K., and Verburg, E. 1996. Muscular function, metabolism, and electrolyte shifts during prolonged repetitive exercise in humans. *Acta Physiol. Scand.* 156: 271–78.

Vollestad, N.K., Wesche, J., and Sejersted, O.M. 1990. Gradual increase in leg oxygen uptake during repeated submaximal contractions in humans. *J. Appl. Physiol.* 68: 1150–6.

Walmsley, B., Hodgson, J.A., and Burke, R.E. 1978. Forces produced by medial gastrocnemius and soleus muscles during locomotion in freely moving cats. *J. Neurophysiol.* 41: 1203–16.

Westerblad, H., and Allen, D.G. 1991. Changes of myoplasmic calcium concentration during fatigue in single mouse muscle fibers. *J. Gen. Physiol.* 98: 615–35.

Westerblad, H., and Allen, D.G. 1992a. Changes of intracellular pH due to repetitive stimulation of single fibers from mouse skeletal muscle. *J. Physiol. (London)* 449: 49–71.

Westerblad, H., and Allen, D.G. 1992b. Myoplasmic free Mg^{2+} concentration during repetitive stimulation of single fibers from mouse skeletal muscle. *J. Physiol. (London)* 453: 413–34.

Westerblad, H., and Allen, D.G. 1993a. The contribution of $[Ca^{2+}]_i$ to the slowing of relaxation in fatigued single fibers from mouse skeletal muscle. *J. Physiol. (London)* 468: 729–740.

Westerblad, H., and Allen, D.G. 1993b. The influence of intracellular pH on contraction, relaxation, and $[Ca^{2+}]_i$ in intact single fibers from mouse muscle. *J. Physiol. (London)* 466: 611–28.

Westerblad, H., and Allen, D.G. 1994a. Mechanisms of slowing of relaxation in fatigue. *J. Muscle Res. Cell. Mot.* 15: 181.

Westerblad, H., and Allen, D.G. 1994b. Relaxation, $[Ca^{2+}]_i$ and $[Mg^{2+}]_i$ during prolonged tetanic stimulation of intact, single fibers from mouse skeletal muscle. *J. Physiol. (London)* 480: 31–43.

Westerblad, H., and Allen, D.G. 1994c. The role of sarcoplasmic reticulum in relaxation of mouse muscle; effects of 2,5-di-(tert-butyl)-1,4-benzohydroquinone. *J. Physiol. (London)* 474: 291–301.

Westerblad, H., and Allen, D.G. 1996. Slowing of relaxation and $[Ca^{2+}]_i$ during prolonged tetanic stimulation of single fibers from *Xenopus* skeletal muscle. *J. Physiol. (London)* 492: 723–36.

Westerblad, H., Bruton, J.D., and Lännergren, J. 1997. The effect of intracellular pH on contractile function of intact single fibers of mouse muscle declines with increasing temperature. *J. Physiol. (London)* 500: 193–204.

Westerblad, H., Duty, S., and Allen, D.G. 1993. Intracellular calcium concentration during low frequency fatigue in isolated single fibers of mouse skeletal muscle. *J. Appl. Physiol.* 75: 382–88.

Westerblad, H., and Lännergren, J. 1991. Slowing of relaxation during fatigue in single mouse muscle fibers. *J. Physiol. (London)* 434: 323–36.

Westerblad, H., and Lännergren, J. 1994. Changes of the force-velocity relation, isometric tension, and relaxation rate during fatigue in intact, single fibers of *Xenopus* skeletal muscle. *J. Muscle Res. Cell. Mot.* 15: 287–98.

Westerblad, H., and Lännergren, J. 1995. Reduced maximum shortening velocity in the absence of phosphocreatine observed in intact fibers of *Xenopus* skeletal muscle. *J. Physiol. (London)* 482: 383–90.

Westerblad, H., Lännergren, J., and Allen, D.G. 1997b. Slowed relaxation in fatigued skeletal muscle fibers of *Xenopus* and mouse: Contribution of $[Ca^{2+}]_i$ and cross bridges. *J. Gen. Physiol.* 109: 385–99.

Westerblad, H., Lee, J.A., Lamb, A.G., Bolsover, S.R., and Allen, D.G. 1990. Spatial gradients of intracellular calcium in skeletal muscle during fatigue. *Pflüg. Archiv.* 415: 734–40.

Westerblad, H., Lee, J.A., Lännergren, J., and Allen, D.G. 1991. Cellular mechanisms of fatigue in skeletal muscle. *Am. J. Physiol.* 261: C195–209.

Wilson, G.J., Dos Remedios, C.G., Stephenson, D.G., and Williams, D.A. 1991. Effects of sulphydryl modification on skinned rat skeletal muscle fibers using 5,5'-dithiobis-(2-nitrobenzoic acid). *J. Physiol. (London)* 437: 409–30.

Yamamoto, T. 1998. The Ca^{2+}, Mg^{2+} dependent ATPase and the uptake of Ca^{2+} by the fragmented sarcoplasmic reticulum. In *Muscle proteins, muscle contraction, and cation transport*, ed. Y. Tonomura, 303–56. Tokyo: University Park Press.

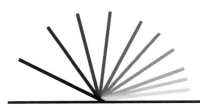

Chapter 20

Stretch-Shortening Cycle Fatigue

P.V. Komi and C. Nicol

Muscular energetics, mechanics, and also fatigue have been studied extensively under conditions of pure isometric, concentric, and eccentric muscle actions. Human movement, however, seldom involves pure forms of isolated concentric, eccentric, or isometric actions (see chapter 3). This is because the body segments are periodically subjected to impact forces, as in running or jumping, or because some external force such as gravity lengthens the muscle. In these phases the muscles are acting eccentrically, and concentric action follows. By definition of eccentric action, the muscles must be active during the stretching phase. The combination of eccentric and concentric actions forms a natural type of muscle function called a stretch-shortening cycle (SSC) (see figure 20.1) (Komi 1984; Norman and Komi 1979).

In SSC the eccentric part influences the performance of the subsequent concentric phase, so that the final (concentric) action can be more powerful than that resulting from the concentric action alone. When direct Achilles tendon forces have been measured during normal SSC activities in human (Komi 1990) or animal (Gregor et al. 1988) experiments, the mechanical (and metabolic) performance of the skeletal muscle has been shown to be considerably modified during the concentric part of the SSC. For example, when the in vivo force-velocity (F-V) curve

Stretch-Shortening Cycle (SSC)

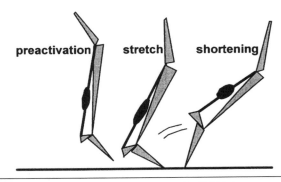

Figure 20.1 Stretch-shortening cycle (SSC) is a natural skeletal muscle function. SSC is shown here schematically for the triceps surae muscle during the contact phase of running. Before landing the muscle is preactivated to be prepared for the stretching phase (eccentric action), which is followed by the shortening (concentric) phase.

Reprinted, by permission, from P.V. Komi, 1984, "Physiological and biomechanical correlates of muscle function: Effects of muscle structure and stretch-shortening cycle on force and speed," *Exercise and sport science reviews* 12: 81-121.

of the SSC is compared to that of the classical F-V curve of the isolated concentric action, SSC situation demonstrates a clear performance enhancement. This performance potentiation of the concentric action following immediately the active prestretch (eccentric action) is a complex phenomenon and can be explained by a variety of elastic and chemomechanical events (Huijing 1992).

In the discussion of force enhancement after stretch, the importance of the reflex influences, and neural activation in general, must also be taken into consideration (Komi 1992). This is especially relevant for two reasons. First, in isolated muscle preparations, the performance enhancement of the concentric part during SSC has been usually studied in conditions where activation levels (electrical stimulation) have been kept constant (e.g., Cavagna, Dusman, and Margaria 1968; Cavagna, Saibene, and Margaria 1965). In normal locomotion (animal and human), the activation of the eccentric part of the SSC can be voluntarily or reflexly regulated to make the power output of the eccentric action match the external load conditions (e.g., impact), as well as the final concentric action (e.g., in submaximal hopping or maximal jump). Second, when an active muscle is put under stretch, the activity from the muscle spindles and the Golgi tendon organs will determine which of these reflexes—facilitatory or inhibitory—will dominate and what the magnitude of the performance potentiation from the reflex sources could be. Hoffer and Andreassen (1981) have documented well that when reflexes are intact the muscle stiffness is greater per same operating force than in an areflexive muscle. Thus, the reflexes may make a net contribution to muscle stiffness during the eccentric part of SSC. Recent experiments with in vivo tendon force measurements (Nicol and Komi 1998) have quantified the importance of stretch reflexes during single passive dorsiflexion stretches in contributing to Achilles tendon force enhancement.

SSC is thus a form of muscle function in which all the major components of performance "sources" are loaded: mechanical, neural, and metabolic. Fatigue during SSC may, however, correspond more clearly to that observed after pure eccentric fatiguing exercise. This would be the case especially from the simple structural point of view. Functionally, however, SSC fatigue appears to be much more complex. The influence of the eccentric phase on the final SSC performance with associated reflex and force enhancements emphasizes that SSC exercise may have more possibilities for fatigue adjustments. The present report makes an attempt to characterize what is currently understood as SSC fatigue and its possible mechanisms of adjustments. The report is a follow-up and updated version of our earlier publications on the topic (e.g., Komi and Gollhofer 1987; Nicol and Komi 1996).

Muscle Fatigue During Isolated Eccentric Actions

It seems evident that the eccentric phase is a dominating part of the SSC. It has been well documented that the maximum force levels during eccentric actions are much higher than in isometric or concentric actions (Asmussen, Hansen, and Lammert 1965; Komi 1973; Levin and Wyman 1927; Singh and Karpovich 1966). The force produced during eccentric activity may be twice that developed during isometric muscle actions, although the total number of strongly bound cross bridges may be only about 10% greater than in isometric contraction (Faulkner, Brooks, and Opiteck 1993). This force enhancement (over, e.g., isometric condition) during stretch (eccentric action) is velocity dependent and has been explored in detail at the muscle fiber/sarcomere level. The recent view of the underlying mechanisms is that a part of the force enhancement during eccentric action is due to increased strain of attached cross bridges, most probably in combination with a slight increase in the number of attached bridges (Edman 1996; Lombardi and Piazzesi 1990; Sugi and Tsuchiya 1988). There is now additional evidence to demonstrate that the cytoskeletal proteins found in the intrasarcomeric region, such as titin, form an important part of the elastic element that can be involved in the force enhancement during eccentric stretch (Edman and Tsuchiya 1996). Although the role of titin as well as of nebulin has been well demonstrated in the passive tension and sarcomere filament assembly of the skeletal muscle (see Patel and Lieber 1997) their role in SSC has not been explored.

It was the concept of high mechanical load in individual muscle fibers and connective tissues that put Asmussen (1956) among the very first ones to quantify the performance and muscle soreness differences between fatiguing concentric (positive) and eccentric (negative) work. He was able to demonstrate for both arm (triceps) and leg (quadriceps) muscles that muscular soreness with palpable muscular changes was always present 1 to 2 days after negative work, but hardly ever after positive work of the same intensity. This finding has formed a basis—although often without reference to Asmussen's work—for the concept of delayed onset muscle soreness (DOMS), typically occurring in relation to unaccustomed exercises, including specifically eccentric actions. Asmussen had later (1975) communicated personally to one of the authors of the present paper (PVK) that he "had come to the conclusion that the connective tissue cells may produce pain-provoking substances, e.g., kinins together with histamines etc. . . . when mechanically overstrained." Even before Asmussen (1956), Hough (1902) had suggested that exercise-induced muscle soreness might be due to some rupture within the muscle, and especially in the connective tissue. Hill (1951) suggested that soreness was due to mechanical injury, distributed microscopically throughout the muscle. As referred to by Asmussen (1956), Bøje (1955) was "inclined to believe that the pains are located in the intramuscular connective tissues, as they are most apt to develop after exercise that have extended the muscles" (p. 110).

These experiments of Asmussen (1956) on the "mechanical stress" concept for muscle soreness stimulated us to design a study where EMG activity and muscle force of the forearm flexors were registered during 40 repeated maximum concentric and eccentric actions of constant velocity in different test situations (Komi and Rusko 1974). Thus, in both conditions the starting maximal EMG activity was the same, but the maximal concentric force was about 60% of the maximal eccentric force (Komi 1973). The results indicated that the muscle tension decreased to 50% in eccentric and to 80% in concentric exercise when the last contractions were compared with the first ones (see figure 20.2). Interestingly, maximal EMG activity of the studied muscles declined continuously during the entire work period, and the pattern and magnitude of this decrease were the same in both types of work (see figure 20.2). In all subjects the eccentric type of fatigue caused severe muscle soreness, which was felt primarily at the myotendinous junction level. Often it was so severe that the subjects could not extend their elbow for a few days after the testing. The recovery from soreness, which was also associated with muscle swelling, lasted for one or sometimes even two weeks. In none of the cases was any substantial soreness felt after concentric fatigue loading. Symptoms of soreness were very much similar to those observed in an earlier study, when severe muscle soreness prevented any positive training adaptation for more than one week. After disappearance of soreness the eccentric training resulted in rapid force increase (Komi and Buskirk 1972). These authors report that the subjects felt severe muscle soreness during the first week of eccentric training when no positive training took place. This was later confirmed in Komi's laboratory (see figure 20.3), when direct coupling was observed between severity of pain and force decrement during early training. Eccentric-induced reduction in performance is therefore usually associated with symptoms of muscle soreness. This can be demonstrated also in figure 20.4, which shows that both the force-time characteristics and EMG/force ratio presented a delayed recovery from eccentric fatigue but not from concentric fatigue (Komi and Viitasalo 1977).

Since these earlier attempts, eccentric exercises have been studied quite extensively, both with animal and human models (for a review, see Kuipers 1994; Waterman-Storer 1991). A common feature for all of these studies is the delayed onset muscle soreness and the progressive damage process, from which muscles will recover at a slow rate. Today the mechanism of muscle damage and symptoms of soreness are much better known, although the final answers are still to be found. MacIntyre, Reid, and McKenzie (1995) have given a comprehensive review on the topic.

Eccentrically Induced Muscle Damage and Soreness

Long lasting, but reversible, ultrastructural muscle damage is the common indicator of injury resulting from exercises involving eccentric muscle actions. It is well documented that the cytoskeletal and myofibrillar abnormalities observed after eccentric muscle actions reach a peak 2 to 3 days after the exercise (Fridén et al. 1983; Fridén, Sjöström, and Ekblom 1983; Newham et al. 1983a). According to Newham and associates (1983a), these lesions would be the precursors of more extensive muscle damage. In this process, Armstrong (1990) differentiates the "initial," the "autogenetic," the "phagocytic," and the "regenerative" stages. The present focus will be on potential mechanisms that may underlie the successive periods shown in figure 20.5.

The "initial stage" (I) includes the events that trigger the whole process. Following eccentric muscle actions, histological studies have reported an extensive disorganization and even disruption of the myofibrillar structures and intermediate filaments, leading to the classically observed Z-line streaming (Fridén, Kjorell, Thornell 1984; Newham, Edwards, and Mills 1994; Waterman-Storer 1991). In pure eccentric work, much evidence exists that the initial local damage results from mechanical rather than from metabolic mechanisms (Brooks, Zerba, and Faulkner 1995; Lieber and Fridén 1993; Newham et al. 1983a; Warren et al. 1993). Because higher tensions can be generated across a given number of recruited fibers under eccentric conditions (Bigland-Ritchie and Woods 1976; Faulkner, Brooks, and Opiteck 1993), it seems that it is not the

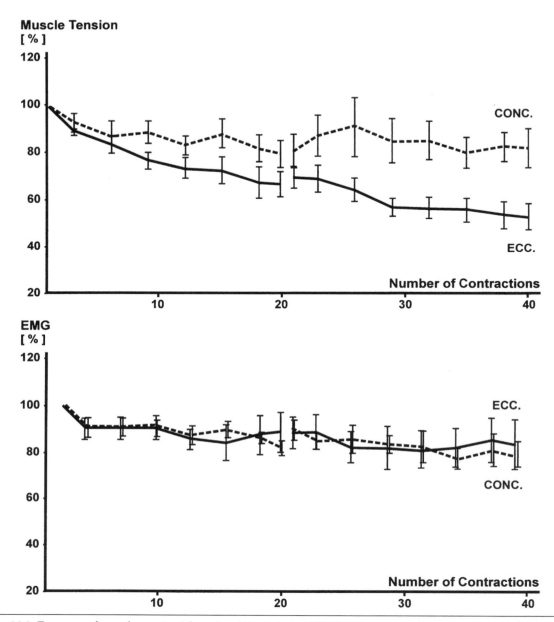

Figure 20.2 Decrease of muscle tension (above) and EMG (below) during the course of 40 repeated maximal contractions of either eccentric or concentric actions. At the start of fatigue the absolute maximal forces were 440 N and 245 N, in eccentric and concentric exercise, respectively. The absolute IEMG values were the same in both types of maximal actions. Note that despite similar reduction in EMG the eccentric fatigue caused much greater reduction in force.

Reprinted, by permission, from P.V. Komi and H. Rusko, 1974, "Quantitative evaluation of mechanical and electrical changes during fatigue loading of eccentric and concentric work," *Scandinavian Journal of Rehabilitative Medicine* 3(suppl.): 121-126.

high stress per se that causes muscle damage, but the magnitude of the active straining (Lieber and Fridén 1993) or the combination of strain and average force (Brooks, Zerba, and Faulkner 1995). Although evidence has been presented that fast twitch fibers would be preferentially recruited during eccentric exercise in man (Nardone, Romano, and Schieppati 1989), there is still controversy whether

eccentric-induced muscle damage is fiber type-specific (for a review, see Fridén and Lieber 1992).

The "autogenetic stage" (II) corresponds to the first 3 to 4 h following the injury and marks the beginning of the degradation process of the membrane structures. According to the reviews of Ebbeling and Clarkson (1989) and Armstrong (1990), the loss of Ca^{2+} homeostasis in the injured fibers

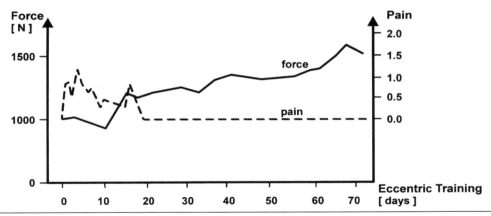

Figure 20.3 Initial increases in strength during high intensity strength training may be prevented by exercise-induced muscle damage and soreness. In this study with leg extensor muscles, the symptoms of pain in the exercising muscles lasted 2 to 3 weeks. Note the reduction in maximal force during the early training period. This finding is essentially the same as that reported by Komi and Buskirk 1972.

From Laine and Louhevaara 1974.

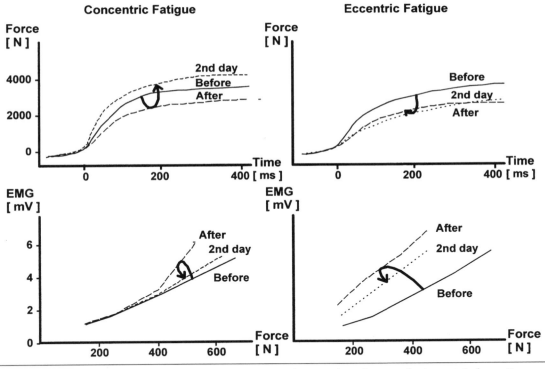

Figure 20.4 Influence of 40 repeated maximum concentric and eccentric actions on the isometric force-time curve of the bilateral leg extension (upper graphs) and on EMG-force relationship in unilateral knee extension (lower graphs). Note the delayed recovery of muscle performance following the eccentric fatigue.

Reprinted, by permission, from P.V. Komi and J.T. Viitasalo, 1977, "Changes in motor unit activity and metabolism in human skeletal muscle during and after repeated eccentric and concentric contractions," *Acta Physiologica Scandinavica* 100: 246-254.

could result in enhanced calcium-activated proteolytic and lipolytic enzyme activity. More investigations are needed, however, to clarify the underlying mechanisms of the increased intracellular Ca^{2+} concentration.

The "phagocytic stage" (III) is characterized by a typical inflammatory response in the tissue that may last for 2 to 4 days or more, with a peak on the third day postexercise (Kihlstrom, Salminen, and Vihko 1984). Inflammation is subclassified into acute

Stages in Ultrastructural Damage

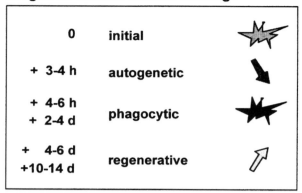

Figure 20.5 Four different stages related to (eccentric) exercise-induced skeletal muscle damage.

Reprinted, by permission, from R.B. Armstrong, 1984, "Mechanisms of exercise-induced delayed onset muscular soreness: A brief review," *Medicine & Science in Sports & Exercise* 16: 529-538.

and chronic. The acute response begins with changes in the vascular wall structure, leading to structural and functional alterations of the basement membrane and to migration of neutrophils and monocytes (Evans and Cannon 1991; Fantone 1993). Mobilization of neutrophils and leucocytes has been reported to be greater following eccentric-type exercise (Fielding et al. 1993; Pizza et al. 1995). Once activated, neutrophils provide a fresh supply of mediators, which may be partly responsible for amplifying and prolonging the inflammation. This is usually associated with redness, swelling, heat, and pain. This stage is also characterized by the presence in the blood of indirect indicators of muscle injury, such as muscular protein metabolites (e.g., troponin I), and increased specific muscle enzyme activity (e.g., creatine kinase, lactate dehydrogenase) (for a review, see Noakes 1987). Referring to the significance of peak CK values, it should be noted that, due to considerable variability of the magnitude of the response of serum enzymes (Clarkson and Ebbeling 1988), the peak values do not reflect the amount of muscle damage and do not correlate well with the functional performance decrements (e.g., Ebbeling and Clarkson 1989; Mair et al. 1995). However, it is suggested that relative changes in CK might be of some relevance to the detection of tissue inflammation and the comprehension of the functional effects of SSC fatigue (as presented later; see figure 20.15, page 396). Finally, Faulkner, Brooks, and Opiteck (1993) have stressed the importance of monocytes and macrophages for the removal of neutrophils and necrotic tissues. There is strong evidence that supports the hypoth-

esis that macrophages are an integral part of the recovery process following exercise-induced muscle injury, especially after eccentric muscle actions (for a review, see MacIntyre, Reid, and McKenzie 1995).

The "regenerative stage" (IV) begins on days 4 to 6 and reflects the regeneration of muscle fibers. In a recent animal study (Lowe et al. 1995), protein synthesis rate was found to increase approximately 48 h after injury, remaining elevated by 83% five days postinjury. By days 10 to 14, muscle protein degradation and synthesis rates had returned to normal, and phagocytic infiltration was not detected. However, muscle mass, protein content, and absolute force production were still lower than before. Similar delays of recovery (Hikida et al. 1983), but also much longer ones (Howell, Chleboun, and Conatser 1993), have been reported in humans following eccentric-induced muscle damage.

Does muscle damage, as discussed above, lead to sensation of pain? Delayed onset muscle soreness (DOMS) is a typical sensation of pain and discomfort that increases in intensity during the first 2 days after unaccustomed or eccentric-type exercises, remaining symptomatic for 1 to 2 more days, and disappearing usually 5 to 7 days after the exercise (for review, see Miles and Clarkson 1994) (see figure 20.6). Sore muscles are often stiff and tender and their ability to produce force is reduced for several days or weeks (Asmussen 1956; Howell, Chleboun, and Conatser 1993; Komi and Buskirk 1972; Komi and Rusko 1974; Komi and Viitasalo 1977; Sherman et al. 1984). At first, tenderness is mostly evident medially, laterally, and at the proximal myotendinous junction of the exercised muscles, but later it spreads throughout the other compartments (Bobbert, Hollander, and Huijing 1986; Howell, Chleboun, and Conatser 1993). In addition, soreness is not constant all the time, being mostly felt when the exercised limbs are extended or fully flexed or when the muscles are palpated deeply (Howell, Chleboun, and Conatser 1993).

In fact, neither the degree nor the timing of ultrastructural damage correlates well with the respective changes in DOMS (Howell, Chleboun, and Conatser 1993; Newham et al. 1983b). Similarly, moderate (Talag 1973) or no relationship (Cleak and Eston 1992) has been reported between DOMS and muscular strength. It is therefore suggested that this delay could reflect the natural course of the inflammatory response to injury (Evans and Cannon 1991). Swelling was also found to be biphasic (Howell, Chleboun, and Conatser 1993). These observations tend to support the early assumption of Hough (1902) that exercise-induced muscle soreness might

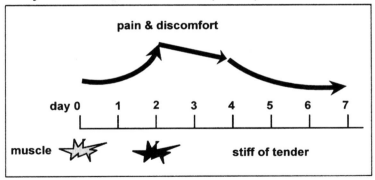

Figure 20.6 Schematic presentation of the subjective feelings of pain and discomfort as they are associated with muscular changes in the course of delayed onset muscle soreness (DOMS).

result from muscle ruptures, especially those affecting the connective tissue.

Although stiffness has not been studied as much as the other parameters associated with muscle soreness, stiffness was found to increase with a delayed time course that parallels that of muscle soreness (Jones, Newham, Clarkson 1987). Several studies (e.g., Howell et al. 1985; Saxton and Donnelly 1995) have shown that following eccentric exercise of the elbow flexors the relaxed elbow takes a more flexed position in standing subjects. This relative flexion is not the result of electrically mediated muscle contraction, but presumably is a consequence of shortening noncontractile elements (Howell et al. 1985; Jones, Newham, Clarkson 1987). According to these authors, this could result instead from an increased passive stiffness of the injured muscles.

When peripheral tissues are damaged, the sensation of pain in response to a given stimulus is enhanced. This phenomenon, termed "hyperalgesia", may involve a lowering of the threshold of nociceptors by the presence of locally released chemicals (for a review, see Jessel and Kelly 1991). Hyperalgesia can occur both at the site of tissue damage and in the surrounding undamaged areas (Fields 1987). The sensation of pain in the skeletal muscles is transmitted by nociceptors that belong to two groups of small diameter group III and IV muscle afferents (Kniffki, Mense, and Schmidt 1978; Mense 1977). These free nerve endings are particularly dense in the regions of connective tissues, but also between intra- and extrafusal muscle fibers as well as near blood vessels, in the Golgi tendon organs, and at the myotendinous junction. It is of interest for the present discussion on the functional effects of muscle damage that small diameter group III and IV muscle afferents have been found to be mostly polymodal, being sensitive to several parameters

associated with either metabolic fatigue or muscle injury (Kniffki, Mense, and Schmidt 1978; Rotto and Kaufman 1988). In case of muscle damage, biochemical substances such as bradykinin and prostaglandins are known to be released, and these increase the spontaneous discharge of both group III and IV mechanoreceptors and nociceptors (Kniffki, Mense, and Schmidt 1978; Mense and Meyer 1988; Rotto and Kaufman 1988). More specifically, group III afferents carry sharp, localized pain, whereas group IV afferents carry dull and diffuse pain. Group IV fibers have been suggested by Armstrong (1984) to be primarily responsible for the sensation of DOMS. Once activated, the nociceptors release neuropeptides, which cause vasodilatation, edema, and release of histamine. These processes lead then to a further and long-lasting activation of some of the sensory endings (Fields 1987).

In addition to peripheral pathways, evidence exists of descending influences on the transmission of the sensory afferent information via the spinocervical tract (Hong et al. 1979; Malmgren and Pierrot-Deseilligny 1987). Variation in receptor types and in the ability to modulate pain at multiple levels in the nervous system could partly explain the intersubject variability in soreness perception. These observations raise the question of potential interactions between muscle damage, muscle activation, and stretch-reflex regulation, as will be discussed later. To the best of our knowledge the literature lacks reports on effects of pure eccentric-induced fatigue on reflex sensitivity. The potential influence of the sensitization of small diameter muscle afferents on the alpha-motoneuron pool activation will be discussed further in the paragraph related to the potential effects of SSC type fatigue on the stretch-reflex sensitivity.

Finally, a beneficial effect of either eccentric (Komi and Buskirk 1972; Mair et al. 1995) or SSC (Byrnes et al. 1985; Schwane, Williams, and Sloan 1987) training has been suggested from the measurement of absence of soreness or diminished plasma CK activity after subsequent repetitions of the same exercise. A single bout of intensive eccentric exercise (Clarkson and Tremblay 1988) has been shown to attenuate the DOMS felt during subsequent bouts of exercise performed within the next six weeks. This phenomenon is known as *repeated bout effect*. Thus, it has been shown that when the same exercise is repeated 1 to 10 weeks after the first one, the recovery of strength and range of motion is faster (e.g., Clarkson and Tremblay 1988), soreness rating is less (e.g., Byrnes et al. 1985), and the presence in the blood of muscle proteins is also clearly reduced (e.g., Mair et al. 1995). These observations confirm the force adaptation curve of eccentric training (Komi and Buskirk 1972; see also figure 20.3, p. 387). According to a recent animal study (McBride, Gorin, and Carlsen 1995), the adaptation associated with a single exposure to eccentric exercise would be reduced in aged muscles. This beneficial mechanism remains, in any case, very complex. From the recent study of Mair and coworkers (1995), it may be seen, for instance, that the repetition of an eccentric exercise bout after 13 days did cause significant reductions in muscular damage, soreness, and functional impairments, whereas in the case of repetition after 4 days, soreness and functional impairments were not abolished. Complete repair does not seem, however, to be a prerequisite for the observed repeated bout effect or rapid training protective effects (for a review, see Ebbeling and Clarkson 1989). From the observations of Nosaka and Clarkson (1995), it can be seen also that repeated bouts of eccentric exercise (every three days) do not necessarily exacerbate or reduce the initially induced damage and soreness, but tend to delay the recovery process. This is in line with our observation of a maintained reflex inhibition when SSC performances were repeated every five days. This effect is discussed further, later in the chapter (see figure 20.14, page 395).

Specific Functional Effects of Fatiguing SSC Exercises

From the preceding discussion it seems evident that fatiguing SSC exercises, due to the important eccentric phase, will also be associated with structural muscle damage and associated soreness. Their intensity and duration are naturally dependent on the type of SSC exercise, as well as familiarity (adapta-

tion) of the subjects to this exercise. There is, however, one major difference between the eccentric phase of the natural SSC exercise and that of the pure eccentric loading as used in many human studies. In SSC the eccentric (braking) phase takes place usually quite rapidly, with relatively high initial stretch velocity (see e.g., Fukashiro et al. 1995). In this regard, it is important to recognize that even a submaximal hopping performed with preferred frequency can load the triceps surae muscle twice more than, for example, maximal squatting jumps (Komi 1990). Thus, effective contribution of the reflex loops becomes essentially important in SSC where stretching loads are high and muscle stiffness must be well regulated to meet the external loading conditions. This is the major reason why natural SSC exercises, with high velocity and short duration eccentric phase, have probably much more functional consequences than the pure eccentric exercises, which seldom, if at all, occur in natural locomotion. In the experiments to be reviewed in the following paragraphs, the impact loads in the SSC exercises were carefully controlled, but also varied in terms of intensity and duration.

Short-Term SSC Fatigue

A special sledge apparatus was used in several of our fatigue studies both for arm (see figures 20.7 and 20.8) and leg SSC exercises (see figure 20.11). The apparatus has been described in detail elsewhere (Kaneko, Komi, and Aura 1984; Komi et al. 1987). It was designed originally to study isolated forms of eccentric and concentric exercises as well as their combinations (see also the more recent paper of Oksanen et al. 1990). The first fatigue study (Gollhofer et al. 1987b) was performed on this apparatus as arm SSC exercises. Healthy men performed 100 submaximal SSCs with both arms, so that they were lying on the sledge facing the force plate, which was attached to the lower end and perpendicularly to the sliding metal bars of the sledge. The results showed that during 100 SSCs the fatigue was characterized by increases in the contact times for both eccentric and concentric phases of the cycle. A more detailed graphic analysis revealed that the force-time curves during contact on the platform were influenced by fatigue. The initial force peak became higher and the subsequent drop of force more pronounced (see figure 20.9). As shown by figure 20.10, however, the reflex contribution to sustain the repeated stretch loads became enhanced, especially when measured during the maximal drop test conditions, immediately after the fatigue loading. Thus, in a nonfatigue state the muscles were

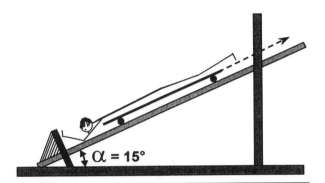

Figure 20.7 Schematic presentation of a subject-sledge system for arm SSC exercise, which consisted of 100 repeated rebound "jumps."

Reprinted, by permission, from A. Gollhofer, P.V. Komi, N. Fujitsuka, and M. Miyashita, 1987b, "Fatigue during stretch-shortening cycle exercises. II: Changes in neuromuscular activation patterns of human skeletal muscle," *International Journal of Sports Medicine* 8(suppl 1): 38-47.

able to damp the impact in SSC by a smooth force increase and by a smooth joint motion. However, repeated damping movements followed by the concentric action may have eventually become so fatiguing that the neuromuscular system changed its "stiffness" regulation. This change was characterized especially by high impact force peak followed by a temporary force decline. However, in the subsequent and highest stretch loading condition, which consisted of maximal falls on the floor (see figure

20.8), all the examined reflex components (short, medium, and long latency components) decreased in amplitude after the fatiguing exercise. Thus, in the submaximal and maximal sledge tests the stretch reflex contribution during fatigue could be interpreted to imply attempts of the nervous system to compensate, by increasing activation, the loss of the muscles' contractile force or ability to resist repeated impact loads. On the other hand, the falls on the floor represented the highest impact load condition, where reflex inhibition seemed logical. In order to examine further the potential adjustments of the neuromuscular system to SSC fatigue, a series of studies was performed in which a reflex test was performed in a more controlled situation—that of passive stretches.

Stretch Reflex Sensitivity After SSC Fatigue

In the basic protocol with the leg extensor muscles, fatigue was induced on the sledge apparatus by performing as many rebounds as possible to a given rising height (70% of the maximal one; see figure 20.11). Exhaustion was reached after 100 to 400 repetitions (Horita et al. 1996; Nicol et al. 1996), which corresponded to 2 to 5 min of intensive work. The potential influence of fatigue on the reflex sensitivity was investigated in the first study immediately before and after fatigue, as well as 2 and 4 or 5 days later. In this study, a vertical drop jump test

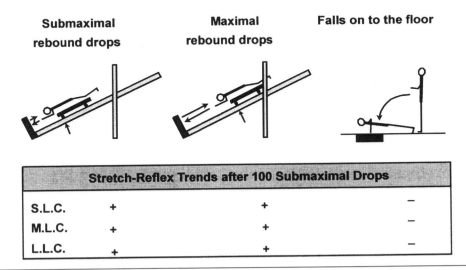

Stretch-Reflex Trends after 100 Submaximal Drops		
S.L.C.	+	+	−
M.L.C.	+	+	−
L.L.C.	+	+	−

Figure 20.8 In addition to the submaximal fatigue loading (upper, left), the other two test loads in the arm SSC exercise included maximal rebound jumps (middle) and forward falls onto the floor (right). The lower panel shows the corresponding increases (+) or decreases (−) in the short, medium, and long latency EMG components, comparing before and after fatigue.

Adapted, by permission, from P.V. Komi and A. Gollhofer, 1987, "Muscular function in exercise and training," *Medicine and Sport Science* 26: 119-127. © Karger Publishers in Basel, Switzerland.

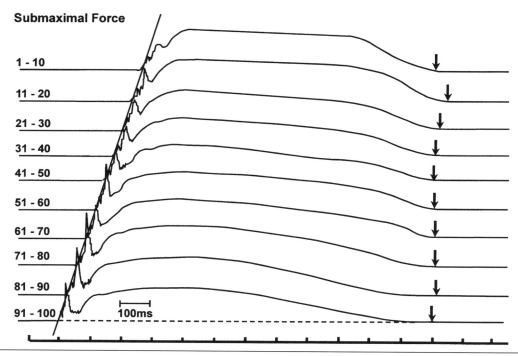

Figure 20.9 The fatiguing arm SSC exercises shown in figure 20.8 demonstrated progressive changes in the reaction force record during the hand contact with the sledge force plate. The records have been averaged for groups of 10 successive force-time curves. Note the increase in the impact force peak with subsequently greater decrease of force when fatigue progressed.

Reprinted, by permission, from A. Gollhofer, P.V. Komi, M. Miyashita, and O. Aura, 1987a. "Fatigue during stretch-shortening cycle exercises. I: Changes in mechanical performance of human skeletal muscle," *International Journal of Sports Medicine* 8: 71-78.

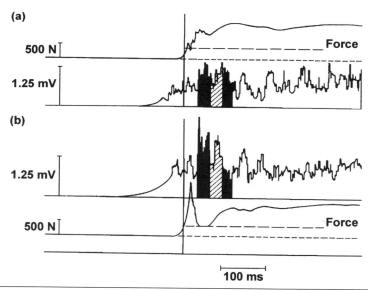

Figure 20.10 Force and EMG records of a subject who performed maximal rebound jumps (see the middle part of figure 20.8) on the sledge before and immediately after 100 exhaustive SSCs. The initial force peak and the subsequent decrease of force increased dramatically due to fatigue. The rectified EMG records suggest augmentations of the short and medium latency reflex components (the first two shaded areas in the EMG records) during the test immediately following the fatiguing SSC exercise. *(a)* Before; *(b)* after.

Reprinted, by permission, from A. Gollhofer, P.V. Komi, N. Fujitsuka, and M. Miyashita, 1987b, "Fatigue during stretch-shortening cycle exercises. II: Changes in neuromuscular activation patterns of human skeletal muscle," *International Journal of Sports Medicine* 8(suppl 1): 38-47.

SSC Exercise

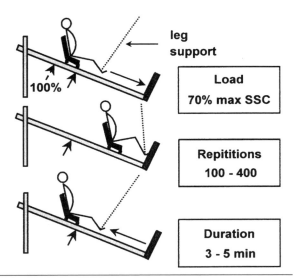

leg support

Load
70% max SSC

Repititions
100 - 400

Duration
3 - 5 min

Figure 20.11 Schematic presentation of the SSC fatigue protocol for leg exercise. The subjects performed repetitive rebound jumps on the sledge from the height corresponding to 70% (heavy arrow) from the maximum drop height. The 100% condition referred to the maximal rebound height they could reach by a single SSC action. During exercise with the 70% load the subjects were instructed to brake in the eccentric phase down to the 90° knee angle and to rebound immediately back to the predetermined height. Depending on the subject, 100 to 400 repetitions were needed for exhaustion.

Reprinted, by permission, from C. Nicol, P.V. Komi, T. Horita, H. Kyrölainen, and T.E.S. Takala, 1996a, "Reduced stretch-reflex sensitivity after exhaustive stretch-shortening cycle exercise," *European Journal of Applied Physiology* 72: 401-409.

from a 50 cm height was also performed immediately after each sledge exercise test. In the second experiment (Nicol, Komi, and Avela 1996), the whole protocol was repeated three times, on days 0, 5, and 10. Each experiment included control sessions, in which subjects performed only the reflex tests. In both experiments, the stretch reflex test of the shank muscles was performed in a sitting position with the knee stabilized at 120° (see figure 20.12). A powerful engine was used to induce stretches of the shank muscles at either 70 or 115° · s^{-1}. This test included a randomized series of 20 mechanical stimuli per velocity. Reflex latency and peak-to-peak amplitude were measured for each recorded reflex and then averaged per velocity. To observe possible changes in the level of spinal excitability and in neuromuscular transmission-propagation the second experiment complemented the reflex tests with H-reflex measurements of the soleus muscle (16 stimuli, 25% maximal M-wave). In this

case, the pedal was fixed so that the ankle angle was kept at 90°. Blood concentrations of lactate and serum creatine kinase activity (CK) were measured at various moments along the protocol.

The analysis of the sledge exercise indicated a significant 5 to 7% increase ($p < .05$) of the contact time toward the time of exhaustion. The EMG analysis revealed a significant $44 \pm 31\%$ reduction ($p < .05$) of the lateral gastrocnemius (LG) preactivation. Time to exhaustion did not differ among the three tests on the sledge in the second experiment. The before-after fatigue comparison of the stretch reflexes revealed an immediate decrease of the peak-to-peak reflex amplitude in both experiments. This was accompanied in the second experiment by a significant reduction ($p < .05$) of the H-reflex/M-wave ratio (see figure 20.13). On the other hand, no significant changes were found in either reflex latency (39 ± 2 ms) or duration. The subsequent analysis of the follow-up period revealed similar trends in both experiments with a significantly delayed recovery of the peak-to-peak stretch reflex amplitude after the sledge exercise. As shown in figure 20.14, the repetition of the sledge exercise on days 5 and 10 further delayed the reflex recovery. In the case of the H-reflex, the stimulus was not mechanical and relied less on muscle spindle sensitivity. The respective changes of the H-reflex and stretch reflex amplitudes were positively related on the 5th day after exercise ($r = .90$ at $70° \cdot$ s^{-1} and $r = .91$ at $115° \cdot$ s^{-1}, $p < .05$). In the first experiment, the decrease in amplitude was accompanied on day 2 by a loss of 20% of the LG responses at the slowest stretching velocity.

From the metabolic analysis, similar mean peak lactate values ($11–15$ mmol $\cdot$ l^{-1}) were found after the three sledge exercises. The serum level of CK presented the characteristic pattern observed after repeated eccentric muscle actions, with an essentially individual, but clearly delayed peak of accumulation, which occurred in each experiment on day 2 postexercise. It should be noted that only some of the subjects returned to their initial CK level on days 4 or 5. The perceived soreness developed usually as soon as 2 h after the exercise and lasted until day 4 in most of the subjects. Furthermore, in the first experiment, the increase in CK activity between the second hour and the second day after exercise was negatively ($r = -.99$, $p < .001$) related to the respective changes in drop jump (DJ) performance (see figure 20.15 left). The drop jump performance decreased significantly immediately after ($p < .05$), remained still reduced ($p < .05$) 2 h and 2 days later, but returned to the prefatigue level on day 4 after the fatiguing exercise. When additional

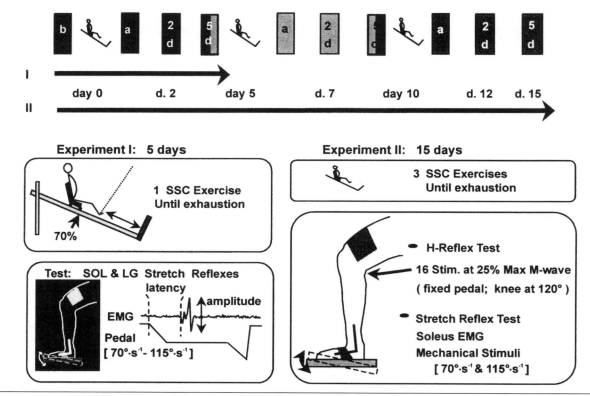

Figure 20.12 Description of the two protocols used in studying SSC fatigue. In experiment I, the effects of one exhaustive SSC exercise were followed for five days. The tests performed before (b), immediately after (a), two days (2d), and five days (5d) later included the double stimulus passive dorsiflexion stretches (left, lower panel) at 70 or 115° · s⁻¹ by a powerful engine. The knee was fixed at 120°. In experiment II the exhaustive SSC exercise was repeated three times: on days 0, 5, and 10. In addition to the passive dorsiflexion test, H-reflex measurements were also performed on the soleus muscle (right, lower panel).

Reprinted, by permission, from C. Nicol, P.V. Komi, T. Horita, H. Kyrölainen, and T.E.S. Takala, 1996a, "Reduced stretch-reflex sensitivity after exhaustive stretch-shortening cycle exercise," *European Journal of Applied Physiology* 72:401-409. C. Nicol, P.V. Komi, and J. Avela, 1996b, "Stretch-shortening cycle fatigue reduces stretch reflex response," Abstract book of the 1996 International Pre-Olympic Scientific Congress. July 10-14, Dallas, Texas. p.108.

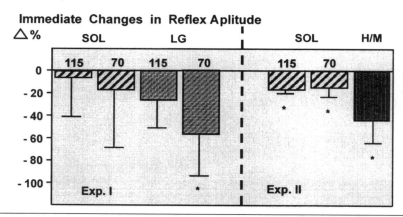

Figure 20.13 Changes in lateral gastrocnemius (LG) and soleus (SOL) stretch reflex amplitudes measured at 2 given shortening velocities (70 and 115° · s⁻¹) immediately after short-term, exhaustive SSC exercise. Experiments I and II refer to the two experiments shown in figure 20.12. In experiment II reduction of the H-reflex/M-wave ratio (H/M) was also apparent.

Reprinted, by permission, from C. Nicol, P.V. Komi, T. Horita, et al, 1996a, "Reduced stretch-reflex sensitivity after exhaustive stretch-shortening cycle exercise," *European Journal of Applied Physiology* 72: 401-409; C. Nicol, P.V. Komi, and J. Avela,1996b, "Stretch-shortening cycle fatigue reduces stretch reflex response," *Abstract book of the July 10-14, 1996 International Pre-Olympic Scientific Congress in Dallas*, 108.

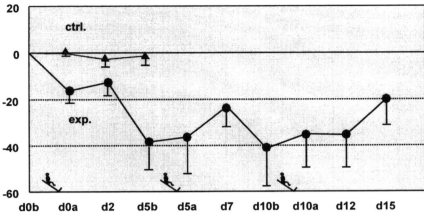

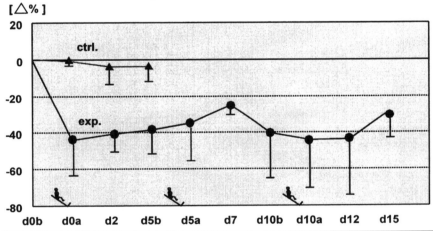

Figure 20.14 Changes in the soleus stretch reflex amplitude (top) and H/M amplitude (bottom) in the course of a 15-day follow-up during which the experimental group repeated the exhaustive short-term SSC exercises on days 0, 5, and 10. From C. Nicol, P.V. Komi, and J. Avela,1996b, "Stretch-shortening cycle fatigue reduces stretch reflex response," *Abstract book of the July 10-14, 1996 International Pre-Olympic Scientific Congress in Dallas,* 108.

subjects were recruited for the project (Horita et al. 1996), the results were qualitatively similar, and statistically slightly stronger. This may imply coupling between performance reduction in SSC and inflammatory processes resulting from muscle injury. This coupling concept may receive further support from the finding that the decrease in the CK activity between days 2 and 4 was related ($p < .05$) to the recovery of the relative peak-to-peak reflex amplitude of LG (see figure 20.15 right). Further analysis of the drop jump test (Horita et al. 1996), revealed that the knee joint stiffness changes were related to changes in short latency (M1 response) reflex components.

These results indicate an immediate reduction and delayed recovery of the stretch reflex loop

sensitivity after exhaustive SSC exercise. This contrasts, however, with the increased reflex sensitivity reported by another SSC fatigue study (Hortobagyi, Lambert, and Kroll 1991). Referring to the study of Gollhofer and coworkers (1987a), it is suggested that the contradictory results of these studies could be attributed to the possibility that the neuromuscular system may adapt differently depending on the fatigue level and requested task. The complexity of the reflex changes with fatigue is also illustrated by the studies of Duchateau and Hainaut (1993), in which fatigue induced by electrostimulation resulted in a reduction in H-reflex amplitude, whereas the long latency reflex amplitude was clearly increased. The reductions observed in our studies

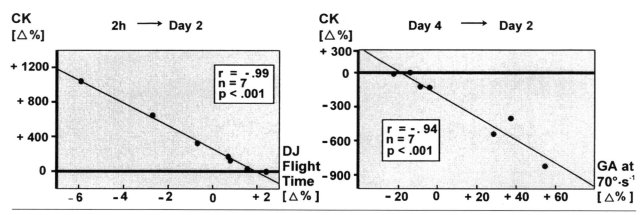

Figure 20.15 Left: Increase in CK activity during the first two days after exhaustive SSC exercise may be associated with decrease in the drop jump performance. Right: The similar association is also possible between the recovery of CK activity and stretch reflex amplitude as measured between days 2 and 4 post-SSC fatigue.

Reprinted, by permission, from C. Nicol, P.V. Komi, T. Horita, H. Kyrölainen, and T.E.S. Takala, 1996a, "Reduced stretch-reflex sensitivity after exhaustive stretch-shortening cycle exercise," *European Journal of Applied Physiology* 72:401-409.

(Avela and Komi 1998a, 1998b; Horita et al. 1996; Nicol et al. 1996) may be expected to result from two major and not mutually exclusive processes: (1) a reduction of the excitatory input (disfacilitation) from the muscle spindles (Macefield et al. 1991) and (2) an increased inhibition originating from the sensitization of small diameter muscle afferents (Garland 1991). Regarding the first hypothesis, it has been observed in mammalian muscle that the spindle activity can be reduced and even suppressed at low intracellular pH (Fukami 1988). Although supported by the observed peak lactate values of 11 to 5 mmol · l⁻¹ (Nicol et al. 1996), this possibility is diminished by the observation of a parallel decrease of the H/M ratio, a reflex loop that may be only partially affected by changes in the afferent signals originating from the muscle spindles. According to Mense and Meyer (1988), a reduced sensitivity of muscle spindle to stretch could also result from some of the chemical agents that are directly released from the damaged muscles.

The hypothesis of a reflex inhibition of the input through the sensitization of small diameter muscle afferents has been suggested by various studies in which fatigue was induced either by isometric work (Balestra, Duchateau, and Hainaut 1992; Woods, Furbush, and Bigland-Ritchie 1987) or by electrostimulation (Bigland-Ritchie et al. 1986; Garland 1991; Garland and McComas 1988). Group III (Mense 1977) and IV (Kniffki, Mense, and Schmidt 1978) nerve endings have been demonstrated to be sensitive to intracellular increases in potassium, phosphate, or lactate, as well as to low pH, which are known to accompany sustained intensive exercises. Kniffki, Mense, and Schmidt (1978) suggested

that the changes induced by normal muscular exercise might not constitute effective stimuli. Muscle damage leads to the release of biochemical substances such as bradykinin and prostaglandins, which enhance the activity of some nociceptors of group III and IV afferents (Kniffki, Mense, and Schmidt 1978; Rotto and Kaufmann 1988). On the perceptual side, there is a prolongation of the pain sensation long after the termination of the stimulus. Most of the present subjects reported long-lasting sensations of muscle pain and discomfort after the exhaustive sledge exercise. It is therefore suggested that the sensitization of small diameter muscle afferents to both chemical and mechanical processes related to muscle damage could partly explain the delayed recovery of the stretch reflex and H-reflex sensitivities (see figure 20.16). This hypothesis is reinforced by the significant relationship previously described between the recovery of the serum CK activity and reflex sensitivity (see figure 20.15, right). This inhibition could partly also explain the loss of 20% of the stretch reflex response at the slow stretching velocity (Nicol et al. 1996). Referring to the significant correlation between the increase in serum CK activity and the decline in the drop jump during the first two days postexercise (see figure 20.15, left), it is suggested that a potential decrease of the stretch reflex sensitivity could have affected the muscle stiffness and, consequently, SSC function. Referring to the potential influence of mechanical active strain (Lieber and Fridén 1993) and mechanical work (Brooks, Zerba, and Faulkner 1995) on the occurrence of muscle damage, it is suggested that some of the testing conditions, such as forward falls (Gollhofer et al. 1987a) or maximal drop jumps (Horita et al. 1996),

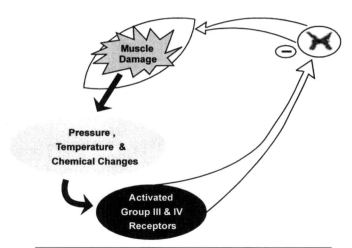

Figure 20.16 Possible coupling between the processes of exercise-induced muscle damage (e.g., after exhaustive SSC exercise) and reflex sensitivity (Bigland-Ritchie et al. 1986; Garland 1991; Garland and McComas 1988; Mense and Meyer 1988; Rotto and Kaufman 1988).

are more likely to reveal a loss of tolerance to impact and a reflex inhibition of the alpha-motoneuron pool activation (e.g., via the activation of GTO). The observed reduction in both stretch reflex and H-reflex sensitivity emphasizes the potential role of a sensitization of group III and IV nerve endings in the long-lasting reflex inhibition.

Long-Term SSC Fatigue

Dramatic effects on the neuromuscular function have also been reported in less intensive, but long-lasting SSC exercises, such as 85 km cross-country skiing (Viitasalo et al. 1982) and marathon running (Avela and Komi 1998a, 1998b; Komi et al. 1986; Nicol, Komi, and Marconnet 1991a, 1991b, 1991c; Pullinen, Leynaert, and Komi 1997; Sherman et al. 1984). These models have been used to investigate more specifically the effect of repetitive submaximal SSCs on the force production capacity and EMG activity of the involved leg extensor muscles.

Viitasalo and coworkers (1982) studied 23 skiers who participated in the Wasa cross-country (X-C) ski marathon with the finishing times ranging from 5.3 to 9.5 h. The measurements performed the day before and 1 to 2 h after the race, demonstrated clear changes in maximal knee extension performance of concentric as well as isometric actions. The maximal torque decreased significantly, but its reduction was much less than that of the maximal rates of torque production and relaxation. As expected, the maximal IEMG levels also decreased significantly and even more than the maximal torque. The leg extensor muscles in X-C skiing are

subjected to smoother impact and stretch loads as compared to running. In running, the duration of the phase including the initial impact and the subsequent braking (eccentric phase) is very short (50–120 ms), and the repeated loading will consequently have greater stretch-induced effects on the stiffness regulation than in X-C skiing. Thus, it would not be surprising to observe that a marathon run, which is usually shorter in duration than an 85 km X-C ski race, could induce more dramatic reductions in the force production. Maximal isometric knee extension has been shown to decrease during a marathon run on average by 26% (Nicol, Komi, and Marconnet 1991a) or 35.5% (Sherman et al. 1984). As in X-C skiing the marathon run was associated with a significant change in the shape of the torque-time curve, indicating that when the race and fatigue progressed the subjects needed more time to reach given absolute and relative force levels (Nicol, Komi, and Marconnet 1991a). In our more recent marathon study (Pullinen, Leynaert, and Komi 1997), the isometric force-time curve for knee extension showed not only a dramatic reduction immediately after the run, but also a slow recovery that was not complete on day 7 postmarathon (see figure 20.17). The maximal isometric force and EMG demonstrated similar patterns with delayed recovery (see figure 20.18). In the study of Nicol, Komi, and Marconnet (1991b) the loss in maximal force of the quadriceps muscle group was accompanied by a clear decline ($39 \pm 9\%$) in the capacity to maintain a 60% submaximal isometric level of force (fatigue test). The parallel analysis of the electromyographic activity revealed a large decrease in maximal activity ($p < .05$), but also the need for an initial increase of the neural activation at submaximal force level, which is likely to indicate a deterioration of the muscle function. These observations support the previous assumptions that various modifications of neural activation may take place to compensate for the exercise-induced contractile fatigue. In line with the hypothesis of Gollhofer et al. (1987a, 1987b), the present results indicate a facilitation at submaximal force level and an inhibition in maximal force condition.

Force platform, kinematic, and electrogoniometer techniques can be used to study what changes take place in the SSC itself when measured either during running or jumping. In this same marathon study (Nicol, Komi, and Marconnet 1991b), the overall performance reduction as judged by a maximal sprint test started to develop after the first 20 km and went down by $15 \pm 10\%$; values that are similar to those observed a few years earlier (Komi et al.

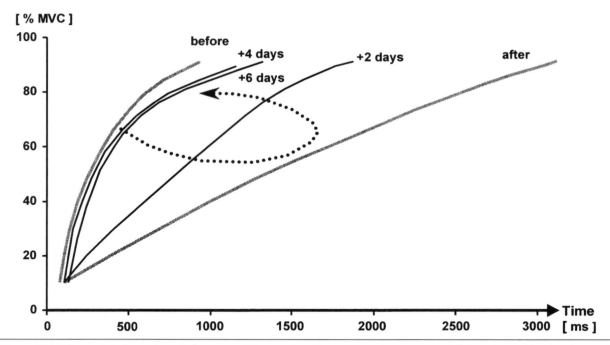

Figure 20.17 Isometric force-time (f-t) curves of the maximal knee extension measured before, immediately after, and at several intervals after the marathon. Note the great shift to the right in the postmarathon f-t curve, which recovered slowly back to the original position.

Data from Pullinen et al. 1997.

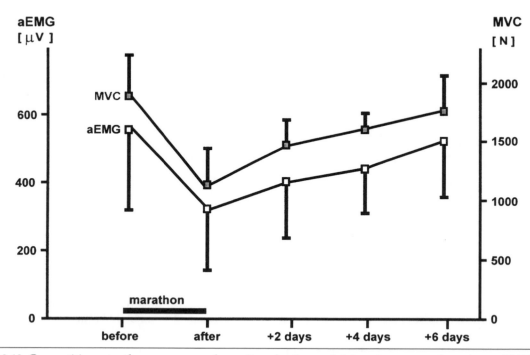

Figure 20.18 Competitive marathon run causes dramatic reduction and delayed recovery of maximum EMG and force of the isometric knee extension.

Data from Pullinen et al. 1997.

1986). In both marathon studies, contractile failure and changes in the neural activation were detected also in the SSC-type performances. In addition to reduction in either sprint run or drop jump tests observed in the second marathon report (Nicol, Komi, and Marconnet 1991b), the before-after marathon comparison of the ground reaction force-time curves revealed a clear drop in the vertical force after the impact peak with a concomitant increase of the contact time (see figure 20.19). In fact, both braking and push-off phases of the contact were longer as the fatigue progressed (for comparison see the short-term sledge exercise, figure 20.9, page 392). A decrease in the average horizontal force during the push-off phase was also demonstrated during the latter half of the marathon race (Nicol, Komi, and Marconnet 1991a). The marathon running resulted also in immediate postfatigue kinematic changes. The contact on the ground was made with more extended leg after marathon (fatigue condition), but with subsequent greater knee flexion (see figure 20.19). The natural consequence would be a longer push-off phase for both drop jump and marathon run tests. These observations are in good agreement with the ground reaction force (Gollhofer et al. 1987a, 1987b) and knee angle (Horita et al. 1996) changes regarding the effects of

short-term SSC exercises. In short-term fatigue, the displacement of the knee joint angle was different in the drop jumps performed before SSC fatigue, immediately after, and two days later (see figure 20.20). It should be noted that the knee joint was deeply flexed when measured two days after, when the muscle soreness was greatest and the peak CK activity at its highest.

These kinematic and ground reaction force changes are a natural consequence of the lost contractile capacity in the muscle. The nervous system makes an attempt to compensate for this loss by changing the muscle activation patterns. In the high stretch load drop jump test this is seen in the increased prelanding EMG (Horita et al. 1996), with subsequent adjustments with the short latency (M1) and medium latency (M2) EMG components. The global EMG changes after a marathon that are shown in figure 20.21 demonstrate that when EMG recordings were made in running at a constant marathon speed, all of the investigated muscles (vastus medialis, vastus lateralis, and gastrocnemius) increased their activity, especially in the push-off phase (Komi et al. 1986). In addition, the EMG/force ratio increased almost twofold for the push-off phase during the course of the marathon. This represents an increased level of muscle activation

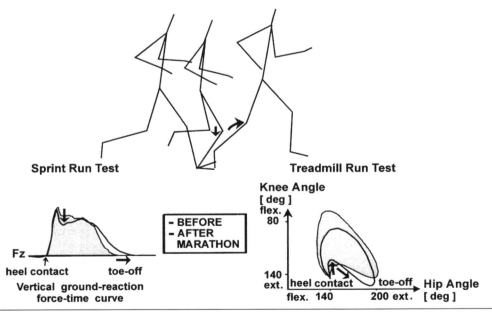

Figure 20.19 Influence of a marathon run on the vertical ground reaction force (left) and on the knee angle/hip angle diagram (right). Note a sharp drop in the peak of the sprint force-time curve (left) after the marathon. The knee angle/hip angle diagram shows a greater knee flexion immediately after the heel contact in a postmarathon situation.

Reprinted, by permission, from C. Nicol, P.V. Komi, and P. Marconnet, 1991a, "Fatigue effects of marathon running on neuromuscular performance. I: Changes in muscle force and stiffness characteristics," *Scandinavian Journal of Medicine and Science in Sports* 1: 10-17; C. Nicol, P.V. Komi, and P. Marconnet, 1991c, "Effects of a marathon fatigue on running kinematics and economy," *Scandinavian Journal of Medicine and Science in Sports* 1: 195-204.

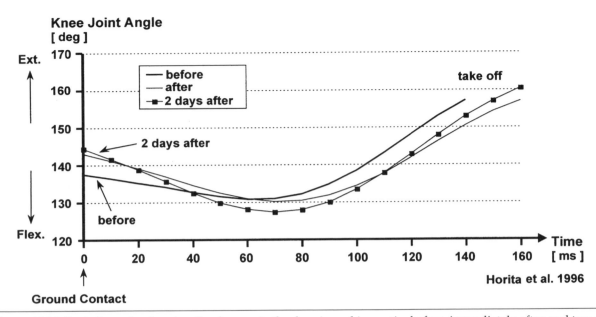

Figure 20.20 Typical graphs showing the changes in the drop jump kinematics before, immediately after, and two days after an exhaustive SSC fatigue on the sledge. It should be noted that in the nonfatigue situation the knee joint was more extended upon ground contact and that it was deeply flexed two days after the fatigue as compared to the two other trials.

Reprinted, by permission, from T. Horita, P.V. Komi, C. Nicol, and H. Kyrölainen, 1996, "Stretch shortening cycle fatigue: Interactions among joint stiffness, reflex, and muscle mechanical performance in the drop jump," *European Journal of Applied Physiology* 73: 393-403.

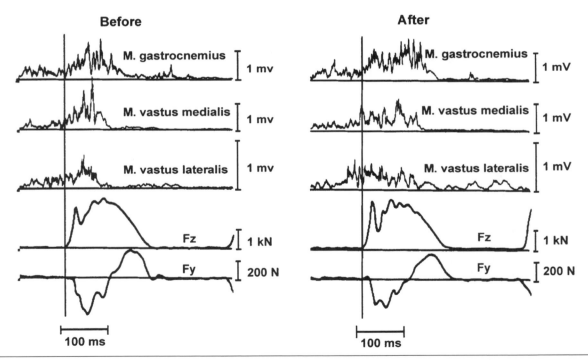

Figure 20.21 Examples of the vertical (Fz) and horizontal (Fy) ground reaction forces as well as of EMG activities of the selected muscles while running at constant submaximal speed before and after a marathon.

Reprinted, by permission, from P.V. Komi, T. Hyvärinen, A. Gollhofer, and A. Mero, 1986, "Man-shoe-surface interaction: Special problems during marathon running," *Acta Universitatis Ouluensis* 179: 69-72.

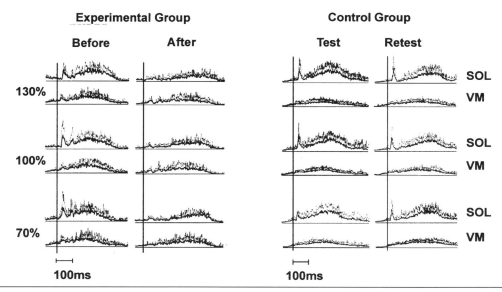

Experimental Group

Before   **After**

Control Group

Test   **Retest**

130%   SOL   VM

100%   SOL   VM

70%   SOL   VM

100ms   100ms

Figure 20.22 Demonstration of reduction (inhibition?) of the postmarathon stretch reflex response. The subjects performed 10 successive rebound jumps on the sledge from three different drop height conditions before and after a marathon run. Especially in the soleus (SOL) muscle, the before marathon records show a clear segmentation peak about 35 ms after the onset of foot contact on the sledge force plate. The time delay of this segmentation peak is consistent with the delay of the short latency stretch reflex component reported by Lee and Tatton (1982). This peak could not be detected after the marathon.

Reprinted, by permission, from J. Avela and P.V. Komi, 1998b, "Reduced stretch reflex sensitivity and muscle stiffness after long-lasting stretch-shortening cycle (SSC) exercise," *European Journal of Applied Physiology* 78:403-410.

during the preactivation and braking phases. The same constant running velocity (= same propelling force) could not have been performed without this dramatic increase in the EMG (muscle activation) of the push-off phase. It is suggested that with the progress of fatigue, there is change in the stiffness characteristics and, hence, a reduction in SSC-type performances. Consistent with this, our more recent marathon study (Avela and Komi 1998a, 1998b) revealed a clear postexercise reflex inhibition while performing maximal SSCs on the sledge apparatus (see figure 20.22). It would be of interest to know the time course of recovery of these reflex components and how this is related to possible interaction from muscle damage and soreness. Thus, it is likely that while the repeated stretch loads caused a reduction in tolerance to stretch with subsequent increase in the braking time, the transfer from stretch to shortening also became longer. Reduced power during the eccentric phase and increased coupling time between the stretch and shortening suggest greater loss of energy by dissipation as heat (Cavagna, Dusman, and Margaria 1968), resulting in less transfer of energy (kinetic to potential to kinetic). The final influence would then be reduced potentiation of the concentric phase in the SSC performances. The longer

push-off phase in the sprint performance test after marathon suggests an unsuccessful attempt of the neuromuscular system to compensate for the induced contractile failure.

Concluding Hypothesis

The major argument of this article is that the naturally occurring SSC, when repeated until exhaustion, will load the neuromuscular system in a more complex (thorough?) way than isolated forms of muscle actions. Evidence has been presented that both short-term and long-duration fatiguing SSC exercises lead to deterioration of neuromuscular function. SSC fatigue results usually in reversible muscle damage and has considerable influence on muscle mechanics, joint and muscle stiffness, as well as on reflex intervention (see figure 20.23). The effects are usually in the direction of reduced performance and can be both acute and delayed in nature. The reduced neural input to the muscle is at least partly of reflex origin. The reduced stretch reflex sensitivity can be demonstrated in association with exhaustive SSC exercise and its reduction is immediate with delayed recovery lasting from a few hours until several days. While the immediate postexercise decrease in stretch reflex sensitivity

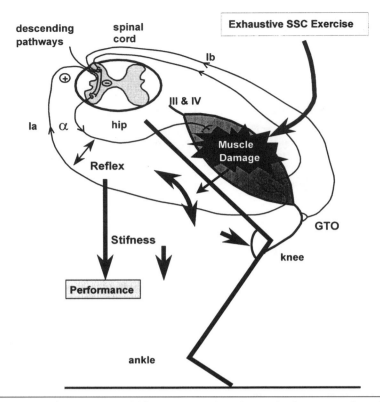

Figure 20.23 Proposed coupling between SSC exercise-induced muscle damage and performance reduction. Muscle damage changes stiffness regulation through changes in the afferent inputs from the muscle spindle, Golgi tendon organ (GTO), and group III and IV afferent nerve endings. The events occur in the following order: (1) Due to muscle damage the stretch reflex sensitivity decreases; (2) muscle (and joint) stiffness regulation becomes disturbed (reduced); (3) efficiency of SSC function (performance) decreases. The proposed mechanism may be even more apparent in the triceps surae muscle as compared to the quadriceps group of the figure.

may have a strong metabolic origin, the delayed recovery could result from the indirect influence of muscle damage, via the development of inflammation and muscle soreness, in the sensitization of small diameter group III and IV afferents. Muscle spindle sensitivity is also considered as tenable, but probably of lesser importance due to parallel decline in the H-reflex response. In addition to the inhibitory signal the progressive withdrawal of the spindle-mediated support may also play a role in modulating the reduction of neural input to the muscle. Direct mechanical damage in the intrafusal muscle fibers could also be a possibility (McBride, Gorin, and Carlsen 1995; Avela and Komi 1986b). Similarly the role of GTO in SSC fatigue needs to be considered.

References

Armstrong, R.B. 1984. Mechanisms of exercise-induced delayed onset muscular soreness: A brief review. *Med. Sci. Sports Exerc.* 16: 529–38.

Armstrong, R.B. 1990. Initial events in exercise-induced muscular injury. *Med. Sci. Sports Exerc.* 22: 429–35.

Asmussen, E. 1956. Observations on experimental muscle soreness. *Acta Rheum. Scand.* 2: 109–16.

Asmussen, E., Hansen, D., and Lammert, O. 1965. The relation between isometric and dynamic muscle strength in man. Communications from the Testing and Observation Institute of the Danish National Association for Infantile Paralysis, No. 20.

Avela, J., and Komi, P.V. 1998a. Interaction between muscle stiffness and stretch reflex sensitivity after long-term stretch-shortening cycle (SSC) exercise. *Muscle & Nerve* 21: 1224–27.

Avela, J., and Komi, P.V. 1998b. Reduced stretch reflex sensitivity and muscle stiffness after long-lasting stretch-shortening cycle (SSC) exercise. *Eur. J. Appl. Physiol.* 78: 403–10.

Balestra, C., Duchateau, J., and Hainaut, K. 1992. Effects of fatigue on the stretch reflex in a human muscle. *Electroencep. Clin. Neurophysiol.* 85: 46–52.

Bigland-Ritchie, B., and Woods J.J. 1976. Integrated EMG and O_2 uptake during positive and negative work. *J. Physiol. (London)* 260: 267–77.

Bigland-Ritchie, B., Dawson, N.J., Johansson, R.S., and Lippold, O.C.J. 1986. Reflex origin for the slowing of motoneuron firing rates in fatigue of human voluntary contractions. *J. Physiol. (London)* 379: 451–59.

Bobbert, M.F., Hollander, A.P., and Huijing, P.A. 1986. Factors in delayed onset muscular soreness of man. *Med. Sci. Sports Exerc.* 18: 75–81.

Bøje, O. 1955. *Bevaegelsestaere, Traening og Øvelsesterapi.* Copenhagen: Fremad.

Brooks, S.V., Zerba, E., and Faulkner, J.A. 1995. Injury to muscle fibers after single stretches of passive and maximally stimulated muscles in mice. *J. Physiol. (London)* 488: 459–69.

Byrnes, W.C., Clarkson, P.M., White, J.S., Hsieh, S.S., Frykman, P.N., and Maughan, R.J. 1985. Delayed onset muscle soreness following repeated bouts of downhill running. *J. Appl. Physiol.* 59: 710–5.

Cavagna, G.A., Dusman, B., and Margaria, R. 1968. Positive work done by a previously stretched muscle. *J. Appl. Physiol.* 24: 21–32.

Cavagna, G.A., Saibene, F.P., and Margaria, R. 1965. Effect of negative work on the amount of positive work performed by an isolated muscle. *J. Appl. Physiol.* 20: 157–58.

Clarkson, P.M., and Ebbeling, C. 1988. Investigation of serum creatine kinase variability after muscle-damaging exercise. *Clin. Sci.* 75: 257–61.

Clarkson, P.M., and Tremblay, I. 1988. Rapid adaptation to exercise-induced muscle damage. *J. Appl. Physiol.* 65: 1–6.

Cleak, M.J., and Eston, R.G. 1992. Muscle soreness, swelling, stiffness, and strength loss after intensive eccentric exercise. *Br. J. Sp. Med.* 26: 267–72.

Duchateau, J., and Hainaut, K. 1993. Behavior of short and long latency reflexes in fatigued human muscles. *J. Physiol. (London)* 471: 787–99.

Ebbeling, C.B., and Clarkson, P.M. 1989. Exercise-induced muscle damage and adaptation. *Sports Med.* 7: 207–34.

Edman, K.A.P. 1996. Fatigue vs. shortening-induced deactivation in striated muscle. *Acta Physiol. Scand.* 156: 183–92.

Edman, K.A.P., and Tsuchiya, T. 1996. Strain of passive elements during force enhancement by stretch in frog muscle fibers. *J. Physiol. (London)* 490(1): 191–205.

Evans, W., and Cannon, J.G. 1991. The metabolic effects of exercise-induced muscle damage. In *Exercise and sport science reviews.* Vol. 19. Ed. J.C. Holloszy, 99–125. Baltimore: Williams & Wilkins.

Fantone, J.C. 1993. Basic concepts in inflammation. In *Sport-induced inflammation: Clinical and basic science concepts,* ed. W.B. Leadbetter, J.A. Buckwalter, and S.L. Gordon, 25–54. Maryland: American Orthopaedic Society of Sports Medicine.

Faulkner, J.A., Brooks, S.V., and Opiteck, J.A. 1993. Injury to skeletal muscle fibers during contractions: Conditions of occurrence and prevention. *Phys. Ther.* 73: 911–21.

Fielding, R.A., Manfredi, T.A., Ding, W., Fiatarone, M.A., Evans, W.J., and Cannon, J.G. 1993. Acute phase response in exercise. III. Neutrophil and IL-1 beta accumulation in skeletal muscle. *Am. J. Physiol.* 265: R166–72.

Fields, H.L. 1987. *Pain.* 35. New York: McGraw-Hill.

Fridén, J., and Lieber, R.L. 1992. Structural and mechanical basis of exercise-induced muscle injury. *Med. Sci. Sports Exerc.* 24: 521–30.

Fridén, J., Seger, J., Sjöström, M., and Ekblom, B. 1983. Adaptive response in human skeletal muscle subjected to prolonged eccentric training. *Int. J. Sports Med.* 4: 177–83.

Fridén, J., Sjöström, M., and Ekblom, B. 1983. Myofibrillar damage following intense eccentric exercise in man. *Int. J. Sports Med.* 4: 170–6.

Fridén, J., Kjorell, U., and Thornell, L.E. 1984. Delayed muscle soreness and cytoskeletal alterations: An immunocytochemical study in man. *Int. J. Sports Med.* 5: 15–18.

Fukami, Y. 1988. The effects of NH_3 and CO_2 on the sensory ending of mammalian muscle spindles: Intracellular pH as a possible mechanism. *Brain Res.* 463: 140–3.

Fukashiro, S., Komi, P.V., Järvinen, M., and Mijashita, M. 1995. Comparison between the directly measured Achilles tendon force and the tendon force calculated from the ankle joint moment during vertical jumps. *Clin. Biomech.* 8: 25–30.

Garland, S.J. 1991. Role of small diameter afferents in reflex inhibition during human muscle fatigue. *J. Physiol. (London)* 435: 547–58.

Garland, S.J., and McComas, A.J. 1988. Reflex inhibition of human soleus muscle during fatigue. *J. Physiol. (London)* 429: 17–27.

Gollhofer, A., Komi, P.V., Fujitsuka, N., and Miyashita, M. 1987a. Fatigue during stretch-shortening cycle exercises. II. Changes in neuromuscular activation patterns of human skeletal muscle. *Int. J. Sports Med.* 8(suppl. 1): 38–47.

Gollhofer, A., Komi, P.V., Miyashita, M., and Aura, O. 1987b. Fatigue during stretch-shortening cycle exercises. I. Changes in mechanical performance of human skeletal muscle. *Int. J. Sports Med.* 8: 71–78.

Gregor, R.J., Roy, R.R., Whiting, W.C., Hodgson, J.A., and Edgerton, V.R. 1988. Force-velocity potentiation in cat soleus muscle during treadmill locomotion. *J. Biomechanics* 218: 721-32.

Hikida, R.S., Staron, R.S., Hagerman, F.C., Sherman, W.M., and Costill, D.L. 1983. Muscle fiber necrosis associated with human marathon runners. *J. Neurol. Sci.* 59: 185–203.

Hill, A.V. 1951. The mechanics of voluntary muscle. *Lancet* 261: 947.

Hoffer, J.A., and Andreassen, S. 1981. Regulation of soleus stiffness in premammillary cats: Intrinsic and reflex components. *J. Neurophysiol.* 45: 267–85.

Hong, S.K., Kniffki, K.D., Mense, S., Schmidt, R.F., and Wendisch, M. 1979. Descending influences on the responses of spinocervical tract neurons to chemical stimulation of fine muscle afferents. *J. Physiol.* 290: 129–40.

Horita, T., Komi, P.V., Nicol, C., and Kyröläinen, H. 1996. Stretch-shortening cycle fatigue: Interactions among joint stiffness, reflex, and muscle mechanical performance in the drop jump. *Eur. J. Appl. Physiol.* 73: 393–403.

Hortobagyi, T., Lambert, N.L., and Kroll, W.P. 1991. Voluntary and reflex responses to fatigue with stretch-shortening cycle exercise. *Can. J. Sports Sci.* 6: 142–50.

Hough, T. 1902. Ergographic studies in muscle soreness. *Am. J. Physiol.* 7: 76–92.

Howell, J.N., Chila, A.G., Ford, G., David, D., and Gates, T. 1985. A electromyographic study of elbow motion during postexercise muscle soreness. *J. Appl. Physiol. (London)* 58: 1713–18.

Howell, J.N., Chleboun, G., and Conatser, R. 1993. Muscle stiffness, strength loss, swelling, and soreness following exercise-induced injury in humans. *J. Physiol. (London)* 464: 183–96.

Huijing, P.A. 1992. Elastic potential of muscle. In *Strength and power in sport*, ed. P.V. Komi, 151–68. Oxford: Blackwell Scientific.

Jessel, T.M., and Kelly, D.D. 1991. Pain and analgesia. In *Principles of neural science*, ed. E.R. Kandel, J.H. Schwartz, and T.M. Jessel, 385–99. New York: Elsevier.

Jones, D.A., Newham, D.J., and Clarkson, P.M. 1987. Skeletal muscle stiffness and pain following eccentric exercise of the elbow flexors. *Pain* 30: 233–42.

Kaneko, M., Komi, P.V., and Aura, O. 1984. Mechanical efficiency of concentric and eccentric exercise performed with medium to fast contraction rates. *Scand. J. Sport Sci.* 6: 15–20.

Kihlstrom, M., Salminen, A., and Vihko, V. 1984. Prednisolone decreases exercise-induced acid hydrolase response in mouse skeletal muscle. *Eur. J. Appl. Physiol.* 53: 53–56.

Kniffki, K.D., Mense, S., and Schmidt, R.F. 1978. Responses of group IV afferent units from skeletal muscle to stretch, contraction, and chemical stimulation. *Exp. Brain Res.* 31: 511–22.

Komi, P.V. 1973. Relationship between muscle tension, EMG, and velocity of contraction under concentric and eccentric work. In *New developments in electromyography and clinical neurophysiology*. Vol. 1. Ed. J.E. Desmedt, 596–606. Basel: Karger.

Komi, P.V. 1984. Physiological and biomechanical correlates of muscle function: Effects of muscle structure and stretch-shortening cycle on force and speed. In *Exercise and sport sciences reviews*. Vol. 12. Ed. R.L. Terjung, 81–121. Lexington, MA: Collamore.

Komi, P.V. 1990. Relevance of in vivo force measurements to human biomechanics. *J. Biomech.* 23(Suppl. 1): 23–34.

Komi, P.V. 1992. Stretch-shortening cycle. In *Strength and power in sport*, ed. P.V. Komi, 169–79. Oxford, UK: Blackwell.

Komi, P.V., and Buskirk, E.R. 1972. Effect of eccentric and concentric conditioning on tension and electrical activity in human muscle. *Ergonomics* 5: 417–31.

Komi, P.V., Fukashiro, S., and Järvinen, M. 1992. Biomechanical loading of Achilles tendon during normal locomotion. *Clin. Sports Med.* 11(3): 521–31.

Komi, P.V., and Gollhofer, A. 1987. Fatigue during stretch-shortening cycle exercise. In *Muscular function in exercise and training*. Vol. 26. Ed. P. Marconnet and P.V. Komi, 119–27. Basel: Karger.

Komi, P.V., Hyvärinen, T., Gollhofer, A., and Mero, A. 1986. Man-shoe-surface interaction: Special problems during marathon running. *Acta Univ. Oulu* 179: 69–72.

Komi, P.V., and Rusko, H. 1974. Quantitative evaluation of mechanical and electrical changes during fatigue loading of eccentric and concentric work. *Scand. J. Rehab. Med.* 3(suppl.): 121–26.

Komi, P.V., Salonen, M., Järvinen, M., and Kokko, O. 1987. In vivo registration of Achilles tendon forces in man. I. Methodological development. *Int. J. Sports Med.* 8(Suppl. 1): 3–8.

Komi, P.V., and Viitasalo, J.T. 1977. Changes in motor unit activity and metabolism in human skeletal muscle during and after repeated eccentric and concentric contractions. *Acta Physiol. Scand.* 100: 246–54.

Kuipers, H. 1994. Exercise-induced muscle damage. *Int. J. Sports Med.* 15: 132–52.

Laine, K., and Louhevaara, V. 1974. [Eccentric strength training and muscle soreness (in Finnish).] Master's thesis, University of Jyväskylä (Finland).

Lee, R.G., and Tatton, W.G. 1978. Long loop reflexes in man: Clinical applications. In *Cerebral motor control in man: Long loop mechanisms*. Vol. 4. Ed. J.E. Desmedt, 320–33. Basel: Karger.

Levin, A., and Wyman, J. 1927. The viscous elastic properties of muscle. *Proc. Roy. Soc. B.* 101: 218–43.

Lieber, R.L., and Fridén, J. 1993. Muscle damage is not a function of muscle force but active muscle strain. *J. Appl. Physiol.* 74: 520–6.

Lombardi, V., and Piazzesi, G. 1990. The contractile response during steady lengthening of stimulated frog muscle fibers. *J. Physiol. (London)* 431: 141–71.

Lowe, D.A., Warren, G.L., Ingalls, C.P., Boorstein, D.B., and Armstrong, R.B. 1995. Muscle function and protein metabolism after initiation of eccentric contraction-induced injury. *J. Appl. Physiol.* 79: 1260–70.

Macefield, G., Hagbarth, K.E., Gorman, R., Gandevia, S.C., and Burke, D. 1991. Decline in spindle support to α-motoneurons during sustained voluntary contractions. *J. Physiol. (London)* 440: 497–512.

MacIntyre, D.L., Reid, W.D., and McKenzie, D.C. 1995. Delayed muscle soreness: The inflammatory response to muscle injury and its clinical implications. *Sports Med.* 20: 24–40.

Mair, J., Mayr, M., Müller, E., Haid, C., Artner-Dworzak, E., Calzolari, C., Larue, C., and Puschendorf, B. 1995. Rapid adaptation to eccentric exercise-induced muscle damage. *Int. J. Sports Med.* 16: 352–56.

Malmgren, K., and Pierrot-Deseilligny, E. 1987. Inhibition of neurons transmitting nonmonosynaptic Ia excitation to human wrist flexor motoneurons. *J. Physiol. (London)* 405: 765–83.

McBride, T.A., Gorin, F.A., and Carlsen, R.C. 1995. Prolonged recovery and reduced adaptation in aged rat muscle following eccentric exercise. *Mech. Age. Rev.* 83: 185–200.

Mense, S. 1977. Nervous outflow from skeletal muscle following chemical noxious stimulation. *J. Neurophysiol.* 267: 75–88.

Mense, S., and Meyer, H. 1988. Bradykinin-induced modulation of the response behavior of different types of feline group III and IV muscle receptors. *J. Physiol. (London)* 398: 49–63.

Miles, M.P., and Clarkson, P.M. 1994. Exercise-induced muscle pain, soreness, and cramps. *J. Sports Med. Phys. Fitness* 34: 203–16.

Nardone, A., Romano, C., and Schieppati, M. 1989. Selective recruitment of high threshold human motor units during voluntary isotonic lengthening of active muscles. *J. Physiol. (London)* 409: 451–71.

Newham, D.J., McPhail, G., Mills, K.R., and Edwards, R.H.T. 1983a. Ultrastructural changes after concentric and eccentric contractions of human muscle. *J. Neurol. Sci.* 61: 109–22.

Newham, D.J., Mills, K.R., Quigley, B.M., and Edwards, R.H.T. 1983b. Pain and fatigue after concentric and eccentric muscle contractions. *Clin. Sci.* 64: 55–62.

Newham, D.J., Edwards, R.H.T., and Mills, K.R. 1994. Skeletal muscle pain. In *Textbook of pain,* ed. P.D. Wall and R. Melzack, 423–40. Edinburgh: Churchill Livingstone.

Nicol, C., and Komi, P.V. 1996. Neuromuscular fatigue in stretch-shortening cycle exercises. In *Human muscular function during dynamic exercise.* Vol. 41. Ed. P. Marconnet, B. Saltin, P.V. Komi, and J. Poortmans, 134–47. Basel: Karger.

Nicol, C., and Komi, P.V. 1998. Significance of passively induced stretch reflexes on Achilles tendon force enhancement. *Muscle & Nerve* 21: 1546–48.

Nicol, C., Komi, P.V., and Avela, J. 1996. Stretch-shortening cycle fatigue reduces stretch reflex response. Abstract book of the 1996 International Pre-Olympic Scientific Congress. July 10–14, Dallas, Texas. p. 108.

Nicol, C., Komi, P.V., Horita, T., Kyröläinen, H., and Takala, T.E.S. 1996a. Reduced stretch reflex sensitivity after exhaustive stretch-shortening cycle exercise. *Eur. J. Appl. Physiol.* 72: 401–9.

Nicol, C., Komi, P.V., and Marconnet, P. 1991a. Fatigue effects of marathon running on neuromuscular performance. I. Changes in muscle force and stiffness characteristics. *Scand. J. Med. Sci. Sports* 1: 10–17.

Nicol, C., Komi, P.V., and Marconnet, P. 1991b. Fatigue effects of marathon running on neuromuscular performance. II. Changes in force, integrated electromyographic activity and endurance capacity. *Scand. J. Med. Sci. Sports* 1: 18–24.

Nicol, C., Komi, P.V., and Marconnet, P. 1991c. Effects of a marathon fatigue on running kinematics and economy. *Scand. J. Med. Sci. Sports* 1: 195–204.

Noakes, T.D. 1987. Effect of exercise on serum enzyme activities in humans. *Sports Med.* 4: 245–47.

Norman, R.W., and Komi, P.V. 1979. Electromechanical delay in skeletal muscle under normal movement condition. *Acta Physiol. Scand.* 106: 241–48.

Nosaka, K., and Clarkson, P.M. 1995. Muscle damage following repeated bouts of high force eccentric exercise. *Med. Sci. Sports Exerc.* 27: 1263–69.

Oksanen, P., Kyröläinen, H., Komi, P.V., and Aura, O. 1990. Estimation of errors in mechanical efficiency. *Eur. J. Appl. Physiol.* 61: 473–78.

Patel, T., and Lieber, R.L. 1997. Force transmission in skeletal muscle: From actomyosin to external tendons. In *Exercise and sport sciences reviews.* Vol. 25. Ed. John O. Holloszy. American College of Sports Medicine Series. Baltimore: Williams and Wilkins.

Pizza, F.X., Mitchell, J.B., Davis, B.H., Starling, R.D., Holtz, R.W., and Bigelow, N. 1995. Exercise-induced muscle damage: Leucocyte and lymphocyte subsets. *Med. Sci. Sports Exerc.* 27: 363–70.

Pullinen, T., Leynaert, M., and Komi, P.V. 1997. Neuromuscular function after marathon. Abstract book of the XVI ISB Congress, August 24–27, 1997, Tokyo.

Rotto, D.M., and Kaufman, M.P. 1988. Effect of metabolic products of muscular contraction on discharge of group III and IV afferents. *J. Appl. Physiol.* 64: 2306–13.

Saxton, J.M., and Donnelly, A.E. 1995. Light concentric exercise during recovery from exercise-induced muscle damage. *Int. J. Sports Med.* 16: 347–51.

Schwane, J.A., Williams, J.S., and Sloan, J.H. 1987. Effects of training on delayed muscle soreness and creatine kinase activity after running. *Med. Sci. Sports Exerc.* 19: 584–90.

Sherman, W.M., Armstrong, L.E., Murray, T.M., Hagerman, F.C., Costill, D.L., Staron, R.C., and Ivy, J.L. 1984. Effect of a 42.2 km footrace and subsequent rest or exercise on muscular strength and work capacity. *J. Appl. Physiol.* 57: 1668–73.

Singh, M., and Karpovich, P.V. 1966. Isotonic and isometric forces of forearm flexors and extensors. *J. Appl. Physiol.* 21: 1435–37.

Sugi, H., and T. Tsuchiya. 1988. Stiffness changes during enhancement and deficit of isometric force by slow length changes in frog skeletal muscle fibers. *J. Physiol. (London)* 407: 215–29.

Talag, T.S. 1973. Residual muscular soreness as influenced by concentric, eccentric, and static contractions. *Res. Quart.* 44: 458–68.

Viitasalo, J.T., Komi, P.V., Jacobs, I., and Karlsson, J. 1982. Effects of a prolonged cross-country skiing on neuromuscular performance. In *Exercise and sport biology.* Vol. 12. Ed. P.V. Komi, 191–98. Champaign, IL: Human Kinetics.

Warren, G.L., Hayes, D.A., Lowe, D.A., and Armstrong, R.B. 1993. Mechanical factors in the initiation of eccentric contraction-induced injury in rat soleus muscle. *J. Physiol. (London)* 464: 457–75.

Waterman-Storer, C.M. 1991. The cytoskeleton of skeletal muscle: Is it affected by exercise? A brief review. *Med. Sci. Sports Exerc.* 23: 1240–9.

Woods, J.J., Furbush, F., and Bigland-Ritchie, B. 1987. Evidence for fatigue-induced reflex inhibition of motoneuron firing rates. *J. Neurol.* 58: 125–37.

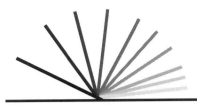

Chapter 21

Muscle Fatigue Monitored by Force, Surface Mechanomyogram, and EMG

C. Orizio

During muscle contraction the mechanical and electrical activities of the recruited fibers can be monitored by different biological signals. Commonly, the mechanical aspect of muscle activity is studied using the output of force or movement transducers, the latter tracking the kinematics of limbs or body segments. Electrodes detecting the cumulative effect of the action potentials of the recruited muscle fibers monitor the electrical aspect of muscle activity. The resulting signal is the electromyogram (EMG). Recently it has been possible to follow the mechanical activity of the motor units (MUs) by means of transducers that detect the muscle surface movement or vibration, which is apparently due to the pressure waves associated with the dimensional changes of the active fibers within the muscle. By analogy with the EMG electrodes these transducers (microphones, accelerometers, piezoelectric contact sensors) summate at their output the muscle surface oscillations due to contraction of the recruited MU. To emphasize the mechanical nature of this signal, independent of detector used at the muscle surface, we will refer to this signal as the surface mechanomyogram (MMG) when new data is presented. In order to

respect the past authors' choices, in the sections regarding the MMG the term indicated in each paper (i.e., muscle sound, acoustic myogram, phonomyogram, vibromyogram, etc.) is reported.

The aim of this chapter is to present and tentatively discuss the reliability of some parameters of the MMG and EMG, resulting from time and frequency domain signal processing, as tools to follow the development of localized muscular fatigue of in situ and in vivo muscle (see chapter 19). The force signal changes are considered only in the stimulation-induced muscle fatigue as a reference signal for the interpretation of MMG changes.

Fatigue has been defined by Enoka and Stuart (1985) as the "progressive increase in the effort required to exert a desired force and the eventual progressive inability to maintain this force in sustained or repeated contractions" (p. 2281). In figure 21.1 some aspects of this definition are presented. In this example, force is maintained while clear changes in the EMG and MMG are present. These changes probably indicate adaptations in the activation pattern of the MU pool to counteract the reduced mechanical performance of active muscle fibers. In

Acknowledgments: The author is grateful to Prof. R. Merletti for the extensive discussion we had about the physiological bases of the changes in the muscular biological signals at fatigue and to Prof. A. Veicsteinas who stimulated and contributed to the research in the MMG and muscular fatigue fields at the Human Physiology Chair of Brescia University. This work was supported by MURST ex-40% grants (1998).

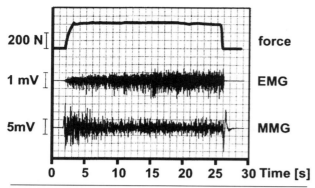

Figure 21.1 Force, EMG, and surface mechanomyogram (MMG) in biceps brachii. Sustained isometric effort at 80% of the maximal voluntary contraction. The different influence of fatigue on changes in motor unit activation pattern and in the mechanical and electrical properties of the fibers on force, EMG, and MMG is evident. Original data from Human Physiology Laboratory, University of Brescia, Italy.

other words, the changes may reflect the increase in the effort required to maintain a constant mechanical output. The different behavior of the signals in figure 21.1 shows how the problem of studying fatigue is complex and also how difficult it is to assess what aspects of the phenomenon are monitored by force, MMG, or EMG changes throughout the exercise.

In an attempt to provide a clear picture of how localized muscle fatigue can be quantified by means of force, MMG, and EMG analysis, this chapter is divided into three sections: fatigue (local muscle fiber changes and muscle motor control alterations), changes in muscle mechanical and electrical signals, and applications.

In order to deal with the most reproducible experimental design and with biological signals as constantly as possible, the data presented in this chapter are preferentially related to isometric contractions.

Fatigue: Local Muscle Fiber Changes and Muscle Motor Control Alterations

Both the sarcolemmal action potential and the tension generation process are influenced by fatigue. They are separately presented.

Influence of Fatigue on Fiber Mechanics and Electrical Activity

The effect of fatigue on muscle mechanics is well represented by changes of the twitch parameters,

(peak twitch force Pt, the peak rate of tension development dP/dt, the contraction time Ct, and the one-half relaxation time 1/2 Rt) as well as by the force-frequency relationship (for detailed description see chapter 19 by MacIntosh and Allen; here only a brief summary of the fatigue effect on fiber mechanics will be provided). Twitch alteration depends "on specific cellular and molecular events associated with the cross-bridge cycle" (Fitts 1994, pp. 51–52). Exhaustive reviews on this topic can be found in the literature (Allen, Lannergren, and Westerblad 1995; Fitts 1994). During sustained activity the muscle fiber accumulates H^+ (due to lactic acid production), Pi (due to phosphocreatine breakdown), and ADP (from ATP breakdown). All these metabolites affect the performance of the contractile machinery (Allen, Lannergren, and Westerblad 1995). At the same time, changes in the ionic gradients across the fiber membrane will affect the shape and conduction velocity of the sarcolemmal action potential.

Reduction of the Force Peak

High intracellular H^+ concentration may determine (1) a reduction of the force generated per cross bridge, (2) an alteration of the myofibrillar sensitivity to Ca^{2+} with a production of less force for a given intracellular [Ca^{2+}], and (3) a reduction of the Ca^{2+} release from the sarcoplasmic reticulum (SR) (Fitts 1994). The accumulation of Pi contributes to the force reduction by inhibition of the transition of the attached cross bridges from the low (before ATP breakdown) to the high force state (after ATP breakdown) (Fitts 1994). Moreover, it seems that Pi acts synergistically with the H^+ to reduce the force per cross bridge and the myofibrillar Ca^{2+} sensitivity (Allen, Lannergren, and Westerblad 1995).

Reduction in Rate of Tension Development

The parameter reflecting the rate of the force generation (dP/dt) reaches a maximum value during the on phase of the twitch. The rate of transition of the cross bridges from weak to strong binding state is reduced by the [H^+], by suboptimal levels of Ca^{2+} released from SR (Fitts 1994), and by an increase of [Pi] (Allen, Lannergren, and Westerblad 1995).

Elongation of the Contraction and Relaxation Processes

In fatigue, the Ca^{2+} transient is prolonged. This is due to the redistribution of Ca^{2+} "from the SR release site to the Ca^{2+} binding protein parvalbumin and the

SR pump" (p. 61), which results in a reduction of the driving force for Ca^{2+} release, and hence a reduced rate of Ca^{2+} release and a greater load placed on the re-uptake processes (Fitts 1994). Moreover, the SR Ca^{2+} pump rate is negatively influenced by the intracellular H^+ increase (Byrd et al. 1989). The consequences of the elongation of the Ca^{2+} transient on the twitch force parameters are reflected in the elongation of CT and 1/2 RT; they can reach 300% and 800%, respectively, of their prefatigue value (Thompson et al. 1992). Also, an impairment of the cross-bridge detachment mechanism (related to the H^+ effect on the myofibrillar ATPase) seems to contribute to slowing the relaxation phase (Allen, Lannergren, and Westerblad 1995).

Reduction in Shortening Velocity

In fatigue, the reduction of the maximum shortening velocity at 0 load (Vmax) is due to the intracellular $[H^+]$ and $[ADP]$ increases, inhibiting the cross-bridge transition from weakly bound (low force) to strongly bound (high force) state and impeding the detachment of cross bridges, respectively (Fitts 1994).

Twitch Potentiation

In the first part of a sustained activity of a motor unit an increase in the mechanical response (single twitch) to the motor command is present. This is true for low frequencies of stimulation (staircase phenomenon) as well as for tetanic frequencies of activation (Krarup 1981; Takamori, Gutman, and Shane 1971). The resulting twitch may also be longer in duration than normal because of an elongation of the "active state" due to a slowed re-uptake of the Ca^{2+} by the sarcoplasmic reticulum (Takamori, Gutman, and Shane 1971).

Fiber Action Potential Changes

One of the outcomes of the fatiguing process is an alteration of the ionic distribution across the fiber membrane. In particular, there is an increase in the intracellular Na^+ and in the extracellular K^+ with a decrease in the intracellular K^+. These changes seem to be related to activation of Ca^{2+} dependent K^+ channels or ATP-sensitive K^+ dependent channels (these are affected by the synergistic action of the ATP reduction and of the H^+ increase; Stephenson et al. 1995), to the increased K^+ efflux, and to a relative inability of the Na^+/K^+ pump to keep the ionic gradients at correct levels for a normal sarcolemmal resting potential. As a consequence the resting potential is less negative and in turn the sarcolemmal action potential is influenced, showing a reduction in amplitude. This may be reflected in a reduction of the amplitude of the MU action potential (Moritani, Muro, and Nagata 1986) and an elongation with a corresponding decrease of the propagation velocity (Fitts 1994; Fuglevand 1995). It seems unlikely that the sarcolemmal action potential changes affect the reduction of the muscle fiber mechanical performance, given that large action potential changes do not impair the muscle fiber activation (Sandow 1952). Considering different muscles and fiber diameters, the propagation velocity of the action potential along the muscle fiber ranges between 1.5 and 6.5 m/s (Trontelj and Stalberg 1995). Within the first minute of activity at fixed stimulation frequency the propagation velocity quickly decreases, reaching half of the initial value at 3 min (Trontelj and Stalberg 1995). Changes in the muscle fiber conduction velocity have also been linked to the extracellular H^+ changes in the muscle (Brody et al. 1991). Lindstrom, Kadefors, and Petersen (1977) and Merletti, Knaflitz, and De Luca (1990) have provided techniques for noninvasive estimation of the average muscle fiber conduction velocity. This last parameter is a weighted combination of the conduction velocities of multiple active muscle fibers.

Concluding Remarks

The sustained activity of the muscle fibers determines changes in both the intracellular and extracellular environments. The related alterations of the muscle mechanics and sarcolemmal action potential listed above should be tracked by the changes of the twitch force parameters and of the force/frequency relationship, as well as of the parameters of the surface mechanomyogram and electromyogram during muscle stimulation.

Influence of Fatigue on Muscle Motor Control

Both EMG and MMG are biological signals in which the electrical and mechanical activities of individual motor units are summated (Basmajian and De Luca 1985; Orizio 1993; Orizio, Liberati et al. 1996). From signal analysis theory we know that the properties of a compound signal are influenced by the shape and the frequency of the trains of the elementary events, as well as by the number of the interfering trains and the degree of their synchronization. In the case of the motor units the number of trains, their individual frequency, and the degree of synchronization are specific features of the motor units' activation pattern. In fact, the number of trains indicates the degree of MU recruitment (REC) while the internal frequency indicates the MU firing rate (FR).

Changes in Motor Unit Firing Rate

The prolongation of relaxation time in fatigue provides a greater degree of fusion for a given FR. It has been demonstrated that the reduction of the stimulation frequency (from 60 Hz to 30 Hz) during electrically elicited contraction is able to sustain a higher force output over time than delivering 60 Hz continuously (Binder-Macleod and Guerin 1990). Jones, Bigland-Ritchie, and Edwards (1979) demonstrated that the force reduction during a sustained MVC could be mimicked with a reduction of the stimulation rate. This tuned change of both the fiber relaxation time (increased) and force (reduced) with the motoneuron discharge rate (decreased) is termed "muscle wisdom" (Marsden, Meadows, and Merton 1983). "It optimises the force and ensures an economical activation of fatiguing muscle by the central nervous system" (Enoka and Stuart 1992, p. 1638). The theory of muscle wisdom seems to be well suited to explain the changes in the muscle motor control during MVC. Indeed it has been well known since the eighties that during MVC the MU FR decreases (Bigland-Ritchie et al. 1983a, 1983b; Grimby 1986; Kukulka, Russell, and Moore 1986; Moritani, Muro, and Nagata 1986). This allows the MU to generate the maximal force with discharge rates just sufficient to meet the changes in their contractile properties (Bellemare et al. 1983). FR reduction in fatigue seems to be due to a sensory feedback able to monitor the elongation of the twitch force and to inhibit the motoneuron discharge (Bigland-Ritchie and Woods 1984; Bigland-Ritchie et al. 1986b). The information about the fatigued muscle may reach the spinal motoneurons via group Ia fibers from muscle spindles (Bongiovanni and Hagbarth 1990; Windhorst and Kokkoroyiannis 1991), inducing a "disfacilitation" of the homonymous motoneurons due to the reduction of the afferent spindle discharge during sustained contraction, or via group III/IV small afferents from mechanoreceptors (Garland 1991; Hayward, Wesselman, and Rymer 1991). This latter group of afferents seems to be sensitive to fiber mechanical changes (more evident for group III) and chemical changes (more evident for group IV) such as alteration in [K$^+$], [H$^+$], [lactate], [bradikinin], and [arachidonic acid]. The large afferents from spindles and the group III small afferents seem to be transiently active at the beginning of effort while the group IV fibers respond, after a latent period, with no decrement over time (Garland and Kaufmann 1995). The intrinsic properties of the motoneuron have also to be considered as a FR modulating factor. It has been clearly demonstrated that during a constant current

stimulation of the motoneuron its discharge rate decreases with time (Kernell and Monster 1982a, 1982b). On this basis Sawczuk, Powers, and Binder (1995) suggested that the so-called "late adaptation" was a contributing factor in reduction of MU firing rate during sustained effort. Different conclusions about the factors inducing the FR reduction during sustained effort came from the study of De Luca, Foley, and Zeynep (1996). These AA authors investigated the first 8 to 15 s of isometric contraction, representing 30 to 80% MVC in tibialis anterior and in first dorsal interosseus, and reported a slower motor unit FR decrease during 80% MVC than during 30% MVC. Given that at high contraction levels "the excitation of peripheral receptors and the accumulation of metabolites would be greater" (p. 1514) and that "late adaptation" is expected to be more evident at 80% than 30% MVC, they concluded that the peripheral sensory feedback and the motoneurons' intrinsic properties are not playing a major role in MU FR reduction in the first 8 to 15 s of effort (De Luca, Foley, and Zeynep 1996).

The complexity of the segmental motor control circuitry regulating the MU activity, on which the negative force feedback during fatigue should converge, has been briefly reviewed by Windhorst and Boorman (1995). The necessity for further studies on the possible inputs and mechanisms of functioning of the MU control system is underlined by the discrepancies concerned with behavior of the FR during sustained effort reported in the literature. Researchers have described the motor unit firing rate during submaximal contraction as increasing during sustained effort (Maton 1981; Maton and Gamet 1989) and during intermittent isometric effort (Bigland-Ritchie, Furbush, and Woods 1986; Dorfman et al. 1990), or decreasing during static contraction (Bigland-Ritchie, Cafarelli, and Vollestad 1986; Person and Kudina 1972). Rather stable discharge rate has also been reported (Bigland-Ritchie, Cafarelli, and Vollestad 1986; Maton and Gamet 1989). The previously cited papers deal primarily with the "motor unit population average rate." In reality when a single motor unit action potential train is followed in time it seems that a more general agreement exists about the FR dynamics. Recently Garland and coworkers (1994) were able to follow the behavior of a single MU in the biceps brachii during sustained submaximal isometric contraction. In most MUs, (32 out of 45), the FR decreased, while in 7 it was constant and in 6 FR was increasing. The increase or the decrease in the FR was related with a low (< 10 Hz) or high (> 20 Hz) initial discharge rate, respectively. The dependence of the FR dynamics on the characteristics of the active MU has been demonstrated by De Luca, Foley, and Zeynep (1996) in tibialis anterior and in first dorsal interosseus. The recorded motor units show clear FR decreases during constant isometric contraction at 30%, 50%, or 80% MVC (the target force was reached with a rate of 10% MVC/s). The slower FR decrease has been reported for the MU with higher threshold and lower average firing rate. Considering the possible MU twitch potentiation in this first part of contraction De Luca, Foley, and Zeynep (1996) suggested that the firing rate reduction during the early stage of contraction may be the result of a fine tuning, operated by the central nervous system, of the motoneuron firing pattern to the motor unit mechanical properties. The dependence of the FR time courses on the MU type was already indicated in biceps brachii by Gydikov and Kosarov (1974). They showed a transient FR increase, followed by a clear decrease, for a MU action potential train recorded from a fast twitch motor unit and a rather stable discharge from slow twitch motor units.

At the present time there are still a few questions unanswered. How may the individual firing patterns of the MU combine to produce the global MU FR? What is the behavior of the global MU firing in the first 15 to 20 s of a steplike (not trapezoid) isometric contraction when the effort level determines the recruitment of all the MUs already from the beginning (for example, at 80% MVC in biceps brachii)? Is the twitch potentiation process (De Luca, Foley, and Zeynep 1996) able to explain the maintained constant force already in the first few seconds? Do we also have to hypothesize a transient increase in the MU firing rate in this time interval? Indeed this last FR modulation mechanism has already been indicated to be the sole tool to increase force in the 80% to 100% MVC range in unfatigued biceps brachii. It could work also for keeping constant the muscle force output in the first few seconds of near maximal efforts.

A more global change in the firing rate pattern occurs with the synchronization and grouping of the MU activity (Broman, Bilotto, and De Luca 1985a; Hermens et al. 1992; Krogh-Lund and Jorgensen 1993; Maton and Gamet 1989). The presence of a "common drive" (De Luca et al. 1982) to a motoneuronal pool determining "unison behaviour of the firing rate of motor units" (Basmajian and De Luca 1985, p. 151) has been recently confirmed by Garland and coworkers (1994) during biceps brachii sustained contraction. The tendency toward grouping of MU activity was early described by Lippold

(1981), and its evidence in the EMG was linked to the existence of a stretch reflex from the active muscle (Gamet and Maton 1989; Matthews and Muir 1980) determining a rhythmic discharge of the MU (Elble and Randall 1976). The tendency to synchronization and grouping increases with fatigue (Gamet and Maton 1989; Hermens et al. 1992; Krogh-Lund and Jorgensen 1993; Lippold, Redfearn, and Vuco 1957).

Changes in Motor Unit Recruitment

During sustained maximal voluntary contraction the motor unit derecruitment, in particular for fast, most fatigable motor units, is a well-known issue (Grimby 1986). During submaximal contraction a recruitment of new motor units, to compensate for already active MU mechanical failure, has been described (Bigland-Ritchie, Cafarelli, and Vollestad 1986; Dorfman, Howard, and McGill 1990; Garland et al. 1994; Maton and Gamet 1989; Moritani, Muro, and Nagata 1986). It seems that one of the effects of fatigue is to reduce the recruitment threshold for the MU (Garland et al. 1994; Maton and Gamet 1989). This may trigger the activation of a given MU at an effort level at which it is usually silent. Recently Fallentin, Jorgensen, and Simonsen (1993) were not able to detect recruitment of new MU during 40% MVC but observed it at 10% MVC sustained effort. In this last paper the sensory feedback, previously described for MU FR decrease, has been invoked as the determinant of the recruitment block.

Concluding Remarks

All the changes in the motor unit recruitment level and firing pattern aimed to tailor the motor commands of the motoneuronal pool to the actual mechanical properties of the peripheral muscle fibers can be considered as the output of an integrative activity of segmental neurons and interneurons. This task is accomplished by spinal neural networks (including I_a and I_b inhibitory interneurons and Renshaw cells) on which the descending motor commands and the sensory afferent information converge (Windhorst and Boorman 1995). In summary, the number and the firing rate of MU throughout sustained contraction are defined by the interaction at the motoneuronal level of the following:

• The reduction of the recruitment threshold of MU. A hypothesis (Kernell 1995) suggests that the background activity of the above-described circuitry may be changed with fatigue, leading to an increase of the recruited MU. This mechanism may act together with the one controlling the firing rate in order to use more MUs at low FR during fatiguing contraction.

- The sensory feedback of the "muscle wisdom" providing a MU discharge rate tailored to the prolonged MU twitch. This optimization of the FR and MU contractile properties exploits the shift of the force frequency relationship to the left, that is, more tension is developed at a given FR because of a more complete summation of the prolonged mechanical response of the MU fibers (Kernell 1995).

- The decline in central drive. A correlation between the discharge of area 4 of the motor cortex cells and EMG was provided (Maton 1991).

It has to be emphasized here that fatigue manifestations at the muscular level are task dependent, and the interpretation of the results of a fatiguing test always has to take into account the exercising conditions (Enoka and Stuart 1992).

The alterations in the recruitment or firing rate of the MU described above should be reflected by changes in the properties of the MMG and EMG recorded during voluntary contraction.

Changes in Muscle Mechanical and Electrical Signals

In this section the changes of the force, surface mechanomyogram, and electromyogram detected during electrical stimulation and voluntary contraction are discussed. The signal detection techniques are considered at the beginning of this section because this will contribute to understanding the changes of the signals throughout contraction.

Force and MMG Recordings

The measurement of the force exerted at the application points is usually done by strain gauges or load cells oriented in the same direction as the force vector. A recent analysis of the measurement techniques for biomechanical parameters has been provided by Nigg (1994).

It is possible to record the MMG with different detectors.

Piezoelectric Contact Sensors

One of the most popular transducers able to detect the muscle surface oscillations due to the motor unit activities is the HP 21050-A contact sensor. Its mass (about 40 g) and its dimensions (diameter 3 cm,

height 2.5 cm) limit its application to large muscles (for example, biceps brachii or quadriceps). As reported by Orizio and coworkers (1992) the force applied for mechanical coupling with the muscle surface should be lower than 200 g. This ensures a flat frequency response of the transducer within the range of interest for MMG (5–150 Hz). This contact sensor can be adapted in order to allow the insertion of the surface electrodes for EMG detection (Orizio 1993). The means of securing the contact sensor to the muscle should avoid relative movements of the detector in respect to the skin and the generation of too much pressure on the piezo element during the bulging of the active muscle. In the author's experience, the use of double-sided sticky tape to secure large dimension transducers to the muscle should be avoided. In fact, this solution allows the detector to oscillate during muscle activity, generating spurious signals. The problem of the changes in piezoelectric crystal response as a function of the contact pressure has been analysed by Smith and Stokes (1993). These authors used three contact pressures and concluded that no differences existed in the acoustic myogram (AMG) amplitude, at the same level of contraction, when using 180 or 790 Pa. A clear increase in AMG amplitude was observed for a contact pressure of 1200 Pa (corresponding to an exerted force of 49.6 g). They concluded that repeated measure on the same subject requires monitoring of the contact pressure in order to be comparable.

Microphones

Several papers report the use of electret condenser microphones (Dalton, Comerford, and Stokes 1992; Dalton and Stokes 1991, 1993; Goldenberg et al. 1991; Petitjean and Bellemare 1994; Petitjean and Maton 1995; Stokes and Dalton 1991). Usually these kinds of devices need a case, into which electrodes for surface EMG can be inserted (Petitjean and Maton 1995), in order to be secured to the muscle by means of double-sided sticky tape or by rubber bands or Velcro straps. In some cases an ultrasound coupling gel has been used to get a better transmission of the muscle surface vibrations to the microphone (Stokes and Dalton 1991), or the microphone has been recessed a few millimeters in a foam case coupled to the muscle skin with surgical cement (Goldenberg et al. 1991). A comparison between the signal obtained from an electret condenser microphone and the HP 21050-A has been done by Bolton and coworkers (1989). It was observed that the "muscle sound" recorded by the two transducers from thenar muscles during electrical stimulation of the median nerve at the wrist were very similar. With the electret condenser microphone, the pressure exerted for the mechanical coupling of the device with the muscle influences the signal amplitude (Bolton et al. 1989).

Accelerometers

The use of this kind of transducer in muscle surface oscillation detection dates back to the seventies (Lammert, Jorgensen, and Einer-Jensen 1976). The mass of these first devices was about 60 g. In contrast, the accelerometers used in the nineties have a mass lower than two g (Barry, Hill, and Dukjin 1992; Herzog et al. 1994a; Orizio, Diemont et al. 1996; Orizio, Esposito et al. 1997; Orizio, Liberati et al. 1996; Zhang et al. 1992). The use of accelerometers should be encouraged because their light mass does not influence the dynamics of the muscle surface oscillations during either voluntary or stimulated contraction. Moreover, the method of fixation of the accelerometer to the skin can be a simple double-sided tape with no additional contact pressure applied to the muscle. Finally, the use of this transducer enables the comparison of the measurements between the different studies because the acceleration is expressed as $m \cdot s^{-2}$ instead of transducer dependent units (mV) (Barry, Hill, and Dukjin 1992).

Optical Methods

In order to monitor the absolute muscle surface movement, our laboratory is introducing a new detection technique. It is based on a laser distance sensor. This technique estimates the actual distance of an object from the head of the laser beam source based on the fact that the reflected laser beam reaches a specific position in the sensor, which depends on the distance of the reflecting surface. Each position in the sensor provides a specific output voltage. Given that the skin surface over the muscle is not a shiny surface the dependence of the measurement accuracy on the orthogonality between the laser head and the reflecting surface is trivial. The advantages of this technique are the following: the muscle is free to move without applied loads (the measuring system has no inertia), the output voltage is convertible to absolute displacement expressed as mm (as a consequence the signal is directly comparable between different studies), the bandwidth is very large (0–10 kHz); the investigated area is highly defined (spot diameter < 1 mm), the accuracy due to the instrument resolution is very good (6 μm). The first recordings done with this method in cat gastrocnemius (Orizio, Baratta et al. 1997) and in humans (see figures 21.4 and 21.5, pages 418 and 419) strongly suggest that laser-detected MMG can

be an absolutely reliable tool to investigate muscle mechanics.

Electromyogram Recordings

The recording of the muscle intrinsic electrical activity can be carried out by different types of electrodes.

Indwelling Electrodes (Needle and Fine Wire)

The indwelling recordings present the advantage of small pick-up area with the possibility of investigating the electrical activity of individual motor unit activity. During the investigation the needle can be easily and precisely repositioned to gain the most meaningful EMG. Moreover the EMG of deep muscles can be investigated with little interference from the electrical activity of adjacent muscles (Herzog, Guimaraes, Zhang 1994). The needle electrode is the most commonly used electrode. All the modern EMG needles are an evolution of the Adrian and Bronk (1929) electrode. They can be monopolar (with only one wire within the cannula) or multipolar with several wires in the cannula (all of them bared at the tips) in order to increase the number of the recording sites. The relative geometrical relationship among the emerging bared tips allows different application of the needle electrodes such as single fiber recordings, single or multiple motor unit recordings, and evaluation of the size of the territory of a motor unit (Basmajan and De Luca 1985). One of the major drawbacks of these electrodes is the pain that can be experienced not only at the insertion but also during the muscle contraction (in particular at high intensity of effort). This is a limiting factor to their use in fatigue studies, as well as in sport medicine and in rehabilitation. The fine wire electrodes represent the indwelling electrode types suited for kinesiological investigation. In fact the presence of 25 to 75 μm diameter wires in the muscle provides only limited discomfort to the subjects during muscle contraction. The positioning of the wires is usually done by means of needle cannulas of 25 or 27 gauge. The cannulas are then withdrawn and the wires are left in place. The recording sites are at the last 1 to 2 mm, bare wire ends. These electrodes detect the activities of a little portion of muscle, and the motor unit action potentials are easily monitored. During activity the pumping action of the muscle may change the position of the wires. It is useful to perform some contraction/relaxation cycles before any measurements are taken. This increases the reliability of measurement (Basmajian and De Luca 1985).

Surface Electrodes

The most important feature, from the application point of view, of the surface EMG is that this is a noninvasive technique. The subject has no discomfort at all: no needle puncture and no pain, because of a foreign body present in the muscle mass during contraction. Usually the electrodes are silver disks (diameter < 1 cm) or bars with a superficial layer of silver chloride. This treatment of the surface electrode is aimed to reduce the degree of the skin-electrode noise and interface polarization, generating an undesired DC voltage. The surface EMG is able to provide a general picture of the electrical activity of a superficial muscle. As a consequence it is useful in monitoring the degree of muscle involvement during specific motor tasks, as a control signal for external prosthesis, in biofeedback techniques, and, finally, as a tool to monitor the development of localized muscle fatigue. One drawback of this technique is the cross talk between the EMGs of the investigated muscle and of adjacent ones.

An important aspect of the surface EMG detection is related to the skin preparation. It is crucial to reduce and match the contact resistance between electrodes and skin. For this purpose the skin can be shaved and abraded with sandpaper or special paste. An acceptable skin resistance could be of the order of a few kilo-ohms (and in any case less than 1/100 of the amplifier input impedance). Conductive gel may be used to further improve the electrical transmission between the skin and the electrodes. The fixation of the electrodes may be done by elastic bands or by conductive or adhesive paste.

General Considerations About EMG Detection

The biophysical basis of EMG generation and detection can be found in textbooks with considerable detail (Basmajan and De Luca 1985) or in more concise chapters of books not exclusively devoted to EMG (Herzog, Guimaraes, Zhang 1994). Here only a brief summary of recording techniques is provided. The muscle electrical activity can be picked up by a monopolar or bipolar detection system. In the first case only one "exploring" electrode, relative to the ground, is placed into the muscle or on the muscle surface. This technique is suitable for single fiber studies. The bipolar detection provides an important improvement in the signal quality. Two "exploring" electrodes are used. All of the electrical events from the recording territory that influences both electrodes are cancelled. In fact the electrodes are connected to the inputs of a "differential" ampli-

fier that subtracts their electrical potentials, relative to the reference ground electrode, and amplifies the result. In conclusion only the difference in electrical activity between the two detection points on the muscle is evident since the common mode noise and disturbance are cancelled. An array of four silver bars (1 mm long, 1 mm diameter, and 1 cm spaced) has been developed (Broman, Bilotto, and De Luca 1985b). This device has been used to measure noninvasively the conduction velocity (estimating the delay of the depolarization wave between the two triplets of electrodes using the "double differential technique") and has been demonstrated to reduce the cross talk from adjacent muscles (De Luca and Merletti 1988). In this kind of array the electrodes have a fixed position because they are included in a rigid plastic case (3.6 cm × 1.5 cm, thickness 4 mm). This is very important because it allows comparison of the EMG data from different persons and facilitates the repositioning of the probe for repeated measures in the same subject. Indeed one of the problems affecting the EMG measures is the influence of the shape, the dimension, and the relative distance of the two electrodes of the most used differential configuration. The use of a probe with constant geometrical parameters is strongly suggested. In order to make it possible to "weight" the difference of the EMG in the different studies, it is always important to report the geometry of the detection system and its location with respect to innervation and termination zones.

From a general point of view the surface EMG electrodes (similar to the indwelling electrodes) monitor the presence of electrical sources within the muscle. When a fiber is active each of its depolarized zones can be considered as a current tripole generated at the neuromuscular junction and extinguished at the tendon insertion (Rosenfalck 1969). The type of representation provided by the EMG electrodes is strongly influenced by the filtering action due to the tissue between the tripole and the detection points. As a function of the tissue thickness (both the muscular and subcutaneous fat tissues should be considered), the filter attenuates the amplitude of the voltage corresponding to the electric field generated by the tripole and smoothes the edges of the voltage changes related to the spatial distribution of the three electrical sources. In other words, the tissue acts as a "spatial filter" attenuating and smoothing the MUAP voltage distribution, which is detected on the surface of the skin. As a consequence of this filtering, two identical MUAPS (two identical tripoles) traveling at the same velocity at different depths will generate surface EMG with different properties. The surface MUAP generated farther away will have lower amplitude and lower frequency content. On the other hand, two identical tripoles, at the same distance from the skin but travelling at different velocities, will generate the same spatial distribution of surface potential but time signals that are scaled with respect to each other by factors related to the conduction velocities. An idea of useful "standard for reporting EMG data" can be found in the *Journal of Electromyography and Kinesiology* (pages iii–iv, volume 6, 1996).

Comparison Between Surface and Indwelling EMG in Fatigue Monitoring

A recent paper by Krivickas and coworkers (1996) deals with the problem of comparison between the fine wire and surface EMG in the detection of electrical manifestation of fatigue. With respect to the initial value the fine wire EMG median frequency presents larger changes than surface EMG median frequency with fatigue. The intersession reliability is poorer for fine wire than surface EMG (it is difficult to find the same location for the two wires, the same interelectrode distance, and depth). Moreover there is a relative discomfort for the subject during electrode positioning and contraction, and it is impossible to know the geometry of the detection apparatus (the final distance between the two bared tips of the wires is unknown). The only important feature of the fine wire EMG for fatigue studies is the possibility to reach deep muscle.

Force and Surface Mechanomyogram During Stimulation

The force changes at fatigue and their consequences on the force/frequency relationship was specifically described in chapter 19 by MacIntosh and Allen. Here the force oscillation and the simultaneously recorded MMG will be compared during single twitches and an increasing stimulation rate before and after fatigue. The aim is to evaluate how the changes in muscle mechanics due to fatigue are reflected in the two signals. The possibility of using the electrically elicited surface mechanomyogram as an index of mechanical fatigue was first verified by Barry, Hill, and Dukjin (1992). These authors studied the first dorsal interosseus muscle surface mechanomyogram by means of an accelerometer (mass = .5 g). Fatigue was induced by voluntary contractions, 25 s on, 5 s off at 30%, 50%, or 70% MVC in different trials. As

illustrated in figure 21.2 the magnitude of the surface oscillation during single twitches elicited by supramaximal stimulation of the ulnar nerve decreased with fatigue. The rate of the decrement was proportional to the effort intensity of the fatiguing contraction. The twitch force amplitude decreased in a similar manner. Barry, Hill, and Dukjin (1992) concluded that the alteration in the acceleration signal "may reflect fatigue-induced changes in contractile properties of the muscle" (p. 309). The recording of the phonomyogram (PMG) at the 8th intercostal space allowed Petitjean and Bellemare (1994) to investigate the presence of fatigue in each hemidiaphragm by single twitches evoked by stimulation of the phrenic nerve. Fatigue was induced by artificially enhancing the airway's resistance. A decrease in the transdiaphragmatic pressure and of the amplitude of PMG was reported in the fatigued diaphragm. We subsequently studied the surface mechanomyogram before and after fatigue in a limb muscle (Orizio, Diemont et al. 1996). The tibialis anterior was percutaneously and supramaximally stimulated at the most proximal motor point before and after a 40 s stimulation at 35 Hz. The stimulation pattern elicited four single twitches (interpulse period 1000 ms) followed by a sweep from 1 to 50 Hz. During the sweep the stimulation rate increased by 1 Hz from one stimulus to the following. The force oscillation and the surface MMG, recorded by an accelerometer (mass = .5 g), were compared. At fatigue the amplitude of MMG and force oscillation during the single twitch decreased by 44% and 40%, respectively. The relationship of both the MMG amplitude and the force oscillation versus the stimulation rate shifted toward the left (see figure 21.3). During a single twitch the MMG duration ranged between 60 and 70 ms. This corresponds closely to the contraction time reported for ankle flexors in humans (Belanger and McComas 1983; Moglia et al. 1995). As a consequence of these measurements, it can be concluded that the accelerometer detects mainly the muscle surface oscillation during the contraction phase. Recently we used the laser distance sensor to pick up the muscle surface displacement in tibialis anterior with the same stimulation pattern. In figures 21.4 and 21.5 the force and laser-detected MMG from one representative subject are compared. It is evident that the muscle surface displacement closely mirrors the force generation process. As it is possible to verify in figure 21.5 all of the single twitch parameters considered in section I changed with fatigue: peak torque (2.8 → 1.97 N · m), peak of

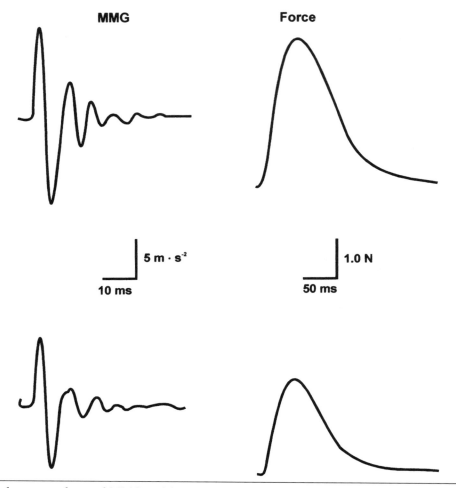

Figure 21.2 Accelerometer-detected MMG and force in fresh (top) and fatigued (bottom) first dorsal interosseus. The force reduction and the slowing of the tension generation process are reflected in the decrease of twitch amplitude and of the acceleration signal amplitude, respectively.

Reprinted, by permission, from D.T. Barry, T. Hill, and I. Dukjin, 1992, "Muscle fatigue measured with evoked muscle vibrations," *Muscle & Nerve* 15: 303-309. Reprinted by permission of John Wiley & Sons, Inc.

displacement (.88 → .47 mm); maximal torque generation rate (18 → 9 N · m/s), maximal displacement rate (3.4 → 1.4 mm/s); 1/2 relaxation time for torque (80 → 130 ms), 1/2 relaxation time for displacement (63 → 94 ms). Moreover, at fatigue the force and displacement at tetanic stimulation frequency are reduced, and the oscillation of both the signals decreases at a higher rate throughout the sweep (see figure 21.4).

In conclusion, the stimulated surface MMG is able to detect the changes in the contractile properties associated with fatigue. In particular when detected by microphones or accelerometers, the signal amplitude reduction during single twitch indicates the slowing of the rate of force generation (dP/dt). The shift to the left of the oscillation amplitude versus stimulation rate indicates the higher degree

of fusion of the mechanical events in fatigued muscle due to the prolongation of the mechanical response. When the laser-detected surface displacement is considered, it seems obvious to conclude that MMG, like force, directly mirrors the muscle fiber mechanical changes at fatigue. Finally, the MMG during stimulation is a reliable adjunct tool to study the alteration of the tension generation process in fatigue. It is specifically of value when the force signal related to the muscle under investigation is difficult or impossible to measure.

Surface Mechanomyogram During Voluntary Contraction

Barry, Geiringer, and Ball (1985) were first to report changes in "acoustic myography" due to fatigue. During slowly developing fatigue (5 sequences of

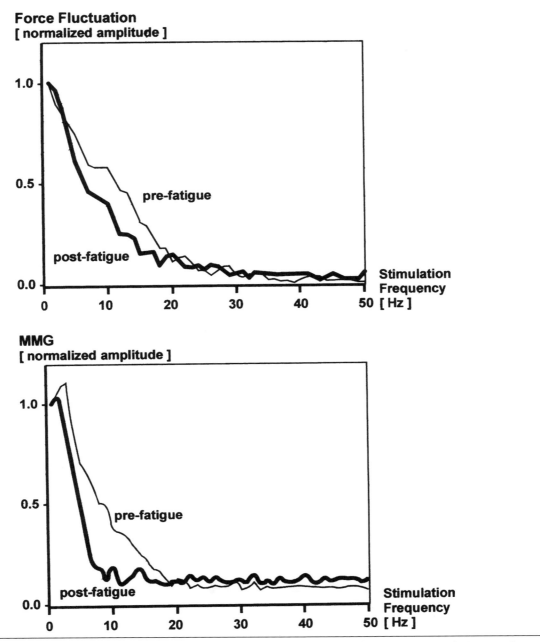

Figure 21.3 Amplitude of force fluctuation (top) and MMG (bottom) as a function of increasing stimulation rate (sweep from 1-50 Hz) in human tibialis anterior. Fatigue reduces the ability of the muscle to provide unfused mechanical response when the time interval between one stimulus and the following shortens. As a consequence the relationship shifts to the left. Original data from Human Physiology Laboratory, University of Brescia, Italy.

increasing weights isometrically sustained for 20 s) the acoustic myogram (AMG) versus force relationship became steeper. During a sustained 75% MVC and the following period in which the force declined to 35% MVC, the AMG and force were described to reduce in parallel while EMG did not change. As a consequence a dissociation of the electrical and mechanical events can be detected by the use of AMG. After this study, many others have addressed the surface mechanomyogram properties of fatigued muscle.

Time Domain Analysis

The surface MMG presents different characteristics throughout sustained isometric effort. Its integrated value or its root mean square (RMS) may increase,

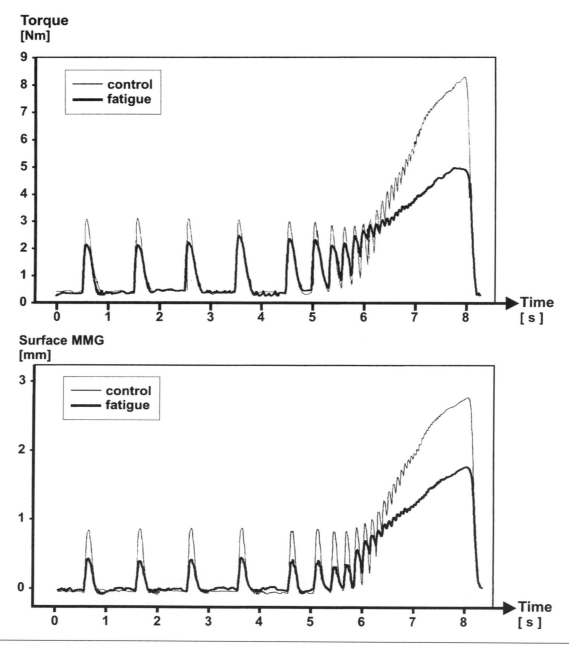

Figure 21.4 Force (top) and laser-detected surface MMG (bottom) during a sequence of 5 single twitches and the 1 to 50 Hz sweep in tibialis anterior. Fatigue influences the mechanical responses for both single twitches (see also figure 21.5) and the increasing stimulation rates. Due to fatigue the force oscillations and MMG present a faster decrease during the sweep (this result confirms the data in figure 21.3). Original data from Human Physiology Laboratory, University of Brescia, Italy.

remain stable, or decrease. An increase in the signal amplitude has been described in biceps brachii at 10% MVC (Keidel and Keidel 1989) and 20% MVC (Orizio, Perini, and Viecsteinas 1989), in abductor digiti minimi at 15 and 25% MVC (Goldenberg et al. 1991), and finally in the quadriceps at 20% and 40% MVC (Rodriquez et al. 1993; Rodriquez et al. 1996).

A stable amplitude of the signal was reported for biceps brachii at 40% MVC (Orizio, Perini, and Viecsteinas 1989) and at 50% MVC (Zwarts and Keidel 1991). A reduction of the MMG with fatigue was evident in biceps brachii at 60 and 80% MVC (see figure 21.1) (Orizio, Perini, and Viecsteinas 1989), and in quadriceps (Rodriquez et al. 1993), or

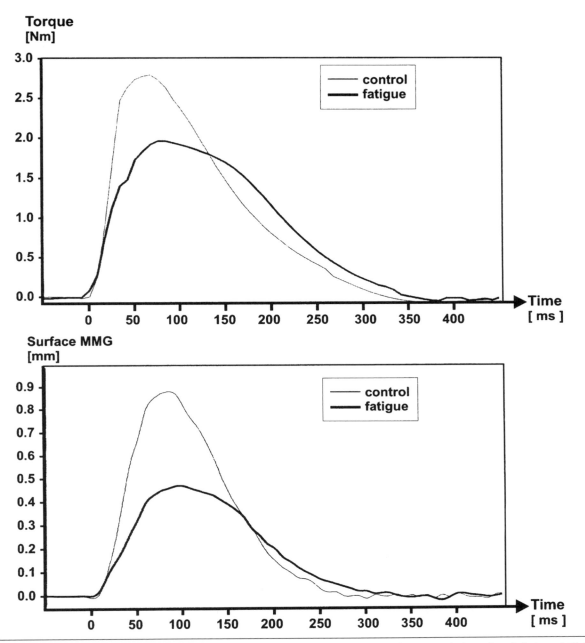

Figure 21.5 Force (top) and laser-detected MMG (bottom) during single twitches in unfatigued and fatigued tibialis anterior. The responses during the first 5 single twitches reported in figure 21.4 have been averaged. The similar behavior of the two signals at fatigue is evident. For twitch parameter values, see text. Original data from Human Physiology Laboratory, University of Brescia, Italy.

in the abductor digiti minimi at 50% MVC and 75% MVC (Goldenberg et al. 1991). With MVC both in the biceps (Orizio 1992) and in the quadriceps (Orizio and Veicsteinas 1992) a MMG reduction was reported. Even if they did not stress it, an MMG reduction was clear also in the paper by Barry, Geiringer, and Ball (1985) (see figure 21.3). The decrease took place within the first 40 s of activity,

when the force was still at the target level (75% MVC).

With reference to the motor unit activation pattern modifications in fatigue (section I) and with data from the literature, the following can be concluded:

1. The increase in the MMG power (low level of contraction) may indicate the recruitment of new

motor units and the increase of the overall MU firing and the developing of MU synchronization and grouping.

2. The decrease in the MMG power (high level of contraction) may indicate the derecruitment of fast fatiguing MU, the decrease in the MU firing rates, the prolongation of the relaxation time reducing the dimensional changes between one motor command and the next (and in turn the related pressure waves generating the muscle surface oscillation), and the increase of the intramuscular pressure (Sadamoto, Bonde-Petersen, and Suzuki 1983) impairing the dimensional changes of the active fibers.

3. The stable MMG power during effort suggests a balanced influence of the mechanisms invoked for behavior 1 and 2, above.

When the quadriceps muscle was fatigued by intermittent isometric contractions (10 s on and 10 s off at 75% MVC until only 40% MVC could be attained), the relationship between MMG power and force before fatigue and after 15 min of recovery was found unaltered from that before fatigue (Stokes and Dalton 1991). This suggested that after 15 min of recovery the possibility to generate the same level of absolute force was completely recovered, or the recruitment and firing rate adaptations were enough to compensate for muscle fiber fatigue (Stokes and Dalton 1991). During the on phases of the same fatiguing protocol the acoustic myogram power decreased together with the force up to 50% MVC. In the last contractions from 50% to 40% MVC the AMG amplitude increased slightly (Dalton, Comerford, and Stokes 1992). The early decrease and the late increase in AMG power may be explained by the factors listed above for sustained contraction (1 and 2).

Frequency Domain Analysis

Holding a 3 kg weight with the elbow in a fixed position provoked an increase in the 8 to 18 and 20 to 30 Hz bands in the vibromyogram (VMG) from biceps (Keidel and Keidel 1987). During 50% MVC, Zwarts and Keidel (1991) reported VMG spectrum changes from bimodal to unimodal. In the same work the spectrum mean frequency has been indicated to decrease during sustained MVC. After truncation of the frequency content below 14 Hz, Goldenberg and coworkers (1991) described a reduction of the AMG mean frequency from 23 to 17 Hz during sustained 50% MVC contraction of the abductor digiti minimi. A more detailed analysis of the sound myogram (SMG) during different levels of isometric fatiguing contraction has been carried

out by Orizio and coworkers (1992). It was shown that during 20% MVC the frequency content of the SMG did not change significantly. During 80% MVC a transient increase of the high frequency content with an accentuation of the bimodality of the spectra were found. Afterwards a continuous shift of the MMG spectrum toward the lower frequencies was found. At 40 and 60% MVC the transient increase was less pronounced. During MVC the high frequency content of the SMG continuously decreased with time (Orizio 1992). At exhaustion the power was distributed always well below 20 Hz (Orizio 1992; Orizio et al. 1992). During sustained MVC in muscle with different fiber typing (vastus lateralis of sprinters, long distance runners, and controls) the shift of the spectra toward the lower frequencies took place in a rapid and slow phase (Orizio and Veicsteinas 1992). The first one always lasted about 30 s and presented the steeper and larger power reduction in the high frequency band of the SMG spectra. This was attributed to the fatiguing process of the fast twitch fiber MU, and it represented a different percentage of the total contraction time in the three groups (70% sprinters, 45% sedentary, 40% long distance runners). The changes in the spectral parameters of the mean frequency always followed closely the morphological changes of the SMG spectrum (Orizio 1992; Orizio et al. 1992). Herzog and coworkers (1994) showed that during sustained 70% MVC the quadriceps vibromyogram power spectrum shifted clearly toward the lower frequencies. The mean frequency has been reported as rather stable while the force was kept close to the target. Indeed, the analysis of figure 21.3 (in Herzog et al. 1994) shows a transient increase in the mean frequency (MF) of a representative subject, followed by a decrease resembling the MF behavior of the SMG from the biceps at 80% MVC (Orizio et al. 1992). When 40% MVC is sustained to failure in quadriceps an increase in the low frequency band (.5–10 Hz) has been reported (Rodriquez et al. 1996). In the quadriceps muscle 15 min after the fatiguing protocol the AMG mean power frequency was not different than in fresh muscle at isometric force output from 10 to 100% MVC (Dalton and Stokes 1993).

From the above reported studies about the spectral changes of the surface mechanomyogram during fatiguing exercises the following conclusion can be made:

1. The reduction in high frequency content of the MMG may reflect the changes in the shape (prolongation) of the twitches summated in MMG, the motor unit firing rate decrease, the derecruitment of

fast twitch motor units, and the synchronization and grouping of the active MU similar to the force tremor in fatigue.

2. The transient increase in high frequency content of the signal spectrum at high effort levels may reflect the increase in the MU firing rate or alternatively a not-yet-demonstrated effect of the potentiation phenomenon on the MU mechanics eliciting faster twitches.

Concluding Remarks

The surface mechanomyogram can be considered a useful tool to evaluate changes in muscle mechanics in fatigue. In fact, during electrical stimulation the alteration of the signal parameters parallel those of the force. During sustained voluntary contraction the properties of the MMG reflect the changes of the shape of the elementary summated events and of the motor unit activation pattern.

Electromyogram

It is well known that the changes in the electrical activity recorded from a fatiguing muscle can be detected much earlier than the mechanical failure point (Basmajian and De Luca 1985) (see also figure 21.1). For this reason the electromyogram "has been used as a guide to fatigue for many years" (Edwards 1981, p. 10). The electromyogram changes are mediated firstly by the influence of fatigue on the electrical properties of the sarcolemma. On this basis this section first deals with muscle fiber conduction velocity alteration and afterward with EMG changes in the time and frequency domains during both voluntary and stimulated contractions.

Muscle Fiber Conduction Velocity

The propagation velocity of the sarcolemmal action potentials can be estimated in the active portion of the muscle as an "average" muscle conduction velocity (CV). For a review on the topic, see Arendt-Nielsen and Zwarts (1989). The CV can be estimated invasively by needle electrodes (Troni, Cantello, and Raniero 1983), or noninvasively by surface electrodes (Merletti, Knaflitz, and De Luca 1990) placed at a known distance. The delay between the electrical events at the two detection points allows calculation of their propagation velocity. It has to be taken into account that surface evaluation usually provides higher CV than an invasive method (Van der Hoeven 1995). When muscle is fresh the increases of CV with force, together with specific alteration in muscle geometry, reflect the recruitment of MU with larger fibers and with corresponding higher

CV (Broman, Bilotto, and De Luca 1985a) and the increase in firing rate (Arendt-Nielsen, Gantchev, and Sinkjaer 1992). The decrease of CV with fatigue is a well-known phenomenon, and it has been verified by means of surface EMG (Linssen et al. 1993; Merletti, Knaflitz, and De Luca 1992; Van der Hoeven and Lange 1994) and by needle EMG (Van der Hoeven, van Weerden, and Zwarts 1993).

In biceps brachii during sustained MVC (Van der Hoeven, van Weerden, and Zwarts 1993), 80% MVC (Broman, Bilotto, and De Luca 1985a, 1985b), and 60% MVC (Eberstein and Beattie 1985) conduction velocity changes from 4.5 to about 3 m/s, from 4.2 to 3.6 m/s, and from about 4.5 to about 3 m/s, respectively. In 20 s efforts of the tibialis anterior the CV changed at 80% MVC from 6 to 4.2 m/s; on the contrary CV was rather stable at 20% MVC (Merletti, Knaflitz, and De Luca 1990). At low level contraction the CV changes in elbow flexors (biceps brachii, BB; brachioradialis, BR) showed the following muscle-specific changes: 10% MVC—no real changes (Krogh-Lund 1993), 15% MVC—significant reduction only for BR (from 4.2 to 3.7 m/s) (Krogh-Lund and Jorgensen 1992), 30% MVC—in BR CV decreased from 4.3 m/s to 2.9 m/s and in BB from 4.6 to 3.9 m/s (Krogh-Lund and Jorgensen 1993), 40% MVC—the changes were similar to 30% MVC (Krogh-Lund 1993).

The length of the muscle strongly influences the degree of the CV alteration with fatigue. Indeed, stretched muscle shows a larger CV decrement for the same fatiguing effort (Arendt-Nielsen, Gantchev, and Sinkjaer 1992). The conduction velocity decreases also during electrical muscle stimulation. This is well documented by the analysis of the evoked myoelectric signal (M-wave). Figure 21.6 presents the changes in the tibialis anterior M-wave, during a 20 Hz stimulation for 1 min at the main motor point. The duration and the shape of the M-wave are altered. During a 20 s low (20 Hz) or high (40 Hz) stimulation rate, the tibialis anterior CV decreased by 10% or 40% from the initial value, respectively (Merletti, Knaflitz, and De Luca 1990). Above 20 Hz, in some cases, the CV variation seems to be dependent on the number of applied stimuli. This suggests that the metabolic alterations at fatigue could be the summation of stepwise changes induced by each stimulus (Merletti, Knaflitz, and De Luca 1992).

In summary, the decrease of the muscle fiber conduction velocity allows detection of (1) the alteration in the sarcolemmal action potential due to its time scaling and its shape alteration at fatigue, both during voluntary and stimulated contraction and

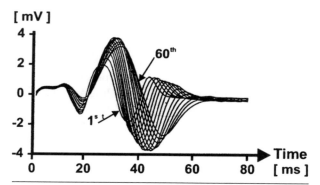

Figure 21.6 Electrically evoked responses (M-waves) from human tibialis anterior during 1 min of stimulation at 20 Hz. Each curve is the average of 20 M-waves. The prolongation and the change of shape of the electrical response to the stimulus with time are clearly demonstrated.

Reprinted, by permission, from R. Merletti and L.R. Lo Conte, 1992, "Electrically evoked surface myoelectric signals," *Functional Neurology* 7 (suppl.): 153-158. Reprinted by permission of CIC Edizioni Interzionali.

(2) the derecruitment of fast fatigable MU (with high conduction velocity), as well as (3) firing rate reduction during voluntary contraction.

EMG Changes in Time Domain

The amplitude variables that are classically used in EMG processing are the root mean square (RMS), the average rectified value (ARV), and the integrated EMG (Basmajian and De Luca 1985). In this section, changes of these variables computed from surface EMG are discussed. This signal is the most appropriate technique for monitoring the alteration of the muscle electrical activity due to fatigue, because a large part of muscle is investigated and no discomfort is imposed on the subject during recording. The increase or the decrease of the signal amplitude variables from the onset to exhaustion, during a sustained isometric contraction, depends on the intensity of the effort. During maximal voluntary contraction the EMG amplitude decreases continuously (Bigland-Ritchie, Jones, and Woods 1979; Kranz, Cassell, and Inbar 1985; Moritani, Muro, and Nagata 1986; Orizio 1992). On the contrary, during submaximal sustained contraction the EMG amplitude increases (De Luca 1985; Gamet and Maton 1989; Krogh-Lund 1993; Krogh-Lund and Jorgensen 1991, 1992, 1993; Orizio 1992; Viitasalo and Komi 1977; West et al. 1995; Zijdewind, Kernell, and Kukulka 1995). This increment is much more evident in the final stages of the effort. The same result is found during repetitive submaximal contractions (Bigland-Ritchie, Furbush, and Woods 1986; Christensen and Fuglsang-Frederiksen 1988; Stokes and Dalton 1991; Sundelin and Hagdberg

1992). During repeated isokinetic contractions of plantarflexors of the foot the EMG presented a steady value or a slight increase only in the first part of the fatiguing trial (Gerdle et al. 1987), at high velocity the quadriceps presented a large increase in amplitude during concentric and not during eccentric contraction (Tesch et al. 1990). In summary, from the literature, the following conclusions can be stated:

1. The EMG amplitude decrease may reflect the MU derecruitment or the MU firing rate decrease and a possible decrease in the MU action potential amplitude.

2. The EMG amplitude increase may reflect changes in the shape of the muscle fiber action potential and decrease in conduction velocity and a slowing of the signal (a slower action potential has a larger integral and RMS value), recruitment of new motor units, and synchronization and grouping of active motor units.

The proof that the large decrease in the EMG during sustained voluntary contraction is not related to alteration in the sarcolemmal action potential but to several concurrent factors can be provided by means of muscle stimulation. Indeed Bigland-Ritchie, Jones, and Wood (1979) recorded practically normal electrically evoked action potentials after 60 s sustained MVC.

The behavior of the evoked compound action potential is more complex when fatigue is induced by electrical stimulation. Fatigue, induced by stimulation at ≥ 80 Hz (Botterman 1995), involves a failure in the excitation of the sarcolemma, possibly due to accumulation of K^+ in the extracellular space (Benzanilla et al. 1972; Bigland-Ritchie, Jones, and Wood 1979). When this occurs, there is a concurrent reduction of the action potential amplitude and of the twitch force; they recover with the same quick rate after stimulation is stopped. During fatigue induced by stimulation at rates < 40 Hz (Botterman 1995), the recovery of the EMG is relatively quick, complete within several minutes (Edwards et al. 1977; Sandercock et al. 1985), and this recovery is completely uncoupled with the twitch force, which may remain depressed for long periods (even for hours). The mechanism of this fatigue (see chapter 19) has been attributed to a disturbance in excitation-contraction coupling. The slow twitch motor units seem to be little affected by this last type of fatigue (Sandercock et al. 1985). The reported uncoupling between force reduction and electrical activity, confirmed also during a fatigue test in single MU (Bevan et al. 1992), leads some authors to indicate that EMG may be an unreliable guide to fatigue under certain conditions (Fitts 1994).

EMG Changes in Frequency Domain

In his contribution to a symposium on fatigue, Edwards (1981) underlined that an approach to muscle fatigue investigation based only on rectified EMG is of limited value and that an alternative "is based on analysis of the power spectrum" (p. 10) of the surface EMG. In reality, from the sixties up to the current time, a huge number of papers have confirmed by mathematical analysis the historic observation of Piper (1912) at the beginning of this century that, during sustained contraction, the frequency content of the EMG decreases (for background on this topic see Basmajian and De Luca 1985). On the basis of the model for the generation of the EMG as the summation of the different MU action potential trains (MUAPTs) presented by Basmajian and De Luca (1985) the spectrum of the surface electromyogram can be represented by the following equation (from Basmajian and De Luca 1985):

$$S_m (\varpi, t, F) = R (\varpi, d) [1/(\nu^2(t,F)) \cdot \qquad 21.1$$
$$G(\varpi d/2 \, \nu \, (t, F))]$$

where

R is the magnitude for the electrode filter function, and

d is the interelectrode distance of a bipolar electrode.

G summarizes the influences of the number of active motor units, the interpulse interval (IPI) statistics (IPI coefficient of variation), the degree of synchronization (i.e., the larger the number of synchronized MUs with stable firing, the larger the peak in the lower part of the spectrum, below 40 Hz), and the shape of the MUAP (the smoother the shape and lower the rate of zero crossings, the greater the low frequency power).

Moreover, other factors can contribute to the EMG spectrum: the mean conduction velocity (υ) (the lower the conduction velocity, with longer MUAPs, the larger the shift toward the lower frequencies); the effort intensity (F); and duration (t). If general agreement exists on the fact that the EMG spectrum shifts toward the lower frequencies with fatigue, a large debate is still open on which spectral parameter can be used to monitor fatigue or which aspect of muscle fatigue is retrievable in median frequency, mean frequency, peak frequency, and

ratio between the power in the high and low frequency bandwidths. The ratio (130–233 Hz power / 20–40 Hz power) decreases dramatically during 60 s MVC (Edwards 1981). It has been demonstrated that this parameter is strongly dependent on the chosen boundaries for the spectrum bandwidths partitioning, and, moreover, its variance is larger than other spectral parameters such as median frequency (MDF) (Stulen and De Luca 1981). The most widely used parameters to represent the spectrum information are the median and the mean (MNF) frequencies. The absolute value of MDF or MNF is dependent on relative geometry and location of the electrode placement. As a consequence it is hard to compare the results of different papers on the topic, and only the relative differences with respect to the initial value can be compared. As was the case for CV, MDF and MNF show specific behavior depending on the muscle and the intensity of contraction. "For low load levels (about 10% MVC) the classical decrease in MDF or MNF over time is rarely reported" (Duchene and Goubel 1993, p. 380). Exceptions to this (i.e., an increase in the spectral parameters) have been reported in the trapezius (Hagberg and Hagberg 1989; Oberg et al. 1992) and in vastus lateralis (Arendt-Nielsen, Mills, and Forster 1989). On the contrary, in the quadriceps at sustained 10% and 20% MVC a decrease, even if not dramatic, of the MNF (Christensen and Fuglsang-Frederiksen 1988) has been reported (in the same study the intermittent contraction of the same muscle at 20% MVC did not result in MNF reduction over 1 h exercise). In biceps brachii at 10% (Krogh-Lund 1993) and 15% MVC (Krogh-Lund and Jorgensen 1992) there was a decrease in MNF. At 25% MVC of the elbow flexors the MNF decreased more in the brachioradialis than in the biceps brachii (Gamet and Maton 1989). At higher contraction level an extensive literature demonstrates that MNF and MDF decrease with fatigue. The higher the intensity of the effort, the greater the rate of reduction of the frequency parameters, as well as the greater the amount of the decrease (Basmajian and De Luca 1985; Merletti, Knaflitz, and De Luca 1990; Merletti and Roy 1996a; Orizio 1992; Orizio et al. 1992). In summary the most typical behavior of the mean and median frequencies during well-maintained isometric contraction as well as during intermittent isometric contractions (Linssen et al. 1993) or dynamic contractions (Gerdle et al. 1987; Sundelin and Hagberg 1992; Takaishi, Yasuda, and Moritani 1994; Tesh et al. 1990) is represented by a reduction with time. Also the MNF of the electrically elicited M-wave decreases its value as the signal's time scaling and shape alteration suggest (see figure 21.6). During electrically evoked activity of the tibialis anterior, increasing the stimulation rate, the asymptotic value and the time constant of the MDF and MNF exponential decrease are lower and shorter, respectively (Merletti, Knaflitz, and De Luca 1990). In the same study it was verified that stimulation rates > 30 Hz provide more dramatic changes of MDF and MNF than voluntary contraction sustained for 20 s at 80% MVC (Merletti, Knaflitz, and De Luca 1990). Intermittent stimulation with periods of recovery generates less fatigue than sustained constant frequency stimulation, both from the electrical and the mechanical point of view (Hainaut and Duchateau 1989).

The EMG spectral changes and the MNF and MDF reduction at fatigue may indicate a reduction in average muscle fiber conduction velocity, alteration in MU synchronization and grouping, and possible derecruitment of the MU with reduction of the interferential pattern of the EMG.

Detailed Analysis of the Changes in CV, MNF, and MDF in Fatigue

The model of Lindstrom, Magnusson, and Petersen (1970) suggested that the EMG spectral shift with fatigue could be explained by means of changes in CV (Basmajian and De Luca 1985). In reality, even if spectral parameters and CV were considered linearly related (Basmajian and De Luca 1985; Eberstein and Beattie 1985) some doubts about a full proportionality between their changes at fatigue were evident already in the late '80s. Broman, Bilotto, and De Luca (1985a), analyzing surface EMG in tibialis anterior (fatigue induced by a sustained 80% MVC effort), showed that the MDF and MNF decreased twice as much as did CV. Zwarts, van Weerden, and Haenen (1987), during 40% MVC sustained contraction of the biceps brachii, observed that MDF changed much more than CV. They concluded that "the shift of the power spectrum to lower frequencies during fatigue cannot be explained by changes in muscle fiber CV alone" (p. 216). Similar results indicating larger changes for MNF than CV were found in studies on tibialis anterior during voluntary contraction at 20 and 80% MVC (Merletti, Knaflitz, and De Luca 1990), in biceps brachii during intermittent isometric contractions (Linssen et al. 1993), and in quadriceps during exhausting 80% MVC (Arendt-Nielsen, Gantchev, and Sinkjaer 1992). The issue of the CV as the sole or the main factor in EMG spectral changes has been investigated extensively by Krogh-Lund and Jorgensen (1991, 1992, 1993), and see also Krogh-Lund (1993), at low level

isometric contraction (from 10–40% MVC) of the elbow flexors. In these last papers it is suggested that during low level isometric contraction the recruitment of new motor units with higher CV may compensate for the CV decrement due to fatigue in the MU active from the beginning. At the same time, the presence of strong synchronization and grouping of the MUs determines the larger decrease of the MNF. This conclusion, indicating a large influence of the MU synchronization and grouping, in addition to the CV variation on the EMG properties, seems to be supported by the fact that, in the second half of the sustained effort, a constant CV coexists with a large peak below 30 Hz in the spectrum and a dramatic increase in the RMS. Over the same time interval the decrease of the MNF is 50% of its total reduction. It can be concluded that at low and intermediate level of effort the uncoupling between the CV and MNF and MDF trends may reflect the changes in recruitment and firing pattern of the active motor units.

The divergence in the trend of CV, MNF, and MDF should not be present during stimulated contraction where no influence of the MU activation pattern is possible. In reality, also in this case the CV decreases at a lower rate than MNF and MDF (Merletti, Knaflitz, and De Luca 1990). Indeed the widening of the M-wave in time, determining the well-known EMG spectral compression, may be due not only to a decrease in the muscle fiber conduction velocity according to the Lindstrom, Magnusson, and Petersen (1970) model, but also to an increase of the length of the depolarization zone or to a nonuniform decrease of muscle fiber CV of the different MUs simultaneously stimulated. This fact would result in an increase of CV dispersion between the active fibers with an alteration of the M-wave shape (Merletti, Knaflitz, and De Luca 1990). The importance of this last factor has been confirmed by Merletti and Roy (1996b) by means of a computer simulated EMG study in which the CV of the different motor units has been decreased at different rates. In summary, also during stimulated contraction CV, MNF, and MDF reflect different aspects of the sarcolemmal electrical fatigue. The spectral parameters are influenced not only by the reduction of the action potential CV, but also by its nonuniform decrement among the different MUs and by the possible lengthening of the depolarization zone.

Concluding Remarks

The analysis of EMG is a crucial tool to follow the development of localized muscle fatigue. In fact, the changes in the motor unit action potential shape and velocity of propagation, the nonuniform CV changes among the different MUs, and the modification of the MU firing rate are well reflected in the muscle fiber CV measure and EMG time and frequency domain analysis. Caution is needed for a correct interpretation of the EMG changes, especially during voluntary contractions (when CV, motor unit firing, and recruitment may change together). Under certain conditions the EMG monitors only specific aspects of the fatigue process. EMG changes can be uncoupled from alterations of the muscle mechanical output (see the spectral compression of EMG with constant output force, or the well-preserved EMG with large force reduction during low frequency stimulation).

Applications

It is useful in many circumstances to be able to detect the presence of fatigue in an objective manner. Measures of force, EMG, and MMG have been used for this purpose. In the following section, two approaches are presented: indexes of muscle fatigue and combined analysis of MMG and EMG in both the time domain and the frequency domain.

Some Indexes of Muscular Fatigue

No doubts exist about the influence of fatigue on force, MMG, and EMG parameters, but no clear indication can be found in the literature about the way to use these parameter changes as a tool to foresee the endurance of a muscle undergoing a certain exercise. The indexes most frequently used in research are the decrement, the initial slope, or the time constant of the lines or curve fitted by least squares method on the time kinetics of the signal parameters (Merletti, Lo Conte, and Orizio 1991; Duchene and Goubel 1993). Recently a new dimensionless index based on a regression free method has been suggested as a tool to estimate fatigue (Merletti, Lo Conte, and Orizio 1991). This is a ratio in which the area of a reference rectangle (height: the first value of the time series of the considered parameter at the beginning of contraction, width: the exercise duration) is the denominator, and the area between the upper side of the rectangle and the relative value of the parameter at each time during contraction time is the numerator. Its value is between 0 and 1 for decreasing variables, negative for increasing variables. All these indexes provide a measure of fatigue. They are able to indicate if one muscle is fatigued more than another one, or if one

exercise is more exhausting than another one. During sustained submaximal contractions of the knee extensors, the rate of the EMG median frequency reduction and the rate of the reduction of the MVC value (produced intermittently throughout the submaximal effort) were found highly correlated (Mannion and Dolan 1996). On this basis the EMG median frequency has been suggested as a tool "to monitor fatigue, where fatigue is defined as the inability to generate the maximum force that can be produced by the muscle in the fresh state" (Mannion and Dolan 1996, p. 411). Recently Merletti and Roy (1996a) were able to demonstrate that the slope of the MDF or MNF of the EMG computed during the first 30 s of sustained isometric exercise (50%, 60%, 70%, or 80% MVC) is predictive of the endurance time for submaximal effort. This is important because it allows the clinicians and rehabilitation therapists to estimate the possible mechanical performance of the muscular group under study.

The Combined Analysis of MMG and EMG in Muscle Fatigue Detection

In this subsection the results of the MMG and EMG analysis already reported in previous parts of the chapter are briefly recalled only to point out the meaning of the comparison between MMG and EMG properties at fatigue.

Time Domain Analysis

The EMG and MMG compared analysis is useful for near maximal sustained isometric effort. In this case the increase in EMG RMS is paralleled by a reduction in MMG RMS. The uncoupling between MMG and EMG RMS indicates that at near maximal isometric effort

- the dimensional changes of the MU fibers are reduced (paralleling the twitch force reduction),
- fatigue tremor (due to MU synchronization and grouping) is not present or is not enough to influence the surface oscillation properties, and
- the MU firing rate reduction by means of the "muscle wisdom" may not be fully proportional to the twitch prolongation of the fatigued fibers.

In conclusion, the divergence of the two signal amplitude parameters may suggest that, despite a well-maintained electrical activity, the muscle fibers generate tension in a "fusionlike" situation.

Dalton, Comerford, and Stokes (1992) underlined that the simultaneous EMG RMS increase and the surface MMG RMS decrease may indicate the presence of peripheral fatigue due to excitation-contraction coupling failure. The analysis of surface MMG together with EMG may be a tool to distinguish, in specific conditions, between central and peripheral fatigue (Dalton, Comerford, and Stokes 1992). At MVC and at low levels of effort the time behavior of the two signals is similar, indicating that EMG and MMG are similarly influenced by the changes in fatigue, the increased fiber mechanical and electrical signals, and the motor unit recruitment and firing pattern.

Frequency Domain Analysis

The comparison of the spectra of simultaneously recorded MMG and EMG (Herzog et al. 1994; Orizio 1992; Zwarts and Keidel 1991) shows that the peaks of the MMG spectra overlap nearly completely the peaks of the EMG spectra below 40 Hz. In this frequency range information about the global motor unit firing pattern is contained in the EMG spectrum. This suggests that a great part of the MMG information is related to the MU firing rate (Goldenberg et al. 1991; Keidel and Keidel 1989; Orizio 1993; Orizio et al. 1990; Orizio, Liberati et al. 1996). This may be due to the longer duration of the elementary events summated in MMG (fiber twitch duration about 100–200 ms), providing a near sinusoidal response even at the lowest firing frequencies of recruitment (about 10 Hz). On this basis the surface MMG can be considered as the summation of more or less distorted sinusoids and not as the summation of trains of narrow waveforms as the EMG. As a consequence, the frequency information (in the 10–40 Hz range) may dominate the MMG signal while the shape (of the motor units action potential) information dominates the EMG signal. On this basis it may be hypothesized that this divergence, when it occurs, may be a tool to evaluate the role of the global firing of the MU in force generation. Finally, considering the MMG and the EMG as compound signals in which the mechanical and electrical MU activities are summated, it has been suggested that with a cross-spectral analysis or coherence (Orizio et al. 1991) analysis between the two signals, data about the global motor unit firing rate may be retrieved. This technique has been rarely applied and further studies with reliable signal models are needed. In fatigue the morphology and the position of the cross-spectrum as well as the trend of its mean frequency with time, fit well with the reported trends of the MU firing rate throughout sustained effort (Orizio 1992).

Concluding Remarks

The combined use of the force signal, MMG, and EMG may contribute to retrieve data on specific aspects of the electrical and mechanical muscle activities as well as about the muscle motor control at fatigue. Moreover, an estimation of the degree of muscle fatigue can be made. During a sustained contraction the compared analysis in the time domain may monitor a possible "fusionlike" situation or "peripheral fatigue," and in the frequency domain may allow inferences about the global motor units firing rate behavior.

References

Adrian, E.D., and Bronk, D.W. 1929. The discharge of impulses in motor nerve fibers: The frequency of discharges in reflex and voluntary contractions. *J. Physiol. (London)* 67: 119–51.

Allen, D.G., Lannergren, J., and Westerblad, H. 1995. Muscle cell function during prolonged activity: Cellular mechanisms of fatigue. *Exp. Physiol.* 80: 497–527.

Arendt-Nielsen, L., Gantchev, N., and Sinkjaer, T. 1992. The influence of muscle length on muscle fiber conduction velocity and development of muscle fatigue. *Electroencephalogr. Clin. Neurophysiol.* 85: 166–72.

Arendt-Nielsen, L., Mills, K.R., and Forster, A. 1989. Changes in muscle fiber conduction velocity, mean power frequency, and mean EMG voltage during prolonged submaximal contractions. *Muscle & Nerve* 12: 493–96.

Arendt-Nielsen, L., and Zwarts, M. 1989. The measurement of muscle fiber conduction velocity in humans: Techniques and applications. *J. Clin. Neurophysiol.* 6: 173–90.

Barry, D.T., Geiringer, S.R., and Ball, R.D. 1985. Acoustic myography: A noninvasive monitor of motor unit fatigue. *Muscle & Nerve* 8: 189–94.

Barry, D.T., Hill, T., and Dukjin, I. 1992. Muscle fatigue measured with evoked muscle vibrations. *Muscle & Nerve* 15: 303–9.

Basmajian, J.V., and De Luca, C.J. 1985. *Muscles alive: Their functions revealed by electromyography.* Baltimore: Williams & Wilkins.

Belanger, A.Y., and McComas, A.J. 1983. Contractile properties of muscles in myotonic dystrophy. *J. Neurol. Neurosurg. Psychiatry* 46: 625–31.

Bellemare, F., Woods, J.J., Johansson, R., and Bigland-Ritchie, B. 1983. Motor unit discharge rates in maximal voluntary contractions of three human muscles. *J. Neurophysiol.* 50: 1380–92.

Benzanilla, F., Caputo, C., Gonzales-Serratos, H., and Venosa, R.A. 1972. Sodium dependence of the inward spread of activation in isolated twitch muscle fibers of the frog. *J. Physiol. (London)* 223: 507–23.

Bevan, L., Laouris, Y., Reinking, R.M., and Stuart, D.G. 1992. The effect of the stimulation pattern on the fatigue of single motor units in adult cats. *J. Physiol. (London)* 449: 85–108.

Bigland-Ritchie, B., Cafarelli, E., and Vollestad, N.K. 1986. Fatigue of submaximal static contractions. *Acta Physiol. Scand.* 128: 137–48.

Bigland-Ritchie, B., Furbush, F., and Woods, J.J. 1986. Fatigue of intermittent submaximal voluntary contractions: Central and peripheral factors. *J. Appl. Physiol.* 61: 421–29.

Bigland-Ritchie, B., Johansson, R., Lippold, O.C.J., Smith, S., and Woods, J.J. 1983a. Changes in motoneuron firing rates during sustained maximal voluntary contractions. *J. Physiol. (London)* 340: 335–46.

Bigland-Ritchie, B., Johansson, R., Lippold, O.C.J., and Woods, J.J. 1983b. Contractile speed and EMG changes during fatigue of sustained maximal voluntary contractions. *J. Neurophysiol.* 50: 313–24.

Bigland-Ritchie, B., Jones, D.A., and Woods, J.J. 1979. Excitation frequency and muscle fatigue: electrical responses during human voluntary and stimulated contractions. *Exp. Neurol.* 64: 414–27.

Bigland-Ritchie, B., and Woods, J.J. 1984. Changes in muscle contractile properties and neural control during human muscular fatigue. *Muscle & Nerve* 7: 691–99.

Binder-Macleod, S.A., and Guerin, T. 1990. Preservation of force output through progressive reduction of stimulation frequency in human quadriceps femoris muscle. *Physical Therapy* 70: 619–25.

Bolton, C.F., Parkes, A., Thompson, T.R., Clark, M.R., and Sterne, C.J. 1989. Recording sound from human skeletal muscle: Technical and physiological aspects. *Muscle & Nerve* 12: 126–34.

Bongiovanni, L.G., and Hagbarth, K.E. 1990. Tonic vibration reflexes elicited during fatigue from maximal voluntary contractions in man. *J. Physiol. (London)* 423: 1–14.

Botterman, B.R. 1995. Task dependent nature of fatigue in single motor units. In *Fatigue: Neural and muscular mechanisms*, ed. S.C. Gandevia, R.M. Enoka, A.J. McComas, D.G. Stuart, and C.K. Thomas, 351–60. New York: Plenum Press.

Brody, L.R., Pollock, M.T., Roy, S.H., De Luca, C.J., and Celli, B. 1991. pH-induced effects on median frequency and conduction velocity of the myoelectric signal. *J. Appl. Physiol.* 71: 1878–85.

Broman, H., Bilotto, G., and De Luca, C.J. 1985a. Myoelectric signal conduction velocity and spectral parameters: Influence of force and time. *J. Appl. Physiol.* 58: 1428–37.

Broman, H., Bilotto, G., and De Luca, C.J. 1985b. A note on the noninvasive estimation of muscle fiber conduction velocity. *IEEE Trans. BME* 32: 341–44.

Byrd, S.K., McCutcheon, L.J., Hodgson, D.R., and Gollnick, P.D. 1989. Altered sarcoplasmic reticulum function after high intensity exercise. *J. Appl. Physiol.* 67: 2072–77.

Christensen, H., and Fuglsang-Frederiksen, A. 1988. Quantitative surface EMG during sustained and intermittent submaximal contractions. *Electroencephalog. Clin. Neurophysiol.* 70: 239–47.

Dalton, P.A., Comerford, M.J., and Stokes, M.J. 1992. Acoustic myography of the human quadriceps muscle during intermittent fatiguing activity. *J. Neurol. Sci.* 109: 56–60.

Dalton, P.A., and Stokes, M.J. 1991. Acoustic myography reflects force changes during dynamic concentric and eccentric contractions of the human biceps brachii muscle. *Eur. J. Appl. Physiol.* 63: 412–416.

Dalton, P.A., and Stokes, M.J. 1993. Frequency of acoustic myography during isometric contraction of fresh and fatigued muscle and during dynamic contractions. *Muscle & Nerve* 16: 255–61.

De Luca, C.J. 1985. Myoelectrical manifestations of localized muscular fatigue in humans. *CRC Crit. Rev. Biomed. Eng.* 11: 251–79.

De Luca, C.J., Foley, P.J., and Zeynep, E. 1996. Motor unit control properties in constant force isometric contractions. *J. Neurophysiol.* 76: 1503–15.

De Luca, C.J., LeFever, R.S., McCue, M.P., and Xenakis, A.P. 1982. Control scheme governing concurrently active human motor units during voluntary contractions. *J. Physiol. (London)* 329: 129–42.

De Luca, C.J., and Merletti, R. 1988. Surface myoelectric signal cross talk among muscles of the leg. *Electroencephalog. Clin. Neurophysiol.* 69: 568–75.

Dorfman, L.J., Howard, J.E., and McGill, K.C. 1990. Triphasic behavioral response of motor units to submaximal fatiguing exercise. *Muscle & Nerve* 13: 621–28.

Duchene, J., and Goubel, F. 1993. Surface electromyogram during voluntary contraction: Processing tools and relation to physiological events. *Crit. Rev. Biomed. Eng.* 21: 313–97.

Eberstein, A., and Beattie, B. 1985. Simultaneous measurement of muscle conduction velocity and EMG power spectrum changes during fatigue. *Muscle & Nerve* 8: 768–73.

Edwards, R.H.T. 1981. Human muscle function and fatigue. In *Human muscle fatigue: Physiological mechanism*, ed. R. Porter and J. Whelan, 1–18. London: Pitman Medical.

Edwards, R.H.T., Young, A., Hosking, G.P., and Jones, D.A. 1977. Human skeletal muscle function: Description of tests and normal values. *Clin. Sci. Mol. Med.* 52: 283–90.

Elble, R.J., and Randall, J.E. 1976. Motor-unit activity responsible for 8 to 12 Hz component of human physiological finger tremor. *J. Neurophysiol.* 39: 370–83.

Enoka, R.M., and Stuart, D.G. 1985. The contribution of neuroscience to exercise studies. *Fed. Proc.* 44: 2279–85.

Enoka, R.M, and Stuart, D.G. 1992. Neurobiology of muscle fatigue. *J. Appl. Physiol.* 72: 1631–48.

Fallentin, N., Jorgensen, K., and Simonsen, E.B. 1993. Motor unit recruitment during prolonged isometric contractions. *Eur. J. Appl. Physiol.* 67: 335–41.

Fitts, R.H., 1994. Cellular mechanisms of muscle fatigue. *Physiol. Rev.* 74: 49–97.

Fuglevand, A.J. 1995. The role of the sarcolemma action potential in fatigue. In *Fatigue: Neural and muscular mechanisms,* ed. S.C. Gandevia, R.M. Enoka, A.J. McComas, D.G. Stuart, and C.K. Thomas, 101–8. New York: Plenum Press.

Gamet, D., and Maton, B. 1989. The fatigability of two agonistic muscles in human isometric voluntary submaximal contraction: An EMG study. I. Assessment of muscular fatigue by means of surface EMG. *Eur. J. Appl. Physiol.* 58: 361–68.

Garland, S.J. 1991. Role of small diameter afferents in reflex inhibition during human muscle fatigue. *J. Physiol. (London)* 435: 547–58.

Garland, S.J., Enoka, R.M., Serrano, L.P., and Robinson, G.A. 1994. Behavior of motor units in human biceps brachii during a submaximal fatiguing contraction. *J. Appl. Physiol.* 76(6): 2411–19.

Garland, S.J., and Kaufman, M.P. 1995. Role of muscle afferents in the inhibition of motoneurons during fatigue. In *Fatigue: Neural and muscular mechanisms,* ed. S.C. Gandevia, R.M. Enoka, A.J. McComas, D.G. Stuart, and C.K. Thomas, 271–78. New York: Plenum Press.

Gerdle, B., Hedberg, R., Jonsson, B., and Fugl-Meyer, A.R. 1987. Mean power frequency and integrated electromyogram of repeated isokinetic plantarflexions. *Acta Physiol. Scand.* 130: 501–6.

Goldenberg, M.S., Yack, H.J., Cerny, F.J., and Burton, H.W. 1991. Acoustic myography as an indicator of force during sustained contractions of a small hand muscle. *J. Appl. Physiol.* 70: 87–91.

Grimby, L. 1986. Single motor unit discharge during voluntary contraction and locomotion. In *Human muscle power,* ed. N.L. Jones, N. McCartney, and A.J. McComas, 111–22. Champaign, IL: Human Kinetics.

Gydikov, A., and Kosarov, D. 1974. Some features of different motor units in human biceps brachii. *Pflügers Arch.* 347: 75–88.

Hagberg, C., and Hagberg, M. 1989. Surface EMG amplitude and frequency dependence on exerted force for the upper trapezius: A comparison between right and left sides. *Eur. J. Appl. Physiol.* 58: 641–45.

Hainaut, K., and Duchateau, J. 1989. Muscle fatigue, effects of training and disuse. *Muscle & Nerve* 12: 660–9.

Hayward, L., Wesselman, U., and Rymer, Z.W. 1991. Effects of muscle fatigue on mechanically sensitive afferents of slow conduction velocity in the cat triceps surae. *J. Neurophysiol.* 65: 360–9.

Hermens, H.J., Bruggen, T.A.M., Baten, C.T.M., Rutten, W.L.C., and Boom, H.B.K. 1992. The median frequency of the surface EMG power spectrum in relation to motor unit firing and action potential properties. *J. Electromyogr. Kinesiol.* 2: 15–25.

Herzog, W., Guimaraes, A.C.S., and Zhang, Y.T. 1994. EMG. In *Biomechanics of the musculoskeletal system,* ed. B.M. Nigg and W. Herzog, 308–36. New York: Wiley.

Herzog, W., Zhang, Y.T., Vaz, M.A., Guimaraes, A.C.S., and Janssen, C. 1994. Assessment of muscular fatigue using vibromyography. *Muscle & Nerve* 17: 1156–61.

Jones, D.A., Bigland-Ritchie, B., and Edwards, R.H.T. 1979. Excitation frequency and muscle fatigue: Mechanical responses during voluntary and stimulated contractions. *Exp. Neurol.* 64: 401–13.

Keidel, M., and Keidel, W.D. 1987. Spectral analysis of muscle vibrations: Computer vibromyography (C-VMG). *Electroencephalogr. Clin. Neurophysiol.* 68: 66–67.

Keidel, M., and Keidel, W.D. 1989. The computer vibromyography as a biometric process in studying muscle function. *Biomed. Technik* 34: 107–16.

Kernell, D., and Monster, A.W. 1982a. Motoneuron properties and motor fatigue: An intracellular study of gastrocnemius motoneurons of the cat. *Exp. Brain Res.* 46: 197–204.

Kernell, D., and Monster, A.W. 1982b. Time course and properties of late adaptation in spinal motoneurons in the cat. *Exp. Brain Res.* 46: 191–96.

Kernell, D. 1995. Neuromuscular frequency coding and fatigue. In *Fatigue: Neural and muscular mechanisms,* ed. S.C. Gandevia, R.M. Enoka, A.J. McComas, D.G. Stuart, and C.K. Thomas, 135–45. New York: Plenum Press.

Kranz, H., Cassell, J.F., and Inbar, G.F. 1985. Relation between electromyogram and force in fatigue. *J. Appl. Physiol.* 59: 821–25.

Krarup, C. 1981. Enhancement and diminution of mechanical tension evoked by staircase and by tetanus in rat muscle. *J. Physiol. (London)* 311: 355–72.

Krivickas, L.S., Nadler, S.F., Davies, M.R., Petroski, G.F., and Feinberg, J.H. 1996. Spectral analysis during fatigue: Surface and fine wire electrode comparison. *Am. J. Phys. Med. & Rehab.* 75: 15–20.

Krogh-Lund, C. 1993. Myoelectric fatigue and force failure from submaximal static elbow flexion sustained to exhaustion. *Eur. J. Appl. Physiol.* 67: 389–401.

Krogh-Lund, C., and Jorgensen, K. 1991. Changes in conduction velocity, median frequency, and root mean square amplitude of the electromyogram during 25% maximal voluntary contraction of the triceps brachii muscle, to limit of endurance. *Eur. J. Appl. Physiol.* 63: 60–9.

Krogh-Lund, C., and Jorgensen, K. 1992. Modification of myoelectric power spectrum in fatigue from 15% maximal voluntary contraction of human elbow flexor muscles, to limit of endurance: Reflection of conduction velocity variation and/or centrally mediated mechanisms. *Eur. J. Appl. Physiol.* 64: 359–70.

Krogh-Lund, C., and Jorgensen, K. 1993. Myoelectric fatigue manifestations revisited: Power spectrum, conduction velocity, and amplitude of human elbow flexor muscles during isolated and repetitive endurance contractions at 30% maximal voluntary contraction. *Eur. J. Appl. Physiol.* 66: 161–73.

Kukulka, C.G., Russell, A.G., and Moore, M.A. 1986. Electrical and mechanical changes in human soleus muscle during sustained maximum isometric contractions. *Brain. Res.* 362: 47–54.

Lammert, O., Jorgensen, F., and Einer-Jensen, N. 1976. Accelerometermyography (AMG). I. Method for measuring mechanical vibrations from isometrically contracted muscles. *Biomechanics.* Vol. A. 152–56. Baltimore: University Park Press.

Lindstrom, L., Kadefors, R., and Petersen, I. 1977. An electromyographic index for localized muscle fatigue. *J. Appl. Physiol.* 43(4): 750–4.

Lindstrom, L., Magnusson, R., and Petersen, R. 1970. Muscle fatigue and action potential conduction velocity changes studied with frequency analysis of EMG signals. *Electromyogr. Clin. Neurophysiol.* 10: 341–56.

Linssen, H.J.P., Stegeman, D.F., Joosten, M.G., van 'T Hof, M.A., Binkhorst, R.A., and Notermans, S.L.H. 1993. Variability and interrelationships of surface EMG parameters during local muscle fatigue. *Muscle & Nerve* 16: 849–56.

Lippold, O.C.J. 1981. The tremor in fatigue. In *Human muscle fatigue: Physiological mechanisms.* 234–48. Ciba Foundation Symposium 82. London: Pitman Medical.

Lippold, O.C.J., Redfearn, J.W.T., and Vuco, J. 1957. The rhythmical activity of groups of motor units in the voluntary contraction of muscle. *J. Physiol. (London)* 137: 473–87.

Mannion, A.F., and Dolan, P. 1996. Relationship between myoelectric and mechanical manifestation of fatigue in quadriceps femoris muscle group. *Eur. J. Appl. Physiol.* 74: 411–19.

Marsden, C.D., Meadows, J.C., and Merton, P.A. 1983. "Muscular wisdom" that minimizes fatigue during prolonged effort in man: Peak rates of motoneuron discharge and slowing of discharge during fatigue. In *Advances in neurology.* Vol. 39. Ed. J.E. Desmedt, 169–212. New York: Raven Press.

Maton, B. 1981. Human motor unit activity during the onset of muscle fatigue in submaximal isometric isotonic contraction. *Eur. J. Appl. Physiol.* 46: 271–81.

Maton, B. 1991. Central nervous changes in fatigue induced by local work. In *Muscle fatigue: Biochemical and physiological aspects,* ed. G. Atlan, L. Beliveau, and P. Bouissou, 207–21. Paris: Masson.

Maton, B., and Gamet, D. 1989. The fatigability of two agonist muscles in human isometric voluntary submaximal contraction: An EMG study. *Eur. J. Appl. Physiol.* 58: 369–74.

Matthews, P.B.C., and Muir, R.B. 1980. Comparison of electromyogram spectra with force spectra during elbow tremor. *J. Physiol. (London)* 302: 427–41.

Merletti, R., Knaflitz, M., and De Luca, C.J. 1990. Myoelectric manifestations of fatigue in voluntary and electrically elicited contractions. *J. Appl. Physiol.* 69: 1810–20.

Merletti, R., Knaflitz, M., and De Luca, C.J. 1992. Electrically evoked myoelectric signals. *CRC Crit. Rev. Biomed. Eng.* 19: 293–340.

Merletti, R., and Lo Conte, L.R. 1992. Electrically evoked surface myoelectric signals. *Func. Neurol.* 7(Suppl.): 153–58.

Merletti, R., Lo Conte, L.R., and Orizio, C. 1991. Indexes of muscle fatigue. *J. Electromyogr. Kinesiol.* 1: 20–33.

Merletti, R., and Roy, S.H. 1996a. Myoelectric and mechanical manifestations of muscle fatigue in voluntary contractions. *J. Orth. Sports Phys. Ther.* 24: 342–53.

Merletti, R., and Roy, S.H. 1996b. Myoelectric manifestations of muscle fatigue: A simulation study. Proc. 11th Int. Congress of ISEK. Enschede. 176–77.

Moglia, A., Alfonsi, E., Piccolo, G., Arrigo, A., Bollani, E., and Malaguti, S. 1995. Twitch response of striated muscle in patients with progressive external ophtalmoplegia, mitochondrial myopathy, and focal cytochrome c-oxidase deficiency. *Ital. J. Neurol. Sci.* 16: 159–66.

Moritani, T., Muro, M., and Nagata, A. 1986. Intramuscular and surface electromyogram changes during muscle fatigue. *J. Appl. Physiol.* 60: 1179–85.

Nigg, B.M. 1994. Acceleration. In *Biomechanics of the musculoskeletal system,* ed. B.M. Nigg and W. Herzog, 237–53. Chichester, UK: Wiley.

Oberg, T., Sandsjo, L., Kadefors, R., and Larsson, S-E. 1992. Electromyographic changes in work-related myalgia of the trapezius muscle. *Eur. J. Appl. Physiol.* 65: 251–57.

Orizio, C. 1992. Sound myogram and EMG cross-spectrum during exhausting isometric contractions in humans. *J. Electromyogr. Kinesiol.* 2: 141–49.

Orizio, C. 1993. Muscle sound: Bases for the introduction of a mechanomyographic signal in muscle studies. *Crit. Rev. in Biomed. Eng.* 21: 201–43.

Orizio, C., Baratta, R., Zhou, B.H., Solomonow, M., and Veicsteinas, A. 1997. Relationship between laser-detected surface displacement and force in stimulated muscle. Proc. Spring Meeting of S.I.F. Firenze 25–27 March, *Pflügers Arch. Eur. J. Physiol.* 434: R-56.

Orizio, C., Diemont, B., Alfonsi, E., Moglia, A., and Veicsteinas, A. 1996. Force fluctuations and surface mechanomyogram during electrically elicited muscle contractions. *J. Muscle Res. Cell. Motil.* Proc. XXIV European Muscle Conference. Firenze 13–16 September 1995. 17: 158–59.

Orizio, C., Diemont, B., Bianchi, A., Liberati, D., Cerutti, S., and Veicsteinas, A. 1991. Coherence analysis between sound myogram and electromyogram. Proc. of the 13th Annual International Conference of the IEEE-MBS, Orlando. 942–43.

Orizio, C., Esposito, F., Paganotti, I., Marino, L., Rossi, B., and Veicsteinas, A. 1997. Electrically-elicited surface mechanomyogram in myotonic dystrophy. *Ital. J. Neurol. Sci.* 18: 185–90.

Orizio, C., Liberati, D., Locatelli, C., De Grandis, D., and Veicsteinas, A. 1996. Surface mechanomyogram reflects muscle fibers twitches summation. *J. Biomech.* 29: 475–81.

Orizio, C., Perini, R., Diemont, B., Maranzana Figini, M., and Veicsteinas, A. 1990. Spectral analysis of muscular sound during isometric contraction of biceps brachii. *J. Appl. Physiol.* 68: 508–12.

Orizio, C., Perini, R., Diemont, B., and Veicsteinas, A. 1992. Muscle sound and electromyogram spectrum analysis during exhausting contractions in man. *Eur. J. Appl. Physiol.* 65: 1–7.

Orizio, C., Perini, R., and Veicsteinas, A. 1989. Changes of muscular sound during sustained isometric contraction up to exhaustion. *J. Appl. Physiol.* 66: 1593–98.

Orizio, C., and Veicsteinas, A. 1992. Sound myogram analysis during sustained maximal voluntary contraction in sprinters and long distance runners. *Int. J. Sports Med.* 13: 594–99.

Person, R.S., and Kudina, L.P. 1972. Discharge frequency and discharge pattern of human motor units during voluntary contraction of muscle. *Electroencephalogr. Clin. Neurophysiol.* 32: 471–83.

Petitjean, M., and Bellemare, F. 1994. Phonomyogram of the diaphragm during unilateral and bilateral phrenic nerve stimulation and changes with fatigue. *Muscle & Nerve* 17: 1201–9.

Petitjean, M., and Maton, B. 1995. Phonomyogram from single motor units during voluntary isometric contraction. *Eur. J. Appl. Physiol.* 71: 215–22.

Piper, H. 1912. *Electrophysiologie Menschlicher Muskeln* Berlin: Springer-Verlag.

Rodriquez, A.A., Agre, J.C., Franke, T.M., Swiggum, E.R., and Curt, J.T. 1996. Acoustic myography during isometric fatigue in postpolio and control subjects. *Muscle & Nerve* 19: 384–87.

Rodriquez, A.A., Agre, J.C., Knudtson, E.R., Franke T.M., and Ng, A.V. 1993. Acoustic myography compared to electromyography during isometric fatigue and recovery. *Muscle & Nerve* 16: 188–92.

Rosenfalck, P. 1969. Intra- and extracellular potential fields of active nerve and muscle fibers. *Acta Physiol. Scand. Suppl.* 321: 168.

Sadamoto, T., Bonde-Petersen, F., and Suzuki, Y. 1983. Skeletal muscle tension, flow, pressure, and EMG during sustained isometric contractions in humans. *Eur. J. Appl. Physiol.* 51: 395–408.

Sandercock, T.G., Faulkner, J.A., Albers, J.W., and Abbrecht, P.H. 1985. Single motor unit and fiber action potentials during fatigue. *J. Appl. Physiol.* 58: 1073–79.

Sandow, A. 1952. Excitation-contraction coupling in muscular response. *Yale J. Biol. Med.* 25: 176–201.

Sawczuk, A., Powers, R.K., and Binder, M.D. 1995. Spike frequency adaptation studied in hypoglossal motoneurons of the rat. *J. Neurophysiol.* 73: 1799–1810.

Smith, T.G., and Stokes, M.J. 1993. Technical aspects of acoustic myography (AMG) of human skeletal muscle, contact pressure and force/AMG relationships. *J. Neurosci. Meth.* 47: 85–92.

Stephenson, D.G., Lamb, G.D., Stephenson, G.M., and Fryer, M.W. 1995. Mechanisms of excitation-contraction coupling relevant to skeletal muscle fatigue. In *Fatigue: Neural and muscular mechanisms,* ed. S.C. Gandevia, R.M. Enoka, A.J. McComas, D.G. Stuart, and C.K. Thomas, 45–56. New York: Plenum Press.

Stokes, M.J., and Dalton, P.A. 1991. Acoustic myography for investigating human skeletal muscle fatigue. *J. Appl. Physiol.* 71: 1422–26.

Stulen, F.B., and De Luca, C.J. 1981. Frequency parameters of the myoelectric signal as a measure of muscle conduction velocity. *IEEE Trans. Biomed. Eng.* 28: 515–23.

Sundelin, G., and Hagberg, M. 1992. Electromyographic signs of shoulder muscle fatigue in repetitive arm work paced by the methods-time measurement system. *Scand. J. Work Environ. Health* 18: 262–68.

Takaishi, T., Yasuda, Y., and Moritani, T. 1994. Neuromuscular fatigue during prolonged pedaling exercise at different pedaling rates. *Eur. J. Appl. Physiol.* 69: 154–58.

Takamori, M., Gutman, L., and Shane, S.R. 1971. Contractile properties of human skeletal muscle: Normal and thyroid disease. *Arch. Neurol.* 25: 535–46.

Tesch, P.A., Dudley, G.A., Duvoisin, M.R., Hather, B.M., and Harris, R.T. 1990. Force and EMG signal patterns during repeated bouts of concentric or eccentric muscle actions. *Acta Physiol. Scand.* 138: 263–71.

Thompson, L.V., Balog, E.M., Riley, D.A., and Fitts, R.H. 1992. Muscle fatigue in frog semitendinosus: Alterations in contractile function. *Am. J. Physiol.* 262: C1500–6.

Troni, W., Cantello, R., and Raniero, I. 1983. Conduction velocity along human fibers in situ. *Neurology* 33: 1453–59.

Trontelj, J.V., and Stalberg, E. 1995. Single fiber electromyography in studies of neuromuscular function. In *Fatigue: Neural and muscular mechanisms,* ed. S.C. Gandevia, R.M. Enoka, A.J. McComas, D.G. Stuart, and C.K. Thomas, 109–19. New York: Plenum Press.

Van der Hoeven, J.H. 1995. Conduction velocity in human muscle fibers—normal values and technical notes: Invasive and surface EMG determination techniques. In *Conduction velocity in human muscle. An EMG study in fatigue and neuromuscular disorders.* Thesis Rijrsuniversiteit Groningen (ISBN 90-367-0551-7). 15–28.

Van der Hoeven, J.H., and Lange, F. 1994. Supernormal muscle fiber conduction velocity during intermittent isometric exercise in human muscle. *J. Appl. Physiol.* 77: 802–6.

Van der Hoeven, J.H., van Weerden, T.W., and Zwarts, M.J. 1993. Long-lasting supernormal conduction velocity after sustained maximal isometric contraction in human muscle. *Muscle & Nerve* 16: 312–20.

Viitasalo, J.H.T., and Komi, P.V. 1977. Signal characteristics of EMG during fatigue. *Eur. J. Appl. Physiol.* 37: 111–21.

West, W., Hicks, A., Clements, L., and Dowling, J. 1995. The relationship between voluntary electromyogram, endurance time and intensity of effort in isometric handgrip exercise. *Eur. J. Appl. Physiol.* 71: 301–5.

Windhorst, U., and Kokkoroyiannis, T. 1991. Interaction of recurrent inhibitory and muscle spindle afferent feedback during muscle fatigue. *Neuroscience* 43: 249–59.

Windhorst, U., and Boorman, G. 1995. Overview: Potential role of segmental motor circuitry in muscle fatigue. In *Fatigue: Neural and muscular mechanisms*, ed. S.C. Gandevia, R.M. Enoka, A.J. McComas, D.G. Stuart, and C.K. Thomas, 241–58. New York: Plenum Press.

Zhang, Y.T., Frank, C.B., Rangayyan, R.M., and Bell, G.D. 1992. A comparative study of simultaneous vibromyography and electromyography with active human quadriceps. *IEEE Trans. BME* 39: 1045–52.

Zijdewind, I., Kernell, D., and Kukulka, C.G. 1995. Spatial differences in fatigue-associated electromyographic behavior of the human first dorsal interosseus muscle. *J. Physiol. (London)* 483(2): 499–509.

Zwarts, M.J., and Keidel, M. 1991. Relationship between electrical and vibratory output of muscle during voluntary contraction and fatigue. *Muscle & Nerve* 14: 756–61.

Zwarts, M.J., van Weerden, T.W., and Haenen, H.T.M. 1987. Relationship between average muscle fiber conduction velocity and EMG power spectra during isometric contraction, recovery, and applied ischemia. *Eur. J. Appl. Physiol.* 56: 212–16.

Fatigue and Exercise Summary

B.R. MacIntosh

Everybody knows what it feels like to experience skeletal muscle fatigue. It is something that affects elite athletes who push the limits of human capability but can touch anybody who exerts him- or herself in physical effort. Fatigue could be the result of repetitious, low-intensity movement patterns with small muscle groups; high-intensity exercise using a substantial portion of the total muscle mass; or any form of exercise between these extremes.

When contractile activity is sustained or repeated over a period of time, it impairs the ability of the muscle to generate tension and perform work. A period of recovery will restore the contractile response to the prefatigue level, though in some cases several hours of recovery is needed. Recovery can take even longer when there is structural damage resulting from the exercise. This is more apt to occur when the exercise included eccentric contractions.

There are specific changes to the contractile response which usually coincide with the attenuated ability to generate force including: the slowing of relaxation; a prolonged twitch contraction time; a greater decrease in force at low frequencies than high ones; and, in some cases, a decrease in maximal force production. When shortening is permitted during the contractions under study, decreases in maximum velocity of shorting is observed. When force-velocity properties are characterized, there is a decrease in the degree of curvature of the force-velocity relationship. The consequence of less curvature is that the decrease in power output is less than would be predicted based on the measured decrease in maximal velocity and isometric force.

Clinical detection of muscle fatigue is an important goal in solving problems of apparent muscle weakness. A great deal of effort has gone into finding ways to detect muscle fatigue, preferably by noninvasive means. The current practice is to use surface electromyography for this purpose. The surface EMG signal has characteristic changes which occur in fatigue including an increase in magnitude of EMG (root mean square) while sustaining a constant submaximal effort, or a decrease in the median frequency obtained with Fourier transform of the data. Recently, considerable advances have been made in understanding the vibration signal emitted by the muscle, and it appears that this signal may be useful for detecting fatigue as well.

In chapter 19, MacIntosh and Allen provide some background information on muscle fatigue. They relate the various ways fatigue is studied, describe the consequences of fatigue in terms of alterations in the contractile response, and review the current understanding of the potential cellular mechanisms of fatigue.

Muscle fatigue can be demonstrated in vivo and studied at this level in terms of characterizing the gross contractile changes associated with fatigue. However, to understand the specific cellular mechanisms which cause fatigue, it is necessary to study muscle at the single cell level. To understand the specific molecular processes altered in fatigue, it may be necessary to study muscle at the subcellular level. This would include skinned fiber preparations as well as myofibrils and isolated molecules. Unfortunately the more precision you have in controlling the environment and measuring the specific response of molecules, the less you know about how it fits into a whole muscle response. An example of inappropriate extrapolation of cellular mechanisms to whole muscle response is the commonly held belief that acidosis results in decreased sensitivity to calcium and therefore contributes to fatigue. The evidence for this conclusion comes from skinned fiber experiments which were conducted on mammalian tissue at room temperature. Recently, comparable experiments have been done with intact and skinned fibers at more physiological temperatures, and the results indicate that acidosis is unlikely to have this direct effect on mammalian muscle at physiological temperatures.

The pursuit of understanding the cellular mechanism of muscle fatigue has been a primary goal of a large number of scientists over the past century (and longer). Several reasons may explain why our understanding of muscle fatigue is less than complete. It is likely that the cellular processes responsible for decreased contractile response vary according to the manner in which the fatigue was

435

induced. Therefore, there is not a unifying theory of muscle fatigue, but rather a series of potential contributing factors that may contribute to a varying extent according to the manner in which fatigue was induced. These factors, which may include just about any step in excitation-contraction coupling, ultimately end up with less calcium released in the cytoplasm of the activated muscle tissue. However, there are several potential factors which could alter the calcium sensitivity, including increased inorganic phosphate. A reduction in calcium release in muscle cells could result from decreased Ca^{2+} stores, altered voltage sensor operation, or impaired channel opening. Increased Mg^{2+} concentration or decreased ATP concentration are potential factors which could alter Ca^{2+} release.

In chapter 20, Komi and Nicol present an interesting perspective on muscle fatigue. They deal specifically with the case where the exercise has involved eccentric contractions, and in particular, the stretch-shortening cycle (SSC). Stretch-shortening cycle contractions typically occur during voluntary movements in daily activities, such as running, cross-country skiing, throwing, and jumping. This type of contraction can result in severe damage, particularly when prior training with eccentric contractions has not been undertaken. There are well-recognized stages of damage and recovery which have been identified:

- *The initial stage:* myofilament disarray and Z-line streaming
- *The autogenic stage:* degradation of membranes, loss of Ca^{2+} homeostasis
- *The phagocytic stage:* inflammation and infiltration
- *The regenerative stage:* marked by net protein synthesis and muscle fiber regeneration

Impaired contractile response of muscle which has undergone eccentric contractions, or SSC, can result from direct damage to contractile protein, neural inhibition of motor pathways, or attenuation of reflex contribution to the activation of the involved muscles.

Regardless of how muscle fatigue has come about or what the cellular mechanism is, the detection of muscle fatigue has been considered an important clinical pursuit. This is not as simple as measuring the contractile output of a muscle because fatigue may be present even when maximal voluntary effort is not changed. Low-frequency fatigue results in attenuated contractile response when stimulation is at a relatively low frequency (incompletely fused tetanic contraction), such as during submaximal effort, but high frequency response (maximal voluntary effort) may not be altered.

Two general approaches are used to detect fatigue under these circumstances. These two approaches include electromyography (EMG) and mechanomyography (MMG) and they are described by Orizio in chapter 21. Electromyography relies on detecting current flow within the muscles, which depends on action potential propagation along membranes of the activated muscle fibers. Mechanical vibration of the muscle is detected in a variety of ways including microphones, piezoelectric contact sensors, and accelerometers. The detectable vibration is thought to be the result of incompletely fused tetanic contractions resulting in pressure waves transmitted to the skin.

Surface EMG (as opposed to indwelling electrodes) is more commonly used, and the signal can be analyzed in a variety of ways: estimation of relative magnitude by calculation of a root mean square or integrated rectified signal; or estimation of frequency components by fast Fourier transform. When a muscle has been previously active to the point when fatigue is present, EMG analysis will detect an increased signal while the subject maintains a given constant isometric force and a decrease in the median frequency in the fast Fourier transform. The increased magnitude of the EMG signal is probably due to increased recruitment of motor units or possibly increased frequency of activation of the active motor units. There is some discrepancy in the literature as to whether there is an increase or a decrease in firing frequency of individual motor units though this may be related to the timing of the measurement. The decrease in median frequency of the fast Fourier transform has been associated with a decrease in the conduction velocity of action potential propagation on the active muscle fibers.

Several characteristic changes to the MMG signal seem to correspond to the known changes in the contractile response of fatiguing muscle. However, there are apparently conflicting observations which can be resolved if consideration is given to the circumstances under which the recordings were obtained (submaximal versus maximal contractions and stimulated versus voluntary activation). The slowing of the contractile response results in a greater fusion between sequential twitch responses and, therefore, a lower magnitude of vibration. This is more apparent during maximal effort. Fast Fourier analysis of the MMG signal reveals a decrease in the median frequency of the vibration signal. This could result from either (or both) a decrease in mean firing frequency of individual motor units or a slowing of the contractile response to each activation (increased twitch contraction time and slower relaxation). Late changes in fatigue while sustaining a submaximal effort, including increased amplitude and decreased frequency of vibration, may be related to greater synchronization of motor unit activation.

Orizio also describes a new, potentially useful technique for indirect detection of muscle force. This technique relies on measuring skin displacement by laser reflection. This technique shows promise but requires further validation before being accepted clinically for evaluation of muscle fatigue.

Fatigue and Exercise Definitions

Biological

electromyogram (EMG): biological signal detectable from the muscle by needle or surface electrodes. It reflects the electrical activities of the recruited motor units. The dimension of the investigated motor unit pool, i.e., the number of MUAPs contributing to the signal, depends on the selectivity of the electrodes.

fatigue: fatigue is understood to mean failure of the ability to sustain a given exercise task, including failure to maintain force or power output. Alternatively, muscle fatigue has been defined as a "reversible state of force depression...". In part IV, fatigue is used to indicate less force or work per contraction for a given stimulation.

> **central fatigue:** attenuation of the force response which is due to failure in the central nervous system, rather than with the steps in muscle activation which lie outside the central nervous system.

> **low and high-frequency fatigue:** following a period of exercise, if the contractile response is more attenuated at low than at high frequencies, then the term low-frequency fatigue is used to describe the nature of the force depression. If the contractile response is more depressed at high frequencies of activation than at low frequencies, then it is considered to be high frequency fatigue. In general, high-frequency fatigue results from a sustained muscle contraction, and recovery from this state of force depression is rapid (seconds to minutes). In contrast, low-frequency fatigue results from a variety of stimulation protocols, including intermittent low or high frequency contractions and recovery is very slow (hours to days).

> **muscle fatigue:** a progressive phenomenon beginning at the onset of contraction. It determines an increasing effort to keep constant the muscle mechanical output. The process involves central and peripheral control systems providing time-related changes in motor unit levels of recruitment and firing rate. The outputs of the control systems take into account the state of the actuator, i.e. the muscle, by means of specific afferents.

> **peripheral fatigue:** fatigue is considered to be peripheral in origin when it results from failure in one of the following: conduction of the motor neuron action potential; neuromuscular junction transmission; muscle fiber action potential generation and propagation; calcium release within the fiber; calcium binding with troponin; or impaired cross-bridge interaction.

integrated signal: an indication of the amplitude of the signal obtained by the calculation of the area beneath the rectified signal. It can be calculated without time window (in this case it increases continuously from the beginning to the end of processing) or with time window (T). In this last case the parameter indicates time dependent signal modifications. The units of this parameter is $V \cdot s$ (submultiples are allowed).

$$I[m(t)] = \int_t^{t+T} |m(t)| \, dt$$

motor unit action potential (MUAP): the summation of all the sarcolemmal action potentials of the muscle fibers of an active motor unit.

motor unit activation pattern: strategy of activation of the motor unit pool. It refers to the chosen interrelation between the number (recruitment) and the firing rate of the active motor units as a tool to control the muscle force output.

muscle fiber conduction velocity: the velocity of propagation of the MUAP along the muscle fibers. It depends on the membrane characteristics. The latter may change with fatigue.

muscle force: the force can be considered as the cause of the movement of a body or of the deformation of the constraints to which it is applied. A more precise analysis of the muscle action recognizes that muscle does not generate force but tension at the tendon. This tension is transmitted via bony structures to the points of applications of the force (the arms of a nut-cracker or the foot contact area on the ground during a jump).

muscle wisdom: the tailoring, by the muscle motor control system, of the motoneuron firing frequency to the actual mechanical properties of the muscle fibers. In practice an increase in the 1/2 relaxation time, monitored by a proprioceptive feedback, induces a decrease of the motoneuron firing rate. This provides an "economical activation of the fatiguing muscle by the central nervous system" (Enoka and Stuart 1992).

power density spectrum: the result of the signal decomposition in basic sinusoids of different frequency. The weight of each sine (its power) in the signal generation is estimated. The power vs. frequency relationship is the spectrum. The most common algorithm to process the signal is the Fast Fourier Transform. Other methods, such as the autoregressive methods, are widely used. In practice they act as a prism on the incident light beam: the contained wavelengths (colours) are displayed at the emerging site. The information contained in the power spectrum is usually compressed in the following frequency parameters: the *mean frequency* (the centroid frequency of the spectrum considered as a geometrical figure) and the *median frequency* (the frequency dividing the spectrum in two halves of equal area). In fatigue the EMG and the MMG spectrum and their frequency parameters shift toward the lower frequencies.

root mean square (RMS): is an indication of the power of a signal calculated over a specific time window (T). The variable is measured in V or its submultiples.

$$\text{RMS}\,[m(t)] = (1/T \int_{t}^{t+T} m^{2}(t)\,dt)^{\frac{1}{2}}$$

single twitch: the transient mechanical output of a single motor stimulus. Its amplitude depends on the number of the involved motor units. The characteristic parameters are: amplitude, contraction time (time to the force peak), maximum derivative of the force during the contraction phase, half relaxation time (time spent to reach the half of the force peak during the relaxation phase). The values of the above listed parameters depend on temperature, species, fiber type, and prior activity.

surface mechanomyogram (MMG): biological signal detectable at the muscle surface by specific contact sensors. It reflects the mechanical activities of the recruited motor units. The synchronous dimensional changes of the active fibers of a motor unit determines pressure waves providing muscle surface oscillations that can be recorded as MMG.

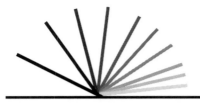

Name Index

A

Abbott, A.V. 133
Abbott, B.C. 76, 280
Abdel-Aziz, Y.I. 32
Abdenour, T.E. 335
Abraham, L.D. 184
Abrahamse, S.K. 277
Acierno, S.P. 343
Adams, W.C. 140
Adrian, E.D. 415
Adrian, R.H. 373
Aerts, P. 24, 25
Agarwal, B.D. 292
Akeson, W.H. 295
Akizuki, S. 295
Ala-Ketola, L. 312
Albrecht, K.H. 201
Aleshinsky, S.Y. 5, 9
Alexander, R.McN. 5, 20f, 21f, 22, 23, 25, 28, 51, 53, 70f, 76
Alhadeff, L. 122
Allen, D.G. 366, 368, 369, 370, 374, 375, 376, 377, 378, 410, 411
Allen, G.M. 366
Allen, L.H. 118
Allison, N. 311
Aloia, J.F. 309, 312
Alt, W. 332, 333, 339
Ambland, B. 225
Ameredes, B.T. 201
Amiel, D. 322
Anderson, R.A. 120
Andreassen, S. 386
Andreasson, G.O. 337, 338
Angulo-Kinzler, R.M. 53
Anstrom, M. 320
Anton, M. 5, 62
Apor, P. 205
Archambault, J.M. 317
Ardigò, L.P. 70, 70f, 73, 77
Arendt-Nielsen, L. 423, 425, 426
Ariano, M.A. 203
Armstrong, L.E. 115

Armstrong, R.B. 88, 203, 388, 391
Arsac, L.M. 208
Asatryan, D.G. 230
Ascenzi, A. 291
Ashley, C.C. 373, 375
Ashman, R.B. 291, 293
Asmussen, E. 5, 31, 386, 387, 390
Assaiante, C. 225
Astrand, P.O. 80
Atha, J. 213, 216
Attarian, D.E. 335
Aubert, X.M. 280
Audu, M.L. 166
Aura, O. 392
Avela, J. 398, 399, 403

B

Baer, E. 301
Baer, J.T. 105
Bagshaw, C.R. 165
Bahler, A.S. 201, 202
Bakels, R. 367
Baker, A.J. 138
Baker, B.E. 342
Baker, J.A. 51
Bakker, A.J. 378
Baldwin, K.M. 310
Balestra, C. 398
Ball, R.D. 418, 421
Balsom, P.D. 114
Bangsbo, J. 144
Bar, E. 295
Baratta, R.V. 195, 203
Barclay, C.J. 138, 372
Barclay, J.K. 136, 137f, 138f, 366
Barker, D. 171
Barnes, W.S. 114
Bar-Or, O. 108, 116, 207
Barry, D.T. 415, 417, 418, 418f, 421
Barstow, T.J. 372
Bartmus, U. 235

Basmajian, J.V. 411, 413, 415, 423, 424, 425, 426
Bassett, D.R. 207
Bates, B. 255
Baudinette, R.V. 27
Baylink, D.J. 311
Beals, K.A. 105
Beattie, B. 423, 426
Beaupré, G.S. 312, 319, 320
Beelen, A. 367
Begley, S. 63
Behiri, J.C. 294
Bélanger, A.Y. 417
Bellemare, F. 412, 414, 417
Belli, A. 208
Below, P.R. 116
Bennet, M. 51, 53
Bennett-Clark, H.C. 5, 22
Bennett, D.J. 170, 171, 172
Bennett, M.B. 22, 24
Benzanilla, F. 424
Bergman, G. 258
Bergström, M. 371
Berns, G.S. 341
Bers, D.M. 136
Bertram, J.E.A. 70f, 311
Best, C.H. 193
Bevan, L. 367, 424
Bevegard, B.S. 80
Beynnon, B.D. 342, 343
Bielinski, R.N. 112
Bier, D.M. 113
Biewener, A.A. 27, 311, 312
Bigland, B. 76
Bigland-Ritchie, B. 372, 387, 398, 412, 413, 424
Bilotto, G. 413, 415, 423, 426
Binder-Macleod, S.A. 412
Binder, M.D. 168, 412
Biolo, G. 112
Biot, M.A. 296
Birch, H.L. 22
Birch, R. 114
Birgit, L. 172
Bischof, M. 338

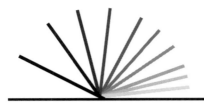

Subject Index

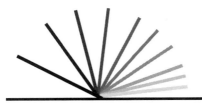

About the Editors

Benno M. Nigg is currently professor of biomechanics at the University of Calgary in Calgary, Alberta, Canada, where he is also director of the Human Performance Laboratory. He received his doctorate in natural science and physics from ETH Zurich. He serves on the editorial boards of the *Journal of Biomechanics*, the *Journal of Sports Sciences*, and several other professional publications and also is a member of numerous professional organizations. He resides in Calgary.

Brian R. MacIntosh is a professor in the Human Performance Laboratory at the University of Calgary, Alberta, Canada. He has training and research experience in human kinetics and physiology and is an associate editor for the *Canadian Journal of Applied Physiology (CJAP)*. Dr. MacIntosh received his doctorate in medical science from the University of Florida. He is a fellow of the American College of Sports Medicine and a member of the Canadian Society for Exercise Physiology. He resides in Calgary.

Joachim Mester is a university professor and the chair for human performance, Institute for Theory and Practice of Training and Movement at the German Sport University in Cologne. He has had training in exercise physiology, sensory physiology, and motor control. Dr. Mester received his doctorate in sport science from the department of sports medicine at the University of Bochum in 1978 and is president of the European College of Sport Science and a member of numerous organizations. He resides in Cologne.

OTHER BOOKS
FROM HUMAN KINETICS

Kinematics of Human Motion
Vladimir M. Zatsiorsky, PhD

1998 • Hardback • 432 pp • ISBN 0-88011-676-5
$49.00 ($73.50 Canadian)

This book is the first major text on the kinematics of human motion and is written by one of the world's leading authorities on the subject. While the book is advanced and assumes knowledge of calculus and matrix algebra, the emphasis is on explaining movement concepts, not mathematical formulae. The text features 23 refreshers of the basic concepts and many practical examples. The book is well illustrated and clearly written as the author skillfully integrates mechanical models with biological experiments.

Neurophysiological Basis of Movement
Mark L. Latash, PhD

1998 • Hardback • 280 pp • ISBN 0-88011-756-7
$45.00 ($67.50 Canadian)

Neurophysiological Basis of Movement is the only contemporary comprehensive textbook on the neurophysiology of voluntary movement. The book also covers relevant information from the study of biomechanics, anatomy, control theory, and motor disorders. It emphasizes neurophysiological mechanisms that apply to the processes of voluntary movements. The text covers a semester's worth of material about the neurophysiological aspects of five major areas: cells, reflexes, structures, behaviors, and disorders.

Advances in Exercise Immunology
Laurel T. Mackinnon, PhD

1999 • Hardback • Approx 376 pp • ISBN 0-88011-562-9
$49.00 ($73.50 Canadian)

The author of the groundbreaking *Exercise and Immunology* returns with *Advances in Exercise Immunology*, a thorough and unparalleled study of the relationship between exercise and immune function. You'll learn why athletes are susceptible to illness during intense training, how various immune system components respond to exercise, how regular exercise may influence disease progression—including cancer and HIV/AIDS—and whether exercise may help restore immune function in the aged and during spaceflight.

To request more information or to order, U.S. customers call 1-800-747-4457,
e-mail us at **humank@hkusa.com**, or visit our Web site at **www.humankinetics.com**.
Persons outside the U.S. can contact us via our Web site or use the appropriate telephone number,
postal address, or e-mail address shown in the front of this book.

Human Kinetics
The Information Leader
in Physical Activity